Mindfulness in Behavioral Health

Mindfulness-based therapy is one of the fastest evolving treatment approaches in psychology and related fields. It has been used to treat many forms of psychological and psychiatric distress and medical conditions as well as to foster health and wellness. Early empirical studies and meta-analyses of current research suggest that mindfulness-based therapies are effective and long lasting, but much more data from research and training studies are needed to fully understand its nature and effective practice. The Mindfulness in Behavioral Health series aims to foster this understanding by aggregating this knowledge in a series of high-quality books that will encourage and enhance dialogue among clinicians, researchers, theorists, philosophers and practitioners in the fields of psychology, medicine, social work, counseling and allied disciplines. The books in the series are appropriate for upper level undergraduate and graduate courses. Each book targets a core audience, but also appeals to others interested in behavior change and personal transformation.

Amy Finlay-Jones • Karen Bluth
Kristin Neff
Editors

Handbook of Self-Compassion

Springer

Editors
Amy Finlay-Jones
Telethon Kids Institute
Nedlands, WA, Australia

Karen Bluth
University of North Carolina at Chapel Hill
Chapel Hill, NC, USA

Kristin Neff
University of Texas at Austin
Austin, TX, USA

ISSN 2195-9579 ISSN 2195-9587 (electronic)
Mindfulness in Behavioral Health
ISBN 978-3-031-22347-1 ISBN 978-3-031-22348-8 (eBook)
https://doi.org/10.1007/978-3-031-22348-8

This Springer imprint is published by the registered company Springer Nature Switzerland AG
The registered company address is: Gewerbestrasse 11, 6330 Cham, Switzerland

To Judy Long, Celedra Gildea, Silvia Fernandez Campos, Maria Paula Jimenez, Petrina Barson, and Janina Scarlet, my dear friends, and wise teachers

Amy Finlay-Jones

To all teens and young adults everywhere – May self-compassion research provide the foundation to alleviate suffering and help navigate the often treacherous waters of this life stage.

Karen Bluth

To Chris Germer and the Center for Mindful Self-Compassion, who helped take self-compassion from an idea to a practice that is now taught to tens of thousands of people around the globe.

Kristin Neff

Preface

Since its emergence as an area of scientific enquiry almost 20 years ago, a mounting body of work attests to the benefits of self-compassion for individual and collective well-being. Alongside this, a global movement has driven communities of practice to support self-compassion cultivation across populations and contexts. Bringing together world-leading researchers, clinicians, and practitioners, the *Handbook of Self-Compassion* provides a deep dive into self-compassion research over the past two decades, consolidating research findings, integrating these with clinical insights, and providing recommendations for refining and extending work in this field. It explores critical questions of how to enhance rigor and precision in self-compassion conceptualization and measurement, how we can understand the development of self-compassion across the life course, how self-compassion is best cultivated, and why it is associated with adaptive outcomes. As we continue to pursue new frontiers in self-compassion science and practice, this handbook is an invaluable resource for researchers and practitioners alike.

Nedlands, WA, Australia — Amy Finlay-Jones
Chapel Hill, NC, USA — Karen Bluth
Austin, TX, USA — Kristin Neff

Contents

About the Editors

Editors

Amy Finlay-Jones, Ph.D., is a senior research fellow and head of the Early Neurodevelopment and Mental Health Team at Telethon Kids Institute. After completing training in clinical psychology, Dr. Finlay-Jones undertook further postgraduate training in health economics. Her research reflects the continuum of clinical to through implementation science, with a focus on translating evidence into policy and practice to improve mental health outcomes for children and families. Dr. Finlay-Jones' personal connection to self-compassion practice came at a young age when she was diagnosed with a chronic illness. Since that time, she has undertaken academic study of self-compassion, including developing the world's first online self-compassion training program during her doctoral studies. Dr. Finlay-Jones is interested in the development of self-compassion in children and adolescents, its application within clinical contexts, and its intersection with self-regulation. She is a trained teacher of the Compassion Cultivation Training program, the Mindful Self-Compassion program, the Mindfulness-Based Compassionate Living program, and the Making Friends with Yourself program, and she has a strong interest in co-designing accessible approaches to self-compassion training with underrepresented groups.

Karen Bluth, Ph.D., is research faculty in the Department of Psychiatry at the University of North Carolina, and a fellow at the University of North Carolina Frank Porter Graham Child Development Institute, where she conducts research on self-compassion and its influences on the emotional well-being of teens. Dr. Bluth is a certified instructor of Mindful Self-Compassion, co-creator of curriculum *Mindful Self-Compassion for Teens (formerly known as Making Friends with Yourself), Embracing Your Life,* the adaptation of Mindful Self-Compassion for young adults, and one of the creators of *Self-Compassion for Educators,* a self-compassion program offered through Mindful Schools. Dr. Bluth is author of the books *The Self-Compassion Workbook for Teens: Mindfulness and Compassion Skills to Overcome Self-Criticism and Embrace Who You Are, The Self-Compassionate Teen: Mindfulness and Compassion Skills to Conquer Your Critical Inner Voice,* co-author of *Mindfulness and Self-Compassion for Teen ADHD: Build Executive Functioning Skills, Increase Motivation, and Improve Self-*

Confidence (New Harbinger Publishers), and the Audible Original *Self-Compassion for Girls: A Guide for Parents, Teachers, and Coaches.* Additionally, Dr. Bluth is a mindfulness practitioner for over 40 years and an educator with 18 years of classroom teaching experience. In addition to teaching and mentoring self-compassion classes for teens and young adults, she trains teachers in *Mindful Self-Compassion for Teens* internationally.

Kristin Neff, Ph.D., is Associate Professor of Educational Psychology at the University of Texas at Austin. During Kristin's last year of graduate school, she became interested in Buddhism and has been practicing meditation in the Insight Meditation tradition ever since. While doing her postdoctoral work, she decided to conduct research on self-compassion – a central construct in Buddhist psychology and one that had not yet been examined empirically. Kristin developed a theory and created a scale to measure the construct over 20 years ago. She has written numerous academic articles on self-compassion and has been recognized as one of the most highly cited and influential scholars in the field of psychology. Kristin is also the author of the bestselling books *Self-Compassion: The Proven Power of Being Kind to Yourself* and *Fierce Self-Compassion: How Women Can Harness Kindness to Speak Up, Claim Their Power, and Thrive*. Along with her colleague Chris Germer, she developed the Mindful Self-Compassion program and co-founded the Center for Mindful Self-Compassion. They co-wrote *The Mindful Self-Compassion Workbook* and *Teaching the Mindful Self-Compassion Program*. She conducts workshops and lectures on self-compassion internationally. For more information go to www.self-compassion.org.

Contributors

Joanna J. Arch, Ph.D., is Associate Professor of Psychology and Neuroscience at the University of Colorado Boulder, Member in Cancer Prevention and Control at the University of Colorado Cancer Center, and a licensed clinical psychologist. Her research focuses on developing and evaluating interventions designed to address anxiety disorders and to improve well-being among adults with cancer, with a focus on mindfulness, compassion, and acceptance-based interventions. She has received funding from the National Institutes of Health, American Cancer Society, and the Templeton Foundation.

Kohki Arimitsu, Ph.D., is Professor of Clinical Psychology in the Department of Psychological Science at Kwansei Gakuin University, Japan. He is currently president of Japan Society for Research on Emotions (JSRE). He has published and edited 14 books, the most recent being *Handbook of Emotion Regulation* which 54 Japanese researchers contributed, and several scientific articles regarding self-compassion. His research focuses on emotion regulation, compassion-based intervention for anxiety related disorders, and cultural differences of self-compassion and well-being.

J. Austin holds a master's degree in medical psychology and is currently in the final stage of pursuing her Ph.D. in the Department of Psychology, Health & Technology at the University of Twente. She is trained in (self-)compassion practices, interventions, theory, and research. Her research interests include: 1) the application, conceptualization, and measurement of (self-compassion and self-criticism; 2) resilience processes (including growth facilitators such as gratitude) in the adaptation to adversities such as cancer and chronic illness; and 3) the role of technology in contributing to a compassionate society and in offering interventions that are rooted in both theory and practice. In her mixed methods doctoral research, Austin develops and evaluates a mobile self-compassion intervention for people with newly diagnosed cancer.

Anna Boggiss is a Ph.D., candidate and pre-intern health psychologist in the Department of Psychological Medicine at the University of Auckland, New Zealand, and a Mindful Self-Compassion for Teens Teacher Trainee. Anna's Ph.D. has focused on developing a clinically usable self-compassion intervention for adolescents with type 1 diabetes. More broadly, she is passionate about partaking in research aimed at improving the wellbeing of young people, especially those with chronic health conditions, and continuing to develop clinically usable coping skill interventions.

E. T. Bohlmeijer, Ph.D., graduated in 2007 on the effects of life-review on depression in older adults. Since 2007 he worked as an associate professor and since 2011 as full professor in the Department of Psychology, Health & Technology at the University of Twente. His research focuses on public mental health: the development and evaluation of interventions aiming to promote well-being and reduce distress in the general population and people with chronic diseases. These, partly eHealth and mHealth, interventions are primarily based on Compassion, Acceptance and Commitment Therapy and Positive Psychology. He developed the model of sustainable mental health to integrate meaning- and strengths-based interventions in mental health care. Since 2010, Professor Bohlmeijer (co-) published over 150 peer-reviewed papers and three scientific books. He is editor of the Dutch Handbook of Positive Psychology.

Christine Brähler, DClinPsy, Ph.D., is a clinical psychologist, psychotherapist, author, and honorary lecturer at the University of Glasgow. She has been teaching self-compassion around the world since 2008. She is a key contributor to the first clinical adaptations and to the non-clinical training program, Mindful Self-Compassion, and its teacher training. Her mission is to guide people to overcome obstacles to gentle and fierce self-compassion to connect with others from a place of empowerment and genuine compassion.

Lydia Brown, Ph.D., is a clinical psychologist and researcher in the fields of healthy aging and women's mental health. She holds a Ph.D. in Clinical Psychology and completed her postdoctoral training at Harvard Medical School and The University of Melbourne. Her research focuses on how psychological resilience factors can help people cope with aging and physical health issues, especially the menopause. After studying meditation in a Sri Lankan Buddhist monastery during her 20s, she developed a lifetime interest in self-compassion and meditation. She remains an active meditation practitioner exploring the Buddhist understanding of suffering and compassion. Alongside her research, she is a clinical psychologist in private practice. She also teaches into the University of Melbourne's Master of Clinical Psychology program, helping train the next generation of psychologists.

Joseph Ciarrochi, Ph.D., is a Research Professor at the Institute of Positive Psychology and Education, Australian Catholic University. His research seeks to help individuals and organizations apply the latest findings from behavioral science to achieve health, well-being and peak performance. Professor Ciarrochi has published over 90 scientific journal articles and many books, including the bestselling, Get out of your mind and into your life teens and The Thriving Adolescent. He has been honored with over 4.5 million dollars in research funding. His work has been discussed on T.V., and in magazines, newspaper articles, and radio. Joseph is ranked in the top 2% of scientists in the world across all disciplines.

Melissa Clepper-Faith, MD, MPH, is a pediatrician with public health training and experience working with diverse populations. After 24 years in clinical practice, including the establishment and management of a private pediatric practice in Hillsborough, NC, she completed a master's in public health at Gillings School of Global Public Health, UNC-Chapel Hill. Her current position as a translational research program and policy coordinator in the FRONTIER program at the Frank Porter Graham Child Development Institute, UNC-Chapel Hill, focuses on neuro-prevention, the translation of neuroscience knowledge into programs and policies that support healthy child development and ameliorate the adverse effects of toxic stress and child maltreatment.

Nathan S. Consedine, Ph.D., is health psychologist in the School of Medicine at the University of Auckland. His training is in the experimental study of emotion and emotion regulation, specifically looking at how such factors may be linked to physical health in diverse groups. Current research foci include compassion in healthcare, disgust in medical contexts, self-compassion, and mindfulness. After graduating from Canterbury in 2000 and spending 10 years working on grants in New York, Nathan returned to traditional academia in New Zealand in 2009. In addition to teaching in health psychology and medical programs, he has supervised 50+ master's and Ph.D. students, including studies evaluating psychosocial interventions in patient populations and testing how patient, physician, clinical and environmental factors impact medical compassion. Nathan has published more than 170 articles and chapters and reviews for numerous international journals and funding agencies. He enjoys fishing, playing tennis with his son, and listening to the sort of music that his colleagues dislike.

C. H. C. Drossaert, Ph.D., is Associate Professor of Health Psychology in the Department of Psychology, Health & Technology at the University of Twente. She teaches health psychology to bachelor's and master's students. In 2002 she graduated on "Psychosocial aspects of participation in Breast Cancer Screening." Her current research interests are in the field of patient participation and patient empowerment. Central in her work is the question: "How can we equip patients and their caregivers with adequate information, skills or tools, to facilitate their active involvement in the management of their disease, or to help them live a happy and meaningful life despite or with the condition?" Dr. Drossaert is especially interested in the role of (internet-) technology herein.

Leah J. Ferguson, Ph.D. (she/her/hers), is an associate professor in the College of Kinesiology at the University of Saskatchewan. A Métis woman, Leah's nationally funded sport psychology research explores how women athletes flourish and reach their potential, and she is particularly interested in understanding the role and impact of self-compassion in the lives of women athletes. Leah translates her sport psychology research into practice through her applied work as a Mental Performance Consultant with the Sport Medicine & Science Council of Saskatchewan, and she is a Professional Member of the

Canadian Sport Psychology Association. She works with athletes, coaches, and parents to facilitate well-being and performance for positive sport experiences. She has worked with athletes and teams at all competitive levels – community, provincial, varsity, national, and international, including athletes competing at the Paralympic Games.

Madeleine I. Fraser, Ph.D., is a qualified clinical psychologist and provides assessment and treatment in a private practice based in Sydney CBD. She is also a full-time clinical psychology lecturer At Australian Catholic University (ACU, Strathfield campus). She supervisors honors and master's research projects in topics related to clinical psychology and with a particular focus on self-compassion. Madeleine lectures in the undergraduate and postgraduate psychology programs and is also an AHPRA board approved supervisor. Dr. Ferrari completed a Master of Philosophy and a Doctor of Clinical Psychology at Macquarie University, and a Ph.D. at the University of Sydney titled "Self-Compassion in Adolescence: A Protective Psychological Framework for Relating to Oneself." She is a member of the Healthy Brain and Mind Research Centre at ACU and maintains an active research profile.

Anna Fitch is a recent graduate of Santa Clara University and has her Master of Arts in Counseling Psychology. During her time in university, she worked closely with renowned mindfulness researcher Dr. Shauna Shapiro, serving as both research and teaching assistant to Dr. Shapiro. Throughout her studies, Anna amassed an extensive knowledge of both theory and implementation of evidence-based mindfulness practices. With a specialty in Positive Psychology, she holds a deep passion for the relationship between mindfulness and wellbeing. Post graduation, Anna began working toward her clinical licensure as a Marriage and Family Therapist along with her certification in Mindfulness-Based Stress Reduction. Currently, she serves as a counselor for diverse populations in the San Francisco Bay Area. Anna implements her backbone in mindfulness and health psychology to offer accessible and approachable practices based on clinical needs. Additionally, Anna utilizes her expertise to consult with start-ups and local organizations interested in incorporating mindfulness and meditation into the workplace.

Christopher Germer, Ph.D., is a clinical psychologist and lecturer on psychiatry (part-time) at Harvard Medical School. He co-developed the *Mindful Self-Compassion (MSC)* program with Kristin Neff in 2010 and they wrote two books, *The Mindful Self-Compassion Workbook* and *Teaching the Mindful Self-Compassion Program.* MSC has been taught to over 200,000 people worldwide. Dr. Germer is also the author of *The Mindful Path to Self-Compassion*; he co-edited two influential volumes on therapy, *Mindfulness and Psychotherapy*, and *Wisdom and Compassion in Psychotherapy.* Dr. Germer is a founding faculty member of the Center for Mindfulness and Compassion, Harvard Medical School, and the Institute for Meditation and Psychotherapy, Cambridge MA, USA. He maintains a small psychotherapy practice in Massachusetts, USA.

Paul Gilbert, Ph.D., is Professor of Clinical Psychology at the University of Derby and visiting professor at the University of Queensland. He has researched evolutionary approaches to mental health alongside clinical work as a Consultant Clinical Psychologist for over 40 years in the NHS. He founded and developed Compassion Focused Therapy (CFT) and established the Compassionate Mind Foundation in 2006, which promotes wellbeing through facilitating the scientific understanding and application of compassion. Professor Gilbert has written over 300 publications including 22 books. He was awarded an OBE in 2011 by the Queen for services to mental health

Aleah Goold is a graduate student in the Master of General Psychology Program, with a focus in Experimental Psychology, at the University of Memphis. She graduated from The University of Tennessee at Martin in 2019 with a B.A. in Psychology and a minor in Sociology. For her thesis, she is currently studying the impact of a self-compassion intervention on intrinsic motivation in college students. She has also been a Teaching Assistant for the General Psychology class at the University of Memphis since Fall 2020, which includes her teaching small sections of the course once a week. Aleah's main drive is to let her research inform her teaching, and her teaching inform her research, with the ultimate goal to help students become more open-minded and eager about their learning. She hopes to continue with her education and earn a Ph.D., and eventually teach Psychology at a university level.

Penelope Hasking, Ph.D., Professor Hasking's work focuses on mental health in adolescents and young adults. Her primary interests are in the social and cognitive factors that initiate and maintain non-suicidal self-injury (NSSI) among youth. She is also interested in the needs of school staff who address NSSI in the school setting, and the views of parents of young people who self-injure. More recently she has focused her work on the "experience" of self-injury, delving into topics such as NSSI stigma, experience of disclosure, meanings of scarring, and approaching NSSI recovery from a person-centered, multifaceted framework.

Katarina L. Huellemann (she/her), M.Sc., is a Ph.D. student in psychology studying body image, gender, adaptive physical activity, and compassion-focused interventions. Her research considers sociocultural and individual forces that may foster stigma towards the self and others, and how this stigma may contribute to worsened mental and physical well-being in various populations (e.g., women, adolescents). Her research has been published in *Body Image*, *Mindfulness*, and the *Canadian Journal of Public Health*. She is supervised by Dr. Rachel Calogero and Dr. Eva Pila at Western University in Canada, and is a member of both the Stigma, Objectification, Bodies, and Resistance Lab (Calogero) and the Body Image and Health Lab (Pila). She is also an assistant editor for *Sex Roles: A Journal of Research*.

Caroline Hunt, Ph.D., is a clinical psychologist, and currently president of the Australian Clinical Psychology Association (ACPA). She is a professor in the School of Psychology at the University of Sydney where she is also the academic lead of the Clinical Psychology Unit. Her teaching key role is to oversee the training of clinical psychology students. Caroline has worked in the field of anxiety disorders for over 20 years. Her expertise and interests are in anxiety disorders, the assessment and prevention of peer victimization in schools, and emotional problems in youth. Caroline has developed and implemented several school-based intervention programs targeting emotional problems, and more recently bullying and victimization.

John Jackson, Ph.D., is a psychologist working primarily in the role of training director for postdoctoral, doctoral, and master's level interns at a university counseling center. John's training interests involve methods of attending to and addressing trainee self-awareness and self-regulation, particularly with attention to attachment style and enactment of self-compassion. In his clinical work with university students, John's interests lie in addressing complex trauma and related health difficulties. John integrates psychodynamic and Emotionally Focused Therapy (EFT) approaches in individual therapy and finds great challenge and reward in an interpersonal process approach to group therapy.

James N. Kirby, Ph.D., is a senior lecturer, clinical psychologist, and co-director of the Compassionate Mind Research Group at the School of Psychology at the University of Queensland. He has two major areas of focus to his research, 1) understanding the decision-making process to how we choose to act compassionately or not, and 2) evaluating the effectiveness of Compassion Focused Therapy. James also maintains an active practice where he delivers Compassion Focused Therapy for those experiencing self-criticism and shame. James also holds a Visiting Fellowship at the Center for Compassion and Altruism Research and Education at Stanford University.

Alex Kirk completed his PhD in clinical psychology and neuroscience at the University of Colorado Boulder. He is a naval medical officer with research and clinical interests spanning anxiety disorders, medical/physical health correlates of mental health concerns, and behavioral mechanisms of change.

Christine Lathren, Ph.D. is a research assistant professor at the University of North Carolina at Chapel Hill within the Program on Integrative Medicine. Her research interests include examining the impact of self-compassion intervention in various caregiving and family contexts using both quantitative and qualitative methods. Recent interests explore if and how learning self-compassion might improve parent-child relationships, parenting behavior, and child socioemotional outcomes in families experiencing high stress. Other work examines the impact of self-compassion training for professional caregivers, including certified nursing assistants in long-term care contexts.

Helena Moreira, Ph.D., is a clinical psychologist and an assistant professor at the Faculty of Psychology and Educational Sciences of the University of Coimbra, Portugal. After obtaining a Ph.D. in Psychology from the University of Coimbra, she did 6 years of postdoctoral studies at the same university, studying the interrelationship between parenting and the parent-child relationship. During her postdoctoral years, she became very interested in understanding the role of self-compassion and mindfulness in parenting. Since then, she has been conducting extensive research on mindful and compassionate parenting at different stages of the life cycle, with a particular emphasis on the post-partum period. Recently, Helena Moreira has also become interested in a transdiagnostic approach to the treatment of children's emotional disorders. She has published over 80 academic articles on topics related to parenting, mindful parenting, self-compassion, attachment, and emotion regulation. She serves as associate editor of the *Journal of Child and Family Studies* and of the *Mindfulness* journal.

Amber D. Mosewich, Ph.D., (she/her/hers) is an associate professor in the Faculty of Kinesiology, Sport, and Recreation at the University of Alberta. Amber's research interests focus on stress, coping, and emotion within the sport domain. A key directive of her research is to understand the psychological skills and resources necessary to promote adaptive responses to stress and emotion and support athletes in attaining their performance potential while maintaining high levels of well-being. One major focus in this line of research is understanding the role of self-compassion in athlete support and development, and the application of self-compassion in sport more broadly. Amber has been and continues to be involved in several research and applied initiatives directed at providing psychological support for athletes involved in grassroots, developmental, national, and international sport programming.

Anaïs Ortiz is a doctoral student at the University of Florida. She studies ways in which healthy relationships with ourselves can help us be kinder and more compassionate with others. Anais is particularly interested in integrating psychological concepts with Eastern roots, such as mindfulness and self-compassion, with Western research on well-being and transformative experiences. She is additionally extending this research to examine the social effects of these concepts on prosocial behavior, self-expansion, and meaning in life. In addition to this work, she enjoys putting research into practice through teaching yoga and meditation.

Alina Pavlova, M.A., M.Sc., is a Ph.D. candidate in the Department of Psychological Medicine at the University of Auckland studying the nature of self- and other-focused compassion in healthcare. By investigating the relationships between physician-, organizational-, and patient-related factors, Alina's biggest commitment is to design multilevel interventions to enhance care at patient, physician, and organizational levels. Adjacent to her Ph.D., Alina is currently clinically training to become a Health Psychologist and is involved in self-harm and suicide prevention research collaborating with international experts and lived-experience researchers worldwide. Before her Ph.D., Alina completed two master's degrees (in Sociology and Economics) from the Erasmus University Rotterdam, where she studied stigma in the context of mental health. Alina is a Yoga Alliance Certified yoga instructor and teaches yoga and mindfulness in the community.

Nicola Petrocchi, Ph.D., is Adjunct Professor of Psychology at John Cabot University in Rome and a compassion-focused therapist in Rome. Dr. Petrocchi founded Compassionate Mind Italia, the Italian association, linked to the Compassionate Mind Foundation (UK), with the aim to deepen research and promote training and good practice of CFT in Italy. Dr. Petrocchi sees patients privately (both in English and Italian) in his studio in the center of Rome. He manly sees patients in individual sessions, and he runs a 12-week Compassionate Mind Trainings for clients struggling with issues of shame and self-criticism.

Trisha L. Raque, Ph.D., is a licensed psychologist and an associate professor in the Department of Counseling Psychology at the University of Denver. She is the Chair of the American Psychological Association Division 17 (Counseling Psychology) Health Section. She serves on the editorial board of the *Journal of Counseling Psychology* and the *Journal of Career Assessment*. Dr. Raque's area of research includes cancer survivorship, self-compassion, the intersection of work and health, and health equities. She applies social justice principles, intersectionality, and the Multicultural Orientation to cancer survivorship scholarship and advocacy.

Baljinder K. Sahdra, Ph.D., is a Senior Lecturer and full-time researcher at the Institute for Positive Psychology and Education, Australian Catholic University. She has previously held positions at the University of California, Davis, and University of Western Sydney, Australia. Dr. Sahdra's research publications reflect her diverse substantive interests in psychological assessment, educational psychology, personality, developmental psychology, and mindfulness related constructs and interventions. They also showcase diverse computational methods, including structural equation modelling, multilevel modelling, network analysis, mixture modelling, longitudinal analysis, text mining, and machine learning advances in psychometrics. She has been awarded with several prestigious awards and competitive grants ($7+ million). Her research is published in top-tier journals, is highly cited, and has been featured in The Sydney Morning Herald, The Sun, Radio Canada International, Boston Globe, New York Daily Post, Huffington Post, New Scientist, The Guardian, ABC Radio, and other major media outlets.

Benjamin J. Sereda, M.Sc. (he/him/his), is a doctoral student in the Faculty of Kinesiology, Sport, and Recreation at the University of Alberta. In addition to his involvement in sport psychology and performance research, he actively coaches and consults with athletes at a variety of ages, competitive levels, and sport contexts. He is passionate about using research to inform efforts to practically support athletes in achieving their sport-related goals while supporting their well-being both inside and outside of sport. Ben's research has focused on how athletes and performers perceive and attempt to manage demands that they experience surrounding training and competition. Further, guided by his applied work and coaching experience, Ben is particularly interested in how athletes attend to, perceive, and respond to setbacks and challenges in sport.

Anna Serlachius, Ph.D., is a health psychologist and senior lecturer in the Department of Psychological Medicine at the University of Auckland. She completed her M.Sc. in Health Psychology at University College London and went on to do a Ph.D. at the University of Melbourne/Murdoch Children's Research Institute. After her Ph.D., she worked on the Cardiovascular Risk in Young Finns Study, based at Helsinki University and later at Columbia University. Dr. Serlachius has published more than 40 journal articles and her research program aims to develop cost-effective and scalable interventions to improve psychological and physical health outcomes, with a focus on youth with chronic health conditions. Her work is increasingly focused on developing and testing digital wellbeing interventions. She is section editor for JMIR Pediatrics and Parenting (*Journal of Medical Internet Research*).

Shauna Shapiro, Ph.D., is a best-selling author, clinical psychologist, and internationally recognized expert in mindfulness and self-compassion. She is a professor at Santa Clara University and has published over 150 papers and three critically acclaimed books, translated into 16 languages. Shauna has presented her research to the King of Thailand, the Danish Government, Bhutan's Gross National Happiness Summit, and the World Council for Psychotherapy, as well as to Fortune 100 Companies including Google, Cisco Systems, and LinkedIn. Her work has been featured in the *Wall Street Journal, Mashable, Wired, USA Today, Dr. Oz, the Huffington Post,* and the *American Psychologist.* Shauna is a *summa cum laude* graduate of Duke University and a Fellow of the Mind and Life Institute, co-founded by the Dalai Lama. Her TEDx Talk, *The Power of Mindfulness*, has been viewed over 2.5 million times

Fuschia M. Sirois, Ph.D., is a professor in the Department of Psychology at the University of Durham in England, and a former Canada Research Chair in Health and Well-being. She has authored over a 120 peer-reviewed journal papers, over 200 conference papers, 19 book chapters, and authored/edited 10 books. Her research has been funded by several national funding agencies including the Social Sciences and Humanities Research Council, Canada, The Economic and Social Research Council (UK), the Engineering and Physical Sciences Research Council (UK), and the Welsh Government. Professor Sirois' research investigates the temporal, affective, cognitive, and behavioral dynamics of personality traits and states that that help or hinder people in their efforts to regulate their emotions, thoughts, and behaviors when dealing with life's challenges. Her research aims to understand ways to enhance resilience and support physical and mental health across a range of populations, including vulnerable populations and individuals living with chronic illness. This research has a particular focus on how positive psychology qualities and interventions, including self-compassion, can support self-regulation and enhance physical health and well-being.

Elizabeth T. Slivjak is a doctoral student in clinical psychology at the University of Colorado Boulder. Her research focuses on the development of self-compassion-based interventions among anxious adults.

Penelope Strauss, Ph.D., (she/her) is a Research Fellow in Youth Suicide Prevention at the Telethon Kids Institute and an adjunct research fellow at the University of Western Australia. Her current research aims to decrease rates of suicidality among LGBTQA+ young people. Penelope received her Master of Public Health and Ph.D. from the University of Western Australia. She led the *Trans Pathways* project, the seminal study conducted on the mental health of trans and gender diverse young people in Australia and barriers experienced when accessing medical and mental health services.

Tracy L. Tylka (she/her), Ph.D., is Professor of Psychology at The Ohio State University. She is also editor-in-chief of *Body Image: An International Journal of Research* and serves on the editorial board of *Eating Disorders: A Journal of Treatment and Prevention*. Her research interests include positive body image, negative body image, intuitive eating, disordered eating, and sociocultural influences on these variables. She edited the *Handbook of Positive Body Image and Embodiment* with Niva Piran and wrote the *Positive Body Image Workbook: A Clinical and Self-Improvement Guide* with Nichole Wood-Barcalow and Casey Judge. Along with colleagues, she has developed several scales to measure these constructs, such as the Body Appreciation Scale-2, Functionality Appreciation Scale, Broad Conceptualization of Beauty Scale, Intuitive Eating Scale-2, Interpersonal Sexual Objectification Scale, Personal Safety Anxiety and Vigilance Scale, and Transgender Congruence Scale. To date, she has written over 100 peer-reviewed articles and book chapters.

Abra Vigna, Ph.D., is an action researcher and evaluator at the University of Wisconsin-Madison Population Health Institute in the School of Medicine and Public Health. Dr. Vigna received her Ph.D. in Human Development and Family Studies from the School of Human Ecology at the University of Wisconsin-Madison. Her research focuses on self-compassion as a resource associated to resilience to adversity and is inspired by over a decade of experience working to advance health equity via direct-service, coalition building, and community-engaged action research.

Jia Wei Zhang, Ph.D., earned a bachelor's degree in Psychology from San Francisco State University and his Ph.D. in Social/Personality Psychology from the University of California, Berkeley. He is currently Assistant Professor of Psychology at the University of Memphis. His scholarship takes a multimethod and reproducible approach to address emotion, emotion regulation, well-being, mental health, resilience, education, ethnic, and diversity topics. For instance, one line of his research has shown that self-compassion, a sympathetic reaction extended toward ourselves when we are faced with difficult experiences, is an adaptive psychological tool that can help people respond to difficult experiences in ways that can protect and enhance their well-being.

Kathryn Ziemer, Ph.D., is a licensed clinical psychologist and the clinical director of Old Town Psychology, a psychology practice that provides evidence-based therapy. She received her Ph.D. in Counseling Psychology from the University of Maryland. She has conducted research at the National Institutes of Mental Health, the Social and Decision Analytics Lab at Virginia Polytechnic Institute and State University, and the research firm Ipsos Public Affairs. Her research interests include attachment, self-compassion, meaning in life, brief psychological interventions, and the promotion of positive psychology. She has over 10 years of experience providing psychological services, including evidence-based therapy, to clients experiencing depression, anxiety, chronic health conditions, infertility, relationship issues, and life transitions. She has previously provided psychological services at the DC Veterans Affairs Medical Center, the University of Maryland Counseling Center, and the George Mason University Cognitive Assessment Program.

1 Self-Compassion: Theory and Measurement

Kristin Neff

Introduction

Self-compassion can be defined as compassion turned inward. In order to understand what self-compassion is, therefore, it helps to consider what occurs in the experience of compassion more generally. Goetz et al. (2010) define compassion as "the feeling that arises when witnessing another's suffering and that motivates a subsequent desire to help" (p. 351). Note that we must be present with the suffering of others, as uncomfortable as it might be, in order for feelings of care and concern to arise. This requires mindfulness so that we can turn toward and be aware of pain rather than avoiding or resisting it. Also central to compassion is a sense of interconnection with others who are suffering (Cassell, 2002). Blum (1980) writes "compassion involves a sense of shared humanity, of regarding the other as a fellow human being" (p. 511). In fact, this is what differentiates compassion from pity, or feeling sorry for someone separate from yourself. The experience of compassion is similar when applied to our own suffering, whether our pain stems from failure, feelings of personal inadequacy, or life distress more generally.

K. Neff (✉)
University of Texas at Austin, Austin, TX, USA
e-mail: kneff@austin.utexas.edu

The Elements of Self-Compassion

According to my model (Neff, 2016), self-compassion forms a bipolar continuum ranging from uncompassionate to compassionate self-responding in moments of suffering. It is comprised of overlapping but conceptually distinct elements that can be loosely organized into three broad domains – how people emotionally respond to suffering (with kindness or judgment), cognitively understand their predicament (as part of the human experience or as isolating), and pay attention to pain (in a mindful or overly identified manner).

Self-Kindness vs. Self-Judgment Most of us try to be kind and supportive toward our friends and loved ones when they feel badly about themselves or experience life challenges. We may voice words of warmth and understanding to let them know we care – perhaps even offering a physical gesture of affection such as putting a hand on their shoulder. We are often much harsher with ourselves, however, saying mean and judgmental things that we would never say to a friend. With self-compassion, however, we turn this around. We take a benevolent and supportive attitude rather than condemning ourselves as worthless. We acknowledge our shortcomings while caring for ourselves regardless. This type of self-acceptance decreases feelings of unworthiness.

A. Finlay-Jones et al. (eds.), *Handbook of Self-Compassion*, Mindfulness in Behavioral Health,
https://doi.org/10.1007/978-3-031-22348-8_1

Self-kindness involves more than merely ending self-criticism, however. It involves actively opening up our hearts to ourselves, showing concern for our distress. We are motivated to try to ease our struggles if we can – not because we're inadequate as we are, but because we care. We often treat ourselves with cold stoicism rather than support when challenged and move straight into problem-solving mode without attending to our emotional needs. With self-kindness, however, we are emotionally available when life becomes difficult. We allow ourselves to be moved by our own pain, stopping to say, "This is really hard right now. How can I care for myself in this moment?" When we respond to ourselves with kindness, we feel validated, supported, and encouraged, similar to how we feel when receiving kindness from another. This helps us to cope with the challenges we face.

Common Humanity vs. Isolation The sense of common humanity that is inherent to self-compassion helps us to feel connected to rather than separate from others. When we fail or make mistakes, we tend to irrationally feel like everyone else is just fine and it's only *me* who has blown it. This isn't a logical process, but an emotional reaction that narrows our understanding and distorts reality. And even when our struggles stem from difficult life circumstances that we don't blame ourselves for, we tend to feel that somehow everyone else is having an easier time of it. We react as if "something has gone wrong" and forget that part of being human means facing challenges and being vulnerable. This feeling of abnormality creates a frightening sense of disconnection and loneliness that greatly exacerbates our suffering.

With self-compassion, however, we recognize that life challenges are part of being human, an experience we all share. In fact, our struggles are what make us card-carrying members of the human race. The element of common humanity also helps to distinguish self-compassion from self-love. While self-love is important, it leaves out an essential factor – other people. Compassion is, by definition, relational. It implies a basic mutuality in the experience of suffering and springs from the acknowledgement that the shared human experience is imperfect. When we're in touch with our humanity, we remember that everyone experiences suffering. The triggers are different, the circumstances are different, the degree of pain is different, but the experience of imperfection is shared. When we remember our common humanity, we feel less isolated and alone.

Mindfulness vs. Overidentification In order to have compassion for ourselves, we need to be willing to turn toward our own pain, to acknowledge it mindfully. Mindfulness is a type of balanced awareness that neither avoids nor exaggerates the discomfort of our present-moment experience (Kabat-Zinn, 1994). We can't show ourselves compassion if we don't acknowledge we're in pain. At the same time, if we fight and resist the fact that we're suffering, our attention may become so absorbed by our pain that we can't step outside ourselves and adopt the perspective needed to care for ourselves. We may become overly identified with our negative thoughts or feelings and be swept away by our aversive reactions. This type of rumination narrows our focus and exaggerates implications for self-worth (Nolen-Hoeksema, 1991). Not only did I fail, "I am a failure." Not only did something terrible happen, "My life is horrific." Overidentification tends to reify our moment-to-moment experience so that we perceive transitory events as definitive and permanent.

With mindfulness, however, we can recognize that our negative thoughts and feelings are just that – thoughts and feelings – which helps us to be less absorbed by and identified with them. We have the perspective necessary to extend compassion for our difficulties. It also provides the space needed to ask "How can I best care for myself right now?" It takes courage to turn toward and directly face our pain, but this act of courage is essential if we are going to alleviate our suffering. For this reason, mindfulness is the pillar on which self-compassion rests.

The Structure of Self-Compassion

When we're self-compassionate, we feel kinder and less judgmental toward ourselves, more connected to humanity, and less isolated and adopt a mindful perspective on our suffering while being less identified with it. The components of self-compassion overlap and interact. For instance, the accepting stance of mindfulness lessens overidentification but also helps to soften self-judgment and provides the wisdom needed to recognize our common humanity. Similarly, self-kindness reduces self-judgment and because it is an affiliative emotion also helps us to feel safer, making it easier to be mindful of our pain. Moreover, realizing that imperfection is part of being human lessens feelings of isolation and helps us to take things less personally so that we are less overidentified with and judgmental of our problems. In this way, self-compassion can be seen as a dynamic system that represents a synergistic state of interaction between its constituent elements (Neff, 2016). Research supports understanding self-compassion as a system: a study by Dreisoerner et al. (2021) found that inducing one element of self-compassion through a writing exercise changed levels of the other elements, suggesting they mutually engender one another.

The construct of self-compassion can also be conceptualized as a bipolar continuum ranging from uncompassionate self-responding (UCS; self-judgment, isolation, and overidentification) to compassionate self-responding (CS; self-kindness, common humanity, and mindfulness). As people become more self-compassionate, they increase in CS and decrease in UCS. Analyses of how the components of self-compassion are configured within individuals finds three basic patterns (Phillips, 2019; Ullich-French & Cox, 2020): low levels of USC and high levels of CS, high levels of CS and low levels of UCS, or moderate levels of both, suggesting that these are not independent dimensions.

Studies show that variation in UCS is a stronger predictor of psychopathology than variation in CS. This makes sense given that self-judgment, isolation, and overidentification directly feed into negative mood states like depression and anxiety. Once people have a moderate level of self-compassion (and are therefore no longer depressed or anxious), becoming even more self-compassionate is unlikely to have a strong effect on how depressed and anxious they are. Some scholars have proposed that because UCS and CS (often referred to as "coldness" and "warmth") have different associations with outcomes, they should be conceptualized as two separate and independent constructs (Brenner et al., 2018; Muris & Petrocchi, 2017; Pfattheicher et al., 2017). In fact, Muris and colleagues (Muris et al., 2019; Muris & Petrocchi, 2017) have argued that reduced self-judgment, isolation, and overidentification should be dropped from the definition and measurement of self-compassion altogether. However, it is common for each end of a bipolar continuum to differentially predict outcomes. Consider the physical analogy of temperature. If one were to conduct a study asking people how warm they are, how cold they are, and how numb their hands are, coldness items would predict numbness more than the warmth items. It would be strange to argue that this finding means coldness and warmth are independent constructs, however, especially given that the way to make your hands less cold is to warm them up! Other bipolar continuums such as wet/dry also evidence differential associations of each end with outcomes. The fact that different poles of a continuum have different associations with outcomes has no bearing on whether or not they can be conceptualized or measured as a unitary construct.

Most criticisms of self-compassion as a unitary construct have been based on cross-sectional findings, but an examination of how the components of self-compassion change in real time using experimental methods can shed more light on the how the construct operates. Research indicates that self-compassion interventions increase the three elements representing CS and decrease the three elements representing UCS simultaneously (Ferrari et al., 2019), suggesting movement along a single continuum. This is true for studies using a wide variety of methodologies such as self-compassion mood inductions (Neff et al., 2021b), self-compassion meditation training (Albertson et al., 2015; Toole & Craighead, 2016;

Wallmark et al., 2012), online psycho-education (Finlay-Jones et al., 2017; Krieger et al., 2016), compassion-focused therapy (Beaumont et al., 2016; Kelly & Carter, 2015; Kelly et al., 2017), or mindful self-compassion training (Finlay-Jones et al., 2018; Neff, 2016). Mindfulness-based interventions also yield a simultaneous increase in CS and decrease in UCS (Birnie et al., 2010; Greeson et al., 2014; Raab et al., 2015; Whitesman & Mash, 2016). A study by Mantzios et al. (2020) examined the effect of targeting CS and UCS separately through an experimental manipulation. Participants were assigned to a brief intervention that either asked them to relate to a difficulty they were having with kindness, a sense of common humanity and mindfulness (CS), or else they were asked to relate to the difficulty without judgment, a sense of isolation or overidentification (reduced UCS). Levels of self-compassion increased for both groups equally. These findings strongly suggest that CS and UCS are not independent constructs but instead form a continuum.

The fact that CS and UCS operate in tandem may be linked to the functioning of the sympathetic and parasympathetic nervous systems (Porges, 2007; see Chap. 17). UCS can be seen to reflect sympathetic activity – the stress response turned inward when our self-concept is threatened (Gilbert, 2000). Self-judgment reflects the fight response in the form of self-criticism and self-attack. Isolation represents the flight response – the desire to flee from others and hide in shame. Overidentification can be viewed as the freeze response – becoming getting stuck in a ruminative cycle of negative thoughts. CS reflects parasympathetic activity, generating feelings of safety that directly counter feelings of threat. Self-kindness involves nurturing and supporting ourselves, counteracting the fight response. Common humanity generates feelings of connection and affiliation, counteracting the flight response. Mindfulness provides a sense of perspective and psychological flexibility that counteracts the freeze response (Creswell, 2015; Tirch et al., 2014). Notably, a recent study by Kim et al. (2020) used fMRI imagery to examine reactions to negative emotional stimuli. They found that UCS increased activity in the anterior insula, anterior cingulate, and the amygdala and that CS suppressed activity in these very same regions, illustrating how warmth counteracts coldness.

Fierce and Tender Self-Compassion

Self-compassion is aimed at the alleviation of suffering, and this occurs through a process of both acceptance and change. Sometimes people view self-compassion as merely going easy on oneself, but to alleviate our suffering, we often need to work hard or take active steps to protect ourselves. Recently I proposed the concept of fierce and tender self-compassion as a useful framework for understanding these two sides of self-compassion (Neff, 2021). Although this framework is used as a metaphor and has not been examined empirically, it is helpful for understanding the expression of self-compassion. Tender self-compassion involves "being with" ourselves in an accepting way. It entails soothing and comforting ourselves, reassuring ourselves that we aren't alone, and being present with and validating our pain. This is the healing power of self-compassion. When we feel hurt or inadequate, we're there for ourselves in a loving manner, acknowledging our pain and embracing ourselves as we are. This nurturing quality allows us to be less concerned with *what* is happening in our experience – whether it's painful, difficult, challenging, or disappointing – and to be more focused on how we are *relating* to it. We learn to be with ourselves in a new way. Rather than being lost in and engulfed by our pain, we're compassionate to ourselves *because* we're in pain. The care and concern we extend ourselves allows us to feel safe and accepted. When we open our hearts to what is, it generates a level of warmth that helps heal our wounds. For instance, self-compassion helps us recover from trauma (Scoglio et al., 2018), promotes self-acceptance (Zhang et al., 2020), and combats shame (Johnson & O'Brien, 2013).

The fierce quality of self-compassion is associated with "acting in the world" to alleviate suffering. It looks different depending on the action required but tends to involve protecting ourselves, providing for our needs, or motivating change. *Protecting* means saying "no" to others who are crossing our boundaries or standing up to injustice. Research shows that self-compassion helps people feel stronger, more empowered, and assertive (Stevenson & Allen, 2017). It enables people to be more comfortable standing up to bullies (Vigna et al., 2017) and confronting others who threaten harm (Allen et al., 2017). *Providing* means taking action to give ourselves what we genuinely need. Self-compassionate people are more likely to engage in self-care behaviors such as physical exercise (Homan & Sirois, 2017). They are more likely to fulfill basic psychological needs for autonomy, competence, and relatedness (Gunnell et al., 2017) and to engage in activities that they truly enjoy and find satisfying (Schellenberg et al., 2016). Self-compassionate people are also more *motivated* to make healthy changes in their lives (see Chaps. 12 and 18), taking actions that promote learning (Hope et al., 2014) and personal growth (Breines & Chen, 2012). Rather than promoting complacency, self-compassion provides the personal initiative needed to take charge and fulfill one's dreams (Dundas et al., 2017).

Like yin and yang (Palmer, 1997), we need access to both tenderness and fierceness for wholeness and well-being. As people aim to alleviate their suffering with compassion, sometimes tender acceptance is called for and at other times fierce action is required. In order for us to support ourselves in a healthy manner, these two ways of being must be balanced and integrated. If not, they are in danger of becoming what's known in Buddhist psychology as a "near enemy." This term refers to a state of mind that appears similar to the desired state – hence it is "near" – but actually undermines it, which is why it's an "enemy." For instance, when acceptance occurs without willingness to take action, it can turn into complacency. Although it's important to love and accept ourselves as we are in the moment, that doesn't mean we want to *stay* as we are in the moment. If our behavior is unhealthy, we don't want to only accept ourselves, but we also want to make a change. At the same time, when the force of protection arises without access to feelings of tender care, it can turn into hostility and aggression toward others. We may start to see a situation as us versus them, I'm right and you're wrong. Self-compassion is empowered, but it's not overpowering. Similarly, trying to meet our needs without sufficient self-acceptance can morph into selfishness, without attention being paid to the needs of others. And the desire to motivate change that is not in balance with acceptance of our human weakness can result in striving or perfectionism. When fierce and tender self-compassion are in balance, we take action to make things better – not because we're unacceptable as we are, but because we care about ourselves. The more secure we feel in this tender self-acceptance, the more energy becomes available to fiercely protect ourselves, fulfill our needs, and achieve our goals.

How Does Self-Compassion Relate to Mindfulness?

Because mindfulness is a core component of self-compassion in my model, it is worth considering how these constructs are similar and how they differ (Neff & Dahm, 2014). First, the type of mindfulness that is part of self-compassion is narrower in scope than mindfulness more generally. The mindfulness component of self-compassion refers specifically to awareness of *suffering*. Mindfulness in general, however, refers to the ability to pay attention to any experience – positive, negative, or neutral – with equanimity. Self-compassion as a total construct is also broader in scope than mindfulness because it includes the elements of self-kindness and common humanity: actively soothing and comforting oneself when painful experiences arise and remembering that such experiences are part of being human. These qualities are not inherently part of mindfulness more narrowly defined. We can be mindfully aware of painful thoughts and feelings without actively soothing and comforting ourselves or remembering that these feelings are part of the shared human experience.

Another distinction between mindfulness and self-compassion lies in their respective targets. Whereas mindfulness is a way of relating to experience, self-compassion is a way of relating to the *experiencer* who is suffering (Germer, 2009; Germer & Barnhofer, 2017). While it's possible to be mindful of eating a raisin, an exercise commonly used to teach mindfulness (Kabat-Zinn, 1990), it wouldn't make sense to give compassion to the raisin, because it doesn't experience pain. Mindfulness involves nonjudgmentally accepting the thoughts, emotions, and sensations that arise in present-moment awareness. Compassion involves the desire for sentient beings to be happy and free from suffering (Salzberg, 1995). These theoretical distinctions between self-compassion and mindfulness should be made lightly, however, because both are experiences that cannot be fully captured by language or logic. At some level, both refer to a state of open heart and mind and cannot be fully disentangled.

How Does Self-Compassion Relate to Self-Esteem?

When I first operationally defined self-compassion (Neff, 2003b), I contrasted it with self-esteem and proposed that it was a healthier alternative (also see Chap. 3). Self-esteem refers to how much one likes or values the self, based on congruence with personal standards or on comparisons with others (Harter, 1999). There is general consensus that self-esteem is essential for good mental health, while the lack of self-esteem undermines well-being, fostering depression, anxiety, and other pathologies (Leary, 1999). There are potential problems with high self-esteem, however, not in terms of having it, but in terms of getting and keeping it (Crocker & Park, 2004).

High self-esteem requires standing out in a crowd – being special and above average (Heine et al., 1999). This is a problem, because it's logically impossible for everyone to be above average at the same time. Self-esteem is often evaluated in comparison with those who are "better" or "worse" than we are. Not surprisingly, the attempt to maintain self-esteem has been associated with narcissism and feelings of superiority (Bushman & Baumeister, 1998), inflated and unrealistic self-views (Sedikides, 1993), prejudice (Aberson et al., 2000), and bullying behavior (Salmivalli et al., 1999). To the extent that the self is evaluated in distinction to others, one may feel the need to derogate others to feel better about oneself.

Moreover, as William James (1890) proposed over a century ago, self-esteem involves evaluating personal performances (how good am I?) in comparison with set standards (what counts as good enough?) in domains of perceived importance (it's important to be good at this). This means that feelings of state self-esteem can be unstable, bouncing up and down according to our latest success or failure (Crocker et al., 2003). Self-esteem is a fair-weather friend, there for us in good times, deserting us when our luck turns.

Self-compassion is different from self-esteem. Although they're both strongly linked to psychological well-being, self-esteem is a positive evaluation of self-worth, while self-compassion isn't a judgment or an evaluation at all. Instead, self-compassion is way of *relating* to the ever-changing landscape of who we are with kindness and acceptance – especially when we fail or feel inadequate. Self-compassion doesn't require feeling better than others, but it simply requires acknowledging the shared human condition of imperfection. This means that we don't have to feel better than others to feel good about ourselves. Self-compassion also offers more emotional stability than self-esteem because it is always there for you – when you're on top of the world and when you fall flat on your face. Self-compassion is a portable friend we can always rely on, in good times and bad.

In many ways, self-compassion can be seen as a healthy way to value oneself. My research suggests that self-compassion and self-esteem are strongly correlated (Neff & Vonk, 2009) but that once their overlap is accounted for, self-compassion is not linked to social comparison, narcissism, and contingent self-worth the way global self-esteem is and offers greater stability in self-worth over time.

Common Misgivings About Self-Compassion

There are many blocks to self-compassion in Western culture, often resulting from misconceptions about its meaning and consequences (Robinson et al., 2016). One common misconception is that self-compassion is *selfish*. Many people assume that spending time and energy being kind and caring toward themselves automatically means spending less time helping others. But research indicates that self-compassion leads to more caring relationship behavior and actually helps us sustain helping others without burning out (see Chaps. 14, 15, and 16). Another common misconception about self-compassion is that it means feeling sorry for yourself – that it's just a dressed-up form of *self-pity*. In fact, self-compassion is an *antidote* to self-pity and reduces the tendency to wallow in suffering. Self-compassion allows us to kindly acknowledge difficult feelings without becoming lost in them. It also reduces self-focus by framing suffering in the context of the shared human experience.

Some people fear that self-compassion will make them *weak* and that harsh self-judgment is needed to be tough. In this case, feelings of compassion are confused with "being nice" all the time. However, compassion can be fierce, taking a strong and resolute stand against anything that causes harm. It also leads to incredible strength and resilience in difficult circumstances (see Chap. 10). Another common misgiving about self-compassion is that it will lead to *self-indulgence*. Doesn't being kind to yourself mean giving yourself whatever you want? It must be remembered that self-compassion has its eye on the prize – the alleviation of suffering. Self-indulgence, on the other hand, involves giving oneself short-term pleasure at the cost of long-term harm. Research shows that self-compassion increases health promoting behaviors (see Chap. 18) rather than self-indulgence.

Perhaps the biggest block to self-compassion is the belief that it will undermine our motivation to improve. We think that self-criticism is necessary to reach our goals. In this case, there is confusion between *harsh self-judgment* and *constructive criticism*. Harsh self-judgment motivates through fear of inadequacy, whereas constructive criticism motivates through care and the desire to learn and grow. Because self-compassion makes it safe to fail, people are more able to learn from their failures and try again. For this reason, self-compassion is a more effective motivator than harsh self-criticism and provides grit and focus as we work toward our goals (see Chap. 12).

The Measurement of Self-Compassion

The majority of studies on self-compassion have been conducted using the Self-Compassion Scale (SCS; Neff, 2003a), which has been translated into at least 22 different languages (Neff & Tóth-Király, 2021). The SCS is a 26-item self-report measure that is designed to measure self-compassion as I have defined it (Neff, 2003b). It's a straightforward assessment of how often people engage in the various thoughts, emotions, and behaviors that align with the different dimensions of self-compassion. It measures self-compassion as a general construct but has six subscales which can be used to individually to examine the constituent components of self-compassion. Sample items are *self-kindness* ("I try to be loving toward myself when I'm feeling emotional pain."), *self-judgment* ("I'm disapproving and judgmental about my own flaws and inadequacies."), *common humanity* ("When things are going badly for me, I see the difficulties as part of life that everyone goes through."), *isolation* ("When I think about my inadequacies, it tends to make me feel more separate and cut off from the rest of the world."), *mindfulness* ("When I'm feeling down, I try to approach my feelings with curiosity and openness."), and *overidentification* ("When something upsets me, I get carried away with my feelings."). Self-judgment, isolation, and overidentification items are reverse coded so that higher scores indicate a relative lack of UCS.

There is a large body of research indicating that scores on the SCS are associated with well-being, constituting construct validity. For example, higher scores on the SCS have been linked to greater levels of happiness, optimism, life satisfaction, body appreciation, perceived competence, and motivation (Hollis-Walker & Colosimo, 2011; Neff et al., 2005, 2007a, b, 2008, 2018a, b) and lower levels of depression, anxiety, stress, rumination, self-criticism, perfectionism, body shame and fear of failure (Breines et al., 2014a, b; Finlay-Jones et al., 2015; Neff, 2003a; Neff et al., 2005, 2018a, b; Raes, 2010), and healthier physiological responses to stress (Breines et al., 2014a; Friis et al., 2016). There is also evidence for predictive validity. Longitudinal studies have found that self-compassion levels predict stress, depression, anxiety, suicidality, and coping over time (Stefan, 2019; Stutts & Blomquist, 2018; Stutts et al., 2018; Zeller et al., 2015; Zhu et al., 2019).

The SCS demonstrates good discriminant validity. First, it is not significantly associated with social desirability (Neff, 2003a). Self-compassion can also be empirically differentiated from self-esteem, and the SCS demonstrates incremental predictive validity with regard to the construct (Krieger et al., 2015; Neff & Vonk, 2009) including in longitudinal research (Marshall et al., 2015). In addition, self-compassion can be differentiated from self-criticism. Although a key feature of self-compassion is the lack of self-judgment, overall SCS scores still negatively predict anxiety and depression when controlling for self-criticism and negative affect (Neff, 2003a; Neff et al., 2007a, b). Neff et al. (2007a, b) found that the SCS predicted significant variance in positive well-being after controlling for all the Big Five personality traits. Moreover, Neff et al. (2018b) established incremental validity with neuroticism in three separate studies, and Stutts et al. (2018) found that self-compassion predicted depression, anxiety, and stress while controlling for neuroticism in a longitudinal study.

The SCS demonstrates known groups validity: undergraduate and community adults have significantly lower scores on the SCS than individuals who practice Buddhist meditation, as would be predicted given the Buddhist roots of the construct (Neff, 2003a; Neff & Pommier, 2013). Similarly, clinical populations have lower levels of self-compassion than nonclinical populations (e.g., Castilho et al., 2015; Werner et al., 2012), which is also to be expected given that a lack of self-compassion is seen as a transdiagnostic feature of clinical populations (Schanche, 2013). The scale demonstrates good convergent validity as well. For instance, therapists' ratings of how "self-compassionate" individuals were (using a single item) after a brief interaction significantly correlated with self-reported SCS scores (Neff et al., 2007a, b), and there was a strong association (.70) between self-reported and partner-reported scores on the SCS among couples in long-term romantic relationships (Neff & Beretvas, 2013). Similarly, high levels of agreement (.77) were found between independent coders using SCS items to rate the level of self-compassion displayed in brief verbal dialogs (Sbarra et al., 2012). These findings suggest that the SCS measures behaviors that are clearly observable by others.

Factor Structure of the SCS

Neff (2003a) originally used confirmatory factor analysis (CFA) to examine the factor structure of the SCS and found adequate fit for a higher-order model and a six-factor correlated model, justifying use of the SCS as a total score or else six subscale scores. Since then, several other validation studies have been carried out on the SCS (for an overview, see Neff et al., 2019). While the six-factor correlated model has generally been replicated, findings of a single higher-order factor have been inconsistent. Some studies have found support for a higher-order model (e.g., Benda & Reichová, 2016; Castilho et al., 2015; Dundas et al., 2016), but others have not (e.g., Brenner et al., 2017; Costa et al., 2015; López et al., 2015; Montero-Marín et al., 2016; Neff et al., 2017; Williams et al., 2014).

As mentioned earlier, certain researchers have argued that the SCS should not be measured with a total score representing a general construct of self-compassion but should instead be measured with two factors representing CS and UCS (e.g., Brenner et al., 2017; Costa et al., 2015; López et al., 2015; Muris & Petrocchi, 2017). Support for a two-factor solution to the SCS has also been inconsistent, however (e.g., Cleare et al., 2018; Neff et al., 2017). We have argued that a bifactor approach combined with Exploratory Structural Equation Modeling (ESEM) offers the most theoretically consistent way to examine the SCS, given that the six components are thought to operate as a dynamic system (Neff & Tóth-Király, 2021). Bifactor ESEM is increasingly popular for analyzing multidimensional constructs (Morin et al., 2016; Morin et al., 2020). The approach allows for the explicit expression of item cross-loadings as opposed to the overly strict confirmatory factor analysis (CFA) which doesn't allow for any cross-loadings. Given that the components of self-compassion mutually engender one another (Dreisoerner et al., 2020), some cross-loadings between factors should be expected. And like CFA, this approach can confirm the fit of a priori models. In a large international collaboration (Neff et al., 2019), we used bifactor ESEM to examine the factor structure of the SCS in 20 diverse samples ($N = 11{,}685$), and support was found in every sample for use of 6 subscale scores or a total score, but support was not found separate scores representing CS and UCS. Moreover, 95% of the reliable variance could be attributed to a general factor. This pattern of results has been replicated in other studies (Neff et al., 2018a, b; Tóth-Király, et al., 2017).

For researchers who are primarily interested in the benefits of self-compassion in terms of implications for intervention, use of a total SCS score is probably most appropriate given that the elements of self-compassion operate in tandem as a system. For those more interested in unpacking the mechanisms of *how* self-compassion enhances well-being, however, it may be useful to examine the six constituent components separately. However, one should use caution if entering the six subscales simultaneously in regression analyses to determine their differential association with outcomes. Given the deep intertwining of the various components in the definition, operation, and measurement of self-compassion, and given that almost all of the reliable variance in item responding on the SCS is explained by a general self-compassion factor, separating out the shared variance of the six subscales could change their meaning in a way that may undermine the interpretability of findings.

Other Measures of Self-Compassion

There are different formats of the SCS that are available to researchers. A 12-item short form of the SCS exists and is frequently used in research given its shorter length and strong correlation with the long form (SCS-SF; Raes et al., 2011). Recently, we have created a 17-item youth version of the SCS designed for middle school-aged populations (Neff et al., 2021a), which can be used to measure total levels of self-compassion or else the six subscales. Additionally, we have created a state Self-Compassion Scale (S-SCS; Neff et al., 2021b) designed to measure feelings of self-compassion in the present moment (e.g., "I'm being supportive toward myself"; "I'm taking a balanced view of this painful situation.") There is an 18-item-long version that can measure state levels of the 6 components of self-compassion, as well as a 6-item short version that measures global levels of state self-compassion only. Although not strictly another format of the SCS, we have also created a 16-item measure of compassion for others with a similar structure to the SCS (Pommier et al., 2020). Compassion was operationalized as experiencing kindness, a sense of common humanity, mindfulness, and lessened indifference toward the suffering of others.

It should be noted that we are in the midst of creating a revised version of the SCS (Neff & Toth-Kiraly, in preparation). Currently the SCS, at 26 items, is too long for many researchers who often need to create shorter surveys for their participants. Also, one potential problem with the SCS is that some items are more focused on general life suffering and others are focused on mistakes and failures, adding a source of variance to items that is independent from self-compassion itself. Items in the revised version mention both personal and general life suffering. The revised SCS will have an 18-item-long version that can measure each of the 6 subscales and a short 6-item version (with 1 item from each subscale) that can be used to measure the general construct of self-compassion. The data we have collected so far provides strong evidence for the psychometric validity of both scales.

Of course, other measures of self-compassion exist in the literature based on different conceptualizations of the construct. For instance, Gilbert and colleagues created the Forms of Self-Criticism and Self-Reassurance Scales (Gilbert et al., 2004) to measure these two ways of relating to oneself. These correspond loosely to the self-kindness and self-judgment subscales of the SCS and are typically used separately. More recently, Gilbert et al. (2017) developed the Compassion Engagement and Action Scales (CEAS), based on the broadly used definition of compassion as sensitivity to suffering with a commitment to try to alleviate it (Goetz et al., 2010). The CEAS includes a self-compassion subscale with items tapping into two elements: engagement with distress (e.g., tolerating and being sensitive to distress) and the motivation to alleviate that distress (e.g., thinking about and taking actions to help). Notably, the CEAS does not include warmth, kindness, concern, or feelings of shared humanity as a feature of compassion. Gu et al. (2020) have created a measure of self-compassion that assesses five elements: recognizing suffering, understanding the universality of suffering, feeling moved by suffering, tolerating uncomfortable feelings aroused in response to suffering, and the motivation to alleviate suffering. Garnefski and Kraaij (2019) have created a measure of self-compassionate coping that focuses mainly on self-kindness. Of all these measures, however, the SCS is most commonly used, and its psychometric validity has been examined most extensively.

Self-Compassion Across Groups

An important question concerns whether there are differences in the prevalence or functioning of self-compassion across groups. Recently, we examined the generalizability of the SCS via tests of measurement invariance across a wide range of populations, including student, community, and clinical samples, and also varying by gender, age, and culture (Tóth-Király & Neff, 2021). Our data set (n = 10,997) included a total of 18 samples from 15 countries (Australia, Brazil, Canada, France, Germany, Greece, Hungary, Iran, Italy, Korea, Norway, Portugal, Spain, United Kingdom, United States) in 12 languages. Overall, findings demonstrated a striking degree of invariance in the reliability and factor structure of the SCS across groups. Although the SCS functioned the same way in all groups, there were differences in mean levels of self-compassion displayed *between* groups.

In terms of gender, females reported slightly lower levels self-compassion than males. This result aligns well with a prior meta-analysis, including a total of 88 study samples, showing that males reported slightly higher levels of self-compassion than females (Yarnell et al., 2015). This gender difference may be because females tend to be more self-critical and judge themselves more negatively than males (Leadbeater et al., 1999). They also feel less entitled to meet their own needs (McGann et al., 2006). However, these gender differences appear to be less a function of sex and more a function of gender role socialization which emphasizes self-sacrifice for females and entitlement for males. Self-compassion involves meeting one's needs in order to alleviate suffering, so norms of self-sacrifice work against this process while norms of entitlement encourage it. Research demonstrates that sex differences are negligible once gender roles are taken into account (Yarnell et al., 2019).

Our results also indicated that people become more self-compassionate as they become older, consistent with past research (Homan, 2016; Neff

& Pommier, 2013; Neff & Vonk, 2009). In fact, the increasing self-acceptance experienced by the elderly (Ardelt, 1997, Erikson, 1968) is probably bidirectionally linked to increasing self-compassion with age (Neff et al., 2007a, b). The wisdom that comes from maturity and experience allows for a kinder and more balanced stance toward oneself that recognizes that shared nature of human suffering and the ability to relate to life difficulties and personal imperfection with mindfulness and compassion enhances life satisfaction and acceptance. The fact that self-compassion increases with age is good news for the elderly, given the many benefits that self-compassion provides in terms of mental and physical well-being (Allen et al., 2012; Perez-Blasco et al., 2016).

Our results also found cultural differences in self-compassion. Among undergraduates, we found that Korean students reported the highest level of self-compassion. This is somewhat surprising, given the influence of Confucianism in East Asian cultures which is thought to promote self-criticism as a means of achievement (Heine, 2003). Of course, caution should be used in interpreting results given that they may have been sample-specific, and a great deal of more research will be needed to determine if these findings replicate. For community adults, it was found that Spanish, Italian, Hungarian, Brazilian, and Australian participants had the highest level of self-compassion. We found that those from the United Kingdom, France, and Greece tended to have the lowest levels, with Americans and Germans in between. Of course, the fact that comparisons were confounded by age composition and gender renders findings somewhat difficult to interpret, and these variables will need to be disentangled in the future. A few things are noteworthy about the findings, however. First, English speaking countries differed in level of self-compassion, indicating that culture may be more important than language in relative impact on self-compassion. Also, there did not appear to be a general trend for more economically prosperous nations to have more self-compassion, suggesting that healthy self-attitudes are not merely a product of standard of living.

Other Methods of Researching Self-Compassion

The majority of research on self-compassion has used the SCS to examine trait levels of self-compassion and its relationship to other psychological traits. This approach, however, limits researchers' ability to make causal inferences. To address this limitation, there is an increasing trend toward experimental research that examines how *change* in self-compassion impacts outcomes (Ferrari et al., 2019). Some scholars have examined the efficacy of self-compassion interventions such as the mindful self-compassion training (Germer & Neff, 2019; Neff & Germer, 2013) or compassion-focused therapy (Gilbert, 2010; Sommers-Spijkerman et al., 2018) to determine how such interventions change well-being over time. Another promising methodology involves inducing a self-compassionate state of mind. One of the first studies to attempt to induce a self-compassionate mind state was conducted by Leary et al. (2007), who asked participants to recall a past event that made them feel badly about themselves and then guided them through a series of writing prompts designed to evoke self-compassion. The study found that compared to control conditions, those in the self-compassionate writing condition experienced a greater decrease in negative affect. Several researchers have used this induction in experimental studies of self-compassion (e.g., Blackie & Kocovski, 2018; Odou & Brinker, 2014). Other researchers have used variations on this writing task or different approaches such as guided meditation (e.g., Breines & Chen, 2012; Kirschner et al., 2019) to determine if changes in state self-compassion impact well-being.

As part of establishing the validity of the state SCS (Neff et al., 2021b), we created a self-compassion mind state induction (SCMI) that is based on a practice known as the Self-Compassion Break found in the MSC program (Germer & Neff, 2019). The induction first asks individuals to think of a situation they are struggling with and then to write mindfully about the feelings evoked by a difficulty, so they can accept and validate their painful feelings.

Examples of self-compassionate language are given such as "this is really hard right now." They are next instructed to write about their common humanity, recognizing that they are not alone in their struggle. Examples are given such as "everyone is imperfect, I'm not alone." They are then instructed to write to themselves with kindness, giving themselves the type of care, understanding, and support they would normally show to a good friend. Examples are given such as "I'm here for you." Finally, participants are invited to reflect on what they have written so that the message of self-compassion can be absorbed and integrated. There is a control condition that asks participants to first write about a difficult situation in a descriptive manner (parallel to mindfulness), then to indicate who was involved in the situation (parallel to common humanity), next to describe any words spoken in the situation (parallel to self-kindness), and finally to reflect on what they had written. The parallel nature of the control condition helps to ensure that participants in both conditions are focused on the difficult situation, with only participants in the self-compassion condition actively changing their responses to it. It is hoped that the availability of this SCMI and also the newly created S-SCS will help facilitate more experimental research on self-compassion in the future.

So far, research on self-compassion using experimental methods has yielded findings that converge with cross-sectional studies using the SCS such as increased happiness, life satisfaction, body appreciation, motivation, and less depression, anxiety, stress, and shame (e.g., Albertson et al., 2015; Breines & Chen, 2012; Diedrich et al., 2014; Johnson & O'Brien, 2013; Leary et al., 2007; Neff & Germer, 2013; Odou & Brinker, 2014; Shapira & Mongrain, 2010; Smeets et al., 2014). Taken as a whole, therefore, findings obtained in the field of self-compassion research appear to be robust.

Conclusion

The field of self-compassion research is growing exponentially, as the chapters in this handbook will attest. Even more important than self-compassion theory, measurement, or research, however, is practice. Self-compassion isn't just a good idea; it's something you *do*. Self-compassion transforms lives. It's a learnable skill that radically improves our ability to cope with painful experiences without becoming overwhelmed. Although it's directly aimed at suffering, the feelings of warmth, connection, and presence entailed by self-compassion provide satisfaction and meaning.

In 2010 my close colleague Chris Germer and I created an eight-week training program to teach self-compassion called Mindful Self-Compassion (MSC; Germer & Neff, 2019; Neff & Germer, 2018). A few years later we founded a nonprofit called the Center for Mindful Self-Compassion (CMSC) to train teachers, and since then MSC has been taught to hundreds of thousands of individuals around the globe, meaning that self-compassion training is scalable. This has far-reaching implications for countries and regions where mental health professionals are in short supply. Although the MSC program isn't therapy, it is highly therapeutic and is an important mental health resource for those needing effective ways to cope with their distress. In addition to the MSC program, CMSC has been working to develop self-compassion training programs for young adults, teens, children, parents, couples, healthcare providers, psychotherapists, businesspeople, and athletes. As researchers continue to examine the benefits of self-compassion in more realms of life, and as more people begin to experience its benefits first-hand, we are likely to see impacts in the larger culture. Instead of being distrustful of self-compassion, hopefully it will be widely recognized as an essential skill for living a healthy life.

References

Aberson, C. L., Healy, M., & Romero, V. (2000). Ingroup bias and self-esteem: A meta-analysis. *Personality and Social Psychology Review, 4*(2), 157–173. https://doi.org/10.1207/S15327957PSPR0402_04

Albertson, E. R., Neff, K. D., & Dill-Shackleford, K. E. (2015). Self-compassion and body dissatisfaction in women: A randomized controlled trial of a brief meditation intervention. *Mindfulness, 6*(3), 444–454. https://doi.org/10.1007/s12671-014-0277-3

Allen, A. B., Goldwasser, E. R., & Leary, M. R. (2012). Self-compassion and well-being among older adults. *Self and Identity, 11*, 428–453. https://doi.org/10.1080/15298868.2011.595082

Allen, A. B., Robertson, E., & Patin, G. A. (2017). Improving emotional and cognitive outcomes for domestic violence survivors: The impact of shelter stay and self-compassion support groups. *Journal of Interpersonal Violence, 36*(1-2), NP598-NP624. https://doi.org/10.1177/0886260517734858

Ardelt, M. (1997). Wisdom and life satisfaction in old age. *The Journals of Gerontology Series B: Psychological Sciences and Social Sciences, 52B*(1), P15–P27. https://doi.org/10.1093/geronb/52B.1.P15

Beaumont, E., Irons, C., Rayner, G., & Dagnall, N. (2016). Does compassion-focused therapy training for health care educators and providers increase self-compassion and reduce self-persecution and self-criticism? *Journal of Continuing Education in the Health Professions, 36*(1), 4–10. https://doi.org/10.1097/ceh.0000000000000023

Benda, J., & Reichová, A. (2016). [Psychometric characteristics of the Czech version of the Self-Compassion Scale] Psychometrické charakteristiky české verze Self-Compassion Scale (SCS-CZ). *Československá Psychologie., 60*(2), 20–36.

Birnie, K., Speca, M., & Carlson, L. E. (2010). Exploring self-compassion and empathy in the context of mindfulness-based stress reduction (MBSR). *Stress and Health, 26*(5), 359–371. https://doi.org/10.1002/smi.1305

Blackie, R. A., & Kocovski, N. L. (2018). Examining the relationships among self-compassion, social anxiety, and post-event processing. *Psychological Reports, 121*(4), 669–689. https://doi.org/10.1177/0033294117740138

Blum, L. (1980). Compassion. In A. O. Rorty (Ed.), *Explaining emotions* (pp. 507–517). University of California Press.

Breines, J. G., & Chen, S. (2012). Self-compassion increases self-improvement motivation. *Personality and Social Psychology Bulletin, 38*(9), 1133–1143. https://doi.org/10.1177/0146167212445599

Breines, J. G., Thoma, M. V., Freeman, T., Wu, Z., & Rohleder, N. (2014a). Self-compassion as a predictor of cortisol responses to repeated acute psychosocial stress. *Psychological Bulletin, 130*(3), 355–391. https://doi.org/10.1016/j.bbi.2013.11.006

Breines, J., Toole, A., Tu, C., & Chen, S. (2014b). Self-compassion, body image, and self-reported disordered eating. *Self and Identity, 13*(4), 432–448. https://doi.org/10.1080/15298868.2013.838992

Brenner, R. E., Heath, P. J., Vogel, D. L., & Credé, M. (2017). Two is more valid than one: Examining the factor structure of the Self-Compassion Scale (SCS). *Journal of Counseling Psychology, 64*(6), 696–707. https://doi.org/10.1037/cou0000211

Brenner, R. E., Vogel, D. L., Lannin, D. G., Engel, K. E., Seidman, A. J., & Heath, P. J. (2018). Do self-compassion and self-coldness distinctly relate to distress and well-being? A theoretical model of self-relating. *Journal of Counseling Psychology, 65*(3), 346. https://doi.org/10.1037/cou0000257

Bushman, B. J., & Baumeister, R. F. (1998). Threatened egotism, narcissism, self-esteem, and direct and displaced aggression: Does self-love or self-hate lead to violence? *Journal of Personality and Social Psychology, 75*(1), 219. https://doi.org/10.1037/0022-3514.75.1.219

Cassell, E. J. (2002). Compassion. In C. R. Snyder & S. J. Lopez (Eds.), *Handbook of positive psychology* (pp. 434–445). Oxford University Press.

Castilho, P., Pinto-Gouveia, J., & Duarte, J. (2015). Evaluating the multifactor structure of the long and short versions of the self- compassion scale in a clinical sample. *Journal of Clinical Psychology, 71*, 856–870. https://doi.org/10.1002/jclp.22187

Cleare, S., Gumley, A., Cleare, C. J., & O'Conner, J. C. (2018). An investigation of the factor structure of the self-compassion scale. *Mindfulness, 9*(2), 618–628. https://doi.org/10.1007/s12671-017-0803-1

Costa, J., Marôco, J., Pinto-Gouveia, J., Ferreira, C., & Castilho, P. (2015). Validation of the psychometric properties of the self-compassion scale. Testing the factorial validity and factorial invariance of the measure among borderline personality disorder, anxiety disorder, eating disorder and general populations. *Clinical Psychology & Psychotherapy, 23*, 460–468. https://doi.org/10.1002/cpp.1974

Creswell, J. D. (2015). Biological pathways linking mindfulness with health. In K. W. Brown, J. D. Creswell, & R. M. Ryan (Eds.), *Handbook of mindfulness: Theory, research, and practice* (pp. 426–440). Guilford Press.

Crocker, J., & Park, L. E. (2004). The costly pursuit of self-esteem. *Psychological Bulletin, 130*, 392–414. https://doi.org/10.1037/0033-2909.130.3.392

Crocker, J., Luhtanen, R. K., Cooper, M. L., & Bouvrette, S. (2003). Contingencies of self-worth in college students: Theory and measurement. *Journal of Personality and Social Psychology, 85*, 894–908. https://doi.org/10.1037/0022-3514.85.5.894

Diedrich, A., Grant, M., Hofmann, S. G., Hiller, W., & Berking, M. (2014). Self-compassion as an emotion regulation strategy in major depressive disorder. *Behaviour Research and Therapy, 58*, 43–51. https://doi.org/10.1016/j.brat.2014.05.006

Dreisoerner, A., Junker, N. M., & van Dick, R. (2020). The relationship among the components of self-compassion: A pilot study using a compassionate writing intervention to enhance self-kindness, common humanity, and mindfulness. *Journal of Happiness Studies, 22*(1), 21–47.

Dreisoerner, A., Junker, N. M., & Van Dick, R. (2021). The relationship among the components of self-compassion: A pilot study using a compassionate writing intervention to enhance self-kindness, common humanity, and mindfulness. *Journal of Happiness Studies, 22*(1), 21–47. https://doi.org/10.1007/s10902-019-00217-4

Dundas, I., Svendsen, J. L., Wiker, A. S., Granli, K. V., & Schanche, E. (2016). Self-compassion and depressive symptoms in a Norwegian student sample. *Nordic Psychology, 68*, 58–72. https://doi.org/10.1080/19012276.2015.1071203

Dundas, I., Binder, P. E., Hansen, T. G., & Stige, S. H. (2017). Does a short self-compassion intervention for students increase healthy self-regulation?: A randomized control trial. *Scandinavian Journal of Psychology, 58*(5), 443–450. https://doi.org/10.1080/19012276.2015.1071203

Erikson, E. H. (1968). *Identity: Youth and Crisis*. Norton.

Ferrari, M., Hunt, C., Harrysunker, A., Abbott, M. J., Beath, A. P., & Einstein, D. A. (2019). Self-compassion interventions and psychosocial outcomes: A meta-analysis of RCTs. *Mindfulness, 10*(8), 1455–1473. https://doi.org/10.1007/s12671-019-01134-6

Finlay-Jones, A. L., Rees, C. S., & Kane, R. T. (2015). Self-compassion, emotion regulation and stress among Australian psychologists: Testing an emotion regulation model of self-compassion using structural equation modeling. *PLoS One, 10*(7). https://doi.org/10.1371/journal.pone.0133481

Finlay-Jones, A., Kane, R., & Rees, C. (2017). Self-compassion online: A pilot study of an Internet-based self-compassion cultivation program for psychology trainees. *Journal of Clinical Psychology, 73*(7), 797–816. https://doi.org/10.1002/jclp.22375

Finlay-Jones, A., Xie, Q., Huang, X., Ma, X., & Guo, X. (2018). A pilot study of the 8-week mindful self-compassion training program in a Chinese community sample. *Mindfulness, 9*(3), 993–1002. https://doi.org/10.1007/s12671-017-0838-3

Friis, A. M., Johnson, M. H., Cutfield, R. G., & Consedine, N. S. (2016). Kindness matters: A randomized controlled trial of a mindful self-compassion intervention improves depression, distress, and HbA1c among patients with diabetes. *Diabetes Care, 39*, 1963–1971. https://doi.org/10.2337/dc16-0416

Garnefski, N., & Kraaij, V. (2019). The self-compassionate coping measure (4 items): Psychometric features and relationships with depression and anxiety in adults. *Advances in Health and Behavior, 2*(2), 75–78. https://doi.org/10.25082/AHB.2019.02.001

Germer, C. (2009). *The mindful path to self-compassion*. Guilford Press.

Germer, C., & Barnhofer, T. (2017). Mindfulness and compassion: Similarities and differences. In P. Gilbert (Ed.), *Compassion: Concepts, research and applications*. Routledge.

Germer, C. K., & Neff, K. D. (2019). *Teaching the mindful self-compassion program: A guide for professionals*. Guilford Press.

Gilbert, P. (2000). Social mentalities: Internal 'social' conflicts and the role of inner warmth and compassion in cognitive therapy. In P. Gilbert & K. G. Bailey (Eds.), *Genes on the Couch: Explorations in evolutionary psychotherapy* (pp. 118–150). Psychology Press.

Gilbert, P. (2010). *Compassion focused therapy: Distinctive features*. Routledge.

Gilbert, P., Clarke, M., Hempel, S., Miles, J. N., & Irons, C. (2004). Criticizing and reassuring oneself: An exploration of forms, styles and reasons in female students. *British Journal of Clinical Psychology, 43*(1), 31–50. https://doi.org/10.1348/014466504772812959

Gilbert, P., Catarino, F., Duarte, C., Matos, M., Kolts, R., Stubbs, J., Ceresatto, L., Duarte, J., Pinto-Gouveia, J., & Basran, J. (2017). The development of compassionate engagement and action scales for self and others. *Journal of Compassionate Health Care, 4*(1), 1–24. https://doi.org/10.1186/s40639-017-0033-3

Goetz, J. L., Keltner, D., & Simon-Thomas, E. (2010). Compassion: An evolutionary analysis and empirical review. *Psychological Bulletin, 136*(3), 351–374. https://doi.org/10.1037/a0018807

Greeson, J. M., Juberg, M. K., Maytan, M., James, K., & Rogers, H. (2014). A randomized controlled trial of Koru: A mindfulness program for college students and other emerging adults. *Journal of American College Health, 62*(4), 222–233. https://doi.org/10.1016/j.paid.2016.11.032

Gu, J., Baer, R., Cavanagh, K., Kuyken, W., & Strauss, C. (2020). Development and psychometric properties of the Sussex-Oxford compassion scales (SOCS). *Assessment, 27*(1), 3–20. https://doi.org/10.1177/1073191119860911

Gunnell, K. E., Mosewich, A. D., McEwen, C. E., Eklund, R. C., & Crocker, P. R. (2017). Don't be so hard on yourself! Changes in self-compassion during the first year of university are associated with changes in well-being. *Personality and Individual Differences, 107*, 43–48. https://doi.org/10.1016/j.paid.2016.11.032

Harter, S. (1999). *The construction of the self: A developmental perspective*. Guilford Press.

Heine, S. J. (2003). An exploration of cultural variation in self-enhancing and self-improving motivations. In V. Murphy-Berman & J. J. Berman (Eds.), *Cross-cultural differences in perspectives on the self* (pp. 118–145). University of Nebraska Press.

Heine, S. J., Lehman, D. R., Markus, H. R., & Kitayama, S. (1999). Is there a universal need for positive self-regard? *Psychological Review, 106*, 766–794. https://doi.org/10.1037/0033-295X.106.4.766

Hollis-Walker, L., & Colosimo, K. (2011). Mindfulness, self-compassion, and happiness in non-meditators: A theoretical and empirical examination. *Personality and Individual Differences, 50*, 222–227. https://doi.org/10.1016/j.paid.2010.09.033

Homan, K. J. (2016). Self-compassion and psychological well-being in older adults. *Journal of Adult Development, 23*, 111–119. https://doi.org/10.1007/s10804-016-9227-8

Homan, K. J., & Sirois, F. M. (2017). Self-compassion and physical health: Exploring the roles of perceived stress and health-promoting behaviors. *Health Psychology Open, 4*(2), 1–9. https://doi.org/10.1177/2055102917729542

Hope, N., Koestner, R., & Milyavskaya, M. (2014). The role of self-compassion in goal pursuit and well-being among university freshmen. *Self and Identity, 13*(5), 579–593. https://doi.org/10.1080/15298868.2014.889032

James, W. (1890). *Principles of psychology*. Encyclopedia Britannica.

Johnson, E. A., & O'Brien, K. A. (2013). Self-compassion soothes the savage ego-threat system: Effects on negative affect, shame, rumination, and depressive symptoms. *Journal of Social and Clinical Psychology, 32*(9), 939–963. https://doi.org/10.1521/jscp.2013.32.9.939

Kabat-Zinn, J. (1990). *Full catastrophe living: Using the wisdom of your body and mind to face stress, pain, and illness*. Dell.

Kabat-Zinn, J. (1994). *Wherever you go, there you are: Mindfulness meditation in everyday life*. New York: Hyperion.

Kelly, A. C., & Carter, J. C. (2015). Self-compassion training for binge eating disorder: A pilot randomized controlled trial. *Psychology and Psychotherapy, 88*, 285–303. https://doi.org/10.1111/papt.12044

Kelly, A. C., Wisniewski, L., Martin-Wagar, C., & Hoffman, E. (2017). Group-based compassion-focused therapy as an adjunct to outpatient treatment for eating disorders: A pilot randomized controlled trial. *Clinical Psychology and Psychotherapy, 24*(2), 475–487. https://doi.org/10.1002/cpp.2018

Kim, J. J., Parker, S. L., Doty, J. R., Cunnington, R., Gilbert, P., & Kirby, J. N. (2020). Neurophysiological and behavioural markers of compassion. *Scientific Reports, 10*, 6789. https://doi.org/10.1038/s41598-020-63846-3

Kirschner, H., Kuyken, W., Wright, K., Roberts, H., Brejcha, C., & Karl, A. (2019). Soothing your heart and feeling connected: A new experimental paradigm to study the benefits of self-compassion. *Clinical Psychological Science, 7*(3), 545–565. https://doi.org/10.1177/2167702618812438

Krieger, T., Hermann, H., Zimmermann, J., & Grosse Holtforth, M. (2015). Associations of self-compassion and global self-esteem with positive and negative affect and stress reactivity in daily life: Findings from a smart phone study. *Personality and Individual Differences, 87*, 288–292. https://doi.org/10.1016/j.paid.2015.08.009

Krieger, T., Martig, D. S., van den Brink, E., & Berger, T. (2016). Working on self-compassion online: A proof of concept and feasibility study. *Internet Interventions, 6*, 64–70. https://doi.org/10.1016/j.invent.2016.10.001

Leadbeater, B. J., Kuperminc, G. P., Blatt, S. J., & Hertzog, C. (1999). A multivariate model of gender differences in adolescents' internalizing and externalizing problems. *Developmental Psychology, 35*(5), 1268. https://doi.org/10.1037/0012-1649.35.5.1268

Leary, M. R. (1999). Making sense of self-esteem. *Current Directions in Psychological Science, 8*, 32–35. https://doi.org/10.1111/1467-8721.00008

Leary, M. R., Tate, E. B., Adams, C. E., Allen, A. B., & Hancock, J. (2007). Self-compassion and reactions to unpleasant self-relevant events: The implications of treating oneself kindly. *Journal of Personality and Social Psychology, 92*, 887–904. https://doi.org/10.1037/0022-3514.92.5.887

López, A., Sanderman, R., Smink, A., Zhang, Y., van Sonderen, E., Ranchor, A., & Schroevers, M. J. (2015). A reconsideration of the self-compassion scale's total score: Self-compassion versus self-criticism. *PloS One, 10*(7). https://doi.org/10.1371/journal.pone.0132940

Mantzios, M., Koneva, A., & Egan, H. (2020). When 'negativity' becomes obstructive: a novel exploration of the two-factor model of the self-compassion scale and a comparison of self-compassion and self-criticism interventions. *Current Issues in Personality Psychology, 8*(4), 289–300. https://doi.org/10.5114/cipp.2020.100791

Marshall, S. L., Parker, P. D., Ciarrochi, J., Sahdra, B., Jackson, C. J., & Heaven, P. C. (2015). Self-compassion protects against the negative effects of low self-esteem: A longitudinal study in a large adolescent sample. *Personality and Individual Differences, 74*, 116–121. https://doi.org/10.5114/cipp.2020.100791

McGann, V. L., Steil, J. M., Worell, J., & Goodheart, C. D. (2006). The sense of entitlement: Implications for gender equality and psychological well-being. In J. Worell & C. D. Goodheart (Eds.), *Handbook of girls' and women's psychological health: Gender and well-being across the lifespan* (pp. 175–182). Oxford University Press.

Montero-Marín, J., Gaete, J., Demarzo, M., Rodero, B., Lopez, L. C. S., & García-Campayo, J. (2016). Self-criticism: A measure of uncompassionate behaviors toward the self, based on the negative components of the self-compassion scale. *Frontiers in Psychology, 7*, 1281. https://doi.org/10.3389/fpsyg.2016.01281

Morin, A. J. S., Arens, A. K., & Marsh, H. W. (2016). A bifactor exploratory structural equation modeling framework for the identification of distinct sources of construct-relevant psychometric multidimensionality. *Structural Equation Modeling: A Multidisciplinary Journal, 23*(1), 116–139. https://doi.org/10.1080/10705511.2014.961800

Morin, A. J., Myers, N. D., & Lee, S. (2020). Modern factor analytic techniques: Bifactor models, exploratory structural equation modeling (ESEM), and bifactor-ESEM. In G. Tenenbaum & R. C. Eklund (Eds.), *Handbook of Sport Psychology* (pp. 1044–1073). Wiley. https://doi.org/10.1002/9781119568124.ch51

Muris, P., & Petrocchi, N. (2017). Protection or vulnerability? A meta-analysis of the relations between the positive and negative components of self-compassion and psychopathology. *Clinical Psychology & Psychotherapy, 24*(2), 373–383. https://doi.org/10.1002/cpp.2005

Muris, P., Otgaar, H., & Pfattheicher, S. (2019). Stripping the forest from the rotten trees: Compassionate self-responding is a way of coping, but reduced uncompassionate self-responding mainly reflects psychopathology. *Mindfulness, 10*(1), 196–199. https://doi.org/10.1007/s12671-018-1030-0

Neff, K. D. (2003a). Development and validation of a scale to measure self-compassion. *Self and Identity, 2*, 223–250. https://doi.org/10.1080/15298860309027

Neff, K. D. (2003b). Self-compassion: An alternative conceptualization of a healthy attitude toward oneself. *Self and Identity, 2*, 85–102. https://doi.org/10.1080/15298860309032

Neff, K. D. (2016). The self-compassion scale is a valid and theoretically coherent measure of self-compassion. *Mindfulness, 7*(1), 264–274. https://doi.org/10.1007/s12671-015-0479-3

Neff, K. D. (2021). *Fierce self-compassion: How women can harness kindness to speak up, claim their power, and thrive*. 374 pages. Harper Wave.

Neff, K. D., & Beretvas, S. N. (2013). The role of self-compassion in romantic relationships. *Self and Identity, 12*(1), 78–98. https://doi.org/10.1080/15298868.2011.639548

Neff, K. D., & Dahm, K. (2014). Self-compassion: What it is, what it does, and how it relates to mindfulness. In M. Robinson, B. Meier, & B. Ostafin (Eds.), *Mindfulness and self-regulation* (pp. 121–140). Springer.

Neff, K. D., & Germer, C. (2013). A pilot study and randomized controlled trial of the mindful self-compassion program. *Journal of Clinical Psychology, 69*(1), 28–44. https://doi.org/10.1002/jclp.21923

Neff, K. D., & Germer, C. (2018). *The mindful self-compassion workbook*. Guilford Press.

Neff, K. D., & Pommier, E. (2013). The relationship between self-compassion and other-focused concern among college undergraduates, community adults, and practicing meditators. *Self and Identity, 12*(2), 160–176. https://doi.org/10.1080/15298868.2011.649546

Neff, K. D., & Tóth-Király, I. (2021). The self-compassion scale. In O. N. Medvedev, C. U. Krägeloh, R. J. Siegert, & N. N. Singh (Eds.), *Handbook of assessment in mindfulness*. Springer.

Neff, K. D., & Vonk, R. (2009). Self-compassion versus global self-esteem: Two different ways of relating to oneself. *Journal of Personality, 77*, 23–50. https://doi.org/10.1111/j.1467-6494.2008.00537.x

Neff, K. D., Hsieh, Y., & Dejitterat, K. (2005). Self-compassion, achievement goals, and coping with academic failure. *Self and Identity, 4*, 263–287. https://doi.org/10.1080/13576500444000317

Neff, K. D., Kirkpatrick, K., & Rude, S. S. (2007a). Self-compassion and adaptive psychological functioning. *Journal of Research in Personality, 41*, 139–154. https://doi.org/10.1016/j.jrp.2006.03.004

Neff, K. D., Rude, S. S., & Kirkpatrick, K. (2007b). An examination of self-compassion in relation to positive psychological functioning and personality traits. *Journal of Research in Personality, 41*, 908–916. https://doi.org/10.1016/j.jrp.2006.08.002

Neff, K. D., Pisitsungkagarn, K., & Hseih, Y. (2008). Self-compassion and self-construal in the United States, Thailand, and Taiwan. *Journal of Cross-Cultural Psychology, 39*(3), 267–285. https://doi.org/10.1177/0022022108314544

Neff, K. D., Whittaker, T., & Karl, A. (2017). Evaluating the factor structure of the Self-Compassion Scale in four distinct populations: Is the use of a total self-compassion score justified? *Journal of Personality Assessment, 99*, 596–607. https://doi.org/10.1080/00223891.2016.1269334

Neff, K. D., Long, P., Knox, M., Davidson, O., Kuchar, A., Costigan, A., et al. (2018a). The forest and the trees: Examining the association of self-compassion and its positive and negative components with psychological functioning. *Self and Identity, 17*(6), 627–645. https://doi.org/10.1080/15298868.2018.1436587

Neff, K. D., Tóth-Király, I., & Colosimo, K. (2018b). Self-compassion is best measured as a global construct and is overlapping with but distinct from neuroticism: A response to Pfattheicher, Geiger, Hartung, Weiss, and Schindler (2017). *European Journal of Personality, 32*(4), 371–392. https://doi.org/10.1002/per.2148

Neff, K. D., Tóth-Király, I., Yarnell, L., Arimitsu, K., Castilho, P., Ghorbani, N., Guo, H. X., Hirsch, J., Hupfeld, J., Hutz, C., Kotsou, I., Lee, W. K., Montero-Marin, J., Sirois, F., de Souza, L., Svendsen, J., Wilkinson, R., & Mantios, M. (2019). Examining the factor structure of the self-compassion scale using exploratory SEM bifactor analysis in 20 diverse samples: Support for use of a total score and six subscale scores. *Psychological Assessment, 31(1), 27–45.* https://doi.org/10.1037/pas0000629

Neff, K. D., Bluth, K., Tóth-Király, I., Davidson, O., Knox, M. C., Williamson, Z., & Costigan, A. (2021a). Development and validation of the self-compassion scale for Youth. *Journal of Personality Assessment, 103*(1), 92–105. https://doi.org/10.1080/00223891.2020.1729774

Neff, K. D., Tóth-Király, I., Knox, M. C., Kuchar, A., & Davidson, O. (2021b). The development and validation of the state self-compassion scale (long-and short form). *Mindfulness, 12*(1), 121–140. https://doi.org/10.1007/s12671-020-01505-4

Nolen-Hoeksema, S. (1991). Responses to depression and their effects on the duration of depressive episodes. *Journal of Abnormal Psychology, 100*, 569–582. https://doi.org/10.1037/0021-843X.100.4.569

Odou, N., & Brinker, J. (2014). Exploring the relationship between rumination, self-compassion, and mood. *Self and Identity, 13*(4), 449–459. https://doi.org/10.1080/15298868.2013.840332

Palmer, M. (1997). *Book of Yin and Yang: Understanding the Chinese philosophy of opposites*. Piatkus Books.

Perez-Blasco, J., Sales, A., Meléndez, J. C., & Mayordomo, T. (2016). The effects of mindfulness and self-compassion on improving the capacity to adapt to stress situations in elderly people living in the community. *Clinical Gerontologist, 39*, 90–103. https://doi.org/10.1080/07317115.2015.1120253

Pfattheicher, S., Geiger, M., Hartung, J., Weiss, S., & Schindler, S. (2017). Old wine in new bottles? The case of self-compassion and neuroticism. *European Journal of Personality, 31*(2), 160–169.

Phillips, W. J. (2019). Self-compassion mindsets: The components of the self-compassion scale operate as a balanced system within individuals. *Current Psychology, 40*(10), 5040–5053. https://doi.org/10.1007/s12144-019-00452-1

Pommier, E., Neff, K. D., & Tóth-Király, I. (2020). The development and validation of the compassion scale. *Assessment, 127*(1), 21–39. https://doi.org/10.1177/1073191119874108

Porges, S. W. (2007). The polyvagal perspective. *Biological Psychology, 74*, 116–143. https://doi.org/10.1016/j.biopsycho.2006.06.009

Raab, K., Sogge, K., Parker, N., & Flament, M. F. (2015). Mindfulness-based stress reduction and self-compassion among mental healthcare professionals: A pilot study. *Mental Health, Religion and Culture, 18*(6), 503–512. https://doi.org/10.1080/13674676.2015.1081588

Raes, F. (2010). Rumination and worry as mediators of the relationship between self-compassion and depression and anxiety. *Personality and Individual Differences, 48*(5), 757–761. https://doi.org/10.1016/j.paid.2010.01.023

Raes, F., Pommier, E., Neff, K. D., & Van Gucht, D. (2011). Construction and factorial validation of a short form of the self-compassion scale. *Clinical Psychology & Psychotherapy, 18*(3), 250–255. https://doi.org/10.1002/cpp.702

Robinson, K. J., Mayer, S., Allen, A. B., Terry, M., Chilton, A., & Leary, M. R. (2016). Resisting self-compassion: Why are some people opposed to being kind to themselves? *Self and Identity, 15*(5), 505–524. https://doi.org/10.1080/15298868.2016.1160952

Salmivalli, C., Kaukiainen, A., Kaistaniemi, L., & Lagerspetz, K. M. (1999). Self-evaluated self-esteem, peer-evaluated self-esteem, and defensive egotism as predictors of adolescents' participation in bullying situations. *Personality and Social Psychology Bulletin, 25*(10), 1268–1278. https://doi.org/10.1177/0146167299258008

Salzberg, S. (1995). *Lovingkindness: The revolutionary art of happiness*. Shambala.

Sbarra, D. A., Smith, H. L., & Mehl, M. R. (2012). When leaving your ex, love yourself: Observational ratings of self-compassion predict the course of emotional recovery following marital separation. *Psychological Science, 23*, 261–269. https://doi.org/10.1177/0956797611429466

Schanche, E. (2013). The transdiagnostic phenomenon of self-criticism. *Psychotherapy, 50*(3), 316.

Schellenberg, B. J., Bailis, D. S., & Mosewich, A. D. (2016). You have passion, but do you have self-compassion? Harmonious passion, obsessive passion, and responses to passion- related failure. *Personality and Individual Differences, 99*, 278–285. https://doi.org/10.1016/j.paid.2016.05.003

Scoglio, A. A., Rudat, D. A., Garvert, D., Jarmolowski, M., Jackson, C., & Herman, J. L. (2018). Self-compassion and responses to trauma: The role of emotion regulation. *Journal of Interpersonal Violence, 33*(13), 2016–2036. https://doi.org/10.1177/0886260515622296

Sedikides, C. (1993). Assessment, enhancement, and verification determinants of the self-evaluation process. *Journal of Personality and Social Psychology, 65*(2), 317. https://doi.org/10.1037/0022-3514.65.2.317

Shapira, L., & Mongrain, L. (2010). The benefits of self-compassion and optimism exercises for individuals vulnerable to depression. *Journal of Positive Psychology, 5*(5), 377–389. https://doi.org/10.1080/17439760.2010.516763

Smeets, E., Neff, K., Alberts, H., & Peters, M. (2014). Meeting suffering with kindness: Effects of a brief self-compassion intervention for female college students. *Journal of Clinical Psychology, 70*(9), 794–807. https://doi.org/10.1002/jclp.22076

Sommers-Spijkerman, M. P. J., Trompetter, H. R., Schreurs, K. M., & Bohlmeijer, E. T. (2018). Compassion-focused therapy as guided self-help for enhancing public mental health: A randomized controlled trial. *Journal of Consulting and Clinical psychology, 86*(2), 101. https://doi.org/10.1037/ccp0000268

Ştefan, C. A. (2019). Self-compassion as mediator between coping and social anxiety in late adolescence: A longitudinal analysis. *Journal of Adolescence, 76*, 120–128. https://doi.org/10.1016/j.adolescence.2019.08.013

Stevenson, O., & Allen, A. B. (2017). Women's empowerment: Finding strength in self-compassion. *Women & Health, 57*(3), 295–310. https://doi.org/10.1080/03630242.2016.1164271

Stutts, L. A., & Blomquist, K. K. (2018). The moderating role of self-compassion on weight and shape concerns and eating pathology: A longitudinal study, *International Journal of Eating Disorders, 51*(8), 879–889. https://doi.org/10.1002/eat.22880

Stutts, L. A., Leary, M. R., Zeveney, A. S., & Hufnagle, A. S. (2018). A longitudinal analysis of the relationship between self-compassion and the psychological effects of perceived stress. *Self and Identity, 17*(6), 609–626.

Tirch, D., Schoendorff, B., & Silberstein, L. R. (2014). *The ACT practitioner's guide to the science of compassion: Tools for fostering psychological flexibility*. New Harbinger.

Toole, A. M., & Craighead, L. W. (2016). Brief self-compassion meditation training for body image distress in young adult women. *Body Image, 19*, 104–112. https://doi.org/10.1016/j.bodyim.2016.09.001

Tóth-Király, I., & Neff, K. D. (2021). Is self-compassion universal? Support for the measurement invariance of the self-compassion Scale across populations. *Assessment, 28*(1), 169–185. https://doi.org/10.1177/1073191120926232

Tóth-Király, I., Bőthe, B., & Orosz, G. (2017). Exploratory structural equation modeling analysis of the self-compassion scale. *Mindfulness, 8*(4), 881–892. https://doi.org/10.1007/s12671-016-0662-1

Ullrich-French, S., & Cox, A. E. (2020). The use of latent profiles to explore the multi-dimensionality of self-compassion. *Mindfulness, 11*, 1483–1499. https://doi.org/10.1007/s12671-020-01365-y

Vigna, A. J., Poehlmann-Tynan, J., & Koenig, B. W. (2017). Does self-compassion facilitate resilience to stigma? A school-based study of sexual and gender minority youth. *Mindfulness, 9*(3), 914–924. https://doi.org/10.1007/s12671-017-0831-x

Wallmark, E., Safarzadeh, K., Daukantaite, D., & Maddux, R. E. (2012). Promoting altruism through meditation: An 8-week randomized controlled pilot study. *Mindfulness, 4*(3), 223–234. https://doi.org/10.1007/s12671-012-0115-4

Werner, K. H., Jazaieri, H., Goldin, P. R., Ziv, M., Heimberg, R. G., & Gross, J. J. (2012). Self-compassion and social anxiety disorder. *Anxiety, Stress & Coping, 25*(5), 543–558. https://doi.org/10.1080/10615806.2011.608842

Whitesman, S., & Mash, R. (2016). Examining the effects of a mindfulness-based professional training module on mindfulness, perceived stress, self-compassion and self-determination. *African Journal of Health Professions Education, 7*(2), 220–223. https://doi.org/10.7196/AJHPE.460

Williams, M. J., Dalgleish, T., Karl, A., & Kuyken, W. (2014). Examining the factor structures of the Five Facet Mindfulness questionnaire and the self-compassion scale. *Psychological Assessment, 26*(2), 407–418. https://doi.org/10.1037/a0035566

Yarnell, L. M., Stafford, R. E., Neff, K. D., Reilly, E. D., Knox, M. C., & Mullarkey, M. (2015). Meta-analysis of gender differences in self-compassion. *Self and Identity, 14*(5), 499–520. https://doi.org/10.1080/15298868.2015.1029966

Yarnell, L. M., Neff, K. D., Davidson, O., & Mullarkey, M. (2019). Gender differences in self-compassion: Examining the role of gender role orientation. *Mindfulness., 10*(6), 1136–1152. https://doi.org/10.1007/s12671-018-1066-1

Zeller, M., Yuval, K., Nitzan-Assayag, Y., & Bernstein, A. (2015). Self-compassion in recovery following potentially traumatic stress: Longitudinal study of at-risk youth. *Journal of Abnormal Child Psychology, 43*(4), 645–653. https://doi.org/10.1007/s10802-014-9937-y

Zhang, J. W., Chen, S., & Tomova Shakur, T. K. (2020). From me to you: Self-compassion predicts acceptance of own and others' imperfections. *Personality and Social Psychology Bulletin, 46*(2), 228–242. https://doi.org/10.1177/0146167219853846

Zhu, L., Yao, J., Wang, J., Wu, L., Gao, Y., Xie, J., Liu, A., Ranchor, A., & Schroevers, M. J. (2019). The predictive role of self-compassion in cancer patients' symptoms of depression, anxiety, and fatigue: A longitudinal study. *Psycho-oncology, 28*(9), 1918–1925. https://doi.org/10.1002/pon.5174

2 Self-Compassion and Mindfulness

Shauna Shapiro and Anna Fitch

Introduction

The word mindfulness—*Sampajañña* in Pali—means clear comprehension. Mindfulness helps us to see clearly, so we can make wise choices and respond to life's challenges effectively and with a clear mind. Awareness of the present moment inhibits the automatic biases that frequently dominate cognitive processing, such as attention to perceived threat or negative thoughts about oneself or the world. In doing so, mindfulness is thought to facilitate more impartial interpretation of reality, giving rise to "clear seeing" or "clear knowing" (Analayo, 2019). From this place of clarity, one is given the opportunity to consciously discern what the present moment requires and respond in a proper and practiced way (Shapiro & Carlson, 2009). Thus, mindfulness can both reduce maladaptive and habitual cognitive, emotional, and behavioral patterns and also mitigate the impact of these patterns on distress and well-being (Verplanken & Fisher, 2014). Although research has demonstrated the myriad benefits of mindfulness (Grossman et al., 2010; Leyland et al., 2019), the recent interest has often led to its oversimplification and overcommodification. As a result, mindfulness bears the danger of becoming a buzz word of modern society with the intent of monetizing wellness and offering mindfulness training without context or nuanced content. This watered-down, amorphous version of mindfulness limits its full potential and can lead to discouragement and, worse, iatrogenic effects.

Self-compassion, at its most basic level, is the ability to show yourself kindness in the face of suffering. Mindfulness is considered foundational to self-compassion: we cannot be kind to ourselves unless we first have the awareness and acknowledge we are in pain. Mindfulness helps us see suffering clearly. Self-compassion adds: "Be kind to yourself in the midst of suffering" (Germer & Neff, 2013, p. 861). Accordingly, self-compassion also has the potential to facilitate and deepen mindfulness, by supporting us to stay present with difficult experiences. Indeed, many scholars and practitioners emphasize that self-compassion is inherent within mindfulness practice. In this way, mindfulness and self-compassion are distinct constructs that mutually engender each other.

This chapter closely examines the relationship between mindfulness and self-compassion, illustrating both the overlap and fundamental differences. We aim to provide a comprehensive and clear understanding of mindfulness, explicitly highlighting its three components: intention, attention, and attitude. As mindfulness and self-compassion are inextricably linked, the purpose of this chapter is to unravel their integral aspects,

S. Shapiro (✉) · A. Fitch
School of Education and Counseling Psychology, Santa Clara University, Santa Clara, CA, USA
e-mail: slshapiro@scu.edu; afitch@scu.edu

A. Finlay-Jones et al. (eds.), *Handbook of Self-Compassion*, Mindfulness in Behavioral Health,
https://doi.org/10.1007/978-3-031-22348-8_2

examine how they interact, and help to articulate the utility of both mindfulness and self-compassion to support health and well-being on an individual and collective level. Throughout this exploration, questions arise: What are the constructs of mindfulness and self-compassion and how are they defined? What are the fundamental components and practices of each? Where are they similar and where do they differ? How can mindfulness and self-compassion be integrated in order to bring about the greatest health and healing, and which populations would this integration be most effective?

Distinguishing Mindfulness and Self-Compassion Practices

To begin, we need to understand mindfulness and self-compassion as separate practices. While mindful awareness is a state of being, mindful practice is the mechanism through which consistent mindfulness is achieved (Shapiro & Carlson, 2009). Just as patterns of cognitive distortion such as worry and rumination can become a "default" through repeated use (McEvoy et al., 2013), mindfulness too can become an integral habit, becoming both a way of being and a formal practice. Focused-attention meditation is one mindfulness practice that specifically aims to cultivate selective attention on a specific object (e.g., the breath). To do so, one must also simultaneously monitor one's attention to identify when it has wandered away from the chosen object, disengage one's attention from any distractions, and return one's attention to the attentional object accordingly (Lutz et al., 2008). Open monitoring meditation builds on focused attention, with the aim of remaining in a state of monitoring anything that occurs in one's present-moment experience, *without* focusing on an attentional object (Lutz et al., 2008).

Traditionally, self-compassion was incorporated in compassion and lovingkindness practices of Buddhism and has been implicitly taught through the practice of mindfulness-based interventions (MBIs; Rudaz et al., 2019). However, with the pioneering work of Neff and Germer (2013), self-compassion practice and the Mindful Self-Compassion (MSC) program were developed to cultivate self-compassion more explicitly. Current research shows the benefits of a direct approach of training in MSC (Neff & Germer, 2013). This program utilizes compassion and lovingkindness meditations, as well as integrating dyadic, group, and individual activities designed to cultivate these constructs. In addition to the focus on one's cognitive, sensory, and affective experience that is characteristic of mindfulness meditation, loving-kindness and compassion practices focus specifically on intra- and interpersonal experiences. In lovingkindness mediation, for example, one practices sending compassion to oneself, loved ones, and all beings who suffer (Hofmann et al., 2011). In compassion meditation, through the awareness of our own suffering or the suffering of others, one can cultivate the inherent compassion that naturally arises. While compassion meditation facilitates recognizing and alleviating suffering, loving-kindness mediation focuses on wishing others unconditional well-being (Hofmann et al., 2011).

In Neff and Shapiro's theoretical framework of general state mindfulness, four distinct elements of mindfulness are identified (Neff & Shapiro, 2019): (1) paying attention to present-moment experiences, (2) how we relate to experiences, (3) how we relate to the experiencer, and (4) the wisdom to see both the experience and experiencer with clarity. Self-compassion predominantly focuses on the third element, how we relate to the experiencer. Additionally, while general state mindfulness can be applied to all life's experiences, mindfulness within self-compassion focuses primarily on how we relate to the experiencer (ourselves) in moments of suffering. When suffering arises, we utilize the practice of mindfulness to become aware of our pain and identify the experience as suffering. When suffering has been identified, we include the explicit practice of self-compassion to actively soothe and support ourselves. The practice of self-compassion is a crucial ancillary to mindfulness in times of suffering; it directly responds to the emotional need that is present with support and comfort.

The Three Pillars of Mindfulness Practice

Whether through focused attention or open monitoring, the practice of mindfulness is made up of three essential pillars: intention, attention, and attitude (Shapiro & Carlson, 2009). Intention underpins *why* we pay attention. It helps to establish the purpose of our practice and keeps our focus on what we wish to achieve through mindfulness, thereby guiding the practice in a deliberate and purposeful direction. Attention helps us train and stabilize our focus in the present moment. Attention permits us to discern *when* and *what* to practice. Since the only constant is change, attention allows us to perceive the current moment with clarity. By utilizing attention, it is possible to become aware of what practice the present moment necessitates. This awareness provides the ability to select the most beneficial practice. Attitude guides *how* we pay attention—specifically, with kindness and curiosity.

This third pillar of mindfulness—attitude—is perhaps the most important and often overlooked. Too often, we practice mindfulness with an attitude of judgment—judging ourselves when our mind wanders, judging ourselves for not doing it right. This "judging" attitude, however, limits the ability to fully benefit from mindful practice. Researchers Heatherton and Wagner (2011) have studied the brain's response to an attitude of self-judgment. Using fMRI scanning, their research purports that shame diminishes the activity of the prefrontal cortex (PFC) and the communication between the amygdala and the PFC, thereby limiting the ability to self-regulate, make informed decisions, and adaptively learn from experience (Heatherton & Wagner, 2011). Further research conducted by Pulcu et al. (2014) found that shame not only dampens activation of the PFC but also increases amygdala activity, which elicits a fear response. Thus, judgment and shame initiate a fear response, which then subsequently reduces the activity in the PFC responsible for acquiring new insights and learning.

Intention: *Why* We Pay Attention Intentions establish an objective and keep the practice of mindfulness active rather than passive. They connect us with our personal visions, aspirations, and motivations. Our intentions keep us on course, reminding us again and again of what is most important. Research on the role of intention in mindfulness is limited; however, an early study found that intentions underpinning meditation practice shift over time and are associated with outcomes of practice (Shapiro, 1992). In their classification framework of mindfulness practices, Levit-Binnun et al. (2021) outline four distinct intentions: (1) to gain insight into the way in which how we relate to our experience influences our well-being or distress; (2) to gain insight into the changing nature of our internal and external world; (3) to gain insight into how our sense of self influences our distress and well-being; and (4) to gain insight into how positive and prosocial mental states influence well-being. Intentions 1–3 are based on the characteristics of experience considered fundamental to understanding and alleviating mental distress in the Buddhist tradition, while the fourth intention reflects the Brahmaviharas (i.e., "loving kindness," "compassion," "sympathetic joy," and "equanimity"). In practice, intentions can take the form of mantras, phrases, or even a single guiding word, such as "see clearly," "respond with kindness," or simply "peace."

Attention: Training and Stabilizing Our Focus in the Present One of the core tenets of mindfulness practice is that we pay attention to our present-moment experiences to truly experience the essence of life. Mindfulness supports the ability to be attentive and aware in the present moment. As mindfulness practice trains and stabilizes attention, we begin to see that which is unfolding before us more clearly. Indeed, research supports the assertion that present-moment attention is associated with more adaptive stress responses over time (Donald et al., 2016), whereas less awareness of the present moment is associated with lower psychological well-being (Stawarczyk et al., 2012). Killingsworth and Gilbert (2010) found that, on average, our minds wander 47% of the time. This means that our attention to the present moment is only fully engaged for approximately half our lives. The phenomenon of the wandering mind has been colloquially termed "monkey mind" because our mind behaves like a monkey, constantly swinging from thought to thought. Swept into future fears or lost in ruminations of the past, we miss the present moment. When our minds wander, we often miss important details of the present moment and construct inaccurate narratives about what is happening, often based on our previous conditioning instead of the reality of the present. Thus, the monkey mind may cling to inaccurate stories which can cause an incorrect forecasting and subsequent behavioral responses in the future.

In support of the assertion that mindfulness is associated with more adaptive attentional outcomes, numerous studies have demonstrated improved attentional processing and differences in associated brain structures related to both trait mindfulness and mindfulness practice (Yordanova et al., 2021; Malinowski, 2013; Allen et al., 2012). In a recent series of meta-analyses, Verhaeghen (2021) examined (a) the effects of mindfulness training, relative to control conditions, on attention outcomes; (b) attentional capacities in long-term meditators compared to those who have never meditated; and (c) the relationship between trait mindfulness and attention. The first meta-analysis demonstrated that mindfulness training had significant, small to medium effects on several components of attention, including inhibition, working memory, attention shifting, and sustained attention, with more training sessions associated with greater effects. The second meta-analysis found that across studies, long-term meditators reported significantly better performance on attention tests than those who have never meditated, again with small to medium effects. Finally, the third meta-analysis documented a small but significant association between trait mindfulness and attention across studies (Verhaeghen, 2021).

Some studies provide insights into the differential impacts of mindfulness- and compassion-based practices on brain structure and function related to attention. For example, Carter et al. (2005) found that Tibetan Buddhist monks performed significantly better on a task of sustained attention following focused-attention meditation, but not following compassion meditation. In a series of studies, Yordanova and colleagues (2020, 2021) found that among experienced meditators, focused attention, open monitoring, and lovingkindness meditation had both shared and distinct neural patterns. Specifically, focused-attention meditation was associated with increased beta connectivity in the right hemisphere of the brain, whereas open monitoring was associated with beta connectivity in the left hemisphere. The authors proposed that these patterns represented differences in the type of attention (i.e., narrow versus wide) or the amount of information attended to (i.e., small versus large).

Attitude: *How* We Pay Attention Of the three pillars of mindfulness, attitude is perhaps the one most often overlooked and the most relevant for this discussion. While intention reminds us *what* is most important and attention stabilizes our mind in the present, our attitude affects *how* we pay attention. *How* we pay attention determines our ability to see clearly, to learn effectively, and to respond wisely and compassionately. To benefit from mindfulness, an attitude of kindness and curiosity is essential (Shapiro et al., 2006). If intention and attention are met with a cold and critical attitude, the practice may have contradictory consequences (Shapiro et al., 2006; Kabat-Zinn, 2003). This could result in a practice that is condemning or judgmental of inner experience, further cultivating neuronal pathways of judgment and shame instead of compassion and acceptance. An attitude of kindness and curiosity enables the learning and information processing areas of the brain to function more effectively, resulting in being able to more objectively evaluate our situation so that we can effectively respond. In fact, the bringing together of these two elements—kindness and presence—is reflected in the Japanese character for mindfulness which is comprised of two interactive figures: one is presence; the other is heart or mind (Santorelli, 1999). Therefore, an equally accurate translation of mindfulness is *heartfulness* (Shapiro & Schwartz, 2000). This underscores the importance of cultivating and including open-hearted qualities during the practice of mindfulness. It is important to note that this "attitude" of kindness is central to the practice of MSC.

One misconception in the current mindfulness "revolution" is that it often neglects this important pillar of attitude (Shapiro, 2020). People often think of "kindness" as a side note, or, worse, they mistakenly believe it will make them soft and cause them to lose their edge. However, the opposite is true; an attitude of kindness and curiosity is directly linked to performance and well-being (Hanson et al., 2021). In support of this assertion, research has found that an attitude of kindness may elicit and strengthen positive well-being. Symeonidou et al. (2019) examined the effects of participant kindness on their subjective well-being. Participants were asked to keep a daily record of their kind actions for a week, while those in the control condition were asked to simply record their typical daily actions. Measurements of subjective well-being were administered pre, post, and at 1- and 2-month follow-ups. A significant small-to-medium positive correlation was found between the number of kind actions undertaken and participant well-being (Symeonidou et al., 2019). What is more, those in the kindness intervention showed increases in well-being during follow-up measurements. This suggests that well-being is not static and can be positively infused through self-recognition of one's kindness.

Additionally, research has found similarly promising results for those who also practice self-kindness. Rowland and Curry (2019) conducted a randomized and blind study to determine the effect of numerous kindness interventions. Participants were split into four kindness groups as well as a control group. Each group directed kindness to a different receiver (family and friends, strangers, self, or observational), except for those in the control who were not directly instructed to engage in acts of kindness. After pre and post measurements for happiness and well-being were analyzed, two important findings emerged. Firstly, each kindness group demonstrated a significant increase in happiness and well-being compared to the control, and the number of kind acts was positively correlated with happiness. This further substantiates the benefits of kindness found by previous research. Secondly, no significant difference was identified between kindness groups. In other words, directing kindness toward oneself was just as beneficial as directing kindness toward others (Rowland & Curry, 2019).

An attitude of kindness and self-kindness has also been found to support personal performance and motivation levels. In one report, researchers conducted two studies to measure the effectiveness of a lovingkindness meditation (LKM) training program on motivational states (Masters-Waage et al., 2021). Using a randomized controlled trial, the first study split participants into three groups (LKM, general mindfulness, and waitlist control). After completion of the program, analysis revealed longitudinal increases in affect and motivation for those in the LKM intervention when compared to general mindfulness and the control condition. Further, the second study reported on a weeklong study to identify the effects of LKM on workplace motivation. Participants were split into three groups (LKM, formal meditation practice, and informal mindfulness practice). Each of these groups was instructed to actively practice mindfulness, but the LKM group was explicitly directed to practice kindness toward themselves and others. Researchers found that those in the LKM condition reported increased daily motivation levels after meditation practice. While workplace performance was not significantly different between groups, this research provides evidence for the efficacy of multiple meditation practices, both formal and informal (Masters-Waage et al., 2021).

Establishing an attitude of kindness and curiosity toward our inner experience is not sugar-coating our emotions or trying to suppress or change them, but rather allowing us to experience our emotions in a safe and courageous way. An attitude of kindness and curiosity activates the relaxation response in the brain, which releases neurotransmitters such as acetylcholine, endorphins, and oxytocin which activate the parasympathetic branch of the autonomic system (Uvnäs-Moberg et al., 2014; Gerritsen & Band, 2018). Rigoni et al. (2015) found that enacting an attitude of kindness during a meditation practice initiates a release of dopamine—one of the brain's neurotransmitters responsible for learning and rewards and associated with feelings of pleasure and motivation.

Mindfulness is the awareness that arises when all three elements—intention, attention, and attitude—synergistically arise to meet the present moment. This mindful awareness allows us to see the present moment clearly and respond productively. By bringing an attitude of kindness and curiosity, we can practice genuine heart-mindfulness. This is where self-compassion and mindfulness overlap, and the two practices work together to improve overall well-being.

Defining Self-Compassion

The formal definition of self-compassion is articulated by Kristin Neff as a dynamic interplay of compassionate self-responding components of mindfulness, self-kindness, and common humanity and uncompassionate self-responding components of overidentification, self-judgment, and isolation (Neff, 2003). By clearly defining the construct of self-compassion, we can more fully understand the need for the explicit practice of self-compassion. The components of self-compassion as defined by Neff (2003) are mindfulness, self-kindness, and common humanity.

Mindfulness Mindfulness is the foundation for self-compassion. We must have the awareness that we are in pain before we can soothe the pain. Within the construct of self-compassion, mindfulness is defined as having a balanced perspective *specifically* when faced with challenges, without exaggerating or evading difficult emotions or potential negative consequences (Neff, 2003). This is contrasted with the more general definition of mindful awareness, which focuses on awareness of *all* experience, both pleasant and unpleasant.

Self-Kindness Kindness is the mechanism used to actively comfort and support when we are suffering. It is important to note that we are not soothing ourselves to make the pain go away or as an act of avoiding the pain. Rather, we are soothing ourselves *because* we are suffering due to pain. The existence of pain, and the resultant experience of suffering, is the reason we call upon the practice of self-compassion. Mindfulness without kindness can lead to the resistance of pain and inadvertently increase suffering. It is necessary here to distinguish the difference between pain and suffering. Pain is a part of life; it exists regardless of our reaction to it. Suffering, however, is optional and is dependent on how strongly we resist the pain. This can be explained by the equation, suffering = pain × resistance (S. Young, personal communication, 2000). Pain is inevitable, and suffering increases depending on the degree to which we resist pain.

In this way, mindfulness, or being aware of pain, does not actively alleviate suffering, but rather it makes the experiencer acutely aware of what they are feeling in relation to the pain. At times, the experiencer may realize that, in fact, pain is not as persistent or intense as previously thought, and therefore there may be some moments of alleviation of the pain. Alternatively, being aware of pain with openness and curiosity may make the experiencer more overwhelmed by the pain when the pain is great. In either case, adding self-kindness *because one is in pain* begins the process of soothing suffering. Mindfulness facilitates awareness of the pain, while self-compassion provides the experiencer with the opportunity to alleviate the pain. When practiced in this manner, self-kindness allows the experiencer to be more intimate with the pain rather than detach from it (Shapiro et al., 2006).

Self-compassion offers an alternative coping mechanism to avoidance. While it may seem intuitive to distance ourselves from pain, research has found that acknowledging pain with an attitude of kindness offers more benefit (Costa & Pinto-Gouveia, 2013). In a study of over 100 patients with chronic pain, Costa and Pinto-Gouveia (2013) analyzed the relationship between experiential avoidance, self-compassion, and psychological distress. Hierarchical regressions were conducted after self-report measures were completed, and the results speak to the benefits of self-compassion in times of suffering. A significant positive correlation was identified between patients who practiced avoidance and psychological distress (depression, anxiety, and stress). Alternatively, those who engaged in self-compassion and approached their pain with kindness showed significantly less psychological distress (Costa & Pinto-Gouveia, 2013). Despite seeming contradictory, when an attitude of self-kindness is present, being with physical pain decreases overall suffering, while detachment from pain may lead to increased psychological suffering.

Of course, being kind to oneself in the face of suffering can be challenging, and it may even feel contradictory to our nature. When things go wrong, we often try to suppress the pain, berate ourselves, or leap into problem-solving mode. Whether through reliance on primitive survival instincts, or cultural influences, this is how we often treat ourselves (Lu et al., 2021). Interestingly, we do not treat others this way; when friends are facing challenges, we readily respond with kindness and compassion. Why would we so readily offer a friend this benevolence yet are so hesitant to extend the same compassion to ourselves? Further, given the aforementioned research and the pitfalls of shame, why would we not cultivate an inner environment that is best able to promote productive growth and meaningful change?

The concept of self-kindness, particularly within the practice of self-compassion, is largely misunderstood in Western culture. Self-kindness is often perceived as indulgent, selfish, or permissive. The misconception asserts that directing kindness to the self is an overall weakness. Primarily, the concern is that self-kindness may result in laziness or could diminish the quantity of kindness we have available for others (Gilbert et al., 2012, 2011). However, as was previously discussed, self-kindness has been shown to increase personal motivation (Masters-Waage et al., 2022). Additionally, those who practice self-kindness have been found to be more emotionally available to offer kindness to others (Waytz & Hofmann, 2020; Hashem & Zeinoun, 2020).

Waytz and Hofmann (2020) examined the prosocial benefits of self-kindness in comparison with kindness shown toward others. In a randomized study, participants were separated into three distinct groups to perform one of three kindness-related activities. The first group was tasked with acting out kindness, such as donating to charity or helping a coworker. The second group was asked to think moral thoughts, such as hoping for someone's success or thinking kindly about a friend. The final group was tasked with engaging in acts of self-kindness, for example, taking time to relax, making a nice meal, or speaking to themselves with nonjudgment. All groups demonstrated an increase in well-being, gratitude, and elevation (feeling inspired to spread kindness). Interestingly, however, the researchers found that those who practiced self-kindness experienced significantly less burnout (Waytz & Hofmann, 2020). People who took the time to bring kindness to themselves were less emotionally exhausted and therefore more available for others. While all forms of kindness were shown to lead to increased kindness toward others, self-kindness demonstrated the unique benefit of protecting against burnout.

These results have been replicated, with researchers finding that healthcare workers who practice self-kindness tend to experience less emotional exhaustion (Hashem & Zeinoun, 2020). While burnout was highly prevalent in the sample, those who measured high in self-kindness demonstrated adaptive coping in times of stress and were readily able to help and offer kindness to patients. Thus, self-kindness has the potential to increase one's desire and availability to be kind to others, making it an important complement to other-focused kindness and compassion.

Common Humanity Common humanity is the understanding that we are not alone in our suffering. When we struggle, we often feel that this is "my" personal problem and that we are the "only one" suffering. This way of perceiving leads to the conclusion that we are alone in our pain, making us feel isolated and separate. Understanding that suffering is part of the human experience allows us to not feel isolated from others, and we may experience an easing of a sense of despair. By reframing our relationship to suffering in terms of a common human experience, we are more able to be in touch with the pain of the present moment. We are able to see suffering not as a result of something that we have done wrong, for example, but as a fundamental and shared aspect of life. It is important to note that although common humanity and interdependence are foundational themes in mindfulness practice, they are not explicitly included in the instructions of mindfulness practice, another important distinction between self-compassion and mindfulness practice.

Unique and Overlapping Benefits of Mindfulness and Self-Compassion

Decades of research demonstrate the link between mindfulness practice and better mental health and overall well-being, including psychological, cognitive, and physical domains (Shapiro & Walsh, 2003), for both clinical and nonclinical populations (Donald et al., 2019; Fjorback et al., 2011; Ni et al., 2020; Lomas et al., 2018). For example, mindfulness has been shown to increase a sense of meaning and life satisfaction (LeBlanc et al., 2021; Aşık & Albayrak, 2021), aid in creativity and problem solving (Henriksen et al., 2020), and improve sleep and chronic pain (Doorley et al., 2021).

Benefits of Mindfulness Training In healthcare, mindfulness interventions have shown a decrease in healthcare worker stress as well as an overall increase in well-being and positive patient evaluations (Lomas et al., 2018; Irving et al., 2009). Likewise, research has exemplified the use of mindfulness in other workplaces to increase employee health and overall performance (Good et al., 2016). For example, mindfulness programs have been utilized in the military, and findings have demonstrated that mindfulness programs help soldiers make wiser choices during high-stress situations (Jha et al., 2015). In another example, mindfulness programs have been shown to result in a decrease of post-traumatic stress symptoms while increasing resiliency among veterans (Davis et al., 2019).

Benefits of Self-Compassion Training The practice of self-compassion—cultivated and taught through various inductions, short programs, as well as eight-session established training programs such as Mindful Self-Compassion developed by Neff and Germer (2013)—has demonstrated positive outcomes in a vast array of domains, including depression, eating behavior, anxiety, stress, and self-criticism (Ferrari et al., 2019). Current research has found comparable benefits to mindfulness practice, with an additional focus on benefits that are specifically targeted by self-compassion practice, such as increased self-worth and psychological well-being (Neff et al., 2005; Baer et al., 2012; Dundas et al., 2017).

Research shows that self-compassion interventions are beneficial, leading to significant improvements in rumination, stress, self-criticism, and anxiety (Ferrari et al., 2019) and there are certain populations for whom this consideration could be especially important. Researchers have found that a lack of self-compassion may be a factor of vulnerability for developing symptoms of depression (Krieger et al., 2016) and could possibly predict psychopathology, such as anxiety and depression (MacBeth & Gumley, 2012). Self-compassion has been shown to elicit adaptive coping skills (emotion regulation, acceptance, and self-soothing) and could act as both a preventative defense to psychopathology and a mediator to recovery (Finlay-Jones, 2017). Regarding the latter, populations that lack self-compassion may considerably benefit from direct self-compassion training. For example, individuals who are overly harsh and critical toward themselves seem to benefit more from explicit self-compassion training (Rudaz et al., 2019). Due to intense feelings of shame, guilt, and unworthiness that individuals with anxiety, depression, and PTSD face, mindfulness practice may be difficult, and beginning practice with self-compassion may be more accessible (Van Dam et al., 2010). In people who are highly self-critical, mindfulness practice may inadvertently deepen the pathways of self-criticism, and explicitly beginning with self-compassion may be a more effective way to intervene (Gilbert & Procter, 2006; Smeets et al., 2014).

Mechanisms of Mindfulness and Self-Compassion

One fundamental factor that relates to how both mindfulness and self-compassion practice achieve positive outcomes is that both these practices depend on neuroplasticity, that is, that our brain structure can be modified through repeated practice. Neuroplasticity refers to the way that the brain and other aspects of the nervous system can change in structure and function in response to internal and external stimuli (Cramer et al., 2011). Contrary to previously held beliefs in neuroscience, brain development does not stop at a certain age with recent findings demonstrating that the brain can change both structurally and functionally across the entire lifespan (Toricelli et al., 2021). Neuroplasticity affirms this capacity for the brain to continue to adapt and grow (Grafman, 2000; Voss et al., 2017).

When Mindfulness and Self-Compassion Meet

Available evidence suggests that self-compassion and mindfulness are overlapping constructs which complement each other (Bluth & Blanton, 2013). At the most basic level, mindfulness supports the present-moment awareness that we are suffering and is therefore the foundation of self-compassion. Self-compassion entails the ability to be aware of suffering (mindfulness), be kind to yourself in the face of this suffering (self-kindness), and recognize that you're not alone in your suffering (common humanity). Without the awareness that mindfulness brings, it is difficult to clearly discern one's emotional needs and is therefore impossible to engage in self-compassion, as self-compassion practice responds to the question "What do I need?" Mindfulness provides the nonjudgmental "witness state" of consciousness that helps us see our suffering clearly. This provides the opportunity to *choose* to comfort and meet suffering with kindness instead of succumbing to habitual reactions of shame or avoidance.

Similarly, self-compassion is integral to the nonjudgmental component of mindfulness. Mindfulness allows us to discern the present moment with clarity, while self-compassion provides us with the resources to respond with kindness. When suffering arises, the need for explicit self-compassion practice is discerned through our mindful awareness. In this way, mindfulness and self-compassion can work synergistically to strength and inform each practice. In summary, while mindfulness and self-compassion overlap, they are two distinct constructs that work together to support the experiencer during times of suffering. Mindfulness provides the clarity of mind to recognize the experience of suffering and approach it with an attitude of nonjudgment. Self-compassion focuses explicitly on the experiencer's reaction to themselves, promoting an active role of kindness and ameliorating feelings of isolation due to an understanding of common humanity.

Research also suggests that self-compassion is an integral component of mindfulness, and great success has been observed by implementing explicit self-compassion practice with a foundation of mindfulness (Neff & Germer, 2013). Although it is a distinct practice, self-compassion builds the quality of kindness, which is essential to the attitude component of mindfulness practice. In fact, some studies suggest that self-compassion is the key mechanism through which mindfulness interventions achieve outcomes (Kuyken et al., 2010; Rowe et al., 2016; Galla, 2016, 2017). Two studies conducted with adolescent participants during weeklong mindfulness meditation retreats reported that while both mindfulness and self-compassion increased well-being, within-person changes in self-compassion predicted beneficial outcomes more constantly than within-person changes in mindfulness (Galla, 2016, 2017). These findings indicate the importance of self-compassion as a potential way in which mindfulness achieves beneficial outcomes.

While mindfulness training teaches the importance of observing nonjudgmentally, it does not explicitly provide an active way to provide comfort to oneself when pain is encountered. It assumes that when observing painful experiences, thoughts, or emotions nonjudgmentally, compassion for the self will naturally emerge, and does so for many practitioners. However, perhaps due to the extensive messaging that we get from our culture about not being "good enough," many are challenged with accessing compassion for the self, and explicit self-compassion practice can be enormously helpful. Teaching self-compassion explicitly provides participants with tools so that they can learn to be kind to themselves in a more direct way. As the behavior of supporting oneself is practiced, neuronal pathways are established and being self-compassionate becomes more of a habit.

Conclusion

While the body of mindfulness and self-compassion research is growing, comparatively little research has been devoted to understanding their unique and overlapping dimensions. Continued exploration is needed with the goal to determine how we can more effectively integrate both in our personal and professional lives. There are many unanswered questions and a substantial amount of work that needs to be done in order to facilitate the growth of a meaningful and illuminative understanding of the relationship between mindfulness and self-compassion.

Methodologically, there is a need for much greater care in the design of studies. This includes such elements as grounding research in theory to guide the development of research expectations; selecting participants in a way that ensures sufficient power and generalizability of findings and/or allows for detailed examination of theoretically relevant mechanisms, processes, and outcomes; utilizing study designs that permit the examination of changes and effects both cross-sectionally and longitudinally; and resisting the tendency of over-relying on a single methodology (e.g., self-report measures, translational imaging technologies).

Rigorously designed RCTs with large sample sizes and diverse populations are needed to elucidate the potential of combining mindfulness and self-compassion practices. In particular, there is very little research on young kids, and future work needs to be done in both clinical and nonclinical youth. Finally, future research needs to determine how to best integrate these practices into complimentary interventions, to produce positive synergistic benefits. The collaboration of mindfulness and self-compassion intervention is still in its infancy, yet it seems clear that if handled skillfully, this integration may prove enriching for both, enabling them to become partners in the understanding, healing, and enhancement of the human mind and heart.

References

Allen, M., Dietz, M., Blair, K. S., van Beek, M., Rees, G., Vestergaard-Poulsen, P., et al. (2012). Cognitive-affective neural plasticity following active-controlled mindfulness intervention. *The Journal of Neuroscience, 32*(44), 15601–15610. https://doi.org/10.1523/JNEUROSCI.2957-12.2012

Analayo, B. (2019). Clear knowing and mindfulness. *Mindfulness, 11*(4), 862–871. https://doi.org/10.1007/s12671-019-01283-8

Aşık, E., & Albayrak, S. (2021). The effect of mindfulness levels on the life satisfaction of nursing students. *Perspectives in Psychiatric Care, 58*(3), 1055–1061. https://doi.org/10.1111/ppc.12898

Baer, R. A., Lykins, E. L. B., & Peters, J. R. (2012). Mindfulness and self-compassion as predictors of psychological wellbeing in long-term meditators and matched nonmeditators. *The Journal of Positive*

Psychology, 7(3), 230–238. https://doi.org/10.1080/17439760.2012.674548

Bluth, K., & Blanton, P. W. (2013). Mindfulness and self-compassion: Exploring pathways to adolescent emotional well-being. *Journal of Child and Family Studies, 23*(7), 1298–1309. https://doi.org/10.1007/s10826-013-9830-2

Carter, O. L., Presti, D. E., Callistemon, C., Ungerer, Y., Liu, G. B., & Pettigrew, J. D. (2005). Meditation alters perceptual rivalry in Tibetan Buddhist monks. *Current Biology, 15*(11), R412–R413. https://doi.org/10.1016/j.cub.2005.05.043

Costa, J., & Pinto-Gouveia, J. (2013). Experiential avoidance and self-compassion in chronic pain. *Journal of Applied Social Psychology, 43*(8), 1578–1591. https://doi.org/10.1111/jasp.12107

Cramer, S. C., Sur, M., Chen, W. G., Cohen, L. G., Decharms, C., Duffy, C. J., et al. (2011). Harnessing neuroplasticity for clinical applications. *Brain, 134*(Pt 6), 1591–1609. https://doi.org/10.1093/brain/awr039

Davis, L. L., Whetsell, C., Hamner, M. B., Carmody, J., Rothbaum, B. O., Allen, R. S., Bartolucci, A., Southwick, S. M., & Bremner, D. (2019). A multi-site randomized controlled trial of mindfulness-based stress reduction in the treatment of posttraumatic stress disorder. *Psychiatric Research and Clinical Practice, 1*(2), 39–48. https://doi.org/10.1176/appi.prcp.20180002

Donald, J. N., Atkins, P. W. B., Parker, P. D., Christie, A. M., & Ryan, R. M. (2016). Daily stress and the benefits of mindfulness: Examining the daily and longitudinal relations between present-moment awareness and stress responses. *Journal of Research in Personality, 65*, 30–37. https://doi.org/10.1016/j.jrp.2016.09.002

Donald, J. N., Sahdra, B. K., Van Zanden, B., Duineveld, J. J., Atkins, P. W. B., Marshall, S. L., et al. (2019). Does your mindfulness benefit others? A systematic review and meta-analysis of the link between mindfulness and prosocial behaviour. *The British Journal of Psychology, 110*(1), 101–125. https://doi.org/10.1111/bjop.12338

Doorley, J. D., Greenberg, J., Stauder, M., & Vranceanu, A.-M. (2021). The role of mindfulness and relaxation in improved sleep quality following a mind–body and activity program for chronic pain. *Mindfulness, 12*(11), 2672–2680. https://doi.org/10.1007/s12671-021-01729-y

Dundas, I., Binder, P. E., Hansen, T. G. B., & Stige, S. H. (2017). Does a short self-compassion intervention for students increase healthy self-regulation? A randomized control trial. *Scandinavian Journal of Psychology, 58*(5), 443–450. https://doi.org/10.1111/sjop.12385

Ferrari, M., Hunt, C., Harrysunker, A., Abbott, M. J., Beath, A. P., & Einstein, D. A. (2019). Self-compassion interventions and psychosocial outcomes: A meta-analysis of RCTs. *Mindfulness, 10*(8), 1455–1473. https://doi.org/10.1007/s12671-019-01134-6

Finlay-Jones, A. L. (2017). The relevance of self-compassion as an intervention target in mood and anxiety disorders: A narrative review based on an emotion regulation framework. *Clinical Psychologist, 21*(2), 90–103. https://doi.org/10.1111/cp.12131

Fjorback, L. O., Arendt, M., Ørnbøl, E., Fink, P., & Walach, H. (2011). Mindfulness-based stress reduction and mindfulness-based cognitive therapy – A systematic review of randomized controlled trials: Systematic review of mindfulness RCTs. *Acta Psychiatrica Scandinavica, 124*(2), 102–119. https://doi.org/10.1111/j.1600-0447.2011.01704.x

Galla, B. M. (2016). Within-person changes in mindfulness and self-compassion predict enhanced emotional well-being in healthy, but stressed adolescents. *Journal of Adolescence, 49*(C), 204–217. https://doi.org/10.1016/j.adolescence.2016.03.016

Galla, B. M. (2017). "Safe in my own mind:" Supporting healthy adolescent development through meditation retreats. *Journal of Applied Developmental Psychology, 53*, 96–107. https://doi.org/10.1016/j.appdev.2017.09.006

Germer, C. K., & Neff, K. D. (2013). Self-compassion in clinical practice. *Journal of Clinical Psychology, 69*(8), 856–867. https://doi.org/10.1002/jclp.22021

Gerritsen, R. J. S., & Band, G. P. H. (2018). Breath of life: The respiratory vagal stimulation model of contemplative activity. *Frontiers in Human Neuroscience, 12*, 397–397. https://doi.org/10.3389/fnhum.2018.00397

Gilbert, P., & Procter, S. (2006). Compassionate mind training for people with high shame and self-criticism: Overview and pilot study of a group therapy approach. *Clinical psychology and psychotherapy, 13*(6), 353–379. https://doi.org/10.1002/cpp.507

Gilbert, P., McEwan, K., Matos, M., & Rivis, A. (2011). Fears of compassion: Development of three self-report measures. *Psychology and Psychotherapy: Theory, Research and Practice, 84*(3), 239–255. https://doi.org/10.1348/147608310X526511

Gilbert, P., McEwan, K., Gibbons, L., Chotai, S., Duarte, J., & Matos, M. (2012). Fears of compassion and happiness in relation to alexithymia, mindfulness, and self-criticism. *Psychology and Psychotherapy: Theory, Research and Practice, 85*(4), 374–390. https://doi.org/10.1111/j.2044-8341.2011.02046.x

Good, D. J., Lyddy, C. J., Glomb, T. M., Bono, J. E., Brown, K. W., Duffy, M. K., Baer, R. A., Brewer, J. A., & Lazar, S. W. (2016). Contemplating mindfulness at work: An integrative review. *Journal of Management, 42*(1), 114–142. https://doi.org/10.1177/0149206315617003

Grafman, J. (2000). Conceptualizing functional neuroplasticity. *Journal of Communication Disorders, 33*(4), 345–355. https://doi.org/10.1016/s0021-9924(00)00030-7

Grossman, P., Niemann, L., Schmidt, S., Walach, H. (2010). Mindfulness-based stress reduction and health benefits: A meta-analysis. *Focus on Alternative and Complementary Therapies, 8*(4), 500–500. https://doi.org/10.1016/S0022-3999(03)00573-7

Hanson, R., Shapiro, S. L., Hutton-Thamm, E., Hagerty, M. R., & Sullivan, K. P. (2021). Learning to learn from positive experiences. *The Journal of Positive Psychology*, 1–12. https://doi.org/10.1080/17439760.2021.2006759

Hashem, Z., & Zeinoun, P. (2020). Self-compassion explains less burnout among healthcare professionals. *Mindfulness, 11*(11), 2542–2551. https://doi.org/10.1007/s12671-020-01469-5

Heatherton, T. F., & Wagner, D. D. (2011). Cognitive neuroscience of self-regulation failure. *Trends in Cognitive Sciences, 15*(3), 132–139. https://doi.org/10.1016/j.tics.2010.12.005

Henriksen, D., Richardson, C., & Shack, K. (2020). Mindfulness and creativity: Implications for thinking and learning. *Thinking Skills and Creativity, 37*, 100689–100689. https://doi.org/10.1016/j.tsc.2020.100689

Hofmann, S. G., Grossman, P., & Hinton, D. E. (2011). Loving-kindness and compassion meditation: potential for psychological interventions. *Clinical psychology review, 31*(7), 1126–1132. https://doi.org/10.1016/j.cpr.2011.07.003

Irving, J. A., Dobkin, P. L., & Park, J. (2009). Cultivating mindfulness in health care professionals: A review of empirical studies of mindfulness-based stress reduction (MBSR). *Complementary Therapies in Clinical Practice, 15*(2), 61–66. https://doi.org/10.1016/j.ctcp.2009.01.002

Jha, A. P., Morrison, A. B., Dainer-Best, J., Parker, S., Rostrup, N., & Stanley, E. A. (2015). Minds "at attention": Mindfulness training curbs attentional lapses in military cohorts. *PLoS One, 10*(2), e0116889–e0116889. https://doi.org/10.1371/journal.pone.0116889

Kabat-Zinn, J. (2003). Mindfulness-based interventions in context: Past, present, and future. *Clinical Psychology: Science and Practice, 10*(2), 144–156. https://doi.org/10.1093/clipsy.bpg016

Krieger, T., Berger, T., & Holtforth, M. G. (2016). The relationship of self-compassion and depression: Cross-lagged panel analyses in depressed patients after outpatient therapy. *Journal of Affective Disorders, 202*, 39–45. https://doi.org/10.1016/j.jad.2016.05.032

Kuyken, W., Watkins, E., Holden, E., White, K., Taylor, R. S., Byford, S., et al. (2010). How does mindfulness-based cognitive therapy work? *Behaviour Research and Therapy, 48*(11), 1105–1112. https://doi.org/10.1016/j.brat.2010.08.003

Killingsworth, M. A., & Gilbert, D. T. (2010). A wandering mind is an unhappy mind. *Science, 330*(6006), 932. https://doi.org/10.1126/science.1192439

LeBlanc, S., Uzun, B., & Aydemir, A. (2021). Structural relationship among mindfulness, reappraisal and life satisfaction: The mediating role of positive affect. *Current Psychology, 40*(9), 4406–4415. https://doi.org/10.1007/s12144-019-00383-x

Leyland, A., Rowse, G., & Emerson, L. M. (2019). Experimental effects of mindfulness inductions on self-regulation: Systematic review and meta-analysis. *Emotion (Washington, D.C.), 19*(1), 108–122. https://doi.org/10.1037/emo0000425

Levit-Binnun, N., Arbel, K., & Dorjee, D. (2021). The mindfulness map: A practical classification framework of mindfulness practices, associated intentions, and experiential understandings. *Frontiers in Psychology, 12*, 727857–727857. https://doi.org/10.3389/fpsyg.2021.727857

Lomas, T., Medina, J. C., Ivtzan, I., Rupprecht, S., & Eiroa-Orosa, F. J. (2018). A systematic review of the impact of mindfulness on the well-being of healthcare professionals. *Journal of Clinical Psychology, 74*(3), 319–355. https://doi.org/10.1002/jclp.22515

Lu, S., Wei, F., & Li, G. (2021). The evolution of the concept of stress and the framework of the stress system. *Cell Stress, 5*(6), 76–85. https://doi.org/10.15698/cst2021.06.250

Lutz, A., Slagter, H. A., Dunne, J. D., & Davidson, R. J. (2008). Attention regulation and monitoring in meditation. *Trends in Cognitive Sciences, 12*(4), 163–169. https://doi.org/10.1016/j.tics.2008.01.005

MacBeth, A., & Gumley, A. (2012). Exploring compassion: A meta-analysis of the association between self-compassion and psychopathology. *Clinical Psychology Review, 32*(6), 545–552. https://doi.org/10.1016/j.cpr.2012.06.003

Malinowski, P. (2013). Neural mechanisms of attentional control in mindfulness meditation. *Frontiers in Neuroscience, 7*, 8–8. https://doi.org/10.3389/fnins.2013.00008

Masters-Waage, T., Reb, J., Tov, W., & Bandara, U. (2021). An initial examination of state and longitudinal effects of loving-kindness practice on affective and motivational states at work. Mindfulness. https://doi.org/10.1007/s12671-021-01781-8

Masters-Waage, T. C., Reb, J., Tov, W., & Bandara, U. (2022). An initial examination of state and longitudinal effects of loving-kindness practice on affective and motivational states at work. *Mindfulness, 13*(1), 174–187. https://doi.org/10.1007/s12671-021-01781-8

McEvoy, P. M., Watson, H., Watkins, E. R., & Nathan, P. (2013). The relationship between worry, rumination, and comorbidity: Evidence for repetitive negative thinking as a transdiagnostic construct. *Journal of Affective Disorders, 151*(1), 313–320. https://doi.org/10.1016/j.jad.2013.06.014

Neff, K. D. (2003). Self-compassion: An alternative conceptualization of a healthy attitude toward oneself. *Self and Identity, 2*(2), 85–101. https://doi.org/10.1080/15298860309032

Neff, K. D., & Germer, C. K. (2013). A pilot study and randomized controlled trial of the mindful self-compassion program. *Journal of Clinical Psychology, 69*(1), 28–44. https://doi.org/10.1002/jclp.21923

Neff, K. D., Hsieh, Y.-P., & Dejitterat, K. (2005). Self-compassion, achievement goals, and coping with academic failure. *Self and Identity, 4*(3), 263–287. https://doi.org/10.1080/13576500444000317

Neff, K. D., & Shapiro, S. L. (2019). The Science of mindfulness and self-compassion: How to build new habits to transform your life. Boulder, CO: Sounds True

Ni, Y., Ma, L., & Li, J. (2020). Effects of mindfulness-based stress reduction and mindfulness-based cognitive therapy in people with diabetes: A systematic review and meta-analysis. *Journal of Nursing Scholarship, 52*(4), 379–388. https://doi.org/10.1111/jnu.12560

Pulcu, E., Lythe, K., Elliott, R., Green, S., Moll, J., Deakin, J. F., Zahn, R. (2014). Increased amygdala response to shame in remitted major depressive disorder. *PLOS One, 9*(1), e86900. https://doi.org/10.1371/journal.pone.0086900

Rigoni, D., Demanet, J., & Sartori, G. (2015). Happiness in action: The impact of positive affect on the time of the conscious intention to act. *Frontiers in Psychology, 6*. https://doi.org/10.3389/fpsyg.2015.01307

Rowe, A. C., Shepstone, L., Carnelley, K. B., Cavanagh, K., & Millings, A. (2016). Attachment security and self-compassion priming increase the likelihood that first-time engagers in mindfulness meditation will continue with mindfulness training. *Mindfulness, 7*(3), 642–650. https://doi.org/10.1007/s12671-016-0499-7

Rowland, L., & Curry, O. S. (2019). A range of kindness activities boost happiness. *The Journal of Social Psychology, 159*(3), 340–343. https://doi.org/10.1080/00224545.2018.1469461

Rudaz, M., Ledermann, T., Twohig, M. P., & Levin, M. E. (2019). Does a brief mindfulness training enhance heartfulness in students? Results of a pilot study. *OBM Integrative and Complementary Medicine, 4*(4), 15. https://doi.org/10.21926/obm.icm.1904059

Santorelli, S. (1999). *Heal thy self: Lessons on mindfulness in medicine*. Random House.

Shapiro, D. H. (1992). A preliminary study of long-term meditators: Goals, effects, religious orientation, cognitions. *The Journal of Transpersonal Psychology, 24*(1), 23–39.

Shapiro, S. L., & Carlson, L. E. (2009). *The art and science of mindfulness: Integrating mindfulness into psychology and the helping professions*. American Psychological Association.

Shapiro, S. L., & Schwartz, G. E. (2000). The role of intention in self-regulation: Toward intentional systemic mindfulness. In M. Boekarts, P. R. Pintrich, & M. Zeidner (Eds.), *Handbook of self-regulation* (pp. 253–273). Academic Press.

Shapiro, S. L., & Walsh, R. (2003). An analysis of recent meditation research and suggestions for future directions. *The Humanistic Psychologist, 31*(2–3), 86–114. https://doi.org/10.1080/08873267.2003.9986927

Shapiro, S. L., Carlson, L. E., Astin, J. A., & Freedman, B. (2006). Mechanisms of mindfulness. *Journal of Clinical Psychology, 4*(4), 373–386. https://doi.org/10.1002/jclp.20237

Smeets, E., Neff, K., Alberts, H., & Peters, M. (2014). Meeting suffering with kindness: Effects of a brief self-compassion intervention for female college students. *Journal of Clinical Psychology, 70*(9), 794–807. https://doi.org/10.1002/jclp.22076

Stawarczyk, D., Majerus, S., Van Der Linden, M., & D'Argembeau, A. (2012). Using the daydreaming frequency scale to investigate the relationships between mind-wandering, psychological well-being, and present-moment awareness. *Frontiers in Psychology, 3*(363). https://doi.org/10.3389/fpsyg.2012.00363

Symeonidou, D., Moraitou, D., Papachristos, C., & Anastassios, S. (2019). Promoting subjective wellbeing through a kindness intervention. *Hellenic Journal of Psychology, 16*(1), 1–21.

Toricelli, M., Pereira, A., Souza Abrao, G., Malerba, H., Maia, J., Buck, H., et al. (2021). Mechanisms of neuroplasticity and brain degeneration: Strategies for protection during the aging process. *Neural Regeneration Research, 16*(1), 58–67. https://doi.org/10.4103/1673-5374.286952

Uvnäs-Moberg, K., Handlin, L., & Petersson, M. (2014). Self-soothing behaviors with particular reference to oxytocin release induced by non-noxious sensory stimulation. *Frontiers in Psychology, 5*, 1529–1529. https://doi.org/10.3389/fpsyg.2014.01529

Van Dam, N. T., Sheppard, S. C., Forsyth, J. P., & Earleywine, M. (2010). Self-compassion is a better predictor than mindfulness of symptom severity and quality of life in mixed anxiety and depression. *Journal of Anxiety Disorders, 25*(1), 123–130. https://doi.org/10.1016/j.janxdis.2010.08.011

Verhaeghen, P. (2021). Mindfulness as attention training: Meta-analyses on the links between attention performance and mindfulness interventions, long-term meditation practice, and trait mindfulness. *Mindfulness, 12*(3), 564–581. https://doi.org/10.1007/s12671-020-01532-1

Verplanken, B., & Fisher, N. (2014). Habitual worrying and benefits of mindfulness. *Mindfulness, 5*(5), 566–573. https://doi.org/10.1007/s12671-013-0211-0

Voss, P., Thomas, M. E., Cisneros-Franco, J. M., & de Villers-Sidani, É. (2017). Dynamic brains and the changing rules of neuroplasticity: Implications for learning and recovery. *Frontiers in Psychology, 8*. https://doi.org/10.3389/fpsyg.2017.01657

Waytz, A., & Hofmann, W. (2020). Nudging the better angels of our nature: A field experiment on morality and well-being. *Emotion, 20*(5), 904–909. https://doi.org/10.1037/emo0000588

Yordanova, J., Kolev, V., Mauro, F., Nicolardi, V., Simione, L., Calabrese, L., Malinowski, P., & Raffone, A. (2020). Common and distinct lateralised patterns of neural coupling during focused attention, open monitoring and loving kindness meditation. *Scientific Reports, 10*(1), 7430–7430. https://doi.org/10.1038/s41598-020-64324-6

Yordanova, J., Kolev, V., Nicolardi, V., Simione, L., Mauro, F., Garberi, P., Malinowski, P., & Raffone, A. (2021). Attentional and cognitive monitoring brain networks in long-term meditators depend on meditation states and expertise. *Scientific Reports, 11*(1), 4909–4909. https://doi.org/10.1038/s41598-021-84325-3

3 To Be Compassionate and Feel Worthy: The Bidirectional Relationship Between Self-Compassion and Self-Esteem

Madeleine I. Fraser, Joseph Ciarrochi, Baljinder K. Sahdra, and Caroline Hunt

Introduction

Some appear to easily adopt a self-compassionate way of relating to themselves, that is, they can extend to themselves the same supportive kindness as they would a good friend. However, many more appear to be more likely to be resistant to accepting the rationale for self-compassion, for example, "it sounds nice, but it's not for me," or they may engage with self-compassion on a superficial level, for example, "I'm hard on myself only when I need to be." Measuring and understanding concerns about self-compassion and a reluctance to adopt this framework are not new; indeed, there is a validated measure of fears of compassion (Gilbert et al., 2011). What has been less closely examined is the potential relationship between difficulties embracing self-compassion and a lack of self-esteem or sense of self-acceptance and self-worth. It may be the case that many internal psychological barriers to self-compassion are related to a belief of being unworthy of such compassion. It is of interest to both self-compassion researchers and clinicians to better understand how these constructs interact and relate to each other.

Self-compassion and self-esteem are interrelated constructs. Self-compassion refers to a tendency to relate to oneself with unconditional support and a desire to help, rather than be self-critical (Gilbert, 2014; Neff, 2003a). Relatedly, self-esteem generally refers to global appraisal of one's self-worth that is positive (Rosenberg, 1965a, b; Rosenberg et al., 1995). When both constructs are high in an individual, they are likely to present in a similar fashion. Illustratively, a psychological flourishing individual, that is, someone who engages in a rich, meaningful, and value-driven life (Ryan & Deci, 2000), is likely to be high in both self-compassion and self-esteem. They can be psychologically flexible and acknowledge and address their own needs with kindness and respect. In addition, they hold a generally positive global view of their identity and sense of self while also acknowledging and accepting imperfections that make them human. Clients seeking psychological support are likely to present to therapy with low self-compassion and low self-esteem. They may treat themselves with harsh self-criticism and be uncomfortable with or even despise the person they see themselves to be. In these clients, the subtle differences

M. I. Fraser (✉)
School of Behavioural Health Sciences, Australian Catholic University – Strathfield Campus, Strathfield, NSW, Australia
e-mail: Madeleine.Ferrari@acu.edu.au

J. Ciarrochi · B. K. Sahdra
Institute of Positive Psychology and Education, Australian Catholic University – North Sydney Campus, North Sydney, NSW, Australia

C. Hunt
School of Psychology, University of Sydney, Camperdown, NSW, Australia

A. Finlay-Jones et al. (eds.), *Handbook of Self-Compassion*, Mindfulness in Behavioral Health,
https://doi.org/10.1007/978-3-031-22348-8_3

and relationships between self-compassion and self-esteem become important and raise several questions. Is there a difference between self-compassion and self-esteem? Is one construct more important than the other? Should these constructs be treated as identical and targeted together through intervention? Does improving one construct through psychological intervention naturally lead to improvement in the other? While much about the relationship between self-compassion and self-esteem remains unknown, a growing body of research offers valuable clinical and research insights.

This chapter seeks to summarize what is currently known about these constructs, why they are important, and the nature of their interactive relationship with each other. Specifically, we explore different versions of self-esteem and how they interact with self-compassion. This chapter will draw together and review a diverse body of studies which has examined self-compassion and self-esteem constructs as targets for psychological intervention. As a result, we will propose that intrinsic self-esteem has a bidirectional relationship with self-compassion. Thus, a sense of self-worth and self-acceptance (intrinsic self-esteem) influences our capacity to relate to oneself with compassion and a motivation to help, not harm (self-compassion), and visa-versa. Extending this description of a bidirectional relationship, this chapter will also propose the two constructs can be thought of as being in an interactive network, with their relationship being highly idiosyncratic and dependent on the individual and context. The implications of the available evidence and some directions for future research will be discussed.

Self-Esteem: Definition, Consequences, and Measurement

Historically, researchers and practitioners have argued almost universally for the value of self-esteem to psychological well-being (Lyubomirsky et al., 2006), yet there remains controversy in how the construct is defined (Levy, 2019; Eromo & Levy, 2017). While self-esteem is generally regarded as a multifaceted construct, at the core of the definition is a judgment of self-worth and self-acceptance (Kernis, 2002; Eromo & Levy, 2017; Deci & Ryan, 1995; Rosenberg 1965a, b). Thus, the self-esteem construct is an affectively laden self-evaluation (Leary & Tangney, 2003), based on one's own values, attributes, and accomplishments. Self-esteem has conceptual similarities to constructs such as self-worth or positive self-regard. Self-esteem is generally conceptualized as a global or *trait* construct (i.e., people's general evaluations of their self-worth), a *state* construct (i.e., more temporary feelings of self-esteem), or a *domain-specific* self-evaluation (i.e., the way that people appraise their performance or worth in a particular domain) (Brown & Marshall, 2006). Further, it has also been proposed that self-esteem can be conceptualized both as a *belief* and as a *motive* (also referred to as conscious and nonconscious self-esteem; Epstein, 2006), with research demonstrating that humans are motivated to create and maintain a positive self-image across the lifespan (James, 1890; Macdonald, 1994).

A considerable body of research supports the psychological benefits of maintaining high self-esteem. Theorists have proposed that high self-esteem may serve several functions, such as maintaining well-being and positive affect, providing feedback about coping efforts, reflecting status in social hierarchies, facilitating self-determination, and providing vital information about eligibility for social inclusion and exclusion (Leary & MacDonald, 2003). High self-esteem is consistently linked with healthy functioning (Leary, 1999) and a range of positive psychological constructs, such as coping with emotional stressors and encouraging development of one's skills and capacities (Pyszczynski et al., 2004). Conversely, low self-esteem predicts psychopathology symptoms (Zeigler-Hill, 2011), including loneliness (Brighi et al., 2012), peer rejection (Ammerman et al., 1993), and suicide ideation (Harter, 1993). Furthermore, low self-esteem has also been consistently linked with poor health behaviors such as cigarette smoking during adolescence (Carters & Byrne, 2013) and illicit substance use (Donnelly et al., 2008). The apparent benefits of having high self-esteem have led to the development of targeted psychological interventions, particularly in schools (Dalgas-Pelish, 2006; Lai et al., 2009; Thijs & Verkuyten, 2017).

Despite the documented benefits of self-esteem, there also appear to be costs to pursuing high self-esteem. Crocker and Park (2004a) reviewed some of the detrimental effects of pursuing high self-esteem, including heightened negative emotions such as shame and sadness when failure is encountered, anxiety at having to "prove" one's self-esteem, and engaging in activities that boost self-esteem yet are ultimately self-defeating (e.g., negative gossip, self-deception). Kernis (2003) explains that high self-esteem can be damaging for one's mental health if characterized by defensiveness, contingence on performance, and instability. Illustratively, high self-esteem may involve positive feelings of self-worth; however, these feelings may also be fragile and highly vulnerable to the environmental context. Engaging in such self-esteem boosting behavior results in interference with activities that satisfy the needs for competence, relatedness, and autonomy that are considered core to well-being (Ryan & Brown, 2003). Ryan and Brown (2003) explain that the pursuit of high self-esteem often leads to behavior which the individual may not value but is seen as worthy by others, thus leading to greater conformity or in some situations greater risk-taking or self-compromising choices.

Different outcomes associated with self-esteem have led researchers to consider whether there are certain facets of self-esteem that are more adaptive than others (Crocker et al., 2003; Kernis, 2003; Kernis et al., 1993). This had led to the identification of several important dimensions of self-esteem, such as *contingent* self-esteem and *intrinsic* self-esteem, that appear to have important implications for psychological well-being.

Contingent Self-Esteem

A core problem with self-esteem is the degree to which one's positive self-evaluations involve comparison with others, successful performance, and perceived success. Contingent self-esteem is a form of self-esteem which emphasizes this problem. Contingent self-esteem refers to the degree that one feels accomplished and worthwhile in relation to others (Harter, 1999) and involves making a judgment of one's place in a social hierarchy. For example, "I have worth, because I am better than you." Contingent self-esteem also involves evaluation of how one is perceived by others. Accordingly, those who pursue contingent self-esteem may be preoccupied with others' opinions and may experience insecurity or worthlessness when others' views of them are unfavorable (Deci & Ryan, 1995). Contingent self-esteem is also referred to as fragile or unstable self-esteem because feelings of self-worth may fluctuate markedly depending on whether or not one is successful and on the feedback received by others. Furthermore, to create a favorable comparison, people with high contingent self-esteem may engage in belittling others (Crocker et al., 1987), defending against negative feedback (Fitch, 1970), and convincing themselves that they are entitled to special treatment. In other words, contingent self-esteem can result in a "zero-sum game" whereby some win and others lose.

Given that contingent self-esteem is based on judgment and comparison of self to others (Eromo & Levy, 2017), the social context is of great importance. This constant comparison to others is problematic, as one's sense of self-worth is almost entirely shaped by who is selected as the comparator. For example, a student attending school X has put in great effort and produced a high-quality piece of work. This same student is told their grade for the work was the highest in the class and, as a result, is likely to feel very proud of their accomplishments and experiences high self-esteem. Let's say that same student completes the same work with the same effort and receives the same grade but is attending school Y with different students and thus is told they ranked 50th in the class. In this scenario, the same student may be at risk of feeling less accomplished and less proud of their work and perhaps not experiencing the same bolstering effect of the feedback on their self-esteem. This hypothetical example illustrates that effort and the quality of work produced can be almost irrelevant to self-esteem. Instead, the comparison to other students in one's immediate context determines the level of pride and subsequent self-esteem felt. In other words, your level of self-esteem depends on whom you are standing next to. Also referred to as the "big fish, little pond" effect, this tendency for a student's academic self-concept to be based on their standing in comparison with their school peers has been measured internationally (Loyalka et al., 2018).

Threats to fragile self-esteem can trigger strong defenses, as if something precious is being taken. For example, attempts to protect one's self-esteem when experiencing social criticism may trigger arrogance or aggression (Walker & Bright, 2009). In other words, a high value is placed on perceived self-esteem, and threats to this are viewed as grave. Sedikides and Alicke (2012) further explain that we all routinely engage in self-enhancement and self-protection behaviors to manage threats to self-esteem. Self-enhancement encompasses efforts to maximize positive views of ourselves, for example, the self-serving bias involves attributing success to our own internal traits and attributing failures to external forces such as an unfair judge, faulty equipment, or poor instruction. Self-protection motives refer to attempts to reduce or minimize negative self-views, for example, the selective self-memory bias may result in systematic failures to recall negative information about oneself. Such processes which seek to preserve one's sense of self-esteem inevitably incur a degree of dishonesty or incorrect information. Yet these same processes can, albeit superficially, maintain self-esteem and thus become heavily reinforced as they serve to support the pursuit of one's goals (Sedikides & Alicke, 2012). Such trends have led to lively debate in the literature questioning whether the benefits of pursuing self-esteem outweigh the costs (Crocker & Park, 2004a, b; Pyszczynski & Cox, 2004). In summary, when contingent self-esteem is high, an individual may be high functioning and experience a positive self-view; however, this form of self-esteem is fragile and encounters many unwanted and unintended consequences when rigidly pursued.

Intrinsic Self-Esteem

In contrast to contingent self-esteem, intrinsic self-esteem refers to a form of self-evaluation that is relatively independent of comparison with others. For example, the statement "I am inherently worthwhile and deserve to be treated fairly" enables one's sense of self-worth to be maintained regardless of environmental context or social comparison. This sense of self-worth is likely to be preserved no matter whom the person compares themselves to. Core to intrinsic self-esteem is an acknowledgement of one's inherent self-worth, independent to accomplishments and comparisons to others, and resulting in self-acceptance and self-liking (Kernis, 2002; Leary, 1999). Self-acceptance is a powerful mindset; to be self-accepting means to be acknowledging and not merely tolerant of but also open and nonresistant to one's flaws. Intrinsic self-esteem also does not incur the same costs as the pursuit of high contingent self-esteem and may serve as a helpful and a strong predictor of psychological well-being (Crocker & Park, 2004a, b; Ryan & Brown, 2003; Kernis, 2002).

Intrinsic self-esteem has long been acknowledged as important. For example, references are made to this concept in the philosophical arguments of the enlightenment period, proposing that all people are born equal and deserving of dignity and respect (Rousseau & May, 2002). Similar concepts to intrinsic self-esteem have been studied using a range of different labels. For example, Deci and Ryan (1995) coined the term "true self-esteem" to reflect an autonomous way of judging oneself which is not a result of achieving outcomes nor social approval. Ryan and Brown (2003) refer to "noncontingent self-esteem" as the experience of oneself as fundamentally worthy of esteem and love. Thus, noncontingent self-esteem serves a protective role, existing distinct from, and thus not dependent on, both successes and failures (Ryan & Brown, 2003). Similarly, Kernis (2003) proposes that distinct from high self-esteem which may be fragile and defensive, "optimal self-esteem" incorporates qualities such as genuine authenticity, stability, and noncontingent self-evaluations. More recently, Eromo and Levy (2017) propose a broader conceptualization of self-appraisal, distinguishing between accurate and distorted forms. This broader proposal enables further acknowledgement of different forms of self-esteem across this continuum of accuracy, for example, positive versus negative forms of self-appraisal (which may have varying levels of accuracy) and stable (consistent across time) or unstable (high fluctuation; Eromo & Levy, 2017). The current chapter focuses on a parsimonious distinction between intrinsic self-esteem, a concept encapsulating these prior definitions of other forms of healthy self-esteem, and more traditional conceptualizations of contingent self-esteem, which are based on a judgment of self-worth based on comparison to others.

Intrinsic self-worth suggests that all people could be considered worthwhile, thus removing the need for a zero-sum game involving a judgment of better versus worst. An additional benefit of intrinsic self-esteem is its stability. A key feature of intrinsic self-esteem is stability in that sense of self-worth across time and context (Kernis, 2005). Intrinsic self-esteem is inherently portable and carried by the individual regardless of who may be available for social comparison or other feedback in the immediate social context. Strong intrinsic self-esteem is more likely to withstand the pressures of daily life and endure instances of failure or shame. Thus, a review of the literature overwhelmingly suggests there are greater benefits and fewer harms to pursuing intrinsic self-esteem in comparison with contingent self-esteem for psychological well-being.

Measurement of Self-Esteem

A diverse range of measures of self-esteem exist (Blascovich & Tomaka, 1991), reflecting the multifaceted nature of this construct. For example, the Self-Esteem Inventory (Coopersmith, 1967) measures positive self-regard in four areas: peers, parents, school, and personal interests. The Social Self-Esteem Scale (Ziller et al., 1969) measures stability of self-esteem, especially when placed under strain. In contrast, the Contingencies of Self-Worth Scale (Crocker et al., 2003) assesses seven sources of contingent self-esteem including academic, appearance, approval from others, competition, family support, God's love, and virtue.

The most commonly used measure of global self-esteem is the brief ten-item Rosenberg Self-Esteem Scale (Rosenberg SES; Rosenberg, 1965a, b). The Rosenberg SES measures one's general feelings of self-worth as a person using items such as "On the whole, I am satisfied with myself" on a four-point Likert scale from *Strongly Agree* to *Strongly Disagree* (Rosenberg, 1965a, b). The Rosenberg SES measures a generalized and global assessment of one's feelings of self-worth as a person. A closer examination of the ten items suggests this scale may reflect intrinsic self-esteem rather than the comparative elements of self-esteem, albeit imperfectly. Items such as "I take a positive attitude toward myself" and "On the whole, I am satisfied with myself" (Rosenberg, 1965a, b) may arguably reflect a sense of self-worth with reduced dependency in comparison with others or context. While there is no specific measure of intrinsic self-esteem, global measures such as the Rosenberg SES may tap into a sense of self-worth that is global, noncontingent on comparison, and enduring, which more closely resembles intrinsic self-esteem than contingent self-esteem.

Self-Compassion: Definition, Consequences, and Measurement

Gilbert (2009) developed a framework of "compassion" based on evolutionary and attachment theory (see Chap. 4). Gilbert's model proposes there are three types of emotion regulation systems which constantly interact: the self-soothing, safe system; the threat and protection system; and the drive, excitement-seeking system (Gilbert, 2009, 2014). Gilbert (2009) conceptualizes both compassion for self and other as part of the self-soothing, affect regulation system. The threat and protection system evolved to enhance our threat-detection abilities and to quickly mobilize us to act with the goal of self-preservation. Strong emotions linked to this system include anxiety, anger, and disgust. The drive and excitement system are activated when engaging in behaviors that strive to accomplish rewards and resources. Based on an evolutionary perspective, these resources can include food, alliances, sexual opportunities, achievements, and validation or territories. Importantly, the self-soothing system is triggered during contentment, when an individual is not perceiving threat or engaging in resource-seeking. This system is associated with feelings of securing, peacefulness, well-being, and safety (Gilbert, 2010).

In a complementary approach, Neff (2003b) developed a framework for self-compassion based on Buddhist concepts. This framework comprises a bipolar continuum ranging from uncompassionate to compassionate self-responding. Compassionate self-responding involves self-kindness or being supportive and caring toward oneself during times of difficulty; common humanity, or an acknowledgement of the imperfect nature of being human; and mindfulness, which in this context refers to a healthy detachment from one's thoughts and feelings. Conversely, uncompassionate self-responding involves self-criticism or a tendency to be harsh and judgmental toward one's perceived flaws; isolation, or feeling separate to others and alone; and overidentification which is characterized by feeling fused and stuck with our thoughts and feelings, especially those which cause distress. Initially framed as a healthy alternative to self-esteem, Neff (2003b) argued that self-compassion does not come with the same costs attached to the pursuit of contingent self-esteem, such as the development of narcissistic traits, self-centeredness, and a lack of concern for others.

The exponential growth in self-compassion research has largely been unified through use of the Self-Compassion Scale (SCS), a 26-item measure of the different facets of self-compassion as conceptualized by Neff (2003a, b). Most self-compassion research utilizes the SCS or its Short Form (SCS-SF; Raes et al., 2011). Newer versions of this scale have also been developed for specific populations including early adolescence (SCS-Y; Neff et al., 2021) and adults diagnosed with diabetes (SCS-D; Tanenbaum et al., 2018). A state-based version of the scale has also been developed to capture self-compassion in the moment, likely to be of interest in experimental studies and when examining changes in self-compassion across time (Neff et al., 2021). The scale has also been translated into many different languages including Japanese (SCS-J; Arimitsu, 2014) and Iranian (Azizi et al., 2013) and Brazilian (de Souza & Hutz, 2016).

Research has consistently found that higher levels of self-compassion are robustly associated with a plethora of benefits across the lifespan, such as greater psychological well-being (Zessin et al., 2015), decreased symptoms of psychopathology (MacBeth & Gumley, 2012; Marsh et al., 2018), and increased health-promoting behaviors (Sirois et al., 2015). Furthermore, meta-analyses have consistently found support for the efficacy of self-compassion-based interventions for improving psychological well-being outcomes. Self-compassion-based interventions can enhance one's ability to be self-compassionate (Kirby et al., 2017; Ferrari et al., 2019) and lead to other beneficial psychological outcomes such as reduced anxiety and depression (Ferrari et al., 2019; Kirby et al., 2017) and improvements in rumination, eating behaviors, mindfulness, and life satisfaction (Ferrari et al., 2019).

Are Self-Compassion and Self-Esteem Distinct Concepts?

Self-compassion and self-esteem, particularly intrinsic self-esteem, are psychological constructs which are closely related and overlap. At high levels of self-compassion and intrinsic self-esteem, the differences between these constructs are less clear. Individuals high in both constructs are likely to present in a similar way: psychologically flourishing with a sense of purpose, meaningful relationships, and a generally positive attitude to their sense of self. Indeed, research suggests there is a strong correlational relationship between self-compassion and global measures of self-esteem, which may more accurately reflect features of intrinsic rather than contingent self-esteem (Souza & Hutz, 2016; Stephenson et al., 2018; Neff & Vonk, 2009; Eller et al., 2014; Pohl et al., 2021; Holas et al., 2021; Thoma et al., 2021). In contrast, contingent self-esteem appears more likely to have a very weak relationship with self-compassion given its outward focus on social comparison (Eromo & Levy, 2017), fragility (Walker & Bright, 2009), and propensity toward artificial self-enhancement (Sedikides & Alicke, 2012).

The following section of this chapter will consider the research comparing self-compassion with different forms of self-esteem. We will argue that contingent self-esteem is not likely to be positively related to self-compassion, but intrinsic self-esteem (meaning an acceptance of self) is likely to have a bidirectional relationship with self-compassion. This section will build toward the proposal that self-compassion and intrinsic self-esteem can be thought of as constructs connected in an interactive network (Ciarrochi et al., 2021), with improvements in one leading to improvements in the other, and vice versa.

Self-Esteem and Self-Compassion Are Correlated

Research suggests there is a meaningful relationship between self-compassion and global measures of self-esteem, such as the Rosenberg SES (Souza & Hutz, 2016; Stephenson et al., 2018; Neff & Vonk, 2009; Eller et al., 2014; Holas et al., 2021; Thoma et al., 2021). As argued above, global measures of self-esteem may more readily reflect features of intrinsic self-esteem than contingent self-esteem and be related to self-compassion. Illustratively, a study of 432 Brazilian citizens reported a moderate positive correlation between the Rosenberg SES and the SCS (Souza & Hutz, 2016). Similarly, in a study involving 184 US undergraduate students, Stephenson et al. (2018) found that self-esteem and self-compassion, using the same Rosenberg SES and SCS-SF measures, were positively moderately correlated. Neff and Vonk (2009, Study 1) surveyed a large adult community population from the Netherlands using a battery of self-report measures of varied psychological outcomes. The authors examined correlations of the SCS total score with different forms of self-esteem in a large cross-sectional community sample. They found significant, positive, and moderate correlations between self-compassion and global self-esteem (measured by Vonk et al.'s (2008) scale, deemed equivalent to Rosenberg SES), which were significantly and negatively correlated with contingent forms of self-esteem including a focus on social approval, appearance, performance, and social comparison.

Theoretically, the negative correlation between self-compassion and contingent self-esteem found in Neff and Vonk's (2009) study can be explained with reference to Neff's (2003a) self-compassion model. Contingent self-esteem contrasts to each of the three components. Instead of self-kindness, contingent self-esteem is associated with aggressive competitiveness (Koivula et al., 2002); instead of mindfulness, it encourages overidentification and fusion with one's self-evaluation (Rohmann et al., 2019); and instead of common humanity, it encourages comparison with – and potentially devaluation of – others (Schütz & Tice, 1997) or devaluation of self (Alfasi, 2019). According to Gilbert's compassion model, high levels of contingent self-esteem may also act as a barrier to self-compassion. This is because the competitive focus of self-esteem, based on comparisons to others, is likely to be related to the threat and drive/excitement systems (Gilbert, 2015). The drive/excitement system may encourage a sense of competitiveness and focus on one's place within a social hierarchy, attempting to rank as superior to others (Gilbert, 2015). Similarly, the threat system may also encourage a sense of competitiveness with others: competition over resource security or safety (Gilbert, 2015). Thus, the sense of comparison which self-esteem often encourages is likely to activate the systems which undermine one's ability to engage in self-compassion and activate the self-soothing system.

Further research has examined the relationship between self-esteem measures and the subscales of the SCS in more detail. Eller et al. (2014) studied self-report data collected from participants diagnosed with HIV across the United States, Puerto Rico, Canada, Namibia, China, and Thailand. The authors found a significant, small, and negative correlation between higher levels of self-esteem (as measured by the Rosenberg SES) and lower levels of self-judgment (negative subscales of the SCS) but reported no significant correlation with self-kindness (positive subscales of the SCS). Thoma et al. (2021) reported in a supplementary table attached to their paper that in adults who had been maltreated as children, self-esteem (as measured by a German version of the Rosenberg SES) did not significantly correlate with the self-kindness nor self-judgment subscales of the SCS (German translation) but positively correlated with the Common Humanity and Overidentified subscales and negatively correlated with Isolation and Mindfulness subscales. Taken together, such cross-sectional research suggests that although self-compassion and self-esteem are related, correlations across different measures and subscales tend to vary, suggesting these constructs are distinct from each other.

Self-Compassion and Self-Esteem Differentially Predict Outcomes

Self-compassion and intrinsic self-esteem are significantly correlated and may be considered related constructs (Souza & Hutz, 2016; Neff & Vonk, 2009; Eller et al., 2014; Thoma et al., 2021), yet clearer differences between these constructs emerge when we examine research findings about their predictive relationship with other psychological outcomes. Some studies support self-compassion as a more stable predictor of self-worth (Neff & Vonk, 2009) and authenticity (Zhang et al., 2019) than self-esteem. Further analyses reported by Neff and Vonk (2009, Study 1) found self-compassion had a stronger predictive relationship compared to global self-esteem with several outcomes including self-esteem stability, global self-esteem contingency, specific areas of self-esteem contingency (social approval, performance, and appearance), and social comparison. Self-esteem stability refers to daily fluctuations in feelings of self-worth, while global self-esteem contingency refers to self-esteem that is contingent on outcomes such as receiving social approval from others, performing to a high standard, and having an appearance which is culturally understood to be attractive. The authors suggest this predictive strength of self-compassion may be a result of self-compassionate individuals embracing all aspects of themselves, the good and the unpleasant, with an open-hearted awareness. In comparison, high self-esteem may encourage a focus on the positive and desirable aspects of self and an avoidance of the undesirable. The exception to this was narcissism, which was predicted by global self-esteem but not self-compassion, potentially supporting the argument that pursuing and maintaining self-esteem may incur unwanted consequences such as extreme self-involvement. Yet in further analyses, Neff and Vonk (2009, Study 2) found self-compassion and self-esteem were equivalent predictors of positive mood states including happiness, optimism, and positive affect.

Authenticity refers to a sense of alignment with one's true or genuine self and is a construct of interest because it is associated with greater psychological well-being, life satisfaction, and positive affect (Kernis & Goldman, 2006; Toor & Ofori, 2009). Across five studies, Zhang et al. (2019) found self-compassion was a stronger predictor of authenticity than self-esteem. Zhang et al. (2019) found a positive correlation between self-compassion (measured by the SCS-SF) and authenticity when controlling for self-esteem in university students (measured by Rosenberg SES; Study 1), which was replicated using a daily diary method of data collection to test the ecological validity of the finding (Study 2). Study 3 experimentally induced a state of self-compassion in participants and found this resulted in higher self-reported authenticity than a self-esteem condition and a control group. This predictive relationship was further replicated across Iranian, Malaysian, Turkish, and American populations and longitudinally (Study 4 and 5). Study 4 and Study 5 also found the link between self-compassion and authenticity could be explained by reductions in fear of negative evaluation and heightened optimism. Taken together, this cumulative series of studies demonstrates that self-compassion is a stronger predictor of authenticity, over and above self-esteem (Zhang et al., 2019).

Additionally, some studies have found that self-compassion shares stronger associations with mental health outcomes than self-esteem. For example, for males living with a positive HIV status, uncompassionate self-responding as measured by the SCS (all negative items summed to create a composite score) was a stronger predictor of depression symptoms than self-esteem, self-efficacy, HIV symptoms, and demographic variables (Eller et al., 2014). Pohl et al. (2021) examined self-compassion and self-esteem in adults with borderline personality disorder (BPD) who had experienced childhood trauma. In this group, self-compassion moderated the positive correlation between childhood trauma and BPD symptom severity, but self-esteem did not. These findings suggest that higher self-compassion may weaken or reduce the negative psychological consequences of childhood trauma, but self-esteem does not offer the same protection. In addition, Stephenson et al. (2018) found that self-esteem did not predict irrational beliefs, such as unrealistic personal standards, after accounting for level of self-compassion, thus suggesting that many benefits of having high self-esteem can be accounted for by self-compassion.

In contrast, a smaller number of studies have found support for self-esteem as a stronger predictor of psychological well-being outcomes, over and above self-compassion. One such study, Thoma et al. (2021) found that self-esteem, but not self-compassion, mediated the effect of childhood abuse on mental health. Of note, this study was based on a sample of Swiss older adults (mean age 70 years) who were identified as having been affected by compulsory child welfare services as children, and the data were collected retrospectively. It is not clear why this study differs in finding support for self-esteem over self-compassion. Potential explanations include the older population assessed by the study, the nature of the challenges they faced as children, a result of the method of retrospective data collection, or a genuine reflection of the importance of self-esteem over self-compassion. Thus, although some research found self-esteem a stronger predictor of mental health symptoms compared to self-compassion (Thoma et al., 2021), a larger body of research supports self-compassion as a stronger predictor over self-esteem for a range of psychological outcomes. Such outcomes are diverse and include self-worth (Neff & Vonk, 2009), authenticity (Zhang et al., 2019), depression (Eller et al., 2014), and BPD symptoms (Pohl et al., 2021).

Self-Compassion May Mediate the Beneficial Impact of Self-Esteem

Neff (2003b) acknowledged there are benefits that come with high self-esteem, however, at times individuals can use unhealthy methods to obtain it, and that self-esteem may have unhealthy consequences. Neff (2003b) proposed that self-compassion can overcome many of the shortcomings associated with the pursuit of self-esteem. For example, the pursuit of high self-esteem may lead to an inflated sense of self which seeks to artificially elevate oneself and disparage others (since one's value is defined in relation to others). Such elevated self-esteem may leave one to be underprepared for times of struggle or failure and unable to have empathy for others. In contrast, self-compassion is not contingent on such artificial judgments. Instead, core to the self-compassion model is an acknowledgement of the natural imperfections and failings of common humanity, encouraging compassion to these aspects of ourselves rather than using denial or suppression to avoid them. Thus, self-compassion may be beneficial for individuals with high self-esteem as they not only have a positive sense of self-worth, but this is based on a more grounded and realistic acceptance of the imperfections of being human. Such individuals may also be better equipped to cope in times of error or embarrassment, being open and accepting to one's flaws while retaining a sense of self-worth.

A recent study which empirically supports the notion that self-compassion may explain the beneficial impacts of self-esteem, especially intrinsic self-esteem, was conducted by Holas et al. (2021). The authors found a moderately strong and significant correlation between self-compassion (SCS total score) and self-esteem (Rosenberg SES) in socially anxious adults. While low self-compassion and low self-esteem significantly predicted more problematic anxiety symptoms, self-esteem was a stronger predictor compared to self-compassion. In addition, self-compassion partially mediated the relationship between self-esteem and social anxiety. The authors interpret these findings to suggest that self-compassion can buffer the negative effects of contingent features of self-esteem, such as a sense of self-worth contingent on competition and appearance. In other words, the presence of self-compassion contributes protective features to one's sense of self-worth. Based on these findings, Holas et al. (2021) propose that self-compassion may involve additional benefits over self-esteem such as facilitating a more balanced and rational stance toward life adversities, resulting in less unhelpful anxiety and greater acceptance that imperfections are part of being human.

Convergent findings by DeLury and Poulin (2018) suggest that self-compassion buffers the effect of a self-esteem threat on academic task performance. The authors conducted an experiment where first year psychology students were randomly allocated to a self-esteem threat, where participants were asked to write in detail about a negative academic event in their past that caused shame or to describe their travel to campus that morning (control). Subsequently, participants were randomly allocated to a neutral writing task or a self-compassion writing task. The self-compassion task prompted participants to consider that everyone has experienced something similar (common humanity), consider how they'd treat a friend in the same situation (kindness), and describe triggered emotions in an accepting fashion (acceptance). Finally, all participants completed a difficult verbal analogy test. As anticipated, those who received the self-compassion induction did not perform poorly on the test after exposure to the threat. Those who received the threat and did not receive the self-compassion induction had a much poorer performance on the difficult test. This experiment supports the role of self-compassion in serving as a buffer against threat effects on performance.

A Comparison of Self-Compassion and Self-Esteem Focused Interventions

The nature and effects of self-compassion and self-esteem are further explored in experimental trials which compare psychological interventions developed to target these constructs. Most such studies have focused on similar outcomes (body image related concerns) and adopt a similar psychological intervention (brief writing tasks). Understanding how self-compassion- and self-esteem-based interventions compare in their effectiveness and whether there are meaningful differences provides valuable insight to the nature and effects of these constructs. A recent meta-analysis of CFT and other compassion-based interventions ($k = 8$) found that, notwithstanding considerable heterogeneity, these interventions had a medium positive effect on self-esteem (Thomason & Moghaddam, 2021). This meta-analysis noted that brief self-compassion-based interventions tended to show little improvement in self-esteem or reported large confidence intervals indicating poor reliability. Thomason and Moghaddam (2021) propose these results reflect the need for sufficient time within an intervention (at least 20 hours) to overcome discomfort and distress reactions to self-compassion, especially when early childhood experiences may lack the opportunity to develop a soothing emotion regulation system. The small number of studies included in the review calls for further research in this area to consolidate and extend on these promising findings. To the best of our knowledge, there has not yet been a review of the effectiveness of self-esteem-based interventions for self-compassion outcomes.

Several papers compare the effectiveness of self-compassion- and self-esteem-based interventions. Moffitt et al. (2018) compared self-esteem, self-compassion, and positive distraction writing tasks and measured their effects on reducing state body dissatisfaction. One hundred forty-nine female undergraduates (mean age 22 years) spent 15 min looking at and rating a series of 16 magazine advertisements on a computer which featured thin, young women. This task was designed to trigger body dissatisfaction in the participants. Subsequently the participants were asked to write a paragraph to themselves for 3 min according to instructions. They were randomly allocated to one of three intervention groups with self-esteem, self-compassion, or positive distraction instructions. The self-esteem group was asked to write a description of their positive qualities such as personal attributes and accomplishments. The self-compassion group was asked to express kindness, compassion, and understanding toward their weight, shape, and appearance. In contrast, the positive distraction group acted as a control comparison and was asked to write about an enjoyable hobby. As anticipated, the self-compassion group reported significantly lower dissatisfaction with weight and appearance, as well as significantly higher self-improvement motivation compared to the self-esteem and positive distraction groups. In addition, participants with high trait body dissatisfaction, independent of the experiment, benefited the most from the self-compassion intervention. Moffitt et al. (2018) concluded that these findings provide robust support for the effectiveness of self-compassion, over and above self-esteem interventions and controls, for body dissatisfaction.

Seekis et al. (2017) used a 15-minute writing task to trigger body image concerns in female university students. Ninety-six female university students (mean age 19 years) were asked to read a hypothetical scenario and imagine they were the protagonists who had unflattering photos of themselves posted on social media by a friend. The researchers then randomly allocated each participant to a self-compassion, self-esteem, or control writing group. The self-compassion group was given writing prompts that were designed to encourage the core elements of self-compassion. For example, to induce self-kindness, participants were instructed "Write a letter to yourself expressing understanding, kindness, and concern. Write in a way you might express concern to a close friend who experienced a similar event." In comparison, the self-esteem group was given prompts such as "Describe why an unflattering photo of yourself does not really indicate anything about the kind of person you are" and asked to list their competent characteristics. The control group was given prompts about irrelevant information such as "List the subjects you did in your final year of high school. Which did you like or dislike and why?" After the intervention, both the self-compassion and self-esteem groups showed higher body satisfaction than the control group. The benefit of self-compassion was evident in the difference in body appreciation scores, which were higher than both the self-esteem and control groups after the intervention. These effects, however, were not sustained at the 2-week follow-up (Seekis et al., 2017). Like the findings of Moffitt et al. (2018), this study provides support for the benefit of targeting self-compassion over self-esteem to improve body satisfaction; however, these benefits appeared to be short term and not sustained at a 2-week follow-up.

Barbeau et al. (2021) also found support for targeting self-compassion to address body appreciation and healthy eating in women. They randomly allocated university and community recruited women (mean age 29 years) to one of three writing groups; each was required to record daily writing activities for 7 consecutive days. The self-compassion and self-esteem groups wrote a journal recount via email on a moment over the last 24 hours when they felt self-conscious about their body, exercise, or eating habits. After this journal recount, they were given either a self-compassionate or self-esteem focused prompt to reflect on the event. The same procedure was carried out for the control group but focused on a particular event, not related to self-consciousness. Comparing the three groups after the intervention revealed the self-compassionate group experienced clinically significant changes in bulimic symptoms while the other two groups did not.

Similarly, Albertson et al. (2015) also found support for self-compassion training in reducing contingent self-esteem based on appearance. Women with body image concerns were recruited through social media ($N = 238$) and randomized to a 3-week self-compassion meditation group ($M_{age} = 38.42$) or a waitlist control ($M_{age} = 36.42$). The intervention included access to podcasts containing a 20-minute self-compassion meditation which participants were asked to practice once a week, such as the Compassionate Body Scan and a variant of the Loving-Kindness Mediation. Participants reported practicing the podcasts 3.6 days each week, demonstrating reasonable intervention adherence. The intervention group demonstrated significantly greater gains in self-compassion (as measured by SCS) and greater reductions in contingent self-worth based on appearance (CSW, Contingent Self-Worth subscale). These results were maintained at a 3-month follow-up, suggesting that self-compassion practice can significantly improve appearance-based self-worth and that such effects are sustained with time.

In addition to body image concerns, a willingness to disclose self-esteem threatening events to others is a behavior which relates to psychological well-being and may increase opportunity for an individual to receive required support or assistance. Dupasquier et al. (2020) found that practicing self-compassion promotes the disclosure of self-esteem threatening events. Disclosing such information which would typically trigger shame and social withdrawal is an important and helpful behavior. Participants were asked to write in detail about an event that occurred in the past 5 years and made them feel bad about themselves at present. Participants were then randomly allocated to an experimental writing manipulation that used writing prompts to promote self-compassion, self-esteem, or free writing (control group). The self-esteem and self-compassion conditions both led to deeper and lengthier disclosures than the control condition which involved writing about the event in a nondirective way. The self-compassion group had a slight advantage in producing lengthier written responses than the self-esteem group, while there was no significant difference in depth of content. Dupasquier et al. (2020) concluded that engaging in a self-compassion exercise and repairing self-esteem are both effective in encouraging disclosure of distressing information, with self-compassion offering a slight advantage through promoting lengthier responses. Such responses allow greater opportunity for individuals to access social support to protect long-term psychological well-being.

Psychological intervention studies which compare self-compassion- with self-esteem-based interventions consistently demonstrated support for self-compassion over self-esteem approaches (Moffit et al., 2018; Seekis et al., 2017; Barbeau et al., 2021, Alberston et al., 2015). There were strong commonalities between these intervention studies in relation to the population studied (young, university female students) and the outcomes targeted (related to body image concerns); therefore, it is not clear whether these findings readily generalize to other populations or psychological processes. Despite these limitations, these intervention studies show growing support for the clinical utility of self-compassion-based interventions. Specifically, self-compassion-based writing tasks were more efficacious compared to self-esteem-based writing tasks for improving body dissatisfaction, self-improvement motivation, and clinical symptoms of bulimia (Barbeau et al., 2021; Moffitt et al., 2018; Seekis et al., 2017). In addition, self-compassion practice also seems to weaken contingent forms of self-esteem (Albertson et al., 2015) and offer a slight advantage to disclosure of distressing information compared to self-esteem (Dupasquier et al., 2020). Self-compassion interventions also tend to contribute to improvements in self-esteem (Thomason & Moghaddam, 2021). Such research confirms that increasing self-compassion compared to increasing self-esteem through psychological intervention results in meaningful psychological differences.

How Self-Compassion and Self-Esteem Interact: Longitudinal Research

Longitudinal research which measures self-compassion and self-esteem over an extended time period provides further insight to the potential bidirectional and causal relationship between these constructs. Using experience sampling via mobile phone, Krieger et al. (2015) examined self-compassion and self-esteem in relation to positive and negative affect over a 2-week period in a community sample (n = 105). They found that both self-compassion (using a German translation of the SCS) and self-esteem (a German translation of the Rosenberg SES) were positively correlated with positive affect and negatively correlated with negative affect and perceived stress. Interestingly, after controlling for the effect of self-esteem, self-compassion continued to predict both positive and negative affect. When the influence of self-compassion was controlled, however, self-esteem no longer predicted these outcomes. This longitudinal study lends further support to the understanding that while both self-compassion and self-esteem are helpful for cultivating good mental health, self-compassion may have a greater effect, independent of self-esteem. In addition, the authors found that beneficial effects of self-esteem on positive and negative affect are largely explained by self-compassion.

A pivotal study which may inform our understanding of self-compassion and self-esteem's bidirectional relationship was a longitudinal study of 2488 high school students conducted by Marshall et al. (2015). The authors collected measures of self-compassion (SCS-SF) and self-esteem (Rosenberg SES) in grade 9 and 10 and measured general mental health in grade 10. The authors used structural equation modeling to assess how the two constructs interacted and predicted general mental health. This study found that high levels of self-esteem in grade 9 were related to better mental health in grade 10, regardless of one's level of self-compassion. Self-compassion became important, however, when an adolescent reported low self-esteem. In these instances, low self-compassion appeared to exacerbate the negative effects of low self-esteem. In contrast, those who were also high in self-compassion experienced a protective buffer which weakened the link between low self-esteem and subsequent poor mental health. This robust study found that self-compassion and self-esteem have independent longitudinal effects on changes to adolescent mental health. In addition, an ability to practice self-compassion and accept personal failings as normal appears to protect against negative self-judgments and thus weaken the negative consequences of low self-esteem. Thus, while similar, the self-compassion construct appears to offer additional psychological benefits which mitigate the harms of low self-esteem.

Examining changes in self-compassion and self-esteem over time can also provide insight into how these constructs might engender each other. To explore this, Donald et al. (2018) collected data from 2809 Australian school students spanning 17 schools (M_{age} = 14.7 years, SD = 0.45). Consistent with prior research (Marshall et al., 2015; Krieger et al., 2015), Donald et al. (2018) used the SCS-Short Form and the Rosenberg SES. Across 4 years, from grade 9 to 12, students completed surveys at the same time each year. Using autoregressive cross-lagged modeling, the study found high self-esteem consistently predicted improvement in self-compassion, yet self-compassion did not predict self-esteem. These findings suggest that developmentally, self-esteem may be an antecedent of the development of self-compassion, but self-compassion is less important as a foundation to develop self-esteem. Based on these findings, Donald et al. (2018) suggested that a positive self-evaluation, or sense of self-worth, may more readily give rise to self-compassionate responses when an individual is faced with difficulty. Thus, the presence of negative self-evaluations, or a lack of a basic sense of self-worth, may encourage the perception of self-compassion as threatening and not deserved. Donald et al. (2018) argue that during the vulnerable time of adolescence, the types of self-evaluations that arise appear to have a critical influence on naturally developing a self-compassionate response.

The longitudinal studies reviewed in this chapter (Donald et al., 2018; Krieger et al., 2015; Marshall et al., 2015) suggest that there is a bidirectional relationship between self-esteem and self-compassion. Krieger et al. (2015) found in community adults that the link between self-esteem and mental health disappeared when controlling for self-compassion. Similarly, Marshall et al. (2015) found that across time, adolescents who were high in self-compassion were protected from the poor mental health effects of low self-esteem, whereas Donald et al. (2018) found that self-esteem predicted self-compassion in adolescents, but not vice versa. Taken together, research suggests that feelings of unworthiness (low self-esteem) often lead to poor mental health, but this link is weakened by high levels of self-compassion (Marshall et al., 2015; Krieger et al., 2015), while in other studies, self-esteem is a predictor of self-compassion but not vice versa (Donald et al., 2018). These findings may suggest the two constructs can be thought of as being in an interactive network, the relationship between constructs changing in an interactive way across different individuals in different contexts.

The Self-Compassion and Self-Esteem Interaction: Application of the Interactive Network Model

This conceptualization of self-esteem and self-compassion as part of an interactive network, rather than distinct and entirely separate constructs, aligns with a bigger movement in psychology to consider mental health diagnoses and psychological interventions as processes rather than distinct packages (Ciarrochi et al., 2021). Traditional psychological models tend to ask which diagnosis best fits the presenting symptoms and which evidence-based treatment best matches the diagnosis. In contrast, an interactive model approach asks, based on this specific client, their circumstances, at this stage of intervention, what biopsychosocial processes should be targeted and how? Illustratively, Ciarrochi et al. (2021) use the example of a depression diagnosis. Rather than conceptualizing depression as a list of distinct symptoms such as depressed mood, significant weight change, sleep disruption, and recurrent thoughts about death, it may be more helpful to think of depression as a series of processes in an interactive network. For example, depression may represent a cyclical interactive network of hopelessness, rumination, sadness, and low behavioral activation which feeds into and amplifies itself. This reframing as a dynamic interactive network rather than a list of symptoms may be more useful for understanding the development, maintenance, and exacerbation of symptoms, in addition to formulating and sequencing a treatment plan.

The interactive model also provides a useful framework for tailoring psychological treatment to the needs of the individual. Continuing with the example of a depression diagnosis, evidence-based treatment packages for depression may include cognitive-behavioral therapy which targets automatic negative thoughts and dysfunctional beliefs, acceptance and commitment therapy which targets psychological flexibility, or compassion-focused therapy which develops the self-soothing emotion regulation system. Such treatment packages are overlapping, often complex, and target multiple processes that will influence different people in different environmental and cultural contexts differently. Adopting an interactive network approach (Ciarrochi et al., 2021) encourages clinicians to focus on processes of change which target the idiosyncratic presentation of the individual rather than treatment packages. Such processes of change may include components of these treatment packages such as cognitive restructuring to target hopelessness, mindfulness to target rumination, emotional acceptance to target sadness, and behavioral activation to target inactivation. Within the interactive network model, clinicians begin with a detailed formulation of the processes involved in client's presenting problem and select processes of change to target each, rather than following a treatment package and tweaking all components to fit the client.

Applied to self-esteem and self-compassion, an interactive network model suggests the practice of self-compassion is likely to also increase one's feelings of worthiness and acceptance of flaws or imperfections. At the same time, a sense of self-worth is likely to make it easier to accept feelings of compassion and acceptance to oneself. While this reciprocal relationship has not directly been tested through intervention studies, this is supported by longitudinal research (Donald et al., 2018; Krieger et al., 2015; Marshall et al., 2015). From a treatment perspective, intervention research reviewed in this chapter overwhelmingly supports self-compassion-based interventions as more effective than those focused on self-worth across a number of psychological outcomes (Barbeau et al., 2021; Moffitt et al., 2018; Seekis et al., 2017; Albertson et al., 2015; Dupasquier et al., 2020). In addition, self-compassion interventions result in modest improvements in self-esteem (Thomason & Moghaddam, 2021). It may be the case that the practice of self-compassion and acceptance of humanity's natural flaws may intuitively encourage an individual to further develop a sense of self-worth. Yet the interactive model suggests that the relative importance or focus on one's self-worth and ability to practice self-compassion are likely to be dynamic and change constantly rather than be fixed. Further intervention research would benefit from examining changes in self-compassion and self-esteem across time and in relation to each other, particularly as longitudinal research (Donald et al., 2018; Krieger et al., 2015; Marshall et al., 2015) suggests these constructs are best understood within an interactive network.

A key advantage of the network approach is that it is naturally ideographic, i.e., it draws attention to the fact that different individuals may have different configurations of the network. As a result, it may not be just a matter of self-compassion and self-esteem influencing each other more or less similarly for most people, but rather the two may be linked to each other directly and/or via other variables differently in different people across different contexts. The studies reviewed in this chapter indicated some variation in the relationship measures between self-esteem and self-compassion, and these differences were present despite analyses being conducted on larger groups of participants; no studies reported data or results on an individual basis. The network approach supports the need for more person-centered and highly tailored interventions for promoting self-compassion rather than one-size-fits-all treatment protocols.

Conclusion

The aim of this chapter was to review research which provides insight to the relationship between self-compassion and self-esteem. This chapter identified a version of self-esteem, intrinsic self-esteem, which is likely to be beneficial to the development and maintenance of self-compassion. Intrinsic self-esteem contrasts to contingent self-esteem, which refers to a focus on thinking positively and boosting your self-worth in comparison with others. Attempts to pursue or maintain contingent self-esteem frequently incur negative consequences and are likely to be unrelated to self-compassion (Crocker et al., 1987; Loyalka et al., 2018; Walker & Bright, 2009). This chapter has reviewed empirical research which suggests that intrinsic forms of self-esteem and self-compassion are significantly correlated and there is likely overlap between these constructs (Souza & Hutz, 2016; Neff & Vonk, 2009; Eller et al., 2014), yet there are also important differences. These differences are reflected by findings that self-compassion is a stronger predictor of psychological outcomes than self-esteem including a more stable predictor of self-worth (Neff & Vonk, 2009), authenticity (Zhang et al., 2019), and BPD symptoms (Pohl et al., 2021). A much smaller body of research found the reverse, that self-esteem was a stronger predictor of mental health outcomes than self-compassion, specifically the effect of involvement of child welfare services on poor mental health later in life (Thoma et al., 2021).

In addition, self-compassion can buffer or mitigate some of the harmful consequences of high self-esteem, such as reducing social anxiety (Holas et al., 2021) and improving academic performance when self-esteem is threatened (DeLury & Poulin, 2018). Extending our knowledge on these relationships further, self-compassion-based interventions also appear to be more effective in promoting psychological well-being (Moffitt et al., 2018; Seekis et al., 2017), reducing bulimia symptoms (Barbeau et al., 2021), weakening harmful contingent self-esteem (Albertson et al., 2015), and increasing disclosure to supportive others of self-esteem threatening events (Dupasquier et al., 2020). Further longitudinal research suggests that feelings of unworthiness often lead to poor mental health, unless accompanied by high levels of self-compassion (Donald et al., 2018; Krieger et al., 2015), while in other studies, self-esteem is a predictor of self-compassion but not vice versa (Marshall et al., 2015).

Drawing together this body of research on the relationship between self-compassion and self-esteem, especially intrinsic self-esteem, it appears this relationship is best understood as bidirectional and connected within an interactive network (Ciarrochi et al., 2021). This means that while distinct, feeling worthy and practicing compassion are closely linked psychological constructs, and improvements in one are likely to lead to improvements in other, and vice versa. Future research could expand our knowledge of these constructs through applying the network approach and using this to guide treatment selection for those who would benefit from cultivating their self-compassion response style and sense of self-worth.

References

Albertson, E. R., Neff, K. D., & Dill-Shackleford, K. E. (2015). Self-compassion and body dissatisfaction in women: A randomized controlled trial of a brief meditation intervention. *Mindfulness, 6*(3), 444–454. https://doi.org/10.1007/s12671-014-0277-3

Alfasi, Y. (2019). The grass is always greener on my Friends' profiles: The effect of Facebook social comparison on state self-esteem and depression. *Personality and Individual Differences, 147*, 111–117.

Ammerman, R. T., Kazdin, A. E., & Van Hasselt, V. B. (1993). Correlates of loneliness in nonreferred and psychiatrically hospitalized children. *Journal of Child and Family Studies, 2*(3), 187–202. https://doi.org/10.1007/BF01321330

Arimitsu, K. (2014). Development and validation of the Japanese version of the self-compassion scale. *Japanese Journal of Psychology*.

Azizi, A., Mohammadkhani, P., Lotfi, S., & Bahramkhani, M. (2013). The validity and reliability of the Iranian version of the Self-Compassion Scale. *Practice in Clinical Psychology, 1*(3), 149–155.

Barbeau, K., Guertin, C., Boileau, K., & Pelletier, L. (2021). The effects of self-compassion and self-esteem writing interventions on women's valuation of weight management goals, body appreciation, and eating behaviors. *Psychology of Women Quarterly, 46*(1), 82–98. https://doi.org/10.1177/03616843211013465

Blascovich, J., & Tomaka, J. (1991). Measures of self-esteem. In J. P. Robinson, P. R. Shaver, & L. S. Wrightsman (Eds.), *Measures of personality and social psychological attitudes* (pp. 115–160). Academic. https://doi.org/10.1016/B978-0-12-590241-0.50008-3

Brighi, A., Melotti, G., Guarini, A., Genta, M. L., Ortega, R., Mora-Merchán, J., et al. (2012). Self-esteem and loneliness in relation to cyberbullying in three European countries. *Cyberbullying in the Global Playground: Research from International Perspectives*, 32–56. https://doi.org/10.1002/9781119954484.ch3

Brown, J. D., & Marshall, M. A. (2006). The three faces of self-esteem. In M. H. Kernis (Ed.), *Self-esteem: Issues and answers* (pp. 4–9). Psychology Press.

Carters, M. A., & Byrne, D. G. (2013). The role of stress and area-specific self-esteem in adolescent smoking. *Australian Journal of Psychology, 65*(3), 180–187. https://doi.org/10.1111/ajpy.12019

Ciarrochi, J., Hayes, S. C., Hayes, L., Sahdra, B., Ferrari, M., Yap, K., & Hofmann, S. G. (2021). From package to process: An evidence-based approach to processes of change in psychotherapy. *Comprehensive Clinical Psychology, 1*, 26–44. https://doi.org/10.1016/B978-0-12-818697-8.00085-6

Coopersmith, S. (1967). *The antecedents of self-esteem*. Freeman.

Crocker, J., Thompson, L. L., McGraw, K. M., & Ingerman, C. (1987). Downward comparison, prejudice, and evaluations of others: effects of self-esteem and threat. *Journal of Personality and Social Psychology, 52*(5), 907.

Crocker, J., & Park, L. E. (2004a). The costly pursuit of self-esteem. *Psychological Bulletin, 130*(3), 392. https://doi.org/10.1037/0033-2909.130.3.392

Crocker, J., & Park, L. E. (2004b). Reaping the benefits of pursuing self-esteem without the costs? Reply to DuBois and Flay (2004), Sheldon (2004), and Pyszczynski and Cox (2004). *Psychological Bulletin, 130*(3), 430–434. https://doi.org/10.1037/0033-2909.130.3.430

Crocker, J., Luhtanen, R. K., Cooper, M. L., & Bouvrette, A. (2003). Contingencies of self-worth in college students: Theory and measurement. *Journal of Personality and Social Psychology, 85*(5), 894–908. https://doi.org/10.1037/0022-3514.85.5.894

Dalgas-Pelish, P. (2006). Effects of a self-esteem intervention program on school-age children. *Pediatric Nursing, 32*(4), 341–348.

Deci, E. L., & Ryan, R. M. (1995). Human Autonomy. In M. H. Kernis (Ed.), *Efficacy, agency, and self-esteem. The Springer series in social clinical psychology*. Springer. https://doi.org/10.1007/978-1-4899-1280-0_3

DeLury, S. S., & Poulin, M. J. (2018). Self-compassion and verbal performance: Evidence for threat-buffering and implicit self-related thoughts. *Self and Identity, 17*(6), 710–722. https://doi.org/10.1080/15298868.2018.1477829

Donald, J. N., Ciarrochi, J., Parker, P. D., Sahdra, B. K., Marshall, S. L., & Guo, J. (2018). A worthy self is a caring self: Examining the developmental relations between self-esteem and self-compassion in adolescents. *Journal of Personality, 86*(4), 619–630. https://doi.org/10.1111/jopy.12340

Donnelly, J., Young, M., Pearson, R., Penhollow, T. M., & Hernandez, A. (2008). Area specific self-esteem, values, and adolescent substance use. *Journal of Drug Education, 38*(4), 389–403. https://doi.org/10.2190/DE.38.4.f

Dupasquier, J. R., Kelly, A. C., Moscovitch, D. A., & Vidovic, V. (2020). Cultivating self-compassion promotes disclosure of experiences that threaten self-esteem. *Cognitive Therapy and Research, 44*(1), 108–119. https://doi.org/10.1007/s10608-019-10050-x

Eller, L. S., Rivero-Mendez, M., Voss, J., Chen, W. T., Chaiphibalsarisdi, P., Iipinge, S., et al. (2014). Depressive symptoms, self-esteem, HIV symptom management self-efficacy and self-compassion in people living with HIV. *AIDS Care, 26*(7), 795–803. https://doi.org/10.1080/09540121.2013.841842

Epstein, S. (2006). Conscious and unconscious self-esteeem from the perspective of cognitive-experiential self-theory. In M. H. Kernis (Ed.), *Self-esteem: Issues and answers* (pp. 69–76). Psychology Press.

Eromo, T. L., & Levy, D. A. (2017). The rise, fall, and resurgence of "self-esteem": A critique, reconceptualization, and recommendations. *North American Journal of Psychology, 19*(2), 255–302.

Fitch, G. (1970). Effects of self-esteem, perceived performance, and choice on causal attributions. *Journal of Personality and Social Psychology, 16*(2), 311.

Ferrari, M., Hunt, C., Harrysunker, A., Abbott, M. J., Beath, A. P., & Einstein, D. A. (2019). Self-compassion interventions and psychosocial outcomes: A meta-analysis of RCTs. *Mindfulness, 10*(8), 1455–1473. https://doi.org/10.1007/s12671-019-01134-6

Gilbert, P. (2009). Introducing compassion-focused therapy. *Advances in Psychiatric Treatment, 15*(3), 199–208. https://doi.org/10.1192/apt.bp.107.005264

Gilbert, P. (2010). An introduction to compassion focused therapy in cognitive behavior therapy. *International Journal of Cognitive Therapy, 3*(2), 97–112. https://doi.org/10.1521/ijct.2010.3.2.97

Gilbert, P. (2014). The origins and nature of compassion focused therapy. *British Journal of Clinical Psychology, 53*(1), 6–41. https://doi.org/10.1111/bjc.12043

Gilbert, P. (2015). The evolution and social dynamics of compassion. *Social and Personality Psychology Compass, 9*(6), 239–254. https://doi.org/10.1111/spc3.12176

Gilbert, P., McEwan, K., Matos, M., & Rivis, A. (2011). Fears of compassion: Development of three self-report measures. *Psychology and Psychotherapy: Theory, Research and Practice, 84*(3), 239–255. https://doi.org/10.1348/147608310X526511

Harter, S. (1993). Causes and consequences of low self-esteem in children and adolescents. In *Self-esteem* (pp. 87–116). Springer.

Harter, S. (1999). *The construction of the self: A developmental perspective*. Guilford Press.

Holas, P., Kowalczyk, M., Krejtz, I., Wisiecka, K., & Jankowski, T. (2021). The relationship between self-esteem and self-compassion in socially anxious. *Current Psychology*, 1–6. https://doi.org/10.1007/s12144-021-02305-2

James, W. (1890). *Principles of psychology*. Encyclopedia Britannica.

Kernis, M. H. (2002). *Self-esteem as a multifaceted construct*. State University of New York Press.

Kernis, M. H. (2003). Toward a conceptualization of optimal self-esteem. *Psychological Inquiry, 14*(1), 1–26. https://doi.org/10.1207/S15327965PLI1401_01

Kernis, M. H. (2005). Measuring self-esteem in context: The importance of stability of self-esteem in psychological functioning. *Journal of Personality, 73*(6), 1569–1605. https://doi.org/10.1111/j.1467-6494.2005.00359.x

Kernis, M. H., & Goldman, B. M. (2006). A multicomponent conceptualization of authenticity: Theory and research. *Advances in Experimental Social Psychology, 38*, 283–357. https://doi.org/10.1016/S0065-2601(06)38006-9

Kernis, M. H., Cornell, D. P., Sun, C.-R., Berry, A., & Harlow, T. (1993). There's more to self-esteem than whether it is high or low: The importance of stability of self-esteem. *Journal of Personality and Social Psychology, 65*(6), 1190–1204. https://doi.org/10.1037/0022-3514.65.6.1190

Kirby, J. N., Tellegen, C. L., & Steindl, S. R. (2017). A meta-analysis of compassion-based interventions: Current state of knowledge and future directions. *Behavior Therapy, 48*(6), 778–792. https://doi.org/10.1016/j.beth.2017.06.003

Koivula, N., Hassmén, P., & Fallby, J. (2002). Self-esteem and perfectionism in elite athletes: Effects on competitive anxiety and self-confidence. *Personality and Individual Differences, 32*(5), 865–875.

Krieger, T., Hermann, H., Zimmermann, J., & Grosse Holtforth, M. (2015). Associations of self-compassion and global self-esteem with positive and negative affect and stress reactivity in daily life: Findings from a smart phone study. *Personality and Individual Differences, 87*, 288–292. https://doi.org/10.1016/j.paid.2015.08.009

Lai, H.-R., Lu, C.-M., Jwo, J.-C., Lee, P.-H., Chou, W.-L., & Wen, W.-Y. (2009). The effects of a self-esteem program incorporated into health and physical education classes. *Journal of Nursing Research, 17*(4), 233–240. https://doi.org/10.1097/JNR.0b013e3181c003c9

Leary, M. R. (1999). The social and psychological importance of self-esteem. In R. M. Kowalski & M. R. Leary (Eds.), *The social psychology of emotional and behavioral problems: Interfaces of social and clinical psychology* (pp. 197–221). American Psychological Association.

Leary, M. R., & Tangney, J. P. (2003). Individual differences in self-esteem: A review and theoretical integration (2ed.). Guilford Publications.

Leary, M. R., & MacDonald, G. (2003). Individual differences in self-esteem: A review and theoretical integration. In M. R. Leary & J. P. Tangney (Eds.), *Handbook of self and identity* (pp. 401–418). The Guilford Press.

Levy, D. A. (2019). The "self-esteem" enigma: A critical analysis. *North American Journal of Psychology, 21*(2), 305–338.

Loyalka, P., Zakharov, A., & Kuzmina, Y. (2018). Catching the big fish in the little pond effect: Evidence from 33 countries and regions. *Comparative Education Review, 62*(4), 542–564. https://doi.org/10.1086/699672

Lyubomirsky, S., Tkach, C., & DiMatteo, M. R. (2006). What are the differences between happiness and self-esteem. *Social Indicators Research, 78*(3), 363–404. https://doi.org/10.1007/s11205-005-0213-y

MacBeth, A., & Gumley, A. (2012). Exploring compassion: A meta-analysis of the association between self-compassion and psychopathology. *Clinical Psychology Review, 32*(6), 545–552. https://doi.org/10.1016/j.cpr.2012.06.003

Macdonald, G. (1994). Self-esteem and the promotion of mental health. In D. Trent & C. Reed (Eds.), *Promotion of mental health* (pp. 19–20). Avebury.

Marsh, I. C., Chan, S. W., & MacBeth, A. (2018). Self-compassion and psychological distress in adolescents – A meta-analysis. *Mindfulness, 9*(4), 1011–1027. https://doi.org/10.1007/s12671-017-0850-7

Marshall, S. L., Parker, P. D., Ciarrochi, J., Sahdra, B., Jackson, C. J., & Heaven, P. C. (2015). Self-compassion protects against the negative effects of low self-esteem: A longitudinal study in a large adolescent sample. *Personality and Individual Differences, 74*, 116–121. https://doi.org/10.1016/j.paid.2014.09.013

Moffitt, R. L., Neumann, D. L., & Williamson, S. P. (2018). Comparing the efficacy of a brief self-esteem and self-compassion intervention for state body dissatisfaction and self-improvement motivation. *Body Image, 27*, 67–76. https://doi.org/10.1016/j.bodyim.2018.08.008

Neff, K. D. (2003a). The development and validation of a scale to measure self-compassion. *Self and Identity, 2*(3), 223–250. https://doi.org/10.1080/15298860309027

Neff, K. D. (2003b). Self-compassion: An alternative conceptualization of a healthy attitude toward oneself. *Self and Identity, 2*(2), 85–101. https://doi.org/10.1080/15298860390129863

Neff, K. D., & Vonk, R. (2009). Self-compassion versus global self-esteem: Two different ways of relating to oneself. *Journal of Personality, 77*(1), 23–50. https://doi.org/10.1111/j.1467-6494.2008.00537.x

Neff, K. D., Tóth-Király, I., Knox, M. C., Kuchar, A., & Davidson, O. (2021). The development and validation of the state self-compassion scale (long-and short form). *Mindfulness, 12*(1), 121–140.

Pohl, S., Steuwe, C., Mainz, V., Driessen, M., & Beblo, T. (2021). Borderline personality disorder and childhood trauma: Exploring the buffering role of self-compassion and self-esteem. *Journal of Clinical Psychology, 77*(3), 837–845. https://doi.org/10.1002/jclp.23070

Pyszczynski, T., & Cox, C. (2004). Can we really do without self-esteem? Comment on Crocker and Park (2004). *Psychological Bulletin, 130*(3), 425–429. https://doi.org/10.1037/0033-2909.130.3.425

Pyszczynski, T., Greenberg, J., Solomon, S., Arndt, J., & Schimel, J. (2004). Why do people need self-esteem? A theoretical and empirical review. *Psychological Bulletin, 130*(3), 435. https://doi.org/10.1037/0033-2909.130.3.435

Raes, F., Pommier, E., Neff, K. D., & Van Gucht, D. (2011). Construction and factorial validation of a short form of the self-compassion scale. *Clinical Psychology & Psychotherapy, 18*(3), 250–255. https://doi.org/10.1002/cpp.702

Rohmann, E., Hanke, S., & Bierhoff, H.-W. (2019). Grandiose and vulnerable narcissism in relation to life satisfaction, self-esteem, and self-construal. *Journal of Individual Differences, 40*(4), 194.

Rosenberg, M. (1965a). *Society and the adolescent self-image*. Princeton University Press.

Rosenberg, M. (1965b). Rosenberg self-esteem scale (RSE). *Acceptance and commitment therapy. Measures package, 61*(52), 18.

Rosenberg, M., Schooler, C., Schoenbach, C., & Rosenberg, F. (1995). Global self-esteem and specific self-esteem: Different concepts, different outcomes. *American Sociological Review, 60*(1), 141–156. https://doi.org/10.2307/2096350

Rousseau, J.-J., & May, G. (2002). *The social contract and the first and second discourses*. Yale University Press.

Ryan, R. M., & Brown, K. W. (2003). Why we don't need self-esteem: On fundamental needs, contingent love, and mindfulness. *Psychological Inquiry, 14*(1), 71–76.

Ryan, R. M., & Deci, E. L. (2000). Self-determination theory and the facilitation of intrinsic motivation, social development, and well-being. *American Psychologist, 55*(1), 68. https://doi.org/10.1037/0003-066X.55.1.68

Schütz, A., & Tice, D. M. (1997). Associative and competitive indirect self-enhancement in close relationships moderated by trait self-esteem. *European Journal of Social Psychology, 27*(3), 257–273.

Sedikides, C., & Alicke, M. D. (2012). Self-enhancement and self-protection motives. In R. M. Ryan & R. M. Ryan (Eds.), *The Oxford handbook of human motivation* (pp. 303–322). Oxford University Press.

Seekis, V., Bradley, G. L., & Duffy, A. (2017). The effectiveness of self-compassion and self-esteem writing tasks in reducing body image concerns. *Body Image, 23*, 206–213. https://doi.org/10.1016/j.bodyim.2017.09.003

Sirois, F. M., Kitner, R., & Hirsch, J. K. (2015). Self-compassion, affect, and health-promoting behaviors. *Health Psychology, 34*(6), 661. https://doi.org/10.1037/hea0000158

Souza, L. K. d., & Hutz, C. S. (2016). Self-compassion in relation to self-esteem, self-efficacy and demographical aspects. *Paidéia, 26*(64), 181–188. https://doi.org/10.1590/1982-43272664201604

Souza, L. K., & d., & Hutz, C. S. (2016). Self-compassion in relation to self-esteem, self-efficacy and demographical aspects. *Paidéia, 26*(64), 181–188. https://doi.org/10.1590/1982-43272664201604

Stephenson, E., Watson, P. J., Chen, Z. J., & Morris, R. J. (2018). Self-compassion, self-esteem, and irrational beliefs. *Current Psychology: A Journal for Diverse Perspectives on Diverse Psychological Issues, 37*(4), 809–815. https://doi.org/10.1007/s12144-017-9563-2

Tanenbaum, M., Adams, R., Gonzalez, J., Hanes, S., & Hood, K. (2018). Adapting and validating a measure of diabetes-specific self-compassion. *Journal of Diabetes and its Complications, 32*(2), 196–202.

Thijs, J., & Verkuyten, M. (2017). Promoting positive self-esteem in ethnic minority students: The role of school and classroom context. In N. J. Cabrera & B. Leyendecker (Eds.), *Handbook on positive development of minority children and youth* (pp. 325–342). Springer Science + Business Media. https://doi.org/10.1007/978-3-319-43645-6_20

Thoma, M. V., Bernays, F., Eising, C. M., Maercker, A., & Rohner, S. L. (2021). Child maltreatment, lifetime trauma, and mental health in Swiss older survivors of enforced child welfare practices: Investigating the mediating role of self-esteem and self-compassion. *Child Abuse & Neglect, 113*, 104925. https://doi.org/10.1016/j.chiabu.2020.104925

Thomason, S., & Moghaddam, N. (2021). Compassion-focused therapies for self-esteem: A systematic review and meta-analysis. *Psychology and Psychotherapy: Theory, Research and Practice, 94*(3), 737–759. https://doi.org/10.1111/papt.12319

Toor, S. U. R., & Ofori, G. (2009). Authenticity and its influence on psychological well-being and contingent self-esteem of leaders in Singapore construction sector. *Construction Management and Economics, 27*(3), 299–313. https://doi.org/10.1080/01446190902729721

Walker, J. S., & Bright, J. A. (2009). False inflated self-esteem and violence: A systematic review and cognitive model. *The Journal of Forensic Psychiatry & Psychology, 20*(1), 1–32. https://doi.org/10.1080/14789940701656808

Zeigler-Hill, V. (2011). The connections between self-esteem and psychopathology. *Journal of Contemporary Psychotherapy, 41*(3), 157–164. https://doi.org/10.1007/s10879-010-9167-8

Zessin, U., Dickhäuser, O., & Garbade, S. (2015). The relationship between self-compassion and well-being: A meta-analysis. *Applied Psychology: Health and Well-Being, 7*(3), 340–364. https://doi.org/10.1111/aphw.12051

Zhang, J. W., Chen, S., Tomova, T. K., Bilgin, B., Chai, W. J., Ramis, T., et al. (2019). A compassionate self is a true self? Self-compassion promotes subjective authenticity. *Personality and Social Psychology Bulletin, 45*(9), 1323–1337. https://doi.org/10.1177/0146167218820914

Ziller, R. C., Hagey, J., Smith, M. D., & Long, B. (1969). Self-esteem: A self-social construct. *Journal of Consulting and Clinical Psychology, 33*(1), 84–95. https://doi.org/10.1037/h0027374

Check for updates

4 Self-Compassion: An Evolutionary, Biopsychosocial, and Social Mentality Approach

Paul Gilbert

Introduction

This chapter offers an evolutionary and biopsychosocial approach to self-compassion. It considers the evolution of compassion as emergent from mammalian caring behavior (Gilbert, 1989/2016, 2000, 2020) regulated via the evolution of a range of human cognitive competencies that enable us to reason, mentalize, and have mindful consciousness of consciousness (Gilbert, 2019, 2020). These competencies are also the basis for self-awareness, which has positive and useful effects, but can also have harmful ones, such that Leary (2004) calls it "the curse of the self."

Before exploring how compassion is applied to "the self," it is important to explore the concept of compassion in general because it has been defined in different ways (Gilbert, 2017; Mascaro et al., 2020). One approach is to explore what clusters of psychological phenomena that characterize compassion. In a major review of the literature, Strauss et al. (2016) suggest compassion is made up of five components:

> recognition of suffering, understanding its universality; feeling sympathy, empathy, or concern for those who are suffering (which we describe as emotional resonance); tolerating the distress associated with the witnessing of suffering; and motivation to act or acting to alleviate the suffering. (Strauss et al., 2016, p. 25)

A different evolutionary approach which is complementary has been to track the phylogeny of motivations like care and compassion and identify their basic *if* A *then* do B algorithm. Algorithms underpin most functioning systems. For example, air-conditioning systems work on the principle that *if* temperature goes below a certain level, *then* the air-conditioning will automatically turn off and *if* the temperature goes above a certain level, *then* the air-conditioning will automatically turn on. All it needs is a temperature *detector* that is linked to an output system. A human body example is *if* temperature goes too low, we *then* shiver; if it goes too high, we *then* perspire. Behavioral algorithms for threat can be as follows: *if* confronted by a threat (stimulus), *then* a defensive behavior (response) becomes activated and engages in fight or flight. For eating, it would be as follows: *if* stimulus indicates food, *then* approach, salivate, and eat. If stimulus indicates sexual opportunity, *then* approach and engage in courting. Clearly, these motive-based algorithms depend on key *feature detectors* that are linked into physiological systems that *then* support specific actions. For example, the amygdala and sympathetic nervous system are

P. Gilbert (✉)
Centre for Compassion Research and Training, College of Health and Social Care Research Centre, University of Derby, Derby, UK

The Compassionate Mind Foundation, Derby, UK
e-mail: p.gilbert@derby.ac.uk

A. Finlay-Jones et al. (eds.), *Handbook of Self-Compassion*, Mindfulness in Behavioral Health,
https://doi.org/10.1007/978-3-031-22348-8_4

important for detecting and responding to threat. Damage to the amygdala can mean that animals may not detect threats in their environment and/or not respond to them if they are detected. These algorithms can also be attuned through learning. In humans, the cognitive interpretation of a stimulus can trigger the threat response (e.g., a rise in heart rate triggers fear of a heart attack). However, it is the linkage of stimulus-detector response that is crucial to the algorithm.

The algorithm for caring is *if* encountering signals of distress and need *then* activate a range of physiological systems that have evolved for caring behavior. Looked at this way, a derivative of caring, compassion, can be seen to have evolved from (among other routes; Kessler, 2020) mammalian caring behavior, friendship formation, and concerns to make a useful contribution to the lives of others (Carter et al., 2017; Mikulincer & Shaver, 2017, 2014; Gilbert, 1989/2016, 2009; Gilbert & Simos, 2022; Mayseless, 2016; Goetz et al., 2010). It is useful to be able to identify the algorithm (the stimulus that needs to be detected and the appropriate response) of the caring motive because then we can identify the physiological infrastructures on which it rests. In fact, a good deal is now known about the evolution of the physiological processes of caring behavior. Especially important ones are hormones such as oxytocin, vasopressin, and endorphins (Brown & Brown, 2015; Carter, 2014; Carter et al., 2017); changes in the autonomic nervous system, especially the vagus nerve (Porges, 2007, 2017); and specific neurocircuits (Kim et al., 2020a, b; Singer & Engert, 2019).

Although the "*if* encountering signals of distress and need (e.g., in offspring), *then* act to alleviate and prevent suffering and distress" appears simple, there are many subroutines within it. For example, a mammalian parent, usually the mother, must be able to identify a particular kind of "distress signal/call and need" to identify the appropriate action that will alleviate the distress. The caring action may be to rescue or protect the infant, keep the infant warm, feed the infant, and, more generally, provide a secure base and safe haven for the infant's development into maturity (Cassidy & Shaver, 2016). In the case of a human mother caring for her infant, she will supply many different resources for that infant, providing not only for their physical needs but also for their psychological developmental needs. For example, she will play with her infant, share positive emotions of joy and love, encourage her infant to take risks, and teach, mentor, and demonstrate so her infant learns life skills (Cassidy & Shaver, 2016). In addition, parental caring involves mentalizing (see below) the child and validating the child's experiences (Luyten et al., 2020). In close states of connectedness, mother and infant can be in physiological synchrony, meaning that their interactions are maturing physiological profiles that underpin the development of compassionate motivations in the infant (Feldman, 2012; Lunkenheimer et al., 2018; McFarland et al., 2020). This is partly because providing care and receiving care operate through similar physiological systems; for example, oxytocin is triggered when we receive care and support from others, but it is also a hormone that supports our interest in being caring (Carter, 2014). What "offspring" require from their mother changes significantly as they mature into a child, proceed into adolescence, and continue into adulthood.

None of this can even begin without a *preexisting motivation to care*. In contrast, callousness can be defined as a lack of motivation and an indifference to the needs and suffering of self and others. Caring then involves being motivated to notice; being sensitive and attuned to signals of distress, need, or suffering; and then switching attention to *what is required* to help the one being cared for. This may be providing nutritious foods, shelter, and warmth; appropriate education; and loving, playful, affectionate interactions that stimulate a range of psychophysiological systems that are conducive to well-being. There is now considerable evidence that the caring we experience in childhood has profound effects on our sense of ourselves and whether we are prone to self-criticism or self-acceptance and self-compassion (Music, 2016; Cassidy & Shaver, 2016). Whether we are loved, neglected, or harmed can have a major impact on our epigenetic profiles (Cowan et al., 2016; Slavich, 2020) and brain maturation (Lippard & Nemeroff, 2020), which in turn have profound impacts on the organization of motives and emotions and sense of self. Hence, when working clinically, helping people develop more accepting, self-reassuring, and compassionate orientations to themselves may require considerable work, as experiencing care and support in this way might be unfamiliar or frightening to them. This may also involve working through early traumas and multiple fears and resistances to compassion (Gilbert, 2022a).

What the mother (i.e., care provider) must also do is to keep track of the infant's development and be alerted if problems are arising which will require a change of action. In essence, she must be alert to signals of distress and also the signals of deviation from expected developmental trajectories which indicate unmet needs. Hence, care-compassion can be defined as *a sensitivity to suffering in self and others with a commitment to try to alleviate and prevent it.* In other words, whether we are talking about compassion to others or being compassionate to ourselves, both are rooted in a motivation to pay attention to suffering (not turn away from, dissociate, or deny) and then find the appropriate wisdom to know what to do and, if not known, then be motivated to find out.

The concept of *prevention of suffering* is implicit in Buddhist definitions of compassion. For example, in a collection of lectures, the Dalai Lama (1995) explains that compassion is the motivation to try to alleviate suffering *and the causes of suffering* in all sentient beings. Hence Gilbert and Choden (2013) highlighted that the element of prevention is essential for the full definition of compassion. In addition, we highlighted that adding the concept of prevention has significant implications for how we think about compassion because it requires us to be forward-thinking and address needs to prevent suffering in the future. When it comes to self-compassion, therefore, understanding what enables us to flourish and how to prevent future suffering must be part of self-compassion. Hence, recognizing that smoking or drinking too much and eating the wrong foods are damaging to one's health, and changing one's behavior, would be examples of self-compassion.

This implies another key principle of compassion which is that we become aware how we may easily and inadvertently cause harm to ourselves and others. We can be callous in pursuing our own needs at the expense of others' needs. For example, we know that many children die every day through lack of clean water, insufficient food, and poor-quality medicines, yet we spend more money on feeding our pets and on our gardens than we do in addressing this tragedy. Vegetarians argue that meat eaters are callous to the suffering of their food. We can also show a certain callousness "as indifference to the suffering we cause ourselves" through our pursuit of pleasures as in smoking, drinking too much alcohol, and eating too many chips and pizzas. We can also be very self-harming, both literally and in the way we think about and treat our own minds. For example, there is a large literature showing that harsh forms of self-criticism are associated with many forms of mental health difficulty (Werner et al., 2019; Gilbert, 2022c).

Therefore, care and compassion require us to be attentive to our potential to cause harm to self and others and adopt a motto of life which is "may I live to be helpful not harmful to self and others." Helpfulness and harmfulness can also be applied to our mental lives, and indeed most discourses on self-compassion tend to focus on mental life. A holistic, biopsychosocial approach to self-compassion requires us to be aware that there are many sources of harmfulness to the self, including, for example, diet, lack of exercise, overworking, and staying in toxic relationships. This is particularly true when we recognize that our diets can have a significant impact on a whole range of physiological systems including the vagus nerve which plays an important role in well-being and compassionate behavior (Breit et al., 2018; Di Bello et al., 2020). In compassion focused therapy (CFT) (Gilbert, 2020), mind awareness is cultivated alongside body awareness and body cultivation. Hence, some of the compassionate mind training exercises that are part of CFT offer insight and guidance into how to train/use the body to support the mind. For example, the importance of developing vagal tone, how to use breathing exercises to settle and ground the body and mind, how to use posture and exercise, and paying attention to diet which can influence the vagal nerve have been developed in compassion focused therapy (CFT) (Gilbert & Simos, 2022) and by Porges and colleagues (Porges & Dana, 2018).

From Caring to Compassion

Although many animals care for their young, and at times each other, compassion is not a term we would typically use to refer to these behaviors. In CFT, compassion requires certain types of cognitive competencies to direct our caring motives and with *knowing intentionality*. These cognitive competencies have been evolving over the past few million years or so on our journey to becoming *Homo sapiens* and have impacted how motives and emotions are activated and regulated. They are also responsible for making us the dominant species that we are. There are at least three types of competencies that underpin compassion: nonsocial reasoning, social reasoning, and consciousness of consciousness (Gilbert, 2019, 2020, 2022b).

Nonsocial Reasoning

Humans have evolved extraordinary competencies for reasoning. Byrne (1995) called humans *the thinking ape* and more recently drew attention to the evolution of *insight* (Byrne, 2016). Baron-Cohen (2020) called us the *pattern seekers*, that is, a fundamental attribute of our intelligent mind is to seek "if-and-then" patterns. We have evolved competencies for language and symbol use and thinking in time (i.e., I can act now because of what I want my future to be; study for a career). The use of internal representations (which can be in the form of imagery) plays a key role in our functioning. Sometimes we experience intrusive imagery as in trauma flashbacks, finding ourselves going over an argument we have had, or worrying. However, we can also knowingly and purposely use imagery to problem solve. We can imagine and run "what if…" and "suppose that…" and "imagine this…" simulations in our mind. We can also run imaginary simulations in our mind deliberately to stimulate physiological systems, as in sexual fantasy. Compassion focused visualizations take advantage of this link between our imagined experiences and our physiological responses (Gilbert & Choden, 2013). Together, with other competencies, these enable us to form complex mental representations of the world. Among these representations are the representations we have of ourselves in relationship to others – particularly whether we are cared for, can contribute, and are valued. This is important, because these interpersonal representations play an important role in our mental well-being. Accordingly, a key component of self-compassion is about addressing the need for interconnectedness.

Cognitive therapists (Beck, 1987; Beck et al., 1985) highlight the importance of understanding how people consciously reason, how they come to the conclusions they do, and how they form expectations and derive attributions. These cognitive dimensions are extremely important for our ability to understand suffering and needs in self and others and, by forming mental representations in our minds, work out what is likely to be helpful, either immediately or at some point in the future. The way we think about the nature of suffering and the causes of suffering can help us deal with suffering or can make it much worse. In the Buddhist traditions, this is called *the sutra of the two arrows*. If we are shot by an arrow, the first pain is from the arrow itself, while the second is our reactions to it, of wondering how this could have happened, who did it, and what does it mean for the future. Crucial to how we deal with life setbacks, disappointments, and failures are the thoughts and feelings we have about ourselves. We may experience shame-based self-criticism. Self-criticism is typically linked to monitoring some standard ideal or outcome we failed to achieve. Even when we are trying to be "mindful," we can have a stream of self-critical thoughts such as that we are not doing it correctly, other people can do it better than us, or we will not be able to achieve what we want to achieve and be rejected. There is a complex relationship between the fear of failing, fear of being rejected/excluded and self-criticism (Gilbert, 2022c). We may also label ourselves as stupid, pathetic, and incompetent. While some people self-criticize because they feel they could and should do better, others have a more pathogenic form which is of self-hatred, commonly linked to early abuse. While both are the opposite of self-compassion, they require different interventions (Gilbert, 2022c).

Social Reasoning

Evolutionary theorists have argued that in addition to using general principles of reasoning, when it comes to social relational events, we have specialist ways of reasoning. For example, how do you come to decisions about whether you are sexually attracted to somebody and if they are sexually attracted to you? How do you make decisions about how dominant or controlling somebody seems the first time you meet them or if this person is somebody you would like to make friends with and cooperate with or not? Although there is clear evidence that several species show elements of understanding the emotions and mind of others of their own kind (De Waal & Preston, 2017), humans have evolved extraordinary competencies for what is called *mentalization*. This is an umbrella term that covers theory of mind, empathy, perspective taking, metacognition, and reflection (Fonagy et al., 2018; Kim, 2015; Luyten et al., 2020). Broader than empathy, mentalization also takes in our scientific understanding about the nature of mind which helps us to understand and think about our minds and the minds of others.

Some people can be extremely talented at reasoning, making them excellent scientists, but not be that good with mentalizing the mind of self or others (Baron-Cohen, 2020). The essence of mentalization is therefore the ability to understand the nature of a mind as motivated, intending and feeling. Caring behavior is greatly advanced in humans because we can mentalize and are able to understand both the nature of mental states (which can include suffering) *and* what will help the one who is suffering. Importantly, self-mentalization is the ability for us to understand our own mental state and minds, to have insight into what motivates us, and why we feel what we feel and think what we think (i.e., metacognition). When helping people with mental health problems develop self-compassion, it sometimes takes a while to help them mentalize themselves and really begin to understand the sources of their suffering. In a way, that is one of the main functions of psychotherapy – to help people understand their mind. Therapists have highlighted the fact that we can dissociate from pain and deny it if and when we feel overwhelmed by it. Examples may be that, rather than trying to heal trauma memories using compassion, we try to avoid thinking about them and dissociate from them. Rather than acknowledge that our mental pain is causing our alcohol problems, we simply deny we have a problem. Clearly, these defenses interfere with our capacities for self-compassion. Self-compassion can also fail when we have not been able to work out what our needs are, because obviously then we can't do what we need to do to sustain ourselves. We may fear compassion because of the pain it touches on, or because we feel we do not deserve it.

Consciousness of Consciousness

Humans are a unique species because we can be conscious that we are conscious. We not only can have a thought, an emotion, or a desire, *but we can know that we are having them*. We can be mindful and observe our mind, and deliberately and purposively direct our attention to observe our mind. Hence, we have self-awareness which comes with the desires for self-identity. Many evolutionists see this as a fundamental distinction between humans and other animals (Leary & Buttermore, 2003). It is not all for good however, and Leary (2004) highlights how self-awareness can be something of a curse because it can drive us to overly self-monitor, self-criticize, feel shame, and experience depression and suicidal thoughts and behaviors. It can also drive us to violence and other defensive and destructive behaviors when we seek to protect ourselves or take vengeance on those who have attacked or shamed our sense of self and identity.

Nonetheless, self-awareness and being conscious of being conscious are partly what drives compassion. For example, we can care for our gardens, cars, and houses, but if they become damaged in some way, we don't have compassion for them. Compassion is reserved for sentient beings with a type of consciousness. If I break my leg, although the pain is in my leg, I don't say my leg is suffering. I say I am suffering because I am experiencing pain. So, compassion relates very much to the dimension of understanding the nature of *conscious experiencing*. Other animals can certainly show concern for each other; for example, there is evidence that rats won't pull a lever for food if they associate it with a cage mate getting an electric shock (De Waal & Preston, 2017). However, this is far from having a *conscious awareness of the suffering* of the cage mate. This awareness of conscious experiencing is also part of mentalization and central for mindfulness.

Knowing Intentionality

These three competencies together give rise to what one can call *knowing intentionality*. One example of this is that lions can intend to chase down and kill their prey. However, they can't do this knowingly; they don't consciously know that they are engaged in hunting behavior or causing suffering, they can't do it mindfully, and they can't suddenly decide that ripping the throat out of another animal to eat it is rather callous and it might be better to become a vegetarian. Nor can they decide to get up in the mornings and go circuit training on the savannahs to become better hunters. The fact that we can make these conscious choices is a game changer in our evolutionary story (Leary, 2004) because it means that we can choose to direct our motives deliberately and purposively. This is depicted in Fig. 4.1.

Complexities of Knowing Intentionality For us to develop empathy for others, we need insight into our own minds; the question is, how do we achieve that? One mechanism is via the activity of mirror neurons, whereby we can stimulate similar circuits in our brain when we observe mental states in others (Corradini & Antonietti, 2013). Another important dimension of knowing intentionality is awareness of what is happening inside our own bodies in relationship to how we are interacting with the environment. For example, attention has been directed to what is called

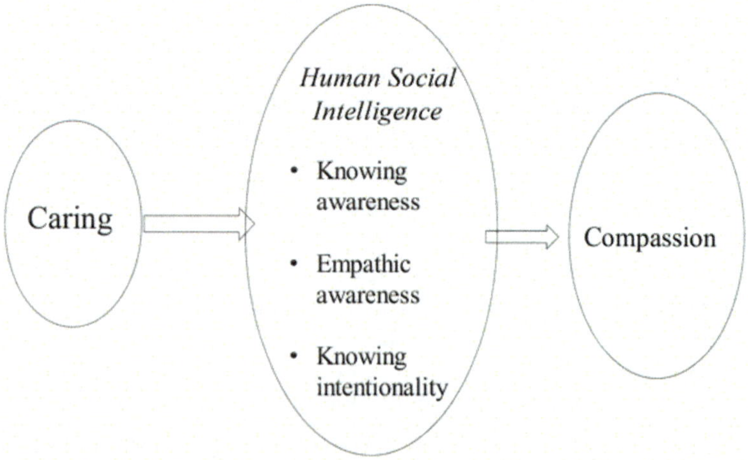

Fig. 4.1 From caring to compassion: From Gilbert (2018). Living Like Crazy. (Reprinted with permission Annwyn House)

interoception (Arnold et al., 2019; Khalsa et al., 2018). Nonsocial forms of interoception relate to detecting hunger, thirst, temperature, and various aches and pains as guides to taking action. In addition, we have awareness of changes in emotional state; examples are noticing our heart rate increases when we become angry or anxious or becoming aware of the feelings in the body when we are happy. In addition, there is the question of how we link our body experiences when in social interactions. For example, in a potential argument, we can be aware of a rising heart rate and facial expression changes and sense of threat that might make us cautious about how we proceed. These different areas indicate the complexity of self-compassion because we can't be compassionate unless we have insight into what we are feeling, thinking, and experiencing. Hence, before some people can develop self-compassion, it will help to have insight into the nature and textures of the mind and their felt experience of different affective states. For example, how can I be compassionate to my depression if I'm dissociating from depression or in denial?

Self-Care Versus Self-Compassion

If we distinguish care and compassion in terms of suffering and having a sentient mind, then we can distinguish between self-care and self-compassion. Self-care is the ability to understand our needs and to address them in a way that is helpful and not harmful. For example, keeping ourselves clean (basic hygiene), keeping fit, eating appropriate foods, resting appropriately, not overworking, and generally *looking after* ourselves are part of self-care. Motivation remains crucial. For example, we may spend a lot of time in the gym not because we want to "care and look after" our bodies but because we want to look attractive or are fearful of being seen as overweight. Nonetheless, individuals can be keen to care for their bodies in relation to their physical health but fail to see the same need for care when it comes to their mental health. People can be harshly self-critical, even driving themselves into anxiety and depression in certain circumstances (Gilbert, 2022c). They may also find being open to the compassionate support and help of others difficult or hard to develop trusting friendships. They may lack insight into what nourishes the mind and promotes well-being, social connectedness, and flourishing.

The Competencies of Human Caring and Compassion

Our new brain competencies offer new ways to approach, activate, and cultivate compassion and the two basic elements of its algorithm of (1) (stimulus) detection and engagement and (2) appropriate (response) action. When using compassion as a guide in therapy, these two elements of compassion can be broken down into their sub-skills that recruit our new brain competencies. Over the years, I have suggested six competencies for engagement and six for appropriate action but only as guides (Gilbert, 2009, 2020). These are given in Fig. 4.2.

Starting our exploration at "9:00 o'clock" on the inner circle, everything begins with concern and care for our own or others' well-being. To engage with the distress in ourselves or others, we must be motivated to be present and aware rather than turning away, dissociating, engaging in denial, or being callous. Then moving around clockwise, we come to sensitivity to the signals of distress and need. This involves some degree of mindfulness. If we are not sensitive, we may not be aware of distress. Being sensitive and noticing distress stimulate a reaction which is called sympathy. Sympathy is related to personal distress and being moved by the plight of self or others (Eisenberg, 2003; Eisenberg et al., 2015). Often, when we are engaging with distress and pain, this can be one of the first exit routes out of compassion, because we can become overwhelmed and close down or pull away. Following logically then, the next competency is called distress tolerance. This is a basic skill that many therapists seek to develop. It typically involves *gradual exposure* toward what is feared or avoided, along with at times cognitive reappraisal. CFT supports these distress tolerance

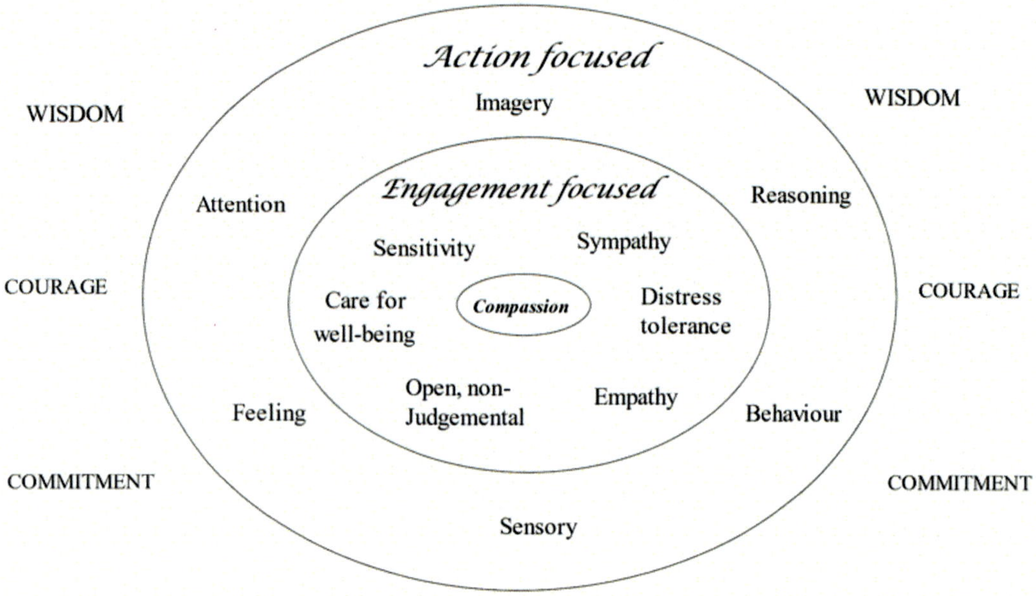

Fig. 4.2 Domains for the therapeutic process. (Adapted from Gilbert (2009). Compassionate Mind. Reprinted with permission from Little Brown)

interventions by teaching people how to access grounding and soothing systems through mindfulness and breathing techniques, among others (Petrocchi & Cheli, 2019). As we develop distress tolerance, we are likely to be able to mentalize (Luyten et al., 2020) and empathically engage with the distress and suffering of self and others. Mentalization enables us to have insight into the nature of our minds and those of others; we can make sense of what is being experienced. In addition, empathy requires us to keep a clear boundary between self and others so that their pain is not mistaken as our pain.

The final competency in the circle is called open non-judgment. It is open because we are not closing down around the nature of our own or other's suffering, and we are non-condemning. Non-judgmental does not mean simple acceptance, because when we recognize distressing feelings in ourselves, we can take steps to change to more helpful ones. In addition, even if we enable acceptance, it does not mean that the *actions* associated with certain brain states are acceptable. In some cases, we will choose not to act on triggered impulses (e.g., vengefulness) but to act against them because we have the mentalizing competency of insight into the consequences or harmful action. Importantly, these six competencies are not linear but support each other. For example, imagine how each of these six competencies might be affected if any of the others were lost.

The outer circle represents the *action-response* qualities of compassion. If we go to hospital with a broken arm and we experience a very empathic doctor who shows all the qualities of the first circle, that will not be enough to help us. We will also want them to have the wisdom to know how to fix the broken arm and take us out of pain. Clearly then, intentionality and the abilities to engage with suffering are not enough; we need wise action. Hence, beginning in the outer circle, we can start by paying attention not to the suffering but to what is likely to be helpful. We can use images and run simulations in our minds and use our new brain reasoning competencies to think through and problem solve. Our capacities for reasoning will also be linked to the knowledge we have about the nature of suffering and how it can be alleviated. So, for example, the doctor can recall his/her training that s/he may have acquired over many years to support compassionate action toward healing the broken arm. When it comes to our own mental states then, again the more we understand about the nature of our brain, and how it creates the desires, emotions, impulses, fears, hopes, joys, ruminations, and plans in the way that it does, the more wisdom we can bring to generating compassionate ways to help our minds with the ups and downs of life. The feelings we have when acting compassionately will depend on what the action is.

At the center of compassion therefore has to be *courage* to engage with pain and difficulties linked to knowing what to do. Wisdom without courage and courage without wisdom are not helpful. However, the nature of the *courage and wisdom* we may need for any particular source of suffering will depend on context. For example, a firefighter about to enter a burning house is likely to be anxious and will need ways to contain their anxiety to pursue the intention to rescue. In addition, they will need particular "wise" skills. In contrast, a counselor counseling a bereaved client might feel sad and be able to contain sadness to provide support; these are very different skills to that of the firefighter. In contrast, somebody fighting injustice will need to work with different themes and different emotions and have different skills. Despite these major differences, what links them is the motivation to address and prevent suffering. This is what makes motivation, rather than any specific emotion, central to compassion. Note, too, that these skills might not be interchangeable because our brave firefighter might not make the most empathic or compassionate parent and our counselor might not make the most courageous firefighter. It is important to see that compassion is not one process and varies with context.

It is the same with self-compassion. What self-compassion will involve will depend on the context of our suffering. Trying to be self-compassionate and supportive when we have just been diagnosed with cancer, and going through treatments, requires a very different set of skills than if we are dealing with depression, coming to terms with lose of loved one, addressing trauma memory of past abuse, learning to be assertive in the face of a bully, leaving a toxic relationship, or making determined efforts to lose weight on health grounds. Clearly then, what links all of these different processes is compassion *motivation* rather than a specific emotion or skill.

Compassion therefore does not have a particular emotion when it is engaged in action. In their paper *Compassion Is Not a Benzo*, Di Bello et al. (2021) highlight the fact that if we are to engage with suffering, then our threat system processes will be activated, and it is how we tolerate and encourage ourselves to behave wisely in those situations that is crucial. The misunderstanding that we need a calm mind to be compassionate is clearly incorrect. When we are engaging with compassion, we are not going on our holidays, but descending into suffering, pain, and difficulty where we need a focused, grounded, courageous mind, not necessarily a calm or fearless one.

Self-compassion is the ability to direct these 12 competencies toward the self. In other words, we can be motivated to be self-compassionate and be attentive and sensitive to sources of our pain and suffering. We can tolerate that distress and begin to empathically attune to understand it and its causes without being harshly self-critical or condemning. When it comes to taking action and how to help ourselves, we can use our new brain competencies to refocus attention, pull on our wisdoms, run imaginary scenarios in our mind, use our abilities to reason and think through things, and maybe discuss with, and seek out, help from friends, mentors or professionals. In any moment of distress, we can be attentive to the body and tolerate feelings that arise as we take action.

There is also increasing evidence that different aspects of self-compassion affect us quite differently. For example, the three positive components of Neff's concept of self-compassion, namely, kindness, mindfulness, and sense of common humanity (Neff, 2011), are significantly linked to measures of well-being and flourishing but less so to measures of mental health difficulty. In contrast, the three negative factors of self-judgment, sense of isolation, and overidentification or absorption are more robustly linked to mental health problems (Gilbert et al., 2017; Pandey et al., 2021). Related to this is that self-reassurance has a greater moderating impact on depression than self-esteem, indicating that all positive self-evaluations do not have the same impact (Petrocchi et al., 2019). Moreover, although measures of self-reassurance and self-compassion are often highly correlated, they are also different. Self-reassurance is focused on a sense of encouragement and the ability to remember one's strength and abilities in times of difficulty (Gilbert et al., 2017). Self-reassurance, like kindness and assertiveness, *is a way of being compassionate* but is not compassion itself.

Wisdom of the Minds We Have

Understanding the nature of our minds is crucial for our ability to become self-compassionate. This can take us into difficult territory of understanding what it is to be a gene-built biological being with a conscious mind. We are, like all other species, a biological creation built to simply survive and reproduce. Considering all other life forms past and present, that is basically all a life form does. Hundreds of millions of years, even before the dinosaurs, it was simply a process of eating and reproducing before being eaten or decay and death. Indeed, no living thing chose that existence or to be what it is. No elephant chose to be an elephant, no rabbit a rabbit, and no human a human. No human chose to be a man or a woman, nor their ethnicity, nor the illnesses they will be vulnerable to, nor where they were born and the families that matured them. We are all running the algorithms that nature has set in our brains and our social experiences have fine-tuned. Clearly, if we have been brought up in a violent background or a very loving background, much of what we become would be different, even our genetic profiles (Cowan et al., 2016).

These are fundamental insights when working with clients. They are the root of compassion because it helps us understand the real meaning of common humanity. We are all biologically created in ways we never chose and socially programmed in ways we never chose. All of us are caught up in various traumas and tragedies of life, some worse than others. All of us have a brief time here, maybe 30,000 days if we are lucky: all of us decay and die, and some of us rather painfully and slowly (e.g., cancers and dementias). All of us become conscious we exist in a body and a brain that have all kinds of impulses and feelings which have been built for us not by us. All of us want to be happy rather than suffer. It is when we personally overidentify with these algorithms and programs that we can get into difficulties. The greatest challenge of all humans is to begin to understand that we are programmed beings, but we can also become "mind aware" of our programming and begin to choose how to live. We can *knowingly and intentionally* try to live to be helpful not harmful, recognizing that we have within us the seeds of great harmfulness. The point of this is to highlight the fact there are many different wisdoms that can orientate us to a compassionate life.

Social Mentalities and the Self

An important aspect of evolution that can throw light on the process of self-compassion is social mentality theory (Gilbert, 1989/2016, 2005, 2017). This highlights the fact that all social motives must coevolve in the context of dynamic, reciprocal interactions and are therefore attuned to such. The evolution of caring behavior is an excellent example of a coevolved social mentality. Before the evolution of attachment and parental investment, reproductive strategies were to produce hundreds, sometimes, thousands, of offspring, very few of which would survive until adulthood. However, a different reproductive strategy evolved with parental investment, whereby very few offspring are produced but they are cared for and provisioned during their maturation into adulthood. This means there is a carer (usually, but not always, the mother) and a cared for, i.e., the infant(s). This is evolving as lock and key because clearly, infants cannot evolve to be dependent on resources from a mother if the mother is not also evolving motives and competencies to be aware of the needs of her infant. The attachment system of mother-infant is thus a clear example of a coevolved, co-regulating process. What is evolving are the physiological infrastructures to be able to engage in these behaviors. Species that produce high numbers of offspring clearly do not have these physiological systems. Importantly, both giving *and* receiving evolved and involve overlapping physiological systems such as the vagus nerve, the hormone oxytocin (Brown & Brown, 2015; Carter et al., 2017; Porges, 2017, 2021), and neurocircuits (Kim et al., 2020a, b).

The Psychological Functions of the Evolved Caring and Attachment Systems

As noted above, parental caring relationships have a range of functions, from feeding, to protecting, thermal regulation, and supporting psychophysiological maturation and regulation (Cassidy & Shaver, 2016). The quality of care we receive also impacts epigenetic development (Cowan et al., 2016). In addition, the evolution of caring behavior provides three main psychological functions for the developing infant. These were identified over 50 years ago by primate researchers, developmental psychologists, and child psychotherapists (Music, 2016; Cassidy & Shaver, 2016). Attachment theorists (Ainsworth, 1969; Bowlby, 1969) highlighted the fact that across species, a caring parent provides key *psychological* resources to the infant that impact its subsequent psychosocial maturation. From an attachment perspective, these include a *secure base*, which in the human context provides multiple inputs to the child including protection, encouragement, support, and signals of warmth, joy, love, and affection along with emotion and behavioral guidance. These enable a child to internalize a sense of his/her own acceptability and lovability in the eyes of others, which builds social confidence and emotion regulation (Music, 2016; Cassidy & Shaver, 2016). Children who do not receive such inputs are more threat sensitive, vigilant, and less trusting, which can compromise well-being (Music, 2016; Cassidy & Shaver, 2016). They are also prone to be more self-critical and less self-compassionate (Gilbert, 2022c).

A second key resource is called a *safe haven*, which is the ability of the parent to act as a soothing object and to help regulate arousal and distress in the infant. There are various caregiving behaviors that can achieve this, including soothing voice tones, stroking, hugs, and facial expressions that convey a genuine sense of empathic connection and concern. As children grow, they can develop transitional objects such as teddy bears that have soothing qualities. In addition, they can develop imaginary friends that can be a source of support (Taylor et al., 2009). As a result of these two core qualities of *secure base*, that is, encouraging and guiding, and *safe haven*, that is, soothing and comforting, the parent and child have what is called *proximity-seeking and maintenance systems*. This third key resource creates motivational orientation such that when the child is distressed, it will seek out those that can guide, comfort, soothe, and support him/her.

Understanding these three core resources of caring is fundamental to understanding the many of the functions of all forms of caring and compassion including to self and to others. It means that when we are being self-compassionate, we can create an inner secure base. We can turn within (proximity-seeking to our inner supports) to find self-reassurance, self-encouragement, and self-support which enables us to face things that are difficult. Indeed, self-compassion is not passively reacting, but actively preparing ourselves for our (heroic) journey through life and to be able to take on its challenges for growth, setbacks and traumas, decay, and death.

We can also function as a safe haven for ourselves. We have within us the competencies for self-regulation and self-soothing. These functions are originally internalized from early attachment relationships, but in some individuals, they may have to be built and cultivated. In addition, if clients have toxic experiences of caring, such as neglect or forms of abuse or criticism, the caring system can require work to detoxify it. That can require grieving for the love and affection that was not present in childhood. In these more complex cases, it means that self-compassion provides the courage to work with one's fractured sense of attachment security. Put another way, we can begin to identify the core functions of compassion and self-compassion as the ability to be encouraging and supportive when facing difficulties and to be soothing and containing when distressed (Gilbert & Simos, 2022).

Compassionate Others and Images

Another key dimension of self-compassion is the ability to reach out to others for help. Hence, some aspects of self-compassion involve being courageous enough to ask for help. Psychotherapy clearly depends on the courage and preparedness of people to seek and be open to help and reveal the nature and extent of their suffering. Indeed, there is growing evidence that being open to the compassion of others is an important buffer against mental health problems in contrast to compulsive self-reliance (Hermanto et al., 2016). Two main reasons that people do not seek out the help of others is due to shame, self-criticism, and social distrust. Given the importance of "receiving compassion" systems in our brain, CFT tries to help people begin to stimulate the experience of receiving by generating images of a compassionate other. There is now very good evidence that imagined forms of relating can have very powerful psychological effects (Gleason, 2013). Just as we have sexual fantasies that involve imagined interactions to stimulate an arousal of specific physiological systems, we can *imagine* interacting with a compassionate other(s) to stimulate compassion systems. Part of the reason for doing this is because compassion is rooted in a social mentality of sending and receiving. Hence, by developing a compassionate image, and imagining dialoguing with the compassionate image, we are activating that sending and receiving process. Importantly, some clients have difficulties generating and responding to their own compassionate images because they struggle to imagine receiving compassion and because compassion feels alien and unfamiliar. For others, it is because compassion connects them to memories of losing compassion connection via shame, rejection, and feeling sad and isolated. These experiences are worked through in the therapy (Gilbert & Simos, 2022).

Nonetheless, rather than (imagining) oneself being compassionate to oneself, some people find it easier to start being self-compassionate by imagining what a compassionate other would say to them or how an imagined compassionate other would be with them. The act of imagining what they would want their compassionate image to be like, and how they would like to relate to it, and be related to by their imagined compassionate other, is part of the work of compassion awareness. This can be a crucial first step into self-compassion, particularly for people who have not received compassionate care in their early lives. These practices can also reveal important fears, blocks, and resistances to compassion.

Self-compassion is not a self-contained process but one that is aware of our interconnectedness, able to feel encouraged, supported, and soothed by others. Indeed, one of the reasons that low self-compassion has become a major issue in the west is partly because of the increase in loneliness and feeling disconnected from caring social communities (Becker et al., 2021; Cacioppo & Patrick, 2008). Self-compassion should not be seen as a process of compulsive self-reliance but one of developing trust in oneself *and* others. Many commentators have noticed that self-criticism is low in supportive communities and hunter-gatherer societies partly because individuals are highly socially connected (Ryan, 2019). There is a strong link between low self-compassion, self-criticism, and a sense of social disconnection (Gilbert, 2022c).

Compassionate Self

There are many strategies that can be used for developing a sense of a compassionate mind and self. One is to help people recognize we are gene-built and socially shaped, and much more of what goes on minds is not of our design. Some of these are based on cognitive techniques, such as inviting individuals to think in certain ways that could offset feelings of self-criticism or loneliness. CFT also seeks to stimulate the caring mentality as it is linked to a range of physiological systems such as the vagus nerve (Porges, 2017, 2021), oxytocin

(Carter et al., 2017), and various neurocircuits (Singer & Engert, 2019). Part of the training is therefore to develop the capacity to use the body to support the mind, through practicing various postures (e.g., yoga), breathing exercises, visualizations, and behavioral practices and how to create compassionate voice tones for self-reassuring thinking. The view is that if you want to run a marathon or climb a mountain, no matter how skilled you are, if you are not physically fit, you will not be able to do it. Similarly, no matter the intention, if people are not able to regulate their autonomic nervous system, for example, and have access to parasympathetic regulation, they may struggle to engage both the secure base and safe haven functions of compassion (Porges, 2021, 2017).

Compassion is based upon courage and wisdom, and both are important when we confront the realities of being a short-lived, vulnerable, biological, gene-built and socially constructed being. As noted above, on the nature of our "tricky brain," clients are guided into important evolved constructs (which offer "compassionate wisdoms on the nature of tricky mind"). Helping people to envision, create, and begin to enact a compassionate sense of self can start by inviting people to imagine particular qualities of a compassionate person and mind or their ideals of compassion, and, then like an actor, imagine becoming them. Asking people to imagine themselves in certain states of minds, such as being self-critical or self-reassuring, is associated with important differences in neurophysiological activation (Longe et al., 2010). Creative imagination of a version of oneself and exploring the impact that has on coping have been investigated in other paradigms. For example, thinking about a life problem and then imagining oneself as one's "best possible self" and exploring it from that perspective are related to emotional change and increased optimism (Meevissen et al., 2011; Peters et al., 2010). Osimo et al. (2015) created a virtual reality scenario where participants raised a personal issue and then offered themselves counseling either as themselves or as (a virtual) Sigmund Freud. Giving oneself counseling "as Sigmund Freud" reduced depressed feelings significantly more than as themselves.

Similarly, practicing positive self-imagery by recalling a time when one felt relaxed and positive was related to higher levels of self-esteem and reduced anxiety in response to anxiety-provoking vignettes such as meeting your partner's parents for the first time (Stopa et al., 2012). Gilbert and Basran (2018) invited people to talk about a minor life difficulty to a partner, who was instructed to simply listen, without saying anything, for 2 or 3 min. The person who shared their difficulty was then asked to reflect on what it was like to be listened to in such a manner and was invited to reflect on how they were thinking about their difficulty, following the experience of being listened to. They then engaged in a compassionate mind induction using the soothing rhythm breathing exercise and then to think about the problem again through their compassionate mind state. For the latter part of the exercise, participants were asked to imagine having all of the characteristics they might have if they were to embody the qualities of compassion. Participants experienced the life difficulty quite differently and had new insights as to how they might be able to work with it. Shifting their focus to a compassionate mind state enabled them to develop more empathy for the problem and imagine different ways of how to cope better. These kinds of projects reveal that switching mental states "on purpose" does create different ways in which we see the world.

Building a Compassionate Mind

To understand any approach to self-compassion requires a clear understanding of that approach's view of compassion and the competencies that support it. Whether compassion is directed toward others or to oneself, it requires a set of competencies that enable a courageous and wise engagement with suffering. Compassionate mind training (CMT) has suggested six competencies for engagement and six competencies for taking action. These were described in Fig. 4.2. There are a range of processes that support these competencies and help build self-compassion. These include the following:

1. *Using the body to support the mind*: These practices are designed to help develop a range of physiological systems that support compassion. These can include posture, breathing, movement, yoga, acupuncture, diet, fitness, grounding, and settling. There are also grounding practices such as safe place and color imagery. These can be also practiced by listening to music or going into nature. Art- and music-based interventions can help people explore different aspects of compassion (Bennett-Levy et al., 2020).
2. *Using new brain competencies*: There are many practices to support mind awareness that include mindfulness, mentalization, and rational thinking. Figure 4.2 outlines six basic competencies for *engaging* with suffering. We also offered six competencies that are focused on the *action* aspect of compassion. Intention without wise action is often not helpful. Together they make up the courage and wisdom elements of compassion. What can help clients engage with these practices is to stimulate a sense of play and allow for curious exploration with pleasure from the process.

As noted throughout this chapter, there are many practices for reflecting on and internalizing core qualities of compassion. Two keys ones are: imagining one's ideal compassionate other and the experiences of relating to that image, and imagining one's own self as having compassionate skills, courage and wisdoms. In regard to the first process, one can imagine compassionate dialogs and interactions designed to stimulate specific physiological systems. The principle is no different than any other imaginary relationship, be it sexual or imagining an argument with somebody; how we imagine interactions influences our physiological state. The second is connecting to and imagining one's own core compassionate qualities, embodying them and regularly, mindfully, remembering to tune into them and practice thinking and acting from that pattern of self (compassionate mind).

3. *Behavioral practice*: Examples of identifying and engaging compassionate ways of thinking and behaving each day with compassionate practice can include making a deliberate effort to walk in the streets with a friendly face and friendly acknowledgements and to deliberately think about how you could be helpful to people and perform helpful acts to yourself and others. Additionally, there are behavioral practices such as compassionate letter writing, listening to one's own compassionate message from a mobile phone, and asking somebody who you think cares about you for help or support.
4. *Addressing the fears, blocks, and resistances*: All psychological processes have fears, blocks, and resistances. For example, there can be certain emotions we are frightened to experience or express. People can be *fearful* of compassion because of what it stimulates in them, such as sadness or anxiety. People may be *blocked* to compassion because they misunderstand it or would like to be compassionate but don't know what to do, or they haven't had a chance to practice. When people are resistant, it is usually because they see it as too costly or not useful. When we are working in psychotherapy, working through the fears, blocks, and resistances to compassion is often *the main focus of the work*, with specific exercises and practices to help clients identify and overcome these barriers.

Conclusion

Compassion emerged from the evolution of caring behavior. There is now a clear science of the physiological systems underpinning caring behavior and how this motivational system can be regulated by more complex cognitive competencies. Indeed, even having a conceptualization of an individualized separate self (which can be compassionate to itself) is a product of certain cognitive competencies. It follows therefore that how we use these competencies to direct motives is fundamental to how we experience ourselves in the world. Self-compassion helps us recognize the nature of personal distress and then develop the courage and wisdom to address it. The nature of the compassion we develop and express depends upon the context and the nature of the suffering we are experiencing. Self-compassion involves many things such as finding the courage to go to hospital if we are hospital phobic or to leave a toxic relationship. It can encourage, support, and guide us when we are trying to engage in health measures such as to lose weight or give up smoking. And it can help us when we are distressed and if we have become self-critical, disappointed in ourselves, or even self-hating and have somehow turned against ourselves. Hence, self-compassion is not one process but is textured by the different feelings and actions associated with those specific episodes of suffering. Nonetheless, despite many different emotions, skills, and behaviors, what links all of them is the motivation to develop the courage and wisdom to recognize, attend to wisely, seek to tolerate, alleviate, and prevent unhelpful personal suffering.

References

Ainsworth, M. D. (1969). Object relations, dependency, and attachment: A theoretical review of the infant-mother relationship. *Child Development, 40*(4), 969–1025. https://doi.org/10.1111/j.1467-8624.1969.tb04561.x

Arnold, A. J., Winkielman, P., & Dobkins, K. (2019). Interoception and social connection. *Frontiers in Psychology, 10*, 2589. https://doi.org/10.3389/fpsyg.2019.02589

Baron-Cohen, S. (2020). *The pattern seekers: A new theory of human invention*. Basic Books.

Beck, A. T. (1987). Cognitive models of depression. *Journal of Cognitive Psychotherapy, 1*(1), 5–37.

Beck, A., Emery, G., & Greenberg, R. (1985). *Anxiety disorders and phobias: A cognitive perspective*. Basic Books.

Becker, J. C., Hartwich, L., & Haslam, S. A. (2021). Neoliberalism can reduce Well-being by promoting a sense of social disconnection, competition, and loneliness. *British Journal of Social Psychology, 60*(3), 947–965. https://doi.org/10.1111/bjso.12438

Bennett-Levy, J., Roxburgh, N., Hibner, L., Bala, S., Edwards, S., Lucre, K., Cohen, G., O'Connor, D., Keogh, S., & Gilbert, P. (2020). Arts-based compassion skills training (ABCST): Channelling compassion focused therapy through visual arts for Australia's indigenous peoples. *Frontiers in Psychology, 11*, 568561. https://doi.org/10.3389/fpsyg.2020.568561

Bowlby, J. (1969). *Attachment and loss*. Basic Books.

Breit, S., Kupferberg, A., Rogler, G., & Hasler, G. (2018). Vagus nerve as modulator of the brain-gut axis in psychiatric and inflammatory disorders. *Frontiers in Psychiatry, 9*, 44. https://doi.org/10.3389/fpsyt.2018.00044

Brown, S. L., & Brown, R. M. (2015). Connecting prosocial behavior to improved physical health: Contributions from the neurobiology of parenting. *Neuroscience and Biobehavioral Reviews, 55*, 1–17. https://doi.org/10.1016/j.neubiorev.2015.04.004

Byrne, R. W. (1995). *The thinking ape: Evolutionary origins of intelligence*. Oxford University Press.

Byrne, R. W. (2016). *Evolving insight: How it is we can think about why things happen*. Oxford University Press.

Cacioppo, J. T., & Patrick, W. (2008). Loneliness: Human nature and the need for social connection. WW Norton & Company.

Carter, S. (2014). Oxytocin pathways and the evolution of human behavior. *Annual Review of Psychology, 65*(1), 17–39. https://doi.org/10.1146/annurev-psych-010213-115110

Carter, S., Bartal, I., & Porges, E. C. (2017). The roots of compassion: An evolutionary and neurobiological perspective. In E. M. Seppälä, E. Simon-Thomas, S. L. Brown, M. C. Worline, D. Cameron, & J. R. Doty (Eds.), *The Oxford handbook of compassion science* (pp. 178–188). Oxford University Press.

Cassidy, J., & Shaver, P. R. (2016). *Handbook of attachment: Theory, research and clinical applications*. Guilford Publications.

Corradini, A., & Antonietti, A. (2013). Mirror neurons and their function in cognitively understood empathy. *Consciousness and Cognition, 22*(3), 1152–1161. https://doi.org/10.1016/j.concog.2013.03.003

Cowan, C. S. M., Callaghan, B. L., Kan, J. M., & Richardson, R. (2016). The lasting impact of early-life adversity on individuals and their descendants: Potential mechanisms and hope for intervention. *Genes, Brain and Behavior, 15*(1), 155–168. https://doi.org/10.1111/gbb.12263

De Waal, F. B., & Preston, S. D. (2017). Mammalian empathy: Behavioural manifestations and neural basis. *Nature Reviews Neuroscience, 18*(8), 498–509. https://doi.org/10.1038/nrn.2017.72

Di Bello, M., Carnevali, L., Petrocchi, N., Thayer, J. F., Gilbert, P., & Ottaviani, C. (2020). The compassionate vagus: A meta-analysis on the connection between compassion and heart rate variability. *Neuroscience & Biobehavioral Reviews, 116*, 21–30. https://doi.org/10.1016/j.neubiorev.2020.06.016

Di Bello, M., Ottaviani, C., & Petrocchi, N. (2021). Compassion is not a benzo: Distinctive associations of heart rate variability with its empathic and action components. *Frontiers in Neuroscience, 15*, 617443. https://doi.org/10.3389/fnins.2021.617443

Eisenberg, N. (2003). Prosocial behavior, empathy, and sympathy. In M. H. Bornstein, L. Davidson, C. L. M. Keyes, & K. A. Moore (Eds.), *Well-being: Positive development across the life course* (pp. 253–265). Lawrence Erlbaum Associates.

Eisenberg, N., VanSchyndel, S. K., & Hofer, C. (2015). The association of maternal socialization in childhood and adolescence with adult offsprings' sympathy/caring. *Developmental Psychology, 51*(1), 7–16. https://doi.org/10.1037/a0038137

Feldman, R. (2012). Parent–infant synchrony: A biobehavioral model of mutual influences in the formation of affiliative bonds. *Monographs of the Society for Research in Child Development, 77*(2), 42–51. https://doi.org/10.1111/j.1540-5834.2011.00660.x

Fonagy, P., Gergely, G., Jurist, E. L., & Target, M. (2018). *Affect regulation, mentalization, and the development of the self.* Routledge.

Gilbert, P. (1989/2016). *Human nature and suffering*. Routledge.

Gilbert, P. (2000). Social mentalities: Internal "social" conflict and the role of inner warmth and compassion in cognitive therapy. In *Genes on the couch: Explorations in evolutionary psychotherapy* (pp. 118–150). Brunner-Routledge.

Gilbert, P. (2005). Social mentalities: A biopsychosocial and evolutionary approach to social relationships. In M. W. Baldwin (Ed.), *Interpersonal cognition* (pp. 299–333). Guilford Press.

Gilbert, P. (2009). *The compassionate mind: A new approach to life's challenges*. Little Brown.

Gilbert, P. (2017). *Compassion: Concepts, research and applications*. Routledge.

Gilbert, P. (2018). Living like crazy. Annwyn House.

Gilbert, P. (2019). Psychotherapy for the 21st century: An integrative, evolutionary, contextual, biopsychosocial approach. *Psychology and Psychotherapy: Theory, Research and Practice, 92*(2), 164–189. https://doi.org/10.1111/papt.12226

Gilbert, P. (2020). Compassion: From its evolution to a psychotherapy. *Frontiers in Psychology, 11*, 586161. https://doi.org/10.3389/fpsyg.2020.586161

Gilbert, P. (2022a). Formulation and fears, blocks and resistances. In P. Gilbert & G. Simos (Eds.), *Compassion focused therapy: Clinical practice and applications*. Routledge.

Gilbert, P. (2022b). Compassion focused therapy as an evolution informed, biopsychosocial science of the mind: History and challenge. In P. Gilbert & G. Simos (Eds.), *Compassion focused therapy: Clinical practice and applications*. Routledge.

Gilbert, P. (2022c). Internal shame and self-disconnection: From hostile self-criticism to compassionate self-correction and guidance. In P. Gilbert & G. Simos (Eds.), *Compassion focused therapy: Clinical practice and applications*. Routledge.

Gilbert, P., & Basran, J. (2018). Imagining one's compassionate self and coping with life difficulties. *EC Psychology and Psychiatry, 7*(12), 971–978.

Gilbert, P., & Choden. (2013). *Mindful compassion*. Little Brown.

Gilbert, P., & Simos, G. (2022). *Compassion focused therapy: Clinical practice and applications*. Routledge.

Gilbert, P., Catarino, F., Duarte, C., Matos, M., Kolts, R., Stubbs, J., et al. (2017). The development of compassionate engagement and action scales for self and others. *Journal of Compassionate Health Care, 4*(1), 1–24. https://doi.org/10.1186/s40639-017-0033-3

Gleason, T. R. (2013). Imaginary relationships. In M. Taylor (Ed.), *The Oxford handbook of the development of imagination* (pp. 251–271). Oxford University Press.

Goetz, J. L., Keltner, D., & Simon-Thomas, E. (2010). Compassion: An evolutionary analysis and empirical review. *Psychological Bulletin, 136*(3), 351. https://doi.org/10.1037/a0018807

Hermanto, N., Zuroff, D. C., Kopala-Sibley, D. C., Kelly, A. C., Matos, M., Gilbert, P., & Koestner, R. (2016). Ability to receive compassion from others buffers the depressogenic effect of self-criticism: A cross-cultural multi-study analysis. *Personality and Individual Differences, 98*, 324–332. https://doi.org/10.1016/j.paid.2016.04.055

His Holiness the Dalai Lama, & Jinpa, T. (1995). *The power of compassion*. HarperCollins.

Kessler, S. E. (2020). Why care: Complex evolutionary history of human healthcare networks. *Frontiers in Psychology, 11*, 199. https://doi.org/10.3389/fpsyg.2020.00199

Khalsa, S. S., Adolphs, R., Cameron, O. G., Critchley, H. D., Davenport, P. W., Feinstein, J. S., et al. (2018). Interoception and mental health: A roadmap. *Biological Psychiatry: Cognitive Neuroscience and Neuroimaging, 3*(6), 501–513. https://doi.org/10.1016/j.bpsc.2017.12.004

Kim, S. (2015). The mind in the making: Developmental and neurobiological origins of mentalizing. *Personality Disorders: Theory, Research, and Treatment, 6*(4), 356. https://doi.org/10.1037/per0000102

Kim, J. J., Cunnington, R., & Kirby, J. N. (2020a). The neurophysiological basis of compassion: An fMRI meta-analysis of compassion and its related neural processes. *Neuroscience & Biobehavioral Reviews, 108*, 112–123. https://doi.org/10.1016/j.neubiorev.2019.10.023

Kim, J. J., Parker, S. L., Doty, J. R., Cunnington, R., Gilbert, P., & Kirby, J. N. (2020b). Neurophysiological and behavioural markers of compassion. *Scientific Reports, 10*(1), 6789. https://doi.org/10.1038/s41598-020-63846-3

Leary, M. R. (2004). *The curse of the self: Self-awareness, egotism, and the quality of human life*. Oxford University Press.

Leary, M. R., & Buttermore, N. R. (2003). The evolution of the human self: Tracing the natural history of self-awareness. *Journal for the Theory of Social Behaviour, 33*(4), 365–404. https://doi.org/10.1046/j.1468-5914.2003.00223.x

Lippard, E. T. C., & Nemeroff, C. B. (2020). The devastating clinical consequences of child abuse and neglect: Increased disease vulnerability and poor treatment response in mood disorders. *The American Journal of Psychiatry, 177*(1), 20–36. https://doi.org/10.1176/appi.ajp.2019.19010020

Longe, O., Maratos, F. A., Gilbert, P., Evans, G., Volker, F., Rockliff, H., et al. (2010). Having a word with yourself: Neural correlates of self-criticism and self-reassurance. *NeuroImage, 49*(2), 1849–1856. https://doi.org/10.1016/j.neuroimage.2009.09.019

Lunkenheimer, E., Tiberio, S. S., Skoranski, A. M., Buss, K. A., & Cole, P. M. (2018). Parent-child coregulation of parasympathetic processes varies by social context and risk for psychopathology. *Psychophysiology, 55*(2), e12985. https://doi.org/10.1111/psyp.12985

Luyten, P., Campbell, C., Allison, E., & Fonagy, P. (2020). The mentalizing approach to psychopathology: State of the art and future directions. *Annual Review of Clinical Psychology, 16*(1), 297–325. https://doi.org/10.1146/annurev-clinpsy-071919-015355

Mascaro, J. S., Florian, M. P., Ash, M. J., Palmer, P. K., Frazier, T., Condon, P., et al. (2020). Ways of knowing compassion: How do we come to know, understand, and measure compassion when we see it? *Frontiers in Psychology, 11*, 547241. https://doi.org/10.3389/fpsyg.2020.547241

Mayseless, O. (2016). *The caring motivation: An integrated theory*. Oxford University Press.

McFarland, D. H., Fortin, A. J., & Polka, L. (2020). Physiological measures of mother–infant interactional synchrony. *Developmental Psychobiology, 62*(1), 50–61. https://doi.org/10.1002/dev.21913

Meevissen, Y. M. C., Peters, M. L., & Alberts, H. J. E. M. (2011). Become more optimistic by imagining a best possible self: Effects of a two week intervention. *Journal of Behavior Therapy and Experimental Psychiatry, 42*(3), 371–378. https://doi.org/10.1016/j.jbtep.2011.02.012

Mikulincer, M., & Shaver, P. R. (2014). Mechanisms of social connection: From brain to group. *American Psychological Association*. https://doi.org/10.1037/14250-000

Mikulincer, M., & Shaver, P. R. (2017). An attachment perspective on compassion and altruism. In P. Gilbert (Ed.), *Compassion: Concepts, research and applications* (pp. 187–202) Routledge/Taylor & Francis Group.

Music, G. (2016). *Nurturing natures: Attachment and children's emotional, sociocultural and brain development*. Routledge.

Neff, K. (2011). *Self compassion: The proven power of being kind to yourself*. Hachette.

Osimo, S. A., Pizarro, R., Spanlang, B., & Slater, M. (2015). Conversations between self and self as Sigmund Freud—A virtual body ownership paradigm for self counselling. *Scientific Reports, 5*(1), 1–14. https://doi.org/10.1038/srep13899

Pandey, R., Tiwari, G. K., Parihar, P., & Rai, P. K. (2021). Positive, not negative, self-compassion mediates the relationship between self-esteem and Well-being. *Psychology and Psychotherapy: Theory, Research and Practice, 94*(1), 1–15. https://doi.org/10.1111/papt.12259

Peters, M. L., Flink, I. K., Boersma, K., & Linton, S. J. (2010). Manipulating optimism: Can imagining a best possible self be used to increase positive future expectancies? *The Journal of Positive Psychology, 5*(3), 204–211. https://doi.org/10.1080/17439761003790963

Petrocchi, N., & Cheli, S. (2019). The social brain and heart rate variability: Implications for psychotherapy. *Psychology and Psychotherapy: Theory, Research and Practice, 92*(2), 208–223. https://doi.org/10.1111/papt.12224

Petrocchi, N., Dentale, F., & Gilbert, P. (2019). Self-reassurance, not self-esteem, serves as a buffer between self-criticism and depressive symptoms. *Psychology and Psychotherapy: Theory, Research and Practice, 92*(3), 394–406. https://doi.org/10.1111/papt.12186

Porges, S. W. (2007). The polyvagal perspective. *Biological Psychology, 74*(2), 116–143. https://doi.org/10.1016/j.biopsycho.2006.06.009

Porges, S. W. (2017). Vagal pathways: Portals to compassion. In E. M. Seppälä, E. Simon-Thomas, S. L. Brown, M. C. Worline, D. Cameron, & J. R. Doty (Eds.), *The Oxford handbook of compassion science* (pp. 189–204). Oxford University Press.

Porges, S. W. (2021). Polyvagal theory: A biobehavioral journey to sociality. *Comprehensive Psychoneuroendocrinology, 7*, 100069. https://doi.org/10.1016/j.cpnec.2021.100069

Porges, S. W., & Dana, D. (2018). Clinical applications of the polyvagal theory: The emergence of polyvagal-informed therapies.. WW Norton & Company.

Ryan, C. (2019). *Civilized to death: The price of progress*. Simon and Schuster.

Singer, T., & Engert, V. (2019). It matters what you practice: Differential training effects on subjective experience, behavior, brain and body in the ReSource project. *Current Opinion in Psychology, 28*, 151–158. https://doi.org/10.1016/j.copsyc.2018.12.005

Slavich, G. M. (2020). Social safety theory: A biologically based evolutionary perspective on life stress, health, and behavior. *Annual Review of Clinical Psychology, 16*(1), 265–295. https://doi.org/10.1146/annurev-clinpsy-032816-045159

Stopa, L., Brown, M. A., & Hirsch, C. R. (2012). The effects of repeated imagery practice on self-concept, anxiety and performance in socially anxious participants. *Journal of Experimental Psychopathology, 3*(2), 223–242. https://doi.org/10.5127/jep.021511

Strauss, C., Lever Taylor, B., Gu, J., Kuyken, W., Baer, R., Jones, F., et al. (2016). What is compassion and how can we measure it? A review of definitions and measures. *Clinical Psychology Review, 47*, 15–27. https://doi.org/10.1016/j.cpr.2016.05.004

Taylor, M., Shawber, A. B., & Mannering, A. M. (2009). Children's imaginary companions: What is it like to have an invisible friend? In K. D. Markman, W. M. P. Klein, & J. A. Suhr (Eds.), *Handbook of imagination and mental simulation* (pp. 211–224). Psychology Press.

Werner, A. M., Tibubos, A. N., Rohrmann, S., & Reiss, N. (2019). The clinical trait self-criticism and its relation to psychopathology: A systematic review – Update. *Journal of Affective Disorders, 246*, 530–547. https://doi.org/10.1016/j.jad.2018.12.069

5 Attachment and Self-Compassion: Associations Across the Lifespan

Trisha L. Raque, Kathryn Ziemer, and John Jackson

Introduction

Self-compassion has been identified as an important self-regulation strategy and a general way of relating to the self that is predictive of a wide range of aspects of interpersonal and intrapersonal functioning (e.g., Zessin et al., 2015). Early life experiences, including the development of attachment to security figures, are key pathways for developing high levels of self-compassion, including how self-compassion relates to self-concept and emotion regulation. A growing body of research is establishing the ways in which attachment and self-compassion relate to each other as well as positively predict a wide range of mental and physical health outcomes across the developmental continuum. This chapter first outlines the construct of self-compassion and then the key tenets of attachment theory before presenting their associations across the lifespan. Beginning with the role of parents in their children's development of attachment styles, we then present information on other salient childhood experiences such as trauma and abuse as they relate to self-compassion, emotion regulation, and early maladaptive schemas. The importance of self-compassion as a mediating pathway between negative early life experiences and well-being will be highlighted. Continuing across the lifespan, this chapter will next cover peer attachment and adult attachment before concluding with a presentation of clinical interventions aimed at increasing secure attachment and self-compassion.

Self-Compassion

Neff and McGehee (2010) define self-compassion as "compassion turned inward" (p. 226). More specifically, self-compassion embodies the ability to treat oneself with kindness when experiencing struggles rather than judging oneself (i.e., self-kindness), to recognize one is not alone in their struggles and imperfections rather than feeling isolated in their distress (i.e., common humanity), and to remain present and observe one's thoughts and feelings rather than overidentifying with, exaggerating, or avoiding distress (i.e., mindfulness; Neff, 2003). Broadly, higher levels of self-compassion have predicted greater optimism, happiness (Neff et al., 2007), social connectedness, self-acceptance (Neff, 2003), coping with stress, resilience (Terry & Leary, 2011), and lower levels of anxiety, depression, self-criticism (Neff, 2003), negative affect (Neff

T. L. Raque (✉)
University of Denver, Denver, CO, USA
e-mail: trisha.raque@du.edu

K. Ziemer
Old Town Psychology, Alexandria, VA, USA

J. Jackson
Central Washington University, Ellensburg, WA, USA

A. Finlay-Jones et al. (eds.), *Handbook of Self-Compassion*, Mindfulness in Behavioral Health,
https://doi.org/10.1007/978-3-031-22348-8_5

et al., 2007), and rumination (Neff, 2009). A meta-analysis by Zessin et al. (2015) on the relationship between self-compassion and well-being reported medium to large effect sizes, with the strongest correlation between self-compassion and psychological well-being ($r = 0.62$, $k = 12$, $n = 1586$), then negative affect ($r = -0.47$, $k = 32$, $n = 5710$), cognitive well-being ($r = 0.47$, $k = 48$, $n = 11, 181$), and finally positive affect ($r = 0.39$, $k = 32$, $n = 5779$).

Neff (2003) has proposed that self-compassion may be a form of emotion regulation (ER) in addition to referring to how one relates to oneself. ER embodies the ways that individuals experience, respond to, and express their reactions to internal and external events (Gross, 1998) and may represent a transdiagnostic risk factor for many mental health issues (Finlay-Jones et al., 2015). In regard to ER, self-compassion has been shown to activate brain regions involved in emotional processing (Hofmann et al., 2011) and to decrease negative emotions (Arimitsu & Hofmann, 2017). Among psychologists experiencing stress, self-compassion was associated with decreased ER difficulties (Finlay-Jones et al., 2015). Finlay-Jones et al. (2015) explain that higher levels of self-compassion may provide greater emotional acceptance and clarity, thereby contributing to healthier ER strategies.

Given the strong associations between self-compassion, emotion regulation, and well-being, it is important to unpack how self-compassion may be developed. Early caregivers and the ways in which children attach to those caregivers may have profound influences on the development of self-compassion; caregivers may influence children's self-concepts and the ways that they learn, both directly and indirectly, to emotionally regulate. Further, experiences with trauma and/or betrayal may impact a person's ability to practice self-kindness, practice mindfulness as a form of emotion regulation, and feel connected to others (e.g., common humanity). Understanding how self-compassion relates to early maladaptive schemas and emotion regulation after trauma may offer insight into how to intervene to increase self-compassion and decrease the potentially deleterious effects of trauma.

Attachment Theory

The need for interpersonal connection is a fundamental human motivation (Baumeister & Leary, 1995), and how we pursue that connection reflects how we relate to the world and ourselves. Attachment theory has been lauded as one of the broadest, most coherent, well-known, and empirically grounded frameworks in the contemporary study of human development and relational functioning (Cassidy & Shaver, 2008). Attachment styles are hypothesized to develop based on early childhood experiences with caregivers and remain relatively consistent into adulthood (Bowlby, 1988), with implications for well-being. Mediators between attachment and well-being have been explored as potential points of intervention, with self-compassion indicated as a potentially key pathway through which attachment relates to mental health (e.g., Raque-Bogdan et al., 2011).

The most fundamental construct of attachment theory is attachment behavior, defined by Bowlby (1969, 1973) as forms of behavior that operate to achieve and/or maintain a child's proximity to a caregiver or attachment figure. Attachment behavior is primarily influenced by the primary caregiver's sensitivity to the child's behavior as well as the quality of interactions that facilitate proximity, help the child feel soothed, and promote a sense of inter- and intrapersonal security. These interactions are in turn influenced by the child's temperament, the caregiver's own attachment history, and environmental factors that can impact caregiving quality, such as family stress. Attachment behaviors support individuals to seek proximity to others in times of danger or potential threat (Bowlby, 1969). Attachment behavior is evolutionarily adaptive and ultimately functions to protect a child from predation and to promote other evolutionarily advantageous behavior, such as feeding, self-regulation, and social interaction (Cassidy & Shaver, 2016).

The quality and form of repeated interactions between a child and an attachment figure give way to the formation of what Bowlby (1973) termed *internal working models*. Internal working models function essentially as cognitive templates regarding oneself and one's external world, including one in relation to others. These structures organize experiences and expectations, shaping the way a child "...perceives events, forecasts the future, and constructs his [*sic*] plans" (Bowlby, 1973, p. 203). When attachment figures are consistently responsive and soothing, a child develops an internal working model of the self as valued and confident and worthy of having their needs met by others. When attachment figures are inconsistently available and/or responsive, or when they are punitive or abusive in their responses to attachment behavior, a child develops an internal working model of the self as unworthy and/or incompetent and not valued enough to have others meet their needs. Thus, experiences with attachment figures throughout childhood promote long-standing, sometimes rigid expectations about how one will be treated by significant others and one's notions of self. Repeated caregiver interactions continue to influence how close relationships are navigated across the lifespan.

During adulthood, internal working models serve as the mental representations of caring and loving relationships that are activated by the attachment system (Mikulincer & Shaver, 2007a, b). Internal working models are characterized by several different ways, often represented as dimensions of anxious, avoidant, and secure, with the taxonomy of fearful, dismissive, and preoccupied built upon these dimensions (Shaver & Fraley, 2008). Anxious attachment is characterized by a positive view of others yet a negative view of the self. Anxious attachment styles manifest in feelings of unworthiness, clinginess, and insecurities over the availability and responsiveness of others. Those with this attachment style may be hypervigilant to signs of abandonment and rejection and seek ongoing reassurance from others, in part because they may be self-critical and doubt whether they are deserving of others' responsiveness to their needs (Wei et al., 2011). Avoidant attachment styles are characterized by a fear of being close to others and holding negative views of others' responsiveness to their needs; thus, they may suppress or deactivate their needs as a form of self-protection (Moreira et al., 2015). They may struggle with intimacy in relationships, may suppress their emotions, may strive for emotional distance from others, and may be rigidly self-reliant when distressed. Withdrawing from others may serve to protect them from anticipated disappointment that others will not meet their needs (Mikulincer et al., 2003a, b). High avoidance and high anxiety comprise fearful attachment, which is characterized by feelings of inadequacy and distrust of others. Together, high avoidance and low anxiety comprise a dismissive attachment style characterized by inflated self-worth and minimization of the value of relationships. The combination of low avoidance and high anxiety constitutes preoccupied attachment, characterized by neediness, dependence on others for validation, and jealousy (Bartholomew & Horowitz, 1991).

Those low on the dimensions of attachment anxiety and avoidance are categorized as securely attached, and this attachment style is characterized by the tendency to hold a positive perspective of oneself and others, higher self-worth, greater capacity for emotion regulation, and higher levels of interpersonal satisfaction and general well-being (Love & Murdock, 2004). Secure attachment captures the expectation that others will be available and respond in a supportive manner to one's needs, with this confidence offering a foundation for positive psychological adjustment (Mikulincer & Shaver, 2003a, b). The majority of the attachment research reviewed in this chapter identifies styles as either secure, anxious, or avoidant rather than as fearful, dismissive, or preoccupied.

For multiple reasons, our overview of attachment theory would be incomplete without attention to the cultural context. Early researchers on attachment, Bowlby (1969/1982) and Ainsworth (1967) drew attention to the importance of context in influencing how attachment and caregiving behaviors are expressed and evaluated. For instance, the cultural context influences how we assess what is considered "responsive," "too responsive," or "not responsive enough" parental behaviors to children's needs for connections (e.g., offering too little/too much comfort and protection; Mirecki & Chou, 2013). Mirecki and

Chou (2013) outline the need to explore the function of a caregivers' behavior within a cultural context and address how adaptive that response is within both the cultural context where that behavior was developed as well as the current cultural context where it is adopted. They noted that "future research directions should examine the effectiveness of clinical applications of attachment theory through a multicultural-driven approach when working with families of various cultures" (Mirecki & Chou, 2013, p. 514).

Additionally, future research should explore the ways in which structural inequities responsible for sexism, racism, heterosexism, poverty, ableism, and other forms of marginalization may influence the development of attachment behavior. For example, being bombarded with societal messages that one's needs are unimportant as a marginalized individual (e.g., woman, racial minority, sexual minority, person with disabilities) may make it challenging not to internalize those messages and not manifest insecure attachment behavior. In a special issue of *Attachment and Human Development* focused on attachment and anti-racism, Stern et al. (2021) outline the importance of asking questions such as, "How do attachment figures teach their children positive racial socialization to help children develop secure internal working models?" and "How might avoidant coping styles be adaptive emotion regulation strategies in the context of predominantly white spaces?" Such inquiries embody the need to account for how power, privilege, and oppression may influence attachment behaviors.

In addition to examining the cultural context in which attachment behaviors are developed, Carr and Battle (2015) emphasize the importance of attending to the cultural context in which attachment behaviors are interpreted as functional or nonfunctional with a social justice lens. In particular, they examine how aspects of attachment avoidance have been interpreted as a strength in a neoliberal society that values self-reliance, independence, hypervigilance for danger, and minimal need for others (e.g., Ein-Dor et al., 2012). After outlining the negative consequences of neoliberalism for psychological well-being, Carr and Battle (2015) express:

> Accordingly, perhaps we, as attachment researchers, are concerned that in the example discussed we have glimpsed how attachment theory and the knowledge it creates might be unwittingly drawn into translating the unacceptable into the acceptable, in the interests of neoliberalism—but at the expense of psychological well-being. Perhaps we are calling for the emergence of a critical consciousness, a critical voice in attachment theory (and beyond) that seeks to interrogate carefully psychological knowledge constructed outside of a value-neutral framework and in the interests of social justice. (p. 172)

They conclude that rather than view attachment avoidance as a way of coping with a neoliberal society, as psychologists, we have a responsibility to recognize the ills in such a society and work to challenge it. As an extension of this view, rather than simply documenting how experiences with marginalization and oppression may make insecure attachment behaviors functional in a dysfunctional society, we have a responsibility to challenge and work to change that dysfunctional society. In short, the cultural context is a critical component in the conceptualization and interpretation of attachment behaviors and how we, as psychologists, advocate for societal change to enhance well-being. With this lens in mind, we now outline how attachment and self-compassion relate over the lifespan.

Connections Between Attachment Styles and Self-Compassion

Among adults, higher levels of attachment security are positively related to a greater ability to engage in self-care and thus self-compassion, whereas those who minimize the value of support provided in relationships (i.e., avoidant attachment style) or exhibit anxiety about whether they are worthy of others' care (i.e., anxious attachment style) are hypothesized to have less access to self-compassion (Neff & McGehee, 2010). Individuals who are clingy or dependent on external sources such as others for validation (i.e., preoccupied) may have less capacity to generate compassion for themselves. Moreover, those who do not trust others and struggle with self-worth (i.e., fearful) may not have a strong foundation for offering themselves acceptance. Mixed findings have been reported for those with dismissive attachment styles and self-compassion, suggesting that those with dismissive attachment may minimize the importance of relationships and thus engage in a form of self-deception that affects their accuracy in identifying their levels of self-compassion (Neff & McGehee, 2010).

Research on compassion training, such as compassion-focused therapy (CFT; Gilbert & Procter, 2006), has found that some people struggle to cultivate compassion for themselves and others, and being prompted to do so triggers avoidance, fear, or grief reactions (Gilbert, 2010). Gilbert (2010) outlines how early life experiences with caregivers influence both attachment styles and affiliative behaviors, with secure attachment resulting in greater capacity for compassion than anxious or avoidant attachment styles. Those who yearned for affection and caring from significant others but failed to receive it may experience self-compassion as increasing their awareness of their unmet needs for close, validating relationships (Gilbert, 2007). Gilbert et al. (2011) developed measures of fear of compassion, including fear of self-compassion, finding that fear of compassion for self was associated with fear of compassion for others, and both were linked to self-coldness, self-criticism, insecure attachment, depression, anxiety, and stress. These findings have been replicated elsewhere with fear of self-compassion relating to lower levels of self-compassion (Joeng & Turner, 2015) and self-compassion and fear of self-compassion predictive of higher levels of anxiety and depression (Joeng et al., 2017).

Early Childhood: Development of Attachment Styles and Self-Compassion

There are multiple frameworks for unpacking how early life experiences may influence the development of attachment styles and self-compassion, including negative life experiences resulting in early maladaptive schemas (EMSs). Conversely, positive early life experiences, such as having experienced mindful parenting, may also influence the development of self-compassion as a positive form of ER and way of relating to oneself. This next section will work through these various perspectives on how early life experiences may affect self-compassion and attachment and the ways in which self-compassion may be protective when facing challenging life events.

Adverse Childhood Experiences

Exposure to adverse childhood experiences may influence how individuals think about and treat themselves. For instance, caregivers who are not emotionally or physically responsive to an infant's cry for help may contribute to the infant's development of an anxious or avoidant attachment style later in life. In turn, attachment style may influence the ability to develop mindfulness and self-compassion and the formation of early maladaptive schemas. Thimm (2017) defines early maladaptive schemas (EMSs) as "negative beliefs about oneself and one's relationship with others that arise from adverse relational experiences in childhood and are associated with a broad range of psychological problems" (p. 3). These beliefs about oneself and others are theorized to develop in childhood and adolescence based on difficulties in having basic needs met (e.g., secure attachment, autonomy) over time and may influence the use of maladaptive coping strategies. Thus, these negative beliefs may become internalized into the individual's sense of self and may form the way the individual sees and interacts with themselves and the world. EMSs emerge from insecure attachment if the belief that takes hold is that one is not worthy of others' consistent care and that others are not dependable for meeting their needs (Cecero et al., 2004; Simard et al., 2011). Further, EMSs may undermine the development of mindfulness and self-compassion; in turn, low awareness of one's experiences in the present moment and self-criticism when experiencing struggles may act to perpetuate the negative impact of EMSs (Thimm, 2017). In a study with 212 undergraduate psychology students (mean age 21.8, *SD* = 4.4 years), participants reported negative associations between EMSs with mindfulness and self-compassion, with self-compassion and mindfulness mediating the pathway between EMSs and psychological distress (Thimm, 2017). Thus, self-compassion and mindfulness represent pathways through which EMSs relate to distress. However, to parse out the causal directions between attachment, mindfulness, self-compassion, and EMSs, longitudinal studies are warranted (Thimm, 2017).

Childhood Abuse Studies examining EMSs after trauma such as childhood abuse have offered further support for the important role that self-compassion may have in mediating between negative early life experiences and psychological distress (e.g., Zeller et al., 2015). Exposure to childhood abuse has been predictive of forming insecure attachments to caregivers, negative appraisals such as "I deserved the abuse" (Barlow et al., 2017), and lower levels of self-compassion (Vettese et al., 2011). For instance, Boyraz et al. (2019) found that participants who reported greater exposure to high betrayal traumas before the age of 18 also reported lower levels of self-compassion and higher post-traumatic stress. Yet, not all who experience trauma develop distress, and studies have revealed the importance of individuals' interpretation of trauma as key to the risk for post-traumatic stress disorder (PTSD), sometimes above and beyond the trauma exposure itself (DePrince et al., 2011). In addition to being predicted by trauma exposure at an early age, self-compassion may play a role in recovery after trauma (Játiva & Cerezo, 2014). Given that self-compassion relates to how one responds to oneself and how one copes during and after challenging moments (Germer & Neff, 2015), theoretically higher self-compassion may help decrease the development of negative trauma appraisals (e.g., shame, self-blame) and PTSD symptoms.

In addition to studying PTSD symptomatology, others have examined how exposure to abuse relates to self-compassion and other forms of psychological distress such as depression and substance use. Wu et al. (2018) report that for college students, self-compassion mediated the pathways between childhood emotional abuse and childhood emotional neglect with depression. Individuals experiencing childhood maltreatment may be less likely to develop self-compassion, potentially leading to higher risk for depression in adulthood (Soffer et al., 2008). Among youth seeking treatment for substance use, Vettese et al. (2011) found that child maltreatment was associated with lower levels of self-compassion and emotion regulation difficulties. Self-compassion predicted emotion dysregulation above and beyond problem substance use, distress, and maltreatment history. Further, self-compassion mediated the association between childhood maltreatment and emotion dysregulation in adolescence.

Bullying Self-compassion may also be an important buffer for negative early life relational experiences outside of the family environment, such as bullying. College students prompted to reflect on their childhood bullying experiences reported that self-compassion served as a significant mediator between attachment and shame as well as between bullying and shame (Beduna & Perrone-McGovern, 2019). Children's attachment styles may make them more susceptible to bullying; specifically, those with insecure attachment who hold cognitive schemas with expectations that others will treat them poorly may be more at risk for being bullied (Walden & Beran, 2010). Thus, self-compassion appears to be helpful for responding to bullying experiences with less shame and self-criticism and instead approaching the effects of bullying with equanimity and self-kindness (Beduna & Perrone-McGovern, 2019). In summary, research on aversive childhood experiences inside and outside the family environment has identified self-compassion as a malleable point of intervention to decrease distress after negative early life experiences. Self-compassion may be especially useful when addressing self-beliefs after trauma (Boyraz et al., 2019), emotion regulation (Vettese et al., 2011), and decreasing shame (e.g., Beduna & Perrone-McGovern, 2019).

The Role of Parents

Mindful Parenting At the other end of the continuum of early life trauma experiences, positive experiences with parents also may represent a key influence on how children develop attachment styles and emotion regulation strategies such as self-compassion. How parents respond to caregiving challenges is connected to their own attachment (Gilbert, 2005). Mindful parenting is characterized by bringing awareness, nonjudgmental acceptance, and compassion into parent-child interactions (Kabat-Zinn & Kabat-Zinn, 1997). It enacts compassion both toward the child and toward parents themselves and enacts self-regulation and emotional awareness that prompts parents to behave in congruence with their values and goals (Duncan et al., 2009). Drawing from attachment theory (Shaver et al., 2017), Moreira et al. (2018) found that mindful parenting may contribute to children's development of secure attachment and self-compassion. In a sample of 563 parent-child dyads (95.6% of which included mothers and 61.5% adolescent girls of mean age of 14 years), Moreira et al. (2018) found that mindful parenting indirectly related to adolescents' self-compassion and mindfulness through adolescents' attachment. Mindful parenting also indirectly related to adolescents' well-being through the pathways of attachment, self-compassion, and mindfulness. Further, adolescents' attachment was indirectly associated with their well-being through the pathways of self-compassion and mindfulness.

Mindful parenting may result in a less reactive approach that creates flexibility in stopping automatic and maladaptive parent-child interactional cycles, thereby cultivating secure relationships with children and creating conditions that foster children's self-compassion and mindfulness. According to Moreira et al.'s (2018) study with adolescents, observational learning and positive family experiences may account for mindful parents' effect on their adolescent's development of self-compassion and mindfulness. By observing their parents practice adaptive coping skills for difficult life situations, thoughts, and emotions, children may learn to relate to themselves in a similarly adaptive manner. Conversely, children who have observed their parents respond to life's challenges with self-criticism, negative emotions, or impulsivity may conclude that any of life's challenges may inevitably create unmanageable and shameful suffering that they alone experience. Children's inner dialogs develop in part because of family experiences; mindful parenting may result in *security-boosting interactions* that foster children and adolescents' secure attachment (Medeiros et al., 2016) thereby fostering the development of a self-compassionate self-soothing system (Moreira et al., 2018).

An interesting finding from this prior work is that parents' ability to listen to their adolescent with complete attention was associated with adolescents' mindfulness. Parents' ability to convey compassion, kindness, sensitivity, and responsiveness to their adolescents' needs indirectly related to adolescents' well-being through attachment. One explanation is that a secure relationship with their parents may allow adolescents to have more cognitive resources available for practicing mindfulness and that experiencing compassionate parenting contributes to the development of attachment security and adaptive emotion regulation strategies. Moreira et al.'s (2018) finding of only an indirect association between mindful parenting and adolescents' mindfulness and of attachment as the mediator indicates the crucial role that attachment plays in how parenting styles may influence their adolescents' mindfulness as well as their well-being.

Parents' Attachment Styles Whether parents engage in mindful parenting may be related to their own attachment styles. Parents' attachment security may affect their caregiving behaviors, including parental sensitivity, responsiveness, supportiveness, parenting stress, and closeness to children (Jones et al., 2015). Parents' reports of their own anxious and avoidant attachments have been associated with negative caregiving emotions, cognitions, and behaviors. Compared to parents with secure attachment, avoidant parents may experience greater ambivalence (Rholes et al., 1995) or less desire for children (Rholes et al., 1997), struggle to adopt caregiving roles (Gillath et al., 2005), feel less confident in their parenting abilities (Rholes et al., 1995), find parenting less meaningful and more stressful (Rholes et al., 2006), and express less support, closeness, and warmth toward their children (Edelstein et al., 2004). Anxiously attached parents may focus more on their own needs (Mikulincer & Shaver, 2007a, b) and experience distress when others need their help, thereby decreasing their sensitivity to their child's needs (Mikulincer et al., 2005). Relative to secure parents, anxious parents have also reported greater parenting stress (Moreira et al., 2015) and more negative attitudes toward parenting (Rholes et al., 1997).

Building upon the research on parenting attachment styles and caregiving behaviors, Moreira et al. (2016) explored self-compassion as the mechanism between mothers' attachment and mindful parenting. Mothers' self-compassion mediated the pathway between maternal attachment anxiety and decreased mindful parenting. Additionally, maternal attachment avoidance directly predicted lower levels of mindful parenting, and self-compassion failed to mediate this pathway. They concluded that attachment avoidant mothers appeared to struggle to practice mindful attention in their caregiving relationships with their child, perhaps instead enacting emotional suppression or inhibition. Having experienced their own parents' model as withholding and emotionally distant, avoidant mothers repeated this pattern with their own children. In contrast, mothers with attachment anxiety appear to experience emotional activation and struggle to implement self-soothing and self-compassion, thereby lowering their likelihood to practice mindful parenting. Moreira et al. (2016) suggest that secure mothers experience less self-criticism about their parenting and greater acceptance of their mothering skills, which is then transposed into how they relate to their children with nonjudgment, caring, and acceptance. Being more aware of, and kind toward, their wide range of emotions and thoughts, rather than ruminating, ignoring, or denying them may allow secure mothers to then self-regulate in their parenting skills.

Other outcomes for children, such as quality of life, also relate to parents' attachment and self-compassion. Moreira et al. (2015) found that in 171 family dyads of children/adolescents aged 8–18 years, the child's quality of life was indirectly predicted by their mother's attachment to their own mother through self-compassion and parenting stress. In other words, mothers' higher levels of attachment-related anxiety and avoidance were connected to worse children's quality of life only through the pathways of mothers' lower self-compassion and higher parenting stress. Interestingly, maternal attachment anxiety and avoidance failed to have direct associations with children's quality of life, although both directly related to parenting stress. These findings indicate that mothers with an anxious or avoidant attachment style have difficulty relating to themselves with warmth and compassion during challenges and experience greater stress with parenting demands. These difficulties with self-compassion and stress are in turn associated with lower quality of life reported by their children. In summary, self-compassion appears to be a critical pathway through which attachment may influence well-being, both for parents and for their children; interventions aimed at increasing self-compassion may help parents break cycles of enactment of anxious or avoidant attachment that may result in lower quality of life for their children and may lead children to internalize insecure attachment styles.

These studies provide further evidence for the importance of attending to individuals' early caregiving schemas and their relationship with attachment, self-compassion, and well-being. Children may develop self-compassionate or self-critical inner dialogs stemming from working models that reflect their family functioning; parents who express caring and warmth may lead children to internalize and provide themselves with caring and warmth (Neff & McGehee, 2010). Other studies have focused on how young adults' recollections of supportive family environments have been associated with adolescents' and young adults' higher levels of self-compassion, whereas others have outlined how attachment and self-compassion enacted in current functioning predicts well-being.

In a college student sample (mean age 21.5 years), Pepping et al. (2015) found that retrospective recall of parenting received in childhood was associated with attachment anxiety, attachment avoidance, and self-compassion. More specifically, attachment anxiety mediated the pathways between parental warmth, parental overprotection, and parental rejection, respectively, with self-compassion; these models explained between 15.3 and 16.2% of the variance in self-compassion. Parental warmth, overprotection, and rejection were correlated with attachment avoidance and self-compassion, but avoidance failed to mediate the associations between poor parenting and self-compassion. For college students, memories of parents as rejecting, critical, and absent of warmth were connected to lower self-compassion through higher levels of attachment anxiety, but not through avoidance. Pepping et al. (2015) suggested that recalling warm parental relationships may help individuals to feel secure and worthy of others' compassion, kindness, and care and not fear abandonment (i.e., low attachment anxiety). In contrast, memories of parents as critical or rejecting may be internalized into negative self-perceptions and self-criticism. Parental overprotection may communicate a lack of confidence in children's ability to cope with life's challenges, hindering opportunities for individuals to learn that they can handle difficult life experiences and cultivate self-compassion in the process (Pepping et al., 2015).

Parental Warmth Rather than examining attachment as the link between parental warmth and self-compassion, Kelly and Dupasquier (2016) identified social safeness and affect as key mediators. Social safeness is characterized as feeling warmth, calmness, and cared for and connected to other people (Gilbert, 2005). After controlling for parental rejection and overprotection, Kelly and Dupasquier (2016) found that recalled parental warmth related to higher self-compassion, lower fear of self-compassion, higher received social support, and lower fear of receiving compassion via social safeness. Thus, perceived parental warmth was associated with feeling connected and safe with others in other relationships, which in turn were associated with individuals' ability to generate compassion toward themselves and receive compassion from others. Although these findings do not point to attachment styles as the key theoretical framework for how parental environments may contribute to the development of self-compassion, they provide further evidence for the importance of parental warmth in how individuals learn to self-soothe and relate to themselves as a key emotion regulation strategy.

Adolescence: The Role of Peer Relationships on Attachment Styles and Self-Compassion

Adolescence represents a developmental stage during which self-compassion may be an especially important protective strategy for emotional regulation. In addition to peer relationships taking on increasing importance, adolescents may hold the *personal fable* cognitive schema in which they view their experiences as unique from that of others. The *personal fable* may lead them to believe that they are isolated in their challenges and that others cannot understand their experiences (Lapsley et al., 1989). Thus, self-compassion could be a helpful tool for breaking through isolation and managing interpersonal challenges with peers during adolescence (Neff & McGehee, 2010).

Prior work supports the hypothesis that secure parental attachments help youth access a self-compassionate response to personal challenges. Peter and Gazelle (2017) report how secure attachment with both parents in early adolescence (mean age 10.65 years) moderated associations between a social withdrawal presentation (characterized by anxiety, solitariness, shyness, and reticence) and self-compassion. Youth who were securely attached to both parents experienced fewer negative associations between a socially anxious interpersonal style and self-compassion. Another study by Jiang et al. (2017) focused on how closeness with parents, likely connected to attachment styles, related to peer relationships and self-compassion. They found that for 658 secondary students in China (mean age 13.58 years), self-compassion mediated associations between closeness with mothers, fathers (partial mediation), and peers to non-suicidal self-injury (NSSI) (Jiang et al., 2017). Self-compassion also mediated the pathway between attachment with peers and NSSI. By internalizing warm, validating, and accepting parental relationships such that they treat themselves similarly, young adolescents may be less likely to engage in NSSI. Likewise, feeling security and warmth in their relationships with peers (with these relationships potentially taking on increasing importance at this stage of development) may help young adolescents practice self-kindness, buffer against feelings of isolation, and reduce risk for NSSI when facing challenges in life.

Adulthood: Link Between Attachment Styles and Self-Compassion

Young Adulthood

As one of the first studies examining attachment and self-compassion in adolescence and young adulthood, Neff and McGehee (2010) found strong relations between higher self-compassion with lower depression, lower anxiety, and higher levels of positive social connection. In addition, family functioning (secure attachment, maternal support) and less endorsement of egocentrism predicted self-compassion. In turn, self-compassion partially mediated associations between family and cognitive variables and well-being. Adolescents (mean age of 15.2 years) and young adults (mean age of 21.1 years) reported similar levels of self-compassion, with self-compassion functioning in the same way in relation to well-being across these two samples. When differentiating the forms that insecure attachment may take, Neff and McGehee (2010) found that those with a preoccupied attachment style (i.e., characterized by neediness or dependency) were related to lower levels of self-compassion but that a dismissive attachment style failed to relate significantly with self-compassion. They explained this finding as individuals with a dismissive attachment style are less likely to consider interpersonal relationships important and potentially less likely to accurately recognize their levels of self-compassion. After controlling for family and cognitive factors, self-compassion continued to explain significant variance in overall well-being (i.e., lower depression and anxiety and higher social connection). Thus, associations between self-compassion and well-being were not confounded by family functioning and egocentrism. Neff and McGehee (2010) note that these findings suggest that self-compassion may be a more accessible point of intervention than attempting to change complicated patterns of family functioning. Conversely, they note that the reverse may also be true such that those higher in well-being may more easily be able to express understanding, compassion, and acceptance of their complicated family dynamics; emotional well-being may result in greater self-compassion toward one's family functioning.

Adulthood

Among adults, self-compassion has partially mediated associations between attachment and outcomes such as mental health (Raque-Bogdan et al., 2011), depression (Øverup et al., 2017) and/or anxiety (Joeng et al., 2017), interpersonal problems (Mackintosh et al., 2017), and subjective well-being (Wei et al., 2011). Multiple studies have reported differential associations between self-compassion and attachment anxiety/avoidance with several studies finding that attachment anxiety held the most predictive power for lower levels of self-compassion (e.g., Joeng et al., 2017; Øverup et al., 2017; Raque-Bogdan et al., 2011). However, most recently, in a sample of adults receiving psychological therapy in primary care for anxiety and/or depression, Mackintosh and colleagues (2017) found that attachment avoidance rather than anxiety played a key role in the connections between attachment, self-compassion, and distress. Attachment avoidance, but not anxiety, along with low self-compassion and high interpersonal problems predicted anxiety and depression, with self-compassion mediating this relationship. They suggest that although individuals with attachment avoidance may maintain a positive self-image, avoidant individuals seeking treatment for anxiety and depression may have difficulty treating themselves with self-kindness, approaching their distress with equanimity, and recognizing that they are not alone in their struggles. Collectively, these studies indicate the importance of examining attachment anxiety and avoidance for adults separately in the context of self-compassion and to be particularly sensitive to assessing for lower levels of self-compassion in individuals with avoidant attachment styles who are experiencing depression and anxiety.

Further, self-compassion has been explored as a mediator along with other positive interpersonal and intrapersonal factors such as mattering (Raque-Bogdan et al., 2011), belonging (Øverup et al., 2017), and empathy (Wei et al., 2011) as well as negative factors such as burdensomeness (Øverup et al., 2017) and fear of self-compassion (Joeng et al., 2017). The connections between self-compassion and variables such as mattering ($r = 0.34$; medium effect size; Raque-Bogdan et al., 2011) and belonging ($r = 0.46$; medium effect size; Øverup et al., 2017) imply that self-compassion not only connects to how individuals feel toward themselves but also relates to how they perceive others' views of them. Longitudinal and experimental studies can help to parse out any causal pathways between self-compassion and positive aspects of social connection such as mattering and belonging.

The Impact of Trauma in Adulthood

For adults reporting trauma, connections between attachment, self-compassion, and stress continue to remain consistent with research on college student and clinical populations. Bistricky et al. (2017) reported that higher experiences with interpersonal trauma (e.g., abuse, assault, rape) predicted higher levels of attachment avoidance and lower levels of self-compassion, which in turn predicted worse interpersonal skills. Interpersonal skills then mediated the pathways between interpersonal trauma, attachment avoidance, and self-compassion with post-traumatic symptoms. They concluded that self-compassion may be a particularly helpful emotion regulation strategy for adults in reducing the distancing, detachment, and rigidity characteristic of attachment avoidance and allowing for affect and thoughts to be integrated into one's self-concept after interpersonal trauma.

The potential for self-compassion to serve as a coping strategy when facing illness has begun to be investigated, specifically in cancer survivorship research. Receiving a cancer diagnosis may represent another form of trauma or threat to safety that may trigger the attachment system to seek care and comfort. Self-compassion has been associated with better mental health outcomes for breast cancer survivors (Pinto-Gouveia et al., 2014), including less depression, anxiety, stress, and body image disturbance (Przezdziecki et al., 2013; Sherman et al., 2017). In a sample with breast cancer survivors of more than 5 years since diagnosis, Arambasic et al. (2019) found that attachment anxiety and avoidance both predicted greater stress, more negative perceived impact of breast cancer on body image, worry, life interference ("I feel like cancer runs my life"), and lower self-compassion. In turn, self-compassion mediated the pathways through which attachment anxiety and avoidance, separately, related to stress and negative perceived impact of cancer. These findings further replicate conclusions drawn about the indirect impact of attachment through self-compassion on a wide range of mental health outcomes for healthy youth, adolescent, student, community, and clinical samples.

Older Adulthood

The benefits of secure attachment and its potential effects on the cultivation of self-compassion may continue into the later stages of life (Homan, 2018). Bowlby (1969/1982) indicated that illness or loss may represent life circumstances in which attachment styles are especially salient. Given that older adults tend to experience more illness and loss, the salience of their attachment styles may be especially relevant for their well-being during this stage of life. A growing body of research has supported the relevance of attachment theory across the life span (e.g., Lopez et al., 2018). Secure attachment in older adulthood has been connected to higher quality of life (Bodner & Cohen-Fridel, 2010) and marital satisfaction (Monin et al., 2014), greater ability to maintain social relationships, fewer depressive symptoms (Gillath et al., 2011), and lower pain perceptions (McWilliams & Bailey, 2010). During the transitions in relationships and physical health status experienced by older adults, secure attachment continues to relate to higher quality of life.

In one of the few studies to examine attachment and self-compassion jointly for older adults, Homan (2018) reports that attachment anxiety and avoidance were both negatively associated with the well-being dimensions of self-acceptance, personal growth, positive relationships, purpose in life, and environmental mastery, and attachment anxiety additionally was associated with lower levels of autonomy. Self-compassion mediated the relations between attachment styles and these aspects of well-being (excluding autonomy) for older adults. In other words, older adults reporting a more secure attachment style also indicated higher levels of self-compassion, and higher self-compassion served as the pathway through which attachment connected to greater acceptance of themselves, increased engagement in activities that promoted their growth, more fulfilling relationships with others, increased sense of purpose in their lives, and greater perceptions that they were living their lives in congruence with their values. To conclude, attachment research with older adults provides evidence that attachment exerts its influence "from the cradle to the grave" (Bowlby, 1979, p. 129) and self-compassion may be one important pathway through which attachment operates on well-being.

Implications for Interventions

As the research presented in this chapter indicates, attachment and self-compassion are intimately connected. Self-compassion is a mechanism through which attachment may influence well-being and other psychological outcomes. Moreover, parenting styles and the family environment may influence attachment as well as self-compassion. As such, early interventions to help parents cultivate secure attachment with their children and foster self-compassion have the potential to make a large difference in children's development. In adulthood, interventions to foster self-compassion also hold potential to correct for insecure attachment styles developed in childhood. This section will first cover interventions focused on parenting in early childhood and then discuss interventions targeting attachment and self-compassion in adulthood.

Mindful parenting, as described earlier in this chapter, has been found to promote positive and secure parent-child attachment and adolescent well-being by fostering self-compassion and mindfulness (Moreira et al., 2018). Early interventions that target parenting styles can help children develop healthy attachment styles and self-compassion from the start. Coatsworth et al. (2010) adapted a Strengthening Families Program: For Parents and Youth 10–14 to incorporate mindful parenting activities with the intent of fostering more positive adolescent-parent relationships and more mindful parenting practices. The adapted program made the mindfulness messages more explicit and included teaching mindfulness practices (e.g., mindful breathing), engaging in mindful reflections (e.g., compassion or loving kindness reflections) at the beginning and end of each session, and providing materials to practice mindfulness at home. The researchers conducted a pilot randomized trial with 65 families comparing the adapted program with the original program and a delayed intervention control group. Results indicate that the families in the adapted program incorporating mindfulness demonstrated greater mindful parenting and higher quality parent-youth relationships as compared to families in the original program.

Mediation analyses found that changes in mindful parenting served as the mechanism through which the benefits of the adapted program operated. This suggests that interventions to help parents engage in mindful parenting practices, including compassion reflections, may be a potential avenue for fostering healthier attachments in children.

Cultural context is important to take into account when considering interventions to improve parenting practices that foster secure attachment and self-compassion. The Family Care Curriculum is a 6-week intervention specifically developed for families experiencing homelessness and includes a culturally sensitive approach to teaching parents positive parenting practices (Sheller et al., 2018). The curriculum involves helping parents understand and provide for their children's emotional, attachment, and developmental needs, as well as understand how their own experiences contribute to their parenting beliefs and patterns. The program also explores the impact of racism, classism, and oppression on the parent-child relationship. Research on the outcomes is ongoing, but the program addresses a much-needed gap in providing a culturally sensitive attachment-based intervention to vulnerable populations.

Individuals who experienced insecure attachments in childhood can still build secure attachment and self-compassion as adults. The following set of interventions are meant to address and rectify insecure attachment styles in adulthood and foster self-compassion. The interventions include attachment-based compassion therapy (ABCT; Garcia-Campayo & Demarzo, 2015; García-Campayo et al., 2016a, b), secure attachment priming (Pepping et al., 2015), and self-compassion priming (Rowe et al., 2016).

Attachment-based compassion therapy (ABCT; Garcia-Campayo & Demarzo, 2015; García-Campayo et al., 2016a, b) was created to address maladaptive attachment styles developed in childhood. It is similar to other compassion interventions in that it includes compassion and mindfulness practices, but it differs in that it incorporates attachment theory. ABCT also incorporates the cultural values of Latin countries. The intervention involves eight sessions, each lasting 2.5 h. During the treatment, participants receive and offer compassion to friends, individuals deemed problematic, unknown individuals, and themselves. They also explore how the parent-child relationship emerged in childhood and address maladaptive attachment styles that may have developed.

In a nonrandomized controlled trial with healthy adults, ABCT demonstrated beneficial outcomes (Navarro-Gil et al., 2018). Compared to the waitlist control group, the participants in the ABCT group significantly increased their level of secure attachment style, self-compassion, and mindfulness and significantly reduced their level of psychological disturbance, anxiety, and avoidance. The increase in secure attachment style was mediated by increases in self-compassion. ABCT has also been found to be effective for adults with fibromyalgia (Montero-Marin et al., 2018). Compared to an active control group (i.e., relaxation exercises), the ABCT group demonstrated significant improvements in their general health status, which was maintained at a 3-month follow-up. ABCT participants also reported improvements in anxiety, depression, quality of life, and psychological flexibility.

Another intervention, secure attachment priming, aims to enhance feelings of security by temporarily activating an individual's mental representations of secure attachment figures (Mikulincer & Shaver, 2007a, b; Pepping et al., 2015). Pepping et al. (2015) examined the effect of secure attachment priming by randomizing undergraduate students to either a 10-min priming exercise or a 10-min interpersonal skills control group. During the secure attachment priming, participants visualized a person who makes them feel comfortable and safe, who is sensitive and responsive to their needs, and who would help them if needed. They were also asked to think about how the person would help them and how it would make the participants feel. Results indicate that security attachment priming increased state attachment security and self-compassion. This offers evidence that even a 10-min intervention can enhance attachment security, regardless of a person's attachment style, and that secure attachment fosters greater self-compassion. Interestingly, attachment anxiety and avoidance did not decrease as a result of the prime, which suggests that the secure attachment priming intervention may need to be modified or enhanced in order to fully address insecure attachment. A separate study found that individuals with higher levels of anxious and avoidant attachment styles felt more comfortable engaging in a compassion-focused imagery exercise after participating in the secure attachment priming intervention (Baldwin et al., 2019). These findings suggest that interventions to enhance attachment security may be needed before self-compassion exercises can be effective for those with insecure attachments.

In addition to secure attachment priming, researchers have also explored self-compassion priming. Rowe et al. (2016) randomized meditation-naïve individuals to 10-min self-compassion priming, secure attachment priming, or neutral priming (i.e., visualizing a shopping trip) conditions prior to engaging participants in a mindfulness meditation exercise. During the self-compassion prime, participants were asked to visualize and write about being completely compassionate and warm toward themselves. The secure attachment prime was similar to that described previously in the Pepping et al. (2015) study. Results indicate that both self-compassion priming and secure attachment priming increased participants' willingness to engage in further mindfulness training as compared to the control group. While secure attachment priming had a direct effect on willingness, the self-compassion prime operated indirectly through increasing state of mindfulness during the meditation exercise. The results of this study indicate that both self-compassion and secure attachment primers can help individuals feel more open to pursuing a challenging exercise.

Aside from fostering secure attachment and self-compassion in healthy adults, certain medical populations may also benefit from attachment-based interventions. A study by Arambasic et al. (2019) found that long-term breast cancer survivors with higher attachment anxiety and avoidance reported greater stress and a larger negative impact from the cancer. Moreover, they found that insecure attachment style was indirectly linked to psychological adjustment via self-compassion, indicating that breast cancer survivors may benefit from self-compassion training. For instance, the My Changed Body intervention (Przezdziecki et al., 2013; Sherman et al., 2017) is a writing intervention that has been specifically developed to foster self-compassion in breast cancer survivors. The intervention has been found to improve psychological distress and body image distress (Sherman et al., 2017). Additional studies are needed to explore whether the intervention affects attachment style. Overall, these studies indicate that interventions to increase self-compassion can foster well-being, especially among those with insecure attachment styles.

Conclusion

Attachment and the development of self-compassion are closely connected. Early life experiences influence the development of secure attachment and self-compassion through parents' own attachment styles, parents' practice of mindful parenting, and individuals' exposure to early life trauma inside or outside of the family. Further, the impact of attachment and self-compassion is felt across the lifespan, with higher levels of self-compassion associated with higher levels of well-being for children, adolescents, adults, and older adults. Promisingly, self-compassion is a malleable point of intervention for parents to potentially influence their children's attachment and self-compassion levels, as well as for adults to increase their ability to be compassionate toward themselves during times of illness (e.g., cancer) and good health.

References

Ainsworth, M. D. S. (1967). *Infancy in Uganda: Infant care and the growth of love*. Johns Hopkins.

Arambasic, J., Sherman, K. A., & Elder, E. (2019). Attachment styles, self-compassion, and psychological adjustment in long-term breast cancer survi-

vors. *Psycho-Oncology, 28*, 1134–1141. https://doi.org/10.1002/pon.5068
Arimitsu, K., & Hofmann, S. G. (2017). Effects of compassionate thinking on negative emotions. *Cognition & Emotions, 31*, 160–167. https://doi.org/10.1080/02699931.2015.1078292
Baldwin, S., Bandarian-Balooch, S., & Adams, R. (2019). Attachment and compassion-threat: Influence of a secure attachment-prime. *Psychology and Psychotherapy-Theory Research and Practice, 193*, 520–536. https://doi.org/10.1111/papt.12244
Barlow, R., Goldsmith Turow, R. E., & Gerhart, J. (2017). Neglect, trauma appraisals, emotion regulation difficulties, and self-compassion predict posttraumatic stress symptoms following childhood abuse. *Child Abuse & Neglect, 65*, 37–47. https://doi.org/10.1016/j.chiabu.2017.01.006
Bartholomew, K., & Horowitz, L. M. (1991). Attachment styles among young adults: A test of a four-category model. *Journal of Personality and Social Psychology, 61*, 226–244. https://doi.org/10.1002/pon.5068
Beduna, K. N., & Perrone-McGovern, K. M. (2019). Recalled childhood bullying victimization and shame in adulthood: The influence of attachment security, self-compassion, and emotion regulation. *Traumatology, 25*, 21–32. https://doi.org/10.1037/trm0000162
Bistricky, S. L., Gallagher, M. W., Roberts, C. M., Ferris, L., Gonzalez, A. J., & Wetterneck, C. T. (2017). Frequency of interpersonal trauma types, avoidant attachment, self-compassion, and interpersonal competence: A model of persisting posttraumatic symptoms. *Journal of Aggression, Maltreatment & Trauma, 26*(6), 608–625. https://doi.org/10.1080/10926771.2017.1322657
Bodner, E., & Cohen-Fridel, S. (2010). Relations between attachment styles, ageism and quality of life in late life. *International Psychogeriatrics, 22*, 1353–1361. https://doi.org/10.1017/S1041610210001249
Bowlby, J. (1969/1982). *Attachment and loss* (Vol. 1. Attachment). Basic Books.
Bowlby, J. (1973). *Attachment and loss, Vol. 2: Separation*. Basic Books.
Bowlby, J. (1979). *The making and breaking of affectional bonds*. Tavistock.
Bowlby, J. (1988). *A secure base: Parent-child attachment and healthy human development*. Basic Books.
Boyraz, G., Ferguson, A. N., Zaken, M. D., Baptiste, B. L., & Kassin, C. (2019). Do dialectical self-beliefs moderate the indirect effect of betrayal traumas on posttraumatic stress through self-compassion? *Child Abuse & Neglect, 96*. https://doi.org/10.1016/j.chiabu.2019.104075
Baumeister, R. F., & Leary, M. R. (1995). The need to belong: Desire for interpersonal attachments as a fundamental human motivation. *Psychological Bulletin, 117*(3), 497–529. https://doi.org/10.1037/0033-2909.117.3.497
Carr, S., & Batelle, I. O. (2015). Attachment theory, neoliberalism, and social conscience *Journal of Theoretical and Philosophical. Psychology, 35*, 160–176. https://doi.org/10.1037/a0038681
Cassidy, J., & Shaver, P. R. (Eds.). (2008). *Handbook of attachment: Theory, research, and clinical applications* (2nd ed.). The Guilford Press.
Cassidy, J., & Shaver, Ph. R. (2016). Handbook of attachment: Theory, research, and clinical applications (3rd ed). New York: The Guilford Press.
Cecero, J. J., Nelson, J. D., & Gillie, J. M. (2004). Tools and tenets of schema therapy: Toward the construct validity of the Early Maladaptive Schema Questionnaire-Research Version (EMSQ-R). *Clinical Psychology & Psychotherapy, 11*, 344–357. https://doi.org/10.1002/cpp.401
Coatsworth, J. D., Duncan, L. G., Greenberg, M. T., & Nix, R. L. (2010). Changing parent's mindfulness, child management skills and relationship quality with their youth: Results from a randomized pilot intervention trial. *Journal of Family Studies, 19*(2), 203–217. https://doi.org/10.1007/s10826-009-9304-8
DePrince, A. P., Chu, A. T., & Pineda, A. S. (2011). Links between specific posttraumatic appraisals and three forms of trauma-related distress. *Psychological Trauma: Theory, Research, Practice, and Policy, 3*, 430–441. https://doi.org/10.1037/a0021576Diedrich
Duncan, L., Coatsworth, J. D., & Greenberg, M. (2009). A model of mindful parenting: Implications for parent child relationships and prevention research. *Clinical Child and Family Psychology Review, 12*(3), 255–270. https://doi.org/10.1007/s10567-009-0046-3
Edelstein, R. S., Alexander, K. W., Shaver, P. R., Schaaf, J. M., Quas, J. A., Lovas, G. S., & Goodman, G. S. (2004). Adult attachment style and parental responsiveness during a stressful event. *Attachment & Human Development, 6*(1), 31–52. https://doi.org/10.1080/146167303100001659584
Ein-Dor, T., Reizer, A., Shaver, P. R., & Dotan, E. (2012). Standoffish perhaps, but successful as well: Evidence that avoidant attachment can be beneficial in professional tennis and computer science. *Journal of Personality, 80*, 749–768. https://doi.org/10.1111/j.1467-6494.2011.00747.x
Finlay-Jones, A. L., Rees, C. S., & Kane, R. T. (2015). Self-compassion, emotion regulation and stress among Australian psychologists: Testing an emotion regulation model of self-compassion using structural equation modeling. PLoS One, 10, e0133481. https://doi.org/10.1037/journal.pone.0133481
Garcia-Campayo, J. & Demarzo, M. (2015). Mindfulness y Compasion: la nueva revolucion. Madrid: Singlatana.
García-Campayo, J., Cebolla, A., & Demarzo, M. M. (2016a). *La ciencia de la compasión: más allá del mindfulness*. Alianza Editorial.
García-Campayo, J., Navarro-Gil, M., & Demarzo, M. (2016b). Attachment-based compassion therapy. *Mindfulness & Compassion, 2*, 68–74. https://doi.org/10.1016/j.mincom.2016.10.004
Germer, C. K., & Neff, K. D. (2015). Cultivating self-compassion in trauma survivors. In V. M. Follette, J. Briere, D. Rozelle, J. W. Hopper, & D. I. Rome

(Eds). Mindfulness-oriented interventions for trauma: Integrating contemplative practices (pp. 43-58). The Guilford Press.

Gilbert, P. (2005). Compassion and cruelty: A biopsychosocial approach. In P. Gilbert (Ed.), *Conceptualizations, research and use in psychotherapy* (pp. 9–74). Routledge.

Gilbert, P. (2007). *Psychotherapy and counselling for depression* (3rd ed.). Sage.

Gilbert, P., & Proctor, S. (2006). Compassionate mind training for people with high shame and self-criticism: Overview and pilot study. *Clinical Psychology and Psychotherapy, 13*, 353–379. https://doi.org/10.1002/cpp.507

Gilbert, P. (2010). *Compassion focused therapy: Distinctive features*. Routledge.

Gilbert, P., McEwan, K., Matos, M., & Rivis, A. (2011). Fears of compassion: Development of three self-report measures. *Psychology and Psychotherapy: Theory, Research and Practice, 84*, 239–255. https://doi.org/10.1348/147608310X526511

Gillath, Oh. Shaver, P. R., & Mikulincer, M. (2005). An attachment-theoretical approach to compassion and altruism. In P. Gilbert (Ed.), Compassion: Its nature and use in psychotherapy (pp. 121–147). London: Brunner-Routledge.

Gillath, O., Johnson, D. K., Selcuk, E., & Teel, C. (2011). Comparing old and young adults as they cope with life transitions: The links between social network management skills and attachment style to depression. *Clinical Gerontologist, 34*, 251–265. https://doi.org/10.1080/07317115.2011.554345

Gross, J. J. (1998). The emerging field of emotion regulation: An integrative review. *Review of General Psychology, 2*, 271–299.

Hofmann, S. G., Grossman, P., & Hinton, D. E. (2011). Loving-kindness and compassion meditation: Potential for psychological interventions. *Clinical Psychology Review, 31*, 1126–1132.

Homan, K. (2018). Secure attachment and eudaimonic well-being in late adulthood: The mediating role of self-compassion. *Aging & Mental Health, 22*(3), 363–370. https://doi.org/10.1080/13607863.2016.1254597

Játiva, R., & Cerezo, M. A. (2014). The mediating role of self-compassion in the relationship between victimization and psychological maladjustment in a sample of adolescents. *Child Abuse & Neglect, 38*, 1180–1190. https://doi.org/10.1016/j.chiabu.2014.04.005

Jiang, Y., You, J., Zheng, X., & Lin, M. (2017). The qualities of attachment with significant others and self-compassion protect adolescents from non-suicidal self-injury. *School Psychology Quarterly, 32*(2), 143–155. https://doi.org/10.1037/spq0000187

Joeng, J. R., Turner, S. L., Kim, E. Y., Choi, S. A., Lee, Y. J., & Kim, J. K. (2017). Insecure attachment and emotional distress: Fear of self-compassion and self-compassion as mediators. *Personality and Individual Differences, 112*, 6–11. https://doi.org/10.1016/j.paid.2017.02.048

Joeng, J. R., & Turner, S. L. (2015). Mediators between self-criticism and depression: Fear of compassion, self-compassion, and importance to others. *Journal of Counseling Psychology, 62*, 453–463. https://doi.org/10.1037/cou0000071

Jones, J. D., Cassidy, J., & Shaver, P. R. (2015). Adult attachment style and parenting. In J. A. Simpson & W. S. Rholes (Eds.), Attachment theory and research: New directions and emerging themes (pp. 234-260). The Guilford Press.

Kabat-Zinn, J., & Kabat-Zinn, M. (1997). *Everyday blessings: The inner working of mindful parenting*. Hyperion.

Kelly, A. C., & Dupasquier, J. (2016). Social safeness mediates the relationship between recalled parental warmth and the capacity for self-compassion and receiving compassion. *Personality and Individual Differences, 89*, 157–161. https://doi.org/10.1016/j.paid.2015.10.017

Lapsley, D. K., FitzGerald, D., Rice, K., & Jackson, S. (1989). Separation-individuation and the "new look" at the imaginary audience and personal fable: A test of an integrative model. *Journal of Adolescent Research, 4*, 483–505. https://doi.org/10.1177/074355488944006

Lopez, F. G., Ramon, K., & Kim, M. (2018). Development and initial validation of a measure of attachment security in late adulthood. *Psychological Assessment, 30*(9), 1214–1225. https://doi.org/10.1037/pas0000568

Love, K. M., & Murdock, T. B. (2004). Attachment to parents and psychological well-being: An examination of young adult college students in intact families and stepfamilies. *Journal of Family Psychology, 18*(4), 60–608. https://doi.org/10.1037/0893-3200.18.4.600

Mackintosh, K., Power, K., Schwannauer, M., & Chan, S. W. Y. (2017). The relationships between self-compassion, attachment and interpersonal problems in clinical patients with mixed anxiety and depression and emotional distress. *Mindfulness, 9*(3), 961–971. https://doi.org/10.1007/s12671-017-0835-6

McWilliams, L. A., & Bailey, S. J. (2010). Associations between adult attachment ratings and health conditions: Evidence from the National Comorbidity Survey Replication. *Health Psychology, 29*, 446–453. https://doi.org/10.1037/a0020061

Medeiros, C., Gouveia, M. J., Canvavarro, M. C., & Moreira, H. (2016). The indirect effect of the mindful parenting of mothers and fathers on the child's perceived well-being through the child's attachment to parents. *Mindfulness, 7*, 916–927. https://doi.org/10.1007/s12671-016-0530-z

Mikulincer, M., & Shaver, P. R. (2003a). The attachment behavioral system in adulthood: Activation, psychodynamics, and interpersonal processes. In M. P. Zanna (Ed.), *Advances in experimental and social psychology* (Vol. 35, pp. 53–152). Academic Press.

Mikulincer, M., & Shaver, P. R. (2003b). Attachment theory and affect regulation: The dynamics, development, and cognitive consequences of attachment-related strategies. *Motivation and Emotion, 27*, 77–102. https://doi.org/10.1023/A:1024515519160

Mikulincer, M., Shaver, P. R., Gillath, O., & Nitzberg, R. A. (2005). Attachment, caregiving, and altruism: Boosting attachment security increases compassion and helping. *Journal of Personality and Social Psychology, 89*(5), 817–839. https://doi.org/10.1037/0022-3514.89.5.817

Mikulincer, M., & Shaver, P. R. (2007a). *Attachment in adulthood: Structure, dynamics, and change.* Guildford.

Mikulincer, M., & Shaver, P. R. (2007b). Boosting attachment security to promote mental health, prosocial values, and inter-group tolerance. *Psychological Inquiry, 18*, 139–156. https://doi.org/10.1080/10478400701512646

Mirecki, R. M., & Chou, J. L. (2013). A multicultural application of attachment theory with immigrant families: Contextual and developmental applications. *Contemporary Family Therapy, 35*, 508–515. https://10.1007/s10591-012-9210-x

Monin, J. K., Zhou, L., & Kershaw, T. (2014). Attachment and psychological health in older couples coping with pain. *GeroPsych, 27*, 115–127. https://doi.org/10.1024/1662-9647/a000110

Montero-Marin, J., Navarro-Gil, M., Puebla-Guedea, M., Luciano, J. V., Van Gordon, W., Shonin, E., & Garcia-Campayo, J. (2018). Efficacy of 'Attachment-Based Compassion Therapy' (ABCT) in the treatment of fibromyalgia: A randomized controlled trial. *Frontiers in Psychiatry, 8*, Article 307. https://doi.org/10.3389/fpsyt.2017.00307

Moreira, H., Carona, C., Silva, N., Nunes, J., & Canavarro, M. C. (2016). Exploring the link between maternal attachment-related anxiety and avoidance and mindful parenting: The mediating role of self-compassion. *Psychology and Psychotherapy: Theory, Research and Practice, 89*, 369–384. https://doi.org/10.1111/papt.12082

Moreira, H., Gouveia, M. J., & Canavarro, M. C. (2018). Is mindful parenting associated with adolescents' well-being in early and middle/late adolescence? The mediating role of adolescents' attachment representations, self-compassion. *Journal of Youth and Adolescence, 47*, 1771–1788. https://doi.org/10.1007/s10964-018-0808-7

Moreira, H., Gouveia, M. J., Carona, C., Silva, N., & Canavarro, M. C. (2015). Maternal attachment and children's quality of life: The mediating role of self-compassion and parenting. *Journal of Child and Family Studies, 24*, 2332–2344. https://doi.org/10.1007/s10826-014-0036-z

Navarro-Gil, M., Lopez-del-Hoyo, Y., Modrego-Alarcon, M., Montero-Marin, J., Van Gordon, W., Shonin, E., & Garcia-Campayho, J. (2018). Effects of attachment-based compassion therapy (ABCT) on self-compassion and attachment style in healthy people. *Mindfulness, 11*, 51–62. https://doi.org/10.1007/s12671-018-0896-1

Neff, K. D. (2003). The development and validation of a scale to measure self-compassion. *Self and Identity, 2*, 223–250. https://doi.org/10.1080/15298860309027

Neff, K. D., Kirkpatrick, K., & Rude, S. S. (2007). Self-compassion and its link to adaptive psychological functioning. *Journal of Research in Personality, 41*, 139–154. https://doi.org/10.1016/j.jrp.2006.03.004

Neff, K. D. (2009). Self-Compassion. In M. R. Leary & R. H. Hoyle (Eds.), *Handbook of Individual Differences in Social Behavior* (pp. 561–573). Guilford Press.

Neff, K. D., & McGehee, P. (2010). Self-compassion and psychological resilience among adolescents and young adults. *Self and Identity, 9*(3), 225–240. https://doi.org/10.1080/15298860902979307

Øverup, C. S., Mclean, E. A., Brunson, J. A., & Coffman, A. D. (2017). Belonging, burdensomeness, and self-compassion as mediators of the association between attachment and depression. *Journal of Social and Clinical Psychology, 36*, 675–703. https://doi.org/10.1521/jscp.2017.36.8.675

Pepping, C. A., Davis, P. J., O'Donovan, A., & Pal, J. (2015). Individual differences in self-compassion: The role of attachment and experiences of parenting in childhood. *Self and Identity, 14*(1), 104–117. https://doi.org/10.1080/15298868.2014.955050

Peter, D., & Gazelle, H. (2017). Anxious solitude and self-compassion and self-criticism trajectories in early adolescence: Attachment security as a moderator. *Child Development, 88*(6), 1834–1848. https://doi.org/10.1111/cdev.12926

Pinto-Gouveia, J., Duarte, C., Matos, M., & Fráguas, S. (2014). The protective role of self-compassion in relation to psychopathology symptoms and quality of life in chronic and in cancer patients. *Clinical Psychology Psychotherapy, 21*(4), 311–323. https://doi.org/10.1002/cpp.1838

Przezdziecki, A., Sherman, K. A., Baillie, A., Taylor, A., Foley, E., & Stalgis-Bilinski, K. (2013). My changed body: Breast cancer, body image, distress and self-compassion. *Psychooncology, 22*(8), 1872–1879. https://doi.org/10.1002/pon.3230

Raque-Bogdan, T. L., Ericson, S. K., Jackson, J., Martin, H. M., & Bryan, N. A. (2011). Attachment and mental and physical health: Self-compassion and mattering as mediators. *Journal of Counseling Psychology, 58*(2), 272–278. https://doi.org/10.1037/a0023041

Rholes, W. S., Simpson, J. A., & Blakely, B. S. (1995). Adult attachment styles and mothers' relationships with their young children. *Personal Relationships, 2*(1), 35–54. https://doi.org/10.1111/j.1475-6811.1995.tb00076.x

Rholes, W. S., Simpson, J. A., Blakely, B. S., Lanigan, L., & Allen, E. A. (1997). Adult attachment styles, the desire to have children, and working models of parenthood. *Journal of Personality, 65*(2), 357–385.

Rholes, W. S., Simpson, J. A., & Friedman, M. (2006). Avoidant attachment and the experience of parenting. *Personality and Social Psychology Bulletin, 32*, 275–285. https://doi.org/10.1111/j.1467-6494.1997.tb00958.x

Rowe, A. C., Shepstone, L., Carnelley, K. B., Cavanagh, K., & Millings, A. (2016). Attachment security and self-compassion priming increase the likelihood that first-time engagers in mindfulness meditation will continue with mindfulness training. *Mindfulness, 7*(3), 642–650. https://doi.org/10.1007/s12671-016-0499-7

Shaver, P. R., & Fraley, R. C. (2008). Attachment, loss, and grief: Bowlby's views and current controversies. In J. Cassidy & P. R. Shaver (Eds.), *Handbook of attachment: Theory, research, and clinical applications* (2nd ed., pp. 48–77). Guilford Press.

Shaver, P. R., Mikulincer, M., Sahdra, B., & Gross, J. (2017). Attachment security as a foundation for kindness toward self and others. In K. W. Brown & M. R. Leary (Eds.), *The Oxford handbook of hypo-egoic phenomena*. Oxford University Press.

Sheller, S. L., Hudson, K. M., Bloch, J. R., Biddle, B., Ewing, E. S. K., & Slaughter-Acey, J. C. (2018). Family Care Curriculum: A Parenting Support Program for Families Experiencing Homelessness. *Maternal and Child Health Journal, 22*, 1247–1254. https://doi.org/10.1007/s10995-018-2561-7

Sherman, K., Woon, S., French, J., & Elder, E. (2017). Body image and psychological distress in nipple-sparing mastectomy: The roles of self-compassion and appearance investment. *Psychooncology, 26*(3), 337–345. https://doi.org/10.1002/pon.4138

Simard, V., Moss, E., & Pascuzzo, K. (2011). Early maladaptive schemas and child and adult attachment: A 15-year longitudinal study. *Psychology and Psychotherapy: Theory, Research and Practice, 84*, 349–366. https://doi.org/10.1111/j.2044-8341.2010.02009.x

Soffer, N., Gilboa-Schechtman, E., & Shahar, G. (2008). The relationship of childhood emotional abuse and neglect to depressive vulnerability and low self-efficacy. *International Journal of Cognitive Therapy, 1*, 151–162. https://doi.org/10.1521/ijct.2008.1.2.151

Stern, J. A., Barbarin, O., & Cassidy, J. (2021). Attachment perspectives on race, prejudice, and anti-racism. *Attachment and Human Development. Ahead of print*, 1–7. https://doi.org/10.1080/14616734.2021.1976920

Terry, M. L., & Leary, M. R. (2011). Self-compassion, self-regulation, and health. *Self and Identity, 10*(3), 352–362. https://doi.org/10.1080/15298868.2011.558404

Thimm, J. C. (2017). Relationships between early maladaptive schemas, mindfulness, self-compassion, and psychological distress. *International Journal of Psychology & Psychological Therapy, 17*, 1–15. https://www.proquest.com/docview/1878050811/citation/71C8C4D5F3A94929PQ/1?accountid=10382

Vettese, L. C., Dyer, C. E., Li, W. L., & Wekerle, C. (2011). Does self-compassion mitigate the association between childhood maltreatment and later emotion regulation difficulties? A preliminary investigation. *International Journal of Mental Health and Addiction, 9*(5), 48–491. https://doi.org/10.1080/15298868.2011.558404

Walden, L. M., & Beran, T. N. (2010). Attachment quality and bullying behavior in school-aged youth. *Canadian Journal of School Psychology, 25*, 5–18. https://doi.org/10.1177/0829573509357046

Wei, M., Liao, K. Y., Ku, T., & Shaffer, P. A. (2011). Attachment, self-compassion, empathy, and subjective well-being among college students and community adults. *Journal of Personality, 79*, 191–221. https://doi.org/10.1111/j.1467-6494.2010.00677.x

Wu, Q., Chi, P., Lin, X., & Du, H. (2018). Child maltreatment and adult depressive symptoms: Roles of self-compassion and gratitude. *Child Abuse & Neglect, 80*, 62–69. https://doi.org/10.1016/j.chiabu.2018.03.013

Zeller, M., Yuval, K., Nitzan-Assayag, Y., & Bernstein, A. (2015). Self-compassion in recovery following potentially traumatic stress: Longitudinal study of at-risk youth. *Journal of Abnormal Child Psychology, 43*, 645–653. https://doi.org/10.1007/s10802-014-9937-y

Zessin, U., Dickhauser, O., & Garbade, S. (2015). The relationship between self-compassion and well-being: A meta-analysis. *Health and Well-Being, 7*, 340–364. https://doi.org/10.1111/aphw.12051

Self-Compassion in Adolescence 6

Karen Bluth and Melissa Clepper-Faith

Introduction

Self-compassion, or a way of self-relating that has been described as *kindness and compassion turned inward* (Neff, 2003), has been associated with better overall emotional well-being in adults in many studies and confirmed in meta-analyses (e.g., MacBeth & Gumley, 2012; Zessin et al., 2015). Furthermore, studies demonstrate a lower physiologic stress response among those who have higher trait self-compassion (Breines et al., 2014; Breines et al., 2015; Svendsen et al., 2016) or who have been primed with self-compassion inductions (Arch et al., 2014); these are discussed in-depth in another chapter in this volume. Overall, self-compassion has been associated with increased happiness, optimism, connectedness to others, and curiosity and inversely associated with mental health disorders, including anxiety and depression in adult clinical and community samples (Neff, 2009).

Although research among adolescent populations is newer and therefore not as extensive as that of adults, studies indicate that self-compassion has similar beneficial effects on the mental health and well-being of adolescents (Marsh et al., 2018). In this chapter, we will first describe why self-compassion is critically important for adolescents and follow with an overview of the adolescent research to date. This chapter does not provide a comprehensive or systematic review but rather includes a diverse sample of the research studies that have been published and suggests ways in which the findings can support healthy adolescent functioning.

K. Bluth (✉) · M. Clepper-Faith
University of North Carolina at Chapel Hill, Chapel Hill, NC, USA
e-mail: karen_bluth@med.unc.edu

The Need for Self-Compassion in Adolescence

Adolescence is a crucial and often tumultuous developmental period, during which the main task of teens is to explore and develop their identity through negotiating social roles, adapting to changing peer and family relationships, and exploring career and vocational goals (Erikson, 1963; Marcia, 1980). While this exploration of identity is taking place, the adolescent brain is going through significant restructuring; myelination of neurons allows the brain's processing speed to increase dramatically, and new neuronal pathways are established while unused pathways are pruned away (Giedd, 2008), allowing for more efficient information processing. Differences in the rate of maturation of brain systems, such as the prefrontal cortex (or attentional control system) and the amygdala (within the limbic system), result in adolescents being more sensitive to emotional highs and lows without always having the advantage of the ability to think clearly about the

A. Finlay-Jones et al. (eds.), *Handbook of Self-Compassion*, Mindfulness in Behavioral Health,
https://doi.org/10.1007/978-3-031-22348-8_6

consequences of their actions (Keating, 2004). This aspect of the adolescent brain has been commonly referred to as "being in a car with a sensitive gas pedal and bad brakes" (Steinberg, 2005). These brain changes are reflected in adolescents' ability to think more abstractly; for the first time, adolescents are able to think about their thinking or *metacognate*. Metacognition, although advantageous in many ways, can also promote greater self-consciousness which can then lead to greater self-criticism. At the same time, changes in sex hormones, in particular, estrogen and testosterone levels in females (Andersen et al., 2022; Copeland et al., 2019), result in increases in depression and anxiety.

In addition to the normative ongoing restructuring of the brain and other physiological changes which take place during this period (Giedd, 2008), this developmental stage may be complicated by environmental challenges that may have long-term deleterious effects on adult health. These challenges often include social stressors such as transitions to different schools and peer and academic pressures (Forbes & Dahl, 2010; Steinberg & Morris, 2001) and may also include in-person and cyberbullying (Patchin & Hinduja, 2013), exposure to violence (Lambert et al., 2005), structural racism (Lambert et al., 2009), physical safety concerns, misogyny (Dosil et al., 2020), inadequate academic resources (Crede et al., 2015), food insecurity (Shanafelt et al., 2016), homelessness (Wang et al., 2019), and lack of future academic and economic opportunities (Johnson et al., 2014). Thus, the complex interaction of various physiological changes taking place at this developmental period, combined with the onslaught of environmental challenges, sets the stage for adolescents to be highly self-critical, anxious, and depressed, particularly for females.

Although adolescence has always been recognized as a time of increased vulnerability to mental health problems, over the last decade, mental health disorders among adolescents have skyrocketed. For example, between 2010 and 2015, high symptoms of mental illness increased by 33% among a US sample of adolescents aged 13–18, and suicides increased by 31%; these changes were mostly driven by females and occurred across all race and ethnicities sampled (Twenge et al., 2018). Mental health disorders in adolescence, particularly when unrecognized or untreated, can establish a maladaptive developmental trajectory and are predictive of substance abuse, poor academic achievement, and mental and physical health disorders in adulthood (Patel et al., 2007). According to the 2018 American Psychological Association survey (American Psychological Association, 2018), US adolescents reported higher stress than any other age group and were least likely to report that their mental health was either excellent or very good. Notably, 75% of teen respondents cited gun violence in general, mass shootings, and school shootings as high on their list of stressors.

Certain subpopulations of adolescents are at higher risk for mental health disorders when compared to their peers and are important groups to support through intervention. For example, studies have shown gender differences exist in how teens respond to stress, with females being at greater risk of developing and maintaining internalizing disorders such as depression and anxiety. Beginning in early adolescence, females report more depressive symptoms than males (Bennik et al., 2014; Nolen-Hoeksema, 2012; Sontag & Graber, 2010), and this global trend continues through adolescence until age 16–19 when females are twice as likely to be depressed as males (Salk et al., 2017). Also, teens who have preexisting health disorders that were diagnosed in childhood are faced with the added normative stressors of adolescence, coupled with the drive for independence and desire to self-direct their medical treatment. For instance, teens with chronic medical illnesses, (e.g., Type 1 diabetes mellitus, eating disorders, cancer) face the added challenges of transitioning to self-care and management of their medications and treatment regimen (Finlay-Jones et al., 2020). Similarly, adolescents with developmental delays may find their health challenges exacerbated by the social and academic stressors of adolescence (North et al., 2013).

Furthermore, a past history of physical, sexual, or emotional abuse is a risk factor for adolescent and adult mental health disorders, including anxiety, depression, and suicidality (Lindert et al., 2014; Lippard & Nemeroff, 2020; Sweeting et al., 2020; Turner et al., 2017). Childhood abuse and other adverse childhood experiences are associated with a higher risk of physical disease in adulthood and adolescence (Felitti et al., 1998; Herrenkohl et al., 2013; Lippard & Nemeroff, 2020). When adverse childhood experiences occur over a prolonged period, the toxic stress which results can cause changes in the developing brain that has lifelong impacts (Kuo et al., 2012; Shonkoff & Garner, 2012). For example, childhood sexual abuse has been linked to both internalizing and externalizing maladaptive behaviors, including depression, and worldwide, prevalence for childhood sexual abuse is 18% for females and 8% for males (Kuehner, 2017). Finding ways to combat the effects of early life diversity is instrumental to disrupting its long-term sequelae.

The stressors faced by marginalized populations, such as Black, Indigenous, and People of Color (BIPOC), exacerbate the challenges of adolescence, and many BIPOC adolescents experience greater psychological, academic, and behavioral challenges than their White counterparts both during and after adolescence (Assari et al., 2017; Hughes et al., 2016; Huynh & Fuligni, 2010; Paradies et al., 2015; Umaña-Taylor & Updegraff, 2007). For example, a 2015 systematic review and meta-analysis found a significant association between experiencing racism and negative mental health outcomes, including depression, stress, and anxiety, particularly for Asian-Americans and Latinx-Americans (Paradies et al., 2015). Additionally, youth who have recently emigrated may experience higher risk of depressive symptoms related to their immigration status and the perception of being regarded as "foreigners" (Davis et al., 2016; Lopez et al., 2016; Sirin et al., 2019). As one example, Latinx youth who recently emigrated to the United States experienced ethnic discrimination in the ninth grade and were more likely to develop depressive symptoms 6 months later (Davis et al., 2016).

Further, sexual/gender minority (SGM) adolescents experience worse mental health outcomes (i.e., greater depression, higher suicide ideation rates) than their sexual majority peers, which is likely a result of higher levels of bias-based bullying, victimization, and social stigma they experience (Gnan et al., 2019; Hatchel et al., 2019; Vigna et al., 2017). In fact, stigmatized SGM youth have two to three times the rates of depression, anxiety, self-injury, and suicide rates relative to their sexual gender majority peers (King et al., 2008; Zaza et al., 2016).

In addition, other subpopulations of adolescents which struggle with nonnormative stressors that exacerbate the normative stressors of adolescence include youth experiencing homelessness or undocumented status (Torres et al., 2018; Zapata et al., 2016), adolescents in the child welfare system (Tanaka et al., 2011), and adolescents who were maltreated or victimized as children (Játiva & Cerezo, 2014; Vettese et al., 2011). Further, youth with multiple marginalized identities, such as those listed above who also struggle with experiences of racial discrimination and are LGBTQIA+, may face an intersectional or compounding effect of stress on their mental health (Vigna et al., 2017).

Importantly, the current social climate, characterized by political instability, racial protests, the global COVID-19 pandemic, and the resulting disruption of school and social structures, has exacerbated many of these risk factors and has increased the stress level of adolescents. One recent study reported that compared to pre-pandemic measures, adolescents in a large pediatric primary care network in the Northeast United States had significant increases in depressive symptoms and suicidal thoughts, particularly among female, non-Hispanic Black, and non-Hispanic White adolescents during the pandemic (Mayne et al., 2021). Female adolescents in this study had a 34% increase in suicidal thoughts compared to before the COVID-19 pandemic, 1 year prior (Mayne et al., 2021). Another study found that the global prevalence of child and adolescent depression and anxiety has doubled in the first year of the COVID-19 pandemic, to current pooled estimate prevalence levels of 25% and 20.5%, respectively (Racine et al., 2021). In fact, Ellis et al. (2020) found that during the COVID-19 pandemic, 43% of adolescents were "very concerned" about the pandemic, as it related to their academic success and relationships with friends, and COVID-related stress was associated with greater loneliness and depression, particularly for those adolescents who spent more time on social media. Further, as connecting with peers and establishing one's place in the social network are critical tasks in adolescent development, and in-person schools and peer interactions have been limited during the pandemic, social media use among adolescents increased during this time, with 48% spending more than 5 hours a day on social media since schools closed (Ellis et al., 2020). Along with the advantages of social contact, online social media use during the pandemic also may have increased unhealthy social media interactions, such as cyberbullying, increasing adolescent mental health challenges (Mayne et al., 2021).

Clearly, there is no magic pill that will alleviate these complex challenges that adolescents face today. However, building inner resources to contend with these stressors is one important strategy for helping to modulate adverse impacts. For example, establishing a sense of identity with a community, finding meaning behind struggles, and acknowledging one's agency (Meyer, 2015) are all internal resources that can be strengthened and have been found to be advantageous in dealing with minority stressors (Meyer, 2015). They are also all aspects of self-compassion; a sense of common humanity emerges from identifying with a community, finding meaning behind struggles is part of growth toward self-kindness (and taught directly in an exercise in the Mindful Self-Compassion program), and acknowledging one's agency and purpose is integral to fierce self-compassion.

As adolescence is a sensitive period for intervention and often a precarious period for emotional stability, providing teens with coping tools such as self-compassion can potentially shift their mental health and behavioral trajectory to a more positive and salubrious pathway. The research providing evidence for links between self-compassion and various dimensions of well-being among adolescents is discussed next.

Associations Between Self-Compassion and Well-Being Outcomes

The first empirical study on self-compassion and adolescence was published in 2010; in this study, Neff and McGehee explored the relationship between self-compassion and psychological resilience in US adolescents and young adults (Neff & McGehee, 2010). Since then, numerous studies have been published, and overall, research findings mirror that of adults. For example, similar to the meta-analysis on adults that shows a large inverse relationship ($r = -0.54$) between self-compassion and psychopathology, defined as stress, anxiety, and depression (MacBeth & Gumley, 2012), Marsh et al.'s (2018) meta-analysis of adolescent samples demonstrated a similar magnitude in the relationship between self-compassion and psychopathology ($r = -0.55$). Supporting this, a systematic review of studies on self-compassion and depressive symptoms in adolescents found that across both longitudinal and cross-sectional studies, self-compassion was inversely related to depressive symptoms (Pullmer et al., 2019a).

Similar associations between self-compassion and various domains of emotional health have been reported. Specifically, Neff and McGehee (2010) found that self-compassion is positively correlated with social connectedness in adolescents; this is noteworthy as establishing healthy peer relationships and stable social networks is critical factor in emotional well-being during adolescence. Additionally, this study also reported positive associations between self-compassion and secure attachment and negative associations between self-compassion and preoccupied and fearful attachment, suggesting that self-compassion is established in an individual through safe and trusting early life relationships. As there has been a paucity of opportunities to develop and cultivate close peer relationships during the time of the pandemic, the ramifications of isolation during this period may influence the emotional well-being of adolescents that may not be apparent for some time.

In understanding the implications of self-compassion for adolescent well-being, it is important to clearly delineate the differences between self-compassion and self-esteem. Due to the many physiological, cognitive, and environmental changes taking place at this developmental stage, many adolescents, particularly females, struggle with low self-esteem; the chapter on self-esteem within this handbook discusses this at length. Although both self-compassion and self-esteem are ways of self-relating, self-esteem, defined as a global evaluation of one's self-worth (Baumeister et al., 2003), differs from self-compassion in that it is acquired and maintained by comparing oneself with others. For example, social media provides a readily accessible way for adolescents to compare themselves to others through number of "likes", "shares," and filtered or curated images posted on social media platforms. Unfortunately, this comparison often has a negative impact on one's self-esteem; in fact, 45% of adolescents aged 15–21 indicated that engaging in social media makes them feel judged, and 38% report that it makes them feel bad about themselves (American Psychological Association, 2018). Yet even if the comparison does not have a negative outcome or even results in a boost in one's self-esteem, the consequence of comparing oneself with others leads to an emotional separation from others at a time in life when what is most needed is connection with others and a sense of belonging. For example, whether one considers oneself above average in a particular trait or below average (i.e., high self-esteem or low self-esteem), an outcome is that one sees oneself as apart from others, rather than feeling accepted and included, or as an integral part of a community. Further, establishing high self-esteem is dependent upon performance and achievement, and inevitably one will not always perform at their best or even fail, and thus one's self-esteem is unstable. In contrast, self-compassion offers a way of self-relating that is stable over time. Through self-compassion, one can always be present to support oneself (Neff & Vonk, 2009).

The relationship between self-compassion and self-esteem was investigated further in a study which used cross-lagged autoregressive structural equation modeling to determine the temporal order of constructs, i.e., which construct predicted the other (Donald et al., 2018). Over 4 years, self-compassion and self-esteem levels of 2809 adolescents were assessed. Results indicated that self-esteem consistently predicted self-compassion levels, whereas self-compassion levels did not predict levels of self-esteem, indicating that among adolescents, it may be necessary to feel valued, worthy, and deserving to give oneself compassion. More research with adolescent populations is needed to replicate these findings.

The different roles of self-compassion and self-esteem as linked to adolescents' psychological well-being are evidenced in a study by Marshall et al. (2015). In this longitudinal study of 2448 high school students in Australia, self-compassion had a buffering effect, protecting against the negative effects of low self-esteem. For grade 9 students who were high in self-compassion, having low self-esteem did not have an effect on their mental health; in grade 10, these students had no significant change in their mental health. However, grade 9 students who were both low in self-compassion and self-esteem had worse mental health when they were assessed in grade 10. Thus, having high self-compassion appeared to protect against the negative repercussions of having low self-esteem (Flett et al., 2003; Hewitt & Flett, 1991; Marshall et al., 2015).

Self-Compassion as a Buffer Against Stress and Trauma

Managing stressful circumstances in a healthy way is an essential skill that adolescents must learn to transition to becoming adults who lead fulfilling and satisfying lives. Being able to cope with stressful circumstances productively and successfully is therefore critical to healthy adolescent development. Among adolescents, self-compassion has been shown to be associated with increased adaptive coping following stressful events. For example, one study found that undergraduates (mean age = 18 years) in Japan who were higher in self-compassion reported 1 month later that they felt more in control of a recent stressful event and felt that it was less threatening to them compared to those who were lower in self-compassion (Chishima et al., 2018). Furthermore, self-compassion was negatively related to avoidance coping; those adolescents who were higher in self-compassion were less likely to engage in maladaptive coping strategies such as denying, disengaging, or distracting themselves from a problem (Chishima et al., 2018).

Self-compassion may also play a buffering role in protecting adolescents from negative outcomes resulting from traumatic early life experiences and chronic stress. For example, among adolescents who had experienced abuse and neglect as children and were part of the Canadian child welfare court system, those who reported greater self-compassion were less likely to experience psychological distress, problem alcohol use, or report a suicide attempt than those with greater self-compassion, even when accounting for early life maltreatment (Tanaka et al., 2011). Among another population of adolescents experiencing chronic stress, Prentice et al. (2021) investigated adolescents and young adults who struggle with chronic physical health problems requiring ongoing medical care. In this study, self-compassion was positively associated with emotional well-being and negatively associated with mental distress. Difficulties in emotion regulation mediated the relationship between self-compassion and distress; this suggests that interventions addressing emotion regulation difficulties and promoting self-compassion may be particularly helpful for teens facing the challenges of chronic medical illness (Prentice et al., 2021).

Being more self-compassionate appears to be beneficial in managing not only long-term chronic stressful events, such as childhood abuse/neglect and health problems, but also acute stressful events. The weeklong Mount Carmel forest fire occurred in northern Israel in December 2010, forcing the evacuation of an educational residential youth village in which high school students lived, 50% of whom had come from homes where they experienced chronic stressors such as poverty, violence, and substance abuse and 20% of whom were orphans. Notably, within 1 month of the fire, 88% of adolescents reported that they feared for their lives when the fire was taking place. Among these youth who were already considered at-risk, both longitudinal and multilevel mediational analyses demonstrated that having greater self-compassion predicted less post-traumatic stress, panic symptoms, depressive symptoms, and suicidality 3 and 6 months later (Zeller et al., 2015).

The potential buffering effect of self-compassion described here is also supported by evidence from physiological responses to stress in an experimental protocol. The Trier Social Stress test is a well-established research protocol that has been used extensively to elicit a physiological stress response that can then be measured in a lab setting. Adolescents were exposed to the Trier Social Stress test in which they were asked to give a speech and perform math computations in front of two neutral-faced lab-coated adults. Adolescents who self-reported greater self-compassion had less of a change in their systolic blood pressure during the lab stressor than those with lower self-compassion (Bluth et al., 2016b), indicating that self-compassionate adolescents may get less activated in stressful situations than adolescents who are less self-compassionate. However, more research using various physiological measures of evaluating stress is needed to substantiate this finding.

In addition to buffering the effects of external stressors (Flett et al., 2003; Hewitt & Flett, 1991; Marshall et al., 2015), self-compassion also appears to protect against the negative impacts of internal characteristics that increase vulnerability to adverse outcomes, such as maladaptive perfectionism. Maladaptive perfectionism is defined by having high personal standards that involve high self-criticism, being overly worried about making mistakes, and fear of being negatively evaluated by others. Many adolescents struggle with maladaptive perfectionism in various domains of their lives, including physical appearance, academics, and their interactions and relationships with peers. This type of perfectionism has been linked to depression (Flett et al., 2003; Hewitt & Flett, 1991) and is considered transdiagnostic in that it affects many aspects of health, including anxiety and eating disorders (Egan et al., 2011). For example, maladaptive perfectionism related to academic achievement among adolescents has been linked to anxiety and depression (Einstein et al., 2000) and negative affect after taking a test (Flett et al., 2009). Ferrari et al. (2018) demonstrated that by treating self-critical thoughts as passing events and as an opportunity to treat oneself with kindness, one is able to interrupt and subvert the pathway to depression. In this study, 541 adolescents in a high school setting completed questionnaires comprising measures of perfectionism, depressive symptoms, and self-compassion. As expected, the results showed a positive correlation between maladaptive perfectionism and depression and a significant moderating or buffering effect of self-compassion on the relationship between maladaptive perfectionism and depression, thus weakening this relationship (Ferrari et al., 2018). In other words, when self-compassion was high in adolescents, the link between maladaptive perfectionism and depression was not as strong, and adolescents who were more self-compassionate were less likely to become depressed.

Many adolescents and young adults struggle with maladaptive perfectionism concerning body image, the subjective assessment or judgment on one's physical appearance. Ntoumanis et al. (2020) studied the association of upward appearance comparison (comparing oneself to another who is perceived as more attractive) on appearance evaluation (the sense of satisfaction with one's own body image) in 396 Greek adolescents. This study investigated appearance-specific self-compassion, i.e., self-compassion that is specific to negative appearance-related thoughts, rather than global self-compassion. For these adolescents, appearance self-compassion moderated, or buffered, the effects of upward comparison on appearance evaluation; those adolescents with high self-compassion were less likely to negatively evaluate their appearance when comparing their image to someone who they considered more attractive (Ntoumanis et al., 2020).

Similarly, Rodgers et al. (2017) found that self-compassion moderated the relationship between perceived overweight status, appearance comparison, and appearance esteem in their study of 232 adolescents, aged 13 to 18 years. Appearance comparison, which is the process of comparing one's physical appearance to others and engaging in mostly unfavorable comparisons, has been identified as a critical mechanism in the maintenance of body image concerns and mediates the relationship between weight status and body dissatisfaction. Perceived overweight status in adolescents is associated with body dissatisfaction and poor self-esteem. The two self-compassion components of common humanity and mindfulness moderated the association between perceived overweight status and appearance comparison among boys, but not among girls. In other words, for those boys who had higher levels of mindfulness and common humanity, the link between perceiving themselves as overweight had less of an effect on negatively evaluating their appearance (Rodgers et al., 2017).

Self-compassion has also been shown to be a moderator in the association between non-suicidal self-injury (NSSI) and depressive symptoms in adolescence (Xavier et al., 2016). NSSI, described as self-injury that is not associated with suicidal intent, is prevalent in adolescents, particularly females. In fact, one study reported that one in four females and one in ten males reported engaging in NSSI over the previous 12 months (Monto et al., 2018). Six hundred forty-three adolescents aged 12–18 years completed questionnaires containing measures of emotional health, risk-taking and self-harm, self-compassion, and daily peer hassles (i.e., conflicts with friends regarding beliefs, opinions, and choices). The results showed that those who reported higher depression scores and daily peer hassle scores were more likely to self-injure. Importantly, self-compassion had a significant modifying or buffering effect on the relationship between depression and NSSI; higher levels of self-compassion were associated with lower levels of NSSI, independent of the adolescents' depression levels (Xavier et al., 2016).

As suicide is the second leading cause of death among adolescents aged 15–19 in the United States and the fourth leading cause of death in China, finding ways to obviate suicide attempts is of critical concern. In one study, self-compassion moderated the association between suicidal ideation and suicidal attempts in a cohort of 520 Chinese adolescents (Sun et al., 2020). Those adolescents with higher self-compassion (measured by high scores when combining the three positive components of self-compassion and low scores when combining the negative components of self-compassion) who had thoughts of suicide were less likely to attempt suicide over a 12-month interval, compared to those with low self-compassion (Sun et al., 2020).

Sex differences in the moderating role of self-compassion have also been found. Results of a cross-sectional study of 1057 adolescents (65% female, mean age 14.5 years) supported the inverse relationship between self-compassion and perceived stress, depression, and anxiety and found that self-compassion moderated the association between perceived stress and anxiety for males only (Lathren et al., 2019) but moderated the association between perceived stress and depression equally between sexes. These findings are consistent with an earlier study showing that self-compassion was a buffer against later onset of anxiety or depression following a traumatic event (Zeller et al., 2015).

Gender and Age as Moderators of Self-Compassion

Similar to findings from a meta-analysis with adults that reported that women had lower levels of self-compassion overall compared to men (Yarnell et al., 2015), most evidence in adolescent studies have also found levels of self-compassion are greater in males (Bluth & Blanton, 2015; Bluth et al., 2016b; Castilho et al., 2017; Pullmer et al., 2019a). However, one study found no difference between male and female adolescents (Neff & McGehee, 2010). Bluth and Blanton (2015) examined differences in self-compassion at different adolescent developmental stages and found that level of self-compassion did not differ between males and females in middle school but differed significantly between males and females in high school, where girls scored lower in self-compassion than their male counterparts, as well as lower in self-compassion than middle school females. Further, the inverse relationship between self-compassion and negative affect was significantly greater among older adolescents compared to that of younger adolescents (Bluth & Blanton, 2015).

Much has been posited about why adolescent females experience greater depressive symptoms and negative affect than males (Nolen-Hoeksema & Girgus, 1994), and likely similar explanations exist for lower self-compassion. First, there may be biological explanations; puberty brings increases of estrogen and testosterone, and these changes, and particularly the amount of fluctuation in these changes, have been linked to depression (Andersen et al., 2022; Copeland et al., 2019). Social changes take place at this time as well; gender intensification theory suggests that in early adolescence, boys and girls take on more "traditional" gender roles for a time, which work in boys' favor; boys tend to dominate and direct conversations, for example, resulting in a loss of agency for girls (Del Giudice, 2015).

Self-compassion levels have also been measured among sexual/gender minority adolescents (i.e., LGBTQIA+) and were found to have lower level of self-compassion than their sexual/gender majority peers, with a medium effect size. Vigna et al. (2017) postulated that this is an effect of the internalization of discrimination and stigma that many SGM experience daily. Indeed, minority stress theory explains that SGM adolescents suffer from substantial proximal and distal stressors, such as discrimination in school, rejection from family, and discomfort with their identity resulting in higher rates of depression, anxiety, and suicidality compared to their non-SGM peers (Goldbach & Gibbs, 2017).

In another study of 238 Canadian high school students, results indicated significantly higher self-compassion in males than in females (Pullmer et al., 2019b); this longitudinal study investigated relationships between baseline self-compassion and the mediating function of psychological distress on eating behaviors and body satisfaction at two time points, 14 to 18 weeks apart. While findings demonstrated that self-compassion was positively associated with body satisfaction, and negatively associated with eating pathology for both males and females at both time points, changes in psychological distress mediated the relationships between self-compassion and the outcomes of body satisfaction and eating pathology for females but not males. The authors suggested that it may be that there are other pathways in which self-compassion influences body satisfaction and eating pathology in males, such as concerns about muscle mass, for example (Pullmer et al., 2019b).

There is some evidence that the response to self-compassion interventions also differs by gender. Bluth and Eisenlohr-Moul (2017) reported a trend for a greater increase in self-compassion in females than males following an 8-week self-compassion program. This is similar to findings from a study of a 6-week mindfulness program in which there was preliminary evidence that females were more engaged than males; for example, only 60% of boys used mindfulness stress-reduction techniques during the post-intervention stress test, compared to 100% of the girls (Bluth et al., 2017). As developmental maturity occurs at different ages for males and females, differences may be related to males' relative lack of maturity and emotional states compared to females of the same age (Steinberg & Morris, 2001). Also, as adolescent males often adopt traditional male hegemonic norms, they may be reluctant to engage in compassion programs that they perceive as being "weak" or "soft" (Kirby & Kirby, 2017); rather, programs that emphasize the cultivation of courage or strength may be more appealing.

Self-compassion levels can also vary according to the age of the adolescent. Stolow et al. (2016) reported that older adolescents (aged 12–16 years) showed higher levels of self-criticism and depressive symptoms, and lower self-esteem and self-compassion, when compared to younger adolescents (aged 9–10 years). Although not measured in this study, previous literature attributes this to the fact that older female adolescents have lower self-compassion, higher negative affect, and less life satisfaction than their younger female adolescent counterparts (Bluth & Blanton, 2015). Further, older adolescents have also shown greater increases in self-compassion scores following an intervention when compared to younger adolescents (Bluth & Eisenlohr-Moul, 2017); this may be because having lower self-compassion levels at baseline affords a greater opportunity to raise level of self-compassion across an intervention. Interestingly, the meta-analysis of studies on self-compassion that examined gender differences found that although men reported higher self-compassion levels than women, this association diminished with increasing age, perhaps due to increased development of the common humanity aspect of self-compassion in older individuals (Yarnell et al., 2015); that is, with age, people are more able to see that experiencing emotions, particularly emotions that are challenging, is part of the experience of being human.

Self-Compassion as a Mediator: The Role of Self-Compassion in Predicting Outcomes

Self-compassion has been shown to be a mechanism which explains the relationship between various risk or protective factors that adolescents are exposed to and their psychological outcomes. For example, Neff and McGehee (2010) reported that self-compassion partially mediated the relationship between early life influences, including maternal support and family functioning and adolescent well-being, defined by depression, anxiety, and connectedness. Adolescents' reports of maternal support and better family functioning led to greater self-compassion, which was then associated with less depression and anxiety and greater sense of connectedness. This suggests that healthy family functioning provides support for establishing and maintaining self-compassion in the individual, which then leads to better mental health during adolescence.

In another study, self-compassion mediated the relationship between shameful memories and traumatic experiences that had been experienced earlier in life and depressive symptoms occurring in adolescence. Shameful memories arise from past experiences in which one is the target of exclusion, rejection, or criticism; the recollection of these experiences later in life can produce intense emotional responses, including feelings of shame, self-criticism, and isolation (Castilho et al., 2017). Adolescents are particularly vulnerable to shame memories because they are engaged in the development of self-identity and social connectedness, and memories of past experiences have a strong effect on the maturation of a positive self-identity (Castilho et al., 2017). Among 1100 adolescents in Portugal, Castilho et al. (2017) found that self-compassion mediated the relationship between adolescents' shame memories and depressive symptoms, such that shame memories were linked with lower self-compassion, which in turn was associated with greater depressive symptoms. Higher levels of self-compassion were associated with higher emotional self-regulation and self-soothing, which may have contributed to the adolescents' ability to use more effective strategies to deal with difficult emotions. One crucial strategy for the healthy functioning of adolescents is the formation of supportive social bonds; this study concluded that self-compassion may facilitate a sense of interpersonal connectedness and of common human experience, which may help adolescents build social bonds and lessen shame and self-criticism, which is then linked with lower depression (Castilho et al., 2017).

Similarly, 109 adolescents with poor school performance who had experienced being victimized (i.e., victimization as a child by a parent or siblings; having experienced assault, robbery, or kidnapping; sexual victimization; indirect victimization such as observing others being victimized; Internet victimization) were assessed (Játiva & Cerezo, 2014). Those who had experienced poly-victimization (i.e., several types of victimization) and directly experienced victimization (as compared to indirect victimization) had worse psychological maladjustment than others, and self-compassion partially mediated the relationship between victimization and psychological maladjustment. In other words, those adolescents who had experienced more victimization and direct victimization had lower levels of self-compassion, which was then associated with higher levels of psychological maladjustment and more internalizing and externalizing problems (Játiva & Cerezo, 2014).

Wu et al. (2019) found that self-compassion mediated the relationship between peer acceptance and non-suicidal self-injury (NSSI) in 816 Chinese adolescents. Increased peer acceptance led to increased self-compassion, which then was associated with decreased depressive symptoms and decreased NSSI. This mediation model was moderated by behavioral impulsivity; for those adolescents with higher levels of impulsivity, increased self-compassion was linked with decreased NSSI, while for adolescents with lower levels of impulsivity, there was no significant relationship between self-compassion and NSSI (Wu et al., 2019). The implications of this study are that for adolescents who are impulsive, such as those with attention deficit hyperactivity disorder (ADHD) who may struggle with being accepted by their peers, cultivating self-compassion may be particularly important for decreasing depressive symptoms and NSSI.

In a study with 1872 US adolescents, Vigna et al. (2017) examined the relationship between bias-based bullying and victimization related to sexual/gender minority (SGM) status. Bias-based bullying is the targeted, stigmatized harassment of those who are perceived as marginalized or "other" in some way by their peers. SGM youth reported higher levels of exposure to adversity, risk-taking behaviors, bias-based bullying, peer victimization, and depression and anxiety, and lower levels of self-compassion, compared to their SGM majority peers, and self-compassion attenuated the relationship between bias-based bullying/peer victimization and both anxiety and depression. The degree to which self-compassion mediated the association of bias-based bullying and anxiety and depression varied according to the adolescents' level of exposure to bias-based bullying; those with higher levels of bias-based bullying reported less of a mediating effect of self-compassion on this association (Vigna et al., 2017). These results suggest that for those with relatively lower levels of bias-based bullying, greater self-compassion could potentially facilitate the pathway to reducing the adverse effects of bias-based bullying and internalization of stigma and thereby improve the mental health outcomes for SGM youth.

Interventions: Cultivating Self-Compassion in Adolescents

Recognizing the beneficial correlates of self-compassion in adolescents, and the positive outcomes of the Mindful Self-Compassion (MSC) program developed by Neff and Germer (2013), an adaptation of MSC was created for adolescents, called Mindful Self-Compassion for Teens (formerly known as Making Friends with Yourself) (MSC-T). MSC-T is currently an eight-session group program involving weekly sessions of 1.5-hour duration that teaches skills of self-compassion. Implementation of MSC-T has demonstrated decreases in depression, anxiety, negative affect, and stress and increases in self-compassion pre- to post-intervention (Bluth et al., 2021; Bluth & Eisenlohr-Moul, 2017; Bluth et al., 2016a; Galla, 2016). Like MSC, MSC-T has a foundation of mindfulness and focuses on teaching skills of self-compassion; in particular, participants in both MSC and MSC-T are taught to recognize that most of us tend to be much harder on ourselves than we are on others and that harsh self-criticism often is detrimental to our mental health. For example, in the MSC-T program, adolescents learn to recognize when their inner critic arises and are taught to then actively take steps to practice being kinder and more compassionate to themselves. Perhaps most importantly, adolescents discover that they are not alone in their struggles and that other adolescents encounter many of the similar emotional pitfalls as do they. Often, this is very eye-opening to adolescents.

In addition to explicitly introducing and teaching self-compassion practices in the MSC-T program, self-compassion is also taught implicitly in mindfulness programs and retreat settings. In these programs, self-compassion is embedded within the context of teaching mindfulness. For example, in the adolescent mindfulness programs such as Learning to BREATHE (Broderick, 2021) and iBme (www.ibme.com), concepts of self-compassion are taught implicitly within the program, and self-compassion practices are included, although the focus of the program is learning mindfulness skills. Galla (2016, 2017) measured change in self-compassion across two studies in two subsequent years in a mindfulness retreat program for adolescents and reported that within-person change in self-compassion was the "driver" behind more of the outcomes than within-person change in mindfulness. Self-compassion predicted decreases in perceived stress, rumination, depressive symptoms, and negative affect and increases in positive affect and life satisfaction, for example (Galla, 2016).

Limitations and Future Directions

As research on adolescents and self-compassion is just over a decade old, there are many gaps in the literature that need to be addressed. First, most studies are cross-sectional which precludes determining a causal link between self-compassion and mental health outcomes. Although Krieger et al. (2016) provided support that self-compassion was antecedent to depression in a cross-lagged panel analysis with an adult sample, longitudinal studies are needed among adolescent populations to support direction of effects. For example, it may be that among adolescents, those with less depressive symptoms are more able to be self-compassionate, rather than self-compassion predicting depressive symptoms, or it may be that these relationships are bidirectional. Additionally, longitudinal studies with at least three time points are needed to determine mediation pathways, as cross-sectional mediation studies have distinct limitations, and should be used only as a preliminary way to conceptualize mediation (MacKinnon, 2008).

Also, there is a lack of standardization of outcome assessments across studies, including standardized measures assessing emotional well-being and mental health. As one example, in their meta-analysis, Pullmer and Chung et al. (2019a) reported that studies utilize numerous measures of depressive symptoms. Using standard measures would allow for comparison across studies and make synthesizing the results of these studies more accurate. Recent debates in the literature over the factor structure of the self-compassion scale (e.g., Muris et al., 2019; Neff et al., 2018) have added variability to the way self-compassion is defined and measured. Hopefully, the publication of the youth self-compassion scale (Neff et al., 2021) will help to standardize self-compassion assessment in youth. In addition, there is lack of standardization of definitions across different age categories (e.g., young teens versus older teens, middle school versus high school), and using agreed-upon definitions of these different stages of development would make it more possible to compare across studies. Similarly, acknowledging multiple genders and establishing criteria for multiple genders within research studies (i.e., providing various gender choices) would better align with current conceptualizations of gender.

There is also a dearth of interventional studies, and the field will benefit from implementing manualized interventions, such as MSC-T, and assessing outcomes in various populations and subpopulations of youth. MSC-T can also be tailored to meet the needs of different populations; for example, Bluth et al. (2021) adapted MSC-T for transgender adolescents with promising findings, and Boggiss et al. (2020) adapted MSC-T to a brief format for adolescents with Type 1 diabetes and disordered eating. Recognizing the differences in self-compassion with regard to gender and age is critical in developing adolescent self-compassion interventions, as it is prudent to tailor interventions to the unique developmental and emotional needs of the participants.

Further, as adolescents are generally adaptable to online learning, and implementing MSC-T online with minor modifications has been shown to be feasible (Bluth et al., 2021), it may be that online implementation of self-compassion interventions is the most effective way to reach adolescents, particularly those in remote areas or those who, due to illness or other reasons, are unable to attend a program in-person (Finlay-Jones et al., 2020). Further, it would be valuable to engage the adolescents as community advisors on how to best tailor the intervention to meet the specific needs of their subpopulation, as well as how to best reach these adolescents and recruit them into the program. For example, in a review of self-compassion as an active component in the prevention of anxiety and depression in adolescents and young adults, Egan et al. (2022) interviewed 20 young people, to ascertain their understanding of self-compassion and how to best offer self-compassion programming to adolescents and young adults. Findings were illuminating; for example, certain terms, such as "intervention" and even "self-compassion" and "self-love," were turnoffs for some, whereas for one participant, appealing to cultivating wisdom and courage was more attractive and would get more "buy-in." Many voiced that they had similar fears of compassion as that of adults; that is, they were afraid of losing motivation to accomplish their goals and expressed that self-compassion sounded "weak" or "lazy." Clearly, it would be important to clarify these misconceptions when teaching self-compassion to adolescents. Additionally, many contributed that designing programs that addressed the diversity which exists among young people is essential, as the "one-size-fits-all" model does not necessarily work. Ideally, these interventional studies need to be randomized and controlled, preferably with an active control group but, at minimum, with a waitlist control or a treatment-as-usual control group. Follow-up assessments (e.g., at 3 months, 6 months, and 1 year) are needed to determine whether outcomes are maintained over time. Lastly, intervention studies need to be replicated by different researchers in different settings to support generalizability.

Further, ascertaining the intervention components that are the most effective in achieving outcomes is needed. Various guided meditations, exercises, didactic instruction, and discussion comprise the interventions. Determining the critical components that are vital in achieving positive outcomes would be advantageous in further tailoring the interventions. Also, it may be that certain components are more efficacious in a particular culture or among a particular subpopulation of adolescents, and other components may be beneficial in other settings. Chio et al. (2021) found differential effects of self-compassion components in dialectical and non-dialectical cultures among adults; there may be similar effects for youth populations as well. Thus far, no research to our knowledge has investigated for whom and under what conditions which components of the interventions are most effective.

In addition, it is important to evaluate the fidelity of the intervention and whether it is being delivered in the way it was intended. Fidelity checks should be included to ensure that all components of the intervention are being taught. Also, the amount of training and experience that the instructors have should be noted, as well as their own practice history with mindfulness and self-compassion. Being able to embody mindfulness and self-compassion, as well as thorough training in teaching contemplative practices in groups, is highly recommended and included as one of the six domains in the Mindfulness-Based Interventions: Teaching Assessment Criteria, a tool created to systematically evaluate the integrity of mindfulness-based interventions (Crane et al., 2013; Griffith et al., 2021). Additionally, experience working with adolescents in an educational or therapeutic capacity should be a requirement, as teaching adolescents has specific challenges unique to their developmental stage.

Future research should use larger and more diverse sample populations to determine whether findings can be generalized to the general population. Self-compassion can be studied in adolescent subpopulations, particularly in marginalized or intersecting groups, including BIPOC, sexual/gender minorities, adolescents experiencing homelessness or undocumented status, those with a history of childhood abuse, those struggling with chronic health conditions, and recent immigrants. The differences between different adolescent genders merit investigation, as it is possible that different interventional approaches and formats may be more effective for different genders. For example, when teaching adolescents who have experienced trauma, it is important to take measures to ensure their emotional safety, providing them with options such as keeping their eyes opened or closed or even disengaging with the exercise completely. For some subpopulations, such as those who have experienced physical or sexual abuse, those with eating disorders, or those who identify as transgender, it may be important to not include the compassionate body scan, as this could be unnecessarily triggering.

Various implementation settings merit study as well, including schools, hospitals, juvenile detention systems, foster care programs, and afterschool programs. Most notably, as most adolescents worldwide attend public or private schools, implementing MSC-T or other self-compassion programs in schools is an important direction to increase the availability and accessibility of such programs. An adaptation of MSC-T for school implementation has been created but has not yet been tested empirically in school settings. The ability to reach all students, not only those whose parents or caregivers have the resources or opportunity to enroll them in self-compassion programs, may have far-reaching implications for the overall mental health of adolescents and, more distally, their long-term academic and behavioral functioning.

Finally, self-compassion and its associations with mental health may vary at different points in the life cycle and at different developmental stages of childhood and adolescence. Ideally, research studies would elucidate the windows of sensitivity in youth development at which the beneficial effects of self-compassion interventions would be most efficacious. For example, it may be that implementing a program to cultivate self-compassion in late childhood for females may help to evade the mental health plummet that is so common among adolescent females. It may be that other genders need a different approach at a different stage of development.

In conclusion, adolescents today struggle with an array of mental health challenges that have implications for lifelong maladaptive trajectories. Self-compassion has been shown to be associated with positive mental health factors, and interventions which cultivate self-compassion have been instrumental in providing coping mechanisms to shift the way in which adolescents respond to these stressors. Implementing self-compassion interventions has tremendous potential for adolescents, and future research is needed to replicate and confirm the current findings.

References

American Psychological Association. (2018). Stress in America: Generation Z. *Stress in America™ Survey*. https://www.apa.org/news/press/releases/stress/2018/stress-gen-z.pdf

Andersen, E., Fiacco, S., Gordon, J., Kozik, R., Baresich, K., Rubinow, D., & Girdler, S. (2022). Methods for characterizing ovarian and adrenal hormone variability and mood relationships in peripubertal females. *Psychoneuroendocrinology, 141*, 105747. https://doi.org/10.1016/j.psyneuen.2022.105747

Arch, J. J., Brown, K. W., Dean, D. J., Landy, L. N., Brown, K. D., & Laudenslager, M. L. (2014). Self-compassion training modulates alpha-amylase, heart rate variability, and subjective responses to social evaluative threat in women. *Psychoneuroendocrinology, 42*, 49–58. https://doi.org/10.1016/j.psyneuen.2013.12.018

Assari, S., Moazen-Zadeh, E., Caldwell, C. H., & Zimmerman, M. A. (2017). Racial discrimination during adolescence predicts mental health deterioration in adulthood: Gender differences among Blacks. *Frontiers in Public Health, 5*, 104. https://doi.org/10.3389/fpubh.2017.00104

Baumeister, R. F., Campbell, J. D., Krueger, J. I., & Vohs, K. D. (2003). Does high self-esteem cause better performance, interpersonal success, happiness, or healthier lifestyles? *Psychological Science in the Public Interest, 4*(1), 1–44. https://doi.org/10.1111/1529-1006.01431

Bennik, E. C., Nederhof, E., Ormel, J., & Oldehinkel, A. J. (2014). Anhedonia and depressed mood in adolescence: Course, stability, and reciprocal relation in the TRAILS study. *European Child & Adolescent Psychiatry, 23*(7), 579–586. https://doi.org/10.1007/s00787-013-0481-z

Bluth, K., & Blanton, P. W. (2015). The influence of self-compassion on emotional well-being among early and older adolescent males and females. *The Journal of Positive Psychology, 10*(3), 219–230. https://doi.org/10.1080/17439760.2014.936967

Bluth, K., & Eisenlohr-Moul, T. A. (2017). Response to a mindful self-compassion intervention in teens: A within-person association of mindfulness, self-compassion, and emotional well-being outcomes. *Journal of Adolescence, 57*, 108–118. https://doi.org/10.1016/j.adolescence.2017.04.001

Bluth, K., Gaylord, S. A., Campo, R. A., Mullarkey, M. C., & Hobbs, L. (2016a). Making friends with yourself: A mixed methods pilot study of a mindful self-compassion program for adolescents. *Mindfulness, 7*(2), 479–492. https://doi.org/10.1007/s12671-015-0476-6

Bluth, K., Roberson, P. N. E., Gaylord, S. A., Faurot, K. R., Grewen, K. M., Arzon, S., & Girdler, S. S. (2016b). Does self-compassion protect adolescents from stress? *Journal of Child and Family Studies, 25*(4), 1098–1109. https://doi.org/10.1007/s10826-015-0307-3

Bluth, K., Roberson, P. N. E., & Girdler, S. S. (2017). Adolescent sex differences in response to a mindfulness intervention: A call for research. *Journal of Child and Family Studies, 26*(7), 1900–1914. https://doi.org/10.1007/s10826-017-0696-6

Bluth, K., Lathren, C., Clepper-Faith, M., Larson, L., Ogunbamowo, D., & Pflum, S. (2021). Improving mental health among transgender adolescents: Implementing Mindful Self-Compassion for teens. *Journal of Adolescent Research, 00*(0), 1–32.10.1177/07435584211062126.

Boggiss, A. L., Consedine, N. S., Schache, K. R., Jefferies, C., Bluth, K., Hofman, P. L., & Serlachius, A. S. (2020). A brief self-compassion intervention for adolescents with Type 1 diabetes and disordered eating: a feasibility study. *Diabetic Medicine, 37*(11), 1854–1860. https://doi.org/10.1111/dme.14352

Breines, J. G., Thoma, M. V., Gianferante, D., Hanlin, L., Chen, X., & Rohleder, N. (2014). Self-compassion as a predictor of interleukin-6 response to acute psychosocial stress. *Brain, Behavior, and Immunity, 37*, 109–114. https://doi.org/10.1016/j.bbi.2013.11.006

Breines, J. G., McInnis, C. M., Kuras, Y. I., Thoma, M. V., Gianferante, D., Hanlin, L., Chen, X., & Rohleder, N. (2015). Self-compassionate young adults show lower salivary alpha-amylase responses to repeated psychosocial stress. *Self and Identity, 14*(4), 390–402. https://doi.org/10.1080/15298868.2015.1005659

Broderick, P. C. (2021). *Learning to breathe: A mindfulness curriculum for adolescents to cultivate emotion regulation, attention, and performance* (2nd ed.). New Harbinger Publications.

Castilho, P., Carvalho, S. A., Marques, S., & Pinto-Gouveia, J. (2017). Self-compassion and emotional intelligence in adolescence: A multigroup mediational study of the impact of shame memories on depressive symptoms. *Journal of Child and Family Studies, 26*(3), 759–768. https://doi.org/10.1007/s10826-016-0613-4

Chio, F. H. N., Mak, W. W. A., & Ben, C. L. (2021). Meta-analytic review on the differential effects of self-compassion components on well-being and psychological distress: The moderating role of dialecticism on self-compassion. *Clinical Psychology Review, 85*, 101986. https://doi.org/10.1016/j.cpr.2021.101986

Chishima, Y., Mizuno, M., Sugawara, D., & Miyagawa, Y. (2018). The influence of self-compassion on cognitive appraisals and coping with stressful events. *Mindfulness, 9*(6), 1–9. https://doi.org/10.1007/s12671-018-0933-0

Copeland, W. E., Worthman, C., Shanahan, L., Costello, E. J., & Angold, A. (2019). Early pubertal timing and testosterone associated with higher levels of adolescent depression in girls. *Journal of the American Academy of Child and Adolescent Psychiatry, 58*(12), 1197–1206. https://doi.org/10.1016/j.jaac.2019.02.007

Crane, R. S., Eames, C., Kuyken, W., Hastings, R. P., Williams, J. M. G., Bartley, T., Evans, A., Silverton, S., Soulsby, J. G., & Surawy, C. (2013). Development and validation of the mindfulness-based interventions – Teaching assessment criteria (MBI:TAC). *Assessment, 20*(6), 681–688. https://doi.org/10.1177/1073191113490790

Crede, J., Wirthwein, L., McElvany, N., & Steinmayr, R. (2015). Adolescents' academic achievement and life satisfaction: the role of parents' education. *Frontiers in Psychology, 6*, 52. https://doi.org/10.3389/fpsyg.2015.00052

Davis, A. N., Carlo, G., Schwartz, S. J., Unger, J. B., Zamboanga, B. L., Lorenzo-Blanco, E. I., Cano, M. A., Baezconde-Garbanati, L., Oshri, A., Streit, C., & Martinez, M. M. (2016). The longitudinal associations between discrimination, depressive symptoms, and prosocial behaviors in U.S. Latino/a recent immigrant adolescents. *Journal of Youth and Adolescence, 45*(3), 457–470. https://doi.org/10.1007/s10964-015-0394-x

Del Giudice, M. (2015). Gender differences in personality and social behavior. In J. D. Wright (Ed.), *International encyclopedia of the social & behavioral sciences* (2nd ed., pp. 750–756). Elsevier. https://doi.org/10.1016/B978-0-08-097086-8.25100-3

Donald, J. N., Ciarrochi, J., Parker, P. D., Sahdra, B. K., Marshall, S. L., & Guo, J. (2018). A worthy self is a caring self: Examining the developmental relations between self-esteem and self-compassion in adolescents. *Journal of Personality, 86*(4), 619–630. https://doi.org/10.1111/jopy.12340

Dosil, M., Jaureguizar, J., Bernaras, E., & Sbicigo, J. B. (2020). Teen dating violence, sexism, and resilience: A multivariate analysis. *International Journal of Environmental Research and Public Health, 17*(8). https://doi.org/10.3390/ijerph17082652

Egan, S. J., Rees, C. S., Delalande, J., Greene, D., Fitzallen, G., Brown, S., Webb, M., & Finlay-Jones, A. (2022). A review of self-compassion as an active ingredient in the prevention and treatment of anxiety and depression in young people. *Administration and Policy in Mental Health, 49*(3), 385–403. https://doi.org/10.1007/s10488-021-01170-2

Egan, S. J., Wade, T. D., & Shafran, R. (2011). Perfectionism as a transdiagnostic process: A clinical review. *Clinical Psychology Review, 31*(2), 203–212. https://doi.org/10.1016/j.cpr.2010.04.009

Einstein, D. A., Lovibond, P. F., & Gaston, J. E. (2000). Relationship between perfectionism and emotional symptoms in an adolescent sample. *Australian Journal of Psychology, 52*(2), 89–93. https://doi.org/10.1080/00049530008255373

Ellis, W. E., Dumas, T. M., & Forbes, L. M. (2020). Physically isolated but socially connected: Psychological adjustment and stress among adolescents during the initial COVID-19 crisis. *Canadian Journal of Behavioural Science / Revue Canadienne des Sciences du Comportement, 52*(3), 177–187. https://doi.org/10.1037/cbs0000215

Erikson, E. H. (1963). *Childhood and society*. W.W. Norton & Company.

Felitti, V. J., Anda, R. F., Nordenberg, D., Williamson, D. F., Spitz, A. M., Edwards, V., & Marks, J. S. (1998). Relationship of childhood abuse and household dysfunction to many of the leading causes of death in adults. The Adverse Childhood Experiences (ACE) Study. *American Journal of Preventive Medicine, 14*(4), 245–258. https://doi.org/10.1016/S0749-3797(98)00017-8

Ferrari, M., Yap, K., Scott, N., Einstein, D. A., & Ciarrochi, J. (2018). Self-compassion moderates the perfectionism and depression link in both adolescence and adulthood. *PLoS One, 13*(2), e0192022. https://doi.org/10.1371/journal.pone.0192022

Finlay-Jones, A., Boyes, M., Perry, Y., Sirois, F., Lee, R., & Rees, C. (2020). Online self-compassion training to improve the wellbeing of youth with chronic medical conditions: Protocol for a randomised control trial. *BMC Public Health, 20*(1), 106. https://doi.org/10.1186/s12889-020-8226-7

Flett, G. L., Besser, A., Davis, R. A., & Hewitt, P. L. (2003). Dimensions of perfectionism, unconditional self-acceptance, and depression. *Journal of Rational-Emotive & Cognitive-Behavior Therapy, 21*, 119–138. https://doi.org/10.1023/A:1025051431957

Flett, G. L., Blankstein, K. R., & Hewitt, P. L. (2009). Perfectionism, performance, and state positive affect and negative affect after a classroom test. *Canadian Journal of School Psychology, 24*(1), 4–18. https://doi.org/10.1177/0829573509332457

Forbes, E. E., & Dahl, R. E. (2010). Pubertal development and behavior: Hormonal activation of social and motivational tendencies. *Brain and Cognition, 72*(1), 66–72. https://doi.org/10.1016/j.bandc.2009.10.007

Galla, B. M. (2016). Within-person changes in mindfulness and self-compassion predict enhanced emotional well-being in healthy, but stressed adolescents. *Journal of Adolescence, 49*, 204–217. https://doi.org/10.1016/j.adolescence.2016.03.016

Galla, B. M. (2017). "Safe in My Own Mind:" Supporting healthy adolescent development through meditation retreats. *Journal of Applied Developmental Psychology, 53*, 96–107. https://doi.org/10.1016/j.appdev.2017.09.006

Giedd, J. N. (2008). The teen brain: Insights from neuroimaging. *The Journal of Adolescent Health, 42*(4), 335–343. https://doi.org/10.1016/j.jadohealth.2008.01.007

Gnan, G. H., Rahman, Q., Ussher, G., Baker, D., West, E., & Rimes, K. A. (2019). General and LGBTQ-specific factors associated with mental health and suicide risk among LGBTQ students. *Journal of Youth Studies, 22*(10), 1393–1408. https://doi.org/10.1080/13676261.2019.1581361

Goldbach, J. T., & Gibbs, J. J. (2017). A developmentally informed adaptation of minority stress for sexual minority adolescents. *Journal of Adolescence, 55*, 36–50. https://doi.org/10.1016/j.adolescence.2016.12.007

Griffith, G. M., Crane, R. S., Baer, R., Fernandez, E., Giommi, F., Herbette, G., & Koerbel, L. (2021). Implementing the Mindfulness-Based Interventions; Teaching Assessment Criteria (MBI:TAC) in mindfulness-based teacher training. *Global Advances in Health and Medicine, 10*, 1–6. https://doi.org/10.1177/2164956121998340

Hatchel, T., Ingram, K. M., Mintz, S., Hartley, C., Valido, A., Espelage, D. L., & Wyman, P. (2019). Predictors of suicidal ideation and attempts among LGBTQ adolescents: The roles of help-seeking beliefs, peer victimization, depressive symptoms, and drug use. *Journal of Child and Family Studies, 28*, 2443–2455. https://doi.org/10.1007/s10826-019-01339-2

Herrenkohl, T. I., Hong, S., Klika, J. B., Herrenkohl, R. C., & Russo, M. J. (2013). Developmental impacts of child abuse and neglect related to adult mental health, substance use, and physical health. *Journal of Family Violence, 28*(2). https://doi.org/10.1007/s10896-012-9474-9

Hewitt, P. L., & Flett, G. L. (1991). Dimensions of perfectionism in unipolar depression. *Journal of Abnormal Psychology, 100*(1), 98–101. https://doi.org/10.1037//0021-843x.100.1.98

Hughes, D., Del Toro, J., Harding, J. F., Way, N., & Rarick, J. R. D. (2016). Trajectories of discrimination across adolescence: Associations with academic, psychological, and behavioral outcomes. *Child Development, 87*(5), 1337–1351. https://doi.org/10.1111/cdev.12591

Huynh, V. W., & Fuligni, A. J. (2010). Discrimination hurts: The academic, psychological, and physical well-being of adolescents. *Journal of Research on Adolescence, 20*(4), 916–941. https://doi.org/10.1111/j.1532-7795.2010.00670.x

Játiva, R., & Cerezo, M. A. (2014). The mediating role of self-compassion in the relationship between victimization and psychological maladjustment in a sample of adolescents. *Child Abuse & Neglect, 38*(7), 1180–1190. https://doi.org/10.1016/j.chiabu.2014.04.005

Johnson, S. R. L., Blum, R. W., & Cheng, T. L. (2014). Future orientation: a construct with implications for adolescent health and wellbeing. *International Journal of Adolescent Medicine and Health, 26*(4), 459–468. https://doi.org/10.1515/ijamh-2013-0333

Keating, D. (2004). Cognitive and brain development. In R. Lerner & L. Steinberg (Eds.), *Handbook of adolescent psychology* (2nd ed., pp. 45–84). Wiley. https://doi.org/10.1002/9780471726746.ch3

King, M., Semlyen, J., Tai, S. S., Killaspy, H., Osborn, D., Popelyuk, D., & Nazareth, I. (2008). A systematic review of mental disorder, suicide, and deliberate self-harm in lesbian, gay and bisexual people. *BMC Psychiatry, 8*, 70. https://doi.org/10.1186/1471-244X-8-70

Kirby, J. N., & Kirby, P. G. (2017). An evolutionary model to conceptualise masculinity and compassion in male teenagers: A unifying framework. *Clinical Psychologist, 21*(2), 74–89. https://doi.org/10.1111/cp.12129

Krieger, T., Berger, T., & Holtforth, M. G. (2016). The relationship of self-compassion and depression: Cross-lagged panel analyses in depressed patients after outpatient therapy. *Journal of Affective Disorders, 202*, 39–45. https://doi.org/10.1016/j.jad.2016.05.032

Kuehner, C. (2017). Why is depression more common among women than among men? *The Lancet Psychiatry, 4*(2), 146–158. https://doi.org/10.1016/S2215-0366(16)30263-2

Kuo, A. A., Etzel, R. A., Chilton, L. A., Watson, C., & Gorski, P. A. (2012). Primary care pediatrics and public health: Meeting the needs of today's children. *American Journal of Public Health, 102*(12), e17–e23. https://doi.org/10.2105/AJPH.2012.301013

Lambert, S. F., Ialongo, N. S., Boyd, R. C., & Cooley, M. R. (2005). Risk factors for community violence exposure in adolescence. *American Journal of Community Psychology, 36*(1–2), 29–48. https://doi.org/10.1007/s10464-005-6231-8

Lambert, S. F., Herman, K. C., Bynum, M. S., & Ialongo, N. S. (2009). Perceptions of racism and depressive symptoms in African American adolescents: The role of perceived academic and social control. *Journal of Youth and Adolescence, 38*(4), 519–531. https://doi.org/10.1007/s10964-009-9393-0

Lathren, C., Bluth, K., & Park, J. (2019). Adolescent self-compassion moderates the relationship between perceived stress and internalizing symptoms. *Personality and Individual Differences, 143*, 36–41. https://doi.org/10.1016/j.paid.2019.02.008

Lindert, J., von Ehrenstein, O. S., Grashow, R., Gal, G., Braehler, E., & Weisskopf, M. G. (2014). Sexual and physical abuse in childhood is associated with depression and anxiety over the life course: Systematic review and meta-analysis. *International Journal of Public Health, 59*(2), 359–372. https://doi.org/10.1007/s00038-013-0519-5

Lippard, E. T. C., & Nemeroff, C. B. (2020). The devastating clinical consequences of child abuse and neglect: increased disease vulnerability and poor treatment response in mood disorders. *The American Journal of Psychiatry, 177*(1), 20–36. https://doi.org/10.1176/appi.ajp.2019.19010020

Lopez, W. D., LeBrón, A. M. W., Graham, L. F., & Grogan-Kaylor, A. (2016). Discrimination and depressive symptoms among Latina/o adolescents of immigrant parents. *International Quarterly of Community*

Health Education, 36(2), 131–140. https://doi.org/10.1177/0272684X16628723

MacBeth, A., & Gumley, A. (2012). Exploring compassion: A meta-analysis of the association between self-compassion and psychopathology. *Clinical Psychology Review, 32*(6), 545–552. https://doi.org/10.1016/j.cpr.2012.06.003

MacKinnon, D. P. (2008). *Introduction to statistical mediation analysis* (1st ed.). Routledge. https://doi.org/10.4324/9780203809556

Marcia, J. E. (1980). Identity in adolescence. In J. Adelson (Ed.), *Handbook of adolescent psychology* (pp. 159–187). Wiley.

Marshall, S. L., Parker, P. D., Ciarrochi, J., Sahdra, B., Jackson, C. J., & Heaven, P. C. L. (2015). Self-compassion protects against the negative effects of low self-esteem: A longitudinal study in a large adolescent sample. *Personality and Individual Differences, 74*, 116–121. https://doi.org/10.1016/j.paid.2014.09.013

Marsh, I. C., Chan, S. W. Y., & MacBeth, A. (2018). Self-compassion and psychological distress in adolescents – A meta-analysis. *Mindfulness, 9*(4), 1011–1027. https://doi.org/10.1007/s12671-017-0850-7

Mayne, S. L., Hannan, C., Davis, M., Young, J. F., Kelly, M. K., Powell, M., Dalembert, G., McPeak, K. E., Jenssen, B. P., & Fiks, A. G. (2021). COVID-19 and adolescent depression and suicide risk screening outcomes. *Pediatrics, 148*(3), e2021051507. https://doi.org/10.1542/peds.2021-051507

Meyer, I. H. (2015). Resilience in the study of minority stress and health of sexual and gender minorities. *Psychology of Sexual Orientation and Gender Diversity, 2*(3), 209–213. https://doi.org/10.1037/sgd0000132

Monto, M. A., McRee, N., & Deryck, F. S. (2018). Nonsuicidal self-injury among a representative sample of US adolescents, 2015. *American Journal of Public Health, 108*(8), 1042–1048. https://doi.org/10.2105/AJPH.2018.304470

Muris, P., Otgaar, H., & Pfattheicher, S. (2019). Stripping the forest from the rotten trees: Compassionate self-responding is a way of coping, but reduced uncompassionate self-responding mainly reflects psychopathology. *Mindfulness, 10*(1), 196–199. https://doi.org/10.1007/s12671-018-1030-0

Neff, K. D. (2003). Self-compassion: An alternative conceptualization of a healthy attitude toward oneself. *Self and Identity, 2*(2), 85–101. https://doi.org/10.1080/15298860309032

Neff, K. D. (2009). The role of self-compassion in development: A healthier way to relate to oneself. *Human Development, 52*(4), 211–214. https://doi.org/10.1159/000215071

Neff, K. D., & Germer, C. (2013). A pilot study and randomized controlled trial of the Mindful Self-Compassion program. *Journal of Clinical Psychology, 69*(1), 28–44. https://doi.org/10.1002/jclp.21923

Neff, K. D., & McGehee, P. (2010). Self-compassion and psychological resilience among adolescents and young adults. *Self and Identity, 9*(3), 225–240. https://doi.org/10.1080/15298860902979307

Neff, K. D., & Vonk, R. (2009). Self-compassion versus global self-esteem: two different ways of relating to oneself. *Journal of Personality, 77*(1), 23–50. https://doi.org/10.1111/j.1467-6494.2008.00537.x

Neff, K. D., Long, P., Knox, M. C., Davidson, O., Kuchar, A., Costigan, A., Williamson, Z., Rohleder, N., Toth-Kiraly, I., & Breines, J. G. (2018). The forest and the trees: Examining the association of self-compassion and its positive and negative components with psychological functioning. *Self and Identity, 17*(6), 627–645. https://doi.org/10.1080/15298868.2018.1436587

Neff, K. D., Bluth, K., Tóth-Király, I., Davidson, O., Knox, M. C., Williamson, Z., & Costigan, A. (2021). Development and validation of the Self-Compassion Scale for Youth. *Journal of Personality Assessment, 103*(1), 92–105. https://doi.org/10.1080/00223891.2020.1729774

Nolen-Hoeksema, S. (2012). Emotion regulation and psychopathology: The role of gender. *Annual Review of Clinical Psychology, 8*, 161–187. https://doi.org/10.1146/annurev-clinpsy-032511-143109

Nolen-Hoeksema, S., & Girgus, J. S. (1994). The emergence of gender differences in depression during adolescence. *Psychological Bulletin, 115*(3), 424–443. https://doi.org/10.1037/0033-2909.115.3.424

North, C. R., Wild, T. C., Zwaigenbaum, L., & Colman, I. (2013). Early neurodevelopment and self-reported adolescent symptoms of depression and anxiety in a national Canadian cohort study. *PLoS One, 8*(2), e56804. https://doi.org/10.1371/journal.pone.0056804

Ntoumanis, N., Stenling, A., Quested, E., Nikitaras, N., Olson, J., & Thøgersen-Ntoumani, C. (2020). Self-compassion and need frustration moderate the effects of upward appearance comparisons on body image discrepancies. *The Journal of Psychology, 154*(4), 292–308. https://doi.org/10.1080/00223980.2020.1716669

Paradies, Y., Ben, J., Denson, N., Elias, A., Priest, N., Pieterse, A., Gupta, A., Kelaher, M., & Gee, G. (2015). Racism as a determinant of health: A systematic review and meta-analysis. *PLoS One, 10*(9), e0138511. https://doi.org/10.1371/journal.pone.0138511

Patchin, J. W., & Hinduja, S. (2013). Cyberbullying among adolescents: implications for empirical research. *The Journal of Adolescent Health, 53*(4), 431–432. https://doi.org/10.1016/j.jadohealth.2013.07.030

Patel, V., Flisher, A. J., Hetrick, S., & McGorry, P. (2007). Mental health of young people: a global public-health challenge. *The Lancet, 369*(9569), 1302–1313. https://doi.org/10.1016/S0140-6736(07)60368-7

Prentice, K., Rees, C., & Finlay-Jones, A. (2021). Self-compassion, wellbeing, and distress in adolescents and young adults with chronic medical conditions: The mediating role of emotion regulation difficulties. *Mindfulness, 2*(9), 2241–2252. https://doi.org/10.1007/s12671-021-01685-7

Pullmer, R., Chung, J., Samson, L., Balanji, S., & Zaitsoff, S. (2019a). A systematic review of the relation between self-compassion and depressive symptoms in adolescents. *Journal of Adolescence, 74*, 210–220. https://doi.org/10.1016/j.adolescence.2019.06.006

Pullmer, R., Coelho, J. S., & Zaitsoff, S. L. (2019b). Kindness begins with yourself: The role of self-compassion in adolescent body satisfaction and eating pathology. *The International Journal of Eating Disorders, 52*(7), 809–816. https://doi.org/10.1002/eat.23081

Racine, N., McArthur, B. A., Cooke, J. E., Eirich, R., Zhu, J., & Madigan, S. (2021). Global prevalence of depressive and anxiety symptoms in children and adolescents during COVID-19: A meta-analysis. *The Journal of the American Medical Association Pediatrics, 175*(11), 1142–1150. https://doi.org/10.1001/jamapediatrics.2021.2482

Rodgers, R. F., Franko, D. L., Donovan, E., Cousineau, T., Yates, K., McGowan, K., Cook, E., & Lowy, A. S. (2017). Body image in emerging adults: The protective role of self-compassion. *Body Image, 22*, 148–155. https://doi.org/10.1016/j.bodyim.2017.07.003

Salk, R. H., Hyde, J. S., & Abramson, L. Y. (2017). Gender differences in depression in representative national samples: Meta-analyses of diagnoses and symptoms. *Psychological Bulletin, 143*(8), 783–822. https://doi.org/10.1037/bul0000102

Shanafelt, A., Hearst, M. O., Wang, Q., & Nanney, M. S. (2016). Food insecurity and rural adolescent personal health, home, and academic environments. *The Journal of School Health, 86*(6), 472–480. https://doi.org/10.1111/josh.12397

Shonkoff, J. P., & Garner, A. S. (2012). The lifelong effects of early childhood adversity and toxic stress. *Pediatrics, 129*(1), e232–e246. https://doi.org/10.1542/peds.2011-2663

Sirin, S. R., Sin, E., Clingain, C., & Rogers-Sirin, L. (2019). Acculturative stress and mental health: Implications for immigrant-origin youth. *Pediatric Clinics of North America, 66*(3), 641–653. https://doi.org/10.1016/j.pcl.2019.02.010

Sontag, L. M., & Graber, J. A. (2010). Coping with perceived peer stress: Gender-specific and common pathways to symptoms of psychopathology. *Developmental Psychology, 46*(6), 1605–1620. https://doi.org/10.1037/a0020617

Steinberg, L., & Morris, A. S. (2001). Adolescent development. *Journal of Cognitive Education and Psychology, 2*(1), 55–87. https://doi.org/10.1891/194589501787383444

Stolow, D., Zuroff, D. C., Young, J. F., Karlin, R. A., & Abela, J. R. Z. (2016). A prospective examination of self-compassion as a predictor of depressive symptoms in children and adolescents. *Journal of Social and Clinical Psychology, 35*(1), 1–20. https://doi.org/10.1521/jscp.2016.35.1.1

Sun, R., Ren, Y., Li, X., Jiang, Y., Liu, S., & You, J. (2020). Self-compassion and family cohesion moderate the association between suicide ideation and suicide attempts in Chinese adolescents. *Journal of Adolescence, 79*, 103–111. https://doi.org/10.1016/j.adolescence.2019.12.010

Svendsen, J. L., Osnes, B., Binder, P.-E., Dundas, I., Visted, E., Nordby, H., Schanche, E., & Sorensen, L. (2016). Trait self-compassion reflects emotional flexibility through an association with high vagally mediated heart rate variability. *Mindfulness, 7*(5), 1103–1113. https://doi.org/10.1007/s12671-016-0549-1

Sweeting, J. A., Garfin, D. R., Holman, E. A., & Silver, R. C. (2020). Associations between exposure to childhood bullying and abuse and adulthood outcomes in a representative national U.S. sample. *Child Abuse & Neglect, 101*, 104048. https://doi.org/10.1016/j.chiabu.2019.104048

Tanaka, M., Wekerle, C., Schmuck, M. L., Paglia-Boak, A., & Research Team, M. A. P. (2011). The linkages among childhood maltreatment, adolescent mental health, and self-compassion in child welfare adolescents. *Child Abuse & Neglect, 35*(10), 887–898. https://doi.org/10.1016/j.chiabu.2011.07.003

Torres, S. A., Santiago, C. D., Walts, K. K., & Richards, M. H. (2018). Immigration policy, practices, and procedures: The impact on the mental health of Mexican and Central American youth and families. *The American Psychologist, 73*(7), 843–854. https://doi.org/10.1037/amp0000184

Turner, S., Taillieu, T., Cheung, K., & Afifi, T. O. (2017). The relationship between childhood sexual abuse and mental health outcomes among males: Results from a nationally representative United States sample. *Child Abuse & Neglect, 66*, 64–72. https://doi.org/10.1016/j.chiabu.2017.01.018

Twenge, J. M., Joiner, T. E., Rogers, M. L., & Martin, G. N. (2018). Increases in depressive symptoms, suicide-related outcomes, and suicide rates among U.S. adolescents after 2010 and links to increased new media screen time. *Clinical Psychological Science, 6*(1), 3–17. https://doi.org/10.1177/2167702617723376

Umaña-Taylor, A. J., & Updegraff, K. A. (2007). Latino adolescents' mental health: exploring the interrelations among discrimination, ethnic identity, cultural orientation, self-esteem, and depressive symptoms. *Journal of Adolescence, 30*(4), 549–567. https://doi.org/10.1016/j.adolescence.2006.08.002

Vettese, L. C., Dyer, C. E., Li, W. L., & Wekerle, C. (2011). Does self-compassion mitigate the association between childhood maltreatment and later emotion regulation difficulties? A preliminary investigation. *International Journal of Mental Health and Addiction, 9*(5), 480–491. https://doi.org/10.1007/s11469-011-9340-7

Vigna, A. J., Poehlmann-Tynan, J., & Koenig, B. W. (2017). Does self-compassion facilitate resilience to stigma? A school-based study of sexual and gender minority youth. *Mindfulness, 9*(3), 1–11. https://doi.org/10.1007/s12671-017-0831-x

Wang, J. Z., Mott, S., Magwood, O., Mathew, C., Mclellan, A., Kpade, V., Gaba, P., Kozloff, N., Pottie, K., & Andermann, A. (2019). The impact of interventions for youth experiencing homelessness on housing, mental health, substance use, and family cohesion: A systematic review. *BMC Public Health, 19*(1), 1528. https://doi.org/10.1186/s12889-019-7856-0

Wu, N., Hou, Y., Chen, P., & You, J. (2019). Peer acceptance and n-suicidal self-injury among Chinese adolescents: A longitudinal moderated mediation model. *Journal of Youth and Adolescence, 48*(9), 1806–1817. https://doi.org/10.1007/s10964-019-01093-0

Xavier, A., Pinto-Gouveia, J., & Cunha, M. (2016). The protective role of self-compassion on risk factors for non-suicidal self-injury in adolescence. *School Mental Health, 8*(4), 476–485. https://doi.org/10.1007/s12310-016-9197-9

Yarnell, L. M., Stafford, R. E., Neff, K. D., Reilly, E. D., Knox, M. C., & Mullarkey, M. (2015). Meta-analysis of gender differences in self-compassion. *Self and Identity, 14*(5), 499–520. https://doi.org/10.1080/15298868.2015.1029966

Zapata Roblyer, M. I., Grzywacz, J. G., Cervantes, R. C., & Merten, M. J. (2016). Stress and alcohol, cigarette, and marijuana use among Latino adolescents in families with undocumented immigrants. *Journal of Child and Family Studies, 25*(2), 475–487. https://doi.org/10.1007/s10826-015-0249-9

Zaza, S., Kann, L., & Barrios, L. C. (2016). Lesbian, gay, and bisexual adolescents: Population estimate and prevalence of health behaviors. *The Journal of the American Medical Association, 316*(22), 2355–2356. https://doi.org/10.1001/jama.2016.11683

Zeller, M., Yuval, K., Nitzan-Assayag, Y., & Bernstein, A. (2015). Self-compassion in recovery following potentially traumatic stress: longitudinal study of at-risk youth. *Journal of Abnormal Child Psychology, 43*(4), 645–653. https://doi.org/10.1007/s10802-014-9937-y

Zessin, U., Dickhäuser, O., & Garbade, S. (2015). The relationship between self-compassion and well-being: A meta-analysis. *Applied Psychology, Health and Well-being, 7*(3), 340–364. https://doi.org/10.1111/aphw.12051

7 Self-Compassion and Positive Aging

Lydia Brown

Introduction

As people age, they experience a wide range of changes that can both positively and negatively influence mental and physical health. Social role transitions are plentiful in later life. Adjustment to an empty nest when children leave home, retirement, and grandparenthood are examples of transitions that can be simultaneously rewarding and yet stressful. While later life is a time of great happiness and fulfillment for many people (Carstensen et al., 2003), there is also considerable heterogeneity in people's ability to adjust to the transitions of later life (Steverink et al., 2001). Some people are better able to maximize the joys of aging while adapting well to inevitable changes, a process known as *positive aging* (Hill, 2011). Others, however, find the transitions of aging more challenging, and their well-being is compromised as a result (Carpentieri et al., 2017). Positive aging is best viewed as a process rather than an outcome (Kunuroglu & Yuzbasi, 2021; Freund & Baltes, 1998; Gergen & Gergen, 2001). Positive aging entails optimizing emotional, physical, and social well-being, and engaging meaningfully in life, despite challenges that may arise such as the emergence of health issues (Gergen & Gergen, 2001).

Self-compassion, with its emphasis on acknowledging and responding kindly to one's own experience of suffering (Neff, 2003), might help explain why some people are better able to optimize well-being and thus experience positive aging. Self-compassion might enable older adults to embrace later life transitions with balanced awareness, kindness, and an underlying sense of common humanity, rather than feeling isolated, overidentified with the negatives of aging, and self-critical of their experience of growing older (Brown et al., 2018a). As visualized in Fig. 7.1, this chapter outlines the literature linking self-compassion to four pillars of positive aging: (1) mental well-being, (2) physical well-being, (3) engagement in activity, and (4) social connectedness (Gergen & Gergen, 2001). Within this discussion, I explore processes that might explain *how* self-compassion facilitates positive aging, via cultivation of a healthy attitude toward aging, acceptance of change, behavioral flexibility, and flexible goal pursuit. First, I explore why self-compassion may be a natural resilience factor, as well as a helpful alternative to self-esteem to cultivate a healthy self-image in later life.

L. Brown (✉)
Melbourne School of Psychological Sciences, University of Melbourne, Parkville, VIC, Australia

Academic and Research Collaborative in Health, La Trobe University, Bundoora, VIC, Australia

Healthscope Hospitals, Melbourne, VIC, Australia
e-mail: lydia.brown@unimelb.edu.au

A. Finlay-Jones et al. (eds.), *Handbook of Self-Compassion*, Mindfulness in Behavioral Health,
https://doi.org/10.1007/978-3-031-22348-8_7

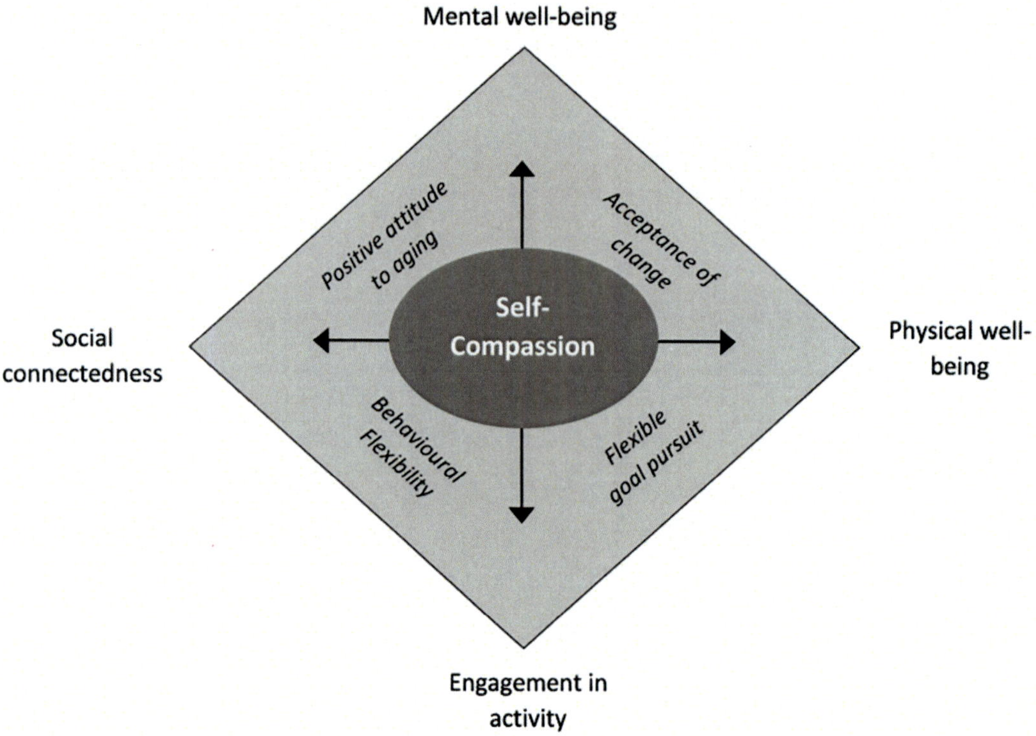

Fig. 7.1 A modified version of Gergan and Gergan's life span diamond model of positive aging. This modified model presents self-compassion as a central ingredient of the four pillars of positive aging: mental well-being, physical well-being, social connectedness, and engagement in activity. Processes that help explain *how* self-compassion facilitates positive aging are in italics

Cultivating a Healthy Self-Image in Later Life

The experience of aging undeniably affects an individual's perception of the self (Hazel, 1991). Key social institutions, including the media and workplaces, proliferate negative views of aging as a time of loss and decline (Australian Human Rights Commission, 2013, 2016). For instance, the Australian Human Rights Commission surveyed over 2000 older adults about their views on media representations of aging, and responses were overwhelmingly negative (Fig. 7.2). Participants believed that the media portrayed older adults as being frail, vulnerable, and burdensome, and furthermore these portrayals negatively influenced participants' attitude toward their own personal views of aging (Australian Human Rights Commission, 2013).

Echoing a cultural context that is predominantly negative about aging, an individual's self-esteem typically follows a sharp downward trajectory in older adulthood. Self-esteem is defined as one's overall sense of self-worth or attitude toward the self (Robins & Trzesniewski, 2005). As seen in Fig. 7.3, representing meta-analytic self-esteem data, self-esteem typically starts off high in childhood then plummets – especially for girls – during adolescence, before gradually rising again throughout early and middle adulthood. Self-esteem then takes a second sharp dive at approximately age 65, and it continues to fall for both men and women during later life, before reaching all-time lows in the eighth decade (Robins & Trzesniewski, 2005). There are substantial individual differences in late life self-esteem trajectories, and those with cognitive decline experience the most pronounced reductions (Wagner et al., 2013). Nevertheless, this

Fig. 7.2 Word cloud of older adult views of media representations of aging. The larger the word font, the more frequently the word was reported by participants. (Reprinted with permission from Australian Human Rights Commission, 2013)

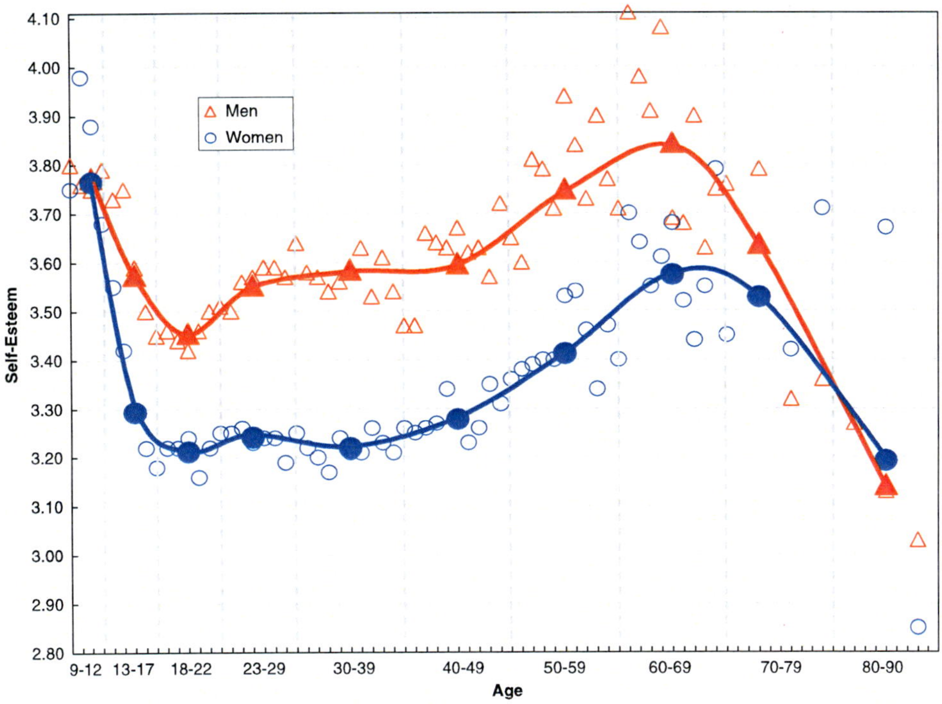

Fig. 7.3 Self-esteem trajectories across the life span. (Reprinted with permission from Robins & Trzesniewski, 2005)

data demonstrates that later life can be a vulnerability factor for low self-esteem. Because attempts to bolster self-esteem can be ineffective (Baumeister et al., 2003), an alternative approach is needed to facilitate healthy self-image in later life.

Unlike self-esteem that declines in older adulthood, evidence demonstrates that self-compassion increases with age (Homan, 2016;

Hwang et al., 2016; Neff & Vonk, 2009). In a study of 2187 adults aged 18–83, Neff and Vonk (2009) found that self-compassion was positively correlated with age, whereas partial correlations revealed that self-esteem significantly decreased as people aged (Neff & Vonk, 2009). Similarly, a Korean study of nearly 2000 younger and midlife adults found evidence of a small but significant positive association between age and self-compassion (r = 0.18) (Hwang et al., 2016). In agreement, in a multigenerational sample of adults aged 18 to 95 (n = 296), Homan (2016) found evidence of a strong positive association between self-compassion and age (r = 0.32). More recently, a large study of community-dwelling adults in the United States (n = 1090) found that the relationship between self-compassion and age followed an inverse U-shaped relationship with age, peaking at approximately 77 years (Lee et al., 2021). While the reasons for this are unclear, it may be that the physical and cognitive decline associated with advanced old age interfere with the capacity for mindfulness or self-kindness or that declining social connections and mobility increase one's sense of isolation.

It is unclear why self-compassion might increase with age, although some researchers have proposed plausible explanations. For example, Homan (2016) suggests that the accumulation of life experiences may lead to people becoming more self-compassionate as they age; as people navigate the challenges of life, they take a more gentle and flexible approach to themselves and others. Self-compassion is strongly related to reflective wisdom (Neff et al., 2007b), which is the capacity to engage in self-examination as a means to attain self-insight, as well as the capacity to gain insight by looking at situations from many different perspectives (Ardelt, 2011). For this reason, Neff and Vonk (2009) suggest that the growth of self-compassion with age might be associated with the development of wisdom (Neff & Vonk, 2009). Reflective wisdom appears to increase in later life, although the association is weak and varies as a function of education (Ardelt et al., 2018). Clearly, more empirical work is needed to understand why people tend to treat themselves more compassionately as they age. However, given that current evidence indicates that self-compassion increases with age, older adults may benefit from using this strength to maximize health and well-being in later life.

Self-Compassion and Mental Well-being in Later Life

An extensive body of research demonstrates that self-compassion is correlated with high mental well-being in younger adults (see Zessin et al., 2015 for a review), and a growing body of research is now finding similar – and in some cases stronger – associations between self-compassion and well-being in older adults. Mental well-being involves two facets: (i) the absence of distressing psychological symptoms such as symptoms of depression, anxiety, and stress and (ii) the presence of positive qualities including happiness, curiosity, and life satisfaction (Huppert & Whittington, 2003). Positive qualities can be further delineated into hedonic well-being, which involves the presence of pleasurable emotions and life satisfaction, and eudaimonic well-being, which is a sense of purpose and meaning in life (Ryan & Deci, 2001). Emerging research shows that self-compassion is correlated with all three of these aspects of mental well-being in later life (Brown et al., 2018a).

Low Levels of Psychological Symptoms Mounting evidence has found that self-compassion is associated with fewer symptoms of depression and anxiety in older adults. In terms of depression, a study by Smith (2015) was one of the first to investigate the association between depression and self-compassion in a sample of American older adults. Smith (2015) recruited 102 independent older adults living in a continuing care retirement community with a mean age of 82 years and found evidence of a strong association between levels of self-compassion and fewer depressive symptoms, whereby self-compassion explained over 50% of the variance in depressive symptoms (Smith,

2015). In agreement, a subsequent US study by Homan (2016) found evidence of a strong correlation between depressive symptoms and self-compassion in a community-based sample of older adults (mean age = 70 years), recruited from a local public library and senior center (Homan, 2016). A Canadian study of patients with chronic obstructive pulmonary disease and healthy controls (mean age = 67 years) echoed this finding, once again demonstrating strong links between self-compassion and depression in older adults (Harrison et al., 2017). This study is important, because it indicates that self-compassion may be protective of depression among older adults with health issues, a topic discussed later in this chapter.

New research is starting to consider the mechanisms that might explain how self-compassion is protective against depressive symptoms in older adults (Hodgetts et al., 2020; Thoma et al., 2021). In a community-based cross-sectional study of 241 Australian adults aged 65 years and older, Hodgetts et al. (2020) investigated the potential role of thinking styles in explaining the relationship between self-compassion and depression. Specifically, the research looked at how ruminative responses, defined as repetitive and passive focus on symptoms of distress, together with a focus on the possible causes and consequences of these symptoms (Nolen-Hoeksema et al., 2008), might account for (i.e., mediate) the relationship between self-compassion and depressive symptoms. The authors found evidence to support a mediation model, whereby older adults higher in self-compassion were less prone to ruminate on their problems, which in turn predicted fewer depressive symptoms. Interestingly, this pattern of associations was stronger for women than men, indicating that self-compassion may be especially protective of depressive symptoms for older women relative to older men. In their discussion, the authors speculate that self-compassion may be more aligned to femininity and therefore may be a more socially appropriate and thus effective strategy to reduce rumination for women relative to men; however, this idea was not tested empirically (Hodgetts et al., 2020; Yarnell et al., 2019). Since levels of self-compassion tend to be slightly lower in women relative to men (Yarnell et al., 2019), an alternate hypothesis is that aging women are in greater need of self-compassion, and thus a unit increase in self-compassion could be associated with a stronger reduction in rumination and depression for women who are starting from a lower self-compassion baseline.

In addition to the cross-sectional associations above, a pilot randomized controlled trial (RCT) has investigated the potential of a ten-session self-compassion-based training program to improve mental well-being and adaptation to stress in a sample of community-dwelling older adults (Perez-Blasco et al., 2016). The authors found that participation in the program was associated with significant reductions in anxiety relative to waitlist control, but there was no significant change in depression. Aside from the small sample size ($n = 45$) which was not adequately powered to detect treatment effects, another concern with this study is that the authors did not use the empirically validated 8-week *Mindful Self-Compassion* (MSC) intervention (Neff & Germer, 2013) and instead developed a study-specific intervention. MSC has been rigorously developed and now trialed in a range of settings, where it has been found to reduce depressive symptomatology (Finlay-Jones et al., 2018; Friis et al., 2016; Neff & Germer, 2013). For ease of between-study comparisons and quality assurance purposes, there is a pressing need to research the efficacy of the standard MSC program in older adult groups, with adequately powered designs that are capable of detecting between-group differences, as well as potential moderators (e.g., gender).

Hedonic Well-being A small group of studies has explored associations between self-compassion and hedonic well-being, including measures of positive affect and satisfaction with life (Allen et al., 2012; Phillips & Ferguson, 2012; Smith, 2015; Kunuroglu & Yuzbasi, 2021). Community-based cross-sectional studies of older adults residing in the United States (Allen et al., 2012) and Turkey (Kunuroglu & Yuzbasi, 2021) have found that self-compassion is positively correlated with life satisfaction. In addition to a bivariate association with life satisfaction, Kunuroglu and Yuzbasi (2021) used path analysis to test a model where self-compassion mediated the relationship between life satisfaction and successful aging. This model was supported, indicating that levels of life satisfaction might help shape self-compassion, which in turn predicts successful aging in older adult groups. However, this study was cross-sectional in nature, and it is plausible that self-compassion may contribute to levels of life satisfaction, rather than the other way around. Indeed, a recent experimental study from Iran adds weight to this idea (Asadi Bijaeyeh et al., 2021). Asadi Bijaeyeh et al. (2021) examined the efficacy of an 8-week self-compassion training program in promoting life satisfaction and resilience among 15 female nursing home residents in Iran, relative to a control group of 15 female residents who did not receive the intervention. The authors found that self-compassion led to improved life satisfaction and resilience post-intervention, with gains maintained at follow-up (Asadi Bijaeyeh et al., 2021). This indicates that self-compassion-based training might help seniors cultivate greater satisfaction with life. More experimental research is needed to see if this finding replicates across cultures and settings.

To date, two studies have found evidence that self-compassion is associated with positive emotionality in later life. Smith (2015) recruited 102 adults residing in a retirement community in the United States (mean age = 82 years) and found that participants with higher self-compassion concurrently reported higher levels of happiness (Smith, 2015). An Australian study by Phillips and Ferguson (2012) examined links between self-compassion and positive affect in a younger sample of community-dwelling older adults with a mean age of 73 (Phillips & Ferguson, 2012). This study also found evidence of a positive association between self-compassion and happiness.

An interesting finding is that the association between self-compassion and well-being might strengthen across the life span. Data to support this idea comes from a large community study of 1813 Korean adults aged 22–61 years (Hwang et al., 2016). In this study, the authors found that the relationship between self-compassion and well-being strengthened with age, with self-compassion being more strongly associated with well-being for midlife compared to younger adults (Hwang et al., 2016). In this study, well-being was measured with the Concise Measure of Subjective Well-Being Scale (Suh & Koo, 2011) which predominantly measures hedonic well-being. The authors draw on developmental theory to help explain their results (Havighurst, 1948). From this perspective, life satisfaction is thought to be derived from increasingly fixed and uncontrollable factors with age. In younger adulthood, there is typically more freedom and opportunity to modify life circumstances to attain satisfaction and happiness. In contrast, as they age, people typically become increasingly established in family units and career pathways that have high barriers to exit. Thus, according to Hwang et al. (2016), the accepting perspective afforded by self-compassion may be increasingly relevant to facilitate adjustment to life circumstances that cannot be changed, because "fixed factors" become increasingly common with age. A caveat, however, is that this study included a sample of younger and midlife adults. In later life, many older adults experience greater flexibility and freedom relative to midlife as children gain independence and leave home. This transition may coincide with fewer financial stressors for some older adults. Older adults often have more flexibility in time, via retirement or reduced working hours, enabling more time to focus on hobbies and meaningful activities. Thus, an alternate hypothesis is that the relevance of self-compassion peaks at midlife and becomes less relevant in later years.

Available evidence, however, does not support this idea. A US study by Greene et al. (2016) surveyed 525 midlife and older adults who identified as being lesbian, bisexual, transgender, intersex, or queer, administering measures of self-compassion and mental and physical health. The authors found that self-compassion was a stronger predictor of mental health among members of the sample who were aged 65 years and over (n = 124) relative to the midlife subsect of the cohort. The authors also found that older adults exhibited higher self-compassion than the midlife group, replicating the finding that self-compassion increases with age (Homan, 2016; Hwang et al., 2016). In their discussion, the authors do not speculate as to why self-compassion was a stronger predictor of mental health in older adulthood relative to midlife. One possibility relates to the notion that self-compassion is a developmental task that deepens with age due to life experiences (Homan, 2016). If older adults struggle to meet this developmental task of building a healthy relationship with themselves as they age, their well-being might suffer as a consequence. Future qualitative work could be helpful to shed light on this intriguing issue by interviewing aging adults about their experiences of self-compassion and well-being, for example.

Eudaimonic Well-being A small number of studies have investigated the association between eudaimonic well-being and self-compassion in older age groups, with unanimous findings that self-compassion is associated with greater eudaimonia (Allen et al., 2012; Homan, 2016; Phillips & Ferguson, 2012; Homan, 2018; Brown et al., 2016). Phillips and Ferguson (2012) developed a path analytic model to investigate the role of self-compassion in predicting two aspects of eudaimonic well-being – meaning in life and ego integrity – in a community sample of 185 older adults. The authors found that self-compassion was nearly twice as strong a predictor of meaning in life compared to positive affect, an index of hedonic well-being.

Phillips and Ferguson (2012) argue that self-compassion might facilitate meaning in life through enabling behavioral flexibility, which is the ability to adapt to find new meaning despite losses and changes. In younger adults, self-compassion seems to help people respond adaptively rather than disengage following a loss or disappointment (Neely et al., 2009; Neff et al., 2007a). In the same way, older adults high in self-compassion might have the flexibility to find new meaningful pursuits following role transitions such as retirement or adapt hobbies to accommodate functional changes rather than disengaging and giving up on them completely (Brandtstädter & Renner, 1990). While the capacity to find new goals is important across the life span, it is known to be particularly central to well-being in later life (Wrosch et al., 2003). As such, self-compassion might be a helpful psychological resource to facilitate behavioral flexibility and new goal setting in later life.

Ego integrity is the last of Eric Erikson's psychosocial stages of life span development (Erikson, 1963). According to this model, the developmental task of later life is to reach a place of fulfillment and a sense of content for a life well lived despite loss and inevitable mortality, rather than falling into despair. In their model, Phillips and Ferguson (2012) found that self-compassion was a positive predictor of ego integrity. This finding has not been replicated to date, and more work is needed to investigate the role of self-compassion in negotiating the developmental tasks of later life.

In sum, a small but compelling body of research has investigated the relationship between self-compassion and mental well-being in later life. Converging research now shows that self-compassion is associated with fewer psychological symptoms and higher hedonic and eudaimonic well-being. While self-esteem typically declines across later adulthood (Robins & Trzesniewski, 2005), self-compassion may be protected or even enhanced in later life and thus could be a natural resource developed through life experience that people can draw on to navigate challenges.

Self-Compassion May Improve Well-being via Promoting a Positive Attitude to Aging

In addition to direct effects, self-compassion may also shape adaptive attitudes to aging which in turn contribute to health, well-being, and functioning in the second half of life. In this way, self-compassion may facilitate a more positive attitude toward aging, which in turn has adaptive ramifications for health (Brown et al., 2016). Brown et al. (2016) developed a structural equation model to investigate relationships between self-compassion, attitudes to aging, and health and well-being outcomes in a sample of 517 midlife women aged 40–60. In this study, women high in self-compassion typically reported a positive view of their personal experience of aging. Specifically, women higher in self-compassion viewed aging as an opportunity for psychological growth, including development of wisdom. They also held more adaptive attitudes about physical aging relative to those lower in self-compassion; for instance, they reported valuing the importance of exercise regardless of age. Finally, they were less likely to feel a sense of age-related psychosocial loss. Taken together, this shows that when aging starts to become personally relevant at midlife, self-compassion appears to help people embrace aging with a more positive attitude, despite the ageist views that are often communicated by social institutions such as the media.

In an earlier study by Allen and Leary (2013), older adults were asked to write about age-related events. The authors found those high in self-compassion embraced a more positive emotional tone when writing about their lived experiences of aging, adding support to the idea that self-compassion can enable people to hold a more adaptive view of age-related experiences. To explore this idea further, future research would do well to investigate if self-compassion inductions and interventions such as the MSC program (Neff & Germer, 2013) might help foster positive attitudes to aging.

Holding a positive view of aging has important benefits for well-being (Bryant et al., 2012) and physical health (Levy et al., 2002). Older adults with a positive view of aging tend to have fewer psychological symptoms and greater satisfaction with life (Bryant et al., 2012). In terms of physical health, a seminal longitudinal study of 660 adults aged 50 years and over found that those with a positive view of aging lived up to 7.5 years longer than those with a negative view of aging (Levy et al., 2002), indicating the significance of age-attitudes. However, many cultures – especially western cultures – hold ageist views (Fig. 7.2. Australian Human Rights Commission, 2013), and this can make it difficult for older adults to feel positively about their experience of aging. Further, empirical work has found that direct attempts to foster a positive attitude to aging can be ineffective (Levy et al., 2014); therefore, indirect pathways to feel positively about aging are needed. In the study by Brown et al. (2016) described above, a model was supported whereby higher self-compassion predicted a positive attitude to aging, which in turn predicted both mental and physical well-being (Brown et al., 2016). From these results, the authors suggest that self-compassion might be an effective pathway to feel more comfortable and accepting of the aging process, and this positive attitude in turn has important ramifications for mental and physical health.

Self-Compassion and Physical Well-being

By age 65, the average adult lives with two or more chronic medical conditions, and these comorbidities increase in prevalence with advancing age (Barnett et al., 2012). There are three pathways by which self-compassion might contribute to good physical health in later life. The first and most established pathway is through facilitating adjustment to illness, meaning self-compassionate people are better able to accept a change in health status. This may serve to lessen the psychological burden of illness. The second pathway is through facilitating healthy behaviors such as exercise and a healthy diet. Finally, but to date least conclusively, self-compassion has potential to exert a direct physiological effect on health in later life, for instance, through immune functioning and anti-inflammatory processes.

Adjustment to Illness A few studies have now considered self-compassion as a moderating factor that attenuates the impact of physical health symptoms on mental health in older adults (Allen et al., 2012; Homan, 2016; Smith, 2015; Herriot

& Wrosch, 2021). For instance, in a community sample of adults aged 67–90 years, Allen et al. (2012) found that self-compassion significantly weakened the association between physical symptoms and well-being. The authors found that for any given level of physical pain, limited mobility or poor self-reported health status, older adults with greater self-compassion experienced higher well-being compared to those with less self-compassion, who were more psychologically affected by their physical health conditions.

Data from studies by Homan (2016) and Smith (2015) largely accord with this idea. Homan (2016) found that self-compassion ameliorated the impact of poor self-rated health on depression (but not anxiety). In a sample of 102 senior residents living in a retirement community, Smith (2015) found no evidence of an overall association between self-reported health and both depression and happiness. However, when self-compassion was included as a moderator, those low on self-compassion evidenced a moderately strong connection between self-reported health and depression and happiness, whereas health was independent of well-being for those with higher self-compassion. This is a particularly interesting finding because it shows that older adults low on self-compassion may be at high risk of experiencing psychological symptoms and reduced well-being in response to health issues. In turn, lower well-being is known to contribute to poorer physical outcomes including morbidity and mortality (Carney & Freedland, 2017; Diener & Chan, 2011), completing a downward self-perpetuating cycle of poor physical and mental health.

A longitudinal study by Herriot and Wrosch (2021) followed health and self-compassion trajectories of 268 older adults with a mean age of 75 years at baseline to further investigate links between self-compassion and physical health. In this study, the authors found that self-compassion was associated with fewer physical symptoms in advanced age (i.e., 83 years, one standard deviation above the mean age of the sample) but not early older adulthood (i.e., 67 years, one standard deviation below the mean age of the sample). Moreover, longitudinal analyses revealed that low self-compassion was associated with a 4-year increase in chronic health issues for participants in advanced old age. In contrast, older participants who had higher self-compassion were protected against a subsequent rise in chronic health issues over the study period. In their discussion of these results, the authors explain that self-compassion may be especially important to health outcomes in advanced old age. Advanced old age is a time where there are often many uncontrollable losses which can contribute to poor health outcomes (Heckhausen et al., 2019), and the capacity to accept these losses with self-kindness, mindfulness, and common humanity may be paramount (Herriot & Wrosch, 2021).

It is plausible that self-compassion training might be helpful for aging adults living with health issues. For instance, Brown et al., 2019 developed a brief, four-session self-compassion-based intervention for midlife and older adults in outpatient hospital treatment for a chronic medical condition. In this small feasibility study (mean age = 64 years), authors found that patients responded well to the training program, rating the sessions as being enjoyable (mean rating 6.8/7) and relevant to daily life (mean rating 6.4/7). The participants also experienced marked reductions in depressive symptoms, as well as trends toward increased self-compassion and positive affect over the course of the program. A limitation of the study was that the intervention was brief; participants provided feedback that they would have preferred more time to practice self-compassion skills to consolidate their learning. While this is a preliminary study, the finding demonstrates that chronically unwell aging adults have an interest in self-compassion-based training, and it may facilitate adaptation to illness. Research to pilot the standard 8-week MSC with older adults is needed to extend this preliminary research and determine whether adaptations are required to tailor the program to the specific needs of older adults.

Health Behaviors Several studies from across the life span have now found that self-compassion is associated with a range of healthy behaviors including exercise, healthy diet, and engagement with medical treatment and high-quality sleep (Dunne et al., 2018; Sirois et al., 2015; Hu et al., 2018; Terry et al., 2013; for reviews see Chaps. 18 and 19 of this Handbook). In a series of studies that included both younger and older adults (age range 18–75 years), Terry et al. (2013) found that self-compassionate people typically felt less guilt or embarrassment about health issues and were more likely to seek medical assistance earlier in the course of illness than those lower on self- compassion. A proactive approach to health may become increasingly important with age, as health issues accumulate. Likewise, the effects of healthy behaviors typically strengthen over time, with behavioral choices earlier in life affecting health and well-being outcomes across the life span (Vaillant, 2008). For this reason, the role that self-compassion might play in facilitating healthy lifestyle choices and a proactive approach to health is of central relevance to later life health and well-being.

Healthy behavior choices made at midlife are thought to be particularly powerful determinants of healthy aging trajectories (Willcox et al., 2006; Kelly et al., 2016). One study by Hallion et al. (2018) looked at self-compassion and physical activity participation in an online survey of 169 midlife adults aged 40 to 65. Contrary to the hypothesis, they found no evidence of a direct link between self-compassion and physical activity. However, the authors found evidence of an association between self-compassion and self-regulatory strategies, which in turn are known to contribute to healthy behaviors such as exercise (e.g., McAuley et al., 2011). This indicates that relationships between self-compassion and health behaviors might be indirect. It should be noted that this was a relatively small study that may have lacked statistical power. Clearly, additional large, well-designed studies are needed to expand on these results by considering links between self-compassion and health behaviors in older adulthood.

Direct Physiological Effects A small but growing body of research has investigated the underlying physiological correlates of trait self-compassion (Svendsen et al., 2016; Breines et al., 2014; Breines et al., 2015; Friis et al., 2015; Herriot et al., 2018), as well as physiological changes caused by self-compassion inductions and training programs (Brown et al., 2019; Friis et al., 2016; Kirby et al., 2017). Herriot et al. (2018) investigated relationships between self-compassion, age-related stressors (including functional disability, life regrets, and physical health issues), and cortisol levels in a community-dwelling group of 233 adults with a mean age of 75 years (Herriot et al., 2018). Cortisol is a hormone that is released in response to stress (Zorn et al., 2017). Findings of the study revealed that self-compassion significantly moderated the relationship between age stressors and diurnal cortisol secretion, whereby self-compassion was increasingly associated with low cortisol secretion in those reporting high regret and health-related stress. This finding demonstrates that self-compassion may dampen the physiological impact of age stressors, in this case through weakening cortisol reactivity to stress.

Physiological benefits of self-compassion have also been observed in younger adult groups (see Chap. 17 for a review), and these findings have implications for older adults. In an experiment of social stress, Breines et al. (2014) asked 41 young adults to give an impromptu speech followed by an arithmetic task in front of a live audience of evaluative judges (Breines et al., 2014). The authors found that self-compassionate participants experienced less stress-induced inflammation (interleukin-6; IL-6) after the stressful encounter compared to those lower on the trait. Stress-induced inflammation can contribute to longer-term health outcomes, increasing the risk of developing cardiovascular disease and diabetes (Juster et al., 2010). IL-6 and IL-6 receptor concentrations have also been found to relate to mental well-being in later life, including relationship quality and purpose in life (Friedman et al., 2007). Therefore, self-compassion may ameliorate inflammatory responses to stress, which in turn may have positive ramifications for health and psychological well-being in later life. It should, however, be noted that evidence linking self-compassion with inflammatory markers is preliminary, and future work is needed to expand on these early results.

Heart rate variability (HRV) is an index of autonomic nervous system functioning associated with both physical and mental health across the life span (Brown et al., 2018b; Thayer et al., 2010; Bhattacharyya et al., 2008). High HRV is indicative of flexible nervous system that adapts efficiently to an ever-changing biopsychosocial environment (Thayer et al., 2012; Appelhans & Luecken, 2006). High resting HRV is associated with emotion regulation (Appelhans & Luecken, 2006), and emotion-based experimental tasks have been used to activate short-term increases in HRV (Smith et al., 2011). Self-compassion is an adaptive emotional regulation skill that helps soothe negative emotional states and is associated with high HRV in college students (Svendsen et al., 2016). Furthermore, self-compassionate inductions and interventions have been reported to increase resting HRV in studies of younger adults (Kirby et al., 2017). Yet little research has explored whether self-compassion training might improve HRV in older adults. In a pilot feasibility study of a self-compassion intervention for aging adults with health issues, Brown et al. (2019) reported a nonsignificant trend toward increased HRV following the training course, but a much larger study is needed to qualify these preliminary results. It should be noted that a recent meta-analysis of randomized controlled trials found no evidence that mindfulness-based training programs lead to improvements in resting-state HRV (Brown et al., 2021). Self-compassion training explicitly teaches people how to soothe negative emotional states, and so self-compassion may be more closely related to HRV compared to general mindfulness training programs (Svendsen et al., 2020).

In sum, there are three pathways by which self-compassion might facilitate healthy physical aging in later life. First, there is relatively strong evidence that self-compassion enables people to adjust to physical health threats, such that mental well-being is buoyed despite the inevitable challenge of being sick. Second, studies of younger adults demonstrate that those high in self-compassion are more likely to take care of themselves by making healthy lifestyle choices and the effects of these choices accumulate over time across the life span and become increasingly prominent with age. Third, self-compassion may lead to direct physiological benefits such as reduced inflammatory response to stress and improved HRV, but large-scale, high-quality research is needed to verify these effects.

Self-Compassion and Engagement in Activity

Later life can bring newfound opportunities for freedom, travel, and the pursuit of hobbies and personal growth. At the same time, role transitions such as retirement and the emergence of health issues can pose challenges for some older adults. With this matrix of gains and losses, flexibility is needed to adapt to change and thus enjoy the pleasures of aging without becoming overwhelmed by the losses.

Selective optimization with compensation (SOC) is a popular developmental model that helps describe how people adjust to inevitable life transitions that occur across the life span and become increasingly common in later life (Freund & Baltes, 1998). Selection involves committing to realistic personal goals and *adjusting* goals when needed based on changing circumstances. An example of this is a tennis player, who, with arthritis and advancing age, might narrow their repertoire and select to play fewer overhead shots in games to avoid pain and care for their joints. Optimization entails allocating resources to achieve higher levels of functioning in selected pursuits, for instance, the older tennis player might spend more time practicing their forehand shots to optimize their performance. Finally, compensation involves finding ways to compensate for loss, so that life priorities can still be pursued, despite any physical decline. To continue with the tennis example, the aging player might elect to play doubles and shorter games. If mobility issues continued to worsen such that playing tennis becomes unrealistic, the player might then compensate by taking up coaching, watching games instead of playing, or fundraising for their local tennis club. In this way, they can continue to engage in the hobby despite physical decline.

Engagement in SOC processes is a strong predictor of well-being in later life (Carpentieri et al., 2017; Freund, 2008). Less research has considered psychological factors that might predict the tendency to engage with SOC processes, versus the alternative which is either the relentless pursuit of unrealistic goals *or* disengagement due to the inability to adapt to change. While a novel area, data from both older and younger adult groups indicates that self-compassion might facilitate adaptive behavior in the form of SOC processes.

Firstly, studies of younger adults indicate that self-compassion is associated with goal regulation (Neff et al., 2005; Neely et al., 2009). Neff et al. (2005) found that college students who self-reported high levels of self-compassion were less likely to embrace outcome-oriented performance goals, in which people are motivated to achieve a goal to defend self-worth. Instead, self-compassionate people were more likely to adopt process-oriented mastery goals, motivated by curiosity and a desire to learn and engage for learning's sake. Research has found that mastery goals are more helpful than performance goals when goal flexibility (i.e., selection, optimization, and compensation) is needed. For instance, mastery is efficacious in the context of challenges such as interpersonal conflict (Darnon et al., 2007) and in the context of age-related symptoms such as memory loss (Hastings & West, 2011). The capacity to hold mastery goals that are more process-oriented may be a great benefit to facilitate flexibility and meaningful engagement with life in older adulthood.

A study by Allen et al. (2012) indicates that self-compassion is associated with goal flexibility in later life. In this study, the authors surveyed 71 older adults on their eagerness to use memory, hearing, and walking aids. The authors found that self-compassionate participants were less bothered about accepting these forms of support (Allen et al., 2012). Thus, in this study, self-compassion appeared to promote flexibility to *compensate* for age-related losses, such that functioning is maintained despite the loss. For instance, a willingness to accept a walking aid when it is justified could enable mobility for a senior to leave their house and engage with their local community. In contrast, those lower on self-compassion who are more bothered about accepting aids could be limited by age-related losses either through refusal to accept the aid or psychological distress caused by using the aid without acceptance. Part of the role of clinicians is to support older adults to see the value of accepting support when it is needed, and helping older adults build self-compassion can facilitate this process of acceptance and adjustment to loss.

There is now some evidence from both younger and older cohorts that self-compassion might enable the *flexible* pursuit of goals through enabling the processes of selection, optimization, and compensation. Self-compassionate people tend to embrace mastery rather than performance goals, and mastery goals are more flexible to accommodate challenges and changes compared to performance goals that are relatively fixed. Further, self-compassionate older adults are more likely to accept aids that can bolster daily life functioning to compensate for age-related losses. Future research is needed to delve more deeply into these issues and to consider the extent to which self-compassion might facilitate flexible pursuit in later life.

Self-Compassion and Connectedness in Later Life

Social isolation and loneliness are powerful predictors of poor health outcomes and risk of premature death (Holt-Lunstad et al., 2015; Nicholson, 2012). Health risks of isolation and loneliness are equivalent in strength to well-established risk factors such as smoking and obesity (Courtin & Knapp, 2017). For this reason, advancing social connection has been identified as a key public health priority (Holt-Lunstad et al., 2017). Unfortunately, those aged 75+ are the age group most at risk of experiencing loneliness (Australian Institute of Health and Welfare, 2019). In turn, lack of connection with others is linked with high rates of suicide that are observed in later life (Van Orden & Conwell, 2011). Given the enormous costs of late life isolation, it is important to consider the extent to which self-compassion might have a role to play in facilitating social connectedness and helping address this major public health crisis. Given that a core component of self-compassion is "common humanity" (i.e., the capacity to recognize that one's difficulties are part of the common human experience), it may be expected that people with higher levels of self-compassion feel more connected to others, particularly during times of difficulty.

Few studies have investigated relationships between self-compassion and connectedness in later life, but early work has showed promising correlational effects (Allen et al., 2012; Homan, 2016). Homan (2016) investigated links between trait self-compassion and positive relationship quality in a sample of 121 older adults and found evidence of a strong relationship between the constructs ($r = 0.56$). Using a different measure, Allen et al. (2012) found that self-compassionate older adults were less likely to report that health issues interfered with their social functioning. While these were cross-sectional studies that limit inferences of causality, this emerging evidence indicates that self-compassion is associated with positive social functioning in later life. More recently, a longitudinal study of 1090 community-dwelling adults (baseline mean age 61.5 and 66 years for women and men, respectively) found that self-compassion was positively associated with mental well-being and inversely associated with loneliness across the life span (Lee et al., 2021).

Studies with younger adults add support to this idea. Self-compassion is associated with high-quality romantic relationships and a felt sense of community and relationship harmony (Neff & Beretvas, 2013; Akın & Akın, 2015; Yang, 2016). While these measures of relationship quality cannot counterbalance the *absence* of relationship that is prevalent in later life, it demonstrates that self-compassion is a skill associated with making the most out of relationships that you do have. Since self-compassion is associated with relationship quality, it is plausible that self-compassionate older adults might be able to derive more psychological benefit from casual social interactions, for instance, a chat with the mailman or local storeowners, relative to those who are less self-compassionate – an interesting hypothesis for future research.

A meta-analysis of interventions designed to prevent social isolation and loneliness in older adults found evidence that group-based programs are more effective than one-to-one programs providing personalized support (Cattan et al., 2005). Moreover, group programs that incorporated educational or training input were the most effective of all in reducing isolation. In their discussion, the authors recommend including training in intrapersonal resources within group programs, such as teaching on self-esteem. Since self-esteem is typically fragile in the face of difficulties and is not easily taught (Neff, 2011), offering group training on self-compassion could be a more effective pathway to social connection for older adults. Modifications of the MSC program to include content that is specifically relevant to older adults could be helpful (Bryant, 2017). In this way, the process of participation in the group would offer a direct route to the experience of social connectedness. Simultaneously, course content on self-compassionate aging might facilitate healthy aging among participants.

Broad Processes Explaining Why Self-Compassion Promotes Positive Aging

As illustrated in Fig. 7.1, self-compassion facilitates positive aging via broad, interrelated processes: attitudes to aging, acceptance of change, behavioral flexibility, and flexible goal pursuit. In turn, these processes enable the four pillars of positive aging which include mental well-being, physical well-being, engagement in activity, and social connectedness. In a society that holds explicit and implicit biases against aging and older adults, embracing a positive view of aging can be challenging, perhaps especially for women who are exposed to negative media representations of aging women (Lemish & Muhlbauer, 2012). Research has found that self-compassion enables individuals to hold more positive views of aging, which in turn has beneficial sequelae for positive aging, especially in the domains of mental and physical health (Brown et al., 2016).

Self-compassion also facilitates adjustment to change. An example of this is converging evidence linking self-compassion to good adjustment to illness in the second half of life (Allen et al., 2012; Homan, 2016; Smith, 2015; Herriot & Wrosch, 2021). Self-compassionate older adults are more likely to maintain high levels of well-being, despite illness and physical symptoms that are increasingly common in later life. Finally, self-compassion enables behavioral flexibility as well as flexible goal pursuit via selection, optimization, and compensation processes (Freund & Baltes, 1998). One example of this is that self-compassionate older adults appear to be less bothered about accepting aids (e.g., for walking and hearing). This type of behavioral flexibility enables continued engagement in life.

Conclusion and Future Directions

Self-compassion is an adaptive psychological resource that facilitates adjustment to challenges across the life span, and it may be especially useful in later life when physical, psychological, and social changes are plentiful. The field of self-compassionate aging is relatively new. The first scientific paper on the topic was published in 2012 (Allen et al., 2012). The past decade has seen a rapid rise in publications, and most study designs have been cross-sectional, investigating associations between trait self-compassion and health and well-being outcomes in older adult groups (Brown et al., 2018a).

From the literature published to date, there is evidence that self-compassion is associated with four pillars of positive aging including (1) mental well-being, (2) physical well-being, (3) engagement in activity, and (4) social well-being (Gergen & Gergen, 2001). First, self-compassion is associated with mental well-being in later life, including reduced risk of psychological symptoms and a greater likelihood of both hedonic and eudaimonic well-being. Self-compassion may also hold benefits for physical health in later life, a second key pillar of positive aging, by facilitating adjustment to health issues, promoting healthy behaviors, and potentially direct physiological pathways such as immune functioning and HRV. Self-compassion is also associated with engagement in activity and social connectedness, the last two pillars of positive aging (Gergen & Gergen, 2001). Self-compassion enables meaningful activity and is associated with high-quality relationships (Homan, 2016), which can protect against loneliness and isolation in later life.

While research is still needed to uncover mechanisms that explain *how* self-compassion facilitates positive aging, there appears to be broad processes that are relevant. Self-compassion works by helping people embrace a positive attitude to aging and adjust to changes that are plentiful in later life, behavioral flexibility, and flexible goal pursuit. These broad processes relate to all four pillars of positive aging. More empirical work is needed to unpack these ideas.

The field of self-compassionate aging is still in its infancy. The field needs to move beyond cross-sectional research to develop and test the efficacy of rigorous self-compassion-based training programs that are modified to suit the needs of older adults (Bryant, 2017). We also need to see more experimental research, to consider if brief self-compassion inductions might be an effective way to prime aging adults to be more self-compassionate toward their experience of aging. In the context of an aging society, there is an urgent need to find cost-effective ways to support aging adults cope with the unprecedented changes and losses of later life (Gurwitz & Pearson, 2019). Self-compassion is a promising psychological resource that may help people respond to these challenges of aging, and more research is needed to expand this field of research.

References

Akın, U., & Akın, A. (2015). Examining the predictive role of self-compassion on sense of community in Turkish adolescents. *Social Indicators Research, 123*(1), 29–38. https://doi.org/10.1007/s11205-014-0724-5

Allen, A. B., & Leary, M. R. (2013). Self-compassionate responses to aging. *The Gerontologist, 54*(2), 190–200. https://doi.org/10.1093/geront/gns204

Allen, A. B., Goldwasser, E. R., & Leary, M. R. (2012). Self-compassion and well-being among older adults. *Self and Identity, 11*(4), 428–453. https://doi.org/10.1080/15298868.2011.595082

Appelhans, B. M., & Luecken, L. J. (2006). Heart rate variability as an index of regulated emotional responding. *Review of General Psychology, 10*(3), 229–240. https://doi.org/10.1037/1089-2680.10.3.229

Ardelt, M. (2011). Wisdom, age, and well-being. In K. W. Schaie & S. L. Willis (Eds.), *Handbook of the psychology of aging* (pp. 279–291). Elsevier.

Ardelt, M., Pridgen, S., & Nutter-Pridgen, K. L. (2018). The relation between age and three-dimensional wisdom: Variations by wisdom dimensions and education. *The Journals of Gerontology: Series B, 73*(8), 1339–1349. https://doi.org/10.1093/geronb/gbx182

Asadi Bijaeyeh, J. S., Amiri Majd, M., Ghamari, M., & Fathi Aghdam, G. (2021). The effectiveness of the self-compassion training on life satisfaction and resilience of the elderly women. *Aging Psychology, 7*(1), 67–55. https://doi.org/10.22126/JAP.2021.6210.1509

Australian Human Rights Commission (2013). Fact or fiction? Stereotypes of older Australians research report. Australian Human Rights Commission (Ed.), Urbis.

Australian Human Rights Commission. (2016). *Willing to work: National inquiry into employment discrimination against older Australians and Australians with disability*. Australian Human Rights Comission.

Australian Institute of Health and Welfare. (2019). *Social isolation and loneliness*. Australian Institute of Health and Welfare.

Barnett, K., Mercer, S. W., Norbury, M., Watt, G., Wyke, S., & Guthrie, B. (2012). Epidemiology of multimorbidity and implications for health care, research, and medical education: A cross-sectional study. *The Lancet, 380*(9836), 37–43. https://doi.org/10.1016/S0140-6736(12)60240-2

Baumeister, R. F., Campbell, J. D., Krueger, J. I., & Vohs, K. D. (2003). Does high self-esteem cause better performance, interpersonal success, happiness, or healthier lifestyles? *Psychological Science in the Public Interest, 4*(1), 1–44. https://doi.org/10.1111/1529-1006.01431

Bhattacharyya, M. R., Whitehead, D. L., Rakhit, R., & Steptoe, A. (2008). Depressed mood, positive affect, and heart rate variability in patients with suspected coronary artery disease. *Psychosomatic Medicine, 70*(9), 1020–1027. https://doi.org/10.1097/PSY.0b013e318189afcc

Brandtstädter, J., & Renner, G. (1990). Tenacious goal pursuit and flexible goal adjustment: Explication and age-related analysis of assimilative and accommodative strategies of coping. *Psychology and Aging, 5*(1), 58. https://doi.org/10.1037//0882-7974.5.1.58

Breines, J. G., Thoma, M. V., Gianferante, D., Hanlin, L., Chen, X., & Rohleder, N. (2014). Self-compassion as a predictor of interleukin-6 response to acute psychosocial stress. *Brain, Behavior, and Immunity, 37*, 109–114. https://doi.org/10.1016/j.bbi.2013.11.006

Breines, J. G., McInnis, C. M., Kuras, Y. I., Thoma, M. V., Gianferante, D., Hanlin, L., Chen, X., & Rohleder, N. (2015). Self-compassionate young adults show lower salivary alpha-amylase responses to repeated psychosocial stress. *Self and Identity, 14*(4), 390–402. https://doi.org/10.1080/15298868.2015.1005659

Brown, L., Bryant, C., Brown, V., Bei, B., & Judd, F. (2016). Self-compassion, attitudes to ageing and indicators of health and well-being among midlife women. *Aging & Mental Health, 20*(10), 1035–1043. https://doi.org/10.1080/13607863.2015.1060946

Brown, L., Huffman, J. C., & Bryant, C. (2018a). Self-compassionate aging: A systematic review. *The Gerontologist, 59*(4), e311–e324. https://doi.org/10.1093/geront/gny108

Brown, L., Karmakar, C., Gray, R., Jindal, R., Lim, T., & Bryant, C. (2018b). Heart rate variability alterations in late life depression: A meta-analysis. *Journal of Affective Disorders, 235*, 456–466. https://doi.org/10.1016/j.jad.2018.04.071

Brown, L., Karmakar, C., Flynn, M., Motin, M. A., Palaniswami, M., Celano, C. M., Huffman, J., & Bryant, C. (2019). A self-compassion group intervention for patients living with chronic medical illness: Treatment development and feasibility study. *The Primary Care Companion for CNS Disorders, 21*(5), 19m02470. https://doi.org/10.4088/PCC.19m02470

Brown, L., Rando, A. A., Eichel, K., Van Dam, N. T., Celano, C. M., Huffman, J. C., & Morris, M. E. (2021). The effects of mindfulness and meditation on vagally mediated heart rate variability: A meta-analysis. *Psychosomatic Medicine, 83*(6), 631–640. https://doi.org/10.1097/PSY.0000000000000900

Bryant, C. (2017). Psychological interventions for older adults: Evidence-based treatments for depression, anxiety, and carer stress. In H. Chiu & K. Shulman (Eds.), *Mental health and illness of the elderly* (pp. 481–514). Springer Science + Business Media. https://doi.org/10.1007/978-981-10-2414-6_21

Bryant, C., Bei, B., Gilson, K., Komiti, A., Jackson, H., & Judd, F. (2012). The relationship between attitudes to aging and physical and mental health in older adults. *International Psychogeriatrics, 24*(10), 1674–1683. https://doi.org/10.1017/S1041610212000774

Carney, R. M., & Freedland, K. E. (2017). Depression and coronary heart disease. *Nature Reviews Cardiology, 14*(3), 145. https://doi.org/10.1038/nrcardio.2016.181

Carpentieri, J., Elliott, J., Brett, C. E., & Deary, I. J. (2017). Adapting to aging: Older people talk about their use of selection, optimization, and compensation to maximize well-being in the context of physical decline. *The Journals of Gerontology: Series B, 72*(2), 351–361. https://doi.org/10.1093/geronb/gbw132

Carstensen, L. L., Fung, H. H., & Charles, S. T. (2003). Socioemotional selectivity theory and the regulation of emotion in the second half of life. *Motivation and Emotion, 27*(2), 103–123. https://doi.org/10.1023/A:1024569803230

Cattan, M., White, M., Bond, J., & Learmouth, A. (2005). Preventing social isolation and loneliness among older people: A systematic review of health promotion interventions. *Ageing & Society, 25*(1), 41–67. https://doi.org/10.1017/S0144686X04002594

Courtin, E., & Knapp, M. (2017). Social isolation, loneliness and health in old age: A scoping review. *Health & Social Care in the Community, 25*(3), 799–812. https://doi.org/10.1111/hsc.12311

Darnon, C., Butera, F., & Harackiewicz, J. M. (2007). Achievement goals in social interactions: Learning with mastery vs. performance goals. *Motivation and Emotion, 31*(1), 61–70. https://doi.org/10.1007/s11031-006-9049-2

Diener, E., & Chan, M. Y. (2011). Happy people live longer: Subjective well-being contributes to health and longevity. *Applied Psychology: Health and Well-Being, 3*(1), 1–43. https://doi.org/10.1111/j.1758-0854.2010.01045.x

Dunne, S., Sheffield, D., & Chilcot, J. (2018). Brief report: Self-compassion, physical health and the mediating role of health-promoting behaviours. *Journal of Health Psychology, 23*(7), 993–999. https://doi.org/10.1177/1359105316643377

Erikson, E. H. (1963). *Childhood and society, revised 2nd ed.* Norton and Co.

Finlay-Jones, A., Xie, Q., Huang, X., Ma, X., & Guo, X. (2018). A pilot study of the 8-week mindful self-compassion training program in a Chinese community sample. *Mindfulness, 9*(3), 993–1002. https://doi.org/10.1007/s12671-017-0838-3

Freund, A. M. (2008). Successful aging as management of resources: The role of selection, optimization, and compensation. *Research in Human Development, 5*(2), 94–106. https://doi.org/10.1080/15427600802034827

Freund, A. M., & Baltes, P. B. (1998). Selection, optimization, and compensation as strategies of life management: Correlations with subjective indicators of successful aging. *Psychology and Aging, 13*(4), 531. https://doi.org/10.1037//0882-7974.13.4.531

Friedman, E. M., Hayney, M., Love, G. D., Singer, B. H., & Ryff, C. D. (2007). Plasma interleukin-6 and soluble IL-6 receptors are associated with psychological well-being in aging women. *Health Psychology, 26*(3), 305. https://doi.org/10.1037/0278-6133.26.3.305

Friis, A., Johnson, M., Cutfield, R., & Consedine, N. (2015). Does kindness matter? Self-compassion buffers the negative impact of diabetes-distress on HbA1c. *Diabetic Medicine, 32*(12), 1634–1640. https://doi.org/10.1111/dme.12774

Friis, A., Johnson, M. H., Cutfield, R. G., & Consedine, N. S. (2016). Kindness matters: A randomized controlled trial of a mindful self-compassion intervention improves depression, distress, and HbA1c among patients with diabetes. *Diabetes Care, 39*(11), 1963–1971. https://doi.org/10.2337/dc16-0416

Gergen, M. M., & Gergen, K. J. (2001). Positive aging: New images for a new age. *Ageing International, 27*(1), 3–23. https://doi.org/10.1007/s12126-001-1013-6

Greene, D. C., Britton, P. J., & Shepherd, J. B. (2016). LGBTQ aging: Mental health at midlife and older adulthood. *Journal of LGBT Issues in Counseling, 10*(4), 180–196. https://doi.org/10.1080/15538605.2016.1233839

Gurwitz, J. H., & Pearson, S. D. (2019). Novel therapies for an aging population: Grappling with price, value, and affordability. *The Journal of the American Medical Association, 321*(16), 1567–1568. https://doi.org/10.1001/jama.2019.2633

Hallion, M., Taylor, A., Roberts, R., & Ashe, M. (2018). Exploring the association between physical activity participation and self-compassion in middle-aged adults. *Sport, Exercise, and Performance Psychology, 8*(3), 305–316. https://doi.org/10.1037/spy0000150

Harrison, S. L., Robertson, N., Goldstein, R. S., & Brooks, D. (2017). Exploring self-conscious emotions in individuals with chronic obstructive pulmonary disease: A mixed-methods study. *Chronic Respiratory Disease, 14*(1), 22–32. https://doi.org/10.1177/1479972316654284

Hastings, E. C., & West, R. L. (2011). Goal orientation and self-efficacy in relation to memory in adulthood. *Aging, Neuropsychology, and Cognition, 18*(4), 471–493. https://doi.org/10.1080/13825585.2011.575926

Havighurst, R. J. (1948). *Developmental tasks and education*. University of Chicago Press.

Hazel, R. (1991). The role of the self-concept in aging. In *Annual review of gerontology and geriatrics, volume 11, 1991: Behavioral Science & Aging* (Vol. 11, p. 110). Springer.

Heckhausen, J., Wrosch, C., & Schulz, R. (2019). Agency and motivation in adulthood and old age. *Annual Review of Psychology, 70*, 191–217. https://doi.org/10.1146/annurev-psych-010418-103043

Herriot, H., & Wrosch, C. (2021). Self-compassion as predictor of daily physical symptoms and chronic illness across older adulthood. *Journal of Health Psychology, 27*(7), 1697–1709. https://doi.org/10.1177/13591053211002326

Herriot, H., Wrosch, C., & Gouin, J.-P. (2018). Self-compassion, chronic age-related stressors, and diurnal cortisol secretion in older adulthood. *Journal of Behavioral Medicine, 41*(6), 850–862. https://doi.org/10.1007/s10865-018-9943-6

Hill, R. D. (2011). A positive aging framework for guiding geropsychology interventions. *Behavior Therapy, 42*(1), 66–77. https://doi.org/10.1016/j.beth.2010.04.006

Hodgetts, J., McLaren, S., Bice, B., & Trezise, A. (2020). The relationships between self-compassion, rumination, and depressive symptoms among older adults: The moderating role of gender. *Aging & Mental*

Health, 25(2), 2337–2346. https://doi.org/10.1080/13607863.2020.1824207

Holt-Lunstad, J., Smith, T. B., Baker, M., Harris, T., & Stephenson, D. (2015). Loneliness and social isolation as risk factors for mortality: A meta-analytic review. *Perspectives on Psychological Science, 10*(2), 227–237. https://doi.org/10.1177/1745691614568352

Holt-Lunstad, J., Robles, T. F., & Sbarra, D. A. (2017). Advancing social connection as a public health priority in the United States. *American Psychologist, 72*(6), 517–530. https://doi.org/10.1037/amp0000103

Homan, K. J. (2016). Self-compassion and psychological well-being in older adults. *Journal of Adult Development, 23*(2), 111–119. https://doi.org/10.1007/s10804-016-9227-8

Homan, K. J. (2018). Secure attachment and eudaimonic well-being in late adulthood: The mediating role of self-compassion. *Aging & Mental Health, 22*(3), 363–370. https://doi.org/10.1080/13607863.2016.1254597

Hu, Y., Wang, Y., Sun, Y., Arteta-Garcia, J., & Purol, S. (2018). Diary study: The protective role of self-compassion on stress-related poor sleep quality. *Mindfulness, 9*(6), 1931–1940. https://doi.org/10.1007/s12671-018-0939-7

Huppert, F. A., & Whittington, J. E. (2003). Evidence for the independence of positive and negative well-being: Implications for quality of life assessment. *British Journal of Health Psychology, 8*(1), 107–122. https://doi.org/10.1348/135910703762879246

Hwang, S., Kim, G., Yang, J. W., & Yang, E. (2016). The moderating effects of age on the relationships of self-compassion, self-esteem, and mental health. *Japanese Psychological Research, 58*(2), 194–205. https://doi.org/10.1111/jpr.12109

Juster, R.-P., McEwen, B. S., & Lupien, S. J. (2010). Allostatic load biomarkers of chronic stress and impact on health and cognition. *Neuroscience & Biobehavioral Reviews, 35*(1), 2–16. https://doi.org/10.1016/j.neubiorev.2009.10.002

Kelly, S., Martin, S., Kuhn, I., Cowan, A., Brayne, C., & Lafortune, L. (2016). Barriers and facilitators to the uptake and maintenance of healthy behaviours by people at mid-life: A rapid systematic review. *PLoS One, 11*(1), e0145074. https://doi.org/10.1371/journal.pone.0145074

Kirby, J. N., Doty, J. R., Petrocchi, N., & Gilbert, P. (2017). The current and future role of heart rate variability for assessing and training compassion. *Frontiers in Public Health, 5*, 40. https://doi.org/10.3389/fpubh.2017.00040

Kunuroglu, F., & Yuzbasi, D. V. (2021). Factors promoting successful aging in Turkish older adults: Self-compassion, psychological resilience, and attitudes towards aging. *Journal of Happiness Studies, 2*, 3663–3678. https://doi.org/10.1007/s10902-021-00388-z

Lee, E. E., Govind, T., Ramsey, M., Wu, T. C., Daly, R., Liu, J., Xin, M. T., Paulus, M. P., Thomas, M. L., & Jeste, D. V. (2021). Compassion toward others and self-compassion predict mental and physical well-being: A 5-year longitudinal study of 1090 community-dwelling adults across the lifespan. *Translational Psychiatry, 11*(1), 397–397. https://doi.org/10.1038/s41398-021-01491-8

Lemish, D., & Muhlbauer, V. (2012). "Can't have it all": Representations of older women in popular culture. *Women & Therapy, 35*(3-4), 165–180. https://doi.org/10.1080/02703149.2012.684541

Levy, B. R., Slade, M. D., Kunkel, S. R., & Kasl, S. V. (2002). Longevity increased by positive self-perceptions of aging. *Journal of Personality and Social Psychology, 83*(2), 261. https://doi.org/10.1037//0022-3514.83.2.26

Levy, B. R., Pilver, C., Chung, P. H., & Slade, M. D. (2014). Subliminal strengthening: Improving older individuals' physical function over time with an implicit-age-stereotype intervention. *Psychological Science, 25*(12), 2127–2135. https://doi.org/10.1177/0956797614551970

McAuley, E., Mullen, S. P., Szabo, A. N., White, S. M., Wójcicki, T. R., Mailey, E. L., Gothe, N. P., Olson, E. A., Voss, M., Erickson, K., Prakash, R., & Kramer, A. F. (2011). Self-regulatory processes and exercise adherence in older adults: Executive function and self-efficacy effects. *American Journal of Preventive Medicine, 41*(3), 284–290. https://doi.org/10.1016/j.amepre.2011.04.014

Neely, M. E., Schallert, D. L., Mohammed, S. S., Roberts, R. M., & Chen, Y.-J. (2009). Self-kindness when facing stress: The role of self-compassion, goal regulation, and support in college students' well-being. *Motivation and Emotion, 33*(1), 88–97. https://doi.org/10.1007/s11031-008-9119-8

Neff, K. D. (2003). Self-compassion: An alternative conceptualization of a healthy attitude toward oneself. *Self and Identity, 2*(2), 85–101. https://doi.org/10.1080/15298860309032

Neff, K. D. (2011). Self-compassion, self-esteem, and Well-being. *Social and Personality Psychology Compass, 5*(1), 1–12. https://doi.org/10.1111/j.1751-9004.2010.00330.x

Neff, K. D., & Beretvas, S. N. (2013). The role of self-compassion in romantic relationships. *Self and Identity, 12*(1), 78–98. https://doi.org/10.1080/15298868.2011.639548

Neff, K. D., & Germer, C. (2013). A pilot study and randomized controlled trial of the mindful self-compassion program. *Journal of Clinical Psychology, 69*(1), 28–44. https://doi.org/10.1002/jclp.21923

Neff, K. D., & Vonk, R. (2009). Self-compassion versus global self-esteem: Two different ways of relating to oneself. *Journal of Personality, 77*(1), 23–50. https://doi.org/10.1111/j.1467-6494.2008.00537.x

Neff, K. D., Hsieh, Y.-P., & Dejitterat, K. (2005). Self-compassion, achievement goals, and coping with academic failure. *Self and Identity, 4*(3), 263–287. https://doi.org/10.1080/13576500444000317

Neff, K. D., Kirkpatrick, K. L., & Rude, S. S. (2007a). Self-compassion and adaptive psychological functioning. *Journal of Research in Personality, 41*(1), 139–154. https://doi.org/10.1016/j.jrp.2006.03.004

Neff, K. D., Rude, S. S., & Kirkpatrick, K. L. (2007b). An examination of self-compassion in relation to positive psychological functioning and personality traits. *Journal of Research in Personality, 41*(4), 908–916. https://doi.org/10.1016/j.jrp.2006.08.002

Nicholson, N. R. (2012). A review of social isolation: An important but underassessed condition in older adults. *The Journal of Primary Prevention, 33*(2-3), 137–152. https://doi.org/10.1007/s10935-012-0271-2

Nolen-Hoeksema, S., Wisco, B. E., & Lyubomirsky, S. (2008). Rethinking rumination. *Perspectives on Psychological Science, 3*(5), 400–424.

Perez-Blasco, J., Sales, A., Meléndez, J. C., & Mayordomo, T. (2016). The effects of mindfulness and self-compassion on improving the capacity to adapt to stress situations in elderly people living in the community. *Clinical Gerontologist, 39*(2), 90–103. https://doi.org/10.1080/07317115.2015.1120253

Phillips, W. J., & Ferguson, S. J. (2012). Self-compassion: A resource for positive aging. *Journals of Gerontology Series B: Psychological Sciences and Social Sciences, 68*(4), 529–539. https://doi.org/10.1093/geronb/gbs091

Robins, R. W., & Trzesniewski, K. H. (2005). Self-esteem development across the lifespan. *Current Directions in Psychological Science, 14*(3), 158–162. https://doi.org/10.1111/j.0963-7214.2005.00353.x

Ryan, R. M., & Deci, E. L. (2001). On happiness and human potentials: A review of research on hedonic and eudaimonic well-being. *Annual Review of Psychology, 52*(1), 141–166. https://doi.org/10.1146/annurev.psych.52.1.141

Sirois, F. M., Kitner, R., & Hirsch, J. K. (2015). Self-compassion, affect, and health-promoting behaviors. *Health Psychology, 34*(6), 661–669. https://doi.org/10.1037/hea0000158

Smith, J. (2015). Self-compassion and resilience in senior living residents. *Seniors Housing and Care Journal, 23*(1), 17–31.

Smith, T. W., Cribbet, M. R., Nealey-Moore, J. B., Uchino, B. N., Williams, P. G., Mackenzie, J., & Thayer, J. F. (2011). Matters of the variable heart: Respiratory sinus arrhythmia response to marital interaction and associations with marital quality. *Journal of Personality and Social Psychology, 100*(1), 103–119. https://doi.org/10.1037/a0021136

Steverink, N., Westerhof, G. J., Bode, C., & Dittmann-Kohli, F. (2001). The personal experience of aging, individual resources, and subjective well-being. *The Journals of Gerontology Series B: Psychological Sciences and Social Sciences, 56*(6), P364–P373. https://doi.org/10.1093/geronb/56.6.p364

Suh, E. K., & Koo, J. S. (2011). A concise measure of subjective well-being (COMOSWB): Scale development and validation. *Korean Journal of Social and Personality Psychology, 25*(1), 95–113.

Svendsen, J. L., Osnes, B., Binder, P. E., Dundas, I., Visted, E., Nordby, H., Schanche, E., & Sørensen, L. (2016). Trait self-compassion reflects emotional flexibility through an association with high vagally mediated heart rate variability. *Mindfulness, 7*(5), 1103–1113. https://doi.org/10.1007/s12671-016-0549-1

Svendsen, J. L., Schanche, E., Osnes, B., Vøllestad, J., Visted, E., Dundas, I., Nordby, H., Binder, P. E., & Sørensen, L. (2020). Is dispositional self-compassion associated with psychophysiological flexibility beyond mindfulness? An exploratory pilot study. *Frontiers in Psychology, 11*, 614. https://doi.org/10.3389/fpsyg.2020.00614

Terry, M. L., Leary, M. R., Mehta, S., & Henderson, K. (2013). Self-compassionate reactions to health threats. *Personality and Social Psychology Bulletin, 39*(7), 911–926. https://doi.org/10.1177/0146167213488213

Thayer, J. F., Yamamoto, S. S., & Brosschot, J. F. (2010). The relationship of autonomic imbalance, heart rate variability and cardiovascular disease risk factors. *International Journal of Cardiology, 141*(2), 122–131. https://doi.org/10.1016/j.ijcard.2009.09.543

Thayer, J. F., Åhs, F., Fredrikson, M., Sollers, J. J., III, & Wager, T. D. (2012). A meta-analysis of heart rate variability and neuroimaging studies: Implications for heart rate variability as a marker of stress and health. *Neuroscience & Biobehavioral Reviews, 36*(2), 747–756. https://doi.org/10.1016/j.neubiorev.2011.11.009

Thoma, M. V., Bernays, F., Eising, C. M., Maercker, A., & Rohner, S. L. (2021). Child maltreatment, lifetime trauma, and mental health in Swiss older survivors of enforced child welfare practices: Investigating the mediating role of self-esteem and self-compassion. *Child Abuse & Neglect, 113*, 104925. https://doi.org/10.1016/j.chiabu.2020.104925

Vaillant, G. E. (2008). *Aging well: Surprising guideposts to a happier life from the landmark study of adult development*. Little, Brown.

Van Orden, K., & Conwell, Y. (2011). Suicides in late life. *Current Psychiatry Reports, 13*(3), 234–241. https://doi.org/10.1007/s11920-011-0193-3

Wagner, J., Gerstorf, D., Hoppmann, C., & Luszcz, M. A. (2013). The nature and correlates of self-esteem trajectories in late life. *Journal of Personality and Social Psychology, 105*(1), 139–153. https://doi.org/10.1037/a0032279

Willcox, B. J., He, Q., Chen, R., Yano, K., Masaki, K. H., Grove, J. S., Donlon, T. A., Willcox, D. C., & Curb, J. D. (2006). Midlife risk factors and healthy survival in men. *The Journal of the American Medical Association, 296*(19), 2343–2350. https://doi.org/10.1001/jama.296.19.2343

Wrosch, C., Scheier, M. F., Miller, G. E., Schulz, R., & Carver, C. S. (2003). Adaptive self-regulation of unattainable goals: Goal disengagement, goal reengagement, and subjective well-being. *Personality and Social Psychology Bulletin, 29*(12), 1494–1508. https://doi.org/10.1177/0146167203256921

Yang, X. (2016). Self-compassion, relationship harmony, versus self-enhancement: Different ways of relating to well-being in Hong Kong Chinese. *Personality and Individual Differences, 89*, 24–27. https://doi.org/10.1016/j.paid.2015.09.006

Yarnell, L. M., Neff, K. D., Davidson, O. A., & Mullarkey, M. (2019). Gender differences in self-compassion: Examining the role of gender role orientation. *Mindfulness, 10*(6), 1136–1152. https://doi.org/10.1007/s12671-018-1066-1

Zessin, U., Dickhäuser, O., & Garbade, S. (2015). The relationship between self-compassion and well-being: A meta-analysis. *Applied Psychology: Health and well-being, 7*(3), 340–364. https://doi.org/10.1111/aphw.12051

Zorn, J. V., Schür, R. R., Boks, M. P., Kahn, R. S., Joëls, M., & Vinkers, C. H. (2017). Cortisol stress reactivity across psychiatric disorders: A systematic review and meta-analysis. *Psychoneuroendocrinology, 77*, 25–36. https://doi.org/10.1016/j.psyneuen.2016.11.036

8 Self-Compassion Across Cultures

Kohki Arimitsu

Introduction

Compassion is a response to others' suffering that involves attunement, empathy, and motivation to relieve that suffering and has been variously defined as an emotion, a virtue or value, and a motivation (Goetz et al., 2010; Strauss et al., 2016). Self-compassion refers to directing these same qualities toward oneself and is characterized by a balanced acceptance of emotions in times of failure or difficulty, recognizing that the experience is shared by others, and channeling kind feelings toward oneself (Neff, 2003b). From an evolutionary perspective, compassion may be conceived as an emotion or motivation that serves an adaptive function across cultures (Goetz et al., 2010). Studies have revealed that behavioral and physiological responses associated with both giving and receiving compassion are common among people from various regions (Gilbert, 2015; Goetz et al., 2010). For example, a gentle touch on the cheek or shoulder of a person in need is a common way of conveying kindness among people in many countries. Similarly, evidence documenting neurophysiological profiles of *self*-compassion suggests that there are biological systems that underpin this phenomenon across cultures (Kim et al., 2020). On the other hand, emotional experiences and their functions are influenced by culture, as the values and social influences that are shared by people living in a certain region influence—among other things—how people relate to themselves and each other.

Neff (2003a) developed the Self-Compassion Scale (SCS), a measure comprising three subscales that measure compassionate self-responding (CS) (self-kindness, common humanity, and mindfulness) and three that measure uncompassionate self-responding (UCS) (self-criticism, isolation, and overidentification). Since their development, several translations of the SCS, its short form (Raes et al., 2011), and youth form have been validated, including translations in Arabic (Alabdulaziz et al., 2020), Chinese, Farsi (Nazari et al., 2022), French (Kotsou and Leys, 2016), German (Hupfeld & Ruffieux, 2011), Greek (Mantzios et al., 2013), Indonesian (Darmawan et al., 2020), Italian (Petrocchi et al., 2014; Veneziani et al., 2017), Japanese (Arimitsu, 2014), Korean (김경의 et al., 2008), Malay (Khatib et al., 2021), Sinhalese (deZoysa et al., 2021), Slovak (Halamová et al., 2018), Slovenian (Uršič et al., 2019), Spanish (Garcia-Campayo et al., 2014), Taiwanese (Chen & Chen, 2019), and Turkish (Deniz et al., 2008, 2022).

An extensive body of research indicates that self-compassion is a modifiable trait and state variable that is reliably associated with adaptive

K. Arimitsu (✉)
Kwansei Gakuin University, Nishinomiya, Japan
e-mail: arimitsu@kwansei.ac.jp

A. Finlay-Jones et al. (eds.), *Handbook of Self-Compassion*, Mindfulness in Behavioral Health,
https://doi.org/10.1007/978-3-031-22348-8_8

physical, psychological, and relational health outcomes (Ewert et al., 2021; MacBeth & Gumley, 2012; Zessin et al., 2015). However, as with most psychological research, most of these studies have been conducted in the United States. This therefore raises the question of whether the construct of self-compassion is understood the same way across different cultures, whether the factor structure of the SCS is culturally influenced, and whether mean scores on the SCS vary according to respondents' cultural background. Studies have found cross-cultural differences in the factor structure of the SCS (Neff et al., 2019; Tóth-Király & Neff, 2020), suggesting that the construct may be experienced or perceived differently across cultures (Montero-Marin et al., 2018). This chapter examines whether cultural factors may impact not only the mean values and factor structure of the SCS but also the relationship between self-compassion and various well-being outcomes. Specifically, this chapter reviews cross-cultural studies to determine whether self-compassion could be affected by the cultural view of the self.

Factor Structure of the SCS Across Cultures

In her original paper describing the factor structure of the SCS, Neff (2003a, b) found support for a higher-order one-factor model, which places one general "self-compassion" factor above the six subscales. This provided a basis for calculating a total self-compassion score (this is done by reverse scoring all self-judgment, isolation, and overidentification items and taking a grand mean of all six subscale means). In a follow-up with four distinct samples, Neff et al. (2017) subsequently found that while the use of a total scale score was justified, this was better explained by a bifactor model, in which each item loads on to both a general factor (i.e., self-compassion) as well as a "group" factor (i.e., one of the six subscales). Neff et al. (2019) then conducted a study of the SCS using 20 samples from 16 countries and 14 languages using both Confirmatory Factory Analysis (CFA) and Exploratory Structural Equation Modeling (ESEM). They tested five different models, including a single-bifactor model, with one general self-compassion factor and six group factors, and a two-bifactor model. The two-bifactor model had two correlated general factors, representing compassionate self-responding (CS) and uncompassionate self-responding (UCS) each with three group factors. Fit statistics using ESEM were excellent for a six-factor correlated model and both bifactor models, although factor loadings indicated that the two general factors were not well specified. Accordingly, the authors recommended the use of the SCS to measure six subscale scores or a total score, but not the CS and UCS factors. Support for the six-factor and bifactor models has been found across countries including Argentina (Cababie & Etchezahar, 2022), France (Kotsou & Leys, 2016), Italy (Petrocchi et al., 2014), and ethnic groups, including African Americans (Zhang et al., 2019).

Measurement Invariance of the SCS

Insight into the generalizability of the SCS across cultures has been provided by Tóth-Király and Neff (2020), who used bifactor Exploratory Structural Equation Modeling (ESEM) to examine measurement invariance of the SCS across groups based on language, population type (student, community, clinical, mixed), gender, and age. Measurement invariance refers to whether the same construct is being measured across groups. Tóth-Király and Neff (2020) tested six dimensions of measurement invariance, using a total of 18 samples collected from 15 countries, representing 12 language groups. They found support for strong invariance across these different language groups, demonstrating that people from different linguistic backgrounds, such as English, Spanish, German, and Greek, conceptualize self-compassion in a similar way. This also suggests that the SCS can be expected to function similarly across linguistic groups, i.e., that associations between self-compassion and outcomes such as depression and well-being will be similar. It should be noted, however, that the sample did not include Chinese and Japanese individuals.

Other studies have explored whether relationships between self-compassion and health outcomes may vary according to ethnicity or culture. Some evidence for cross-cultural differences in the strengths of the association between self-compassion and health outcomes has been found. For example, Arimitsu et al. (2019) demonstrated that self-compassion was more likely to enhance hedonic well-being in individualistic cultures than in collectivistic cultures and that compassion for others was associated with enhanced eudaemonic well-being in collectivistic cultures. In their study of college students, Boyraz et al. (2020) reported that ethnicity moderated the link between self-criticism and perceived health, with negative associations observed between self-criticism and perceived health among Hispanic/Latinx and European American participants, but not among Asian American participants. Interestingly, they also found that after adjusting for self-criticism, self-compassion was positively associated with perceived health among Asian Americans and European Americans but not Hispanic/Latinx participants; further, this relationship was stronger for Asian Americans. However, in a meta-analysis of 168 studies across 27 unique cultures—the largest synthesis of self-compassion studies to date—Chio et al. (2021) found that there were no cross-cultural differences in the strength of the relationships between self-compassion and well-being or psychological distress.

Mean Differences in SCS Across Cultures

In an early study, Neff et al. (2008) compared mean scores on the SCS among participants in Thailand, Taiwan, and the United States. They found significant differences in self-compassion scores: Thai participants had the highest scores, Taiwanese participants had the lowest scores, and Americans were somewhere in between. Additionally, as part of their 2020 study, Tóth-Király and Neff (2020) examined mean differences in self-compassion across different linguistic groups within different populations (students, community members, and clinical samples). Among students, Koreans were found to have the highest self-compassion scores, followed by Iranians, with lower scores found among the Canadian, American, and Norwegian groups. As for community members, Spanish, Italian, Hungarian, and Brazilian samples scored high, while Australian, American, and German samples scored lower and Greeks and British had the lowest scores. In contrast, other studies have found no statistically significant differences in total self-compassion scores across countries. For example, there was no difference between students in China and the United States (Birkett, 2013) or between HIV patients in Canada, China, Namibia, Puerto Rico, and the United States (Kemppainen et al., 2013). However, the aforementioned samples comprised a small number of students and HIV patients, which may have resulted in sampling bias, reflecting an influence other than culture.

Cultivation of Self-Compassion Across Cultures

Another dimension of understanding self-compassion across cultures relates to whether and how it can be cultivated with different cultural or ethnic groups. Self-compassion intervention studies have been conducted in several countries, including China (Huang et al., 2021; Guo et al., 2020), Slovakia (Halamova et al., 2020), Japan (Arimitsu, 2016), and Iran (Rezapour-Mirsaleh et al., 2021). While many of these studies use novel intervention protocols, studies of standardized self-compassion training protocols, such as the Mindful Self-Compassion (MSC) program, provide the opportunity to compare outcomes across cultures. Pilot studies in China of both online (Yeung et al., 2021) and face-to-face versions of MSC (Finlay-Jones et al., 2017) found comparable results to trial outcomes reported in Western cultures (e.g., Neff & Germer, 2013). Furthermore, a meta-analysis of studies aimed at improving mental health by increasing self-compassion has revealed moderate effects on depression, anxiety, and well-being (Kirby et al., 2017). It includes findings not only from the United States and Western countries but also from East Asia, including Japan and China. These studies very clearly show that self-compassion is amenable to cultivation across cultures and that effective cultivation is associated with a range of health and social benefits.

Cultural Differences and Their Influence on Self-Compassion

In sum, research to date suggests that while in many cases the factor structure of the SCS is similar across cultures, and measurement invariance has been demonstrated, this is not a universal finding. Further various studies have shown that there are cultural differences in the SCS means, although self-compassion interventions appear feasible and acceptable across cultures and are associated with similar benefits in Eastern and Western countries. To gain further insight into these findings and consider why self-compassion may or may not vary across cultures, several dimensions of cultural influence must be explored. For example, a robust body of work has explored the ways in which Eastern and Western cultures differ along dimensions of affective expression, self-construal, and beliefs about the world, which are in turn influenced by several factors, including social norms, values, and religious beliefs. In the following section, these dimensions are discussed and research exploring how they intersect with self-compassion is explored.

Self-Construal and Dialecticism

Self-construal theory posits that cultural differences in affect, cognitions, and behavior are influenced by the degree to which a person considers themselves fundamentally connected to or separate from others (Markus & Kitayama, 1991). According to this view, those in *individualistic* cultures—those which focus on the individual rights and needs of each person—are more likely to endorse an *independent* self-construal, in which oneself is perceived as relatively separate from others. Typically, such self-construal is considered characteristic of Western countries. This is contrasted with *collectivistic* cultures, which emphasize the needs of the community, and in which the *interdependent* view of the self prevails. Interdependent self-construal emphasizes the inseparability of self and others and are considered characteristic of Asian countries. There is some evidence that those in collectivistic cultures might be more likely to endorse the "common humanity" facet of self-compassion (Akin and Eroglu, 2013). This has implications for self-compassion, in part because in Western cultures, one of the primary misgivings about self-compassion is that it will undermine motivation (Robinson et al., 2016). In contrast, self-criticism, on the other hand, is often considered necessary to maintain motivation and personal standards, despite self-criticism increasing vulnerability to mental health problems (Schanche, 2013). In Eastern countries, self-criticism may also be viewed positively, but for different reasons, namely, for its role in helping to promote social harmony and maintain positive relationships (Yamaguchi et al., 2014). Interestingly, research has suggested that stronger endorsements of *either* independent or collectivist construals are associated with greater self-criticism, depending on the dominant culture (Yamaguchi et al., 2014). Yamaguchi et al. (2014) found that US participants who reported higher levels of independent self-construal reported more self-criticism, while in Japan, those who reported more interdependent self-construal were also more likely to criticize themselves.

There is some suggestion in the literature that the different functions of self-criticism across Eastern and Western countries may have different implications for mental health. For example, there is evidence demonstrating that in collectivist contexts, interdependent self-construal reduces the detrimental impact of self-criticism on mental health problems (Aruta et al., 2021). Aruta et al. (2021) proposed that when self-criticism occurs in the context of interdependence, the benefits gained by fulfilling social norms and preserving relationships counter any negative impacts of self-criticism, thereby promoting better mental health outcomes. In more independent countries, it might be assumed that criticizing oneself in order to maintain motivation to achieve personal goals does not serve this same affiliative purpose; moreover, self-criticism for competitive motives may increase separation from others and increase vulnerability to depression (Gilbert & Woodyatt, 2017). However, the adaptive function of self-criticism in Eastern cultures may depend on the degree to which an individual feels aligned with cultural values (Aruta et al., 2021). Further, when self-criticism is harsh, and when it co-occurs with feelings of isolation and overidentification, it is likely to be detrimental for mental health, regardless of one's cultural background. Additionally, recent findings from Boyraz et al. (2020) suggest that the impact of culture and ethnicity on the relationships between self-compassion, self-criticism, and health outcomes are more complex and cannot be explained by single measures of cultural value or beliefs.

Dialectical Thinking Dialectical thinking refers to a technique of accepting and integrating two seemingly contradictory and opposing things to create a better idea. The tendency toward dialectical thinking and feeling varies across cultures and may influence the experience of compassion (Chio et al., 2021). People in dialectical cultures tend to experience positive and negative emotions simultaneously because they accept the ambivalence of things (Schimmack et al., 2002). For example, Japanese people are reported to have a dialectical emotional style in which they experience both positive and negative emotions with moderate frequency compared to Americans (Miyamoto & Ryff, 2011). Moreover, people with a moderate dialectical emotional style were found to experience fewer physical symptoms in Japan than in the United States (Miyamoto & Ryff, 2011). These results suggest that a balanced experience of emotions, rather than experiencing more positive and fewer negative emotions, may lead to better well-being among East Asians.

Applying the dialectical thinking to self-compassion, East Asian cultures may tend to have ambivalent emotional experiences, resulting in experiencing equal amounts of self-kindness and self-criticism simultaneously, albeit at more moderate levels than Western cultures. This response pattern could be quite different from Western non-dialectical cultures. If the pattern is evident, it would also be necessary to consider that the factor structure of self-compassion and its relationship with well-being may differ depending on the differences in dialectical thinking among cultures. Chio et al.'s (2021) meta-analysis of data from 27 cultures examined whether dialectical thinking might impact the correlation between CS and UCS components of the SCS. The results suggested that across cultures, there were moderate associations between CS and UCS, but dialectical culture moderated the correlation. In other words, correlations between conflicting constructs, such as self-kindness and self-criticism, were found to be lower in cultures that endorsed more dialectical thinking. However, as the SCS generalizes across different events, the extent to which it can provide insight into whether CS and UCS occur simultaneously is limited. Accordingly, it is important to test this hypothesis using the *state* rather than *trait* Self-Compassion Scale.

A cross-cultural study examined the effects of the CS and UCS on well-being and psychological distress among students in Hong Kong and the United States (Fung et al., 2021). The results revealed that both were associated with well-being and psychological distress only among students in Hong Kong. For American students, UCS were related to well-being, depression, and anxiety, while CS were not related to depression and anxiety. This result is consistent with research on American students and adults (Brenner et al., 2018). Fung et al. (2021) discussed that these cultural differences were found because of the tendency for individuals from collectivist cultures toward dialectical thinking. They proposed that a greater ability and experience to possess seemingly contradictory statements and emotions at the same time led to both UCS and CS uniquely affecting well-being as well as depression and anxiety. They also argued that self-compassion might mitigate the impact of negative traits, such as self-criticism, on mental health issues because of the correlation between CS and UCS. Self-compassion is, however, characterized by an increase in CS and a decrease in UCS, such as being kind to oneself instead of being self-critical. Again, this line of inquiry requires the use of the state Self-Compassion Scale (Neff et al., 2021) to determine profiles of state CS and UCS across cultures and investigate their relationships with health outcomes. Further, it should be noted that even in cultures like Japan, who tend to report higher levels of dialectical thinking, experimental studies have found that CS and UCS of the SCS changed in tandem as a unitary construct following self-compassion intervention (Arimitsu, 2016).

Affective Expression

Another way that culture may influence self-compassion is via the socialization of emotion. For example, each culture has a variety of valued daily-life emotions (Mesquita, 2003), ideal emotions (Tsai, 2007), and behaviors. In a study comparing ideal emotions between the United States and China (Tsai, 2007), it was found that people in the United States tended to idealize high-arousal, positive emotional states (e.g., excitement, enthusiasm) while people in China tended to idealize low-arousal, positive emotional states (e.g., calmness, serenity). These differences in self-views indicate that in collectivistic cultures, other-focused emotions such as friendliness and guilt are likely to be more salient. Conversely, in individualistic cultures, affective responses that emphasize separateness from others—such as pride and anger—may be more salient (Kitayama et al., 2006). This suggests that adaptive emotions themselves differ due to differences in cultural views of the self and relationship with others. Accordingly, the expression of self-compassion, and its intensity and relationship with well-being, might differ between independent and interdependent cultures.

The cultural value or emphasis placed on affective and motivational states such as compassion likely influences their linguistic and behavioral expression in that culture. For example, if a culture emphasizes compassion, it should be relatively easier for helping behaviors to take root in that culture (Koopmann-Holm & Tsai, 2017). Furthermore, the vocabulary related to compassion should be more extensive and more distinct from other words related to emotion. Shaver et al. (1992) found differences in the English, Italian, and Chinese emotion lexicon for compassion (Shaver et al., 1992), suggesting that the way in which compassion is understood and expressed may differ depending on cultural background. Qualitative research to understand the lexicon associated with self-compassion across cultures is an important direction for future research.

There are two types of psychological well-being—hedonic and eudaemonic—which have been shown to vary in degree and relationship to compassion across cultures. High levels of life satisfaction and positive emotions and the absence of negative emotions are referred to as hedonic well-being, which is experienced when one has achieved pleasure and avoided pain. Arimitsu et al. (2019) conducted a comparative study of the relationship between compassion for self and others and hedonic and eudaemonic well-being and psychopathology between the United States, which has an independent view of the self, and Japan, with an interdependent view of the self. They hypothesized that self-compassion, which motivates people to move forward despite failures and difficulties, would maintain and improve hedonic well-being, such as positive emotions and life satisfaction, in independent cultures more than in interdependent cultures. The study revealed that self-compassion was a factor in improving positive affect and life satisfaction in both countries, but the explained variance for positive affect was higher in the United States than in Japan.

On the other hand, in interdependent cultures, it is adaptive to be able to meet the wishes and expectations of those around oneself to the greatest extent possible, and it is necessary to acquire the flexibility to change one's thoughts and actions appropriately depending on others. Therefore, eudemonic well-being is more likely to be enhanced when people have a high level of compassion for others. Among the measures of eudemonic well-being, interpersonal happiness is one that may be enhanced by compassion for others in interdependent cultures. Hitokoto and Uchida (2015) developed a scale that measures interdependent happiness, which serves as a comprehensive assessment of whether one has harmonious relationships with others, makes others happy, is on par with others, and is in a peaceful emotional state. The study found that in Japan, being in harmony with others, rather than one's own superiority over others, leads to life satisfaction and positive emotions. The moderation effect of culture on interdependent happiness was found, with higher compassion associated with higher interdependent happiness only in Japan, an interdependent culture.

Culture and self-construals may also impact the relationship between compassion for others and negative emotions related to others (Arimitsu et al., 2019). Previous cross-cultural studies showed that social anxiety disorder (SAD) is characterized by self-focused attention that increases not only the belief that one is behaving in a socially inappropriate manner but also negative thoughts and feelings about oneself. However, Taijin-Kyofu-Sho (TKS) features the other-focused cognition of fearing that one's inappropriate behavior will make others uncomfortable. Patients with TKS fear offending others by emitting offensive odors, blushing, staring inappropriately, and presenting an improper facial expression or physical deformity (Hofmann & Hinton, 2014). TKS symptoms tend to be more prevalent in interdependent cultures than in independent cultures because the former tends to be more attentive to the feelings of others (Norasakkunkit et al., 2012). Since people with SAD and TKS focus their attention on the self and others, Arimitsu et al. (2019) proposed that their relationship with compassion for self and others may differ, and the relationship might also differ depending on the cultural view of self. In other words, it was predicted that in a self-focused, independent culture, self-compassion would be more robustly associated with lower SAD symptoms than in an interdependent culture, while in an interdependent, other-focused culture, compassion for others would be more likely to be associated with lower TKS symptoms than in an independent culture. The results revealed no evidence of the predicted culture moderation effects. Self-compassion was associated with lower SAD symptoms, while compassion for others was associated with lower TKS symptoms across cultures, but there was no effect of cultural self-construals on the relationship between self-compassion and psychiatric symptoms. It should be noted, however, that the study included only two different cultural groups—Japan and the United States—and had limited indicators for well-being and psychopathological symptoms.

Values and Social Norms

Hofstede et al.'s (2001, 2010) Cultural Dimensions Theory proposes six dimensions of cultural values that are thought to influence behavior. In addition to individualism-collectivism, these dimensions are masculinity-femininity (i.e., the extent to which one is focused on tasks and competitive achievements versus people and relationships), power distance (the degree to which unequal power distribution in a society is accepted by the less powerful), long-term orientation (the extent to which individuals prepare for the future), uncertainty avoidance (the degree of comfort that society members have with ambiguity), and indulgence-restraint (the extent to which individuals endorse hedonism and impulsive behavior). Montero-Marin et al. (2018) examined the relationship between the six dimensions of Hofstede's cultural values and CS and UCS of the SCS in 11 countries and found that the higher the cultural value of long-term orientation, the higher the CS scores. This may be because both self-compassion and long-term orientation reflect a higher level of self-regulation (Biber & Ellis, 2019). Similarly, indulgence—the opposite of self-control—was also associated with lower CS scores. Furthermore, the higher the cultural value of individualism, the higher the UCS scores. Furthermore, in the Korean and Japanese samples, where long-term orientation, uncertainty avoidance, and self-control were high, the correlation between CS and UCS of the SCS was low, the factor loadings of CS items were high, and the factor loadings of the UCS items were low.

A cross-cultural study examined self-compassion with work engagement and mental health problems among Dutch and Japanese workers (Kotera et al., 2020). The results revealed that mental health problems were significantly inversely associated with self-compassion among Japanese workers and with work engagement among Dutch workers. With reference to Hofstede's cultural value dimensions, Japan has a higher tendency toward power inequality, masculinity, and uncertainty avoidance than does the Netherlands, while the Netherlands has a higher tendency toward individualism and indulgence than does Japan. Although the comparison between these two countries is minimal, it suggests that self-compassion has no effect on mental health problems among Dutch people. Other studies, however, revealed contrary findings of work engagement predicting the onset of depression in Japan (Imamura et al., 2016) and self-compassion being associated with lower depression in the Netherlands (Kreemers et al., 2020). Further studies are needed to determine whether these cultural differences are consistently observed not only in such cross-sectional studies but also experimental ones.

Other Factors

Most of the cultural differences reviewed in this chapter were related to individualism-collectivism and cultural self-view. There are, however, other factors and psychological indicators that may impact self-compassion. One such cultural factor is *simpatia* or the tendency to avoid conflict and outright negativity in favor of warm, positive social interactions. This is most prevalent in Latin countries, in which individuals are more likely to help strangers than they are in countries without such traditions of *simpatia* (Levine et al., 2001). The results of a field experiment in 23 major cities around the world revealed large cross-cultural differences in helping behaviors toward strangers, ranging from 93% in Rio de Janeiro to 40% in Kuala Lumpur. A potential direction for future work is to determine whether *simpatia*'s characteristics of warmth and positive relationship building would have a positive impact on self-compassion and whether this varies across cultures.

Limitations and Future Prospects

As the current chapter demonstrates, cultural differences have been examined in the mean and factor structure of SCS, and in the relationship between SCS, well-being, and psychopathology. One of the limitations of this research is the lack of experimental studies. Although several studies have shown that self-compassion could reduce experimentally induced distress, few have investigated this phenomenon cross-culturally using the same protocol in equivalent samples. Large-scale studies and collaborative efforts across countries can support standardization of trial protocols to generate more robust insights into the process and outcomes of self-compassion training across cultures.

In the literature reviewed in this chapter, there is suggestion that the effects of self-criticism and self-compassion on emotions and motivation may vary across collectivist and individualistic countries. An experiment might clarify whether individualism increases the effect of CS on motivation and well-being because of the desire for self-actualization and individual uniqueness or whether collectivism weakens the effect of CS because of their belief of self-improvement effect of self-criticism. It may also be interesting to test whether the effects of self-compassion training are impacted by "priming" participants with a stimulus that reinforces the role of self-compassion within their cultural value system (e.g., priming participants from individualist cultures by telling them that self-compassion will increase their motivation for personal goals or telling participants from collectivist countries that self-compassion will increase group harmony and interpersonal relationships).

Future intervention studies and qualitative work could also focus on whether resistance to self-compassion is influenced by culture and the implications of this for intervention engagement. For example, depending on the religious background of the participants, it may be offensive to emphasize that self-compassion and its meditation techniques are derived from Buddhism. Since compassion is an important part of the doctrines of not only Buddhism but also Christianity, Hinduism, and many other religions, it may not be necessary to inform participants about Buddhist doctrines. Moreover, participants from Buddhist countries who are well versed with Buddhism may feel uncomfortable practicing something slightly different from what they are familiar with in their own culture. For example, one of the practices of Zen Buddhism in Japan is *zazen*, which is different from mindfulness meditation in both doctrine and method. In contrast to some "mental training" meditation approaches, Zazen focuses on "simply" sitting with awareness of body and mind, without the explicit goal of improving mental health or concentration. For participants who are familiar with such traditional religious practices, being told that a meditation method or program has been scientifically proven to be effective may seem awkward. Such instances of knowledge getting in the way of practice are likely to occur in all countries, cultures, and religions as the concepts of mindfulness and compassion and their meditation methods spread.

In order to account for the cultural background of the participants, it should be common practice to match the cultural background of the facilitator as well as the language used or to adapt the method to the culture. Mindfulness-based interventions applied to Hispanic populations have demonstrated that cultural adaptation can be used to improve engagement and implementation without compromising the quality of the research (Castellanos et al., 2020). Programs to improve self-compassion are currently becoming widespread globally. It will be necessary to gather extensive practical knowledge to apply the various findings across cultures and to clarify what kind of innovations are needed in self-compassion interventions.

The second limitation is that most cross-cultural comparisons are based on the self-construal dimensions of collectivism-individualism. Further, cross-cultural studies tend to make comparisons between cultures at the extremes of only one dimension, without accounting for the impact of other dimensions. Given that cultural dimensions are multifaceted and complex, larger samples are required to enable multivariate analyses. Such studies could test whether the influence of self-compassion on emotions and motivation across cultures depends on situational factors, such as whether success or failure is experienced in an in-group or in interactions with an out-group.

The third limitation is the translated version of the SCS; although the one bifactor plus six-factor model of the SCS has a good fit across cultures, it inevitably has exceptions. The results of a meta-analysis (Chio et al., 2021) revealed that cultural differences in the correlations between CS and UCS of the SCS are influenced by dialectical thinking tendency, which allows for the coexistence of each aspect in dialectical cultures. To test this hypothesis, researchers should measure the dialectical tendency across cultures and test the fit of the model.

The fourth limitation is the lack of studies that use behavioral indicators. Although this is related to the lack of experimental studies, it is better to examine how behavior in certain situations differs among cultures to avoid verbal bias. For example, a cultural difference in the form of more sharing behaviors among Asian children has been reported (e.g., Stewart & McBride-Chang, 2000). While individualistic cultures seek self-actualization and individual uniqueness, in collectivistic cultures, parents train their children from an early age to view shared behavior as part of family approval and identity and to be seen as "givers" by others. For self-compassionate behaviors, Gilbert et al. (2017) developed a scale consisting of items such as "Pay attention to what might be helpful to you" and "Create a state of mind that is supportive, helpful, and encouraging to you." If individualistic cultures are more prone to self-actualization and self-enhancement, then the propensity for these behaviors may also be higher in these cultures compared to collectivistic cultures.

The fifth limitation is that the measures of well-being used in previous studies are few. Self-compassion has been examined in relation to psychological well-being, life purpose, and self-acceptance, in addition to hedonic well-being, such as positive emotions and life satisfaction (Zessin et al., 2015). However, few cross-cultural studies have included eudaemonic well-being (e.g., Arimitsu et al., 2019), and further investigation is warranted. In addition, because collectivistic cultures aim to promote interpersonal harmony, interpersonal satisfaction between parents and children or couples might be appropriate indicators. Self-compassion has also been found to lead to maintenance of romantic relationships (Jacobson et al., 2018), and future examination of how self-compassion supports interpersonal functioning across cultures is needed.

Conclusion

In this chapter, cultural differences in self-compassion were reviewed, focusing on differences in the mean values, factor structure, and well-being. The factor structure of self-compassion was found to fit a one bifactor plus six-factor model across cultures. The associations between self-compassion and both well-being and psychological distress are also evident across cultures. However, cross-cultural studies suggest that cultural self-construals and values might impact the correlation between CS and UCS of the SCS in interdependent and dialectical cultures. Some studies have also identified the moderating effects of culture, such as the stronger ability of self-compassion to improve well-being in independent cultures. However, a research question remains as to whether the cultural factors might impact the beneficial effects of self-compassion on motivation and well-being. More experimental studies using the state SCS should be conducted across more cultures in the future.

References

Akin, A., & Eroglu, Y. (2013). Self-compassion and relational-interdependent self-construal. *Studia Psychologica, 55*(2), 111–121. https://doi.org/10.21909/sp.2013.02.629

Alabdulaziz, H., Alquwez, N., Almazan, J. U., Albougami, A., Alshammari, F., & Cruz, J. P. (2020). The Self-Compassion Scale Arabic version for baccalaureate nursing students: A validation study. *Nurse Education Today, 89*, 104420–104420. https://doi.org/10.1016/j.nedt.2020.104420

Arimitsu, K. (2014). Development and validation of the Japanese version of the Self-Compassion Scale. *The Japanese Journal of Psychology, 85*, 50–59. https://doi.org/10.4992/jjpsy.85.50

Arimitsu, K. (2016). The effects of a program to enhance self-compassion in Japanese individuals: A randomized controlled pilot study. *The Journal of Positive Psychology, 11*(6), 559–571. https://doi.org/10.1080/17439760.2016.1152593

Arimitsu, K., Hitokoto, H., Kind, S., & Hofmann, S. G. (2019). Differences in compassion, well-being, and social anxiety between Japan and the USA. *Mindfulness, 10*(5), 854–862. https://doi.org/10.1007/s12671-018-1045-6

Aruta, J. J. B. R., Antazo, B. G., & Paceño, J. L. (2021). Self-stigma is associated with depression and anxiety in a collectivistic context: The adaptive cultural function of self-criticism. *The Journal of Psychology, 155*(2), 238–256. https://doi.org/10.1080/00223980.2021.1876620

Biber, D. D., & Ellis, R. (2019). The effect of self-compassion on the self-regulation of health behaviors: A systematic review. *Journal of Health Psychology, 24*(14), 2060–2071. https://doi.org/10.1177/1359105317713361

Birkett, M. (2013). Self-compassion and empathy across cultures: Comparison of young adults in China and the United States. *International Journal of Research Studies in Psychology, 3*, 25–34. https://doi.org/10.5861/ijrsp.2013.551

Boyraz, G., Legros, D. N., & Berger, W. B. (2020). Self-criticism, self-compassion, and perceived health: Moderating effect of ethnicity. *Journal of General Psychology, 148*, 149–167. https://doi.org/10.1080/00221309.2020.1746232

Brenner, R. E., Vogel, D. L., Lannin, D. G., Engel, K. E., Seidman, A. J., & Heath, P. J. (2018). Do self-compassion and self-coldness distinctly relate to distress and well-being? A theoretical model of self-relating. *Journal of Counseling Psychology, 65*, 346–357.

Cababie, M., & Etchezahar, E. (2022). Adaptation and validation of the Self-Compassion Scale (SCS) in an argentine context. *Current Psychology*. https://doi.org/10.1007/s12144-021-02505-w

Castellanos, R., Spinel, M. Y., Phan, V., Orengo-Aguayo, R. E., Humphreys, K., & Flory, K. (2020). A systematic review and meta-analysis of cultural adaptations of mindfulness-based interventions for Hispanic populations. *Mindfulness, 11*(2), 317–332. https://doi.org/10.1007/s12671-019-01210-x

Chen, Y.-J., & Chen, S.-H. (2019). The Taiwanese version of the Self-Compassion Scale: Psychometric properties, implications for psychological health and self-compassion across multiple generations. *Chinese Journal of Psychology, 61*(1). https://doi.org/10.6129/CJP.201903_61(1).0003

Chio, F. H., Mak, W., & Yu, B. C. (2021). Meta-analytic review on the differential effects of self-compassion components on well-being and psychological distress: The moderating role of dialecticism on self-compassion. *Clinical Psychology Review, 85*, 101986. https://doi.org/10.1016/j.cpr.2021.101986

Darmawan, M., Taufik Akbar Rizqi, Y., Annisa Zaenab Nur, F., Amanda Meuthia, R., & Giofanny Filadelfia, L. (2020). Properti psikometri Self-Compassion Scale versi Indonesia: Struktur faktor, reliabilitas, dan validitas kriteria. *Persona, 9*(2), 189–208. https://doi.org/10.30996/persona.v9i2.3944

Deniz, M. E., Kesici, Ş., & Sümer, A. S. (2008). The validity and reliability of the Turkish version of the Self-Compassion Scale. *Social Behavior and Personality, 36*(9), 1151–1160. https://doi.org/10.2224/sbp.2008.36.9.1151

Deniz, M. E., Satici, S. A., Doenyas, C., & Caglar, A. (2022). Self-Compassion Scale for Youth: Turkish adaptation and exploration of the relationship with resilience, depression, and well-being. *Child Indicators Research, 15*, 1255–1267. https://doi.org/10.1007/s12187-022-09915-7

deZoysa, P., Kumar, S., Amarasuriya, S. D., & Mendis, N. S. R. (2021). Cultural validation of the Self-Compassion Scale and the Subjective Happiness Scale and the influence of gender on self-compassion and subjective happiness in a Sri Lankan undergraduate population. *Asian Journal of Social Psychology, 25*(3), 421. https://doi.org/10.1111/ajsp.12505

Ewert, C., Vater, A., & Schröder-Abé, M. (2021). Self-compassion and coping: A meta-analysis. *Mindfulness, 12*(5), 1063–1077. https://doi.org/10.1007/s12671-020-01563-8

Finlay-Jones, A., Xie, Q., Huang, X., Ma, X., & Guo, X. (2017). A pilot study of the 8-week mindful self-compassion training program in a Chinese community sample. *Mindfulness, 9*(1), 993–1002. https://doi.org/10.1007/s12671-017-0838-3

Fung, J., Chen, G., Kim, J., & Tracy, L. (2021). The relations between self-compassion, self-coldness, and psychological functioning among North American and Hong Kong college students. *Mindfulness, 2*, 2161–2172. https://doi.org/10.1007/s12671-021-01670-0

Garcia-Campayo, J., Navarro-Gil, M., Andrés, E., Montero-Marin, J., López-Artal, L., & Demarzo, M. M. P. (2014). Validation of the Spanish versions of the long (26 items) and short (12 items) forms of the Self-Compassion Scale (SCS). *Health and Quality of Life Outcomes, 12*(1), 4–4. https://doi.org/10.1186/1477-7525-12-4

Gilbert, P. (2015). The evolution and social dynamics of compassion. *Social and Personality Psychology Compass, 9*(6), 239–254. https://doi.org/10.1111/spc3.12176

Gilbert, P., & Woodyatt, L. (2017). An evolutionary approach to shame-based self-criticism, self-forgiveness, and compassion. In L. Woodyatt, E. L. Worthington, & M. Wenzel (Eds.), *Handbook of the psychology of self-forgiveness*. Springer. https://doi.org/10.1007/978-3-319-60573-9

Gilbert, P., Catarino, F., Duarte, C., Matos, M., Kolts, R. L., Stubbs, J., Ceresatto, L., Duarte, J., Pinto-Gouveia, J., & Basran, J. (2017). The development of compassionate engagement and action scales for self and others. *Journal of Compassionate Health Care, 4*, 1–24. https://doi.org/10.1186/s40639-017-0033-3

Goetz, J. L., Keltner, D., & Simon-Thomas, E. R. (2010). Compassion: An evolutionary analysis and empirical review. *Psychological Bulletin, 136*(3), 351–374. https://doi.org/10.1037/a0018807

Guo, L., Zhang, J., Mu, L., & Ye, Z. (2020). Preventing postpartum depression with Mindful Self-Compassion intervention: A randomized control study. *The Journal of Nervous and Mental Disease, 208*(2), 101–107. https://doi.org/10.1097/NMD.0000000000001096

Halamová, J., Kanovský, M., & Pacúchová, M. (2018). Self-Compassion Scale: IRT psychometric analysis, validation, and factor structure – Slovak translation. *Psychologica Belgica, 57*(4), 190–209. https://doi.org/10.5334/pb.398

Halamova, J., Kanovsky, M., Jakubcova, K., & Kupeli, N. (2020). Short online compassionate intervention based on Mindful Self-Compassion program. *Československá Psychologie, 64*(2), 236–250.

Hitokoto, H., & Uchida, Y. (2015). Interdependent happiness: Theoretical importance and measurement validity. *Journal of Happiness Studies, 16*(1), 211–239. https://doi.org/10.1007/s10902-014-9505-8

Hofmann, S., & Hinton, D. (2014). Cross-cultural aspects of anxiety disorders. *Current Psychiatry Reports, 16*, 450. https://doi.org/10.1007/s11920-014-0450-3

Hofstede, G. (2001). *Culture's consequences: Comparing values, behaviors, institutions, and organizations across nations*. (2nd ed.). Thousand Oaks, CA: Sage Publications.

Hofstede, G., Hofstede, G. J., & Minkov, M. (2010). *Cultures and organizations: Software of the mind* (Revised and Expanded 3rd ed.). McGraw-Hill.

Huang, J., Lin, K., Fan, L., Qiao, S., & Wang, Y. (2021). The effects of a self-compassion intervention on future-oriented coping and psychological well-being: A randomized controlled trial in Chinese college students. *Mindfulness, 12*(6), 1451–1458. https://doi.org/10.1007/s12671-021-01614-8

Hupfeld, J., & Ruffieux, N. (2011). Validierung einer deutschen Version der Self-Compassion Scale (SCS-D). *Zeitschrift für Klinische Psychologie und Psychotherapie, 40*(2), 115–123. https://doi.org/10.1026/1616-3443/a000088

Imamura, K., Kawakami, N., Inoue, A., Shimazu, A., Tsutsumi, A., Takahashi, M., & Totsuzaki, T. (2016). Work engagement as a predictor of onset of major depressive episode (MDE) among workers, independent of psychological distress: A 3-year prospective cohort study. *PLoS One, 11*(2), e0148157. https://doi.org/10.1371/journal.pone.0148157

Jacobson, E., Wilson, K., Kurz, A. S., & Kellum, K. (2018). Examining self-compassion in romantic relationships. *Journal of Contextual Behavioral Science, 8*, 69–73. https://doi.org/10.1016/j.jcbs.2018.04.003

Kemppainen, J., Johnson, M. O., Phillips, J. C., Sullivan, K., Corless, I., Reid, P., Iipinge, S., Chaiphibalsarisdi, P., Sefcik, E., Chen, W., Kirksey, K., Voss, J., Rivero-Méndez, M., Tyer-Viola, L., Rose, C. D., Webel, A., Nokes, K., Portillo, C., Holzemer, W., Eller, L., Nicholas, P., Wantland, D., Brion, J., & Beamon, E. (2013). A multinational study of self-compassion and human immunodeficiency virus-related anxiety. *International Nursing Review, 60*(4), 477–486. https://doi.org/10.1111/inr.12056

Khatib, N. A. M., Roseliza-Murni, A., Hoesni, S. M., & Manap, J. (2021). Exploratory factor analysis of the Self-Compassion Scale-Malay Version: Its reliability among adolescents. *International Journal of Academic Research in Business and Social Sciences, 11*(3). https://doi.org/10.6007/IJARBSS/v11-i3/8916

Kim, J. J., Parker, S. L., Doty, J. R., Cunnington, R., Gilbert, P., & Kirby, J. N. (2020). Neurophysiological and behavioural markers of compassion. *Scientific Reports, 10*(1), 6789–6789. https://doi.org/10.1038/s41598-020-63846-3

Kirby, J. N., Tellegen, C. L., & Steindl, S. (2017). A meta-analysis of compassion-based interventions: Current state of knowledge and future directions. *Behavior Therapy, 48*(6), 778–792. https://doi.org/10.1016/j.beth.2017.06.003

Kitayama, S., Mesquita, B., & Karasawa, M. (2006). Cultural affordances and emotional experience: Socially engaging and disengaging emotions in Japan and the United States. *Journal of Personality and Social Psychology, 91*(5), 890–903. https://doi.org/10.1037/0022-3514.91.5.890

Koopmann-Holm, B., & Tsai, J. L. (2017). The cultural shaping of compassion. In E. M. Seppäla, E. Simon-Thomas, S. L. Brown, M. C. Worline, C. D. Cameron, & J. R. Doty (Eds.), *The Oxford handbook of compassion science* (pp. 273–285). Oxford University Press.

Kotera, Y., Van Laethem, M., & Ohshima, R. (2020). Cross-cultural comparison of mental health between Japanese and Dutch workers: Relationships with mental health shame, self-compassion, work engagement and motivation. *Cross Cultural & Strategic Management, 27*(3), 511–530. https://doi.org/10.1108/CCSM-02-2020-0055

Kotsou, I., & Leys, C. (2016). Self-Compassion Scale (SCS): Psychometric properties of the French translation and its relations with psychological well-being, affect and depression. *PLoS One, 11*(4), e0152880–e0152880. https://doi.org/10.1371/journal.pone.0152880

Kreemers, L., Hooft, E. A., Vianen, A. E., & Zilwa, S. C. (2020). Testing a self-compassion intervention among job seekers: Self-compassion beneficially impacts affect through reduced self-criticism. *Frontiers in Psychology, 11*, 1371. https://doi.org/10.3389/fpsyg.2020.01371

Levine, R., Norenzayan, A., & Philbrick, K. (2001). Cross-cultural differences in helping strangers. *Journal of Cross-Cultural Psychology, 32*(5), 543–560. https://doi.org/10.1177/0022022101032005002

MacBeth, A., & Gumley, A. (2012). Exploring compassion: A meta-analysis of the association between self-compassion and psychopathology. *Clinical Psychology Review, 32*(6), 545–552. https://doi.org/10.1016/j.cpr.2012.06.003

Mantzios, M., Wilson, J. C., & Giannou, K. (2013). Psychometric properties of the Greek versions of the self-compassion and mindful attention and awareness scales. *Mindfulness, 6*(1), 123–132. https://doi.org/10.1007/s12671-013-0237-3

Markus, H., & Kitayama, S. (1991). Culture and the self: Implications for cognition, emotion, and motivation. *Psychological Review, 98*(2), 224–253. https://doi.org/10.1037/0033-295X.98.2.224

Mesquita, B. (2003). Emotions as dynamic cultural phenomena. In R. Davidson, H. Goldsmith, & K. R. Scherer (Eds.), *The handbook of the affective sciences* (pp. 871–890). New York: Oxford University Press.

Miyamoto, Y., & Ryff, C. (2011). Cultural differences in the dialectical and non-dialectical emotional styles and their implications for health. *Cognition and Emotion, 25*, 22–39.

Montero-Marin, J., Kuyken, W., Crane, C., Gu, J., Baer, R., Al-Awamleh, A. A., Akutsu, S., Araya-Véliz, C., Ghorbani, N., Chen, Z. J., Kim, M.-S., Mantzios, M., Rolim Dos Santos, D. N., Serramo López, L. C., Teleb, A. A., Watson, P. J., Yamaguchi, A., Yang, E., & García-Campayo, J. (2018). Self-compassion and cultural values: A cross-cultural study of self-compassion using a multitrait-multimethod (MTMM) analytical procedure. *Frontiers in Psychology, 9*, 2638–2638. https://doi.org/10.3389/fpsyg.2018.02638

Nazari, N., Hernández, R. M., Ocaña-Fernandez, Y., & Griffiths, M. D. (2022). Psychometric validation of the Persian self-compassion scale youth version. *Mindfulness, 13*(2), 385–397. https://doi.org/10.1007/s12671-021-01801-7

Neff, K. D. (2003a). The development and validation of a scale to measure self-compassion. *Self and Identity, 2*(3), 223–250. https://doi.org/10.1080/15298860309027

Neff, K. D. (2003b). Self-compassion: An alternative conceptualization of a healthy attitude toward oneself. *Self and Identity, 2*(2), 85–101. https://doi.org/10.1080/15298860309032

Neff, K. D., & Germer, C. K. (2013). A pilot study and randomized controlled trial of the mindful self-compassion program. *Journal of Clinical Psychology, 69*(1), 28–44.

Neff, K. D., Pisitsungkagarn, K., & Hsieh, Y. P. (2008). Self-compassion and self-construal in the United States, Thailand, and Taiwan. *Journal of Cross-Cultural Psychology, 39*(3), 267–285.

Neff, K. D., Whittaker, T. A., & Karl, A. (2017). Examining the factor structure of the Self-Compassion Scale in four distinct populations: Is the use of a total scale score justified? *Journal of Personality Assessment, 99*(6), 596–607. https://doi.org/10.1080/00223891.2016.1269334

Neff, K. D., Tóth-Király, I., Yarnell, L. M., Arimitsu, K., Castilho, P., Ghorbani, N., Guo, H. X., Hirsch, J. K., Hupfeld, J., Hutz, C. S., Kotsou, I., Lee, W. K., Montero-Marin, J., Sirois, F. M., de Souza, L. K., Svendsen, J. L., Wilkinson, R. B., & Mantzios, M. (2019). Examining the factor structure of the Self-Compassion Scale in 20 diverse samples: Support for use of a total score and six subscale scores. *Psychological Assessment, 31*(1), 27–45. https://doi.org/10.1037/pas0000629

Neff, K. D., Tóth-Király, I., Knox, M. C., Kuchar, A., & Davidson, O. (2021). The development and validation of the State Self-Compassion Scale (long and short form). *Mindfulness, 12*(1), 121–140. https://doi.org/10.1007/s12671-020-01505-4

Norasakkunkit, V., Kitayama, S., & Uchida, Y. (2012). Social anxiety and holistic cognition: Self-focused social anxiety in the United States and other-focused social anxiety in Japan. *Journal of Cross-Cultural Psychology, 43*(5), 742–757. https://doi.org/10.1177/0022022111405658

Petrocchi, N., Ottaviani, C., & Couyoumdjian, A. (2014). Dimensionality of self-compassion: Translation and construct validation of the self-compassion scale in an Italian sample. *Journal of Mental Health, 23*(2), 72–77. https://doi.org/10.3109/09638237.2013.841869

Raes, F., Pommier, E., Neff, K. D., & Van Gucht, D. (2011). Construction and factorial validation of a short form of the Self-Compassion Scale. *Clinical Psychology and Psychotherapy, 18*(3), 250–255. https://doi.org/10.1002/cpp.702

Rezapour-Mirsaleh, Y., Shafizadeh, R., Shomali, M., & Sedaghat, R. (2021). Effectiveness of self-compassion intervention on criminal thinking in male prisoners. *International Journal of Offender Therapy and Comparative Criminology, 65*(1), 100–116. https://doi.org/10.1177/0306624X20936192

Robinson, K. J., Mayer, S., Allen, A. B., Terry, M., Chilton, A., & Leary, M. R. (2016). Resisting self-compassion: Why are some people opposed to being kind to themselves? *Self and Identity, 15*(5), 505–524. https://doi.org/10.1080/15298868.2016.1160952

Schanche, E. (2013). The transdiagnostic phenomenon of self-criticism. *Psychotherapy, 50*(3), 316–321. https://doi.org/10.1037/a0032163

Schimmack, U., Oishi, S., & Diener, E. (2002). Cultural influences on the relation between pleasant emotions and unpleasant emotions: Asian dialectic philosophies or individualism-collectivism? *Cognition and Emotion, 16*(2), 705–719. https://doi.org/10.1080/02699930143000590

Shaver, P., Wu, S., & Schwartz, J. (1992). Cross-cultural similarities and differences in emotion and its representation. In M. S. Clark (Ed.), *Review of personality and social psychology* (Vol. 13, pp. 175–212). Sage Publications.

Stewart, S., & McBride-Chang, C. (2000). Influences on children's sharing in a multicultural setting. *Journal of Cross-Cultural Psychology, 31*(3), 333–348. https://doi.org/10.1177/0022022100031003003

Strauss, C., Lever Taylor, B., Gu, J., Kuyken, W., Baer, R., Jones, F., & Cavanagh, K. (2016). What is compassion and how can we measure it? A review of definitions and measures. *Clinical Psychology Review, 47*, 15–27. https://doi.org/10.1016/j.cpr.2016.05.004

Tóth-Király, I., & Neff, K. D. (2020). Is self-compassion universal? Support for the measurement invariance of the Self-Compassion Scale across populations. *Assessment, 28*(1), 169–185. https://doi.org/10.1177/1073191120926232

Tsai, J. (2007). Ideal affect: Cultural causes and behavioral consequences. *Perspectives on Psychological Science, 2*(3), 242–259. https://doi.org/10.1111/j.1745-6916.2007.00043.x

Uršič, N., Kocjančič, D., & Žvelc, G. (2019). Psychometric properties of the Slovenian long and short version of the self-compassion scale. *Psihologija, 52*(2), 107–125. https://doi.org/10.2298/PSI180408029U

Veneziani, C. A., Fuochi, G., & Voci, A. (2017). Self-compassion as a healthy attitude toward the self: Factorial and construct validity in an Italian sample. *Personality and Individual Differences, 119*, 60–68. https://doi.org/10.1016/j.paid.2017.06.028

Yamaguchi, A., Kim, M.-S., & Akutsu, S. (2014). The effects of self-construals, self-criticism, and self-compassion on depressive symptoms. *Personality and Individual Differences, 68*, 65–70. https://doi.org/10.1016/j.paid.2014.03.013

Yeung, A., Xie, Q., Huang, X., Hoeppner, B., Jain, F. A., Tan, E. K., Mai, X., Mischoulon, D., & Guo, X. (2021). Effectiveness of Mindful Self-Compassion training dupported by online peer groups in China: A pilot study. *Alternative Therapies in Health and Medicine*, AT6940. Advance online publication.

Zessin, U., Dickhäuser, O., & Garbade, S. (2015). The relationship between self-compassion and well-being: A meta-analysis. *Applied Psychology: Health and Well-being, 7*(3), 340–364. https://doi.org/10.1111/aphw.12051

Zhang, H., Dong, L., Watson-Singleton, N. N., Tarantino, N., Carr, E. R., Niles-Carnes, L. V., et al. (2019). Psychometric properties of the Self-Compassion Scale (SCS) in an African American clinical sample. *Mindfulness, 10*(7), 1395–1405. https://doi.org/10.1007/s12671-019-01099-6

김경의, 이우경, 채숙희, 이금단, & Yongrae, C. (2008). The validation study of the Korean version of the Self-Compassion Scale. *Korean Journal of Health Psychology, 13*(4), 1023–1044. https://doi.org/10.17315/kjhp.2008.13.4.012

9 Self-Compassion Among Sexual and Gender Minorities: The Importance of Self-Kindness in a Frequently Unkind World

Abra Vigna and Penelope Strauss

Caring for myself is not self-indulgence, it is self-preservation, and that is an act of political warfare.

Audre Lorde

Introduction

The ability to predict the distribution of health, hope, morbidity, and mortality based upon social identity is a clear marker of an unjust society. In most Western nations, people who identify as something other than heterosexual or cisgender are often considered abnormal and predictably experience a disproportionate burden of ill health. For example, although most sexual- and gender-minority (SGM) individuals demonstrate robust mental and physical health (Saewyc, 2011), meta-analyses of population-level data show that suicide attempts are nearly 2.5 times higher in lesbian, gay, and bisexual (LGB) adults (King et al., 2008) and 3 times higher in LGB youth (Marshal et al., 2011) compared to their heterosexual peers. Rates among gender minorities appear to be on par, or higher than sexual minorities (Reisner et al., 2015b). Global research suggests that almost one in two young trans people has attempted suicide (Bradlow et al., 2017; Strauss et al., 2020a). These disproportionate rates of mental health struggles correlate with higher lifetime rates of harassment, discrimination, and victimization experiences that are specific to individuals perceived as having a marginalized identity (Katz-Wise & Hyde, 2012). Up to 94% of SGM youth report having experienced some form of sexual-orientation-based victimization (Mustanski et al., 2011).

The most well-supported theory explaining this phenomenon is a variation of stigma theory called the minority-stress model (Meyer, 2003a, b). The minority-stress model posits that in addition to managing the chronic stress of diminished access to resources and opportunities, minorities are subject to acute stressors in the form of rejection, violence, and harassment that at times result in the internalization of stigma. Together, these forces result in reliably high rates of mortality and morbidity relative to those experienced by nonminority individuals. In this chapter, we examine the data indicating that self-compassion (SC) buffers the impact of these stressors and interrupts internalization of stigma, thereby func-

A. Vigna (✉)
UW Population Health Institute, University of Wisconsin-Madison, Madison, WI, USA
e-mail: abra.vigna@wisc.edu

P. Strauss
Telethon Kids Institute, Nedlands, WA, Australia

A. Finlay-Jones et al. (eds.), *Handbook of Self-Compassion*, Mindfulness in Behavioral Health,
https://doi.org/10.1007/978-3-031-22348-8_9

tioning as a protective factor for minoritized identity groups. Our focus is on the emerging data regarding sexual and gender minorities and includes a summary of the evidence regarding distinctions between singly and multiply marginalized groups, such as between SGM youth who identify as white and between SGM youth who also identify as racial minorities.

A Rose by Any Other Name: LGBA, TNG, TGD, or Sexual- and Gender-Minority Populations

The terms people use to indicate their sexuality and gender identities are ever evolving. Common self-ascribed terms indicating membership in these populations include *lesbian, gay, bisexual, queer, questioning, asexual* (LGBQA), and *trans/transgender, nonbinary, and gender expansive/nonconforming* (TNG). Subcommunities and age cohorts often affirm their own distinct terminology reflective of racial, regional, and temporal specificity (Cohler & Hammack, 2007). Additionally, it has become increasingly common for younger individuals to resist identifying with any label or category (Olive, 2012). Preferred terms used to indicate SGM populations differ between countries, cultures, and individuals.

In academia, the term *sexual and gender minority* (SGM) applies to all individuals who do not fulfill hegemonic expectations of either gender or sexuality. Gender-minority individuals, thus, are persons who do not identify with the gender assigned them at birth (e.g., TNG individuals) and/or do not conduct their lives or express themselves in what are considered gender-conforming ways relative to their context (e.g., gender-expansive or gender-nonconforming individuals) (Toomey et al., 2010). Separately, individuals may be defined as sexual minorities if they do not experience exclusively heterosexual sexual or romantic attractions, do not favor exclusively heterosexual partnering, or do not identify as heterosexual; such individuals may use terms such as *gay*, *lesbian*, *bisexual*, *pansexual*, *demisexual*, and *asexual*, among others, to define their sexuality. In this chapter, we respectfully use the term *SGM* to encompass the diversity of identities of individuals who are not exclusively heterosexual or not fully gender conforming or identify as something other than cisgender. We recognize, however, that *SGM* is not universally used.

Gender identity and sexuality are independent facets of one's identity (Igartua et al., 2009; Rieger & Savin-Williams, 2012). Sexual identity, sexual arousal, and sexual behavior display considerable independence from one another and, for some people, are independently fluid over the life course (Coker et al., 2010; Mock & Eibach, 2012; Rosario et al., 2011). As a result, an individual's current proclaimed sexual identity is not sufficient information to determine their sexual history, the content of their fantasies, their gender presentation, or their gender identity. People who engage in same-sex sexual activity may not self-identify with a sexual-minority identity (i.e., not heterosexual), and not all gender-nonconforming individuals identify as TNG (or any of that term's synonyms) (Mustanski et al., 2014). This may be particularly true for members of age cohorts during whose adolescence nonconformity was heavily policed and for individuals presently in adolescence and thus in a developmental period of identity exploration.

Due to the emergent nature of identity and the shifting vogue of identity descriptors, it is difficult to estimate the proportion of the population that identifies or would qualify as SGM. Additionally, the pressure of cisheteronormativity and the threat of the consequences of deviation from it delay many individuals from affirming their nonheterosexual or gender-nonconforming inclinations to themselves or on a survey. *Cistheteronormativity* refers to the expectation that it is "normal" for each person to affirm the gender legally assigned to them at birth (i.e., are cisgender) and to desire only heterosexual relationships (Gordon & Meyer, 2008; Schilt & Westbrook, 2009). It also refers to the dominant expectation that genitalia are classifiable as indicative of one of two gender categories (i.e., *girl* or *boy*) and that those categories align with certain interests. An example is the expectation that all who are born with vulvas identify as girls/women and both prefer and are better at caretaking activities such as cooking and providing emotional support or childcare/eldercare. Cisheteronormativity also insists that it is unnatural for people to desire sexual and romantic activities with others who have similar genitalia (e.g., *girls have vulvas and should like boys who have penises* and vice versa). While cisheteronormative identities remain dominant, an increasing number of people outwardly identify as diverse in their sexuality and/or gender and at increasingly younger ages. Current estimates for the proportion of adults who are trans or gender diverse are between 0.5% and 1.2% (Flores et al., 2017; Winter et al., 2016), and for high-school-aged people, current available estimates range from 2.3% to 2.7% (Rider et al., 2018; Fisher et al., 2019). Since gender identity and sexuality are typically not included in population-based data collection methods (such as censuses), it is difficult to know how these numbers differ from the true numbers of SGM individuals in the general population. Consequently, existing estimates of the proportion of SGM individuals in the population are likely underestimates.

Stigma and Other Punishments for Violating Social Norms

Stigma is the phenomenon of ascribing less value to some people in a place (Link & Phelan, 2001). Stigma identifies out-group members in order to create an in-group to idealize, privilege, and protect. Stigma ideologies, such as cisheteronormativity and white supremacy, create cultural mythologies regarding what is considered natural or civilized. Such myths in turn justify interpersonal and structural discrimination against, status loss for, and stereotyping of persons or groups with certain characteristics (Hatzenbuehler et al., 2013; Martin, 1990; Sidanius & Pratto, 2012). Stigma ideologies of otherness have a long history in societies with steep social gradients. For centuries, utilizing stigma stories or propaganda has been a key tactic for empires to justify policies and practices to harass, exclude, and eliminate indigenous populations during colonization processes (Tyler, 2018).

In addition to rationalizing unjust social practices such as disproportionate police contact, incarceration, or denial of the full rights of citizenship (e.g., legal marriage or adoption), stigma messages and practices are intended to instill shame in those who violate cisheteronormative expectations. Content analyses of stigma messages directed at SGM populations suggest that the same mythological linkages between legally assigned sex/presumed gender at birth and conformity to gender norms and sexuality are used to instill shame in SGM people (Ioerger et al., 2015; Kimmel, 2004; Nadal et al., 2011; Worthen, 2016; Schilt & Westbrook, 2009). When people believe these myths and experience shame for who they are, the resulting psychological distress can be considerable. As will be described in greater detail below, an abundance of evidence indicates that stigma drives population-level health inequities by increasing the number and intensity of stressors people face, decreasing access to the personal and institutional resources needed to cope with them, and eroding physical and emotional well-being via the internalization of stigma (Hatzenbuehler et al., 2013; Toomey et al., 2010; White et al., 2015).

Empirical Investigations of Stigma: The Minority-Stress Model

Meyer (1995, 2003a, b) and others (Brooks, 1981; Hendricks & Testa, 2012) have delineated three key assumptions of the minority-stress model. The first is that minority stressors are distinct from and additive to the stressors that all individuals within the same culture tend to experience. For example, in addition to dealing with a global pandemic or the loss of a loved one, minorities are less likely to receive adequate medical care when seeking help and more likely to lose their jobs during a social lockdown than are majority individuals. The second assumption is that a state of minority stress is chronic due to relatively stable structures of marginalization. Thirdly, the minority-stress model assumes that minority stress is a consequence of existing in a discriminatory context and not endemic to the individual. Minority stressors are assumed to impact the well-being of an individual via the following three mechanisms: observable discrimination, an individual's response to discrimination, and internalization of stigma.

Mechanism #1: Observable Discrimination The first mechanism is the context of discrimination. This includes interpersonal and/or institutional-level discrimination, violence, and threats to physical and/or material safety or security. Examples of institutional-level discrimination range from policies banning medical doctors from providing gender-affirming medical care to individuals who are transgender, to school policies that ban SGM-inclusive sexual education curricula or prohibit transgender athletes from participating on teams aligned with their affirmed gender, to disproportionate school pushout of SGM youth (Mitchum & Moodie-Mills, 2014). SGM individuals also encounter employment, housing, and health-care discrimination at greater rates than their non-SGM peers (Bachmann & Gooch, 2018; Grant et al., 2011; Strauss et al., 2020a; Strauss et al., 2020b). For example, a very limited subset of the medical workforce provides gender-affirming medical interventions, which are strongly associated with decreased psychological distress and suicidality in SGM populations from youth through adulthood (Dolan et al., 2020; Mahfouda et al., 2017, 2018; Wernick et al., 2019).

Interpersonal discrimination is observed more often among SGM youth than among their non-SGM peers. Common examples include familial rejection or abuse, bias-based harassment by peers and teachers, and exposure to a greater number of adverse childhood experiences (Andersen & Blosnich, 2013; Grant et al., 2011; Grossman & D'Augelli, 2006; Kosciw et al., 2008; McLaughlin et al., 2012; Zaza et al., 2016).

Mechanism #2: Expectation and Concealment Individual reactions and responses are the second mechanism through which stigma impacts health and well-being. This includes the nonconscious autonomic stress response, as well as the conscious decisions people make to keep themselves safe. For example, one common bodily response to discrimination that erodes one's health and well-being is to remain alert for future threat. Hypervigilance is a chronic state of autonomic arousal that prioritizes energy for fight or flight and delays rest, digestion, and healing processes. Hypervigilance is often accompanied by decisions to hide one's identity to avoid potential rejection or discrimination. Since many minorities either conceal their identity or have been denied opportunities to access health care, a common available option for dampening hyperarousal is self-medication via alcohol or unprescribed drugs (Beemyn & Rankin, 2011; Brubaker et al., 2009; Rood et al, 2016; Jerome & Halkitis, 2009). The prevalence of this pattern of responding is reflected in higher rates of substance use and abuse among SGM individuals than among their cisgender, heterosexual peers (Nadal, 2013; Reisner et al., 2015a).

Mechanism #3: Internalization of Stigma The third pathway by which stigma impacts health and well-being is through believing the stigma messages are true. Understood as the internalization of homophobia, transphobia, or stigma, the process is typically unconscious and contributes to the fear or belief that one deserves one's mistreatment for one's deviations from so-called normality (Newcomb & Mustanski, 2010; Puckett et al., 2015). This mechanism is considered the most damaging, as it is the process by which one becomes one's own bully and remains chronically burdened by managing the stress of shame (Russell, 2007). Shame increases one's inclination for self-isolation as well by reducing the likelihood that one will seek external supports, even among individuals who have affirmed a minoritized identity. For example, a common self-policing sentiment that perpetuates isolation among SGM populations is the fear that one is not "queer enough" to seek LGBTIQ-specific mental health services. Internalized homophobia has been one of the strongest predictors of posttraumatic stress symptoms among SGM populations (Dragowski et al., 2011).

In simple terms, the minority-stress model suggests that stigma experiences cause the health disparities seen in minorities (i.e., SGM status → stigma stressors → health inequities). Stigmatized people in societies have their need for belonging unmet in many circumstances; are more likely to be treated with violence by persons, organizations, and the government; have limited recourse when they are mistreated; and are routinely denied access to resources for meeting various needs, such as employment, upward mobility, shelter, and training. The long-term results of increased stressors and reduced resources include physical and mental health challenges and post-traumatic stress symptomatology (Mustanski et al., 2016).

Getting into the Weeds of Mechanism #3: The Psychological Mediation Framework Since the internalization of stigma is pernicious, Hatzenbuehler (2009) formulated the psychological mediation framework to tease apart component variables that contribute to the process of internalization and thus identify potential intervention points. The psychological mediation framework delineates a path analysis by which mechanism one travels through mechanisms two and three—either simultaneously or sequentially—of the minority-stress model to arrive at a psychopathology. This framework identifies the sub-mechanisms of internalization, including emotional (dys)regulation processes such as hypervigilance or rumination (i.e., mechanism one), maladaptive social coping processes such as physical or psychological self-isolation or concealment (i.e., mechanism two), and the tenor of the cognitive meaning-making that accompanies awareness of stigma (see Fig. 9.1). In theory, stigma "gets under the skin" via increased levels of emotional dysregulation, self-isolation or concealment, and negative self-stories, resulting in psychopathology (Hatzenbuehler, 2009).

Tests of this theory are thus far supportive. Individuals who respond to stigma by suppressing their emotions (Hatzenbuehler et al., 2009) or ruminating and thereby overidentifying with their stigma experiences are more likely to report psychopathology (Szymanski et al., 2014; Yadavaia & Hayes, 2012). So too are those who have increased expectations of rejection (Denton et al., 2014), fear of being a burden to others (Baams et al., 2015), or difficulty detaching from negative thoughts (Hatzenbuehler et al., 2009).

Self-Compassion May Support Resilience to Minority Stress for SGM Populations

Resilience is the manifestation of adaptive functioning that supports one's innate tendency to make sense of hardship and integrate it into how one functions. Described as "ordinary magic" because of its near universality, resilience can be seen in the ability to move forward from an adversity rather than regress to the level of functioning experienced before that adversity (Masten, 2001). A close cousin of the study of stress and coping, resilience science focuses on uncovering the constellation of individual, social, and ecological factors that contribute to defying the increased odds of mortality and morbidity that accompany a history of facing adversity (Masten, 2001). In resilience science, factors that reduce the number, or lessen the severity, of the potential negative impacts of adversity are described as protective (Fergus & Zimmerman, 2005; Luthar, 2006).

Although the minority-stress model primarily explains the route to psychopathology, it also provides a framework for explaining how an individual's emotional, cognitive, and social coping strategies can be protective, and thus result in resilience (Meyer, 2015; see Fig. 9.2). For example, bringing present-moment awareness to the distress that arises from experiencing discrimination or stigma and treating oneself with kindness and comfort rather than self-judgment could be an adaptive emotional coping approach. Similarly, seeking out a supportive community to help carry the burden of distress, rather than self-isolating and shouldering it alone, could act as a

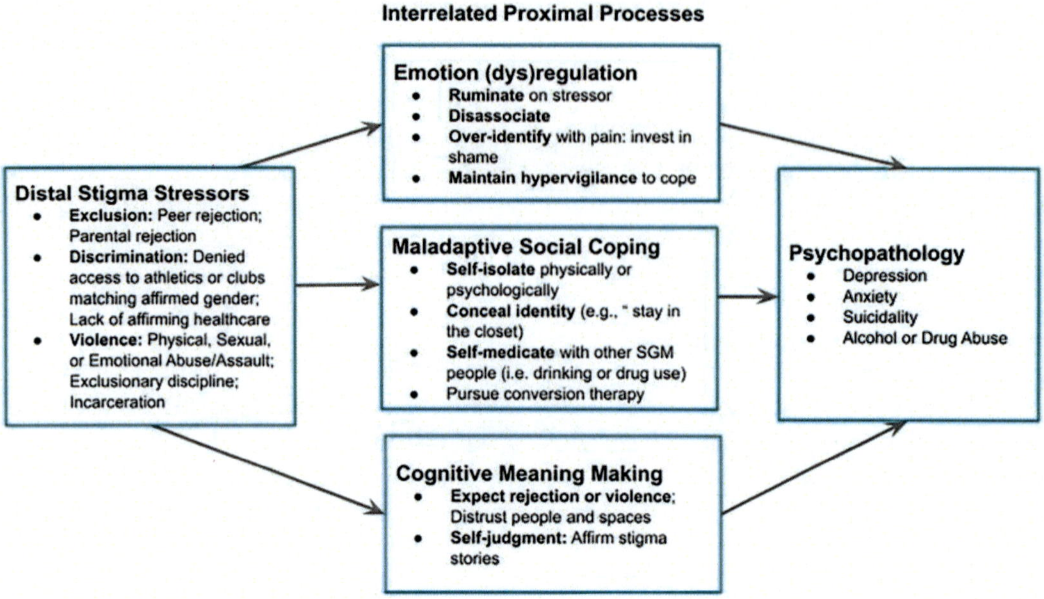

Fig. 9.1 Adaptation of the psychological mediation framework. (Hatzenbuehler, 2009)

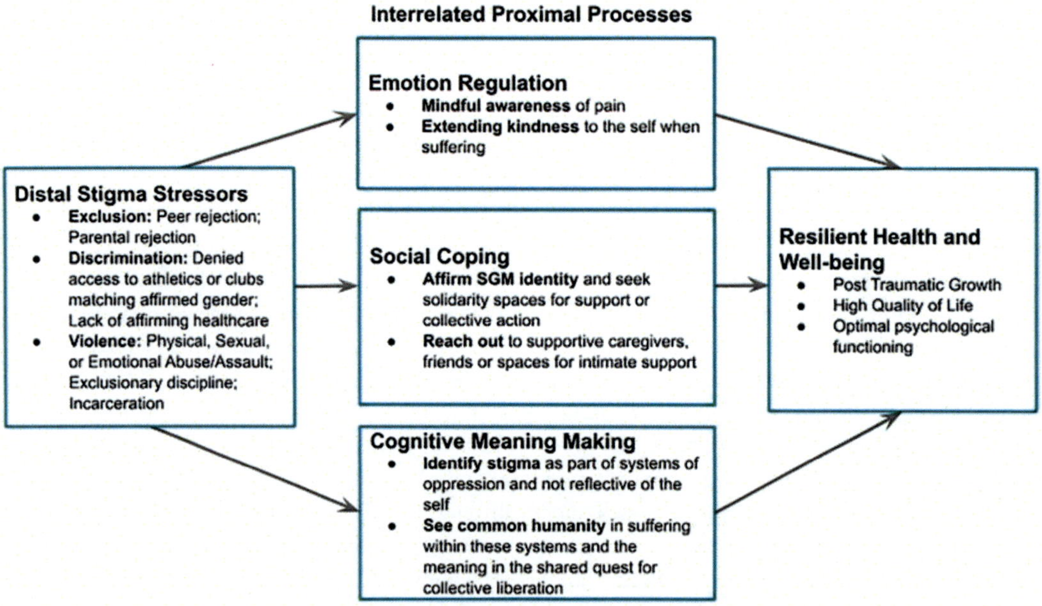

Fig. 9.2 Resilience-focused adaptation of the psychological mediation framework. (Hatzenbuehler, 2009)

buffer against adverse experiences. Finally, making sense of the stress as an experience of common humanity could be a route for adaptive cognitive processes. For example, the prevailing myth of meritocracy suggests that all pain and suffering is a reflection of your value as a human being. In contrast by choosing to see one's experiences as reflective of a common struggle and a normative human response to stigma, one can feel supported by a sense of connection to all suffering and potentially the shared goal of collective liberation.

As a trait-level quality of responding to the self with warmth, acceptance, a sense of collective struggle, and a nonjudgmental presence, compassionate self-responding would theoretically facilitate resilience in the face of minority stress. We will use the remaining portion of this chapter to explore evidence that supports the theory that self-compassion facilitates resilience in the face of stigma among SGM individuals, concluding with recommendations for future research.

Compassionate Self-Responding Is an Adaptive Emotional and Cognitive Coping Response Responding to oneself with compassion entails presence, warmth, and understanding and engenders a sense of a "loving, connected presence" (McGehee et al., 2017; p. 280). Neff's (2003a) theory of self-compassion (SC) outlines three main components (i.e., common humanity, kindness, and mindful attunement) that can be assessed according to whether they are more or less aligned with compassionate self-responding. In contrast, stigma focuses on personal defect, exclusion, and avoidance. Unlike self-esteem, which hinges upon a positive view of a uniquely special self in comparison with others and is thus vulnerable to stigma messages (which are borne of social comparison), SC theory posits that a healthy self-attitude arises from de-emphasizing a separate self (Neff, 2003a, b).

Compassionate self-responding entails offering kindness to the suffering self, seeing suffering as a source of connection with all of humanity, and remaining attuned to the feelings and thoughts that arise from suffering and then dissipate. Uncompassionate self-responding entails judging the self harshly because of an experience of suffering or its cause, interpreting the experience or its cause as evidence of one's unique fallibility, and fixating on, or identifying with, the experience of suffering on a personal (i.e., ego) level, such as in the case of rumination or the personal fable (Alberts et al., 2007). The 26-item Self-Compassion Scale developed by Neff is used to calculate a total score of a systems-level balance of a trait tendency to respond to the self with more kindness and less judgment, with a sense of connection via common humanity rather than withdrawing and self-isolating, and with mindful awareness rather than an overidentification with one's pain. Significant empirical evidence regarding how the autonomic nervous system functions supports the premise that overall SC is a balance of these six components (see Neff et al., 2018, and Neff et al., 2019, for detail).

Self-Compassion Is Associated with Reduced Psychopathology and Increased Well-being Data consistently suggest that people who maintain a more compassionate stance toward themselves are less likely to report symptoms of depression or anxiety; this finding has emerged repeatedly and with large effect sizes (MacBeth & Gumley, 2012; Zessin et al., 2015). People who are more likely to respond to their own suffering with compassion are also less likely to experience public self-consciousness, engage in social comparison or ruminative thought, hold a contingent or unstable sense of self-worth, or engage in thought suppression as a self-regulation strategy (Krieger et al., 2013; Leary et al., 2007; Neff, 2003a, b; Neff & Vonk, 2009; Neff et al., 2007; Raes, 2011). Experimental investigations have shown that people with greater dispositional self-compassion more accurately appraise their own performance, and test subjects who are induced to feel self-compassion experience fewer negative emotions and accept greater personal responsibility for past failures than do subjects induced to have higher self-esteem (Leary et al., 2007).

Inductions of Self-Compassion Appear to Reduce the Sense of Isolation and Thus Reduced Social Resources One hypothesis regarding how self-compassion enables adaptive emotion regulation is that it causes one to feel supported in distress. Specifically, self-compassion inductions appear to trigger a calming hormonal cascade (Swain et al., 2012). Such a calming hormonal cascade is also observed when an individual is in the presence of a close companion during pain-induction experiments

(Brown et al., 2003). According to several experimental designs, activating SC reduces the duration and magnitude of inflammatory responses during social-evaluative stress inductions (Breines et al., 2014, 2015; Bluth et al., 2016a, b). Similarly, researchers have found evidence that SC is associated with a decreased threat response and an increased reliance upon reassuring and kind thoughts during social-evaluative stress-induction procedures (Arch et al., 2014). Finally, data from SC training interventions demonstrate that increased SC is directly associated with reductions in symptoms of depression and anxiety across adolescence and adulthood (Bluth & Eisenlohr-Moul, 2017; Diedrich et al., 2016; Galla, 2016; Neff & Germer, 2013; Shapira & Mongrain, 2010).

Self-Compassion May Also Confer Resilience in the Face of Adversity, Such as Minority Stress As Leary et al. (2007) noted, "self-compassionate people more readily accept [socially] undesirable aspects of their character and behavior without obsessing about them, becoming defensive, or behaving badly than do people low in self-compassion" (p. 901). With the inclination to de-identify with thoughts and to rest in a sense of common humanity, SC may help SGM individuals to view stigma messages as mere messages and not statements of truth about themselves. This may create the space to see that stigma messages have an explicit purpose: to maintain and protect inequitable access to resources as defined by a social hierarchy. The process of acknowledging one's suffering nonjudgmentally and refusing to create an ongoing negative story about oneself has obvious applications for disrupting the internalization of stigma messages. In one of the few investigations of the subscales of the full 26-item Self-Compassion Scale, researchers found that reduced rates of self-judgment and self-isolation account for most of the variance in predicting lower rates of depression and anxiety symptomatology (Van Dam et al., 2011).

SC appears to be protective against the long-term negative impacts of abusive family relationships. Longitudinal research on adolescents who experienced legally documented abuse or neglect suggests that individuals with higher rates of dispositional SC are less likely to report common maltreatment-related impairments, such as psychological distress, substance abuse, and serious suicide attempts (Tanaka et al., 2011). Other research has found that among adolescents with a history of childhood maltreatment variance in compassionate self-responding accounts for more of the variance in emotion dysregulation than do a history of childhood maltreatment, current rates of psychological distress, and problem substance use combined (Vettese et al., 2011). Taken together, these data suggest compassionate self-responding may protect us from the deleterious effects of stressors in much the same way that the internalization of secure attachment serves as a protective factor in the face of acute adversity.

Self-Compassion in SGM Populations

Although the study of SC in SGM populations is still emerging, preliminary correlational data support the theory that compassionate—and uncompassionate—self-responding may partially mediate the presence of well-being disparities in these populations. For example, SC was found to mediate the relationship between early memories of warmth and safeness, and happiness in adulthood among LGB adults (Greene & Britton, 2015). SC also consistently demonstrates a significant inverse relationship with peer victimization and suicidality and a positive relationship with subjective well-being among LGBQ adolescents (Hatchel et al., 2019) and gay men (Beard et al., 2017). And, as with non-SGM populations, it appears that self-kindness (versus self-judgment) and self-isolation (versus common humanity) account for much of the variance in subjective well-being (Beard et al., 2017).

In the first published investigation of SC rates among SGM adolescents, Vigna et al. (2018a) found that SGM youth in a school-based sample report lower levels of SC than their sexual- and gender-majority counterparts, with a medium effect size. Furthermore, mediation analysis suggested that SC may account for more of the variance in depression, suicidality, and anxiety than do adverse childhood experiences, bias-based bullying, and non-bias-based bullying combined. Also, high SC rates appear to have a negative relationship with psychological distress (Vigna et al., 2018a, b).

Self-Compassion, Discrimination, and Self-Stigma In accordance with the psychological mediation framework of the minority-stress model, structural equation modeling of data from a national sample of LGB adults suggests that experiences of discrimination "get under the skin" via increased expectations of rejection, anger, and rumination and reduced SC (Liao et al., 2015). One study found that higher rates of SC, particularly a sense of common humanity, were associated with positive LGB identity development during the process of coming out (i.e., internally and externally affirming one's stigmatized identity), as reflected upon by LGB adults (Crews, 2012). Another study, involving in-depth qualitative analyses of 16 emerging adolescents and described in the same dissertation found that moving from uncompassionate self-responding to compassionate self-responding was correlated with a movement from resistance and fear to self-acceptance and well-being. For many subjects of this study, the key pivot point was the recognition of an experience of solidarity, or common humanity, with others also marginalized by stigma leading to a sense of belonging (Crews, 2012). This same pattern has emerged elsewhere (DiFulvio, 2011).

In one of the few studies involving comparisons among sub-identities of an SGM sample, greater SC when encountering stigma was associated with less proneness to disordered eating in gay men and TNG individuals (Bell et al., 2019). Additionally, path analyses demonstrated that the effects of moderators of internalized stigma on subjective well-being SC significantly moderated the effect on subjective well-being, whereas mindfulness alone did not, among SGM men living with HIV in Hong Kong (Yang & Mak, 2017). Finally, in the only study on the relationship between self-stigma, or the internalization of stigma, and SC, Fredrick et al. (2020) found that public stigma impacts quality of life indirectly first via self-stigma, which then impacts SC among SGM adults. However, since the study was correlational, the mediational relationship remains somewhat speculative.

Self-Compassion and Gender Nonconformity Violating dominant expectations of gender assigned at birth is at the heart of stigma messaging among SGM populations (Gordon & Meyer, 2008). Accordingly, much research has found an inverse relationship between gender nonconformity and mental health, independent of sexual orientation, although the strength of the relationship may differ based upon which gender was legally assigned at birth (Alanko et al., 2009; Rieger & Savin-Williams, 2012; Lippa, 2008; Skidmore et al., 2006; Toomey et al., 2010). One study of gender-nonconforming adults in Singapore found that SC may moderate the relationship between gender nonconformity and subjective well-being via the degree of self-judgment, overidentification, and self-isolation a person is plagued by. However, in this investigation, SC did not moderate the relationship between gender nonconformity and depression (Keng & Liew, 2017). Notably, the statistical model employed lacked a measure of adversity, relying instead upon a measure of gender nonconformity, and although gender nonconformity is frequently associated with significant harassment, this is not always the case. Nevertheless, this study demonstrated that SC attenuates the negative association between gender nonconformity and subjective well-being and thus is relevant to SGM populations and the question of SC's utility in attenuating the impact of stigma on well-being.

Self-Compassion and Parental Rejection The impact of parental acceptance or rejection of a young SGM person is profound (Katz-Wise et al., 2016; Khaleque & Rohner, 2002). In one of the first investigations of the health impacts of degree of family rejection on SGM youth, Ryan et al. (2009) found that young LGB adults who reported higher levels of family rejection were 8.4 times more likely to report having attempted suicide, and 5.9 times more likely to report high levels of depression, compared with peers from families that reported no or low levels of family rejection. Similarly, higher levels of family rejection were associated with 3.4 times increased likelihood of using illegal drugs and 3.4 times increased likelihood of reporting engaging in unprotected sexual intercourse (Ryan et al., 2009). Accordingly, investigations of factors that might moderate the negative impact of parental rejection are common. For example, several researchers have modeled the role of compassionate or uncompassionate self-responding in mediating the relationship between parental rejection or support and psychological distress. Results of these investigations suggest that SC fully mediates the relationship between parental support, parental acceptance at the time of coming out, and present-day levels of internalized homophobia and subjective well-being among LGB adults (Beard, et al., 2017; Gertler, 2014; Toplu-Demirtaş et al., 2018).

Emerging research supports the premise that whether one tends to respond to oneself with compassion is particularly important in explaining the relationship between one's stigma-related parental rejection and depression. For example, one cross-sectional study found that SC mediated the relationship between shame memories and current functioning for gay men, but not for heterosexual men. However, the ability to remain mindfully aware of distress and behave in accordance with one's values (i.e., psychologically flexible) mediated the relationship between shame memories and current depressive symptoms among all participants of the study. One important distinction between the two groups was that the shame memories of the gay men were related to their fathers, whereas the shame memories of the heterosexual men were about nonparental adults. The authors of the study thus speculate that SC is more relevant for sexual-minority men who are already managing ongoing stressors of global belonging as stigmatized people. The authors speculate that SC, under this ongoing thwarting of the need for belonging, either exacerbates the impact of parental rejection when low or compensates for parental rejection when high (Matos et al., 2017).

Self-Compassion and the Intersection of Multiple Marginalized Identities

The minority-stress model is predicated on the notion that social hierarchies define the opportunities, challenges, and resources available to a person according to that person's identity. However, there is no single SGM experience, nor is there a single experience of being a member of any stigmatized group (Cyrus, 2017; Mueller et al., 2015). SGM individuals have racial identities, citizenship statuses, native languages, and unique relationships with colonization practices and globalization patterns, all of which contribute to a diversity of experiences of minority stress and access to the resources needed to keep stress levels manageable (Holley et al., 2012; White Hughto et al., 2015). At its root, intersectionality theory requires that we consider how our various identities combine, or interact, to produce experiences reflective of the pressures placed on our various socially assigned identities, such as those defined according to sexual and gender minority or majority and racial minority or majority (Crenshaw, 1991). For example, a Black queer woman's total experience of oppression is not reducible to an experience of racism or cisheterosexism or sexism; rather, the intersection of these pressures in her life makes for a unique experience. Among other things, intersectionality calls us to recognize that a white queer woman's experience of oppression is distinct from a Black queer woman's experience.

Intersectionality theory thus cautions researchers to resist the urge to apply one-size-fits-all repairs to issues of injustice, such as health inequities (Bowleg, 2012; Moradi et al., 2010). Moreover, as resilience science notes, the more external the adversity one contends with in the form of structural discrimination and violence, the less effective intraindividual resilience factors will be in protecting against the deleterious impacts of stigma-laden contexts (Masten, 2001). This conclusion is particularly germane to inquiry into SC as a resilience factor for SGM individuals. One would expect that the greater the structural discrimination SGM individuals face, the less effective an individual-level resilience factor such as SC would be at buffering the impacts of stigma on well-being.

Investigation into the conditions in which SC serves as a resilience factor among SGM is emergent. The one investigation the authors of this chapter could find suggests that although BIPOC (Black, Indigenous, and People of Color) who are SGM may report two to three times the frequency of exposure to structural discrimination as that experienced by white SGM students, white SGM students report the highest rates of mental health concerns. In this study, levels of structural discrimination were determined according to factors of economic hardship, having an incarcerated parent, not trusting the police, and the assignment of specialized education plans and/or experiencing exclusionary discipline. Despite reporting higher rates of structural discrimination, BIPOC students reported the highest rates of SC, while white SGM students reported the lowest, with a small-to-moderate effect size (Vigna et al., 2018b, 2020). Additionally, within racial groupings, comparisons of SC scores across sexual and gender categories always favored the less marginalized group, with sexual- and gender-majority students having more compassion for the self than their racially matched SGM counterparts, to a moderate and large effect for BIPOC students. In contrast with the authors' hypothesis, the largest difference in SC scores was found between sexual- and gender-*majority* and white SGM, to a large effect size (Vigna et al., 2018b).

Given the wide variability in experiences of marginalization, it is possible that protective factors such as SC function differently at the intersection of racialization, and gender- and sexuality-socialization, experiences. Conditional process analyses of the effect of identity on SC, as moderated by rates of victimization from bias-based bullying, revealed a significant interaction effect for white SGM youth, but not for SoC. This finding suggests that experiencing discrimination in the form of interpersonal bias-based bullying has a stronger relationship with compassionate self-responding for white SGM youth than it does for BIPOC SGM (Vigna et al., 2018b). Similarly, in conditional process analyses of the effect of intersectional identity on depressive symptoms as moderated by SC, white SGM students report the highest rates of depressive symptomatology at average and below-average rates of SC of all identity groups (Vigna et al., 2020).

The higher rates of depression symptomatology, anxiety symptomatology, and suicidality among white SGM students compared with BIPOC SGM students mirror the rates of depression and anxiety among white versus Black adults in the United States (Keyes, 2007, 2009; McGuire & Miranda, 2008). Indeed, a strong pattern of internalization challenges across racialization experiences is well documented. For the past century, white people have consistently had the highest rate of suicide of all the racial groups in the United States, despite benefiting from the greatest structural privileges (National Center for Health Statistics, 2016). It is possible that racialized socialization into the myth of meritocracy discourages white SGM students from seeing their stigma as artificially designed to maintain inequitable power distributions (Bañales et al., 2020). Instead of finding solidarity in the experience of being oppressed, white SGM may be more inclined to believe that there is something wrong with who they are and thus they deserve exclusion and shame. Coupled with the finding that BIPOC students reported the highest rates of SC, this hypothesis suggests that racialization experiences that support the depersonalization of stigma are relevant for consideration in studies of the relationship between SC and mental health.

Existing literature provides preliminary support for the hypothesis that racialization experiences related to the development of critical awareness of social power structures and systems may impact well-being via depersonalization. For example, research comparing internalization rates and mental health between white and BIPOC SGM adults suggests that BIPOC SGM adults are less inclined to internalize SGM-focused stigma and subsequently experience the same patterns in psychological distress. In other words, BIPOC SGM may be less inclined to ascribe their stigma experiences to moral failings or individual flaws associated with being SGM (Moradi et al., 2010). Additionally, longitudinal research has found that system-justifying beliefs, such as that "the system is fair for everyone," were associated with stronger declines in well-being among early-adolescent BIPOC (Godfrey et al., 2019). If SC can support the depersonalization of stigma messages and experiences for all SGM people, it may support the shift toward a collective experience rooted in the quest for liberation that is related to psychological well-being among SGM folks and racial minorities (DiFulvio, 2011; Godfrey et al., 2019).

Self-Compassion Is Not the Sole Answer to Dealing with Stigma

Stigma, and the minority stress that results from it, creates real internal and external challenges. As a root cause of population health inequities, stigma drives and justifies the creation and perpetuation of conditions that exclude, exploit, or violate minorities based on one or many of their identities. Although one cannot *compassion* oneself out of incarceration or poverty, data do suggest that SC is a promising tool for reframing one's understanding of one's material conditions and potentially supporting a shift toward collective action aimed at eliminating stigma structures.

The research summarized in this chapter supports the hypothesis that SC as a style of self-responding that is promotive and protective of psychological functioning during difficulty may help preserve individual functioning and promote engagement with similarly stigmatized others. Given that SC has a dampening effect on the autonomic stress response and encourages the individual to affirm kindness for the self, SC may be a robust intervention for staving off the burnout that often removes activists from structural-change work (Breslow et al., 2015). Additionally, if paired with a system-level understanding of stigma as a tool of oppression, a focus on solidarity with all of humanity may have the effect of resourcing ongoing engagement in the emotionally challenging cross-group efforts for collective liberation from oppressive structures.

Eliminating external sources of stigma is the most efficient and effective way to eliminate the impacts of internalized stigma (Puckett et al., 2016; Russell, 2007). Many of the intergroup efforts to do so focus on improving community resilience, on improving the conditions of communities to support and sustain individual well-being (Hatzenbuehler & Pachankis, 2016; Hall & Zautra, 2010). Aspects of community resilience have been found to promote well-being despite social stigma across levels of social ecology. Evidence-based conditions of community resilience specific to SGM stigma include parental acceptance (Ryan et al., 2010; van Beusekom, et al., 2015), gay-straight alliances in public schools (Hatzenbuehler, 2011; Poteat et al., 2012; Toomey et al., 2012), inclusive school policies (Hatzenbuehler, 2011; Hatzenbuehler et al., 2011), community connectedness (Kertzner et al., 2009), visible SGM communities (Hatzenbuehler, 2011; Hatzenbuehler et al., 2012), and legislation that affirms identities and reinstates the rights of citizenship or provides protections for LGBTQ individuals (Hatzenbuehler et al., 2009a, b, 2010).

Notably, macro-level legislation that reduces social stigma by affirming equality for SGM people has demonstrated promotive impacts on psychological well-being for *both* SGM and non-SGM individuals (Hatzenbuehler et al., 2009a, b, 2010). It is important to emphasize that in all communities, SGM identities are marginalized, so the above conditions of resilience are truly supportive of everyone's resilience only if they reflect and impact the diversity of SGM experiences. This includes having visible, culturally diverse SGM communities and GSAs (gay-straight, or gender-sexuality, alliances), adopting inclusive school policies that teach the truth of a country's history (i.e., critical race theory), and legislation that benefits all SGM people and not just those with the money to hire lawyers to enforce or access protections or benefits, and so on.

Limitations

Overall, there have been few investigations of SC as a resilience factor in SGM populations, despite the likeliness of suitability in these populations. What literature does exist is cross-sectional and relies upon self-reporting and is thus subject to common-method bias and the impossibility of inferring causality (Podsakoff et al., 2003). Additionally, reliance upon school-based or population-based surveys to examine patterns among SGM individuals is problematic as there is evidence to suggest that, at least among adolescents, some SGM individuals are unwilling to "out" themselves for a survey, and thus are counted as either heterosexual or cisgender when they may not be, thereby obscuring the true relationships between SGM identity and various health outcomes (Macapagal et al., 2017).

While we can make some assumptions from the experimental and longitudinal research not explicitly including SGM individuals, data suggest that SC functioning is amplified among stigmatized populations. Further, although few studies with the power to detect differences among subgroups of the SGM category have been published, evidence suggests that significant health differences exist among distinct identities of SGM. For example, individuals who select *bisexual* or write in their sexual orientation on a list of forced-choice options consistently report higher anxiety rates than those who select *heterosexual*, or *gay* or *lesbian* (Wadesworth & Hayes-Skelton, 2015). Other studies have found that young people who select the *bisexual* identifier report higher rates of suicidality and depressive symptomatology than young people who select *gay* or *lesbian* identifiers, while young women who select *lesbian* or *bisexual* report higher rates than their male gay and bisexual counterparts (Gilbey et al., 2019). This distinction implies that there are differences in the experience of stigma, minority stress, resilience, and/or adverse events (e.g., discrimination) among SGM subgroups.

Future Directions

In addition to pursuing longitudinal and experimental designs, future research on SC among SGM populations should include an intersectional approach that prioritizes inquiry into practical applications for reducing health inequities. Although there is limited research documenting the burden of health inequities falling upon transgender and gender-nonconforming BIPOC people—particularly those who present as feminine of center—what exists is alarming. As such, efforts that broaden our collective understanding of how to address and reduce these inequities via multilevel interventions should be prioritized.

Racialization appears to be particularly germane to investigations into SC as a practice for disrupting the internalization of stigma. For example, ample evidence suggests that white racialization conditions lower dispositional SC by heightening a sense of individualization via competition and a fear of others (Kwate & Meyer, 2010; Jost et al., 2004; Okun, n.d.; Sidanius & Pratto, 2012). Conversely, racialization experiences within BIPOC communities may preserve dispositional SC by explicitly countering the hegemonic meritocratic narrative that obscures the organizing influence of whiteness by affirming the influence of oppressive structures in our lives (Neighbors et al., 1996). Limited data suggest that the process of reframing the cause of stigma from the self to society can help reaffirm one's identity and preserve psychological health (Frost, 2011; Godfrey et al., 2017; DiFulvio, 2011). Since extant literature indicates that dominant narratives constructing racial categories play a significant role in predicting mortality and morbidity, future researchers would do well to account for the influence of racial conditioning on SC, as omitting it has likely contributed to the variance's remaining unaccounted for in the research.

Logical next steps for quantitative investigations include enacting purposive sampling strategies to recruit sample groups large enough to detect real differences among identities more precise than *white*, *BIPOC*, *SGM*, and *non-SGM*, as a true intersectional approach recommends. Additionally, future statistical models would benefit from datasets that include variables on multiple levels of the social ecological model to more accurately test the multilevel influences of intersectionality on health outcomes (Seng et al., 2012). However, there is also much to be gained from qualitative inquiries into the topic of intraindividual resilience factors such as SC. For example, we know little of how individuals make sense of what helps them to flourish within stigmatizing contexts. Given that SC appears to play a facilitative role, asking people about the role that SC plays in their flourishing despite adversity would provide important data on construct validity.

Beyond that, there is a dearth of literature on SC interventions among SGM individuals. Further examination of SC therapies with SGM individuals is needed and currently underway (Finlay-Jones et al., 2021). Although this chapter refers generally to people who are SGM, the emerging literature on intersectional identities and SC indicates that socialization experiences regarding identities are important to consider in the research on SC interventions for SGM individuals. For example, given the reliable differences in dispositional self-compassion across gender binaries, gender socialization seems particularly relevant for adapting SC trainings for SGM populations. A recent investigation into the relationship between gender roles, gender identity, and SC showed that the impact of self-identified gender on SC was consistently smaller than the impact of masculine gender-role orientation, suggesting that socialization plays a strong role. Notably, those subjects high in both femininity and masculinity (or instrumental and expressive qualities) tended to have the highest levels of SC (Yarnell et al., 2019). These findings suggest that SC interventions may be most successful if their content and format are adapted to acknowledge and address toxic gender, sexuality, and racial identity messages and experiences.

For several reasons, adolescence is a promising time to introduce SC and online offerings are a promising medium for connecting with SGM youth. First, adolescence is the developmental period during which emotion-regulation strategies are solidified (Steinberg et al., 2006). Second, realizations regarding one's sexuality or gender identity during this time can become stressful and lead to the solidification of habitual responses of managing stigma (Eliason, 1996). Third, SGM adolescents are not concentrated in many geographic areas or may have difficulty physically accessing SGM-affirming services or support groups where such interventions would normally be offered. Fourth, it may be unsafe for SGM adolescents to seek out SGM-specific services or interventions as many live in homes or communities where their physical safety may be compromised if they were to affirm these aspects of their identity. Finally, contemporary adolescent cohorts are accustomed to engaging with one another via virtual spaces (McDermott et al., 2013, 2016).

Moreover, it is imperative that all mental health practitioners working with SGM populations understand the potential impacts of stigma and self-stigma on the individual. Some practitioners even argue that a central tenet of psychological care for SGM populations should be the reduction of stigma and internalized stigma, starting with an individualized approach to clients' conceptualization of stigma (Puckett & Levitt, 2015). Incorporating SC interventions into generalized mental health practice could be useful in helping SGM and questioning individuals to acquire immediate support, as SGM individuals may face difficulty in accessing limited SGM-specific services in a timely manner.

In summary, preliminary findings and the theoretical underpinnings discussed here demonstrate the potential utility of SC among SGM populations; however, further research is needed to determine whether a) SC functions as expected according to those whom it may theoretically benefit, (b) aspects of SC training need to be tailored to specific subgroups, (c) the effectiveness of SC is similar across age groups, and (d) specific SC training formats are more suitable than others (e.g., online versus face-to-face formats). Finally, research with activists engaged in collective liberation in general, and in intersectional spaces specifically, is needed to test the hypothesis that SC supports long-term engagement in structural-change efforts designed to eliminate stigma.

References

Alanko, K., Santtila, P., Witting, K., Varjonen, M., Jern, P., Johansson, A., van der Pahlen, B., & Sandnabba, N. K. (2009). Psychiatric symptoms and same-sex sexual attraction and behavior in light of childhood gender atypical behavior and parental relationships. *Journal of Sex Research, 46*, 494–504. https://doi.org/10.1080/00224490902846487

Alberts, A., Elkind, D., & Ginsberg, S. (2007). The personal fable and risk-taking in early adolescence. *Journal of Youth and Adolescence, 36*(1), 71–76. https://doi.org/10.1007/s10964-006-9144-4

Andersen, J. P., & Blosnich, J. (2013). Disparities in adverse childhood experiences among sexual minority and heterosexual adults: Results from a multi-state probability-based sample. *PLoS One, 8*(1), e54691. https://doi.org/10.1371/journal.pone.0054691

Arch, J. J., Brown, K. W., Dean, D. J., Landy, L. N., Brown, K. D., & Laudenslager, M. L. (2014). Self-compassion training modulates alpha-amylase, heart rate variability, and subjective responses to social evaluative threat in women. *Psychoneuroendocrinology, 42*, 49–58. https://doi.org/10.1016/j.psyneuen.2013.12.018

Baams, L., Grossman, A. H., & Russell, S. T. (2015). Minority stress and mechanisms of risk for depression and suicidal ideation among lesbian, gay, and bisexual youth. *Developmental Psychology, 51*(5), 688–696. https://doi.org/10.1037/a0038994

Bachmann, C. L., & Gooch, B. (2018). *LGBT in Britain: Trans report*. Stonewall & YouGov. https://www.stonewall.org.uk/lgbt-britain-trans-report

Bañales, J., Marchand, A. D., Skinner, O. D., Anyiwo, N., Rowley, S. J., & Kurtz-Costes, B. (2020). Black adolescents' critical reflection development: Parents' racial socialization and attributions about race achievement gaps. *Journal of Research on Adolescence, 30*, 403–417. https://doi.org/10.1111/jora.12485

Beard, K., Eames, C., & Withers, P. (2017). The role of self-compassion in the Well-being of self-identifying gay men. *Journal of Gay & Lesbian Mental Health, 21*(1), 77–96. https://doi.org/10.1080/19359705.2016.1233163

Beemyn, G., & Rankin, S. (2011). *The lives of transgender people*. Columbia University Press.

Bell, K., Rieger, E., & Hirsch, J. K. (2019). Eating disorder symptoms and proneness in gay men, lesbian women, and transgender and gender non-conforming adults: Comparative levels and a proposed mediational model. *Frontiers in Psychology, 9*, 2692. https://doi.org/10.3389/fpsyg.2018.02692

Bluth, K., & Eisenlohr-Moul, T. A. (2017). Response to a mindful self-compassion intervention in teens: A within-person association of mindfulness, self-compassion, and emotional Well-being outcomes. *Journal of Adolescence, 57*, 108–118. https://doi.org/10.1016/j.adolescence.2017.04.001

Bluth, K., Gaylord, S. A., Campo, R. A., Mullarkey, M. C., & Hobbs, L. (2016a). Making friends with yourself: A mixed methods pilot study of a mindful self-compassion program for adolescents. *Mindfulness, 7*(2), 479–492. https://doi.org/10.1007/s12671-015-0476-6

Bluth, K., Roberson, P. N., Gaylord, S. A., Faurot, K. R., Grewen, K. M., Arzon, S., & Girdler, S. S. (2016b). Does self-compassion protect adolescents from stress? *Journal of Child and Family Studies, 25*(4), 1098–1109. https://doi.org/10.1007/s10826-015-0307-3

Bowleg, L. (2012). The problem with the phrase women and minorities: Intersectionality—An important theoretical framework for public health. *American Journal of Public Health, 102*(7), 1267–1273. https://doi.org/10.2105/AJPH.2012.300750

Bradlow, J., Bartram, F., Guasp, A., & Jadva, V. (2017). *School report: The experiences of lesbian, gay, bi and trans young people in Britain's schools in 2017*. Stonewall & University of Cambridge Centre for Family Research. https://www.stonewall.org.uk/sites/default/files/the_school_report_2017.pdf

Breines, J. G., Thoma, M. V., Gianferante, D., Hanlin, L., Chen, X., & Rohleder, N. (2014). Self-compassion as a predictor of interleukin-6 response to acute psychosocial stress. *Brain, Behavior, and Immunity, 37*, 109–114. https://doi.org/10.1016/j.bbi.2013.11.006

Breines, J. G., McInnis, C. M., Kuras, Y., Thoma, M. V., Gianferante, D., Hanlin, L., & Chen & Rohleder, N. (2015). Self-compassionate young adults show lower salivary alpha-amylase responses to repeated psychosocial stress. *Self and Identity, 14*(4), 390–402. https://doi.org/10.1080/15298868.2015.1005659

Breslow, A. S., Brewster, M. E., Velez, B. L., Wong, S., Geiger, E., & Soderstrom, B. (2015). Resilience and collective action: Exploring buffers against minority stress for transgender individuals. *Psychology of Sexual Orientation and Gender Diversity, 2*(3), 253–265. https://doi.org/10.1037/sgd000011

Brooks, V. R. (1981). *Minority stress and lesbian women*. Lexington Books.

Brown, J. L., Sheffield, D., Leary, M. R., & Robinson, M. E. (2003). Social support and experimental pain. *Psychosomatic Medicine, 65*(2), 276–283. https://doi.org/10.1097/01.PSY.0000030388.62434.46

Brubaker, M. D., Garrett, M. T., & Dew, B. J. (2009). Examining the relationship between internalized heterosexism and substance abuse among lesbian, gay, and bisexual individuals: A critical review. *Journal of LGBT Issues in Counseling, 3*, 62–89. https://doi.org/10.1080/15538600902754494

Cohler, B. J., & Hammack, P. L. (2007). The psychological world of the gay teenager: Social change, narrative, and 'normality'. *Journal of Youth and Adolescence, 36*(1), 47–59. https://doi.org/10.1007/s10964-006-9110-1

Coker, T. R., Austin, S. B., & Schuster, M. A. (2010). The health and health care of lesbian, gay, and bisexual adolescents. *Annual Review of Public Health, 31*, 457–477. https://doi.org/10.1146/annurev.publhealth.012809.103636

Crenshaw, K. (1991). Mapping the margins: Intersectionality, identity politics, and violence against women of color. *Stanford Law Review, 43*(6), 1241. https://doi.org/10.2307/1229039

Crews, D. A. (2012). *Exploring self-compassion with lesbian, gay, and bisexual persons*. [Dissertation, University of Utah]. ProQuest Dissertations Publishing. https://socialwork.utah.edu/_resources/documents/phd/recent-phd-graduates/2013/Crews-Douglas-Alan.pdf

Cyrus, K. (2017). Multiple minorities as multiply marginalized: Applying the minority stress theory to LGBTQ people of color. *Journal of Gay & Lesbian Mental Health, 21*(3), 194–202. https://doi.org/10.1080/19359705.2017.1320739

Denton, F. N., Rostosky, S. S., & Danner, F. (2014). Stigma-related stressors, coping self-efficacy, and physical health in lesbian, gay, and bisexual individuals. *Journal of Counseling Psychology, 61*(3), 383. https://doi.org/10.1037/a0036707

Diedrich, A., Hofmann, S. G., Cuijpers, P., & Berking, M. (2016). Self-compassion enhances the efficacy of explicit cognitive reappraisal as an emotion regulation strategy in individuals with major depressive disorder. *Behaviour Research and Therapy, 82*, 1–10. https://doi.org/10.1016/j.brat.2016.04.003

DiFulvio, G. T. (2011). Sexual minority youth, social connection and resilience: From personal struggle to collective identity. *Social Science & Medicine, 72*(10), 1611–1617. https://doi.org/10.1016/j.socscimed.2011.02.045

Dolan, I. J., Strauss, P., Winter, S., & Lin, A. (2020). Misgendering and experiences of stigma within health-care settings for trans individuals in Western Australia: Implications of structural policies and impacts on delivery of care. *Medical Journal of Australia, 12*(4), 150–151. https://doi.org/10.5694/mja2.50497

Dragowski, E. A., Halkitis, P. N., Grossman, A. H., & D'Augelli, A. R. (2011). Sexual orientation victimization and posttraumatic stress symptoms among lesbian, gay, and bisexual youth. *Journal of Gay & Lesbian Social Services, 23*(2), 226–249.

Eliason, M. J. (1996). Identity formation for lesbian, bisexual, and gay persons: Beyond a 'minoritizing' view. *Journal of Homosexuality, 30*, 31–58. https://doi.org/10.1300/J082v30n03_03

Fergus, S., & Zimmerman, M. A. (2005). Adolescent resilience: A framework for understanding healthy development in the face of risk. *Annual Review of Public Health, 26*, 399–419. https://doi.org/10.1146/annurev.publhealth.26.021304.144357

Finlay-Jones, A., Strauss, P., Perry, Y., Waters, Z., Gilbey, D., Windred, M., et al. (2021). Group mindful self-compassion training to improve mental health outcomes for LGBTQIA+ young adults: Rationale and protocol for a randomised controlled trial. *Contemporary Clinical Trials, 102*, 1–7. https://doi.org/10.1016/j.cct.2021.106268

Fisher, C. M., Waling, A., Kerr, L., Bellamy, R., Ezer, P., Mikolajczak, G., Carmen, M., & Lucke, J. (2019). *6th National Survey of Australian secondary students and sexual health 2018* (ARCSHS monograph series no. 113). Australian Research Centre in Sex, Health & Society, La Trobe University. https://doi.org/10.26181/5c80777f6c35e

Flores, A. R., Herman, J. L., Brown, T. N. T., & Conron, K. J. (2017). *Age of individuals who identify as transgender in the United States*. The Williams Institute. https://williamsinstitute.law.ucla.edu/wp-content/uploads/Age-Trans-Individuals-Jan-2017.pdf

Fredrick, E. G., LaDuke, S. L., & Williams, S. L. (2020). Sexual minority quality of life: The indirect effect of public stigma through self-compassion, authenticity, and internalized stigma. *Stigma and Health, 5*(1), 79. https://doi.org/10.1037/sah0000176

Frost, D. M. (2011). Social stigma and its consequences for the socially stigmatized. *Social and Personality Psychology Compass, 5*, 824–839. https://doi.org/10.1111/j.1751-9004.2011.00394.x

Galla, B. M. (2016). Within-person changes in mindfulness and self-compassion predict enhanced emotional well-being in healthy, but stressed adolescents. *Journal of Adolescence, 49*, 204–217. https://doi.org/10.1016/j.adolescence.2016.03.016

Gertler, L. M. (2014). *The coming out experience, internalized homophobia, and self-compassion in LGBQ young adults*. [Dissertation, The Wright Institute]. ProQuest Dissertations Publishing.

Gilbey, D., Mahfouda, S., Ohan, J., Lin, A., & Perry, Y. (2019). Trajectories of mental health difficulties in young people who are attracted to the same gender: A systematic review. *Adolescent Research Review, 5*(3), 281–293. https://doi.org/10.1007/s40894-019-00128-8

Godfrey, E. B., Santos, C. E., & Burson, E. (2017). For better or worse? System-justifying beliefs in sixth-grade predict trajectories of self-esteem and behavior across early adolescence. *Child Development, 90*(1), 180–195. https://doi.org/10.1111/cdev.12854

Godfrey, E. B., Santos, C. E., & Burson, E. (2019). For better or worse? System-justifying beliefs in sixth-grade predict trajectories of self-esteem and behavior across early adolescence. *Child Development, 90*(1), 180–195.

Gordon, A. R., & Meyer, I. H. (2008). Gender nonconformity as a target of prejudice, discrimination, and violence against LGB individuals. *Journal of LGBT Health Research, 3*(3), 55–71. https://doi.org/10.1080/15574090802093562

Grant, J. M., Mottet, L. A., Tanis, J., Harrison, J., Herman, J. L., & Keisling, M. (2011). *Injustice at every turn: A report of the national transgender discrimination survey*. National Center for Transgender Equality and National Gay and Lesbian Task Force. https://www.

thetaskforce.org/injustice-every-turn-report-national-transgender-discrimination-survey/
Greene, D. C., & Britton, P. J. (2015). Predicting adult LGBTQ happiness: Impact of childhood affirmation, self-compassion, and personal mastery. *Journal of LGBT Issues in Counseling, 9*(3), 158–179. https://doi.org/10.1080/15538605.2015.1068143
Grossman, A. H., & D'Augelli, A. R. (2006). Transgender youth: Invisible and vulnerable. *Journal of Homosexuality, 51*(1), 111–128. https://doi.org/10.1300/J082v51n01_06
Hall, J. S., & Zautra, A. J. (2010). Indicators of community resilience: What are they, why bother? In J. W. Reich, A. J. Zautra, & J. S. Hall (Eds.), *Handbook of adult resilience* (pp. 350–371). The Guilford Press.
Hatchel, T., Merrin, G. J., & Espelage, A. D. (2019). Peer victimization and suicidality among LGBTQ youth: The roles of school belonging, self-compassion, and parental support. *Journal of LGBT Youth, 16*(2), 134–156. https://doi.org/10.1080/19361653.2018.1543036
Hatzenbuehler, M. L. (2009). How does sexual minority stigma get under the skin? A psychological mediation framework. *Psychological Bulletin, 135*(5), 707–730. https://doi.org/10.1037/a0016441
Hatzenbuehler, M. L. (2011). The social environment and suicide attempts in lesbian, gay, and bisexual youth. *Pediatrics, 127*(5), 896–903. https://doi.org/10.1542/peds.2010-3020
Hatzenbuehler, M. L., & Pachankis, J. E. (2016). Stigma and minority stress as social determinants of health among lesbian, gay, bisexual, and transgender youth: Research evidence and clinical implications. *Pediatric Clinics, 63*(6), 985–997. https://doi.org/10.1016/j.pcl.2016.07.003
Hatzenbuehler, M. L., Keyes, K. M., & Hasin, D. S. (2009). State-level policies and psychiatric morbidity in lesbian, gay, and bisexual populations. *American Journal of Public Health, 99*(12), 2275–2281.
Hatzenbuehler, M. L., Dovidio, J. F., Nolen-Hoeksema, S., & Phills, C. E. (2009a). An implicit measure of anti-gay attitudes: Prospective associations with emotion regulation strategies and psychological distress. *Journal of Experimental Social Psychology, 45*(6), 1316–1320. https://doi.org/10.1016/j.jesp.2009.08.005
Hatzenbuehler, M. L., Keyes, K. M., & Hasin, D. S. (2009b). State-level policies and psychiatric morbidity in lesbian, gay, and bisexual populations. *American Journal of Public Health, 99*(12), 2275–2281. https://doi.org/10.2105/AJPH.2008.153510
Hatzenbuehler, M. L., McLaughlin, K. A., Keyes, K. M., & Hasin, D. S. (2010). The impact of institutional discrimination on psychiatric disorders in lesbian, gay, and bisexual populations: A prospective study. *American Journal of Public Health, 100*(3), 452–459. https://doi.org/10.2105/AJPH.2009.168815
Hatzenbuehler, M. L., Wieringa, N. F., & Keyes, K. M. (2011). Community-level determinants of tobacco use disparities in lesbian, gay, and bisexual youth: Results from a population-based study. *Archives of Pediatrics & Adolescent Medicine, 165*(6), 527–532. https://doi.org/10.1001/archpediatrics.2011.64
Hatzenbuehler, M. L., O'Cleirigh, C., Grasso, C., Mayer, K., Safren, S., & Bradford, J. (2012). Effect of same-sex marriage laws on health care use and expenditures in sexual minority men: A quasi-natural experiment. *American Journal of Public Health, 102*(2), 285–291. https://doi.org/10.2105/AJPH.2011.30038
Hatzenbuehler, M. L., Phelan, J. C., & Link, B. G. (2013). Stigma as a fundamental cause of population health inequalities. *American Journal of Public Health, 103*(5), 813–821. https://doi.org/10.2105/AJPH.2012.301069
Hendricks, M. L., & Testa, R. J. (2012). A conceptual framework for clinical work with transgender and gender nonconforming clients: An adaptation of the Minority Stress Model. *Professional Psychology: Research and Practice, 43*(5), 460–467. https://doi.org/10.1037/a0029597
Holley, M. L., Stromwall, L. K., & Bashor, H. (2012). Reconceptualizing stigma: Toward a critical anti-oppression paradigm. *Stigma Research and Action, 2*(2), 51–61. https://doi.org/10.5463/sra.v1i1
Igartua, K., Thombs, B. D., Burgos, G., & Montoro, R. (2009). Concordance and discrepancy in sexual identity, attraction, and behavior among adolescents. *The Journal of Adolescent Health, 45*(6), 602–608. https://doi.org/10.1016/j.jadohealth.2009.03.019
Ioerger, M., Henry, K. L., Chen, P. Y., Cigularov, K. P., & Tomazic, R. G. (2015). Beyond same-sex attraction: Gender-variant-based victimization is associated with suicidal behavior and substance use for other-sex attracted adolescents. *PLoS One, 10*(6), e0129976. https://doi.org/10.1371/journal.pone.0129976
Jerome, R. C., & Halkitis, P. N. (2009). Stigmatization, stress, and the search for belonging in Black men who have sex with men who use methamphetamine. *Journal of Black Psychology, 35*(3), 343–365. https://doi.org/10.1177/0095798409333620
Jost, J. T., Banaji, M. R., & Nosek, B. A. (2004). A decade of system justification theory: Accumulated evidence of conscious and unconscious bolstering of the status quo. *Political Psychology, 25*(6), 881–919. https://doi.org/10.1111/j.1467-9221.2004.00402.x
Katz-Wise, S., & Hyde, J. (2012). Victimization experiences of lesbian, gay, and bisexual individuals: A meta-analysis. *Journal of Sex Research, 49*(2–3), 142–167. https://doi.org/10.1080/00224499.2011.637247
Katz-Wise, S. L., Rosario, M., & Tsappis, M. (2016). LGBT youth and family acceptance. *Pediatric Clinics of North America, 63*(6), 1011–1025. https://doi.org/10.1016/j.pcl.2016.07.005
Keng, S. L., & Liew, K. W. L. (2017). Trait mindfulness and self-compassion as moderators of the association between gender nonconformity and psychological health. *Mindfulness, 8*(3), 615–626. https://doi.org/10.1007/s12671-016-0639-0
Kertzner, R. M., Meyer, I. H., Frost, D. M., & Stirratt, M. J. (2009). Social and psychological well-being in lesbians, gay men, and bisexuals: The effects of race,

gender, age, and sexual identity. *American Journal of Orthopsychiatry, 79*(4), 500–510. https://doi.org/10.1037/a0016848

Keyes, C. L. (2007). Promoting and protecting mental health as flourishing: A complementary strategy for improving national mental health. *American Psychologist, 62*(2), 95–108. https://doi.org/10.1037/0003-066X.62.2.95

Keyes, C. L. (2009). The Black–White paradox in health: Flourishing in the face of social inequality and discrimination. *Journal of Personality, 77*(6), 1677–1706. https://doi.org/10.1111/j.1467-6494.2009.00597.x

Khaleque, A., & Rohner, R. P. (2002). Perceived parental acceptance-rejection and psychological adjustment: A meta-analysis of cross-cultural and intracultural studies. *Journal of Marriage and Family, 64*(1), 54–64. https://doi.org/10.1111/j.1741-3737.2002.00054.x

Kimmel, M. S. (2004). Masculinity as homophobia: Fear, shame, and silence in the construction of gender identity. In P. S. Rothenberg (Ed.), *Race, class, and gender in the United States: An integrated study* (pp. 81–93). Worth.

King, M., Semlyen, J., Tai, S. S., Killaspy, H., Osborn, D., Popelyuk, D., & Nazareth, I. (2008). A systematic review of mental disorder, suicide, and deliberate self harm in lesbian, gay and bisexual people. *BMC Psychiatry, 8*(1), 1–17. https://doi.org/10.1186/1471-244X-8-70

Kosciw, J. G., Diaz, E. A., & Greytak, E. A. (2008). *The 2007 national school climate survey: The experiences of lesbian, gay, bisexual and transgender youth in our nation's schools*. GLSEN. https://www.glsen.org/research/2007-national-school-climate-survey

Krieger, T., Altenstein, D., Baettig, I., Doerig, N., & Holtforth, M. G. (2013). Self-compassion in depression: Associations with depressive symptoms, rumination, and avoidance in depressed outpatients. *Behavior Therapy, 44*(3), 501–513. https://doi.org/10.1016/j.beth.2013.04.004

Kwate, N. O. A., & Meyer, I. H. (2010). The myth of meritocracy and African American health. *American Journal of Public Health, 100*(10), 1831–1834. https://doi.org/10.2105/AJPH.2009.186445

Leary, M. R., Tate, E. B., Adams, C. E., Allen, A. B., & Hancock, J. (2007). Self-compassion and reactions to unpleasant self-relevant events: The implications of treating oneself kindly. *Journal of Personality and Social Psychology, 92*(5), 887–904. https://doi.org/10.1037/0022-3514.92.5.887

Liao, K. Y. H., Kashubeck-West, S., Weng, C. Y., & Deitz, C. (2015). Testing a mediation framework for the link between perceived discrimination and psychological distress among sexual minority individuals. *Journal of Counseling Psychology, 62*(2), 226–241. https://doi.org/10.1037/cou0000064

Link, B. G., & Phelan, J. C. (2001). Conceptualizing stigma. *Annual Review of Sociology, 27*(1), 363–385. https://www.jstor.org/stable/2678626

Lippa, R. A. (2008). The relation between childhood gender nonconformity and adult masculinity-femininity and anxiety in heterosexual and homosexual men and women. *Sex Roles, 59*, 684–693. https://doi.org/10.1007/s11199-008-9476-5

Luthar, S. S. (2006). Resilience in development: A synthesis of research across five decades. In D. Cicchetti, D. J. Cohen, & D. J. Cohen (Eds.), *Developmental psychopathology: Risk, disorder, and adaptation* (Vol. 3, 2nd ed.). Wiley.

Macapagal, K., Coventry, R., Arbeit, M. R., Fisher, C. B., & Mustanski, B. (2017). 'I won't out myself just to do a survey': Sexual and gender minority adolescents' perspectives on the risks and benefits of sex research. *Archives of Sexual Behavior, 46*(5), 1393–1409. https://doi.org/10.1007/s10508-016-0784-5

MacBeth, A., & Gumley, A. (2012). Exploring compassion: A meta-analysis of the association between self-compassion and psychopathology. *Clinical Psychology Review, 32*(6), 545–552. https://doi.org/10.1016/j.cpr.2012.06.003

Mahfouda, S., Moore, J. K., Siafarikas, A., Zepf, F. D., & Lin, A. (2017). Puberty suppression in transgender children and adolescents. *The Lancet Diabetes & Endocrinology, 5*(10), 816–826. https://doi.org/10.1016/S2213-8587(17)30099-2

Mahfouda, S., Moore, J. K., Siafarikas, A., Hewitt, T., Ganti, U., Lin, A., et al. (2018). Gender-affirming hormones and surgery in transgender children and adolescents. *The Lancet Diabetes & Endocrinology, 7*(6), 484–498. https://doi.org/10.1016/S2213-8587(18)30305-X

Marshal, M. P., Dietz, L. J., Friedman, M. S., Stall, R., Smith, H. A., McGinley, J., Thoma, B. C., Murray, P. J., D'Augelli, A. R., & Brent, D. A. (2011). Suicidality and depression disparities between sexual minority and heterosexual youth: A meta-analytic review. *Journal of Adolescent Health, 49*(2), 115–123. https://doi.org/10.1016/j.jadohealth.2011.02.005

Martin, C. L. (1990). Attitudes and expectations about children with nontraditional and traditional gender roles. *Sex Roles, 22*(3–4), 151–166. https://doi.org/10.1007/BF00288188

Masten, A. S. (2001). Ordinary magic: Resilience processes in development. *American Psychologist, 56*(3), 227–238. https://doi.org/10.1037//0003-066X.56.3.227

Matos, M., Carvalho, S. A., Cunha, M., Galhardo, A., & Sepodes, C. (2017). Psychological flexibility and self-compassion in gay and heterosexual men: How they relate to childhood memories, shame, and depressive symptoms. *Journal of LGBT Issues in Counseling, 11*(2), 88–105. https://doi.org/10.1080/15538605.2017.1310007

McDermott, E., Roen, K., & Piela, A. (2013). Hard-to-reach youth online: Methodological advances in self-harm Research. *Sexuality Research and Social Policy, 10*(2), 125–134. https://doi.org/10.1007/s13178-012-0108-z

McDermott, E., Hughes, E., & Rawlings, V. E. (2016, June 30). *Queer future final report: Understanding Lesbian, Gay, Bisexual and Trans (LGBT) Adolescents' suicide, self-harm, and help-seeking behaviour*. http://www.queerfutures.co.uk/wp-content/uploads/2016/06/Queer-Futures-Final-Report.pdf

McGehee, P., Germer, C., & Neff, K. D. (2017). Core values in mindful self-compassion. In L. M. Monteiro, R. F. Musten, & J. C. Compson (Eds.), *A Practitioner's guide to mindfulness and ethics* (pp. 279–294). Springer.

McGuire, T. G., & Miranda, J. (2008). New evidence regarding racial and ethnic disparities in mental health: Policy implications. *Health Affairs, 27*(2), 393–403. https://doi.org/10.1377/hlthaff.27.2.393

McLaughlin, K. A., Hatzenbuehler, M. L., Xuan, Z., & Conron, K. J. (2012). Disproportionate exposure to early-life adversity and sexual orientation disparities in psychiatric morbidity. *Child Abuse & Neglect, 36*(9), 645–655. https://doi.org/10.1016/j.chiabu.2012.07.004

Meyer, I. H. (1995). Minority stress and mental health in gay men. *Journal of Health and Social Behavior, 36*(1), 38–56. https://doi.org/10.2307/2137286

Meyer, I. H. (2003a). Prejudice, social stress, and mental health in lesbian, gay, and bisexual populations: Conceptual issues and research evidence. *Psychological Bulletin, 129*(5), 674–697. https://doi.org/10.1037/0033-2909.129.5.674

Meyer, I. H. (2003b). Prejudice as stress: Conceptual and measurement problems. *American Journal of Public Health, 93*(2), 262–265. https://doi.org/10.2105/AJPH.93.2.262

Meyer, I. H. (2015). Resilience in the study of minority stress and health of sexual and gender minorities. *Psychology of Sexual Orientation and Gender Diversity, 2*(3), 209–213. https://doi.org/10.1037/sgd0000132

Mitchum, P., & Moodie-Mills, A. C. (2014). *Beyond bullying: How hostile school climate perpetuates the school-to-prison pipeline for LGBT youth*. Office of Justice Programs: Center for American Progress. https://www.ojp.gov/ncjrs/virtual-library/abstracts/beyond-bullying-how-hostile-school-climate-perpetuates-school

Mock, S. E., & Eibach, R. P. (2012). Stability and change in sexual orientation identity over a 10-year period in adulthood. *Archives of Sexual Behavior, 41*(3), 641–648. https://doi.org/10.1007/s10508-011-9761-1

Moradi, B., Wiseman, M. C., DeBlaere, C., Goodman, M. B., Sarkees, A., Brewster, M. E., & Huang, Y. (2010). LGB of color and white individuals' perceptions of heterosexist stigma, internalized homophobia, and outness: Comparisons of levels and links. *The Counseling Psychologist, 38*(3), 397–424. https://doi.org/10.1177/0011000009335263

Mueller, A. S., James, W., Abrutyn, S., & Levin, M. L. (2015). Suicide ideation and bullying among US adolescents: Examining the intersections of sexual orientation, gender, and race/ethnicity. *American Journal of Public Health, 105*(5), 980–985. https://doi.org/10.2105/AJPH.2014.302391

Mustanski, B., Newcomb, M. E., & Garofalo, R. (2011). Mental health of lesbian, gay, and bisexual youths: A developmental resiliency perspective. *Journal of Gay & Lesbian Social Services, 23*(2), 204–225. https://doi.org/10.1080/10538720.2011.561474

Mustanski, B., Birkett, M., Greene, G. J., Rosario, M., Bostwick, W., & Everett, B. G. (2014). The association between sexual orientation identity and behavior across race/ethnicity, sex, and age in a probability sample of high school students. *American Journal of Public Health, 104*(2), 237–244. https://doi.org/10.2105/AJPH.2013.301451

Mustanski, B., Andrews, R., & Puckett, J. A. (2016). The effects of cumulative victimization on mental health among lesbian, gay, bisexual, and transgender adolescents and young adults. *American Journal of Public Health, 106*(3), 527–533. https://doi.org/10.2105/AJPH.2015.302976

Nadal, K. L. (2013). *That's so gay! Microaggressions and the lesbian, gay, bisexual, and transgender community*. American Psychological Association. https://doi.org/10.1037/14093-000

Nadal, K. L., Issa, M. A., Leon, J., Meterko, V., Wideman, M., & Wong, Y. (2011). Sexual orientation microaggressions: 'Death by a thousand cuts' for lesbian, gay, and bisexual youth. *Journal of LGBT Youth, 8*(3), 234–259. https://doi.org/10.1080/19361653.2011.584204

National Center for Health Statistics. (2016). *Health, United States, 2015: With special feature on racial and ethnic health disparities*. US Government Printing Office. https://stacks.cdc.gov/view/cdc/39108

Neff, K. (2003a). Self-compassion: An alternative conceptualization of a healthy attitude toward oneself. *Self and Identity, 2*(2), 85–101. https://doi.org/10.1080/15298860309032

Neff, K. D. (2003b). The development and validation of a scale to measure self-compassion. *Self and Identity, 2*(3), 223–250. https://doi.org/10.1080/15298860309027

Neff, K. D., & Germer, C. K. (2013). A pilot study and randomized controlled trial of the mindful self-compassion program. *Journal of Clinical Psychology, 69*(1), 28–44. https://doi.org/10.1002/jclp.21923

Neff, K. D., & Vonk, R. (2009). Self-compassion versus global self-esteem: Two different ways of relating to oneself. *Journal of Personality, 77*(1), 23–50. https://doi.org/10.1111/j.1467-6494.2008.00537.x

Neff, K. D., Rude, S. S., & Kirkpatrick, K. L. (2007). An examination of self-compassion in relation to positive psychological functioning and personality traits. *Journal of Research in Personality, 41*(4), 908–916. https://doi.org/10.1016/j.jrp.2006.08.002

Neff, K. D., Long, P., Knox, M., Davidson, O., Kuchar, A., Costigan, A., Williamson, Z., Rohleder, N., Tóth-Király, I., & Breines, J. (2018). The forest and the trees: Examining the association of self-compassion and its positive and negative components with psychological functioning. *Self and Identity, 17*(6), 627–645. https://doi.org/10.1080/15298868.2018.1436587

Neff, K. D., Tóth-Király, I., Yarnell, L. M., Arimitsu, K., Castilho, P., Ghorbani, N., Guo, H. X., Hirsch, J. K., Hupfeld, J., Hutz, C. S., Kotsou, I., Lee, W. K., Montero-Marin, J., Sirois, F. M., de Souza, L. K., Svendsen, J. L., Wilkinson, R. B., & Mantzios, M. (2019). Examining the factor structure of the self-compassion scale in 20 diverse samples: Support for use of a total score and six subscale scores. *Psychological Assessment, 31*(1), 27–45. https://doi.org/10.1037/pas0000629

Neighbors, H. W., Jackson, J. S., Broman, C., & Thompson, E. (1996). Racism and the mental health of African Americans: The role of self and system blame. *Ethnicity & Disease, 6*(1–2), 167–175. https://www.jstor.org/stable/45409645

Newcomb, M. E., & Mustanski, B. (2010). Internalized homophobia and internalizing mental health problems: A meta-analytic review. *Clinical Psychology Review, 30*(8), 1019–1029. https://doi.org/10.1016/j.cpr.2010.07.003

Okun, T. (n.d.). *(Divorcing) White supremacy culture.* http://www.dismantlingracism.org/white-supremacy-culture.html

Olive, J. L. (2012). Reflections on the life histories of today's LGBQ postsecondary students. *Journal of LGBT Youth, 9*(3), 247–265. https://doi.org/10.1080/19361653.2012.652854

Podsakoff, P. M., MacKenzie, S. B., Lee, J. Y., & Podsakoff, N. P. (2003). Common method biases in behavioral research: A critical review of the literature and recommended remedies. *Journal of Applied Psychology, 88*(5), 879–903. https://doi.org/10.1037/0021-9010

Poteat, V. P., Sinclair, K. O., DiGiovanni, C. D., Koenig, B. W., & Russell, S. T. (2012). Gay-straight alliances are associated with student health: A multi-school comparison of LGBTQ and heterosexual youth. *Journal of Research on Adolescence, 23*(2), 319–330. https://doi.org/10.1111/j.1532-7795.2012.00832

Puckett, J. A., & Levitt, H. M. (2015). Internalized stigma within sexual and gender minorities: Change strategies and clinical implications. *Journal of LGBT Issues in Counseling, 9*(4), 329–349. https://doi.org/10.1080/15538605.2015.1112336

Puckett, J. A., Levitt, H. M., Horne, S. G., & Hayes-Skelton, S. A. (2015). Internalized heterosexism and psychological distress: The mediating roles of self-criticism and community connectedness. *Psychology of Sexual Orientation and Gender Diversity, 2*(4), 426–435. https://doi.org/10.1037/sgd0000123

Puckett, J. A., Newcomb, M. E., Garofalo, R., & Mustanski, B. (2016). The impact of victimization and neuroticism on mental health in young men who have sex with men: Internalized homophobia as an underlying mechanism. *Sexuality Research & Social Policy, 13*(3), 193–201. https://doi.org/10.1007/s13178-016-0239-8

Raes, F. (2011). The effect of self-compassion on the development of depression symptoms in a non-clinical sample. *Mindfulness, 2*(1), 33–36. https://doi.org/10.1007/s12671-011-0040-y

Reisner, S. L., Greytak, E. A., Parsons, J. T., & Ybarra, M. L. (2015a). Gender minority social stress in adolescence: Disparities in adolescent bullying and substance use by gender identity. *The Journal of Sex Research, 52*(3), 243–256. https://doi.org/10.1080/00224499.2014.88632

Reisner, S. L., Vetters, R., Leclerc, M., Zaslow, S., Wolfrum, S., Shumer, D., & Mimiaga, M. J. (2015b). Mental health of transgender youth in care at an adolescent urban community health center: A matched retrospective cohort study. *The Journal of Adolescent Health, 56*(3), 274–279. https://doi.org/10.1016/j.jadohealth.2014.10.264

Rider, G. N., McMorris, B. J., Gower, A. L., Coleman, E., & Eisenberg, M. E. (2018). Health and care utilization of transgender and gender nonconforming youth: A population-based study. *Pediatrics, 141*(3), e20171683. https://doi.org/10.1542/peds.2017-1683

Rieger, G., & Savin-Williams, R. C. (2012). Gender nonconformity, sexual orientation, and psychological Well-being. *Archives of Sexual Behavior, 41*(3), 611–621. https://doi.org/10.1007/s10508-011-9738-0

Rood, B., Reisner, S., Surace, F., Puckett, J., Maroney, M., & Pantalone, D. (2016). Expecting rejection: Understanding the minority stress experiences of transgender and gender-nonconforming individuals. *Transgender Health, 1*(1), 151–164. https://doi.org/10.1089/trgh.2016.0012

Rosario, M., Schrimshaw, E. W., & Hunter, J. (2011). Different patterns of sexual identity development over time: Implications for the psychological adjustment of lesbian, gay, and bisexual youths. *Journal of Sex Research, 48*(1), 3–15. https://doi.org/10.1080/00224490903331067

Russell, G. M. (2007). Internalized homophobia: Lessons from the mobius strip. In C. Brown & T. Augusta-Scott (Eds.), *Narrative therapy: Making meaning, making lives* (pp. 151–176). Sage Publications.

Ryan, C., Huebner, D., Diaz, R. M., & Sanchez, J. (2009). Family rejection as a predictor of negative health outcomes in white and Latino lesbian, gay, and bisexual young adults. *Pediatrics, 123*(1), 346–352. https://doi.org/10.1542/peds.2007-352

Ryan, C., Russell, S. T., Huebner, D., Diaz, R., & Sanchez, J. (2010). Family acceptance in adolescence and the health of LGBT young adults. *Journal of Child and Adolescent Psychiatric Nursing, 23*(4), 205–213. https://doi.org/10.1111/j.1744-6171.2010.00246.x

Saewyc, E. M. (2011). Research on adolescent sexual orientation: Development, health disparities, stigma, and resilience. *Journal of Research on Adolescence, 21*(1), 256–272. https://doi.org/10.1111/j.1532-7795.2010.00727.x

Schilt, K., & Westbrook, L. (2009). Doing gender, doing heteronormativity: 'Gender normals,' transgender people, and the social maintenance of heterosexuality. *Gender & Society, 23*(4), 440–464. https://doi.org/10.1177/0891243209340034

Seng, J. S., Lopez, W. D., Sperlich, M., Hamama, L., & Reed Meldrum, C. D. (2012). Marginalized identities, discrimination burden, and mental health: Empirical exploration of an interpersonal-level approach to modeling intersectionality. *Social Science & Medicine, 75*(12), 2437–2445. https://doi.org/10.1016/j.socscimed.2012.09.023

Shapira, L. B., & Mongrain, M. (2010). The benefits of self-compassion and optimism exercises for individuals vulnerable to depression. *The Journal of Positive Psychology, 5*(5), 377–389. https://doi.org/10.1080/17439760.2010.516763

Sidanius, J., & Pratto, F. (2012). Social dominance theory. In P. A. M. Van Lange, A. W. Kruglanski, & E. T. Higgins (Eds.), *Handbook of theories of social psychology* (pp. 418–438). Sage.

Skidmore, W. C., Linsenmeier, J. A., & Bailey, J. M. (2006). Gender nonconformity and psychological distress in lesbians and gay men. *Archives of Sexual Behavior, 35*(6), 685–697.

Steinberg, L., Dahl, R., Keating, D., Kupfer, D. J., Masten, A., & Pine, D. (2006). The study of developmental psychopathology in adolescence: Integrating affective neuroscience with the study of context. In D. Cicchetti & D. Cohen (Eds.), *Developmental psychopathology: Developmental neuroscience* (pp. 710–741). Wiley.

Strauss, P., Cook, A., Winter, S., Watson, V., Wright Toussaint, D., & Lin, A. (2020a). Associations between negative life experiences and the mental health of trans and gender diverse young people in Australia: Findings from trans pathways. *Psychological Medicine, 50*(5), 808–817. https://doi.org/10.1017/S0033291719000643

Strauss, P., Cook, A., Winter, S., Watson, V., Wright Toussaint, D., & Lin, A. (2020b). Mental health issues and complex experiences of abuse among trans and gender diverse young people: Findings from trans pathways. *LGBT Health, 7*(3), 128–136. https://doi.org/10.1089/lgbt.2019.0232

Swain, J. E., Konrath, S., Brown, S. L., Finegood, E. D., Akce, L. B., Dayton, C. J., & Ho, S. S. (2012). Parenting and beyond: Common neurocircuits underlying parental and altruistic caregiving. *Parenting, 12*(2–3), 115–123.

Szymanski, D. M., Dunn, T. L., & Ikizler, A. S. (2014). Multiple minority stressors and psychological distress among sexual minority women: The roles of rumination and maladaptive coping. *Psychology of Sexual Orientation and Gender Diversity, 1*(4), 412–421. https://doi.org/10.1037/sgd0000066

Tanaka, M., Wekerle, C., Schmuck, M. L., Paglia-Boak, A., & Research Team, M. A. P. (2011). The linkages among childhood maltreatment, adolescent mental health, and self-compassion in child welfare adolescents. *Child Abuse & Neglect, 35*(10), 887–898. https://doi.org/10.1016/j.chiabu.2011.07.003

Toomey, R. B., Ryan, C., Diaz, R. M., Card, N. A., & Russell, S. T. (2010). Gender-nonconforming lesbian, gay, bisexual, and transgender youth: School victimization and young adult psychosocial adjustment. *Developmental Psychology, 46*(6), 1580–1589. https://doi.org/10.1037/a0020705

Toomey, R. B., McGuire, J. K., & Russell, S. T. (2012). Heteronormativity, school climates, and perceived safety for gender nonconforming peers. *Journal of Adolescence, 35*(1), 187–196. https://doi.org/10.1016/j.adolescence.2011.03.001

Toplu-Demirtaş, E., Kemer, G., Pope, A. L., & Moe, J. L. (2018). Self-compassion matters: The relationships between perceived social support, self-compassion, and subjective well-being among LGB individuals in Turkey. *Journal of Counseling Psychology, 65*(3), 372–382. https://doi.org/10.1037/cou0000261

Tyler, I. (2018). The hieroglyphics of the border: Racial stigma in neoliberal Europe. *Ethnic and Racial Studies, 41*(10), 1783–1801. https://doi.org/10.1080/01419870.2017.1361542

Van Beusekom, G. V. B., Bos, M. W., Overbeek, G., & Sandfort, T. G. M. (2015). Same-sex attraction, gender nonconformity, and mental health: The protective role of parental acceptance. *Psychology of Sexual Orientation and Gender Diversity, 2*(3), 307–312. https://doi.org/10.1037/sgd000011

Van Dam, N. T., Sheppard, S. C., Forsyth, J. P., & Earleywine, M. (2011). Self-compassion is a better predictor than mindfulness of symptom severity and quality of life in mixed anxiety and depression. *Journal of Anxiety Disorders, 25*(1), 123–130. https://doi.org/10.1016/j.janxdis.2010.08.011

Vettese, L. C., Dyer, C. E., Li, W. L., & Wekerle, C. (2011). Does self-compassion mitigate the association between childhood maltreatment and later emotion regulation difficulties? A preliminary investigation. *International Journal of Mental Health and Addiction, 9*(5), 480–491. https://doi.org/10.1007/s11469-011-9340-7

Vigna, A. B., Poehlmann-Tynan, J., & Koenig, B. (2018a). Does self-compassion facilitate resilience to stigma? A school-based study of sexual and gender minority youth. *Mindfulness, 9*(3), 914–924. https://doi.org/10.1007/s12671-017-0831-x

Vigna, A. B., Poehlmann-Tynan, J., & Koenig, B. (2018b). Does self-compassion covary with minority stress? Examining group differences at the intersection of marginalized identities. *Self and Identity, 17*(6), 687–709. https://doi.org/10.1080/15298868.2018.1457566

Vigna, A. J., Poehlmann-Tynan, J., & Koenig, B. (2020). Is self-compassion protective among sexual and gender-minority adolescents across racial groups? *Mindfulness, 11*(3), 800–815. https://doi.org/10.1007/s12671-019-01294-5

Wadsworth, L. P., & Hayes-Skelton, S. A. (2015). Differences among lesbian, gay, bisexual, heterosexual individuals, and those who reported an other identity on an open-ended response on levels of social anxiety. *Psychology of Sexual Orientation and Gender Diversity, 2*(2), 181–187. https://doi.org/10.1037/sgd0000092

Wernick, J. A., Busa, S., Matouk, K., Nicholson, J., & Janssen, A. (2019). A systematic review of the psychological benefits of gender-affirming surgery. *Urologic*

Clinics, 46(4), 474–486. https://doi.org/10.1016/j.ucl.2019.07.002

White Hughto, J. M., Reisner, S. L., & Pachankis, J. E. (2015). Transgender stigma and health: A critical review of stigma determinants, mechanisms, and interventions. *Social Science & Medicine, 147*, 222–231. https://doi.org/10.1016/j.socscimed.2015.11.010

Winter, S., Diamond, M., Green, J., Karasic, D., Reed, T., Whittle, S., & Wylie, K. (2016). Transgender people: Health at the margins of society. *The Lancet, 388*(10042), 390–400. https://doi.org/10.1016/S0140-6736(16)00683-8

Worthen, M. G. F. (2016). Hetero-cis–normativity and the gendering of transphobia. *International Journal of Transgenderism, 17*(1), 31–57. https://doi.org/10.1080/15532739.2016.1149538

Yadavaia, J. E., & Hayes, S. C. (2012). Acceptance and commitment therapy for self-stigma around sexual orientation: A multiple baseline evaluation. *Cognitive and Behavioral Practice, 19*(4), 545–559. https://doi.org/10.1016/j.cbpra.2011.09.002

Yang, X., & Mak, W. W. S. (2017). The differential moderating roles of self-compassion and mindfulness in self-stigma and well-being among people living with mental illness or HIV. *Mindfulness, 8*(3), 595–602. https://doi.org/10.1007/s12671-016-0635-4

Yarnell, L. M., Neff, K. D., Davidson, O. A., & Mullarkey, M. (2019). Gender differences in self-compassion: Examining the role of gender role orientation. *Mindfulness, 10*(6), 1136–1152.

Zaza, S., Kann, L., & Barrios, L. C. (2016). Lesbian, gay, and bisexual adolescents: Population estimate and prevalence of health behaviors. *Journal of the American Medical Association, 316*(22), 2355–2356. https://doi.org/10.1001/jama.2016.11683

Zessin, U., Dickhäuser, O., & Garbade, S. (2015). The relationship between self-compassion and Well-being: A meta-analysis. *Applied Psychology: Health and Well-Being, 7*(3), 340–364. https://doi.org/10.1111/aphw.12051

10 Self-Compassion as a Resource of Resilience

J. Austin, C. H. C. Drossaert, and E. T. Bohlmeijer

From Risk to Resilience

For a long time, the field of clinical and developmental psychology has been grounded in a disease model of stress and coping and focused almost exclusively on pathology, mental illness, social problems, and risk factors (Zautra & Reich, 2012). Clinical psychology has been predominantly concerned with maladaptive functioning and hardly with the promotion of adaptive functioning (Bohlmeijer & Westerhof, 2021). Well-known risk factors such as unemployment and social isolation were expected to result in diminished functioning and health, assuming a linear relation between exposure to risk factors and resulting poor health outcomes. The underlying premise was that people tend to get lost in despair at times of adversity. However, rates of severe distress and posttraumatic stress symptoms following traumatic experiences have been found lower than anticipated (Zautra & Reich, 2012), and a wealth of research demonstrated positive and adaptive changes following adversity that were unaccounted for by known risk factors (e.g., Garmezy, 1991; Luthar et al., 1993; Rutter, 1993; Werner & Johnson, 2002). It is now firmly established that resilience is a common and ordinary response to experiencing adversity (Christopher, 2004; Masten, 2001; Richardson, 2002; Zautra et al., 2010). In this chapter we will explore self-compassion as a resource of resilience. First, we will discuss how resilience is defined and how it can be related to both mental distress and mental well-being. We will then investigate how self-compassion contributes to resilience using examples of experiencing war and cancer.

J. Austin (✉) · C. H. C. Drossaert · E. T. Bohlmeijer
Department of Psychology, Health & Technology, University of Twente, Enschede, The Netherlands
e-mail: j.austin@utwente.nl

What Is Resilience?

There are varying definitions of resilience. Some emphasize individual traits that contribute to recovery from adversity, such as optimism, agency, and the ability to make meaning of experiences. Others focus more on social and environmental processes that facilitate individual resilience, such as supportive family influences and community cohesion (Ungar, 2012; Zautra et al., 2010). Resilience may in fact be seen as a metatheory, encompassing many fields of inquiry (e.g., social, cognitive, and evolutionary) (Joseph & Linley, 2006; Richardson, 2002). The consensus seems to be that resilience is best defined as *an adaptive response to adversity* (Richardson, 2002; Zautra et al., 2010). A common approach is to infer resilience based on the individual variations in outcomes of people who experienced adversity (Ungar, 2012). According to Zautra et al. (2010), these outcomes can be categorized into recovery, sustainability, and growth. First, *recovery* pertains to the return to baseline func-

A. Finlay-Jones et al. (eds.), *Handbook of Self-Compassion*, Mindfulness in Behavioral Health,
https://doi.org/10.1007/978-3-031-22348-8_10

tioning, or the "bouncing back" emotionally after adversity. This does not necessarily imply recovery to previous parameters, since resiliency refers to maintaining *any* healthy, relatively stable equilibrium of psychological and physical functioning, including a newly established one (Bonanno, 2004). It has even been posited that recovery to pre-trauma functioning leaves the individual vulnerable to future traumatization, since their world view has not been accommodated to encompass the disruptive experience (Joseph & Linley, 2006). Thus, sometimes recovery is about "bouncing forward" (Walsh, 2020). Second, while meeting the challenges of the stressors at hand, resilience is fostered by *sustainability* of approach motivations toward personal values and goals (Zautra et al., 2010). Being able to maintain sources of positive emotions and engagement with meaningful goal-directed activities contributes to a sense of well-being and can be seen as a fundamental aspect of resilience (Tugade et al., 2004). Third, in addition to recovering from adversity and sustaining purposeful living, a disruptive event encompasses a potential for (adversarial or posttraumatic) *growth*. The disruption of normalcy and the individual's worldview during times of adversity asks for a reintegration and meaning-making of life events (Joseph & Linley, 2006; Richardson, 2002; Tedeschi & Calhoun, 2003; Zautra et al., 2010). When experiences are accommodated in a positive direction, growth may occur in self-views (e.g., greater acceptance of vulnerabilities), life orientation (e.g., renegotiating priorities), and social relationships (e.g., increased closeness with others) (Joseph & Linley, 2006). Adversarial growth (or posttraumatic growth, benefit finding) has been linked to increased well-being and reduced distress (Helgeson et al., 2006) and increased social connection and support (Petrie et al., 1999; Rzeszutek & Gruszczyńska, 2018). While it is now established that resilience, consisting of recovery, sustainability, and growth processes, is a normal and beneficial response to adversity, the extent to which resilience is accounted for in the broader mental health context (e.g., models, assessments, and interventions of mental health) is less clear.

Resilience in the Context of Mental Health

Similar to the risk and resilience literature, there has been a tendency to automatically equate mental health with an absence of mental illness or symptoms of psychopathology. However, mental health has two dimensions: mental illness and mental well-being (Westerhof & Keyes, 2010). The World Health Organization (WHO) defines mental well-being as: "A state of well-being in which the individual realizes [their] own abilities, can cope with the normal stresses of life, can work productively and fruitfully and is able to make a contribution to [their] community" (2005, p. 12). In concordance with this definition, mental well-being can be divided into three components. The first component is *emotional* or subjective well-being and comprises satisfaction with life and the presence of positive affect (Diener et al., 1999). The second component is *psychological* well-being and comprises aspects of positive, individual functioning such as autonomy, self-acceptance, and having meaningful goals and positive relationships (Ryff, 1989). The third component is *social* well-being and can be defined as optimal social functioning in terms of social engagement and societal functioning (Keyes, 1998). In this way, mental well-being comprises the presence of both emotional well-being as an indicator of *feeling* well along with psychological and social well-being as indicators of *living* well. When emotional, psychological, and social well-being are all high, this is defined as flourishing (Keyes, 2002). Sustainable mental health can be realized when both the reduction of maladaptive and the promotion of adaptive cognitions, emotions, and behavior are focused on in interventions (Bohlmeijer & Westerhof, 2021; Wood & Tarrier, 2010).

In the context of clinical psychology, the possibility of increased positive functioning as an outcome of treatment of mental illness or distress has long been overlooked (Rottenberg et al., 2018). Yet clients particularly value personal growth as an outcome of treatment (Zimmerman et al., 2006). For example, in a systematic review of qualitative studies on clients' perspectives of recovery among persons recovered from eating disorders, it was found that clinical indices for recovery were considered less important than aspects of recovery related to psychological well-being (De Vos et al., 2017). Also, it has been demonstrated that the presence of mental well-being cannot be taken for granted when mental illness is absent. Mental illness and mental well-being function as two related yet distinct phenomena: one continuum represents the presence or absence of mental well-being, the other the presence or absence of mental illness. This two-related-factor model of mental well-being and illness has demonstrated superiority over one-factor models in large representative surveys of American, English, and Dutch adults (Keyes, 2005; Lamers et al., 2011; Schotanus-Dijkstra et al., 2016; Weich et al., 2011; Westerhof & Keyes, 2010) and also recently in clinical samples (Franken et al., 2018). The two-continua model explains why some people with mental illness may still be able to flourish (De Vos et al., 2017; Westerhof & Keyes, 2010) and clients may reliably change on either distress or mental well-being (Trompetter et al., 2017).

Resilience and the Two-Continua Model of Mental Health

The two-continua model of mental health is also relevant to the conceptualization and research of resilience. One dimension of resilience is related to coping with stress-related difficult emotions and cognitions (Livneh, 2001) (i.e., recovery). Regulating these, often short-term, responses with adaptive coping and emotion regulation strategies is important for restoring previous levels of functioning or bouncing back or forward to an earlier or new equilibrium. Successful coping with negative emotions and cognitions may primarily influence distress and symptoms of mental illness. However, there is also a growing body of literature focusing on a second dimension of resilience, i.e., the adaptive role of positive processes in the aftermath of negative life-events (i.e., sustainability and growth). In reaction to adversity, people may become increasingly aware of and develop positive responses such as positive emotions, gratitude, strengths, virtues, positive relationships, renewed meaning, and values (Peterson & Seligman, 2003; Tedeschi & Calhoun, 1995; Zautra et al., 2010). Successful regulation of these, often longer-term, positive responses may primarily result in personal growth and higher levels of mental well-being. However, although recovery from mental distress may often pertain to short-term processes and sustainability and growth of mental well-being may often pertain to long-term processes, this distinction is of course not absolute. Awareness of positive events and emotions may promote adaptation in the short term (e.g., experiencing gratitude), and negative emotions such as sadness and anger may be experienced in the long term and warrant continued coping. Below we will argue that self-compassion can be related to both distress-reducing and well-being-promoting dimensions of resilience and mental health.

Self-Compassion as an Adaptation and Resilience Resource

Self-compassion refers to a warm, wise, and kind attitude in times of difficulty and the ability to be sensitive to personal suffering (Neff et al., 2007). Self-compassion can be described as an adaptive way of relating to the self when experiencing personal inadequacies or adversity in life (Gilbert, 2010; Neff & McGehee, 2010). Many definitions of (self-)compassion exist, all stating that compassion consists of an awareness of suffering, being moved by the suffering, and acting or being motivated to act to alleviate suffering (Strauss et al., 2016). Variations in the definitions pertain to additional elements of the ability to tolerate uncomfortable feelings in response to perceived suffering and a recognition of commonality with other suffering beings, as well as a focus on compassion for/from others or for oneself (Strauss et al., 2016). For example, Gilbert (2014) describes three flows of compassion: having compassion for the self, having compassion for others, and (being open to and capable of) receiving compassion from others. Focusing on compassion toward the self, Neff (2003) proposed three elements that make up self-compassion: (1) *mindfulness* or holding one's present-moment experience in balanced perspective rather than getting lost in surrounding narratives; (2) *self-kindness*, or treating oneself with care and understanding rather than with harsh self-judgment; and (3) *common humanity*, or the acknowledgment that imperfection is a shared aspect among all humans rather than a sign of personal failure. Together, these elements form a self-compassionate frame of mind. Self-compassion is positively related to resilience in various general and clinical populations (Alizadeh et al., 2018; Bluth et al., 2018; Hayter & Dorstyn, 2014; Nery-Hurwit et al., 2018) and can promote resilience in a number of ways. We will now explore self-compassion as an adaptation resource in the context of recovery (mental distress) and sustainability and growth (mental well-being) processes of resilience, as illustrated in Fig. 10.1.

Recovery from Mental Distress with Self-Compassion

Given that self-compassion is often measured on a spectrum with uncompassionate self-responding at one end (self-judgment, isolation, over-identification) and compassionate self-responding at the other end (kindness, common humanity, and mindfulness), one of the reasons self-compassion may reduce distress is because it facilitates less judgment, isolation and over-identification.

For example, one way in which self-compassion may serve as a resource of recovery and resilience in general is through regulating overwhelming emotions with *mindfulness*. Through mindfulness, experiences are held in balanced awareness without resisting, avoiding, or exaggerating them (Germer & Neff, 2019). Mindfulness facilitates awareness and clarity regarding emotional experiences, which are helpful for active coping (Eckland & Berenbaum, 2020). People high in self-compassion are more able to maintain openness and equanimity in the face of stressors, thereby modifying the context in which these negative experiences occur (Trompetter et al., 2017). In addition, people high in self-compassion seem to exhibit less maladaptive emotion regulation strategies such as experiential avoidance (Allen & Leary, 2010; Eichholz et al., 2020; Finlay-Jones et al., 2015; Raes, 2010; Scoglio et al., 2018; Trompetter et al., 2017). This nonreactive, nonjudgmental stance toward the stressors at hand is linked to higher

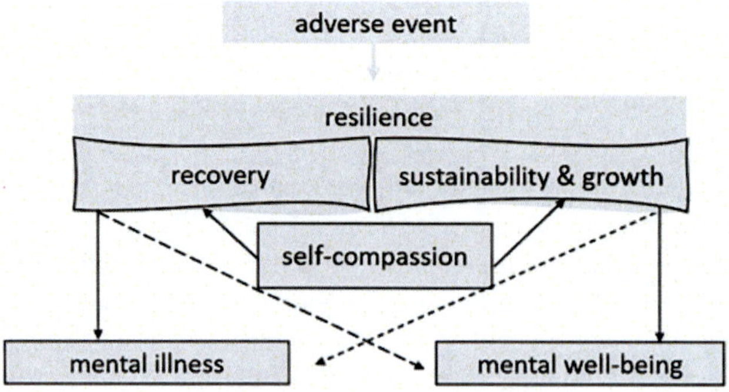

Fig. 10.1 Model of self-compassion (including mindfulness, self-kindness, and common humanity aspects) as a resource of resilience in the context of mental illness and well-being. An adverse event is depicted as a precedent to resilience. This model does not include other (non-resilient) responses to adversity that may occur. Recovery predominantly involves reduction of mental distress/illness (solid arrow) but may also involve boosting well-being (dashed arrow). Sustainability and growth predominantly involve increasing mental well-being (solid arrow) but may also pertain to reducing mental illness (dashed arrow). Self-compassion aids both the distress-reducing and well-being-promoting dimensions of resilience.

resilience in clinical and nonclinical populations (Chien-Chung et al., 2020; Freligh & Debb, 2019; Montero-Marin et al., 2015; Roemer et al., 2015; Zarotti et al., 2020). Gilbert (2014) describes three main emotion regulation systems: one focused on abilities to notice and respond to threat appropriately (threat system), one focused on seeking out and acquiring resources (drive system), and one focused on caring and resting (soothing system). The soothing system represents an evolved mammalian caregiving system, allowing mammals, in contrast to other lifeforms, to protect, nurture, and soothe their immaturely born young. Through mindfulness, the soothing system is accessed, allowing for a non-striving, accepting, and being-in-the-moment experience rather than a state based on activation (e.g., protecting or achieving) (Gilbert, 2014). When recovering from adverse life events, mindfulness thus supports the individual to not get lost in surrounding narratives and stay rooted in present-moment experiences.

In addition to being present, *self-kindness* can be evoked in order to comfort and care for ourselves in the midst of emotional turmoil, thereby alleviating distress (Germer & Neff, 2019). This is related to the notion that (self-)compassion involves the motivation to alleviate suffering and that by practicing self-compassion, we are treating ourselves with the same kindness and understanding that we would treat another during times of adversity (Gilbert, 2014). While mindfulness alone is about the *experience* and *being mode*, self-compassion is about the *experiencer* and *doing mode*. In other words, mindfulness is focused on our relationship with our experiences (how we perceive and respond to external and internal stimuli), while self-compassion focuses on how we relate to ourselves (e.g., as a person having a difficult experience). In the context of recovery, self-kindness may take the form of nurturing, reassuring, or self-supportive thoughts or behaviors, rather than self-critical thinking or destructive behaviors (Neff, 2003). Being critical toward one's own role in causing, exacerbating, or dealing with an adverse event can be seen as an internal threat to our self-concept, thereby activating threat-based emotions such as disappointment, anger, or contempt. In contrast, self-kindness can activate self-soothing emotions (Gilbert, 2014). Indeed, individuals with PTSD with higher levels of shame engage more in self-critical and less in self-reassuring ways of thinking (Harman & Lee, 2010). Self-kindness may also take the form of self-care behaviors to facilitate recovery, for example, by taking rest, self-soothe, lower productivity, guard boundaries, adhere to medical/psychological treatment, and conduct healthy lifestyle behaviors. Evidently, self-kindness not only pertains to being kind and self-soothing (i.e., "yin self-compassion") but may just as well involve firm action such as setting boundaries or undergoing difficult treatments (i.e., "yang self-compassion") (Neff & Germer, 2018). The objective is to alleviate suffering, and whether that involves soothing or firm action likely differs from situation to situation and from person to person.

Finally, self-compassion may foster experiences of *common humanity*, belonging, and social support (Alizadeh et al., 2018; Neff, 2003; Wilson et al., 2020). Through the awareness that suffering is part of life and experiencing suffering is not a personal failing, experiences can be de-shamed, and feelings of self-blame are diminished (Gilbert, 2014; Neff, 2003). This means that common humanity evokes an understanding that we haven't necessarily done something wrong and that emotional challenges are simply part of the human experience. In contrast, in response to adverse events, people often tend to feel as if everyone else has it easier and that the personal suffering is abnormal rather than a part of being human (Germer & Neff, 2019). Self-compassion is related to greater feelings of being supported (Alizadeh et al., 2018; Wilson et al., 2020) and thus may be helpful in counteracting these feelings of isolation.

Sustainability and Growth of Mental Well-being with Self-Compassion

Beyond initial recovery, confrontation with adversity can ignite a process of revaluation of values and goals in life. Self-compassion could foster sustainability of approach motivations and growth, thereby contributing to mental well-being. Commonly, discrepancies between a previous world view (e.g., a just and fair world) and the threat to this worldview instigated by the adversity (e.g., this suffering is unfair) are attempted to be bridged in a process of meaning-making (Park & Ai, 2006). Self-compassion can facilitate the process of meaning-making, for example, by evaluating the adversity in a balanced way (Yela et al., 2020). Research shows that past the initial shock or survival mode following adverse events, rumination and overidentification with the adverse event often persist (García et al., 2015; Im & Follette, 2016; Kim et al., 2017; Szabo et al., 2017). Through *mindfulness*, adverse experiences can continue to be held in balanced awareness (Tubbs et al., 2019; Vujanovic et al., 2009), allowing flexibility to process experiences and to create space for making sustainable and growth-promoting choices. In conjunction, *self-kindness* helps to facilitate proactive action, which could promote well-being in the long term (Akın, 2014). These fierce aspects of self-compassion, such as encouraging growth and drawing boundaries, facilitate getting back to a meaningful life after dealing with adversity (Germer & Neff, 2019). In the context of Gilbert's three emotion regulation systems, this implies a reactivation of the drive system, involving the motivation to engage in valued activities once again. This engagement may take the form of committed action steps (e.g., developing assertiveness skills, refraining from addictive behaviors, targeted exposure to fears) that serve personal values such as taking good care of oneself and others (Tirch et al., 2014). While integrating the adverse experiences into a new narrative, awareness of *common humanity* enables seeing adversity as part of being human, rather than a unique and isolating personal failure or a case of bad luck (Neff, 2003). This may allow for a growth-promoting narrative, with more flexibility in moving onward in life. Past adversity and trauma pose a risk for experiencing loneliness (Hensley et al., 2012; Hyland et al., 2019; Kearney et al., 2018; Zahava Solomon et al., 2015; Zeligman et al., 2017). Loneliness in turn hinders the opportunity to experience growth (Lee et al., 2019; Zeligman et al., 2017). In contrast, self-compassion promotes resilience by facilitating decreased loneliness (Akın, 2014).

Indeed, research shows that self-compassion is related to higher resilience and mental well-being in the context of various types of adversity, such as low well-being (Sommers-Spijkerman et al., 2018), work- and study-related stress and anxiety (Kemper et al., 2015; Lefebvre et al., 2020; Tang, 2019), chronic illness (Baker et al., 2019; Hayter & Dorstyn, 2014; Nery-Hurwit et al., 2018), interpersonal violence (Scoglio et al., 2018), divorce (Masumeh et al., 2019), and other types of trauma such as natural disaster and traffic accidents (Shebuski et al., 2020). More recently, a burgeoning body of research has explored the protective role of self-compassion in buffering against stressors arising as a result of the COVID-19 pandemic (Lau et al., 2020). For example, studies have demonstrated that self-compassion is associated with less pandemic-related stress and greater resilience in populations who have experienced extreme disruption and occupational challenges, including health professionals (Kotera et al., 2021), teachers (Chen, 2022), and parents (Davis et al., 2021).

Despite many decades of research on adversity, very little is known about differences between various kinds of adversity and how they may affect resilience. However, distinctions have been made between chronic adversity or single-incident trauma (Bonanno & Diminich, 2013) and internal or external sources of threat (Sumalla et al., 2009). To explore the ways in which self-compassion may contribute to resilience in depth, we will use two different types of adversity as case studies: going through war (i.e., an external threat) and receiving a cancer diagnosis (i.e., an internal threat). These contrasting types of hardship will aid the exploration of self-compassion as a resource of resilience across adversity types.

Self-Compassion and Resilience in the Context of War Veterans

The Adversity of Going Through War

"Difference exaggerated, invented, or politicized in the extreme can explode into large-scale armed conflict between groups that find others so 'other' that they must be killed" (Sylvester, 2011, p. 1), in other words, war. Exposure to the atrocities of war may involve committing, witnessing, or failing to prevent acts of violence; experiencing betrayal, loss, and sexual misconduct; and observing grotesque mutilations, injury, and death (Forkus et al., 2019; Lueger-Schuster et al., 2012; Snyder, 2014). War affects individuals in myriad ways, as combatant, victim of abuse, family member, service or health professional, and many more (Sylvester, 2011). For this chapter, we will focus on veterans of war and the psychological consequences of surviving a war after serving in the military. Other research addresses mental illness and well-being in military recruits and prisoners of war (e.g., Mantzios, 2014; Solomon et al., 2009).

Veterans are at increased risk for posttraumatic stress symptoms and posttraumatic stress disorder (PTSD). PTSD is a mental health disorder resulting from exposure to trauma, characterized by intrusive reexperiencing of the event, avoidance, negative cognitions, and emotional arousal, among other symptoms (Diagnostic and Statistical Manual of Mental Disorders – Fifth Edition: DSM-5, 2013). The occurrence of PTSD among veterans has been estimated at around 23% (Bryan et al., 2013; Fulton et al., 2015), and a greater proportion may experience posttraumatic stress symptoms (Melvin et al., 2012; Schreuder et al., 2000). Depression (Gadermann et al., 2012), suicidal ideation (Craig et al., 2015), deliberate self-harm (Bryan & Bryan, 2014), elevated levels of anger (Renshaw & Kiddie, 2012; Wilk et al., 2015), and drug and alcohol abuse (Burnett-Zeigler et al., 2011; Jeffery et al., 2013; Seal et al., 2011) are common among veterans. Fundamental views about the self (e.g., that one has control over their experiences), the world (e.g., that it is a safe place), and other people (e.g., that others are fundamentally benevolent) may be challenged (Creamer & Forbes, 2004). Wartime experiences during combat and other missions pose the risk of the so-called moral injury, in which one's deeply held personal beliefs are transgressed or violated. Cognitive dissonance then results from the discrepancy between personal beliefs about one's goodness and the goodness of the world and the disruptive wartime experiences (Litz et al., 2009). Moral injury is in turn associated with increased distress and suicidality (Forkus et al., 2019; Kelley et al., 2019).

Problems in social and occupational functioning, such as unemployment, marital issues, and homelessness, are prevalent among veterans (Held & Owens, 2015; Prigerson et al., 2001). Furthermore, the devastating effects of war are not limited to veterans themselves. Their parents, children, partners, and health professionals may be affected by secondary trauma (Bramsen et al., 2002; Gibbons et al., 2012; Gliske et al., 2019; Johnson et al., 2014; Melvin et al., 2012; Vasterling et al., 2015). Although the adversity of war is clear, the above also implies that if 23% of veterans experience PTSD, that means that 77% of veterans do not experience this level of continued distress. In a longitudinal study that monitored symptoms of PTSD, depression, and anxiety over the course of 2 years, 68% of veterans were characterized as resilient (Isaacs et al., 2017). What helps these veterans to be resilient after wartime, and in particular, what role does self-compassion play in this?

Promoting Resilience in Veterans with Self-Compassion

Research shows that self-compassion is related to decreased PTSD symptomology in veterans (Dahm et al., 2015; Forkus et al., 2019; Hiraoka et al., 2015; Meyer et al., 2019; Rabon et al., 2019) and that self-compassion interventions can help reduce PTSD symptomology (Lang et al., 2019; Steen et al., 2021). Since avoidance is a key part of PTSD, the *mindfulness* component of self-compassion may be helpful to tolerate, be with, or engage with difficult experiences (Kelley et al., 2019). Indeed, mindfulness interventions have been found acceptable (Bravo et al., 2019) and effective (Bremner et al., 2017; King et al., 2013; Polusny et al., 2015) in reducing PTSD symptomology for veterans. The mindfulness component of self-compassion allows for a non-evaluative and accepting stance toward difficult experiences. This approach-oriented attitude could also support veterans to refrain from using substances as an avoidant coping strategy (Forkus et al., 2019).

In the case of moral injury, viewing oneself as a bad person and engaging in self-punishing cognitions or behaviors are typical (Litz et al., 2009). Viewing or participating in morally transgressive events challenges not only one's moral compass but also perceptions of the self as a capable and just individual (Forkus et al., 2019). In contrast, some facets of self-compassion moderate the link between moral injury and adverse outcomes (e.g., suicide ideation), suggesting that the impact of moral injury can be attenuated by self-compassion (Kelley et al., 2019). *Self-kindness* may be particularly helpful to be more understanding toward oneself, counteracting harsh self-criticism and stimulating self-care (Gilbert et al., 2006; Neff, 2003). This may further facilitate breaking the cycle of negative cognitions in trauma-related guilt (Held & Owens, 2015) and depression (Forkus et al., 2019). Similarly, *common humanity* may help to integrate transgressive experiences into a sense of self, in which the veteran is simply an imperfect human being who had to make difficult decisions in a tough situation (Forkus et al., 2019). This could facilitate a sense of belonging as well as connecting with others, especially after going through an isolating experience of war and then having to reintegrate into society.

In addition to recovery, this reintegration requires finding new goals and calibrating meaning in post-war life, in order to facilitate mental well-being. Posttraumatic growth may involve positive perceptions of new possibilities, relations to others, personal strengths, and a new appreciation of life (Cann et al., 2010). Increased emotional stability through mindfulness may be an important resource, as emotional stability facilitates growth in veterans (Heppner et al., 2015; Tsai & Pietrzak, 2017). Behaviors stemming from kinder styles of self-relating may contribute to personal goals and purpose, and purpose in life is in turn related to growth in veterans (Isaacs et al., 2017; Tsai & Pietrzak, 2017). Noteworthy, in the context of Gilbert's drive system (which focuses on acquiring (coping) resources), striving to overcome worthlessness through maladaptive overachieving or addictive behaviors may occur, as is common in veterans. At the same time, the drive system is an important source of vitality, positive emotions, and motivation (Irons & Lad, 2017). While traditional masculinity norms in military culture are related to worse mental health outcomes overall, the masculinity aspect of "success dedication" is linked to greater quality of life and mindfulness in veterans (Ramon et al., 2019). This dedication to success inherent to military culture could be a potential catalyst for facilitating compassionate goals and self-care behaviors, as this drive for success may be a source of vitality to be used for other goals as well. Furthermore, perceived social support (Staugaard et al., 2015; Tsai & Pietrzak, 2017) as well as altruism (Isaacs et al., 2017; Tsai et al., 2016) predicts growth in veterans. Self-compassion could facilitate this growth via common humanity through increased social connection (Germer & Neff, 2019).

Evidence for self-compassion interventions to facilitate resilience in veterans is still limited. Interventions specifically aimed at training self-compassion seem promising (Serpa et al., 2021; Alliger-Horn et al., 2016; Grodin et al., 2019; Lee, 2009), while other types of interventions may increase self-compassion indirectly (e.g., Bergen-Cico et al., 2018). Compassion-based interventions typically include psychoeducation about emotions and various exercises to cultivate compassionate skills and attitudes, which may be adapted to the needs of veterans (Grodin et al., 2019). Lee (2009) describes how the cultivation of compassionate resilience can be helpful for veterans with PTSD. Compassionate resilience, as trained with compassion-based interventions such as compassion-focused therapy, enables the development of self-soothing capabilities and feelings of safeness in the face of confronting memories (Lee, 2009). This may take the form of developing an image of a compassionate self who cares for and self-soothes a traumatized part of the self (Alliger-Horn et al., 2016). Using this imagery to learn compassionate self-talk can break the cycle of self-criticism that is maintaining feelings of threat. Furthermore, being able to access feelings of self-compassion facilitates the development of new perspectives on the meaning of the traumatic event, which may be actively addressed with compassionate rescripting. This can be useful to work though trauma stories and develop more helpful inter- and intrapersonal relationships (Lee, 2009).

Self-Compassion and Resilience in the Context of Cancer

The Adversity of Going Through Cancer

What does it entail to have cancer? Many cancer patients describe receiving a diagnosis of cancer as an event that turns their entire life upside down and demolishes all sense of certainty (Austin et al., 2021). Cancer, a term that describes a group of diseases in which abnormal and uncontrollable cell growth occurs (WHO, 2020), involves profound physical, functional, psychological, and social changes. It is estimated that, worldwide, more than 18 million people each year are diagnosed with cancer, and the physical, emotional, and financial burden of cancer continues to grow globally (WHO, 2020; International Agency for Research on Cancer, 2019). Regarding the physical burdens of a cancer diagnosis, patients face side effects of their treatment (e.g., nausea due to chemotherapy), fatigue, pain, and functional limitations (e.g., decreased mobility). They are at an increased risk for depression and anxiety (Trindade et al., 2018; Zabora et al., 2001), and many cancer patients struggle with distress, negative body image, and self-blame (Callebaut et al., 2017; Przezdziecki et al., 2013; Zabora et al., 2001). In addition, cancer often challenges social roles both within close relationships (e.g., from partner to caregiver or care recipient) and within the societal and employment context (e.g., from employee to being exempted from work). Some cancer patients report feelings of loneliness, for example, because despite abundant social support, the illness is something they only go through by themselves. They may also experience feelings of uselessness because of not being able to contribute to others (Austin et al., 2021). Despite all this, it is estimated that around 65% of people with cancer do not experience prolonged distress (Herschbach et al., 2020; Wang et al., 2016; Zabora et al., 2001). What helps these patients to be resilient after receiving a cancer diagnosis, and what role does self-compassion play in this?

Promoting Resilience in People with Cancer with Self-Compassion

Resilience in the context of cancer pertains to adaptation to a cancer diagnosis and may involve attributes such as meaning-making, positive emotions, social support, and cognitive flexibility (Seiler & Jenewein, 2019). Recovery from cancer may involve different challenges than other (acute) types of trauma, given that the source of the threat is internal, the stressors are widespread, the threat is ongoing in the future, and there is greater perceived control over the threat (Sumalla et al., 2009). Accordingly, individuals with cancer may experience a greater sense of personal failure if disease progression worsens. Seiler and Jenewein (2019) posit that resilience in the context of cancer occurs both directly, via individual coping abilities and personality traits, and indirectly, via redefinition of the individual's self, post-diagnosis. This is akin to recovery (direct) and sustainability and growth (indirect) processes. As previously discussed, self-compassion addresses both of these pathways. Self-compassion appears a relevant adaptation resource in the context of cancer, since higher self-compassion in cancer patients has been associated with lower depression (Pinto-Gouveia et al., 2014; Todorov et al., 2019; Zhu et al.,

2019), anxiety (Todorov et al., 2019; Zhu et al., 2019), and distress (Pinto-Gouveia et al., 2014; Todorov et al., 2019), lower body-image distress (Todorov et al., 2019), and higher treatment adherence (Sirois & Hirsch, 2019) and is directly related to higher resilience (Alizadeh et al., 2018).

To learn more about self-compassion as part of an adaptation process in response to a cancer diagnosis, we conducted an interview study with 26 cancer patients about their experiences of self-compassion (Austin et al., 2021). Patients were asked to familiarize themselves with the concept of self-compassion by trying out various self-compassion exercises for 2 weeks, after which patients were asked about their ideas, experiences, and examples regarding self-compassion in the context of their diagnosis. Various self-compassionate actions and cognitions were described, originating from participants' personal experiences. Related to mindfulness, participants described allowing emotions to arise, as well as releasing them as they come up. This also included taking a balanced perspective, instead of getting lost in negative thoughts. Many different forms of self-kindness were described. In the context of self-kind rather than self-critical thinking, participants described acknowledging that you are going through a hard time; realizing that things aren't your fault; encouraging yourself; acceptance of the condition, negative emotions, and functional limitations; and paying attention to positive things in life.

Further, participants described self-kind behaviors, such as guarding social and physical boundaries, undertaking pleasurable activities, asking for and accepting help, and taking responsibility for their health. While participants did not explicitly describe their experience in terms of common humanity (e.g., realizing that adverse experiences and failings are human and part of life), they described feeling connected to other patients, close family, and friends. Patients acknowledged the importance of self-compassion in the challenging time after receiving a cancer diagnosis. These experiences mostly pertain to the recovery aspect of resilience, since participants were interviewed about their experiences shortly after receiving a cancer diagnosis. However, another qualitative interview study with people with breast cancer found similar aspects in the context of (sustainability and) growth, describing experiences of renewed attitudes toward the self, relationships, and life in general (Barthakur et al., 2016).

Similar to veterans, compassion-based interventions are increasingly available for people with cancer (Austin et al., 2020). While these interventions are promising, attending self-compassion training face-to-face is not always feasible for people with cancer (and, perhaps, other populations who experienced adversity). The often already high load of medical appointments of cancer patients constitutes a substantial burden of care. Lathren et al. (2018) offered a Mindful Self-Compassion training entirely via videoconferencing, thereby addressing a population of young adult cancer survivors at a distance. Mobile technology offers benefits in terms of addressing issues of accessibility and availability (Gemert-Pijnen et al., 2018; Kelders & Howard, 2018) in general and in times of crisis and may offer additional benefits regarding integration of newly learned skills into daily life (Jones et al., 2015; Williams et al., 2007) and interactive and personalization features. In our current work, we are co-designing and evaluating a self-compassion mobile intervention for people with newly diagnosed cancer with cancer patients and oncology nurses (Austin et al., 2022). We expect that participation in mobile interventions such as this one will help cancer patients relate to themselves and their cancer experience with compassion and that this in turn will facilitate their resilience in the face of the diagnosis. This work is an important step in the direction of offering feasible low-threshold self-compassion and resilience interventions for people with limited capacity due to experienced adversity.

Conclusion

In this chapter we have discussed resilience as the ability to adapt to adversity and difficult life circumstances, particularly in the context of surviving military service and surviving cancer. There is growing evidence that self-compassion is an inner resource that promotes resilience and the ability to adapt in two major ways. First, it supports the willingness and ability to acknowledge and process the emotional and cognitive suffering that is created by serious negative life events. This facilitates the recovery aspect of resilience, especially in the context of mental illness. Second, it can be supportive in creating a context in which humans become more aware of important values in their lives and reengage in meaningful activities. This facilitates sustainability and growth aspects of resilience, especially in the context of mental well-being. Self-compassion promotes a sense of care, fierceness, and courage to be resilient in the face of adversity. We recommend future studies to further investigate the different ways in which self-compassion can promote resilience, taking into account both mental distress and mental well-being outcomes.

References

Akın, U. (2014). Self-compassion as a predictor of proactivity. *International Online Journal of Educational Sciences, 6*, 103–111. https://doi.org/10.15345/iojes.2014.01.011

Alizadeh, S., Khanahmadi, S., Vedadhir, A., & Barjasteh, S. (2018). The relationship between resilience with self- compassion, social support and sense of belonging in women with breast cancer. *Asian Pacific Journal of Cancer Prevention, 19*(9), 2469–2474. https://doi.org/10.22034/apjcp.2018.19.9.2469

Allen, A. B., & Leary, M. R. (2010). Self-compassion, stress, and coping. *Social and Personality Psychology Compass, 4*(2), 107–118. https://doi.org/10.1111/j.1751-9004.2009.00246.x

Alliger-Horn, C., Zimmermann, P., & Schmucker, M. (2016). Guilt, shame and compassionate imagery in war: Traumatized german soldiers with ptsd, a pilot study. *Journal of Clinical Medicine, 5*, 90. https://doi.org/10.3390/jcm5100090

Austin, J., Drossaert, C. H. C., Schroevers, M. J., Sanderman, R., Kirby, J. N., & Bohlmeijer, E. T. (2020). Compassion-based interventions for people with long-term physical conditions: A mixed methods systematic review. *Psychology & Health*, 1–27. https://doi.org/10.1080/08870446.2019.1699090

Austin, J., Drossaert, C. H. C., Schroevers, M. J., Sanderman, R., & Bohlmeijer, E. T. (2021). Experiences of self-criticism and self-compassion in people diagnosed with cancer: A multimethod qualitative study. *Frontiers in Psychology, 12*. https://doi.org/10.3389/fpsyg.2021.737725

Austin, J., Drossaert, C. H. C., Van Dijk, J., Sanderman, R., Børøsund, E., Mirkovic, J., et al. (2022). Integrating top-down and bottom-up requirements in eHealth development: The case of a mobile self-compassion intervention for people with newly diagnosed cancer. *JMIR Cancer*. doi:24/05/2022:37502

Baker, D. A., Caswell, H. L., & Eccles, F. J. R. (2019). Self-compassion and depression, anxiety, and resilience in adults with epilepsy. *Epilepsy & Behavior, 90*, 154–161. https://doi.org/10.1016/j.yebeh.2018.11.025

Barthakur, M. S., Sharma, M. P., Chaturvedi, S. K., & Manjunath, S. K. (2016). Posttraumatic growth in women survivors of breast cancer. *Indian Journal of Palliative Care, 22*(2), 157–162. https://doi.org/10.4103/0973-1075.179609

Bergen-Cico, D., Smith, Y., Wolford, K., Gooley, C., Hannon, K., Woodruff, R., et al. (2018). Dog ownership and training reduces post-traumatic stress symptoms and increases self-compassion among veterans: Results of a longitudinal control study. *Journal of Alternative and Complementary Medicine, 24*(12), 1166–1175. https://doi.org/10.1089/acm.2018.0179

Bluth, K., Mullarkey, M., & Lathren, C. (2018). Self-compassion: A potential path to adolescent resilience and positive exploration. *Journal of Child and Family Studies, 27*, 1–11. https://doi.org/10.1007/s10826-018-1125-1

Bohlmeijer, E. T., & Westerhof, G. J. (2021). A new model for sustainable mental health: Integrating well-being into psychological treatment. In J. N. Kirby & P. Gilbert (Eds.), *Making an impact on mental health: The applications of psychological research* (pp. 153–188). Routledge/Taylor & Francis Group. https://doi.org/10.4324/9780429244551-7

Bonanno, G. A. (2004). Loss, trauma, and human resilience: Have we underestimated the human capacity to thrive after extremely aversive events? *American Psychologist Journal, 59*(1), 20–28. https://doi.org/10.1037/0003-066x.59.1.20

Bonanno, G. A., & Diminich, E. D. (2013). Annual research review: Positive adjustment to adversity – Trajectories of minimal–impact resilience and emergent resilience. *Journal of Child Psychology and Psychiatry, 54*(4), 378–401. https://doi.org/10.1111/jcpp.12021

Bramsen, I., van der Ploeg, H. M., & Twisk, J. W. (2002). Secondary traumatization in Dutch couples of World War II survivors. *Journal of Consulting and Clinical Psychology, 70*(1), 241–245.

Bravo, A. J., Witkiewitz, K., Kelley, M. L., & Redman, J. C. (2019). Prevalence of mental health problems and willingness to participate in a mindfulness treatment: An examination among veterans injured in combat. *Mindfulness, 10*(5), 953–963. https://doi.org/10.1007/s12671-018-1047-4

Bremner, J. D., Mishra, S., Campanella, C., Shah, M., Kasher, N., Evans, S., et al. (2017). A pilot study of the effects of mindfulness-based stress reduction on post-traumatic stress disorder symptoms and brain response to traumatic reminders of combat in operation enduring freedom/operation Iraqi freedom combat veterans with post-traumatic stress disorder. *Frontiers in Psychiatry, 8*, 157. https://doi.org/10.3389/fpsyt.2017.00157

Bryan, C., & Bryan, A. (2014). Nonsuicidal self-injury among a sample of United States military personnel and veterans enrolled in college classes. *Journal of Clinical Psychology, 70*(9), 874–885.

Bryan, C. J., Morrow, C. E., Etienne, N., & Ray-Sannerud, B. (2013). Guilt, shame, and suicidal ideation in a military outpatient clinical sample. *Depression and Anxiety, 30*(1), 55–60. https://doi.org/10.1002/da.22002

Burnett-Zeigler, I., Ilgen, M., Valenstein, M., Zivin, K., Gorman, L., Blow, A., et al. (2011). Prevalence and correlates of alcohol misuse among returning Afghanistan and Iraq veterans. *Addictive Behaviors, 36*(8), 801–806. https://doi.org/10.1016/j.addbeh.2010.12.032

Callebaut, L., Molyneux, P., & Alexander, T. (2017). The relationship between self-blame for the onset of a chronic physical health condition and emotional distress: A systematic literature review. *Clinical Psychology and Psychotherapy, 24*(4), 965–986. https://doi.org/10.1002/cpp.2061

Cann, A., Calhoun, L. G., Tedeschi, R. G., Taku, K., Vishnevsky, T., Triplett, K. N., & Danhauer, S. C. (2010). A short form of the posttraumatic growth inventory. *Anxiety, Stress & Coping, 23*(2), 127–137. https://doi.org/10.1080/10615800903094273

Chen, J. J. (2022). Self-compassion as key to stress resilience among first-year early childhood teachers during COVID-19: An interpretative phenomenological analysis. *Teaching and Teacher Education, 111*, 103627. https://doi.org/10.1016/j.tate.2021.103627

Chien-Chung, H., Yafan, C., Huiying, J., Marci, S., Chuwei, L., & Cailee, O. (2020). Mindfulness, life skills, resilience, and emotional and behavioral problems for gifted low-income adolescents in China. *Frontiers in Psychology, 11*. https://doi.org/10.3389/fpsyg.2020.00594

Christopher, M. (2004). A broader view of trauma: A biopsychosocial-evolutionary view of the role of the traumatic stress response in the emergence of pathology and/or growth. *Clinical Psychology Review, 24*(1), 75–98. https://doi.org/10.1016/j.cpr.2003.12.003

Craig, J. B., Rudd, M. D., Evelyn, W., Stacey, Y.-M., & Alan, P. (2015). Nonsuicidal self-injury as a prospective predictor of suicide attempts in a clinical sample of military personnel. *Comprehensive Psychiatry, 59*, 1–7. https://doi.org/10.1016/j.comppsych.2014.07.009

Creamer, M., & Forbes, D. (2004). Treatment of posttraumatic stress disorder in military and veteran populations. *Psychotherapy: Theory, Research, Practice, Training, 41*(4), 388–398. https://doi.org/10.1037/0033-3204.41.4.388

Dahm, K. A., Meyer, E. C., Neff, K. D., Kimbrel, N. A., Gulliver, S. B., & Morissette, S. B. (2015). Mindfulness, self-compassion, posttraumatic stress disorder symptoms, and functional disability in U.S. Iraq and Afghanistan war veterans self-compassion and mindfulness in U.S. war veterans. *Journal of Traumatic Stress, 28*(5), 460–464. https://doi.org/10.1002/jts.22045

Davis, J. A., Gibson, L. Y., Bear, N. L., Finlay-Jones, A. L., Ohan, J. L., Silva, D. T., & Prescott, S. L. (2021). Can positive mindsets be protective against stress and isolation experienced during the COVID-19 pandemic? A mixed methods approach to understanding emotional health and wellbeing needs of perinatal women. *International Journal of Environmental Research and Public Health, 18*(13), 6958. https://doi.org/10.3390/ijerph18136958

De Vos, J. A., LaMarre, A., Radstaak, M., Bijkerk, C. A., Bohlmeijer, E. T., & Westerhof, G. J. (2017). Identifying fundamental criteria for eating disorder recovery: A systematic review and qualitative meta-analysis. *Journal of Eating Disorders, 5*(1), 1–14. https://doi.org/10.1186/s40337-017-0164-0

Diagnostic and statistical manual of mental disorders: DSM-5 (2013). American Psychiatric Association.

Diener, E., Suh, E. M., Lucas, R. E., & Smith, H. L. (1999). Subjective well-being: Three decades of progress. *Psychological Bulletin, 125*(2), 276–302. https://doi.org/10.1037//0033-2909.125.2.276

Eckland, N. S., & Berenbaum, H. (2020). Emotional awareness in daily life: Exploring its potential role in repetitive thinking and healthy coping. *Behavior Therapy*. https://doi.org/10.1016/j.beth.2020.04.010

Eichholz, A., Schwartz, C., Meule, A., Heese, J., Neumüller, J., & Voderholzer, U. (2020). Self-compassion and emotion regulation difficulties in obsessive-compulsive disorder. *Clinical Psychology & Psychotherapy., 27*, 630. https://doi.org/10.1002/cpp.2451

Finlay-Jones, A. L., Rees, C. S., & Kane, R. T. (2015). Self-compassion, emotion regulation and stress among australian psychologists: Testing an emotion regulation model of self-compassion using structural equation modeling. *PLoS One, 10*(7), e0133481. https://doi.org/10.1371/journal.pone.0133481

Forkus, S. R., Breines, J. G., & Weiss, N. H. (2019). Morally injurious experiences and mental health: The moderating role of self-compassion. *Psychological Trauma: Theory, Research, Practice and Policy, 11*(6), 630–638. https://doi.org/10.1037/tra0000446

Franken, K., Lamers, S. M. A., Ten Klooster, P. M., Bohlmeijer, E. T., & Westerhof, G. J. (2018). Validation of the mental health continuum-short form and the dual continua model of well-being and psychopathology in an adult mental health setting. *Journal of Clinical Psychology, 74*(12), 2187–2202. https://doi.org/10.1002/jclp.22659

Freligh, C. B., & Debb, S. M. (2019). Nonreactivity and resilience to stress: Gauging the mindfulness of African American college students. *Mindfulness, 10*(11), 2302–2311. https://doi.org/10.1007/s12671-019-01203-w

Fulton, J. J., Calhoun, P. S., Wagner, H. R., Schry, A. R., Hair, L. P., Feeling, N., et al. (2015). The prevalence of posttraumatic stress disorder in operation enduring freedom/operation Iraqi freedom (OEF/OIF) veterans: A meta-analysis. *Journal of Anxiety Disorders, 31*, 98–107. https://doi.org/10.1016/j.janxdis.2015.02.003

Gadermann, A. M., Engel, C. C., Naifeh, J. A., Nock, M. K., Petukhova, M., Santiago, P. N., et al. (2012). Prevalence of DSM-IV major depression among u.s. military personnel: Meta-analysis and simulation. *Military Medicine, 177*, 47–59. https://doi.org/10.7205/MILMED-D-12-00103

García, F. E., Cova, F., Rincón, P., & Vázquez, C. (2015). Trauma or growth after a natural disaster? The mediating role of rumination processes. *European Journal of Psychotraumatology, 6*(1), 26557. https://doi.org/10.3402/ejpt.v6.26557

Garmezy, N. (1991). Resiliency and vulnerability to adverse developmental outcomes associated with poverty. *American Behavioral Scientist, 34*(4), 416–430. https://doi.org/10.1177/0002764291034004003

Gemert-Pijnen, J. E. W. C. V., Kip, H., & Kelders, S. M. (2018). Introducing eHealth. In S. M. Kelders, J. E. W. C. Van Gemert-Pijnen, H. Kip, & R. Sanderman (Eds.), *eHealth research, theory and development: A multidisciplinary approach* (1st ed., pp. 3–26). Routledge.

Germer, C., & Neff, K. (2019). *Teaching the mindful self-compassion program: A guide for professionals*. Guilford Publications.

Gibbons, S. W., Hickling, E. J., & Watts, D. D. (2012). Combat stressors and post-traumatic stress in deployed military healthcare professionals: An integrative review. *Journal of Advanced Nursing, 68*(1), 3–21. https://doi.org/10.1111/j.1365-2648.2011.05708.x

Gilbert, P. (2010). *The compassionate mind: A new approach to life's challenges*. New Harbinger Publications.

Gilbert, P. (2014). The origins and nature of compassion focused therapy. *British Journal of Clinical Psychology, 53*(1), 6–41.

Gilbert, P., Baldwin, M., Irons, C., Baker, J., & Palmer, M. (2006). Self-criticism and self-warmth: An imagery study exploring their relation to depression. *Journal of Cognitive Psychotherapy, 20*, 183–200. https://doi.org/10.1891/088983906780639817

Gliske, K., Richmond, A., Smischney, T., & Borden, L. M. (2019). Mindfulness strategies: Supporting military parents during reintegration. *Mindfulness, 10*(9), 1721–1729. https://doi.org/10.1007/s12671-019-01156-0

Grodin, J., Clark, J. L., Kolts, R., & Lovejoy, T. I. (2019). Compassion focused therapy for anger: A pilot study of a group intervention for veterans with PTSD. *Journal of Contextual Behavioral Science, 13*, 27–33. https://doi.org/10.1016/j.jcbs.2019.06.004

Harman, R., & Lee, D. (2010). The role of shame and self-critical thinking in the development and maintenance of current threat in post-traumatic stress disorder. *Clinical Psychology & Psychotherapy, 17*(1), 13–24. https://doi.org/10.1002/cpp.636

Hayter, M. R., & Dorstyn, D. S. (2014). Resilience, self-esteem and self-compassion in adults with spina bifida. *Spinal Cord, 52*(2), 167–171. https://doi.org/10.1038/sc.2013.152

Held, P., & Owens, G. P. (2015). Effects of self-compassion workbook training on trauma-related guilt in a sample of homeless veterans: A pilot study. *Journal of Clinical Psychology, 71*(6), 513–526.

Helgeson, V. S., Reynolds, K. A., & Tomich, P. L. (2006). A meta-analytic review of benefit finding and growth. *Journal of Consulting and Clinical Psychology, 74*(5), 797–816.

Hensley, B., Martin, P., Margrett, J. A., MacDonald, M., Siegler, I. C., Poon, L. W., & The Georgia Centenarian Study. (2012). Life events and personality predicting loneliness among centenarians: Findings from the Georgia centenarian study. *The Journal of Psychology, 146*(1–2), 173–188. https://doi.org/10.1080/00223980.2011.613874

Heppner, W. L., Spears, C. A., Vidrine, J. I., & Wetter, D. W. (2015). Mindfulness and emotion regulation. In B. D. Ostafin, M. D. Robinson, & B. P. Meier (Eds.), *Handbook of mindfulness and self-regulation* (pp. 107–120). Springer.

Herschbach, P., Britzelmeir, I., Dinkel, A., Giesler, J. R. M., Herkommer, K., Nest, A., et al. (2020). Distress in cancer patients: Who are the main groups at risk? *Psycho-Oncology, 29*(4), 703–710. https://doi.org/10.1002/pon.5321

Hiraoka, R., Meyer, E. C., Kimbrel, N. A., DeBeer, B. B., Gulliver, S. B., & Morissette, S. B. (2015). Self-compassion as a prospective predictor of PTSD symptom severity among trauma-exposed U.S. Iraq and Afghanistan war veterans. *Journal of Traumatic Stress, 28*(2), 127–133. https://doi.org/10.1002/jts.21995

Hyland, P., Shevlin, M., Cloitre, M., Karatzias, T., Vallières, F., McGinty, G., et al. (2019). Quality not quantity: Loneliness subtypes, psychological trauma, and mental health in the US adult population. *Social Psychiatry and Psychiatric Epidemiology, 54*(9), 1089–1099. https://doi.org/10.1007/s00127-018-1597-8

Im, S., & Follette, V. M. (2016). Rumination and mindfulness related to multiple types of trauma exposure. *Translational Issues in Psychological Science, 2*(4), 395–407. https://doi.org/10.1037/tps0000090

International Agency for Research on Cancer (2019). Ci5plus: Cancer incidence in five continents time trends. Retrieved from http://ci5.iarc.fr/CI5plus/Default.aspx

Irons, C., & Lad, S. (2017). Using compassion focused therapy to work with shame and self-criticism in complex trauma. *Australian Clinical Psychologist, 3*(1).

Isaacs, K., Mota, N. P., Tsai, J., Harpaz-Rotem, I., Cook, J. M., Kirwin, P. D., et al. (2017). Psychological resilience in U.S. military veterans: A 2-year, nationally representative prospective cohort study. *Journal of Psychiatry Research, 84*, 301–309. https://doi.org/10.1016/j.jpsychires.2016.10.017

Jeffery, D. D., Babeu, L. A., Nelson, L. E., Kloc, M., & Klette, K. (2013). Prescription drug misuse among U.S. active duty military personnel: A secondary analysis of the 2008 DoD survey of health related behaviors. *Military Medicine, 178*(2), 180–195.

Johnson, W. B., Bertschinger, M., Snell, A. K., & Wilson, A. (2014). Secondary trauma and ethical obligations for military psychologists: Preserving compassion and competence in the crucible of combat. *Psychological Services, 11*(1), 68–74. https://doi.org/10.1037/a0033913

Jones, D. J., Anton, M., Gonzalez, M., Honeycutt, A., Khavjou, O., Forehand, R., & Parent, J. (2015). Incorporating mobile phone technologies to expand evidence-based care. *Cognitive Behavioral Practice, 22*(3), 281–290. https://doi.org/10.1016/j.cbpra.2014.06.002

Joseph, S., & Linley, P. A. (2006). Growth following adversity: Theoretical perspectives and implications for clinical practice. *Clinical Psychology Review, 26*(8), 1041–1053. https://doi.org/10.1016/j.cpr.2005.12.006

Kearney, M. A., Zeligman, M., Brack, J. L., & Payne, E. (2018). Trauma and dissociation: Predictors of loneliness in students at an urban university. *Journal of College Counseling, 21*(2), 165–179. https://doi.org/10.1002/jocc.12095

Kelders, S. M., & Howard, M. (2018). Opportunities of technology to promote health and well-being. In S. M. Kelders, J. E. W. C. Van Gemert-Pijnen, H. Kip, & R. Sanderman (Eds.), *eHealth research, theory and development: A multidisciplinary approach* (1st ed., pp. 48–68). Routledge.

Kelley, M. L., Bravo, A. J., Davies, R. L., Hamrick, H. C., Vinci, C., & Redman, J. C. (2019). Moral injury and suicidality among combat-wounded veterans: The moderating effects of social connectedness and self-compassion. *Psychological Trauma: Theory, Research, Practice and Policy, 11*(6), 621–629. https://doi.org/10.1037/tra0000447

Kemper, K. J., Mo, X., & Khayat, R. (2015). Are mindfulness and self-compassion associated with sleep and resilience in health professionals? *Journal of Alternative and Complementary Medicine, 21*(8), 496–503. https://doi.org/10.1089/acm.2014.0281

Keyes, C. L. M. (1998). Social well-being. *Social Psychology Quarterly, 61*(2), 121–140.

Keyes, C. L. M. (2002). The mental health continuum: From languishing to flourishing in life. *Journal of Health and Social Behavior, 43*(2), 207–222.

Keyes, C. L. M. (2005). Mental illness and/or mental health? Investigating axioms of the complete state model of health. *Journal of Consulting and Clinical Psychology, 73*(3), 539–548.

Kim, J. S., Jin, M. J., Jung, W., Hahn, S. W., & Lee, S. H. (2017). Rumination as a mediator between childhood trauma and adulthood depression/anxiety in non-clinical participants. *Frontiers in Psychology, 8*, 1597. https://doi.org/10.3389/fpsyg.2017.01597

King, A. P., Erickson, T. M., Giardino, N. D., Favorite, T., Rauch, S. A. M., Robinson, E., et al. (2013). A pilot study of group mindfulness-based cognitive therapy (MBCT) for combat veterans with posttraumatic stress disorder (PTSD). *Depression and Anxiety, 30*(7), 638–645. https://doi.org/10.1002/da.22104

Kotera, Y., Ozaki, A., Miyatake, H., Tsunetoshi, C., Nishikawa, Y., & Tanimoto, T. (2021). Mental health of medical workers in Japan during COVID-19: Relationships with loneliness, hope and self-compassion. *Current Psychology, 40*(12), 6271–6274. https://doi.org/10.1007/s12144-021-01514-z

Lamers, S. M. A., Westerhof, G. J., Bohlmeijer, E. T., ten Klooster, P. M., & Keyes, C. L. M. (2011). Evaluating the psychometric properties of the mental health continuum-short form (MHC-SF). *Journal of Clinical Psychology, 67*(1), 99–110. https://doi.org/10.1002/jclp.20741

Lang, A. J., Malaktaris, A. L., Casmar, P., Baca, S. A., Golshan, S., Harrison, T., & Negi, L. (2019). Compassion meditation for posttraumatic stress disorder in veterans: A randomized proof of concept study. *Journal of Traumatic Stress, 32*(2), 299–309. https://doi.org/10.1002/jts.22397

Lathren, C., Bluth, K., Campo, R., Tan, W., & Futch, W. (2018). Young adult cancer survivors' experiences with a mindful self-compassion (MSC) video-chat intervention: A qualitative analysis. *Self and Identity, 17*(6), 646–665. https://doi.org/10.1080/15298868.2018.1451363

Lau, B. H.-P., Chan, C. L.-W., & Ng, S.-M. (2020). Self-compassion buffers the adverse mental health impacts of COVID-19-related threats: Results from a cross-sectional survey at the first peak of Hong Kong's outbreak. *Frontiers in Psychiatry, 11*. https://doi.org/10.3389/fpsyt.2020.585270

Lee, D. (2009). Compassion-focused cognitive therapy for shame-based trauma memories and flashbacks in post-traumatic stress disorder. In N. Grey (Ed.), *A casebook of cognitive therapy for traumatic stress reactions* (pp. 230–247). Routledge.

Lee, J., Blackmon, B. J., Lee, J. Y., Cochran, D. M., & Rehner, T. A. (2019). An exploration of posttraumatic growth, loneliness, depression, resilience, and social capital among survivors of Hurricane Katrina and the Deepwater Horizon Oil Spill. *Journal of Community Psychology, 47*(2), 356–370. https://doi.org/10.1002/jcop.22125

Lefebvre, J. I., Montani, F., & Courcy, F. (2020). Self-compassion and resilience at work: A practice-oriented review. *Advances in Developing Human Resources, 22*(4), 437–452. https://doi.org/10.1177/1523422320949145

Litz, B. T., Stein, N., Delaney, E., Lebowitz, L., Nash, W. P., Silva, C., & Maguen, S. (2009). Moral injury and moral repair in war veterans: A preliminary model and intervention strategy. *Clinical Psychology Review, 29*(8), 695–706. https://doi.org/10.1016/j.cpr.2009.07.003

Livneh, H. (2001). Psychosocial adaptation to chronic illness and disability. *Rehabilitation Counseling Bulletin, 44*(3), 151–160.

Lueger-Schuster, B., Glück, T. M., Tran, U. S., & Zeilinger, E. L. (2012). Sexual violence by occupational forces during and after World War II: Influence of experiencing and witnessing of sexual violence on current mental health in a sample of elderly Austrians. *International Psychogeriatrics, 24*(8), 1354–1358. https://doi.org/10.1017/S104161021200021X

Luthar, S. S., Doernberger, C. H., & Zigler, E. (1993). Resilience is not a unidimensional construct: Insights from a prospective study of inner-city adolescents. *Development and Psychopathology, 5*(4), 703–717. https://doi.org/10.1017/s0954579400006246

Mantzios, M. (2014). Exploring the relationship between worry and impulsivity in military recruits: The role of mindfulness and self-compassion as potential mediators. *Stress and Health, 30*(5), 397–404.

Masten, A. S. (2001). Ordinary magic. Resilience processes in development. *American Psychologist Journal, 56*(3), 227–238. https://doi.org/10.1037//0003-066x.56.3.227

Masumeh, S., Shirin, A., & Hadees, H. (2019). The effectiveness of self-compassion training on loneliness and resilience in destitute women. *Shenakt Journal of Psychology and Psychiatry, 5*(6), 71–84. https://doi.org/10.29252/shenakht.5.6.71

Melvin, K. C., Gross, D., Hayat, M. J., Jennings, B. M., & Campbell, J. C. (2012). Couple functioning and post-traumatic stress symptoms in US Army couples: The role of resilience. *Research in Nursing & Health, 35*(2), 164–177. https://doi.org/10.1002/nur.21459

Meyer, E. C., Szabo, Y. Z., Frankfurt, S. B., Kimbrel, N. A., DeBeer, B. B., & Morissette, S. B. (2019). Predictors of recovery from post-deployment posttraumatic stress disorder symptoms in war veterans: The contributions of psychological flexibility, mindfulness, and self-compassion. *Behaviour Research and Therapy, 114*, 7–14. https://doi.org/10.1016/j.brat.2019.01.002

Montero-Marin, J., Tops, M., Manzanera, R., Piva Demarzo, M. M., Álvarez de Mon, M., & García-Campayo, J. (2015). Mindfulness, resilience, and burnout subtypes in primary care physicians: The possible mediating role of positive and negative affect. *Frontiers in Psychology, 6*, 1895. https://doi.org/10.3389/fpsyg.2015.01895

Neff, K. (2003). Self-compassion: An alternative conceptualization of a healthy attitude toward oneself. *Self and Identity, 2*, 85–101. https://doi.org/10.1080/15298860309032

Neff, K., & Germer, C. (2018). *The mindful self-compassion workbook: A proven way to accept yourself, build inner strength, and thrive*. Guilford Publications.

Neff, K. D., & McGehee, P. (2010). Self-compassion and psychological resilience among adolescents and young adults. *Self and Identity, 9*(3), 225–240. https://doi.org/10.1080/15298860902979307

Neff, K. D., Kirkpatrick, K. L., & Rude, S. S. (2007). Self-compassion and adaptive psychological functioning. *Journal of Research in Personality, 41*(1), 139–154. https://doi.org/10.1016/j.jrp.2006.03.004

Nery-Hurwit, M., Yun, J., & Ebbeck, V. (2018). Examining the roles of self-compassion and resilience on health-related quality of life for individuals with Multiple Sclerosis. *Disability and Health Journal, 11*(2), 256–261. https://doi.org/10.1016/j.dhjo.2017.10.010

Park, C. L., & Ai, A. L. (2006). Meaning making and growth: New directions for research on survivors of trauma. *Journal of Loss and Trauma, 11*(5), 389–407. https://doi.org/10.1080/15325020600685295

Peterson, C., & Seligman, M. E. P. (2003). Character strengths before and after September 11. *Psychological Science, 14*(4), 381–384.

Petrie, K. J., Buick, D. L., Weinman, J., & Booth, R. J. (1999). Positive effects of illness reported by myocardial infarction and breast cancer patients. *Journal of Psychosomatic Research, 47*(6), 537–543. https://doi.org/10.1016/S0022-3999(99)00054-9

Pinto-Gouveia, J., Duarte, C., Matos, M., & Fráguas, S. (2014). The protective role of self-compassion in relation to psychopathology symptoms and quality of life in chronic and in cancer patients. *Clinical Psychology & Psychotherapy, 21*(4), 311–323. https://doi.org/10.1002/cpp.1838

Polusny, M. A., Erbes, C. R., Thuras, P., Moran, A., Lamberty, G. J., Collins, R. C., et al. (2015). Mindfulness-based stress reduction for posttraumatic stress disorder among veterans a randomized clinical trial. *Journal of the American Medical Association, 314*(5), 456–465. https://doi.org/10.1001/jama.2015.8361

Prigerson, H. G., Maciejewski, P. K., & Rosenheck, R. A. (2001). Combat trauma: Trauma with highest risk of delayed onset and unresolved posttraumatic stress disorder symptoms, unemployment, and abuse among men. *The Journal of Nervous and Mental Disease, 189*(2), 99–108.

Przezdziecki, A., Sherman, K. A., Baillie, A., Taylor, A., Foley, E., & Stalgis-Bilinski, K. (2013). My changed body: Breast cancer, body image, distress and self-compassion. *Psychooncology, 22*(8), 1872–1879. https://doi.org/10.1002/pon.3230

Rabon, J. K., Hirsch, J. K., Kaniuka, A. R., Sirois, F., Brooks, B. D., & Neff, K. (2019). Self-compassion and suicide risk in veterans: When the going gets tough, do the tough benefit more from self-compassion? *Mindfulness, 10*(12), 2544–2554. https://doi.org/10.1007/s12671-019-01221-8

Raes, F. (2010). Rumination and worry as mediators of the relationship between self-compassion and depression and anxiety. *Personality and Individual Differences, 48*(6), 757–761. https://doi.org/10.1016/j.paid.2010.01.023

Ramon, A. E., Guthrie, L., & Rochester, N. K. (2019). Role of masculinity in relationships between mindfulness, self-compassion, and well-being in military veterans. *Psychology of Men & Masculinities, 21*, 357. https://doi.org/10.1037/men0000240

Renshaw, K. D., & Kiddie, N. S. (2012). Internal anger and external expressions of aggression in OEF/OIF veterans. *Military Psychology, 24*(3), 221–235. https://doi.org/10.1080/08995605.2012.678197

Richardson, G. E. (2002). The metatheory of resilience and resiliency. *Journal of Clinical Psychology, 58*(3), 307–321. https://doi.org/10.1002/jclp.10020

Roemer, L., Williston, S. K., & Rollins, L. G. (2015). Mindfulness and emotion regulation. *Current Opinion in Psychology, 3*, 52–57. https://doi.org/10.1016/j.copsyc.2015.02.006

Rottenberg, J., Devendorf, A. R., Kashdan, T. B., & Disabato, D. J. (2018). The curious neglect of high functioning after psychopathology: The case of depression. *Perspectives on Psychological Science: a journal of the Association for Psychological Science, 13*(5), 549–566. https://doi.org/10.1177/1745691618769868

Rutter, M. (1993). Resilience: some conceptual considerations. *Journal of Adolescent Health, 14*(8), 626–631, 690-626. https://doi.org/10.1016/1054-139x(93)90196-v

Ryff, C. D. (1989). Happiness is everything, or is it? Explorations on the meaning of psychological well-being. *Journal of Personality and Social Psychology, 57*(6), 1069–1081. https://doi.org/10.1037//0022-3514.57.6.1069

Rzeszutek, M., & Gruszczyńska, E. (2018). Posttraumatic growth among people living with HIV: A systematic review. *Journal of Psychosomatic Research, 114*, 81–91. https://doi.org/10.1016/j.jpsychores.2018.09.006

Schotanus-Dijkstra, M., Pieterse, M. E., Drossaert, C. H. C., Westerhof, G. J., de Graaf, R., ten Have, M., et al. (2016). What factors are associated with flourishing? Results from a large representative national sample. *Journal of Happiness Studies, 17*(4), 1351–1370.

Schreuder, B. J. N., Kleijn, W. C., & Rooijmans, H. G. M. (2000). Nocturnal re-experiencing more than forty years after war trauma. *Journal of Traumatic Stress, 13*(3), 453–463. https://doi.org/10.1023/A:1007733324351

Scoglio, A. A. J., Rudat, D. A., Garvert, D., Jarmolowski, M., Jackson, C., & Herman, J. L. (2018). Self-compassion and responses to trauma: The role of emotion regulation. *Journal of Interpersonal Violence, 33*(13), 2016–2036. https://doi.org/10.1177/0886260515622296

Seal, K. H., Cohen, G., Waldrop, A., Cohen, B. E., Maguen, S., & Ren, L. (2011). Substance use disorders in Iraq and Afghanistan veterans in VA healthcare, 2001-2010: Implications for screening, diagnosis and treatment. *Drug and Alcohol Dependence, 116*(1–3), 93–101. https://doi.org/10.1016/j.drugalcdep.2010.11.027

Seiler, A., & Jenewein, J. (2019). Resilience in cancer patients. *Frontiers in Psychiatry, 10*, 208–208. https://doi.org/10.3389/fpsyt.2019.00208

Serpa, J. G., Bourey, C. P., Adjaoute, G. N., & Pieczynski, J. M. (2021). Mindful self-compassion (MSC) with veterans: A program evaluation. *Mindfulness, 12*(1), 153–161. https://doi.org/10.1007/s12671-020-01508-1

Shebuski, K., Bowie, J.-A., & Ashby, J. S. (2020). Self-compassion, trait resilience, and trauma exposure in undergraduate students. *Journal of College Counseling, 23*(1), 2–14. https://doi.org/10.1002/jocc.12145

Sirois, F. M., & Hirsch, J. K. (2019). Self-compassion and adherence in five medical samples: The role of stress. *Mindfulness, 10*(1), 46–54. https://doi.org/10.1007/s12671-018-0945-9

Snyder, J. (2014). "Blood, guts, and gore galore": bodies, moral pollution, and combat trauma. *Symbolic Interaction, 37*(4), 524–540. https://doi.org/10.1002/symb.116

Solomon, Z., Dekel, R., Zerach, G., & Horesh, D. (2009). Differentiation of the self and posttraumatic symptomatology among ex-POWs and their wives. *Journal of Marital and Family Therapy, 35*(1), 60–73. https://doi.org/10.1111/j.1752-0606.2008.00102.x

Solomon, Z., Bensimon, M., Greene, T., Horesh, D., & Ein-Dor, T. (2015). Loneliness trajectories: The role of posttraumatic symptoms and social support. *Journal of Loss and Trauma, 20*(1), 1–21. https://doi.org/10.1080/15325024.2013.815055

Sommers-Spijkerman, M. P. J., Trompetter, H. R., Schreurs, K. M. G., & Bohlmeijer, E. T. (2018). Compassion-focused therapy as guided self-help for enhancing public mental health: A randomized controlled trial. *Journal of Consulting and Clinical Psychology, 86*(2), 101–115. https://doi.org/10.1037/ccp0000268

Staugaard, S. R., Johannessen, K. B., Thomsen, Y. D., Bertelsen, M., & Berntsen, D. (2015). Centrality of positive and negative deployment memories predicts posttraumatic growth in danish veterans. *Journal of Clinical Psychology, 71*(4), 362–377. https://doi.org/10.1002/jclp.22142

Steen, M. P., Di Lemma, L., Finnegan, A., Wepa, D., & McGhee, S. (2021). Self-compassion and veteran's health: A scoping review. *Journal of Veterans Studies, 7*(1), 86–130. https://doi.org/10.21061/jvs.v7i1.219

Strauss, C., Lever Taylor, B., Gu, J., Kuyken, W., Baer, R., Jones, F., & Cavanagh, K. (2016). What is compassion and how can we measure it? A review of definitions and measures. *Clinical Psychology Review, 47*, 15–27. https://doi.org/10.1016/j.cpr.2016.05.004

Sumalla, E. C., Ochoa, C., & Blanco, I. (2009). Posttraumatic growth in cancer: Reality or illusion? *Clinical Psychology Review, 29*(1), 24–33. https://doi.org/10.1016/j.cpr.2008.09.006

Sylvester, C. (2011). Experiencing war: An introduction. In C. Sylvester (Ed.), *Experiencing war*. Routledge.

Szabo, Y. Z., Warnecke, A. J., Newton, T. L., & Valentine, J. C. (2017). Rumination and posttraumatic stress symptoms in trauma-exposed adults: A systematic review and meta-analysis. *Anxiety, Stress & Coping, 30*(4), 396–414.

Tang, W. K. (2019). Resilience and self-compassion related with achievement emotions, test anxiety, intolerance of uncertainty, and academic achievement. *Psychological Studies, 64*(1), 92–102. https://doi.org/10.1007/s12646-019-00482-6

Tedeschi, R. G., & Calhoun, L. G. (1995). *Trauma & transformation: Growing in the aftermath of suffering*. Sage.

Tedeschi, R. G., & Calhoun, L. G. (2003). Routes to posttraumatic growth through cognitive processing. In *Promoting capabilities to manage posttraumatic stress: Perspectives on resilience* (pp. 12–26). Charles C Thomas Publisher.

Tirch, D., Schoendorff, B., & Silberstein, L. R. (2014). Compassion and psychological flexibility. In D. Tirch, B. Schoendorff, L. R. Silberstein, P. Gilbert, & S. C. Hayes (Eds.), *The ACT practitioner's guide to the science of compassion: Tools for fostering psychological flexibility*. New Harbinger Publications.

Todorov, N., Sherman, K. A., & Kilby, C. J. (2019). Self-compassion and hope in the context of body image disturbance and distress in breast cancer survivors. *Psycho-Oncology, 28*(10), 2025–2032. https://doi.org/10.1002/pon.5187

Trindade, I. A., Ferreira, C., Borrego, M., Ponte, A., Carvalho, C., & Pinto-Gouveia, J. (2018). Going beyond social support: Fear of receiving compassion from others predicts depression symptoms in breast cancer patients. *Journal of Psychosocial Oncology, 36*(4), 520–528. https://doi.org/10.1080/07347332.2018.1440275

Trompetter, H. R., Kleine, E., & Bohlmeijer, E. T. (2017). Why does positive mental health buffer against psychopathology? An exploratory study on self-compassion as a resilience mechanism and adaptive emotion regulation strategy. *Cognitive Therapy and Research, 41*(3), 459–468.

Tsai, J., & Pietrzak, R. H. (2017). Trajectories of posttraumatic growth among US military veterans: A 4-year nationally representative, prospective cohort study. *Acta Psychiatrica Scandinavica, 136*(5), 483–492. https://doi.org/10.1111/acps.12800

Tsai, J., Sippel, L. M., Mota, N., Southwick, S. M., & Pietrzak, R. H. (2016). Longitudinal course of posttraumatic growth among U.S. military veterans: Results from the national health and resilience in veterans study. *Depression and Anxiety, 33*(1), 9–18. https://doi.org/10.1002/da.22371

Tubbs, J. D., Savage, J. E., Adkins, A. E., Amstadter, A. B., & Dick, D. M. (2019). Mindfulness moderates the relation between trauma and anxiety symptoms in college students. *Journal of American College Health, 67*(3), 235–245. https://doi.org/10.1080/07448481.2018.1477782

Tugade, M. M., Fredrickson, B. L., & Barrett, L. F. (2004). Psychological resilience and positive emotional granularity: Examining the benefits of positive emotions on coping and health. *Journal of Personality, 72*(6), 1161–1190. https://doi.org/10.1111/j.1467-6494.2004.00294.x

Ungar, M. (2012). *The social ecology of resilience: A handbook of theory and practice* [1 online resource (xv, 463 pages): Illustrations]. https://doi.org/10.1007/978-1-4614-0586-3

Vasterling, J. J., Taft, C. T., Proctor, S. P., Macdonald, H. Z., Lawrence, A., Kalill, K., et al. (2015). Establishing a methodology to examine the effects of war-zone PTSD on the family: The family foundations study. *International Journal of Methods in Psychiatric Research, 24*(2), 143–155. https://doi.org/10.1002/mpr.1464

Vujanovic, A. A., Youngwirth, N. E., Johnson, K. A., & Zvolensky, M. J. (2009). Mindfulness-based acceptance and posttraumatic stress symptoms among trauma-exposed adults without axis I psychopathology. *Journal of Anxiety Disorders, 23*(2), 297–303. https://doi.org/10.1016/j.janxdis.2008.08.005

Walsh, F. (2020). Loss and resilience in the time of COVID-19: Meaning making, hope, and transcendence. *Family Process, 59*(3), 898–911. https://doi.org/10.1111/famp.12588

Wang, G.-L., Cheng, C.-T., Feng, A.-C., Hsu, S.-H., Hou, Y.-C., & Chiu, C.-Y. (2016). Prevalence, risk factors, and the desire for help of distressed newly diagnosed cancer patients: A large-sample study. *Palliative and Supportive Care, 15*, 1–10. https://doi.org/10.1017/S1478951516000717

Weich, S., Brugha, T., King, M., McManus, S., Bebbington, P., Jenkins, R., et al. (2011). Mental well-being and mental illness: Findings from the adult psychiatric morbidity survey for England 2007. *The British Journal of Psychiatry: the Journal of Mental Science, 199*(1), 23–28. https://doi.org/10.1192/bjp.bp.111.091496

Werner, E. E., & Johnson, J. L. (2002). Can we apply resilience? In M. D. Glantz & J. L. Johnson (Eds.), *Resilience and development: Positive life adaptations* (pp. 259–268). Springer.

Westerhof, G. J., & Keyes, C. L. M. (2010). Mental illness and mental health: The two continua model across the lifespan. *Journal of Adult Development, 17*(2), 110–119. https://doi.org/10.1007/s10804-009-9082-y

Wilk, J. E., Quartana, P. J., Clarke-Walper, K., Kok, B. C., & Riviere, L. A. (2015). Aggression in US soldiers post-deployment: Associations with combat exposure and PTSD and the moderating role of trait anger. *Aggressive Behavior, 41*(6), 556–565. https://doi.org/10.1002/ab.21595

Williams, G. C., Lynch, M., & Glasgow, R. E. (2007). Computer-assisted intervention improves patient-centered diabetes care by increasing autonomy support. *Health Psychology, 26*(6), 728–734. https://doi.org/10.1037/0278-6133.26.6.728

Wilson, J. M., Weiss, A., & Shook, N. J. (2020). Mindfulness, self-compassion, and savoring: Factors that explain the relation between perceived social support and well-being. *Personality and Individual Differences, 152*. https://doi.org/10.1016/j.paid.2019.109568

Wood, A. M., & Tarrier, N. (2010). Positive clinical psychology: A new vision and strategy for integrated research and practice. *Clinical Psychology Review, 30*(7), 819–829. https://doi.org/10.1016/j.cpr.2010.06.003

World Health Organization: WHO. (2005). *Promoting mental health: Concepts, emerging evidence, practice*. WHO.

World Health Organization: WHO. (2020). *Factsheet about cancer*. Retrieved from https://www.who.int/news-room/fact-sheets/detail/cancer

Yela, J. R., Crego, A., Gómez-Martínez, M., & Jiménez, L. (2020). Self-compassion, meaning in life, and experiential avoidance explain the relationship between meditation and positive mental health outcomes. *Journal of Clinical Psychology, 76*(9), 1631–1652. https://doi.org/10.1002/jclp.22932

Zabora, J., BrintzenhofeSzoc, K., Curbow, B., Hooker, C., & Piantadosi, S. (2001). The prevalence of psychological distress by cancer site. *Psychooncology, 10*(1), 19–28. https://doi.org/10.1002/1099--1611(200101/02)10:1<19::aid-pon501>3.0.co;2-6

Zarotti, N., Povah, C., & Simpson, J. (2020). Mindfulness mediates the relationship between cognitive reappraisal and resilience in higher education students. *Personality and Individual Differences*, 156. https://doi.org/10.1016/j.paid.2019.109795

Zautra, A., & Reich, J. (2012). Resilience: The meanings, methods, and measures of a fundamental characteristic of human adaptation. *The Oxford Handbook of Stress, Health, and Coping*. https://doi.org/10.1093/oxfordhb/9780195375343.013.0009

Zautra, A. J., Arewasikporn, A., & Davis, M. C. (2010). Resilience: Promoting well-being through recovery, sustainability, and growth. *Research in Human Development, 7*(3), 221–238. https://doi.org/10.1080/15427609.2010.504431

Zeligman, M., Bialo, J. A., Brack, J. L., & Kearney, M. A. (2017). Loneliness as moderator between trauma and posttraumatic growth. *Journal of Counseling and Development, 95*(4), 435–444.

Zhu, L., Yao, J., Wang, J., Wu, L., Gao, Y., Xie, J., et al. (2019). The predictive role of self-compassion in cancer patients' symptoms of depression, anxiety, and fatigue: A longitudinal study. *Psycho-Oncology, 28*(9), 1918.

Zimmerman, M., McGlinchey, J. B., Posternak, M. A., Friedman, M., Attiullah, N., & Boerescu, D. (2006). How should remission from depression be defined? The depressed patient's perspective. *American Journal of Psychiatry, 163*(1), 148–150. https://doi.org/10.1176/appi.ajp.163.1.148

11 Self-Compassion and Body Image

Tracy L. Tylka and Katarina L. Huellemann

Introduction

Self-compassion occurs when people are mindful, understanding, and nurturing toward themselves during situations where they experience distressing feelings and concerns about their adequacy (Neff, 2003b). Many situations hold the potential to bring forth distressing feelings and concerns that are specific to the body and appearance. These situations may include a person being told (or receiving a hint) to go on a diet, not being able to fit into a piece of clothing that used to fit comfortably, being rejected by a love interest, realizing age-related changes in their appearance, and seeing an unflattering photograph of their body. Situations such as these are referred to as body image threats (Cash & Williams, 2005). Body image represents a person's "inside view" of their body–that is, their feelings, perceptions, thoughts, and beliefs about their body that impact how they behave toward it (Cash, 2004, p. 1). Body image threats, then, may provoke body-related distress and shame as well as negative thoughts and beliefs (e.g., "I'm so unattractive"), which could encourage negative body-related behaviors and harmful coping strategies (e.g., disordered eating, excessive exercise). Fortunately, generating self-nurturance, understanding, and mindfulness rather than self-judgment and criticism may subvert this deleterious process.

The study of self-compassion holds great relevance for body image theory, research, and practice. During body image threats, a person can activate self-compassion, which facilitates feelings of being cared for, connected to others, and emotionally calm, to offset their distress (Gilbert, 2005). For example, rather than judging their body as deficient, a person high in self-compassion might react in kind, warm-hearted ways toward their body that nurture their self-care (self-kindness) (Neff, 2003b). Rather than feeling alone in their experience of body distress, a person high in self-compassion is able to understand that societal ideals prompt most people to feel badly toward their body at times (common humanity). Rather than ruminating about their negative feelings toward their body or trying to avoid their emotional reactions, a person high in self-compassion is aware of their feelings in a more balanced way (mindfulness). Therefore, self-compassion may be an important protective factor that helps build and maintain positive body image and counteract the development of body dissatisfaction, especially when confronted with body image threats (Tylka & Kroon Van Diest, 2015).

T. L. Tylka (✉)
Department of Psychology, The Ohio State University, Marion, OH, USA
e-mail: tylka.2@osu.edu

K. L. Huellemann
Western University, London, ON, Canada

A. Finlay-Jones et al. (eds.), *Handbook of Self-Compassion*, Mindfulness in Behavioral Health,
https://doi.org/10.1007/978-3-031-22348-8_11

Within the last decade, there have been many advances in understanding the connection between self-compassion and body image. Several positive and negative body image variables have been included in this research. Of note, positive body image is not simply body satisfaction (Tylka & Wood-Barcalow, 2015b). While body satisfaction typically refers to a person's positive evaluation and perception of their appearance, positive body image is more complex, reflecting a person's appreciation, acceptance, respect, and love for their body regardless of their appearance (Wood-Barcalow et al., 2010). Several key positive body image variables include body appreciation (Tylka & Wood-Barcalow, 2015a), functionality appreciation (Alleva et al., 2017), body image flexibility (Sandoz et al., 2013), body compassion (Altman et al., 2017), and broadly conceptualizing beauty (Tylka & Iannantuono, 2016). Table 11.1 provides an overview of key body image variables (and their corresponding measures) included in self-compassion research. Most of this research has assessed self-compassion using the 26-item Self-Compassion Scale (SCS; Neff, 2003a) or the 12-item short-form of the Self-Compassion Scale (SCS-SF; Raes et al., 2011).

Research supports the inverse connection between self-compassion and negative body image (e.g., body dissatisfaction, weight and shape concerns) and positive connection between self-compassion and positive body image (e.g., body appreciation) (Alleva et al., 2017; Tylka & Iannantuono, 2016; Wasylkiw et al., 2012). A recent meta-analysis revealed a strong positive relationship between self-compassion and positive body image ($r = 0.52$) across 20 studies, and a moderate inverse relationship between self-compassion and negative body image ($r = -0.45$) across 21 studies (Turk & Waller, 2020).

This meta-analysis also revealed that individuals receiving self-compassion-related interventions experienced improved body image compared to those in control groups, with the degree of difference being small to moderate in strength ($g = 0.39$) across 13 studies.

Table 11.1 Body image variables often examined in self-compassion research

Variable	Definition	Scale
Body (dis)satisfaction (includes appearance satisfaction, appearance evaluation, weight/shape concerns)	(Dis)Satisfaction with overall body shape, size, and/or weight as well as the shape and size of specific body areas	Many scales have been developed to assess this construct.
Body appreciation	Love and respect for, acceptance and appreciation of, and comfort with the body, including its unique characteristics	Body Appreciation Scale-2 (BAS-2) (Tylka & Wood-Barcalow, 2015a)
Functionality appreciation	Being grateful for, respecting, and honoring the body for what it is capable of doing: Its physical capacities (e.g., walking), internal processes (e.g., healing from a cold), sensory and perceptual abilities (e.g., self-soothing), creative endeavors (e.g., drawing), communication with others (e.g., laughter), and self-care (e.g., showering)	Functionality Appreciation Scale (FAS) (Alleva et al., 2017)
Body image flexibility	A compassionate response to embrace (rather than avoid, escape, or change) negative body-related thoughts and feelings using mindfulness and acceptance while pursuing meaningful and valued behaviors	Body Image-Acceptance and Action Questionnaire (BI-AAQ) (Sandoz et al., 2013)
Body compassion	Regarding one's body with mindfulness, kindness, and awareness of common humanity in terms of its appearance, health, and competence	Body Compassion Scale (BCS) (Altman et al., 2017)
Broadly conceptualizing beauty	Perceiving a wide range of physical appearances as beautiful (e.g., body shape, weight, personal style) as well as drawing from inner characteristics (e.g., confidence, self-acceptance) when defining beauty	Broad Conceptualization of Beauty Scale (BCBS) (Tylka & Iannantuono, 2016)

Self-compassion appears to both prevent and treat body dissatisfaction as well as increase and maintain positive body image (Braun et al., 2016).

In this chapter, we first review the existing theory and research investigating the complex connection between self-compassion and body image while acknowledging that theoretical and empirical advancements in this area continue to burgeon. Scholars have used meta-analytic, cross-sectional, prospective, and diary-based designs to study this connection. We then turn to examining the effects of self-compassion interventions on body image and end with discussing opportunities for the next generation of research on self-compassion and body image.

Theoretical Frameworks and Corresponding Research

From a theoretical standpoint, attitudes toward the self have long been considered in the context of body image. One popular self-view that has been examined in the context of body image is *self-esteem,* which involves how a person perceives their level of competence and self-worth in areas that are personally (and possibly socially) important to them (James, 1890). Simply put, high self-esteem involves a positive evaluation of oneself that has been associated with meaningful benefits including happiness (Lucas et al., 1996) and life satisfaction (Diener & Diener, 1995). Given that self-esteem is influenced by our perceptions of *how we think we appear to others* (Harter, 1999), and that pursuing self-esteem often involves comparing ourselves to others (Neff, 2011), having low self-esteem may lower a person's positive (and heighten their negative) attitudes and feelings about their body specifically (Grossbard et al., 2009). This link may be especially relevant in Western societies where, for some women, self-esteem has been found to be contingent on the degree they evaluate themselves as meeting societal standards of beauty (Crocker et al., 2003). There are several problems linked to pursuing self-esteem. Trying to "keep up" with evaluating oneself positively in comparison to others may lead a person to overlook rather than confront and cope with their personal inadequacies, which is linked to narcissism, bullying, and prejudice (Neff & Vonk, 2009). Further, self-esteem is often contingent on peer approval and success, and deserts people when they fail (e.g., experience a body image threat), which is when they need emotional support the most (Neff, 2011).

Other limitations of self-esteem are that it emerges early in life, remains stable over time, and is highly resistant to intervention (Josephs et al., 2003). Some research has even shown that experimental efforts to enhance self-esteem have no effects or even increase negative body image (Alleva et al., 2015). In relation to body image and appearance, cultivating self-compassion instead of self-esteem may instead allow a person to become more comfortable with their unique physical features that are inconsistent with societal appearance ideals. Further, self-compassion may encourage greater awareness and acceptance of the emotions associated with not meeting societal appearance ideals rather than trying to avoid these by inflating appearance-related self-worth. Thus, self-compassion represents an alternative way of coping with stressful body image threats, experiences, behaviors, and feelings of appearance-based inadequacy.

Body image researchers have proposed that self-compassion is protective of positive body image in several ways (Braun et al., 2016; Tylka & Kroon Van Diest, 2015). First, self-compassion may directly promote positive body image, as it helps people remain kind to their bodies during body image threats. Second, having high self-compassion may make it less likely that people would engage in thoughts or behaviors associated with negative body image (e.g., comparing their appearance to others). Third, self-compassion may help explain how an environment that offers body acceptance can help cultivate a person's body appreciation, thus acting as a mediator. Finally, self-compassion may act as a moderator, changing the strength and/or direction of a relationship between a predictor of negative body image (e.g., how much a person compares their appearance to others) and well-being (e.g., body appreciation). For example, women high in self-compassion have been found to still appreciate their bodies even when comparing their appearance to others, whereas individuals low in self-compassion report lower body appreciation when comparing their bodies to others (Homan & Tylka, 2015; Siegel et al., 2020). While theoretical conceptualizations of self-compassion have shown how it may act as a direct predictor, mediator, or moderator, self-compassion likely operates through multiple pathways when it comes to body image (Tylka et al., 2015). Sociocultural theories, which position media and interpersonal contributions as threats to body image, have provided the framework for researchers to integrate self-compassion and investigate the varied ways it could be connected to body image.

The Tripartite Influence Model

The tripartite influence model postulates that a person's body dissatisfaction can be both directly and indirectly affected by three social agents (peers, parents, and the media) as well as the extent to which they internalize (or idealize) societal appearance ideals and compare their appearance to others (to evaluate whether or not they "meet" appearance ideals in relation to the person or image they are comparing themselves to) (Thompson et al., 1999). When a person internalizes appearance ideals and discovers that they do not meet these appearance ideals (via comparing their appearance with others), they may experience negative outcomes such as lower satisfaction with, and appreciation of, their bodies (Homan & Tylka, 2015; Schaefer & Thompson, 2014). Although the tripartite influence model has mostly been tested in samples of women, research has shown support for it in samples of heterosexual and gay men (Tylka, 2011; Tylka & Andorka, 2012) and bisexual women (Hazzard et al., 2019).

Support for Self-Compassion as a Moderator in the Tripartite Influence Model Self-compassionate responding has been described as a way to offset the self-critical and judgmental nature of internalizing appearance ideals and engaging in appearance-based social comparison (Rodgers et al., 2017). Some research has found support for this theory. For example, both Homan and Tylka (2015) and Siegel et al. (2020) found that women who frequently compared their bodies to others had lower body appreciation, but this relationship was attenuated for women who were high in self-compassion. These two studies also found that self-compassion weakened the link between appearance-contingent self-worth (how much a person evaluates their overall sense of self-worth based on whether or not they meet cultural standards for physical attractiveness) and body appreciation, demonstrating that self-compassion can protect positive body image even in the face of threats related to one's appearance (Homan & Tylka, 2015; Siegel et al., 2020). Another study found that self-compassion moderated the extent that media pressure to lose weight was related to women's disordered eating and thin-ideal internalization (how much they idealize a thin body type) (Tylka et al., 2015). When women high in self-compassion felt pressure from the media to pursue thinness, they did not internalize the thin ideal or engage in disordered eating, yet when women low in self-compassion felt media pressure to be thin, they reported higher disordered eating and thin-ideal internalization.

In their study using an adolescent sample, Rodgers et al. (2017) found that two dimensions of self-compassion (i.e., mindfulness and common humanity) moderated the mediated relationship between perceived weight status and appearance satisfaction through appearance comparison (how often a person compares their appearance to others) for boys. Their findings demonstrated that boys who perceived themselves as "overweight" compared their bodies more and had lower appearance satisfaction, but this relationship was weakened for boys who had high levels of common humanity and mindfulness. Although the self-compassion dimensions were correlated as expected for girls (i.e., positively associated with appearance satisfaction and negatively associated with appearance comparison), none of the self-compassion dimensions buffered the mediated relationship between perceived weight status and appearance satisfaction through appearance comparison for girls. In addition, none of the self-compassion dimensions moderated the relationship between appearance comparison and appearance satisfaction for boys or girls. The total self-compassion score was not assessed in this study. The authors concluded that (a) the moderation findings for girls may have not emerged because they reported lower levels of self-compassion compared to boys, and (b) self-kindness may simply not be potent enough to combat the self-critical nature of appearance comparison in adolescents.

These cross-sectional studies provide some evidence that self-compassion can buffer the negative outcomes that are associated with body image threats. Interestingly, one study to date has examined the moderating effect of self-compassion at the within-person and the between-person level in the context of the tripartite influence model. Kelly et al. (2016) recruited Canadian college women to report on their daily social interactions, affect, self-compassion, eating behaviors, and body image for seven consecutive nights. The authors found that on days when women reported lower self-compassion, interactions with body-focused others (i.e., people who make self-disparaging comments about their body, weight, or eating; people who diet) were related to higher body image concerns as well as lower intuitive eating (an adaptive way of eating that involves relying on internal hunger and fullness cues rather than external cues such as dieting advice; Tylka, 2006). However, this relationship was absent on days when women reported higher self-compassion. At the between-persons level, women low in self-compassion who had more interactions with body-focused others over the course of the week had lower mean levels of body appreciation and intuitive eating, but these relationships were not observed for women high in self-compassion. Thus, fostering both state (temporary shifts in) and trait (stable levels of) self-compassion may have downstream implications for how women cope when immersed in an appearance-focused culture. Given that it is likely impossible for most women to completely avoid contact with body-focused others (some of these people may be close friends or family), aiming to cultivate trait self-compassion as well as daily self-compassionate thoughts may help protect women from body image concerns.

Support for Self-Compassion as a Predictor within the Tripartite Influence Model There has also been some support for self-compassion's role as a predictor of appearance comparison and thin-ideal internalization. Having high self-compassion directly impacts the importance a person places on appearance-related information (i.e., by engaging in less appearance comparison and being less likely to internalize societal appearance ideals) (Andrew et al., 2016). In a sample of Australian women, data analysis using structural equation modeling revealed that women high in self-compassion experienced lower levels of comparing their appearance to others and internalization of societal appearance ideals, which then predicted body appreciation (Andrew et al., 2016). However, the conclusions that can be drawn from this study are limited since a cross-sectional design was used to test mediation, and thus, causal inferences cannot be made (Maxwell & Cole, 2007).

To examine the within-person fluctuations and between-person associations of self-compassion and appearance comparison as predictors of women's body image and related variables (appearance anxiety, drive to be thin, and body dissatisfaction), a second study employed a diary methodology whereby women responded to brief online surveys three times per day every alternate day for 1 week (Thøgersen-Ntoumani et al., 2017). Using a mixed linear modeling approach, the results revealed that appearance anxiety, motivation to be thin, and body dissatisfaction were negatively predicted by daily self-compassionate thoughts and positively predicted by upward appearance comparison (comparing oneself to others perceived as more physically attractive). Thus, although it may be important to cultivate a stable level of high self-compassion to moderate the prolonged impact of appearance-related information from parents, peers, and the media, promoting self-compassionate thoughts daily can also help to reduce feelings of body dissatisfaction.

Objectification Theory

Objectification theory posits that experiencing and witnessing sexual objectification may lead a person to consistently think about their body from an outsider's perspective and consider themselves and their appearance as an object for others' approval and satisfaction, a process known as *self-objectification* (Fredrickson & Roberts, 1997). When a person self-objectifies, they are more vulnerable to negative body-related outcomes such as body shame, body guilt, and eating disorder symptoms (Calogero & Pina, 2011). Although objectification theory has been contextualized and applied to women's experiences (Calogero et al., 2011; Fredrickson & Roberts, 1997), research has revealed that other populations such as gay, bisexual, or transgender men (Wiseman & Moradi, 2010) and transgender women (Comiskey et al., 2020) do experience sexual and self-objectification and its negative downstream consequences to body image, disordered eating, and psychological well-being.

Given that self-compassion inherently contrasts the self-policing nature of self-objectification, it may help to combat self-objectification and its body-related consequences. A person who is high in self-objectification engages in body surveillance (habitually monitors and checks their body) in order to anticipate how another person may evaluate them, which fosters self-disparaging thoughts and feelings and efforts to manipulate their appearance (Calogero & Pina, 2011). In contrast, a person who is high in self-compassion directs their attention inward, cultivating a more balanced and kind awareness of their thoughts and emotions. This may support them to welcome their body as is, not for how it appears to others. These next sections will discuss the various ways that self-compassion has been integrated into objectification theory to illuminate whether self-compassion can reduce self-objectification and mitigate the link between self-objectification and negative body image.

Support for Objectification Theory with Self-Compassion as a Moderator Some evidence suggests that self-compassion may have a moderating effect on the relationship between self-objectification and its negative body-related outcomes. In support of this theory, Liss and Erchull (2015) categorized women into a low self-compassion or a high self-compassion group and tested a model linking self-objectification (conceptualized as body surveillance) to body shame, depression, and disordered eating for each group. Women in the high self-compassion group reported lower levels of body surveillance, body shame, depression, and disordered eating compared to women in the low self-compassion group. Moreover, the pathways between body surveillance and body shame, depression, and disordered eating were not as strong in the high self-compassion group as they were in the low self-compassion group, suggesting that self-compassion moderated these pathways. Daye et al. (2014) found that among college women, having a caregiver that was critical of eating habits was related to higher body surveillance and body shame. However, at high levels of self-compassion, body surveillance and body shame were diminished. These findings suggested that women high in self-compassion were less likely to experience body shame or surveillance despite receiving disapproving messages about their eating habits as children.

Support for Self-Compassion as a Predictor of Objectification Theory Constructs Having high self-compassion may also predict lower levels of objectification theory constructs (e.g., self-objectification, body surveillance). Several researchers have examined this assertion. First, in their model, Andrew et al. (2016) found that high self-compassion predicted reduced self-objectifying thoughts and behaviors in college women. Second, Cox et al. (2019) explored the relationship between women's self-compassion and intrinsic motivation for physical activity with body surveillance and body appreciation as mediators in a cross-sectional design (Sample 1) and examined the prospective relationships between these variables throughout women's participation in a 16-week yoga course (Sample 2). Analysis using structural equation modeling suggested that for Sample 1, self-compassion predicted higher body appreciation, and higher body appreciation predicted greater intrinsic motivation for physical activity. In Sample 2, latent growth curve analyses revealed that changes in self-compassion over the 16-week course predicted higher body appreciation and lower body surveillance, while higher body appreciation predicted greater intrinsic motivation for physical activity. Although the cross-sectional findings did not find any evidence that self-compassion uniquely contributed to lower body surveillance, an examination of these relationships over time suggested that the yoga environment fostered improvements in self-compassion, body surveillance, body appreciation, and intrinsic motivation. Yoga may therefore offer a unique context to practice viewing and appreciating the body more compassionately.

Self-compassion has also been considered as a predictor of lower body surveillance and body shame among female athletes. Mosewich et al. (2011; see also Chap. 13) proposed that self-compassion may be a valuable resource for young women athletes to harness, given that sport contexts emphasize social comparison, self-evaluation, and being evaluated by others. Athletes continuously monitor and evaluate their bodies, which can foster negative self-conscious emotions (e.g., body shame and guilt), but having high self-compassion may be able to mitigate self-conscious emotions. In support of their theory of self-compassion's important role in promoting lower body surveillance and self-conscious emotions, self-compassion explained more variance than self-esteem in shame, guilt, objectified body consciousness (including body surveillance), and fear of failure and negative evaluation (Mosewich et al., 2011).

Social Media, Self-Compassion, and Objectification Theory Constructs Given the millions of photos that display attractive people with thin and fit physiques on social media platforms, it is unsurprising that prior research has linked greater social media use to more body concerns, thin-ideal internalization, and self-objectification (Fardouly & Vartanian, 2015; Trekels et al., 2018). Self-compassion may be able to help people cope with the inherent problem arising from comparing their bodies with others via social media, thereby protecting themselves from harmful body-related mental health outcomes.

To date, two studies have examined this proposition. Among a sample of community women, Modica (2019) found that Facebook appearance exposure (engaging in photo-related activity compared to all types of activity on Facebook) and Facebook appearance comparison (comparing one's body to others' bodies on Facebook) were significantly related to greater body surveillance, and high self-compassion was significantly associated with lower body surveillance. Nevertheless, self-compassion did not moderate the relationship between Facebook appearance comparison and body surveillance, suggesting that having high self-compassion was not enough to attenuate the link between Facebook appearance comparison and body surveillance. Lonergan et al. (2019) revealed that editing/posting selfies (photo manipulation) and being invested emotionally in others' responses to the selfies were related to greater body dissatisfaction in men and women, but self-compassion did not moderate these relationships even though greater self-compassion was associated with lower photo manipulation and emotional investment in others' responses to selfies. Thus, self-compassion appears to be inversely related to constructs that indicate self-objectification when using social media (i.e., body surveillance, manipulating one's appearance in photos), but there is currently no evidence that it has a buffering effect in this context.

To date, no other examined variable (e.g., deriving self-worth from appearance) has been found to moderate (weaken or strengthen) the links between social media constructs, self-objectification, and body dissatisfaction (Modica, 2019). It is possible that the high level of appearance exposure on social media among women, the internalization of these appearance ideals, and the self-objectification that stems from viewing these images are so potent that self-compassion and other variables are not strong enough to reduce women's appearance comparison or self-objectification. An alternative may be for women to cultivate self-compassion and diversify their exposure to different types of images on social media (e.g., diverse body types, nature, motivational quotes), but this has not been empirically tested. Exploring social media use on other image-focused social media platforms (e.g., Instagram, Snapchat, TikTok) and aiming to replicate findings are important next steps to verify whether self-compassion is protective against body surveillance after social media use.

Social Mentalities Theory

Social mentalities theory is derived from evolutionary biology, neurobiology, and attachment theory, and it suggests that being capable of self-soothing (e.g., self-compassion) is dependent on prior caregiving and care-seeking experiences (Gilbert & Irons, 2005; see also Chap. 4). For example, low self-compassion and high fear of self-compassion (believing that one is undeserving of self-compassion and that engaging in self-compassion will make one weak) are thought to emerge from anxious and avoidant attachment styles, or personality traits that stem from unstable or unaffectionate caregiving experiences (Gilbert et al., 2011; also see Chap. 5). As such, a person may engage in self-criticism when their threat system is activated (e.g., from a body image threat) but may also have the ability to activate self-compassion when their self-soothing system is activated (Gilbert, 2015). A few studies have examined the role of fear of self-compassion in relation to body image outcomes. For example, in a sample of Portuguese women, Dias et al. (2020) found that high fear of self-compassion had an indirect effect on eating disorder symptoms through feeling ashamed of their body, suggesting that fear of self-compassion drives negative emotions toward the body. Similarly, in a sample of Canadian women, Huellemann and Calogero (2020) found that low self-compassion and high fear of self-compassion were independently related to body checking behavior (continuously critiquing the size, shape, and appearance of one's body) through stigmatizing oneself in the forms of self-objectification, experiencing body shame, and internalizing weight stigma (believing that negative societal stereotypes about weight apply to oneself). This preliminary research suggests that it may be a fruitful next step for clinicians to develop body image interventions that cultivate self-compassion and reduce fear of self-compassion simultaneously, especially for people who are high in self-criticism or who have experienced unaffectionate childhoods.

Additionally, self-critical personality traits, such as maladaptive perfectionism, may reflect what occurs when a person's threat system has been continuously activated (e.g., from continuous appearance comparison to social media appearance ideals), and being high in perfectionism may adversely impact a person's body image. To test this theory, Barnett and Sharp (2016) conducted two studies. In Study 1, the self-kindness and self-judgment components of self-compassion mediated the inverse association between maladaptive perfectionism and body satisfaction in college women. These findings were replicated with another sample of college women in Study 2, which also revealed that self-judgment mediated the positive link between perfectionism and eating disorder symptoms. The evidence indicates that self-compassion likely acts as a mediator in the context of self-critical personality traits, body image, and eating disorder symptoms.

Furthermore, people adopt various social mentalities when interacting with others, with each triggering different patterns of thinking, feeling, and behaving (Gilbert, 2005). For instance, a competitive-based mentality orients people to appraise their rank relative to others (e.g., engage in appearance comparisons) and to base their behavior on their perceptions of who is superior or inferior. In contrast, a compassion-based mentality supports connecting with others rather than competing with them and is linked to many positive health and well-being effects (Salzberg, 1995). Vimalakanthan et al. (2018) explored the effects of cultivating a caregiving versus competitive mentality whenever participants found themselves comparing their appearance to others. Female undergraduate students were randomly assigned to use one of three practices when they found themselves engaging in appearance comparisons: caregiving (cultivating compassion and loving-kindness toward the person they are comparing their appearance to), downward competition (comparing themselves favorably to the person in various ways except for appearance), and distraction (distracting themselves with a counting task). After learning about their particular practice in a lab-based setting, participants committed to engaging in their practice over a period of 48 h whenever they noticed that they were comparing their appearance to others. Results indicated that the caregiving practice improved body dissatisfaction among women who compared their appearance to others frequently; however, the downward competition practice improved body dissatisfaction among women who compared their appearance to others less frequently. Therefore, cultivating compassion toward others can especially help women who have body dissatisfaction and frequently compare their appearance to others.

Weight Stigma Theories

People living in larger bodies are more likely to experience overt weight stigma (e.g., weight-based discrimination, weight-based teasing, and bullying) and may avoid certain situations where weight-based discrimination would be more likely to occur (e.g., physical activity and health-care settings) (Mensinger et al., 2018; Vartanian & Shaprow, 2008). As such, people with a higher BMI may have lower body image flexibility, which refers to the degree a person is still able to engage in meaningful and valued life activities even when experiencing negative body image-related thoughts or feelings (see Table 11.1; Sandoz et al., 2013). Self-compassion may be able to buffer the effect of having a higher BMI on body image flexibility since it involves interpreting incoming information about one's body with openness and kindness. Kelly et al. (2014a, b) tested this hypothesis in a cross-sectional sample of female undergraduates and found that self-compassion moderated the relationship between high BMI and body image flexibility as well as between high BMI and eating disorder symptoms. Women with a higher BMI were more likely to report lower body image flexibility and eating disorder symptoms, but these relationships were dampened at higher levels of self-compassion.

Webb et al. (2016) examined the relationship between anti-fat attitudes (placing personal blame on individuals for being fat, disliking higher weight people, and fear of weight gain) and fat talk (conversations expressing a personal desire to lose weight or that degrade oneself for their weight) with self-compassion as a moderator. Because Western culture values thinness and lean muscularity, people of all sizes may develop anti-fat attitudes. Fat talk is problematic because it reinforces the belief that slimmer bodies are superior to larger ones. Undergraduate women completed a self-report questionnaire with measures of anti-fat attitudes, fat talk, body shame, and self-compassion. Controlling for BMI, Webb et al. (2016) found that anti-fat attitudes were indirectly associated with fat talk via body shame but that this mediated relationship was attenuated at high levels of self-compassion. In other words, women with higher anti-fat attitudes reported greater body shame and were more likely to engage in fat talk; however, these relationships were weaker for women higher in self-compassion. It is important to recognize that alongside encouraging self-compassion cultivation, macro-level changes must take place that publicly discourage weight-based discrimination and promote body inclusivity.

Within-Person Variability in Self-Compassion, Body Image, and Eating Behavior

Although self-compassion is traditionally studied as a stable characteristic (trait) that differs between people, daily diary studies are an effective method to elucidate within-person variability in self-compassion and explore how shifts in self-compassion contribute to a person's daily body image and eating behavior. To explore the within and between-persons relations between self-compassion, negative body image, and eating disorder symptoms, college women participated in a 4-day daily diary study (Breines et al., 2014). Participants completed multiple online surveys each day on their level of appearance-based self-compassion (perceiving body flaws in a self-compassionate way), eating disorder symptoms, and self-esteem. Controlling for self-esteem, the findings revealed that on days when participants reported higher appearance-based self-compassion, they tended to also report fewer eating disorder symptoms. Of note, participants in a second sample who, in a lab setting, reflected on a perceived body flaw in a self-compassionate way through a writing task reported lower subsequent body shame (Breines et al., 2014).

In another study, female college students completed brief surveys for 7 days on self-compassion, self-esteem, body appreciation, body satisfaction, intuitive eating, and eating disorder symptoms (Kelly & Stephen, 2016). Using multilevel modeling, the results revealed that daily fluctuations in self-compassion predicted daily fluctuations in body image and eating behavior. That is, higher daily self-compassion was related to greater daily body appreciation and intuitive eating as well as lower daily levels of restrained eating at the within-persons level. At the between-persons level, higher average self-compassion scores across the week were associated with more adaptive body image and eating behavior. These findings were held even when controlling for self-esteem. Collectively, Kelly and Stephen demonstrated that self-compassion has the potential to make a meaningful impact on the daily (state) level even if a person is not high in stable (trait) self-compassion. Therefore, interventions that increase state- and trait-based levels of self-compassion may be useful for improving positive body image and decreasing negative body image.

Interventions to Improve Body Image

Given the theoretical and empirical connections between self-compassion and body image, several research teams have developed and investigated self-compassion-based interventions to determine whether these interventions can increase positive body image (e.g., body appreciation, body image flexibility) and reduce negative body image (e.g., body dissatisfaction). Findings from recent reviews (Guest et al., 2019) and meta-analyses (Turk & Waller, 2020) of self-compassion-based interventions suggest that these interventions do improve participants' body image, with several studies reporting lasting effects. Below, we discuss these interventions and the studies that have evaluated their impact on body image.

Self-Compassion Meditations

Certain self-compassion interventions have used Neff and Germer's (2013) 20-min guided self-compassion meditations from their Mindful Self-Compassion program as podcasts participants can listen to daily. As the first study using this method, Albertson et al. (2015) randomized women with body concerns to a self-compassion intervention or a waitlist control group. Women in the intervention group were instructed to listen to the guided self-compassion meditations once a day for 3 weeks. They received a link to one podcast per week; the first included a compassionate body scan, the second included affectionate breathing, and the third focused on loving-kindness directed toward the body. Compared to the control group, intervention participants (who reported listening to the podcast 3.6 days a week on average) experienced greater increases in body appreciation and decreases in body dissatisfaction, body shame, and self-worth based on their appearance, and these improvements were maintained when measured 3 months after the intervention. However, the high attrition (48%) from the initial assessment to follow-up tempered the conclusions.

Exploring whether a more condensed version of Albertson et al.'s (2015) program would produce lower attrition yet similar benefits, Toole and Craighead (2016) randomized undergraduate women to either participate in a self-compassion-based meditation training or a waitlist control group for a 1-week period. They provided the initial training session in a lab to standardize the first exposure. Once a day, participants in the intervention group were emailed a link to Neff and Germer's (2013) audio recordings of a compassionate body scan (days 1–3), affectionate breathing exercise (days 4–5), or loving-kindness body meditation (days 6–7). The shorter time frame (compared to Albertson et al., 2015) did not increase participants' willingness to listen to the podcast on their own (total weekly exposure ranged from 20 to 90 min). Nevertheless, at posttest, participants in the intervention group reported higher body appreciation and lower self-worth based on appearance and body surveillance compared to the waitlist control group (body shame and body dissatisfaction did not differ between groups).

In sum, self-compassion meditation appears to be an effective means of improving body image that can be flexibly delivered using podcasts. Listening to these podcasts have positive effects on women's body image, which may be sustained, at least over the short term. Further, relatively brief exposure may be powerful enough to have an impact. Yet, additional research is needed to determine whether these podcasts would help diverse groups of individuals and whether they can be used to prevent negative body image.

Writing Tasks

In general, writing exercises are an effective mode of delivering both compassion-based and non-compassion-based body image interventions (Guest et al., 2019). Seekis et al. (2017) investigated whether a single-session self-compassion writing task would improve body image concerns among undergraduate women. After reading a negative body image scenario to induce body concerns (participants imagined themselves viewing unflattering pictures of themselves posted on social media by a friend), they completed measures to gauge their current (state) level of body appreciation, body satisfaction, and appearance anxiety. They were randomly assigned to one of three writing groups: self-compassion (designed to induce self-kindness, common humanity, and mindfulness), self-esteem (designed to induce feelings of self-competence and value), or control (writing about academic topics). They followed their condition's prompts and wrote for 15 min. Participants completed the state-based body image measures again immediately after the writing task and again at a 2-week follow-up. The self-compassion writing group experienced higher state body appreciation compared to the self-esteem and control groups at posttest, as well as higher state body appreciation compared to the control group at follow-up. Both self-compassion and self-esteem writing groups reported higher body satisfaction compared to the control group at posttest and follow-up. However, groups did not differ on appearance anxiety. Therefore, brief self-compassion-based writing interventions may help improve women's body appreciation and satisfaction in the face of body image threats, but it may not reduce their appearance anxiety.

Similarly, Moffitt et al. (2018) compared the efficacy of self-compassion and self-esteem writing tasks for reducing state body dissatisfaction. College women viewed 16 magazine images of young, thin, female models for 8 min to induce body dissatisfaction and were encouraged to compare their bodies with the women in the images. Women were then randomized to one of three 3-min writing interventions:

self-compassion (expressing compassion toward their weight, appearance, and body shape), self-esteem (describing their positive qualities, such as personal attributes and accomplishments), or positive distraction (describing a hobby they enjoy). Findings revealed that those in the self-compassion and self-esteem groups had greater reductions in state body dissatisfaction. Participants' trait body dissatisfaction moderated the intervention efficacy: the benefit of the self-compassion intervention became evident at moderate levels of body dissatisfaction and was most apparent at high levels of body dissatisfaction.

Self-compassion intervention also appears beneficial for women undergoing appearance changes due to illness. For example, one study found that women with breast cancer benefited from writing exercises designed to foster self-compassion toward their body-related changes as a result of cancer treatment. Przezdziecki and Sherman (2016) randomly assigned female breast cancer survivors to a self-compassionate writing assignment or an unstructured writing assignment whereby they wrote about an event related to their posttreatment body changes. Women in the self-compassion condition were prompted, in writing, to show understanding and kindness toward their bodies, explore their feelings by putting space between the event and their reactions, and express how other survivors may experience events where they feel uncomfortable about their body. Last, they wrote a self-compassionate letter to themselves. The control condition was simply asked to provide details about the event (i.e., what led up to it, who was present, and how they felt and behaved). The writing length was standardized to four pages for each condition. Although body image was not explored as an outcome, the authors found that the self-compassion intervention reduced negative affect and increased self-compassion.

Overall, writing tasks that focus on taking a self-compassionate stance toward the body after an imagined or real body image threat help increase state body appreciation and body satisfaction, and these positive effects may continue after the intervention. Self-compassion interventions seem to be the most beneficial for women with elevated trait body dissatisfaction and may hold some advantages over self-esteem-based interventions for improving body appreciation.

Mobile Applications

Thus far, one self-compassion-based application has been developed and studied in its ability to improve body image. *BodiMojo* is a mobile application-based 6-week intervention to promote positive body image through self-compassion. The app contains intervention messages (delivered twice daily in the form of an affirmation, behavioral tip, or psychoeducation), mood tracking and emotional regulation, and gratitude journaling. Intervention messages are focused on building self-compassion, addressing body image-related content (media literacy, challenging fat talk, appearance teasing, and appearance comparison), and providing healthy lifestyle-related content (mindful eating, sleep, hygiene, and physical activity). Rodgers et al. (2018) randomized female and male adolescents to *BodiMojo* or an assessment-only control group; both groups completed measures of appearance satisfaction and body image flexibility at baseline, post-intervention, and 6-weeks follow up. Compared to the control group, adolescents in *BodiMojo* increased their appearance satisfaction, but not body image flexibility, over time. While Rodgers et al.'s study yields promising results for improving appearance-related body image, more research is needed to explore additional body image benefits of self-compassion-based mobile apps, such as body appreciation.

Group-Based Delivery

Grounded in self-compassion and cognitive dissonance principles, *Bodies in Motion* (Voelker et al., 2019) is a group-based program that addresses the unique body-related experiences of female athletes and teaches them to respond in more functional ways to sociocultural appearance pressures and sports-related body pressures.

In addition to actively challenging unrealistic appearance ideals, female athletes also learn how to be kinder and more understanding in how they evaluate and respond to themselves when exposed to these ideals while also staying present-focused in their self-awareness and nonjudgmental of their thoughts and feelings. The small group-based program includes an introductory session, followed by four interactive and experiential sessions led by trained program leaders. In their evaluation study, Voelker et al. randomly assigned female athletes from nine NCAA athletic departments to *Bodies in Motion* or a waitlist control group. At posttest, a stringent approach to control family-wise error revealed that groups differed on thin-ideal internalization; however, trends were noted for improved body appreciation, body satisfaction, and shape and weight concerns. In other words, athletes who complete *Bodies in Motion* are likely to experience reduced thin-ideal internalization and may also experience increased body appreciation and satisfaction and decreased weight and shape concerns. *Bodies in Motion* raises important questions about the unique effects of self-compassion to the program (above cognitive dissonance principles) for future research to disentangle.

Conclusions and Future Research Directions

A robust literature exists on the link between self-compassion and body image. This link is complex and multifaceted, with self-compassion often moderating (or buffering) the relationship between negative environmental influences (e.g., appearance-related pressures, sexual objectification) and body image, mediating the link between both modifiable (e.g., meditation exposure, yoga practice) and less modifiable (e.g., attachment) experiences and body image, and contributing to improved body image via established interventions. Yet, there is much more work ahead. First, the vast majority of the research on self-compassion and body image has been conducted with samples of predominantly young adult, White, heterosexual, able-bodied, cisgender women without visible differences and who tend to fall in the middle of the weight spectrum. Much more research needs to be conducted in the area of self-compassion and body image with groups who are not represented in this research, such as men, transgender and nonbinary folx, children and older adults, as well as those with visible differences, limited abilities, higher weights, and of various racial and ethnic cultural groups. While it is worthwhile to determine whether the findings of the reviewed research replicate with diverse groups, it is also important to consider experiences (e.g., weight stigma, appearance-based stigma, racial discrimination, transphobia) that likely impact many of these individuals within the study design and analyses.

Second, researchers could investigate whether integrating self-compassion themes within social media could impact body image. Slater et al. (2017) experimentally examined the impact of exposure to self-compassion quotes on Instagram images, and they found that women who viewed these quotes reported greater body satisfaction, body appreciation, self-compassion, and reduced negative mood compared to women who viewed neutral images. In addition, when women viewed self-compassion quotes alongside images that reflect the fit ideal, they experienced more positive outcomes compared to women who viewed only the fit ideal images. These findings suggest that self-compassion messages within social media have potential to be beneficial to women's body image and offset the negative impact of exposure to thin- and fit-ideal media. More research is needed to replicate and extend these findings as well as determine who may benefit more from these messages.

Third, it is important to both investigate and create interventions to reduce individuals' resistance to and fears of self-compassion. Those who misunderstand what self-compassion involves may believe that it is self-indulgent and could undermine their personal responsibility for their problems, thereby interfering with their motivation and success (Robinson et al., 2016). Indeed, when interviewed about their attitudes toward physical changes in their body as they age, some physically active women aged 65 and older reported that having self-compassion for their aging body was "idealistic" and found it difficult to respond to their body in a self-compassionate manner due to being critical of their appearance and functionality (Bennett et al., 2017). In addition, self-compassion may be a frightening experience for some individuals who feel that they are undeserving of compassion (Gilbert et al., 2011). While the fear of self-compassion has been studied in research on eating disorders (e.g., Kelly et al., 2014a, b), researchers have largely neglected to explore its impact on body image as well as its influence on the success of self-compassion interventions to improve body image. The resistance to and fear of self-compassion could prevent engagement in self-compassion interventions as well as lessen any positive effect of these interventions on body image.

Fourth, longitudinal data exploring the associations between self-compassion and body image are lacking to date. In the one study we did find, Stutts and Blomquist (2018) used a longitudinal design to examine whether self-compassion moderated the relationship of weight and shape concern on disordered eating among a large sample of college women. For the first year, weight/shape concern was more strongly linked to disordered eating for those lower in self-compassion compared to those higher in self-compassion. However, this pattern did not emerge during the second year. Researchers would benefit from exploring longitudinal associations between self-compassion and different dimensions of body image. Embedded in this research, it would be valuable to explore "upward spirals," or combinations of protective factors that trigger self-perpetuating cycles and trajectories of positive psychological growth (Garland et al., 2010), both within intervention-based and nonintervention-based prospective research. For example, self-compassion may predict increased body appreciation, which could then predict even higher self-compassion and additional psychological resources.

Fifth, researchers need to explore the variables that may cultivate self-compassion within the context of enhancing positive body image. While researchers have explored secure attachment (Raque-Bogdan et al., 2016) in this role, other variables may be more amenable to change, such as exposure to nature. Nature has a salutogenic effect on overall health and well-being (Bowler et al., 2010) and body appreciation more specifically (Swami et al., 2018), as being engaged in nature (e.g., hiking, biking, smelling leaves, watching birds) helps them appreciate the various ways their body functions. Nature also promotes opportunities for self-compassion, as it facilitates the stillness needed for mindfulness and an emotional state focused on self- and other-kindness. Among a sample of British women and men, Swami et al. (2019) found that self-kindness and common humanity, but not mindfulness, partially mediated nature exposure's positive associations with both functionality appreciation (appreciating all the things that the body is capable of doing; Alleva et al., 2017) and body appreciation. Additional interventions, such as one that promoted gratitude, may cultivate self-compassion, as well as positive body image through self-compassion (Homan & Tylka, 2018), and thus are worthy to explore in prospective and experimental research.

Sixth, researchers could study whether self-compassion aids the maintenance of positive body image by strengthening individuals' resistance to internalizing the negative messages within an image-focused culture. Those with positive body image interpret incoming information in a body-protective manner whereby most positive information is internalized, and most body image threats are rejected or reframed (Wood-Barcalow et al., 2010). This process is referred to as "protective filtering" (p. 112). To date, we know very little about this process, and thus, it would be worthy to explore whether self-compassion helps individuals strengthen their protective filter to ward against body-related threats.

We are confident that research within these areas will provide novel contributions to the already rich literature exploring the connection between self-compassion and body image covered in this chapter. The way that individuals treat themselves (with kindness, humanity, and attention versus criticism, isolation, and overidentification) maps on to the way they feel about and treat their bodies (with appreciation and respect versus shame and condemnation). Therefore, continuing to study the complex connections between self-compassion and body image is essential to foster psychological health and well-being among diverse individuals.

References

Albertson, E. R., Neff, K. D., & Dill-Shackleford, K. E. (2015). Self-compassion and body dissatisfaction in women: A randomized controlled trial of a brief meditation intervention. *Mindfulness, 6*(3), 444–454. https://doi.org/10.1007/s12671-014-0277-3

Alleva, J. M., Sheeran, P., Webb, T. L., Martijn, C., & Miles, E. (2015). A meta-analytic review of stand-alone interventions to improve body image. *PLoS One, 10*(9), e0139177. https://doi.org/10.1371/journal.pone.0139177

Alleva, J. M., Tylka, T. L., & Kroon Van Diest, A. M. (2017). The Functionality Appreciation Scale (FAS): Development and psychometric evaluation in U.S. community women and men. *Body Image, 23*, 28–44. https://doi.org/10.1016/j.bodyim.2017.07.008

Altman, J. K., Linfield, K., Salmon, P. G., & Beacham, A. O. (2017). The body compassion scale: Development and initial validation. *Journal of Health Psychology*. Advance online publication. https://doi.org/10.1177/2F1359105317718924

Andrew, R., Tiggemann, M., & Clark, L. (2016). Predicting body appreciation in young women: An integrated model of positive body image. *Body Image, 18*, 34–42. https://doi.org/10.1016/j.bodyim.2016.04.003

Barnett, M. D., & Sharp, K. J. (2016). Maladaptive perfectionism, body image satisfaction, and disordered eating behaviors among U.S. college women: The mediating role of self-compassion. *Personality and Individual Differences, 99*, 335–234. https://doi.org/10.1016/j.paid.2016.05.004

Bennett, E. V., Clarke, L. H., Kowalski, K. C., & Crocker, P. R. E. (2017). "I'll do anything to maintain my health": How women aged 65-94 perceive, experience, and cope with their aging bodies. *Body Image, 21*, 71–80. https://doi.org/10.1016/j.bodyim.2017.03.002

Bowler, D. E., Buyung-Ali, L. M., Knight, T. M., & Pullin, A. S. (2010). A systematic review of the evidence for the added benefits to health of exposure to natural environments. *BMC Public Health, 10*(1), 456. https://doi.org/10.1186/1471-2458-10-456

Braun, T. D., Park, C. L., & Gorin, A. (2016). Self-compassion, body image, and disordered eating: A review of the literature. *Body Image, 17*, 117–131. https://doi.org/10.1016/j.bodyim.2016.03.003

Breines, J., Toole, A., Tu, C., & Chen, S. (2014). Self-compassion, body image, and self-reported disordered eating. *Self and Identity, 13*(4), 432–448. https://doi.org/10.1080/15298868.2013.838992

Calogero, R. M., & Pina, A. (2011). Body guilt: Preliminary evidence for a further subjective experience of self-objectification. *Psychology of Women Quarterly, 35*(3), 428–440. https://doi.org/10.1177/0361684311408564

Calogero, R. M., Tantleff-Dunn, S., & Thompson, J. K. (2011). *Self-objectification in women: Causes, consequences, and counteractions*. American Psychological Association.

Cash, T. F. (2004). Body image: Past, present, and future. *Body Image, 1*(1), 1–5. https://doi.org/10.1016/S1740-1445(03)00011-1

Cash, T. F., & Williams, E. F. (2005). Coping with body-image threats and challenges: Validation of the Body Image Coping Strategies Inventory. *Journal of Psychosomatic Research, 58*(2), 190–199. https://doi.org/10.1016/j.jpsychores.2004.07.008

Comiskey, A., Parent, M. C., & Tebbe, E. A. (2020). An inhospitable world: Exploring a model of objectification theory with trans women. *Psychology of Women Quarterly, 44*(1), 105–116. https://doi.org/10.1177/2F0361684319889595

Cox, A. E., Ullrich-French, S., Tylka, T. L., & McMahon, A. K. (2019). The roles of self-compassion, body surveillance, and body appreciation in predicting intrinsic motivation for physical activity: Cross-sectional association, and prospective changes within a yoga context. *Body Image, 29*, 110–117. https://doi.org/10.1016/j.bodyim.2019.03.002

Crocker, J., Luhtanen, R. K., Cooper, M. L., & Bouvrette, A. (2003). Contingencies of self-worth in college students: Theory and measurement. *Journal of Personality and Social Psychology, 85*(5), 894–908. https://doi.org/10.1037/0022-3514.85.5.894

Daye, C. A., Webb, J. B., & Jafari, N. (2014). Exploring self-compassion as a refuge against recalling the body-related shaming of caregiver eating messages on dimensions of objectified body consciousness in college women. *Body Image, 11*(4), 547–556. https://doi.org/10.1016/j.bodyim.2014.08.001

Dias, B. S., Ferreira, C., & Trindade, I. A. (2020). Influence of fears of compassion on body image shame and disordered eating. *Eating and Weight Disorders, 25*(1), 99–106. https://doi.org/10.1007/s40519-018-0523-0

Diener, E., & Diener, M. (1995). Cross-cultural correlates of life satisfaction and self-esteem. *Journal of Personality and Social Psychology, 68*(4), 653–663. https://doi.org/10.1037//0022-3514.68.4.653

Fardouly, J., & Vartanian, L. R. (2015). Negative comparisons about one's appearance mediate the relationship between Facebook usage and body image concerns. *Body Image, 12*(1), 82–88. https://doi.org/10.1016/j.bodyim.2014.10.004

Fredrickson, B. L., & Roberts, T.-A. (1997). Objectification theory: Toward understanding women's lived experiences and mental health risks.

Psychology of Women Quarterly, 21(2), 173–206. https://doi.org/10.1111/j.1471-6402.1997.tb00108.x

Garland, E. L., Fredrickson, B., Kring, A. M., Johnson, D. P., Meyer, P. S., & Penn, D. L. (2010). Upward spirals of positive emotions counter downward spirals of negativity: Insights from the broaden-and-build theory and affective neuroscience on the treatment of emotion dysfunctions and deficits in psychopathology. *Clinical Psychology Review, 30*(7), 849–864. https://doi.org/10.1016/2Fj.cpr.2010.03.002

Gilbert, P. (2005). Compassion and cruelty: A biopsychosocial approach. In P. Gilbert (Ed.), *Compassion: Conceptualisations, research and use in psychotherapy* (pp. 9–74). Routledge.

Gilbert, P. (2015). An evolutionary approach to emotion in mental health with a focus on affiliative emotions. *Emotion Review, 7*(3), 230–237. https://doi.org/10.1177/1754073915576552

Gilbert, P., & Irons, C. (2005). Focused therapies and compassionate mind training for shame and self-attacking. In P. Gilbert (Ed.), *Compassion: Conceptualisations, research and use in psychotherapy* (pp. 263–325). Routledge.

Gilbert, P., McEwan, K., Matos, M., & Rivis, A. (2011). Fears of self-compassion: Development of three self-report measures. *Psychology and Psychotherapy, 84*(3), 239–255. https://doi.org/10.1348/147608310X526511

Grossbard, J. R., Lee, C. M., Neighbors, C., & Larimer, M. E. (2009). Body image concerns and contingent self-esteem in male and female college students. *Sex Roles, 60*(3–4), 198–207. https://doi.org/10.1007/s11199-008-9535-y

Guest, E., Costa, B., Williamson, H., Meyrick, J., Halliwell, E., & Harcourt, D. (2019). The effectiveness of interventions aiming to promote positive body image in adults: A systematic review. *Body Image, 30*, 10–25. https://doi.org/10.1016/j.bodyim.2019.04.002

Harter, S. (1999). *The construction of the self: A developmental perspective*. Guilford Press.

Hazzard, V. M., Schaefer, L. M., Schaumberg, K., Bardone-Cone, A. M., Frederick, D. A., Klump, K. L., Anderson, D. A., & Thompson, J. K. (2019). Testing the tripartite influence model among heterosexual, bisexual, and lesbian women. *Body Image, 30*, 145–149. https://doi.org/10.1016/j.bodyim.2019.07.001

Homan, K. J., & Tylka, T. L. (2015). Self-compassion moderates body comparison and appearance self-worth's inverse relationships with body appreciation. *Body Image, 15*, 1–7. https://doi.org/10.1016/j.bodyim.2015.04.007

Homan, K. J., & Tylka, T. L. (2018). Development and exploration of the gratitude model of body appreciation in women. *Body Image, 25*, 14–22. https://doi.org/10.1016/j.bodyim.2018.01.008

Huellemann, K. L., & Calogero, R. M. (2020). Self-compassion and body checking among women: The mediating role of stigmatizing self-perceptions. *Mindfulness, 11*(9), 2121–2130. https://doi.org/10.1007/s12671-020-01420-8

James, W. (1890). *Principles of psychology*. Encyclopedia Britannica.

Josephs, R. A., Bosson, J. K., & Jacobs, C. G. (2003). Self-esteem maintenance processes: Why low self-esteem may be resistant to change. *Personality and Social Psychology, 29*(7), 920–933. https://doi.org/10.1177/2F0146167203029007010

Kelly, A. C., & Stephen, E. (2016). A daily diary study of self-compassion, body image, and eating behavior in female college students. *Body Image, 17*, 152–160. https://doi.org/10.1016/j.bodyim.2016.03.006

Kelly, A. C., Vimalakanthan, K., & Carter, J. C. (2014a). Understanding the roles of self-esteem, self-compassion, and fear of self-compassion in eating disorder pathology: An examination of female students and eating disorder patients. *Eating Behaviors, 15*(3), 388–391. https://doi.org/10.1016/j.eatbeh.2014.04.008

Kelly, A. C., Vimalakanthan, K., & Miller, K. E. (2014b). Self-compassion moderates the relationship between body mass index and both eating disorder pathology and body image flexibility. *Body Image, 11*(4), 446–453. https://doi.org/10.1016/j.bodyim.2014.07.005

Kelly, A. C., Miller, K. E., & Stephen, E. (2016). The benefits of being self-compassionate on days when interactions with body-focused others are frequent. *Body Image, 19*, 195–203. https://doi.org/10.1016/j.bodyim.2016.10.005

Liss, M., & Erchull, M. J. (2015). Not hating what you see: Self-compassion may protect against negative mental health variables connected to self-objectification in college women. *Body Image, 14*, 5–12. https://doi.org/10.1016/j.bodyim.2015.02.006

Lonergan, A. R., Bussey, K., Mond, J., Brown, O., Griffiths, S., Murray, S. B., & Mitchison, D. (2019). Me, my selfie, and I: The relationship between editing and posting selfies and body dissatisfaction in men and women. *Body Image, 28*, 39–43. https://doi.org/10.1016/j.bodyim.2018.12.001

Lucas, R. E., Diener, E., & Suh, E. (1996). Discriminant validity of well-being measures. *Journal of Personality and Social Psychology, 71*(3), 616–628. https://doi.org/10.1037//0022-3514.71.3.616

Maxwell, S. E., & Cole, D. A. (2007). Bias in cross-sectional analyses of longitudinal mediation. *Psychological Methods, 12*(1), 23–44. https://doi.org/10.1037/1082-989X.12.1.23

Mensinger, J. L., Tylka, T. L., & Calamari, M. E. (2018). Mechanisms underlying weight status and healthcare avoidance in women: A study of weight stigma, body-related shame and guilt, and healthcare stress. *Body Image, 25*, 139–147. https://doi.org/10.1016/j.bodyim.2018.03.001

Modica, C. (2019). Facebook, body esteem, and body surveillance in adult women: The moderating role of self-compassion and appearance-contingent self-worth. *Body Image, 29*, 17–30. https://doi.org/10.1016/j.bodyim.2019.02.002

Moffitt, R. L., Neumann, D. L., & Williamson, S. P. (2018). Comparing the efficacy of a brief self-esteem and self-compassion intervention for state body dissatisfaction and self-improvement motivation. *Body Image, 27*, 67–76. https://doi.org/10.1016/j.bodyim.2018.08.008

Mosewich, A. D., Kowalski, K. C., Sabiston, C. M., Sedgwick, W. A., & Tracy, J. L. (2011). Self-compassion: A potential resource for young women athletes. *Journal of Sport & Exercise Psychology, 33*(1), 103–123. https://doi.org/10.1123/jsep.33.1.103

Neff, K. (2003a). Development and validation of a scale to measure self-compassion. *Self and Identity, 2*(3), 223–250. https://doi.org/10.1080/15298860309027

Neff, K. (2003b). Self-compassion: An alternative conceptualization of a healthy attitude toward oneself. *Self and Identity, 2*(3), 85–101. https://doi.org/10.1080/15298860309032

Neff, K. D. (2011). Self-compassion, self-esteem, and well-being. *Social and Personality Psychology Compass, 5*(1), 1–12. https://doi.org/10.1111/j.1751-9004.2010.00330.x

Neff, K. D., & Germer, C. K. (2013). A pilot study and randomized controlled trial of the mindful self-compassion program. *Journal of Clinical Psychology, 69*(1), 28–44. https://doi.org/10.1002/jclp.21923

Neff, K. D., & Vonk, R. (2009). Self-compassion versus global self-esteem: Two different ways of relating to oneself. *Journal of Personality, 77*(1), 23–50. https://doi.org/10.1111/j.1467-6494.2008.00537.x

Przezdziecki, A., & Sherman, K. A. (2016). Modifying affective and cognitive responses regarding body image difficulties in breast cancer survivors using a self-compassion-based writing intervention. *Mindfulness, 7*(5), 1142–1155. https://doi.org/10.1007/s12671-016-0557-1

Raes, F., Pommier, E., Neff, K. D., & Van Gucht, D. (2011). Construction and factorial validation of a short form of the Self-Compassion Scale. *Clinical Psychology & Psychotherapy, 18*(3), 250–255. https://doi.org/10.1002/cpp.702

Raque-Bogdan, T. L., Piontkowski, S., Hui, K., Schaefer Ziemer, K., & Garriott, P. O. (2016). Self-compassion as a mediator between attachment anxiety and body appreciation: An exploratory model. *Body Image, 19*, 28–36. https://doi.org/10.1016/j.bodyim.2016.08.001

Robinson, K. J., Mayer, S., Allen, A. B., Terry, M., Chilton, A., & Leary, M. R. (2016). Resisting self-compassion: Why are some people opposed to being kind to themselves? *Self and Identity, 15*(5), 505–524. https://doi.org/10.1080/15298868.2016.1160952

Rodgers, R. F., Franko, D. L., Donovan, E., Cousineau, T., Yates, K., McGowan, K., Cook, E., & Lowy, A. S. (2017). Body image in emerging adults: The protective role of self-compassion. *Body Image, 22*, 148–155. https://doi.org/10.1016/j.bodyim.2017.07.003

Rodgers, R. F., Donovan, E., Cousineau, T., Yates, K., McGowan, K., Cook, E., Lowy, A. S., & Franko, D. L. (2018). *BodiMojo*: Efficacy of a mobile-based intervention in improving body image and self-compassion among adolescents. *Journal of Youth and Adolescence, 47*(7), 1363–1372. https://doi.org/10.1007/s10964-017-0804-3

Salzberg, S. (1995). *Loving-kindness: The revolutionary art of happiness*. Shambala Publications.

Sandoz, E. K., Wilson, K. G., Merwin, R. M., & Kellum, K. K. (2013). Assessment of body image flexibility: The Body Image-Acceptance and Action Questionnaire. *Journal of Contextual Behavioral Science, 2*(1–2), 39–48. https://doi.org/10.1016/j.jcbs.2013.03.002

Schaefer, L. M., & Thompson, J. K. (2014). The development and validation of the Physical Appearance Comparison Scale-Revised (PACS-R). *Eating Behaviors, 15*(2), 209–217. https://doi.org/10.1016/j.eatbeh.2014.01.001

Seekis, V., Bradley, G. L., & Duffy, A. (2017). The effectiveness of self-compassion and self-esteem writing tasks in reducing body image concerns. *Body Image, 23*, 206–213. https://doi.org/10.1016/j.bodyim.2017.09.003

Siegel, J. A., Huellemann, K. L., Hillier, C. C., & Campbell, L. (2020). The protective role of self-compassion for women's positive body image: An open replication and extension. *Body Image, 32*, 136–144. https://doi.org/10.1016/j.bodyim.2019.12.003

Slater, A., Varsani, N., & Diedrichs, P. C. (2017). #fitspo or #loveyourself? The impact of fitspiration and self-compassion Instagram images on women's body image, self-compassion, and mood. *Body Image, 22*, 87–96. https://doi.org/10.1016/j.bodyim.2017.06.004

Stutts, L. A., & Blomquist, K. K. (2018). The moderating role of self-compassion on weight and shape concerns and eating pathology: A longitudinal study. *International Journal of Eating Disorders, 51*(8), 879–889. https://doi.org/10.1002/eat.22880

Swami, V., Barron, D., & Furnham, A. (2018). Exposure to natural environments, and photographs of natural environments, promotes more positive body image. *Body Image, 24*, 82–94. https://doi.org/10.1016/j.bodyim.2017.12.006

Swami, V., Barron, D., Hari, R., Grover, S., Smith, L., & Furnham, A. (2019). The nature of positive body image: Examining associations between nature exposure, self-compassion, functionality appreciation, and body appreciation. *Ecopsychology, 11*(4), 243–253. https://doi.org/10.1089/eco.2019.0019

Thøgersen-Ntoumani, C., Dodos, L., Chatzisarantis, N., & Ntoumanis, N. (2017). A diary study of self-compassion, upward social comparisons, and body image-related outcomes. *Health and Well-Being, 9*(2), 242–258. https://doi.org/10.1111/aphw.12089

Thompson, J. K., Heinberg, L. J., Altabe, M., & Tantleff-Dunn, S. (1999). *Exacting beauty: Theory, assessment, and treatment of body image disturbance*. American Psychological Association.

Toole, A. M., & Craighead, L. W. (2016). Brief self-compassion meditation training for body image distress in young adult women. *Body Image, 19*, 104–112. https://doi.org/10.1016/j.bodyim.2016.09.001

Trekels, J., Ward, L. M., & Eggermont, S. (2018). I "like" the way you look: How appearance- focused and overall Facebook use contribute to adolescents' self-sexualization. *Computers in Human Behavior, 81*, 198–208. https://doi.org/10.1016/j.chb.2017.12.020

Turk, F., & Waller, G. (2020). Is self-compassion relevant to the pathology and treatment of eating and body image concerns? A systematic review and meta-analysis. *Clinical Psychology Review, 79*, 101856. https://doi.org/10.1016/j.cpr.2020.101856

Tylka, T. L. (2006). Development and psychometric evaluation of a measure of intuitive eating. *Journal of Counseling Psychology, 53*(2), 226–240. https://doi.org/10.1037/0022-0167.53.2.226

Tylka, T. L. (2011). Refinement of the tripartite influence model for men: Dual body image pathways to body change behaviors. *Body Image, 8*(3), 199–207. https://doi.org/10.1016/j.bodyim.2011.04.008

Tylka, T. L., & Andorka, M. J. (2012). Support for an expanded tripartite influence model with gay men. *Body Image, 9*(1), 57–67. https://doi.org/10.1016/j.bodyim.2011.09.006

Tylka, T. L., & Iannantuono, A. C. (2016). Perceiving beauty in all women: Psychometric evaluation of the Broad Conceptualization of Beauty Scale. *Body Image, 17*, 67–81. https://doi.org/10.1016/j.bodyim.2016.02.005

Tylka, T. L., & Kroon Van Diest, A. M. (2015). Protective factors in the development of eating disorders. In L. Smolak & M. P. Levine (Eds.), *The Wiley-Blackwell handbook of eating disorders* (pp. 430–444). Wiley.

Tylka, T. L., & Wood-Barcalow, N. L. (2015a). The Body Appreciation Scale-2: Item refinement and psychometric evaluation. *Body Image, 12*(1), 53–67. https://doi.org/10.1016/j.bodyim.2014.09.006

Tylka, T. L., & Wood-Barcalow, N. L. (2015b). What is and what is not positive body image? Conceptual foundations and construct definition. *Body Image, 14*, 118–129. https://doi.org/10.1016/j.bodyim.2015.04.001

Tylka, T. L., Russell, H. L., & Neal, A. A. (2015). Self-compassion as a moderator of thinness-related pressures' associations with thin-ideal internalization and disordered eating. *Eating Behaviors, 17*, 23–26. https://doi.org/10.1016/j.eatbeh.2014.12.009

Vartanian, L. R., & Shaprow, J. G. (2008). Effects of weight stigma on exercise motivation and behavior. *Journal of Health Psychology, 13*(1), 131–138. https://doi.org/10.1177/1359105307084318

Vimalakanthan, K., Kelly, A. C., & Trac, S. (2018). From competition to compassion: A caregiving approach to intervening with appearance comparisons. *Body Image, 25*, 148–162. https://doi.org/10.1016/j.bodyim.2018.03.003

Voelker, D. K., Petrie, T. A., Huang, Q., & Chandran, A. (2019). Bodies in motion: An empirical evaluation of a program to support positive body image in female collegiate athletes. *Body Image, 28*, 149–158. https://doi.org/10.1016/j.bodyim.2019.01.008

Wasylkiw, L., MacKinnon, A. L., & MacLellan, A. M. (2012). Exploring the link between self-compassion and body image in university women. *Body Image, 9*(2), 236–245. https://doi.org/10.1016/j.bodyim.2012.01.007

Webb, J. B., Fiery, M. F., & Jafari, N. (2016). "You better not leave me shaming!" Conditional indirect effect analyses of anti-fat attitudes, body shame, and fat talk as a function of self-compassion in college women. *Body Image, 18*, 5–13. https://doi.org/10.1016/j.bodyim.2016.04.009

Wiseman, M. C., & Moradi, B. (2010). Body image and eating disorder symptoms in sexual minority men: A test and extension of objectification theory. *Journal of Counseling Psychology, 57*(2), 154–166. https://doi.org/10.1037/a0018937

Wood-Barcalow, N. L., Tylka, T. L., & Augustus-Horvath, C. L. (2010). But I like my body: Positive body image characteristics and a holistic model for young-adult women. *Body Image, 7*(2), 106–116. https://doi.org/10.1016/j.bodyim.2010.01.001

12 Self-Compassion, Personal Improvement, and Motivation

Anaïs Ortiz, Aleah Goold, and Jia Wei Zhang

Introduction

Motivation is the internal desire and drive to create change in one's own self or environment (Baumeister, 2015). Theoretical perspectives on motivated or goal-directed behaviors (e.g., Carver & Scheier, 1998; Azjen, 1991) suggest that an individual's goals are shaped by two primary parameters: the desirability, or attractiveness of goal attainment, and the feasibility, or likelihood that one's goal will be attained. In turn, feasibility can be influenced by both pragmatic factors and individual differences variables such as optimism (Scheier & Carver, 1987), self-efficacy, and outcome expectancies (Bandura, 1977), while the desirability of a particular goal is influenced by its alignment with the individual's values and their external context. Motivational theories also suggest that the process and outcomes of motivated behaviors are influenced by the degree to which goals are intrinsically or extrinsically determined. For example, Ryan and Deci (2000) contrast *controlled motivation* – which is driven by the need or desire to satisfy "external demand or reward contingenc[ies]" (p. 72) with *self-determined motivation*, which is seen to have both intrinsic drivers and external drivers that align with the values of the individual.

Motivated behavior can be observed in many contexts, from day-to-day goals to decade-long pursuits, and there are a variety of factors that can motivate behavioral change, such as experiencing a failure, transgression, or suffering. For example, some people can be motivated to study harder by receiving a bad grade, or exercise more after being diagnosed with unfavorable health conditions, or an athlete may train longer and harder after losing at a sporting event. What internal factors motivate some people to improve more than others? Recent research suggests that self-compassion may be one such factor that drives people to adaptively engage in improvement-related inclinations. In addition to influencing the types of goals one sets for oneself, evidence also suggests that self-compassion influences how people respond to challenges and setbacks that may impede goal attainment.

Self-compassion is rooted in empathy extended toward the self when one is faced with a difficult experience (Neff, 2011). Self-compassion has three interrelated components: (1) self-kindness, a tendency to apply a caring and tender, rather than judgmental, attitude toward one's difficult experiences; (2) common humanity, the recognition that it is only "human" to make mistakes and that one's suffering is shared by others; and (3) mindfulness or facing

A. Ortiz
Department of Psychology, University of Florida, Gainesville, FL, USA

A. Goold · J. W. Zhang (✉)
Department of Psychology, University of Memphis, Memphis, TN, USA
e-mail: jzhang11@memphis.edu

A. Finlay-Jones et al. (eds.), *Handbook of Self-Compassion*, Mindfulness in Behavioral Health,
https://doi.org/10.1007/978-3-031-22348-8_12

one's failure and observing one's pain with equanimity (Neff, 2011). Thus, self-compassionate people are aware of the emotions related to experiences (both positive and negative), recognize that others share these emotions, and handle setbacks and failures with more calmness and acceptance than those who have less self-compassion. While self-compassion may be incorrectly viewed as going easy on oneself, recent work by Neff (2021) has highlighted both the "tender" and "fierce" sides of self-compassion. Fierce self-compassion, with its focus on assertiveness, determination, and goal-directed behavior, demonstrates that self-determined motivation is an important way that self-compassion manifests. Indeed, a study of female athletes described associations between self-compassion and several facets of mental toughness, including taking perspective, adequate preparation, perseverance, and staying present, which together helped athletes cope with adversity (Wilson et al., 2019).

In both its fierce and tender forms, self-compassion offers an adaptive way of relating to oneself that can be contrasted with less healthy forms of self-relation, such as self-criticism. While self-criticism can motivate behavior intended to maintain or protect one's self-image (e.g., when task performance is driven by avoiding failure or negative evaluation; Mongrain & Zuroff, 1995), studies have routinely demonstrated that self-criticism can undermine autonomous motivation and thwart goal progress (Powers et al., 2007, 2009; Vandenkerckhove et al., 2019). In contrast, self-compassion is associated with several important motivation-related constructs: it is positively related to grit, personal growth initiative, self-efficacy, and goal engagement, and inversely associated with fear of failure and maladaptive perfectionism (Neff et al., 2018). In this chapter, we bring together recent scientific studies of self-compassion and motivation by summarizing the latest research on self-compassion and its impact on goal setting, pursuit, and persistence. Further, we discuss the reasons why self-compassion might be associated with more adaptive motivational processes and outcomes and highlight key directions for future research.

Self-Compassion and Self-Improvement

One of the frequent concerns about self-compassion is that it could lead to complacency or self-indulgence (Kirby et al., 2019). Thus, people typically do not associate self-compassion with efforts to improve themselves. However, research has shown a clear link between self-compassion and the motivation for personal improvement. For example, one study found that participants who were led to write about their moral transgression from a self-compassionate perspective reported greater motivation to redeem themselves and avoid repeating the same transgression in the future compared to participants who used a self-esteem perspective and participants who received no additional reflection instructions (Breines & Chen, 2012). Follow-up experiments found that participants who were led to think about a failed test with self-compassion, compared to self-esteem and control, studied longer for a subsequent test and were more willing to interact with someone who shared a similar weakness but overcame it (Breines & Chen, 2012). Additionally, participants who were induced to talk about their regret with self-compassion, compared to a self-esteem group and a control group with no manipulation, reported more self-improvement motivation. Another study found that the connection between self-compassion and self-improvement motivation was explained by greater acceptance of regret (Zhang & Chen, 2016). In this series of three studies, the authors first coded descriptions of regret posted on a website. Those who were coded as describing their experience with greater self-compassion were more likely to also be coded as demonstrating more self-improvement. This was supported by the second study, in which self-compassion was associated with both self-reported and observer-coded personal improvement following the recall of a regret experience. The third study suggested that the link between self-compassion and personal improvement was mediated by increased acceptances of the regret experience (Zhang & Chen, 2016).

In the context of relationships, self-compassion has been closely associated with motivation to correct mistakes. For example, one study demonstrated that self-compassion was associated with greater motivation to correct interpersonal mistakes among women, although among men this relationship was only found for men high in conscientiousness (Baker & McNulty, 2011). In the same study, self-compassion was also associated with observations of more constructive problem-solving behaviors among men high in conscientiousness. More recently, Zhang and Chen (2017) conducted a study of individuals who attributed personal responsibility for a recent breakup, and compared outcomes across three different conditions in which participants were assigned to a self-compassion or self-esteem induction, or no-induction control (Zhang & Chen, 2017). It was found that the self-compassion group reported greater self-improvement motivation and intended appreciation for a future romantic partner than either the self-esteem or no-induction control groups (Zhang & Chen, 2017).

Self-Compassion and Health

Another good indicator of self-compassion's role in motivating people toward self-improvement rests in the emerging connection between self-compassion and health outcomes, as well as health-related outlooks. In one study, participants were instructed to recall a recent time where they experienced an exercise setback (e.g., not meeting a goal of going to the gym regularly; Semenchuk et al., 2018). The authors found that self-compassion inversely predicted negative emotional experiences related to recalling the abandoned goal, such that those with higher self-compassion were less likely to experience rumination or negative affect. They also found that self-compassion was positively associated with greater goal reengagement (e.g., setting a new, more attainable exercise goal; Semenchuk et al., 2018) and that these relationships were maintained after controlling for self-esteem.

More recently, Biber (2020) found that self-compassion was significantly associated with exercise-related self-regulation in a sample of US adults. Self-regulation refers to the degree to which individuals can flexibly adapt their cognitions, emotions, and behaviors to achieve their goals. Self-regulatory capacities are therefore critical to motivating behavior, underpinning one's capacity to both monitor progress toward a goal, and adjusting one's behavior in the face of challenges and setbacks (Baumeister & Heatherton, 1996; Ryan & Deci, 2000). Biber (2020) also found that self-compassion predicted less anxiety, depression, and self-blame related to health, in addition to greater health satisfaction. Moreover, self-compassionate people reported fewer negative emotions to mild illness and were more willing to see a medical provider to address their symptoms sooner than less self-compassionate counterparts (Terry et al., 2013). Lastly, Terry et al. (2013) demonstrated that self-compassion was associated with less negative emotional reactions to serious illness via positive thoughts and the intention to treat oneself kindly. Also, self-compassionate people were more likely to follow doctors' recommendations for the treatment of their illness compared to their less self-compassionate counterparts (Terry et al., 2013). In short, this evidence suggests that self-compassion enables people to engage in healthier self-regulation that can, in turn, lead to more consistent and persistent exercise habits, as well as more adaptive coping and adherence (i.e., following a doctor's orders) in the face of unfavorable health diagnoses. Evidence supporting the relationship between self-compassion and health behaviors is described in greater detail in Chap. 18.

Self-Compassion and Academic Motivation

Another line of research points to self-compassion's role in academic achievement and motivational outcomes that are relevant in academic contexts, such as procrastination, test anxiety, and academic self-regulation. Academic environments are frequently characterized by high levels of external pressure, including recurrent deadlines, achievement expectations, and the message that future opportunities are contingent on current performance. Responses to such environmental pressures are influenced by the implicit goals that individuals hold regarding their achievement in goal-oriented settings (Elliot & McGregor, 2001). Implicit goals may be categorized along two main dimensions: performance oriented, in which individuals compare their performance to others, and mastery oriented, in which individuals focus on change in their own personal competencies, regardless of others' performance. These dimensions can be further classified into *approach* and *avoidance* dimensions, to create four types of goals: (i) performance approach, in which individuals aim to demonstrate their comparative competence; (ii) performance avoidance, in which individuals aim to avoid appearing incompetent relative to others; (iii) mastery approach, in which individuals aim to develop their own competence; and (iv) mastery avoidance, in which individuals aim to avoid performing worse than they have previously (Elliot & McGregor, 2001). These goals can be characterized as a type of coping strategy for dealing with challenges, which determine an individual's motivation and behavior (Babenko et al., 2018), and align with *intrinsic* (i.e., mastery-oriented) and *extrinsic* (i.e., performance-oriented) models of goal attainment (Korn & Elliot, 2016).

Prior research has found that while both types of approach goals are associated with academic achievement, mastery-approach goals are more likely to be linked to adaptive outcomes, such as curiosity, engagement, and academic self-regulation (Senko et al., 2011). Similarly, while intrinsic and extrinsic motivations are not incompatible, intrinsic motivations have been linked more consistently to greater well-being and higher retention in academic settings. Conversely, there are concerns that performance-oriented goals may increase competitive and dishonest behavior, such as cheating. In an early study, Neff et al. (2005) found that self-compassion was associated with perceived competence and mastery goals, and inversely associated with fear of failure in a sample of college students. This was supported by Kotera and Ting's (2019) findings that people who are more self-compassionate are more intrinsically motivated. In another example, trait self-compassion predicted greater score improvement on a difficult test regardless of whether people were induced with a mastery or performance goal (Shimizu et al., 2015). More recently, a study found that self-compassion moderates the pathway from extrinsic to intrinsic motivation, indicating that those with greater self-compassion are more able to transfer extrinsic goals to intrinsic ones (Kotera et al., 2021). Together, this evidence suggests self-compassion promotes dedication to pursuing intrinsic academic goals and to aligning one's internal motivation with external demands.

Self-compassion has been linked to making more meaningful goals and being more persistent in the face of failure, as well as less concern with peer perceptions (Neely et al., 2009). In a study of college students, Neely et al. (2009) found that self-compassion was positively associated with goal reengagement, and together, these variables were associated with student well-being. Goal reengagement is a useful coping strategy in the face of failure or disappointment, and these findings align with other work demonstrating that self-compassion is associated with more adaptive responses to difficulties in an academic context. For example, one study found that self-compassion predicted adaptive coping techniques (i.e., acceptance and positive reframing) in students who performed poorly in a previous exam (Neff et al., 2005). Other studies showed that self-compassion was associated with the use of cognitively demanding *desirable difficulties*, or self-imposed challenges used during the learning process to optimize comprehension and memory encoding and retrieval (Wagner et al., 2017). Moreover, self-compassion predicted less procrastination in three separate student samples (Sirois, 2014).

Pathways Underpinning the Relationship Between Self-Compassion and Motivation

There are several, interconnected reasons why self-compassion may support more adaptive motivational processes and outcomes, including more adaptive goal setting and greater well-being in the face of blocked goals. Potential mechanisms include greater capacity to admit responsibility for mistakes, higher self-efficacy, less fear of failure, better self-regulatory capacity, and more adaptive affective responses to stress. These pathways are likely interconnected and share dynamic relationships with motivated behaviors over time. However, most of the available literature involves cross-sectional observational research or performance-associated mood inductions (e.g., recalling a failure, receiving negative feedback on a performance). Further, few studies have examined these potential pathways simultaneously. Nevertheless, exploration of the supporting evidence provides guidance for path analysis in future longitudinal and experimental work.

Taking Responsibility for Performance Outcomes

One of the reasons why self-compassion leads to self-improvement motivation may be attributed to enhanced willingness to take responsibility for performance-related outcomes, including under circumstances of failure. In one early study, participants were asked to describe a personal humiliation, failure, or rejection in detail (Leary et al., 2007). Participants were then randomly assigned to respond to their personal humiliation, failure, or rejection from a self-compassionate perspective, induced self-esteem perspective (e.g., led to feel good about themselves), through expressive writing (e.g., exploring their deepest emotions), or with no follow-up response (i.e., control). The results showed that the self-compassion induction led participants to report greater willingness to admit and accept responsibility for the negative event compared to participants in the other three conditions (Leary et al., 2007). In short, self-compassion successfully led people to acknowledge that they were the person who made the mistake. It is possible that those who were induced to be more self-compassionate were more able to acknowledge their mistake because they saw themselves as a human being who, like all human beings, are fallible. The mistake was reparable and reflected on their *actions* rather than their intrinsic value as a person. By reducing self-criticism about one's mistakes (which is generally experienced as aversive), self-compassion may also reduce the likelihood that people avoid thinking about or engaging with their errors.

Another study examined the relationship between self-compassion and self-handicapping and sandbagging, two behaviors designed to alter performance expectations and shift responsibility for inadequate performance away from the self (Peterson, 2014). Self-handicapping involves engaging in behaviors (such as failing to study for an exam) that can be used to provide an explanation for one's failures and therefore protect self-esteem (Jones & Berglas, 1978). Sandbagging is aimed at lowering expectations by feigning inability or predicting subpar performance (Gibson & Sachau, 2000). Like self-handicapping, sandbagging is designed to protect self-esteem but can also result in performance being undermined (Gibson et al., 2002). Among a student sample, Peterson (2014) found that self-compassion was inversely correlated with both self-handicapping and sandbagging, suggesting that people with self-compassion are more likely to take a realistic perspective of their own performance and less likely to try and manipulate others' perceptions of them.

Self-Efficacy, Performance Appraisals, and Failure Construals

A second potential pathway underpinning the relationship between self-compassion and goal maintenance is self-efficacy. A recent meta-analysis found that self-compassion is moderately associated with self-efficacy (Liao et al.,

2021), including across educational (Iskender, 2009) and occupational (Latorre et al., 2021) settings. Accordingly, people with greater self-compassion may feel more confident in their ability to achieve their goals, thereby motivating them to take the required action to do so. Similarly, self-compassion may promote more adaptive motivation in the face of setbacks by changing the way that individuals appraise their performance. The way that people appraise or interpret stressors, such as failures or poor performance, is an integral part of their coping response. In a recent study, Miyagawa et al. (2020) found that self-compassionate people were more likely to report seeing personal failures as a learning opportunity and have less maladaptive failure beliefs. In a follow-up experiment, the same authors showed that people who were led to talk about a personal failure with self-compassion, compared to self-esteem and control, reported less likely to perceive failure as aversive or to be avoided. In this way, individuals with greater self-compassion are more likely to be able to recover from performance setbacks and may experience less self-doubt than those who are less self-compassionate.

Closely linked to these outcomes is the relationship between self-compassion, performance anxiety, and perfectionism. When coupled with goal flexibility and tolerance for mistakes, adaptive perfectionism can support healthy striving toward goals while also feeling accomplished with one's achievements (Lo & Abbott, 2013). Conversely, maladaptive perfectionism entails excessively high standards, low tolerance for errors, and an inability to feel satisfied with performance, regardless of outcome (Lo & Abbott, 2013). The links between maladaptive perfectionism and performance expectations, such as self-efficacy, fear of failure, and performance anxiety are complex (Flett et al., 1992). However, there are some facets of fear of failure – such as fear of shame and embarrassment (Conroy, 2001) – that appear closely linked to maladaptive perfectionism (Conroy, 2004; Sagar & Stoeber, 2009). Fear of failure and performance anxiety have been associated with the tendency to equate personal worth with performance. Accordingly, taking a self-compassionate rather than self-evaluative approach may confer benefits for well-being as well as for goal-related behaviors such as procrastination (for more on the difference between self-compassion and self-evaluation, see Chap. 3).

Available evidence demonstrates inverse relationships between self-compassion and maladaptive perfectionism (Neff et al., 2018; Tobin & Dunkley, 2021; Wei et al., 2020), and there is some evidence that self-compassion can protect against its negative psychosocial consequences. In a study of undergraduate students, Ong et al. (2021) examined psychological flexibility, self-compassion, distress, and symptom impairment in groups of students who had low, average, or high perfectionism scores. They found that while self-compassion appeared to confer protective benefits among individuals with average perfectionism, those with high perfectionism benefited more from psychological flexibility. This suggests that at higher levels of perfectionism, self-compassion alone may not be enough to protect against its deleterious effects. Nevertheless, given the close relationship between self-compassion and psychological flexibility (e.g., Marshall & Brockman, 2016), future research may wish to explore whether self-compassion intervention leads to changes in psychological flexibility among those with different levels of perfectionism.

In one of the few studies to examine multiple mediators of the relationship between self-compassion and goal-related outcomes, Neff et al. (2005) found that among undergraduates, the relationship between self-compassion and achievement goals was mediated both by fear of failure and perceived competence. More recently, the self-compassion-fear of failure link has been tested experimentally. In a study of entrepreneurs, Engel et al. (2021) randomized participants to engage in loving-kindness meditation or to listen to recorded talks about meditation, without engaging in meditation practice. Entrepreneurs are a group for whom fear of failure is a highly salient outcome given the uncertain and high-pressure environment that entrepreneurs operate in. Engel et al. (2021) found that, as hypothesized, participants randomized to the loving-kindness condition reported significant improvements in self-compassion relative to controls, with parallel reductions in fear of failure. Further, they found that the effect of loving-kindness meditation on fear of failure was almost entirely meditated by self-compassion.

Self-Regulation and Adaptive Affective Responses to Challenges

Across contexts, having unattainable or thwarted goals can harm well-being (Wrosch et al., 2013), however, some people are more able to recover from these setbacks than others (Carver & Scheier, 2016). When individuals fixate on unattained goals or when progress toward goals is misaligned with internal expectations, they may be more likely to experience negative affect stress, distress, or burnout. Conversely, individuals who are better able to regulate their emotions and behaviors may be flexible in the face of setbacks and less likely to experience negative well-being outcomes. It follows that another way in which self-compassion might promote more adaptive motivational processes and outcomes is via its association with enhanced self-regulatory capacities. As reviewed by Terry and Leary (2011) in the context of health behaviors, self-compassion is associated with several processes underpinning adaptive self-regulation toward health behaviors, including setting safe, realistic goals, seeking necessary support, and reducing negative feelings, such a shame or anger that can interfere with one's capacity for self-regulation. They also propose that because the focus of self-compassion is on alleviating one's suffering, self-compassion is more likely to promote disengagement with harmful and unproductive goals, and a heightened focus on goals that are more likely to promote well-being (Terry & Leary, 2011).

Self-regulation models of self-compassion are supported by evidence that self-compassion is associated with more adaptive emotional and behavioral regulation in the face of stress (see Ewert et al., 2021; Finlay-Jones, 2017; Inwood & Ferrari, 2018, for reviews). Procrastination is one behavior that is proposed to arise because of self-regulation failure (Steel, 2007), is linked to negative affective outcomes (Klibert et al., 2011), and that may represent an ineffective emotion regulation strategy (Sirois et al., 2019). Accordingly, procrastination may be implicated in a vicious cycle, in which low mood leads to procrastination, which exacerbates negative affect. Sirois et al. (2019) described bedtime procrastination as one example of how procrastination might be enacted to regulate negative mood. In support of a self-regulation model of self-compassion and motivation, they found that one of the reasons individuals with higher self-compassion were less likely to procrastinate was because they had greater access to adaptive emotion regulation strategies and were less likely to experience negative affect. Similarly, other studies with student samples have reported inverse associations between self-compassion and negative affective responses to stress (Zhang et al., 2016), as well as procrastination and performance anxiety (Williams et al., 2008; Zhang et al., 2021a, b). Further, in a sequential mediation model, Yang et al. (2021) found that the link between social support and lower procrastination in college students was mediated first by self-compassion and then by negative affect.

Arguably, by reducing negative affect in the face of stress, adaptive emotion regulation can also lead to more reasoned decision-making (e.g., setting feasible, health-promoting goals) and promote goal engagement in the face of setbacks. Miyagawa et al. (2018) proposed that students with greater self-compassion would be more likely to disengage from an unattained goal and to reengage with an alternative goal, thereby reducing negative affect following goal non-attainment. They found support for this hypothesis; in an undergraduate student sample, students were instructed to recall a failure to attain their goals. Findings showed that self-compassion was associated with both goal disengagement, and reengagement with an alternative goal. Further, the authors found that self-compassion was inversely associated with negative affect after recalling the failed goal, as well as indirectly related to negative affect through goal disengagement.

These findings align with those of Hope et al. (2014) who found that college students who reported more daily self-compassion were less vulnerable to the affective consequences of thwarted daily goal progress. The authors also found that the relationship between self-determined goal motivation and negative affect was moderated by trait self-compassion such that highly self-compassionate students, compared to students with low self-compassion, showed low negative affect fluctuations in the progress of self-determined personally meaningful goals. More recently, daily self-compassion predicted daily subjective well-being despite COVID-19 discrimination experience among Asian Americans. Daily self-compassion predicted increased COVID-19 protective behaviors on days Asian Americans experienced COVID-19 discrimination. Daily acceptance, but not daily reappraisal, explained the link between daily self-compassion and daily subjective well-being (Zhang et al., 2021a, b).

Summary

One of the biggest concerns, or misconceptions, that many people have of self-compassion is that it leads to complacency. Our brief review of the literature across domains suggests the opposite, showing that people who are self-compassionate adapt to failures and are motivated to change. In sum, these studies document evidence that stands in sharp contrast to the argument that self-compassion causes complacency. Instead, self-compassion, independent of self-esteem, functions as a proactive coping strategy that guides people to approach, rather than avoid, difficult life experiences. More research on self-compassion and self-improvement motivation in different domains of life and in various age groups, including elder and end-of-life populations, is needed to further illuminate the impact of self-compassion on goals of various magnitude and length. More information is necessary on the longitudinal benefits of self-compassion in long-term goal regulation and in the face of regret. Nonetheless, we hope future research will continue to illuminate and solidify the connection between self-compassion and motivation toward self-improvement.

References

Azjen, I. (1991). The theory of planned behavior. *Organizational Behavior and Human Decision Processes, 50*(2), 179–211. https://doi.org/10.0116/0749-5978(91)90020-T

Babenko, O., Mosewich, A., Abraham, J., & Lai, H. (2018). Contributions of psychological needs, self-compassion, leisure-time exercise, and achievement goals to academic engagement and exhaustion in Canadian medical students. *Journal of Educational Evaluation for Health Professions, 15*, 2. https://doi.org/10.3352/jeehp.2018.15.2

Baker, L. R., & McNulty, J. K. (2011). Self-compassion and relationship maintenance: The moderating roles of conscientiousness and gender. *Journal of Personality and Social Psychology, 100*(5), 853–873. https://doi.org/10.1037/a0021884

Bandura, A. (1977). Self-efficacy: Toward a unifying theory of behavioral change. *Psychological Review, 84*(2), 191–215. https://doi.org/10.1037/0033-295X.84.2.191

Baumeister. (2015). Toward a general theory of motivation: Problems, challenges, opportunities, and the big picture. *Motivation and Emotion, 40*(1), 1–10. https://doi.org/10.1007/s11031-015-9521-y

Baumeister, R. F., & Heatherton, T. F. (1996). Self-regulation failure: An overview. *Psychological Inquiry, 7*(1), 1–15. https://doi.org/10.1207/s15327965pli0701_1

Biber, D. D. (2020). Exercise identity, self-regulatory efficacy, and self-compassion prepared for psychological studies. *Psychological Studies, 65*(3), 261–269. https://doi.org/10.1007/s12646-020-00556-w

Breines, J. G., & Chen, S. (2012). Self-compassion increases self-improvement motivation. *Personality and Social Psychology Bulletin, 38*(9), 1133–1143. https://doi.org/10.1177/0146167212445599

Carver, C. S., & Scheier, M. F. (1998). *On the self-regulation of behavior*. Cambridge University Press.

Carver, C. S., & Scheier, M. F. (2016). Self-regulation of action and affect. In K. D. Vohs & R. E. Baumeister (Eds.), *Handbook of self-regulation: Research, theory, and application* (pp. 3–23). The Guildford Press.

Conroy, D. E. (2001). Progress in the development of a multidimensional measure of fear of failure: The Performance Failure Appraisal Inventory (PFAI). *Anxiety, Stress, & Coping, 14*(4), 431–452.10.1080/10615800108248365.

Conroy, D. E. (2004). The unique psychological meanings of multidimensional fears of failing. *Journal of Sport and Exercise Psychology, 26*(3), 484–491. https://doi.org/10.1123/jsep.26.3.484

Elliot, A. J., & McGregor, H. A. (2001). A 2 x 2 achievement goal framework. *Journal of Personality and Social Psychology, 80*(3), 501–519. https://doi.org/10.1037//0022-3514.80.3.501

Engel, Y., Noordijk, S., Spoelder, A., & van Gelderen, M. (2021). Self-compassion when coping with venture obstacles: Loving-kindness meditation and entrepreneurial fear of failure. *Entrepreneurship Theory and Practice, 45*(2), 263–290. https://doi.org/10.1177/1042258719890991

Ewert, C., Vater, A., & Schröder-Abé, M. (2021). Self-compassion and coping: A meta-analysis. *Mindfulness, 12*(5), 1063–1077. https://doi.org/10.1007/s12671-020-01563-8

Finlay-Jones, A. (2017). The relevance of self-compassion as an intervention target in mood and anxiety disorders: A narrative review based on an emotion regulation framework. *Clinical Psychologist, 21*(2), 90–103. https://doi.org/10.1111/cp.12131

Flett, G. L., Blankstein, K. R., Hewitt, P. L., & Koledin, S. (1992). Components of perfectionism and procrastination in college students. *Social Behavior and Personality, 20*(2), 85–94. https://doi.org/10.2224/sbp.1992.20.2.85

Gibson, B., & Sachau, D. A. (2000). Sandbagging as a self-presentational strategy: Claiming to be less than you are. *Personality and Social Psychology Bulletin, 26*(1), 56–70. https://doi.org/10.1177/0146167200261006

Gibson, B., Sachau, D. A., Doll, R., & Shumate, R. (2002). Sandbagging in competition: Responding to the pressure of being the favorite. *Personality and Social Psychology Bulletin, 28*(8), 1119–1130. https://doi.org/10.1177/01461672022811010

Hope, N., Koestner, R., & Milyavskaya, M. (2014). The role of self-compassion in goal pursuit and well-being among university freshmen. *Self and Identity, 13*(5), 579–593. https://doi.org/10.1080/15298868.2014.889032

Inwood, E., & Ferrari, M. (2018). Mechanisms of change in the relationship between self-compassion, emotion regulation, and mental health: A systematic review. *Applied Psychology: Health and Well-Being, 10*(2), 215–235. https://doi.org/10.1111/aphw.12127

Iskender, M. (2009). The relationship between self-compassion, self-efficacy, and control belief about learning in turkish university students. *Social Behavior and Personality, 37*(5), 711–720. https://doi.org/10.2224/sbp.2009.37.5.711

Jones, E. E., & Berglas, S. (1978). Control of attributions about the self through self-handicapping strategies: The appeal of alcohol and the role of underachievement. *Personality and Social Psychology Bulletin, 4*(2), 200–206. https://doi.org/10.1177/014616727800400205

Kirby, J. N., Day, J., & Sagar, V. (2019). The 'flow' of compassion: A meta-analysis of the fears of compassion scales and psychological functioning. *Clinical Psychology Review, 70*, 26–39. https://doi.org/10.1016/j.cpr.2019.03.001

Klibert, J., Langhinrichsen-Rohling, J., Luna, A., & Robichaux, M. (2011). Suicide proneness in college students: Relationships with gender, procrastination, and achievement motivation. *Death Studies, 35*(7), 625–645. https://doi.org/10.1080/07481187.2011.553311

Korn, R. M., & Elliot, A. J. (2016). The 2 × 2 standpoints model of achievement goals. *Frontiers in Psychology, 7*, 742. https://doi.org/10.3389/fpsyg.2016.00742

Kotera, Y., & Ting, S.-H. (2019). Positive psychology of Malaysian university students: Impacts of engagement, motivation, self-compassion, and well-being on mental health. *International Journal of Mental Health and Addiction, 19*(1), 227–239. https://doi.org/10.1007/s11469-019-00169-z

Kotera, Y., Taylor, E., Fido, D., Williams, D., & Tsuda-McCaie, F. (2021). Motivation of UK graduate students in education: self-compassion moderates pathway from extrinsic motivation to intrinsic motivation. *Current Psychology*, 1–14. https://doi.org/10.1007/s12144-021-02301-6

Latorre, C., Leppma, M., Platt, L. F., Shook, N., & Daniels, J. (2021). The relationship between mindfulness and self-compassion for self-assessed competency and self-efficacy of psychologists-in-training. *Training and Education in Professional Psychology*. Advance online publication. https://doi.org/10.1037/tep0000395.

Leary, M. R., Tate, E. B., Adams, C. E., Batts Allen, A., & Hancock, J. (2007). Self-compassion and reactions to unpleasant self-relevant events: The implications of treating oneself kindly. *Journal of Personality and Social Psychology, 92*(5), 887–904. https://doi.org/10.1037/0022-3514.92.5.887

Liao, K. Y.-H., Stead, G. B., & Liao, C.-Y. (2021). A meta-analysis of the relation between self-compassion and self-efficacy. *Mindfulness, 12*(8), 1878–1891. https://doi.org/10.1007/s12671-021-01626-4

Lo, A., & Abbott, M. J. (2013). Review of the theoretical, empirical, and clinical status of adaptive and maladaptive perfectionism. *Behaviour Change, 30*(2), 96–116. https://doi.org/10.1017/bec.2013.9

Marshall, E. J., & Brockman, R. N. (2016). The relationships between psychological flexibility, self-compassion, and emotional well-being. *Journal of Cognitive Psychotherapy, 30*(1), 60–72. https://doi.org/10.1891/0889-8391.30.1.60

Miyagawa, Y., Taniguchi, J., & Niiya, Y. (2018). Can self-compassion help people regulate unattained goals and emotional reactions toward setbacks? *Personality and Individual Differences, 134*, 239–244. https://doi.org/10.1016/j.paid.2018.06.029

Miyagawa, Y., Niiya, Y., & Taniguchi, J. (2020). When life gives you lemons, make lemonade: Self-compassion increases adaptive beliefs about failure. *Journal of Happiness Studies, 21*(6), 2051–2068. https://doi.org/10.1007/s10902-019-00172-0

Mongrain, M., & Zuroff, D. C. (1995). Motivational and affective correlates of dependency and self-criticism. *Personality and Individual Differences, 18*(3), 347–354. https://doi.org/10.1016/0191-8869(94)00139-J

Neely, M. E., Schallert, D. L., Mohammed, S. S., Roberts, R. M., & Chen, Y.-J. (2009). Self-kindness when facing stress: The role of self-compassion, goal regulation, and support in college students' well-being. *Motivation and Emotion, 33*(1), 88–97. https://doi.org/10.1007/s11031-008-9119-8

Neff, K. D. (2011). Self-compassion, self-esteem, and well-being. *Social & Personality Psychology Compass, 5*(1), 1–12. https://doi.org/10.1111/j.1751-9004.2010.00330.x

Neff, K. D. (2021). *Fierce self-compassion: How women can harness kindness to speak up, claim their power, and thrive*. Harper Wave.

Neff, K. D., Hsieh, Y.-P., & Dejitterat, K. (2005). Self-compassion, achievement goals, and coping with academic failure. *Self and Identity, 4*(3), 263–287. https://doi.org/10.1080/13576500444000317

Neff, K. D., Long, P., Knox, M. C., Davidson, O., Kuchar, A., Costigan, A., et al. (2018). The forest and the trees: Examining the association of self-compassion and its positive and negative components with psychological functioning. *Self and Identity, 17*(6), 627–645. https://doi.org/10.1080/15298868.2018.1436587

Ong, C. W., Lee, E. B., Petersen, J. M., Levin, M. E., & Twohig, M. P. (2021). Is perfectionism always unhealthy? Examining the moderating effects of psychological flexibility and self-compassion. *Journal of Clinical Psychology, 77*(11), 2576–2591. https://doi.org/10.1002/jclp.23187

Petersen, L.-E. (2014). Self-compassion and self-protection strategies: The impact of self-compassion on the use of self-handicapping and sandbagging. *Personality and Individual Differences, 56*(1), 133–138. https://doi.org/10.1016/j.paid.2013.08.036

Powers, T. A., Koestner, R., & Zuroff, D. C. (2007). Self-criticism, goal motivation, and goal progress. *Journal of Social and Clinical Psychology, 26*(7), 826–840. https://doi.org/10.1521/jscp.2007.26.7.826

Powers, T. A., Koestner, R., Lacaille, N., Kwan, L., & Zuroff, D. C. (2009). Self-criticism, motivation, and goal progress of athletes and musicians: A prospective study. *Personality and Individual Differences, 47*(4), 279–283. https://doi.org/10.1016/j.paid.2009.03.012

Ryan, R. M., & Deci, E. L. (2000). Self-Determination Theory and the facilitation of intrinsic motivation, social development, and well-being. *The American Psychologist, 55*(1), 68–78. https://doi.org/10.1037/0003-066X.55.1.68

Sagar, S. S., & Stoeber, J. (2009). Perfectionism, fear of failure, and affective responses to success and failure: the central role of fear of experiencing shame and embarrassment. *Journal of Sport and Exercise Psychology, 31*(5), 602–627. https://doi.org/10.1123/jsep.31.5.602

Scheier, M. F., & Carver, C. S. (1987). Dispositional optimism and physical well-being: The influence of generalized outcome expectancies on health. *Journal of Personality, 55*(2), 169–210. https://doi.org/10.1111/j.1467-6494.1987.tb00434.x

Semenchuk, B. N., Strachan, S. M., & Fortier, M. (2018). Self-compassion and the self-regulation of exercise: Reactions to recalled exercise setbacks. *Journal of Sport and Exercise Psychology, 40*(1), 31–39. https://doi.org/10.1123/jsep.2017-0242

Senko, C., Hulleman, C. S., & Harackiewicz, J. M. (2011). Achievement Goal Theory at the crossroads: Old controversies, current challenges, and new directions. *Educational Psychologist, 46*(1), 26–47. https://doi.org/10.1080/00461520.2011.538646

Shimizu, M., Niiya, Y., & Shigemasu, E. (2015). Achievement goals and improvement following failure: moderating roles of self-compassion and contingency of self-worth. *Self and Identity, 15*(1), 107–115. https://doi.org/10.1080/15298868.2015.1084371

Sirois, F. M. (2014). Procrastination and stress: Exploring the role of self-compassion. *Self and Identity, 13*(2), 128–145. https://doi.org/10.1080/15298868.2013.763404

Sirois, F. M., Nauts, S., & Molnar, D. S. (2019). Self-compassion and bedtime procrastination: an emotion regulation perspective. *Mindfulness, 10*(3), 434–445. https://doi.org/10.1007/s12671-018-0983-3

Steel, P. (2007). The nature of procrastination: A meta-analytic and theoretical review of quintessential self-regulatory failure. *Psychological Bulletin, 133*(1), 65–94. https://doi.org/10.1037/0033-2909.133.1.65

Terry, M. L., & Leary, M. R. (2011). Self-compassion, self-regulation, and health. *Self and Identity, 10*(3), 352–362. https://doi.org/10.1080/15298868.2011.558404

Terry, M. L., Leary, M., Mehta, S., & Henderson, K. (2013). Self-compassionate reactions to health threats. *Personality and Social Psychology Bulletin, 39*(7), 911–926. https://doi.org/10.1177/0146167213488213

Tobin, R., & Dunkley, D. M. (2021). Self-critical perfectionism and lower mindfulness and self-compassion predict anxious and depressive symptoms over two years. *Behaviour Research and Therapy, 136*, 103780. https://doi.org/10.1016/j.brat.2020.103780

Vandenkerckhove, B., Soenens, B., Van der Kaap-Deeder, J., Brenning, K., Luyten, P., & Vansteenkiste, M. (2019). The role of weekly need-based experiences and self-criticism in predicting weekly academic (mal) adjustment. *Learning and Individual Differences, 69*, 69–83. https://doi.org/10.1016/j.lindif.2018.11.009

Wagner, L., Schindler, S., & Reinhard, M. A. (2017). The positive facet of self-compassion predicts self-reported use of and attitudes toward desirable difficulties in learning. *Frontiers in Psychology, 8*, 1353. https://doi.org/10.3389/fpsyg.2017.01353

Wei, S., Li, L., Shi, J., Liang, H., & Yang, X. (2020). Self-compassion mediates the perfectionism and depression link on Chinese undergraduates. *Annals of Palliative Medicine, 10*(2), 1950–1960. https://doi.org/10.21037/apm-20-1582

Williams, J. G., Stark, S. K., & Foster, E. E. (2008). Start today or the very last day? The relationships among self-compassion, motivation, and procrastination.

American Journal of Psychological Research, 4(1), 37–44.

Wilson, D., Bennett, E. V., Mosewich, A. D., Faulkner, G. E., & Crocker, P. R. E. (2019). "The zipper effect": Exploring the interrelationship of mental toughness and self-compassion among Canadian elite women athletes. *Psychology of Sport and Exercise, 40*, 61–70. https://doi.org/10.1016/j.psychsport.2018.09.006

Wrosch, C., Scheier, M. F., & Miller, G. E. (2013). Goal adjustment capacities, subjective well-being, and physical health. *Social and Personality Psychology Compass, 7*(12), 847–860. https://doi.org/10.1111/spc3.12074

Yang, X., Zhu, J., & Hu, P. (2021). Perceived social support and procrastination in college students: A sequential mediation model of self-compassion and negative emotions. *Current Psychology*. https://doi.org/10.1007/s12144-021-01920-3

Zhang, J. W., & Chen, S. (2016). Self-compassion promotes personal improvement from regret experiences via acceptance. *Personality and Social Psychology Bulletin, 42*(2), 244–258. https://doi.org/10.1177/0146167215623271

Zhang, J. W., & Chen, S. (2017). Self-compassion promotes positive adjustment for people who attribute responsibility of a romantic breakup to themselves. *Self and Identity., 16*(6), 731–759. https://doi.org/10.1080/15298868.2017.1305985

Zhang, Y., Luo, X., Che, X., & Duan, W. (2016). Protective effect of self-compassion to emotional response among students with chronic academic stress. *Frontiers in Psychology, 7*, 1802. https://doi.org/10.3389/fpsyg.2016.01802

Zhang, J. W., Bui, V., Snell, A. N., Howell, R. T., & Bailis, D. (2021a). Daily self-compassion protects Asian Americans/Canadians after experiences of COVID-19 discrimination: Implications for subjective well-being and health behaviors. *Self and Identity, (ahead-of-print)*, 1–23. https://doi.org/10.1080/15298868.2021.2012511

Zhang, J. W., Kessler, E., & Braasch, J. L. G. (2021b). Self-compassion mindsets can predict statistics course performance via intelligence mindsets and statistics anxiety. *Learning and Individual Differences, 90*, 102047. https://doi.org/10.1016/j.lindif.2021.102047

13 Self-Compassion in Competitive Sport

Amber D. Mosewich, Leah J. Ferguson, and Benjamin J. Sereda

The Case for Self-Compassion in Sport

Sport is replete with instances of evaluation and comparison, successes and failures, uncertainties, and sacrifices. Sport can elicit a range of emotions – positive and negative – as a by-product of participation. Through the positives and negatives of sport, and everything in between, we find athletes, and others involved in sport, navigating a range of situations, including consequences of the past and the goals they are working toward. While many involved in sport are privy to traditional mental skills and have adopted individualized practices in self-talk, goal setting, performance planning, and other strategies, there are still recognized gaps in athletes' abilities to manage the emotional, physical, and social demands of sport (e.g., Mosewich et al., 2014). Tales of punishing self-criticism and overwhelming self-evaluation are all too common. The consequences often manifest in burnout, poor performance, injury, and poor mental health and well-being, among others. Self-compassion has been introduced as a potential resource or approach[1] to help manage the demands of the sport domain, support sport performance, and foster well-being in sport. Self-compassion can be soothing, offering support, and understanding, while "fierce self-compassion" enables active and direct responses directed toward one's best interest in pursuit of potential (Neff, 2021).

Current and ongoing events have brought media and, by extension, public attention to the challenges faced by many athletes when it comes to their mental health. Athletes from all types of sport, at all levels of play, and at various stages of their careers have spoken about their personal mental health experiences. It would be remiss to assume these experiences are similar, as they are not; each is complex, unique, and multifaceted. However, the desire to offer support to those in need remains universal. In a domain where high

A. D. Mosewich (✉) · B. J. Sereda
Faculty of Kinesiology, Sport, and Recreation, University of Alberta, Edmonton, Canada
e-mail: amber.mosewich@ualberta.ca

L. J. Ferguson
College of Kinesiology, University of Saskatchewan, Saskatoon, Canada

[1] While initially conceptualized from a trait perspective (Neff, 2003a, b), there is evidence to support that self-compassion can also be prompted or applied as a resource or strategy (e.g., Mosewich et al., 2013). Some people engender self-compassion automatically as a function of their disposition, while others adopt or engage in the approach when coping with difficult events and pursuing goals (Ingstrup et al., 2017). Most research in sport positions self-compassion as a resource or an approach, and such a conceptualization is adopted in the present chapter.

A. Finlay-Jones et al. (eds.), *Handbook of Self-Compassion*, Mindfulness in Behavioral Health,
https://doi.org/10.1007/978-3-031-22348-8_13

performance is the focus, the importance of well-being cannot be overlooked. Additionally, it must be remembered that those involved in sport are not immune to mental health concerns. A recent study by Åkesdotter et al. (2020) indicates more than 50% of athletes will experience a mental health issue in their lifetime. What's more, symptoms often manifest at a young age and recurrent episodes are common. There is a need to support athlete well-being alongside their performance, and self-compassion may be one approach to foster both of those aims.

The present chapter will overview self-compassion research and practice in sport. Although self-compassion is relevant in other areas of physical activity and human movement, such as exercise and physical education, this chapter will focus on the domain of competitive sport. We will begin with addressing what we feel often needs to be discussed up front when working in a sport domain: acknowledging the challenges of presenting self-compassion to the culture of sport, which is one that often possesses and values the norm of self-criticism. We will then overview how self-compassion has been assessed in the sport domain, as this framing is key to the conceptualization of the sport self-compassion literature. What will follow is a summary of the empirical findings. We provide an overview of cognitions, emotions, behaviors, and outcomes associated with self-compassion, and the accompanying implications for well-being and performance. We then present self-compassion development and intervention research that is specific to the sport domain and consider its use in applied sport psychology practice. The chapter concludes with considerations regarding the adoption of self-compassion in sport and areas for future inquiry.

Challenges for Self-Compassion in Sport

Despite the relevance of self-compassion within the sport domain, not everyone is receptive to the approach. Researchers have identified that at least some individuals involved in sport are hesitant to embrace self-compassion (Mosewich et al., 2019a). For example, some athletes have expressed concern that being *too* self-compassionate may lead to settling for mediocrity or complacency in sport (Ferguson et al., 2014; Sutherland et al., 2014). Fear of self-compassion – characterized by being fearful of or resistant to extending compassion toward oneself (Gilbert et al., 2011) – is often marked by high self-criticism and associated with psychological distress in athletes (Ceccarelli et al., 2019; Ferguson et al., 2015; Walton et al., 2020). Fear of self-compassion has also been found to explain additional variance beyond self-compassion on some maladaptive thoughts and behaviors in an athlete sample (Röthlin & Leiggene, 2021). This speaks to the importance of addressing fear of self-compassion among athletes, particularly where there are concerns that self-compassion may undermine performance. Inadequate understanding of self-compassion presents a challenge to its use among athletes and others involved in sport; accordingly, a key goal is "convincing" those involved in sport that being self-compassionate is beneficial.

Fear of self-compassion in the sport domain may be exacerbated by a perceived language barrier. That is, the idea – and corresponding language – of self-compassion may not "fit" within sport participants' typical conceptual models of sport and achievement. The competitive sport environment is replete with extreme training demands, performance expectations, and setbacks, which may appear incompatible with a kind, connected, and balanced self-attitude. While some athletes believe they need to be hard on themselves to achieve their goals, it is important to clearly differentiate between harsh or debilitating self-criticism and constructive evaluation of the self (Mosewich, 2020). Wilson et al. (2019) found that an elite group of women athletes identified the merits of both mental toughness and self-compassion, indicating that the two can even be compatible processes that work in tandem with one another in the pursuit of athletic achievement. It is therefore important to explore effective ways to introduce and talk about self-compassion with those involved in a sport context. Reis et al. (2022) flagged the need for adequate education about self-compassion from knowledgeable sources to help sport participants accurately learn what self-compassion is and how to apply it in their sport experiences. Doing so in a way that stays true to the tenets of self-compassion and is meaningful to athletes and others involved in sport is paramount.

Measurement of Self-Compassion in Sport

Neff (2003b) originally conceptualized self-compassion as trait-like, suggesting that individuals have general tendencies to be more or less compassionate toward themselves. Accordingly, Neff developed a trait-like, domain-general measure of the construct, the Self-Compassion Scale (SCS; Neff, 2003a), intended to assess stable individual differences in self-compassion. While other measures have been developed (e.g., the Sussex-Oxford Compassion for the Self Scale; Gu et al., 2020; The Compassion Motivation and Action Scales; Steindl et al., 2021), most self-compassion in sport research has used Neff's measure or a short form of the scale (i.e., SCS-SF; Raes et al., 2011; see Röthlin et al., 2019, for a review).

Within the self-compassion in sport literature, many researchers have positioned self-compassion as a coping *resource* that can be used after a situation has been appraised, and responding self-compassionately is an option to manage the athlete's experience. In line with this conceptualization, some researchers have assessed the extent to which individuals respond to a given situation (e.g., "being responsible for losing an athletic competition for your team") with *state* self-compassion (e.g., "I tried to be kind to myself"; e.g., Leary et al., 2007; Reis et al., 2015). Similarly, self-compassion has also been positioned as a mental performance *skill* (or collection of skills) or an *approach* that can be taught to athletes to use when experiencing a setback (e.g., Ingstrup et al., 2017; Mosewich et al., 2013). Given these conceptualizations, it may be important to assess self-compassion at a state-level to capture shifts in an individual's self-compassionate mindset across different situations or over time. Neff et al. (2021) recently developed and validated long and short forms of state-specific measures of self-compassion that are positioned to assess changes in self-compassion. It is important for researchers to include Neff et al.'s (2021) state-specific measures in future self-compassion in sport research, particularly when examining self-compassion's malleability through inductions and interventions.

There are at least two critical questions for researchers to consider when measuring self-compassion in sport. First, is self-compassion domain specific? Some researchers have attempted to enhance the relevance of their self-compassion measures to the sport context by priming participants to think about their self-compassionate tendencies in sport. Both Killham et al. (2018) and Lizmore et al. (2017) modified the language of domain-general self-compassion measures (i.e., SCS and SCS-SF, respectively) so the items referred specifically to the sport context, rather than life in general. Mosewich et al. (2021) recently investigated whether a sport-specific measure of self-compassion provides insight beyond that obtained from a domain-general measure of self-compassion (i.e., SCS-SF; Raes et al., 2011). Composite self-compassion scores did not differ between the sport-specific measure and domain-general measure. As a result, there is currently no evidence to support adopting a sport-specific measure of self-compassion over the domain-general measures of self-compassion, but given the limited research, further attention is warranted.

The second important question to consider is whether the language of self-compassion aligns with the competitive sport context. Researchers want to ensure that the language in their measures accurately reflects the underlying constructs they are studying. At the same time, it is important that researchers use language that resonates with and is appropriate for their participants and research contexts. Researchers need to be mindful of participants' reception to the concept, measurement, and application of self-compassion in sport. There is the risk that some sport participants may perceive the language of self-compassion as "too soft" or incompatible with the competitive sport environment (Ferguson et al., 2014; Sutherland et al., 2014). Researchers need to constructively consider the self-compassion measures used in their research, including, but not limited to, issues around language and domain specificity, to strive for appropriate and optimal assessment.

Cognitions, Emotions, Behaviors, and Outcomes Associated with Self-Compassion

The sport domain requires athletes, and others involved in sport, to navigate many types of experiences. Given the difficulty in managing evaluation, comparison, and setbacks, many researchers have focused on the potential of self-compassion as a resource or approach to address sport-related challenges to better support performance and well-being in sport. The connection of self-compassion to performance and well-being, as well as to cognitions, emotions, and behaviors that are viewed as both directly and indirectly related to performance and well-being, are accumulating in the sport domain.

Collectively, research has been used to build a case for self-compassion as a resource to help athletes manage difficult cognitions and emotions, as well as to promote adaptive thoughts, feelings, and behaviors, particularly in the face of challenges. Adaptive perfectionistic strivings (aspects of perfectionism associated with a self-oriented high achievement striving and high personal performance standards; see Stoeber, 2011), behavioral equanimity (remaining calm and composed in responses), positive affect, authentic pride, happiness, persistence of effort, self-determined motivation, positive attitudes toward help-seeking, cognitive reappraisal, and use of emotion- and problem-focused coping approaches are typical of self-compassionate athletes (Barczak & Eklund, 2020; Ceccarelli et al., 2019; Doorley et al., 2022; Jeon et al., 2016; Lizmore et al., 2017; Mosewich et al., 2019b; Mosewich et al., 2021; Reis et al., 2015; Tarasoff et al., 2017; Tingaz et al., 2022; Wasylkiw & Clairo, 2016). Sereda et al. (2022) found that athletes high in self-compassion can adaptively appraise the unexpected stressors they encounter, through balanced perspectives and logical analysis, and effectively cope with them. In contrast, athletes with lower self-compassion report more burnout, concern over mistakes, shame, interpersonal and emotional exhaustion, psychological distress, somatic anxiety, state self-criticism, fear of failure, fear of negative evaluation, maladaptive perfectionistic concerns, worry, rumination, pessimism, and negative affect along with externally regulated motivation and use of avoidance-focused coping approaches than athletes high in self-compassion (Amemiya & Sakairi, 2020; Barczak & Eklund, 2020; Casali et al., 2022; Ceccarelli et al., 2019; Fontana et al., 2017; Huysmans & Clement, 2017; Jeon et al., 2016; Lizmore et al., 2017; Mosewich et al., 2011; Mosewich et al., 2019b; Reis et al., 2019; Tarasoff et al., 2017; Walton et al., 2020; Walker, 2021). Overall, when it comes to cognitions, emotions, and behaviors among sport participants, self-compassion appears to have potential to not only attenuate the negative but also promote the positive.

Self-compassion also appears to foster positive body image among athletes (see Chap. 11 for a complete discussion of self-compassion and body image literature). Athletes with higher self-compassion have greater body appreciation, adaptive eating attitudes and behaviors, and lower social physique anxiety, self-conscious emotions, objectified body consciousness, compulsive exercise tendencies, and disordered eating behaviors than those lower in self-compassion (Adam et al., 2021b; Mosewich et al., 2011; Pila et al., 2022). Mosewich et al. (2009) suggested that the promotion of a "self-compassionate muscularity" (p. 113; i.e., approaching one's muscularity with self-compassion) could attenuate the expectations and evaluation many women athletes encounter surrounding their bodies through enabling women to acknowledge and respect the body and what it can and cannot do, bypassing harsh self-criticism. Body self-compassion among women athletes can promote respect for and satisfaction with one's body, supporting emotion regulation and adaptive perceptions around the body and performance (Eke et al., 2020). Similar sentiments were present among a group of men with spinal cord injury who were former athletes (Smith, 2013). Self-compassion surrounding the body, including appreciating what the body could do, motivated leisure time physical activity, and attenuated harsh self-criticism connected to limitations of the body (Smith, 2013). Thus, these findings suggest that self-compassion promotes acceptance and appreciation regarding the body's appearance and functionality among sport participants. Further, Pila et al. (2022) suggested that self-compassion intervention may provide protection from negative body-related emotional experiences in sport through decreasing outcomes associated with negative body image, as well as preventing body image concerns.

Self-compassion also appears to support well-being among sport participants. For example, Ferguson et al. (2014) found that athletes with greater self-compassion exhibited greater eudaemonic well-being (i.e., autonomy, environmental mastery, personal growth, positive relatedness, purpose in life, and self-acceptance). During adversity, being self-compassionate may support athletes' psychological functioning and development through increased positivity and perseverance, as well as reduced passivity, in reaction to difficult sport experiences (Ferguson et al., 2015). Furthermore, athletes with greater self-compassion at the start of their competitive season have greater autonomy, mastery, purpose, and self-acceptance at the end of their competitive season (Ferguson et al., 2022a). These findings support the contribution of self-compassion to psychological flourishing throughout a competitive sport season. Huysmans and Clement (2017) suggested that athletes high in self-compassion should cope effectively with stressors in sport and life, resulting in a reduction of the magnitude of the stress response (e.g., attention disruption, maladaptive muscle tension) that could increase the risk of injury. While there were no significant findings pertaining to self-compassion and attenuation of injury risk, self-compassion was associated with adaptive responses to stress, specifically a reduction in somatic anxiety, and the accompanying worry and concentration disruption, and avoidance-coping strategies (Huysmans & Clement, 2017). Additionally, the adaptive cognitions, emotions, and behaviors reviewed earlier both relate and contribute to high well-being.

While adaptive cognitions, emotions, and behaviors and well-being contribute to positive performance, those in the sport domain are understandably interested in the direct relationship between self-compassion and physical performance outcomes. Although performance outcomes are complex and impacted by a variety of factors, researchers are starting to examine more explicit performance variables. In a group of women varsity athletes, self-compassion was a significant prospective predictor of goal progress (Mosewich et al., 2019b). Researchers have also found positive associations between self-compassion and perceptions of sport performance (Barczak & Eklund, 2020; Killham et al., 2018). Additionally, some athletes have identified self-compassion as an "energizing internal force" (p. 274) that facilitates their performance through reframing criticism and maintaining a determined approach (Adam et al., 2021a). Research examining the direct links between self-compassion and performance, as well as further establishing self-compassion as an indirect facilitator of performance in sport, will continue.

An important extension of the psychological correlates of self-compassion is the integration of physiological variables into research. Ceccarelli et al. (2019) examined the relationship between self-compassion and athletes' physiological and psychological responses to a recalled sport failure. Athletes with greater self-compassion showed dampened physiological reactivity in the form of blunted heart rate variability withdrawal (i.e., the self-compassionate athletes were better able to regulate their physiological response). Dysregulation of the parasympathetic system during a stressor is associated with slower reaction times and decreased accuracy (Williams et al., 2016), both of which have implications for sport performance. Thus, this study supports self-compassion as a protective factor for athletes' parasympathetic reactivity during a stressor, suggesting that self-compassionate athletes are better able to regulate and balance their physiological state (Ceccarelli et al., 2019).

The Development of Self-Compassion

Given the potential benefits associated with self-compassion in the sport context, there has been increased interest centered on understanding how self-compassion may be developed among sport participants. Beginning to engage with and foster a self-compassionate approach may occur in childhood and adolescence (Ingstrup et al., 2017), as well as during the late teen and young adult years (Frentz et al., 2020), leading to the heightened likelihood of typically responding with self-compassion across the life span. Based on interviews with women athletes high in self-compassion, Ingstrup et al. (2017) identified the roles of important others (e.g., parents, coaches, sport psychologists, peers, siblings) in combination with athletes' personal awareness and self-reflection as key factors in development of self-compassion. The authors concluded that "self-compassion appeared to be a set of skills or resources that were learned through personal reflection and interactions with others" (p. 327).

Although athletes noted the contribution of a variety of significant others in fostering their self-compassion, Ingstrup et al. (2017) highlighted the role of parents as particularly important in its development. Parents appeared to foster an environment conducive to athletes feeling comfortable seeking and receiving support rather than managing sport-related difficulty in isolation (i.e., common humanity). Further, Ingstrup and colleagues noted that parents taught and encouraged the athletes to demonstrate kindness toward the self, particularly during times of failure. Illustrations of mindfulness appeared as parents assisted athletes in putting events into perspective. As such, Ingstrup and colleagues suggested that parents can support the development of self-compassion in athletes through being available to them when they experience adversity in sport, by teaching them to be kind to themselves, and to help them put experiences into perspective.

In addition to parents, peers also seem important to the development of self-compassion. Athletes appeared to learn from and engage in self-compassionate behaviors (e.g., balanced awareness) modeled by teammates when experiencing adversity (Ingstrup et al., 2017). In support of Ingstrup et al.'s (2017) initial qualitative findings, Crozier et al. (2019) found that the more that athletes believed that their teammates were self-compassionate, the more likely it was that the athletes reported being self-compassionate themselves. Further, Jeon et al. (2016) stated that social support had a positive impact on self-compassion, suggesting that social support from others in sport can help athletes to accept and endorse a self-compassionate approach. Similarly, men athletes in Reis et al.'s (2022) study noted that seeing their teammates exhibit self-compassionate behaviors increased their own capacity to engage in a self-compassionate approach to sport-related challenges. Frentz et al. (2020) also noted that teammates' social support and affirmations facilitated more self-compassionate responding.

Coaches, mental skills consultants, and sport psychologists may also play a role in the development of self-compassion (Ingstrup et al., 2017). Athletes in Ingstrup et al.'s (2017) study appeared to develop self-compassion through skills learned in sport psychology sessions. Similarly, Frentz et al. (2020) suggested that support offered by sport psychology consultants can aid athletes in adopting a more self-compassionate approach. Mental performance consultants also attest to the usefulness of self-compassion in supporting masters-level[2] athletes (Makepeace & Young, 2021). Additionally, Frentz et al. (2020) suggested that coaches can significantly influence the development of self-compassion, limiting athlete self-criticism and aiding athletes in taking a new perspective. Interactions with a range of significant others is important in sport

[2]Masters-level athletes are typically over the age of 35 and compete in events designed for older participants.

participants' learning and engagement with a self-compassionate approach.

Although the influence of others is sizeable in the cultivation of self-compassion, there is also an internal component to its development (Frentz et al., 2020; Ingstrup et al., 2017). Both Frentz et al. (2020) and Ingstrup et al. (2017) highlighted self-awareness as a key factor in becoming more self-compassionate. Being able to reflect on previous negative experiences whilst avoiding engaging in excessive self-criticism permitted more self-awareness and self-compassion (Ingstrup et al., 2017). Being aware of, and acknowledging, other athletes going through similar challenges fostered not only common humanity but also the ability to abandon self-criticism for self-compassion (Frentz et al., 2020; Ingstrup et al., 2017). Therefore, internal (i.e., self-awareness and reflection) and external (i.e., learning from significant others) factors, in combination, contribute toward the development of self-compassion among those involved in sport. As such, athletes should be encouraged to reflect on their experiences in a kind and mindful manner and also share their challenges with others – the sharing can facilitate common humanity, and through sharing, the athlete stands to not only learn from others but also teach or model self-compassion for others.

Self-Compassion Intervention

Although it is important to understand how individuals develop self-compassion through organic interactions with peers and family members and through other life experiences, not all sport participants are provided the opportunity and support to develop self-compassion during their formative years. Fortunately, there is evidence to suggest that intervention efforts can successfully foster self-compassion in individuals who may not inherently engage in self-compassionate responding. Given the unique demands of sport, intervention efforts tailored specifically for athletes have been developed, both guided by and contributing to self-compassion intervention development in other areas.

Mosewich et al. (2013) were the first to develop and empirically assess a self-compassion intervention in the sport context. They examined the effectiveness of a 7-day psychoeducation and writing-based intervention in increasing self-compassion and decreasing levels of self-criticism, rumination, and excessive concern over mistakes in women varsity athletes who identified as being highly self-critical in a way that was "less than constructive" (p. 516). The intervention began with a brief psychoeducation presentation overviewing stress and coping in sport and introducing the concept of self-compassion. The athletes were presented with evidence that adopting a self-compassionate approach does not promote passivity and/or complacency, rather that it can nurture goal progress and a more effective focus on sport-related tasks. Such an inclusion was particularly important given some athletes may have had apprehension in adopting a self-compassionate approach (Ferguson et al., 2014; Sutherland et al., 2014). As noted in the introduction, addressing concerns about self-compassion is important in intervention efforts, as is presenting the construct in a manner that will resonate with athletes (Mosewich et al., 2019a; Röthlin et al., 2019).

Following the psychoeducation presentation, the athletes participated in a self-compassion writing exercise pertaining to a sport setback and were then provided instructions surrounding the five remaining writing tasks that they were to complete over the next 7 days (Mosewich et al., 2013). Writing tasks were modeled after Leary et al.'s (2007) prompts designed to encourage athletes to think about a negative event in a self-compassionate way (Mosewich et al., 2013). Reflecting on and processing previous negative sport experiences and setbacks can be valuable in promoting self-compassion (Frentz et al., 2020; Ingstrup et al., 2017). Athletes identified a personally significant sport setback as their negative event. To promote common humanity, athletes were instructed to "list ways in which other people experience similar events." To evoke self-kindness, athletes were asked to "write a paragraph expressing understanding, kindness, and concern to yourself." To assist the athletes in taking a self-kind perspective, they were further told to "write as if you are communicating to a close friend in the same situation." To encourage mindfulness, athletes were asked to "describe the event in an objective and unemotional manner" to acknowledge the event without overidentifying with the negative aspects of the experience.

Mosewich et al. (2013) found that the self-compassion intervention was effective for improving self-compassion and state self-criticism, state rumination, and concern over mistakes compared to an attention-training control group. These findings remained one week and one month after the completion of the intervention. Given that self-compassion interventions typically require more time involvement from participants (e.g., 6 weeks, Bluth et al., 2016; 8 weeks, Neff & Germer, 2013), the format of Mosewich et al.'s (2013) intervention was a notable strength. Given the time demands and constraints faced by athletes and others involved in sport, the relatively brief, flexible, and accessible nature of the intervention may be particularly appealing to sport participants interested in incorporating the new strategies into existing mental performance training and routines.

Building on the work of Mosewich et al. (2013), Reis et al. (2015) proposed that a brief self-compassion induction, if effective, may be more practical and manageable for athletes and researchers relative to the 7-day period in the Mosewich et al. (2013) intervention. Brief self-compassion inductions have been successful outside of the sport domain (e.g., DeLury & Poulin, 2018; Leary et al., 2007). Thus, Reis et al. (2015) examined the viability of a brief self-compassion induction in influencing athletes' reactions, thoughts, and emotions following a hypothetical scenario (i.e., being responsible for losing an athletic competition for their team). Like Mosewich and colleagues, Reis et al. (2015) used writing prompts to induce self-compassion informed by Leary et al. (2007). At a single timepoint, women athletes were asked to write down the ways that other people also experience similar events (common humanity); write a paragraph expressing understanding, kindness, and concern to themselves in the same way that they may express concern to a friend who had experienced something similar (self-kindness); and write about the event in an objective, unemotional manner (mindfulness). Although self-compassion was positively associated with behavioral equanimity and negatively associated with negative affect, catastrophizing thoughts, and personal thoughts, Reis and colleagues did not observe a significant effect of the self-compassion induction on how athletes responded to the hypothetical scenario. The researchers noted that although self-compassion predicts adaptive responses to emotionally difficult sport situations in women athletes (Ferguson et al., 2014, 2015), it may be challenging to quickly induce self-compassion among athletes, particularly when relying on a sole, brief procedure. They further speculated that as the scenario they presented was hypothetical rather than personal, there might have been a lack of personal salience, which could influence the response.

Recently, Röthlin and Leiggene (2021) adapted the psychoeducation and writing intervention presented by Mosewich et al. (2013) for a group of climbers using a pretest and posttest assessment and a wait-list control group. The psychoeducation component consisted of a handout describing the three affect regulation systems (soothing, threat, and protection) proposed by Gilbert (2010) and applied the content to the climbing context. Neff's (2003b) definition of self-compassion was introduced, along with explaining how self-compassion activates the soothing system and subsequently stands to deactivate the threat and protection system. The climbers completed the same five writing tasks presented by Mosewich et al. (2013), this time focusing on a climbing experience. All intervention materials were delivered via email. The intervention led to an increase in self-compassion and a decrease in somatic, but not cognitive, performance anxiety (Röthlin & Leiggene, 2021) in the self-compassion group, compared to a wait-list control. The researchers posited that the reduction of somatic performance anxiety might be due to self-compassion activating the emotional contentment and soothing system, which then attenuates the physical effects of the threat and protection system (Gilbert, 2010). Thus, self-compassion may be a strategy to help athletes reduce physiological arousal and somatic anxiety (Röthlin & Leiggene, 2021). While the lack of significant reduction in cognitive performance anxiety was unexpected, the researchers suggest that an intervention that places more emphasis on cognitive processes, such as systematic mindfulness practice, might be necessary for significant results (Röthlin & Leiggene, 2021).

This new intervention adaption by Röthlin and Leiggene (2021) provides some important novel contributions. Interestingly, the increase in self-compassion appeared to be primarily due to a strengthening of the positive facets of self-compassion rather than to a decrease in its negative facets (Röthlin & Leiggene, 2021). Statistically assessing change in the positive and the negative facets of self-compassion may be relevant for future research. Additionally, the findings of this study lead to the suggestion that even athletes who do not describe themselves as self-critical (recall that the athletes in the Mosewich et al. (2013) intervention identified themselves as self-critical) can increase their self-compassion in a similarly short intervention, delivered remotely.

There are also instances of self-compassion being integrated into existing intervention and therapeutic approaches. *Bodies in Motion* is an intervention rooted in Cognitive Dissonance Theory that introduces self-compassion principles to NCAA women athletes to attenuate body pressures (Voelker et al., 2019). Using self-compassion strategies, athletes build self-awareness regarding their cognitive and emotional responses to appearance ideals and cultivate effective coping skills to navigating such standards. The intervention aimed to have athletes learn to be more mindful, self-aware, nonjudgmental of their thoughts and feelings, and kind and understanding and less self-critical in their evaluations and responses to themselves while navigating body image messaging. Five sessions address origins of appearance ideals, environmental triggers, consequences of ideals, and mindfulness and self-compassion practice in relation to body evaluation. Between sessions, there are self-directed activities and support via social media.

The intervention was successful, as the experimental group showed significantly higher self-compassion and mindfulness after the program compared to baseline, while the control group showed no change. The experimental group also reported less thin-ideal internalization compared to the control group after completing the program. Voelker et al. (2021) interviewed women athletes who had participated in the intervention and found that upon program completion, participants reported increased self-awareness and awareness of their relationships with their bodies, understanding of their body's functionality, appreciation of common humanity with other athletes, acknowledgement of the social constructions of beauty in sport and society, recognition of the negative effects of self-criticism, and ability to apply self-compassion strategies. Positive changes in body attitudes, such as acceptance and gratitude, were also reported.

The MBSoccerP program is another example of self-compassion being integrated into an existing framework (Carraça et al., 2019). The intervention was based on Mindfulness-Based Stress Reduction, Acceptance Commitment Therapy, and Compassionate Mind Training. The 8-week, 9-session pilot study involved male soccer players and incorporated self-compassion strategies, including compassionate imagery, writing, and psychoeducation. Mindfulness, self-compassion, performance evaluation, and flow increased, and experiential avoidance decreased, from pre- to post-intervention, leading Carraça et al. (2019) to attest that self-compassion was an important addition to traditional mindfulness training approaches in the sport domain.

In addition to ongoing research focused on fostering self-compassion, individuals in applied roles have also documented their insight surrounding practically integrating self-compassion in a sport context. Rodriguez and Ebbeck (2015) positioned self-compassion as a potentially valuable resource in helping women gymnasts to attenuate the negative impact of mistakes and setbacks. They described a combination of activities designed to instill a self-compassionate perspective into both practice and competition contexts. The strategies adopted from Neff (2003b) included a writing exercise that highlighted discrepancies between how one treats oneself and how one would treat a teammate during hardship, generating and integrating "self-compassion cues," and creating more caring "motivators" to replace harshly self-critical ones. Additionally, Rodriguez and Ebbeck (2015) created novel strategies tailored to the needs of the athletes that they were working with, including a bead transfer exercise to foster awareness and being mindful of one's positive and negative self-talk, sport-specific physical examples to counter the perception that self-compassion is self-coddling (i.e., "Would you withhold water from yourself during your 4-hour practice to be tough on yourself?"), and collaborative, proactive planning to prevent adverse outcomes. Also informed by her applied work, Baltzell (2016) shared that excessive and harsh self-criticism inherent in the sport domain might be successfully attenuated with compassionate attention, reasoning, behavior, imagery, and scripting. As research demonstrating the value of self-compassion in sport continues to grow, there will be increased need for researchers and stakeholders to provide current, empirically informed recommendations for effective self-compassion integration in applied sport settings.

To continue to develop guidelines and strategies for successful self-compassion intervention and development in sport, there are many important considerations for researchers and practitioners. Mosewich et al. (2019a) asserted that efficacy and effectiveness of sport-based interventions must be thoroughly considered (e.g., replication, participant adherence and engagement, assessing maintenance of change over time). Consistent with Kirby et al.'s (2017) recommendations for compassion-based intervention research, Mosewich et al. (2019a) highlighted the need for established normative data and corresponding clinical cutoffs to assess the effectiveness and efficacy of interventions in sport. As interventions focused on self-compassion continue to grow in number, larger-scale randomized control trials with attention to intervention fidelity are required to advance our understanding of intervention effectiveness (Mosewich, 2020; Mosewich et al., 2019a).

Mosewich and colleagues (2019a) also stressed the importance of thoughtful engagement with intervention design, including intervention length and activities (e.g., writing, verbal psychoeducation sessions, strategies embedded in the sport environment), delivery format (e.g., in-person versus remote, self-directed versus instructor-led), and accessibility (e.g., online availability, degree of access restrictions) in supporting deeper understanding of this area. Since athletes with greater self-compassion at the start of their competitive season have been found to have greater well-being at the end of their season, embedding self-compassion intervention into pre-season activities may have merit (Ferguson et al., 2022a). When designing a self-compassion intervention, consideration should also be given to potential facilitators and barriers to the adoption of self-compassion. For example, given the impact that significant others in sport (e.g., coaches, peers, parents; Crozier et al., 2019; Frentz et al., 2020; Ingstrup et al., 2017; Reis et al., 2022) can have on the development of, and engagement with, self-compassion, Mosewich et al. (2019a) suggested that involvement of such significant others may strengthen intervention design and efforts. Further, researchers should thoroughly consider context (e.g., social support, teammates' self-compassion; Crozier et al., 2019; Frentz et al., 2020; Reis et al., 2022; Röthlin et al., 2019) and athletic characteristics (e.g., competitive level, gender, age, culture, past experience with mental skills; Mosewich et al., 2013, 2019a; Reis et al., 2015) to design robust self-compassion interventions in the sport context.

Although existing self-compassion interventions and inductions in sport have involved writing exercises and psychoeducation sessions, a variety of intervention activities and modalities could be included in intervention procedures. For example, self-reflection and self-awareness exercises, modeling, smartphone-guided meditation and journaling, compassionate imagery, cues, affectionate breathing, group-based activities, proactive planning, soothing touch, and social media groups have been positioned as potential approaches within self-compassion interventions (e.g., Baltzell, 2016; Ebbeck & Austin, 2018; Hägglund et al., 2022; Ingstrup et al., 2017; Rodriguez & Ebbeck, 2015; Schnepper et al., 2020; Seekis et al., 2020; Voelker et al., 2019). Given the potentially unique nature of environments, norms, and cultures within and across sport types and participation categories, researchers interested in self-compassion interventions should consider past research and recommendations and be innovative in their efforts to effectively and appropriately foster self-compassion in their sporting environments.

Adoption of Self-Compassion in Sport

Self-Compassion Promotion Versus Intervention

In considering the adoption of self-compassion in the sport context, it would be remiss not to discuss the unknowns regarding the appropriateness of self-compassion promotion versus intervention. That is, should the aim be widespread promotion or targeted intervention, or is there merit in both? There is a need for further examination of when – and for whom – self-compassion might be relevant and effective (Mosewich, 2020; Mosewich et al., 2019a, b). Such a process needs to examine potential drawbacks of self-compassion, as it cannot be assumed that self-compassion is an adaptive approach for all involved in sport and across all sport contexts (Mosewich, 2020). In terms of self-compassion promotion, accessibility and effectiveness must be a focus, with avoidance of any possible downsides of promotion efforts by engaging in appropriate monitoring. For intervention, we must identify who should be targeted, how they will be identified, the temporal patterning of intervention efforts, and best practice in delivery and monitoring. Continual empirical evaluation is necessary, as is systematic refinement based on current research knowledge.

Connection to Resilience and Motivation

Self-compassion has important implications for two psychological constructs that have great relevance in sport – resilience and motivation. The desire for sport participants to embody resilience and motivation is readily apparent in most, if not all, sport contexts. Importantly, resilience and adaptive motivational tendencies have the potential to support both performance and well-being in sport. Additionally, connecting self-compassion with the desired attributes of resilience and motivation may also help foster acceptance and integration of self-compassion into the sport culture.

Given the focus of self-support and engaging in actions that support what is best for the self, it is not surprising that self-compassion is associated with adaptive motivation. Self-compassion is positively associated with self-determined motivation, personal initiative, and responsibility and negatively related to passivity, which provides evidence supporting positive motivational tendencies in sport (Ferguson et al., 2014). Outside of sport, a notable finding by Breines and Chen (2012) was that self-compassion was associated with self-improvement motivation, which bodes well for continued personal development and goal progress in sport.

Many of the adaptive cognitions, emotions, and behaviors associated with self-compassion (see Mosewich, 2020) could be promotive factors for resilience (see Fergus & Zimmerman, 2005, and Chap. 10 of this handbook). To offer a few specific examples, the positive relationship between self-compassion and optimism (Lizmore et al., 2017), grit (Mosewich et al., 2021), mental toughness (Stamatis et al., 2020), and help-seeking (Wasylkiw & Clairo, 2016) makes a pointed statement regarding self-compassion and resilience in sport. Research outside of sport has explicitly associated self-compassion with resilience (Bluth et al., 2018). In addition, resilience is positively related to personal factors such as focus and concentration (Sarkar & Fletcher, 2014), adaptive coping strategies (Nicholls et al., 2016; Secades et al., 2016), and effective emotion regulation strategies (Tugade & Fredrickson, 2007), which exist among self-compassionate athletes (see Mosewich, 2020, for a review). Though resilience is often thought of in terms of personal factors, it is also impacted by the environment (Fergus & Zimmerman, 2005). Sport environments that exude a positive, task-focused motivational climate support resilience (Vitali et al., 2015), and a self-compassionate sport culture is likely to support such an environment.

Sport-Specific Needs and Factors

Adoption of self-compassion in the sport domain requires careful consideration of the unique nature of the sport context. At the present time, there remain some gaps in our knowledge. For instance, an important variable in sport is competition level. Most sport self-compassion studies combine athletes participating at various competitive levels of sport, from regional to international (e.g., Adam et al., 2021b; Reis et al., 2015). Only select few studies have focused on particular levels of competition, such as recreational (e.g., Fontana et al., 2017), varsity (i.e., competing for university or college teams; e.g., Lizmore et al., 2017; Sereda et al., 2022), and international (e.g., Wilson et al., 2019). Studies comparing levels of self-compassion at different competitive levels are scarce. While no significant differences in self-compassion were found across NCAA Divisions I, II, and III student-athletes (Stamatis et al., 2020), athletes competing at a local level reported higher self-compassion than those competing provincially, nationally, and internationally (Ferguson et al., 2022b). Given this limited examination, exploration of key transitional periods, such as retirement from sport, deselection, and change in level of competition is needed. Exploration of self-compassion at different levels of sport and during key periods in a sport career trajectory will help to delineate potential needs and use in different circumstances and inform promotion and intervention efforts.

A variety of sport experiences exist as a function of gender identity. While we do acknowledge an increase in research involving men in recent times, much of the body of research on self-compassion in sport has focused on women athletes (see Mosewich, 2020). There is limited and equivocal research pertaining to gender differences in the levels of self-compassion in a sport context. Jansen et al. (2021) found women had higher values in the negative scale of self-compassion compared to men. Stamatis et al. (2020) reported men NCAA student-athletes as being higher in self-compassion than their women counterparts, while Hilliard et al. (2019) and Ferguson et al. (2022b) found no differences in self-compassion between men and women athletes. A meta-analysis by Yarnell et al. (2015) found that women in general had slightly lower self-compassion than men. Regardless, it seems prudent to consider issues of gender identity when integrating self-compassion into a sport context. For example, there may be a complex relationship between masculinity and self-compassion among men involved in sport, with men athletes who strongly endorse traditional masculine norms experiencing challenges in adopting self-compassion (Reis et al., 2019). As such, an intervention involving inclusive masculinity (i.e., a version of masculinity that is centered on acceptance of varying enactments of masculinities and a view that versions of masculinity should not be hierarchical, but equal; Anderson, 2009) would be valuable in assessing self-compassion in men's sport (Reis et al., 2019).

Research on sport type is limited and equivocal but stands to be considered as an important sport-specific variable. Jansen et al. (2021) reported lower self-compassion among individual sport athletes and handball players, while the soccer players in their sample reflected higher levels of self-compassion. In a study by Ferguson et al. (2022b), aesthetic sport athletes had lower self-compassion than non-aesthetic sport athletes, but there were no differences between team and individual sport athletes. While it remains important to consider factors such as competition level, gender identity, and sport type, the need for individualized promotion and intervention attempts for those involved in sport should not be overlooked. Traditional mental skills training recognizes the need for individualization, and the justification – that athlete strengths, needs, and environments differ – holds true for use of self-compassion as a resource as well. Adoption of self-compassion in sport is likely to be impacted by several personal and environmental factors.

Although the research on self-compassion in sport has centered on athletes, it is not a stretch to suggest self-compassion would also be useful for others involved in sport. Coaches, sport parents, mental performance coaches, and other members of an integrated support team/performance enhancement team, as well as those working in the administrative side of sport, could also benefit from self-compassion. Initial research provides support for such an assertion. Hägglund et al. (2022) found that high performance coaches who undertook their mindful self-reflection intervention reported behavior changes indicative of self-compassion, such as learning from mistakes and avoiding excessive rumination and harsh self-criticism. Thus, Hägglund et al. (2022) suggest self-compassion may be useful in supporting coach well-being. Teaching coaches, mental performance coaches, sport parents, and other sport personnel about self-compassion and strategies to apply it could position them to support themselves, and the athletes with whom they work. Essentially, we should encourage a self-compassion sport culture – one that promotes self-compassion within oneself and others.

Successfully embedding self-compassion in the way those in sport train, compete, and approach life in general requires examination of the inherent sport culture of the team or training group. There is the potential for others to promote or hinder the development of self-compassion through modeling. Crozier et al. (2019) found that athletes who perceived their teammates as being self-compassionate were more likely to exude self-compassion themselves. Parents, coaches, teammates, sport psychology professionals, and teammates not only form a significant part of the sport environment but play a major role in the sport culture. Further, they can have a notable impact whether self-compassion is modeled, supported, or hindered via cultivation of self-critical norms often (unnecessarily) engrained in the minds of sport participants as a required characteristic for success (Frentz et al., 2020; Ingstrup et al., 2017). Thus, it is important to think of the sport environment when promoting, adopting, and sustaining self-compassion. Including coaches, psychologists, mental skills consultants, parents, and others in supporting athletes to learn and be self-compassionate is critical – as is their own adoption of self-compassion. It is in this way that the broader culture of sport can shift to a more self-compassionate approach.

Full adoption of self-compassion requires buy-in, learning, and integration into the sport participant's unique training and competition context, both at home and when away for training and competition. While likely to be initially effortful and conscious (and applied like a resource), the aim is for the approach to become an automatic response. A self-compassionate sport environment promotes constructive self-reflection as opposed to harsh self-criticism, views challenges as an instance for support and growth, emulates a balanced perspective, and acknowledges each sport participant's background and their physical, mental, spiritual, emotional, social, and cultural needs. It also honors the interrelatedness between performance in sport and well-being in sport.

Additional Areas for Future Research

While the previous section has identified several future research directions in discussing the adoption of self-compassion in sport, there are some additional suggestions to be made. These suggestions further accentuate the need for diverse and representative samples and innovative research designs to better guide appropriate, tailored, and relevant efforts to foster self-compassion in a variety of sport participants across different sport contexts.

Much of the research on self-compassion in sport has focused on adolescents (e.g., Eke et al., 2020; Mosewich et al., 2011; Pila et al., 2022) and young adults (e.g., Adam et al., 2021a; Ingstrup et al., 2017; Mosewich et al., 2019b). There is a need for more research on younger child and youth cohorts, as well as older adult and masters-level athletes. Mental performance consultants who work with masters-level athletes have attested to their use of self-compassion with this population (Makepeace & Young, 2021). The mental performance consultants employed self-compassion strategies with their masters-level athlete clients to help them listen to and accept their bodies and how they function. As such, Makepeace and Young (2021) suggest self-compassion can help masters-level athletes accept and cope with age-related performance changes and limitations as they navigate their unique sport experiences and other life pursuits. Future research needs to examine self-compassion across the sporting life span.

Most studies have adopted a cross-sectional design (e.g., Hilliard et al., 2019; Mosewich et al., 2021), and with the exception of some prospective designs (e.g., Mosewich et al., 2019b) and follow-up qualitative interviews (e.g., Eke et al., 2020; Sutherland et al., 2014), most longitudinal research involves some form of pretest/posttest intervention design (e.g., Mosewich et al., 2013; Röthlin & Leiggener, 2021). Thus, there is a lack of longitudinal tracking of self-compassion among sport participants, which means we know little about how self-compassion may change over time and with experience. Pila et al. (2022) examined self-compassion scores among adolescent girls participating in organized sport over a 3-year period. Self-compassion scores did not change over 3 years across the sample; however, participants differed in their rate of change over time. Thus, Pila et al.'s (2022) research speaks to the importance of considering *within-person* variability or how athletes fluctuate on an individual level over time. Self-compassion is often considered as an individual difference construct, but it is important that future research consider the individual variations as well (Pila et al., 2022).

Conclusion

The collective work of researchers and self-compassion practitioners has established self-compassion as a useful resource in sport. Researchers will continue to build this case, as well as further inform how to foster self-compassion in sport. Self-compassion is associated with adaptive cognitions, emotions, behaviors, and outcomes that stand to support both well-being and performance in sport. As we strive for high performance alongside positive experiences and well-being, self-compassion can support the resilience of those involved in sport and motivate participation across the life span in various sporting roles. Self-compassion involves both evading the negative (i.e., self-judgment, overidentification, isolation) and enhancing the positive (i.e., self-kindness, common humanity, and mindfulness; Neff, 2003b). It involves self-soothing and support, as well as an active "fierce self-compassion" to allow us to take care of ourselves and help us reach our potential (Neff, 2021). In sport, we must strive to manage our difficulties and challenges and harness our strengths. Self-compassion emulates such an approach.

References

Adam, M. E. K., Eke, A. O., & Ferguson, L. J. (2021a). "Know that you're not just settling": Exploring women athletes' self-compassion, sport performance perceptions, and well-being around important competitive events. *Journal of Sport and Exercise Psychology, 43*(3), 268–278. https://doi.org/10.1123/jsep.2020-0196

Adam, M. E. K., Kowalski, K. C., Duckham, R. L., Ferguson, L. J., & Mosewich, A. D. (2021b). Self-compassion plays a role in Canadian women athletes' body appreciation and intuitive eating: A mixed methods approach. *International Journal of Sport Psychology, 52*(4), 287–309. https://doi.org/10.7352/IJSP.2021.52.287

Åkesdotter, C., Kenttä, G., Eloranta, S., & Franck, J. (2020). The prevalence of mental health problems in elite athletes. *Journal of Science and Medicine in Sport, 23*(4), 329–335. https://doi.org/10.1016/j.jsams.2019.10.022

Amemiya, R., & Sakairi, Y. (2020). The role of self-compassion in athlete mindfulness and burnout: Examination of the effects of gender differences.

Personality and Individual Differences, 166, 110167. https://doi.org/10.1016/j.paid.2020.110167

Anderson, E. (2009). *Inclusive masculinity: The changing nature of masculinities*. Routledge.

Baltzell, A. (2016). Self-compassion, distress tolerance, and mindfulness in performance. In A. Baltzell (Ed.), *Mindfulness and performance: Current perspectives in social and behavioral sciences* (pp. 53–77). Cambridge.

Barczak, N., & Eklund, R. C. (2020). The moderating effect of self-compassion on relationships between performance and subsequent coping and motivation. *International Journal of Sport and Exercise Psychology, 18*(2), 256–268. https://doi.org/10.1080/1612197X.2018.1511620

Bluth, K., Gaylord, S. A., Campo, R. A., Mullarkey, M. C., & Hobbs, L. (2016). Making friends with yourself: A mixed methods pilot study of a mindful self-compassion program for adolescents. *Mindfulness, 7*(2), 479–492. https://doi.org/10.1007/s12671-015-0476-6

Bluth, K., Mullarkey, M., & Lathren, C. (2018). Self-compassion: A potential path to adolescent resilience and positive exploration. *Journal of Child and Family Studies, 27*, 3037–3047. https://doi.org/10.1007/s10826-018-1125-1

Breines, J. G., & Chen, S. (2012). Self-compassion increases self-improvement motivation. *Personality and Social Psychology Bulletin, 38*(9), 1133–1143. https://doi.org/10.1177/0146167212445599

Carrança, B., Serpa, S., Rosado, A., & Guerrero, J. P. (2019). A pilot study of a mindfulness-based program (MBSoccerP): The potential role of mindfulness, self-compassion and psychological flexibility on flow and elite performance in soccer athletes. *Revista Iberoamericana de Psicologia del Ejercicio y el Deporte, 14*(1), 34–40.

Casali, N., Ghisi, M., Jansen, P., Feraco, T., & Meneghetti, C. (2022). What can affect competition anxiety in athletes? The role of self-compassion and repetitive negative thinking. *Psychological Reports, 125(*4), 2009–2028. https://doi.org/10.1177/00332941211017258.

Ceccarelli, L. A., Giuliano, R. J., Glazebrook, C. M., & Strachan, S. M. (2019). Self-compassion and psycho-physiological recovery from recalled sport failure. *Frontiers in Psychology, 10*, 1564. https://doi.org/10.3389/fpsyg.2019.01564

Crozier, A. J., Mosewich, A. D., & Ferguson, L. J. (2019). The company we keep: Exploring the relationship between perceived teammate self-compassion and athlete self-compassion. *Psychology of Sport and Exercise, 40*, 152–155. https://doi.org/10.1016/j.psychsport.2018.10.005

DeLury, S. S., & Poulin, M. J. (2018). Self-compassion and verbal performance: Evidence for threat-buffering and implicit self-related thoughts. *Self and Identity, 17*(4), 710–722. https://doi.org/10.1080/15298868.2018.1477829

Doorley, J. D., Kashdan, T. B., Weppner, C. H., & Glass, C. R. (2022). The effects of self-compassion on daily emotion regulation and performance rebound among college athletes: Comparisons with confidence, grit, and hope. *Psychology of Sport and Exercise, 58*, 102081. https://doi.org/10.1016/j.psychsport.2021.102081

Ebbeck, V., & Austin, S. (2018). Burning off the fat oppression: Self-compassion exercises for personal trainers. *Fat Studies, 7*(1), 81–92. https://doi.org/10.1080/21604851.2017.1360670

Eke, A. O., Adam, M. E. K., Kowalski, K. C., & Ferguson, L. J. (2020). Narratives of adolescent women athletes' body self-compassion, performance and emotional well-being. *Qualitative Research in Sport, Exercise and Health, 12*(2), 175–191. https://doi.org/10.1080/2159676X.2019.1628805

Fergus, S., & Zimmerman, M. A. (2005). Adolescent resilience: A framework for understanding healthy development in the face of risk. *Annual Review Public Health, 26*(1), 399–419. https://doi.org/10.1146/annurev.publhealth.26.021304.144357

Ferguson, L. J., Kowalski, K. C., Mack, D. E., & Sabiston, C. M. (2014). Exploring self-compassion and eudaimonic well-being in young women athletes. *Journal of Sport & Exercise Psychology, 36*(2), 203–216. https://doi.org/10.1123/jsep.2013-0096

Ferguson, L. J., Kowalski, K. C., Mack, D. E., & Sabiston, C. M. (2015). Self-compassion and eudaimonic well-being during emotionally difficult times in sport. *Journal of Happiness Studies, 16*(5), 1263–1280. https://doi.org/10.1007/s10902-014-9558-8

Ferguson, L. J., Adam, M. E. K., Gunnell, K. E., Kowalski, K. C., Mack, D. E., Mosewich, A. D., & Murphy, N. (2022a). Self-compassion or self-criticism? Predicting women athletes' psychological flourishing in sport in Canada. *Journal of Happiness Studies, 23*, 1923–1939. https://doi.org/10.1007/s10902-021-00483-1.

Ferguson, L. J., Saini, S., & Adam, M. E. K. (2022b). Safe space or high stakes environments: Comparing self-compassion in differing sport contexts in Canada. *International Journal of Sport Psychology, 53*(1), 1–24. https://doi.org/10.7352/IJSP.2022.53.001

Fontana, M. S., Fry, M. D., & Cramer, E. (2017). Exploring the relationship between athletes' perceptions of the motivational climate to their compassion, self-compassion, shame, and pride in adult recreational sport. *Measurement in Physical Education and Exercise Science, 21*(2), 101–111. https://doi.org/10.1080/1091367X.2017.1278698

Frentz, D. M., McHugh, T.-L. F., & Mosewich, A. D. (2020). Athletes' experiences of shifting from self-critical to self-compassionate approaches within high-performance sport. *Journal of Applied Sport Psychology, 32*(6), 565–584. https://doi.org/10.1080/10413200.2019.1608332

Gilbert, P. (2010). *Compassion focused therapy*. Routledge.

Gilbert, P., McEwan, K., Matos, M., & Rivis, A. (2011). Fears of compassion: Development of three self-report measures. *Psychology and Psychotherapy: Theory, Research and Practice, 84*(3), 239–255. https://doi.org/10.1348/147608310X526511

Gu, J., Baer, R., Cavanagh, K., Kuyken, W., & Strauss, C. (2020). Development and psychometric properties of the Sussex-Oxford Compassion Scales (SOCS). *Assessment, 27*(1), 3–20. https://doi.org/10.1177/1073191119860911

Hägglund, K., Kenttä, G., Thelwell, R. & Wagstaff, C. R. D. (2022). Mindful self-reflection to support sustainable high-performance coaching: A process evaluation of a novel method development in elite sport. *Journal of Applied Sport Psychology, 34*(6), 1125–1148. https://doi.org/10.1080/10413200.2021.1925782.

Hilliard, R. C., Redmond, L. A., & Watson, J. C. (2019). The relationships among self-compassion, stigma, and attitudes toward counseling in student-athletes. *Journal of Clinical Sport Psychology, 13*(3), 374–389. https://doi.org/10.1123/jcsp.2018-0027

Huysmans, Z., & Clement, D. (2017). A preliminary exploration of the application of self-compassion within the context of sport injury. *Journal of Sport and Exercise Psychology, 39*(1), 56–66. https://doi.org/10.1123/jsep.2016-0144

Ingstrup, M. S., Mosewich, A. D., & Holt, N. L. (2017). The development of self-compassion among women varsity athletes. *The Sport Psychologist, 31*(4), 317–331. https://doi.org/10.1123/tsp.2016-0147

Jansen, P., Hoja, S., & Meneghetti, C. (2021). Does repetitive thinking mediate the relationship between self-compassion and competition anxiety in athletes? *Cogent Psychology, 8*(1), 1909243. https://doi.org/10.1080/23311908.2021.1909243

Jeon, H., Lee, K., & Kwon, S. (2016). Investigation of the structural relationships between social support, self-compassion, and subjective well-being in Korean elite student athletes. *Psychological Reports, 119*(1), 39–54. https://doi.org/10.1177/0033294116658226

Killham, M. E., Mosewich, A. D., Mack, D. E., Gunnell, K. E., & Ferguson, L. J. (2018). Women athletes' self-compassion, self-criticism, and perceived sport performance. *Sport, Exercise, and Performance Psychology, 7*(3), 297–307. https://doi.org/10.1037/spy0000127

Kirby, J. N., Tellegen, C. L., & Steindl, S. R. (2017). A meta-analysis of compassion-based interventions: Current state of knowledge and future directions. *Behavior Therapy, 48*(6), 778–792. https://doi.org/10.1016/j.beth.2017.06.003

Leary, M. R., Tate, E. B., Adams, C. E., Batts Allen, A., & Hancock, J. (2007). Self-compassion and reactions to unpleasant self-relevant events: The implications of treating oneself kindly. *Journal of Personality and Social Psychology, 92*(5), 887–904. https://doi.org/10.1037/0022-3514.92.5.887

Lizmore, M. R., Dunn, J. G. H., & Dunn, J. C. (2017). Perfectionistic strivings, perfectionistic concerns, and reactions to poor personal performances among intercollegiate athletes. *Psychology of Sport and Exercise, 33*, 75–84. https://doi.org/10.1016/j.psychsport.2017.07.010

Makepeace, T., & Young, B. W. (2021). Mental performance consultants perspectives on content and delivery of sport psychology services to masters athletes. *Journal of Aging and Physical Activity, 35*(3), 200–212. https://doi.org/10.1123/japa.2021-0141.

Mosewich, A. D. (2020). Self-compassion in sport and exercise. In G. Tenenbaum & R. C. Eklund (Eds.), *Handbook of sport psychology* (4th ed., pp. 382–401). Wiley.

Mosewich, A. D., Vangool, A. B., Kowalski, K. C., & McHugh, T. F. (2009). Exploring women track and field athletes' meanings of muscularity. *Journal of Applied Sport Psychology, 21*(1), 99–115. https://doi.org/10.1080/10413200802575742

Mosewich, A. D., Kowalski, K. C., Sabiston, C. M., Sedgwick, W. A., & Tracy, J. L. (2011). Self-compassion: A potential resource for young women athletes. *Journal of Sport and Exercise Psychology, 33*(1), 103–123. https://doi.org/10.1123/jsep.33.1.103

Mosewich, A. D., Crocker, P. R. E., Kowalski, K. C., & DeLongis, A. (2013). Applying self-compassion in sport: An intervention with women athletes. *Journal of Sport & Exercise Psychology, 35*(5), 514–524. https://doi.org/10.1123/jsep.35.5.514

Mosewich, A. D., Crocker, P. R. E., & Kowalski, K. C. (2014). Managing injury and other setbacks in sport: Experiences of (and resources for) high-performance women athletes. *Qualitative Research in Sport, Exercise and Health, 6*(2), 182–204. https://doi.org/10.1080/2159676X.2013.766810

Mosewich, A. D., Ferguson, L. J., McHugh, T. L. F., & Kowalski, K. C. (2019a). Enhancing capacity: Integrating self-compassion in sport. *Journal of Sport Psychology in Action, 10*(4), 1–9. https://doi.org/10.1080/21520704.2018.1557774

Mosewich, A. D., Sabiston, C. M., Kowalski, K. C., Gaudreau, P., & Crocker, P. R. E. (2019b). Self-compassion in the stress process in women athletes. *The Sport Psychologist, 33*(1), 23–34. https://doi.org/10.1123/tsp.2017-0094

Mosewich, A. D., Dunn, J. G. H., Causgrove Dunn, J., & Wright, K. S. (2021). Domain-specific grit, identity, and self-compassion in intercollegiate athletes. *Sport, Exercise, and Performance Psychology, 10*(2), 257–272. https://doi.org/10.1037/spy0000267

Neff, K. D. (2003a). The development and validation of a scale to measure self-compassion. *Self and Identity, 2*(3), 223–250. https://doi.org/10.1080/15298860390209035

Neff, K. D. (2003b). Self-compassion: An alternative conceptualization of a healthy attitude toward oneself. *Self and Identity, 2*(2), 85–101. https://doi.org/10.1080/15298860390129863

Neff, K. (2021). *Fierce self-compassion: How women can harness kindness to speak up, claim their power, and thrive*. Harper Collins.

Neff, K. D., & Germer, C. K. (2013). A pilot study and randomized controlled trial of the mindful self-compassion program. *Journal of Clinical Psychology, 69*(1), 28–44. https://doi.org/10.1002/jclp.21923

Neff, K. D., Tóth-Király, I., Knox, M. C., Kuchar, A., & Davidson, O. (2021). The development and validation of the State Self-Compassion Scale (long-and short form). *Mindfulness, 12*(1), 121–140. https://doi.org/10.1007/s12671-020-01505-4

Nicholls, A. R., Morley, D., & Perry, J. L. (2016). The model of motivational dynamics in sport: Resistance to peer influence, behavioral engagement and disaffection, dispositional coping, and resilience. *Frontiers in Psychology, 6*, 1–9. https://doi.org/10.3389/fpsyg.2015.02010

Pila, E., Gilchrist, J. D., Kowalski, K., & Sabiston, C. M. (2022). Self-compassion and body-related self-conscious emotions: Examining within-and between-person variation among adolescent girls in sport. *Psychology of Sport and Exercise, 58*, 102083. https://doi.org/10.1016/j.psychsport.2021.102083

Raes, F., Pommier, E., Neff, K. D., & Van Gucht, D. (2011). Construction and factorial validation of a short form of the Self-Compassion Scale. *Clinical Psychology and Psychotherapy, 18*(3), 250–255. https://doi.org/10.1002/cpp.702

Reis, N. A., Kowalski, K. C., Ferguson, L. J., Sabiston, C. M., Sedgwick, W. A., & Crocker, P. R. E. (2015). Self-compassion and women athletes' responses to emotionally difficult sport situations: An evaluation of a brief induction. *Psychology of Sport and Exercise, 16*(3), 18–25. https://doi.org/10.1016/j.psychsport.2014.08.011

Reis, N. A., Kowalski, K. C., Mosewich, A. D., & Ferguson, L. J. (2019). Exploring self-compassion and versions of masculinity in men athletes. *Journal of Sport and Exercise Psychology, 41*(6), 368–379. https://doi.org/10.1123/jsep.2019-0061

Reis, N. A., Kowalski, K. C., Mosewich, A. D., & Ferguson, L. J. (2022). "That's how I am dealing with it – that *is* dealing with it": Exploring men athletes' self-compassion through the lens of masculinity. *Qualitative Research in Sport, Exercise and Health, 14*(2), 245–267. https://doi.org/10.1080/2159676X.2021.1920455.

Rodriguez, M., & Ebbeck, V. (2015). Implementing self-compassion strategies with female college gymnasts. *Journal of Sport Psychology in Action, 6*(1), 44–53. https://doi.org/10.1080/21520704.2014.991052

Röthlin, P., & Leiggener, R. (2021). Self-compassion to decrease performance anxiety in climbers: A randomized control trial. *Current Issues in Sport Science, 6*, 4. https://doi.org/10.36950/2021ciss004

Röthlin, P., Horvath, S., & Birrer, D. (2019). Go soft or go home? A scoping review of empirical studies on the role of self-compassion in the competitive sport setting. *Current Issues in Sport Science, 4*. https://doi.org/10.15203/CISS_2019.013

Sarkar, M., & Fletcher, D. (2014). Psychological resilience in sport performers: A review of stressors and protective factors. *Journal of Sports Sciences, 32*(15), 1419–1434. https://doi.org/10.1080/02640414.2014.901551

Schnepper, R., Reichenberger, J., & Blechert, J. (2020). Being my own companion in times of social isolation–A 14-day mobile self-compassion intervention improves stress levels and eating behavior. *Frontiers in Psychology, 11*, 1–9. https://doi.org/10.3389/fpsyg.2020.595806

Secades, X. G., Molinero, O., Salguero, A., Barquín, R. R., de la Vega, R., & Márquez, S. (2016). Relationship between resilience and coping strategies in competitive sport. *Perceptual and Motor Skills, 122*(1), 336–349. https://doi.org/10.1177/0031512516631056

Seekis, V., Bradley, G. L., & Duffy, A. L. (2020). Does a Facebook-enhanced mindful self-compassion intervention improve body image? An evaluation study. *Body Image, 34*, 259–269. https://doi.org/10.1016/j.bodyim.2020.07.006

Sereda, B. J., Holt, N. L., & Mosewich, A. D. (2022). How women varsity athletes high in self-compassion experience unexpected stressors. *Journal of Applied Sport Psychology, 34*(6), 1149–1169. https://doi.org/10.1080/10413200.2021.1897900.

Smith, B. (2013). Disability, sport and men's narratives of health: A qualitative study. *Health Psychology, 32*(1), 110–119. https://doi.org/10.1037/a0029187

Stamatis, A., Deal, P. J., Morgan, G. B., Forsse, J. S., Papadakis, Z., McKinley-Barnard, S., Scudamore, E. M., & Koutakis, P. (2020). Can athletes be tough yet compassionate to themselves? Practical implications for NCAA mental health best practice no. 4. *PLoS One, 15*(12), e0244579. https://doi.org/10.1371/journal.pone.0244579

Steindl, S. R., Tellegen, C. L., Filus, A., Seppälä, E., Doty, J. R., & Kirby, J. N. (2021). The Compassion Motivation and Action Scales: A self-report measure of compassionate and self-compassionate behaviours. *Australian Psychologist, 56*(2), 93–110. https://doi.org/10.1080/00050067.2021.1893110

Stoeber, J. (2011). The dual nature of perfectionism in sports: Relationships with emotion, motivation, and performance. *International Review of Sport and Exercise Psychology, 4*(2), 128–145. https://doi.org/10.1080/1750984X.2011.604789

Sutherland, L. M., Kowalski, K. C., Ferguson, L. J., Sabiston, C. M., Sedgwick, W. A., & Crocker, P. R. E. (2014). Narratives of young women athletes' experiences of emotional pain and self-compassion. *Qualitative Research in Sport, Exercise and Health, 6*(4), 499–516. https://doi.org/10.1080/2159676X.2014.88858

Tarasoff, L., Ferguson, L., & Kowalski, K. (2017). Self-compassion in an evaluative dance environment. *University of Saskatchewan Undergraduate Research Journal, 3*(1), 1–11.

Tingaz, E. O., Solmaz, S., Ekiz, M. A., & Guvendi, B. (2022). The relationship between mindfulness and happiness in student-athletes: The role of self-compassion – mediator or moderator? *Journal of*

Rational-Emotive & Cognitive-Behavior Therapy, 40, 75–85. https://doi.org/10.1007/s10942-021-00397-0.

Tugade, M. M., & Fredrickson, B. L. (2007). Regulation of positive emotions: Emotion regulation strategies that promote resilience. *Journal of Happiness Studies, 8*(3), 311–333. https://doi.org/10.1007/s10902-006-9015-4

Vitali, F., Bortoli, L., Bertinato, L., Robazza, C., & Schena, F. (2015). Motivational climate, resilience, and burnout in youth sport. *Sport Sciences for Health, 11*, 103–108. https://doi.org/10.1007/s11332-014-0214-9

Voelker, D. K., Petrie, T. A., Huang, Q., & Chandran, A. (2019). Bodies in Motion: An empirical evaluation of a program to support positive body image in female collegiate athletes. *Body Image, 28*, 149–158. https://doi.org/10.1016/j.bodyim.2019.01.008

Voelker, D. K., Petrie, T. A., Fairhurst, K., & Casanave, K. (2021). "My body loves me, so I should love it back": A qualitative evaluation of the bodies in motion program with female collegiate athletes. *Sport, Exercise, and Performance Psychology, 10*(1), 43–58. https://doi.org/10.1037/spy0000211

Walker, S. P. (2021). Self-compassion mediates the relationship between dispositional mindfulness and athlete burnout among adolescent squash players in South Africa. *South African Journal of Sports Medicine, 33*(1), 1–6. https://doi.org/10.17159/2078-516X/2021/v33i1a11877

Walton, C. C., Baranoff, J., Gilbert, P., & Kirby, J. (2020). Self-compassion, social rank, and psychological distress in athletes of varying competitive levels. *Psychology of Sport and Exercise, 50*, 101733. https://doi.org/10.1016/j.psychsport.2020.101733

Wasylkiw, L., & Clairo, J. (2016). Help seeking in men: When masculinity and self-compassion collide. *Psychology of Men & Masculinity, 19*(2), 234–242. https://doi.org/10.1037/men0000086

Williams, D., Thayer, J., & Koenig, J. (2016). Resting cardiac vagal tone predicts intraindividual reaction time variability during an attention task in a sample of young and healthy adults. *Psychophysiology, 53*(12), 1843–1851. https://doi.org/10.1111/psyp.12739

Wilson, D., Bennett, E. V., Mosewich, A. D., Faulkner, G. E., & Crocker, P. R. E. (2019). "The zipper effect": Exploring the interrelationship of mental toughness and self-compassion among Canadian elite women athletes. *Psychology of Sport & Exercise, 40*, 61–70. https://doi.org/10.1016/j.psychsport.2018.09.006

Yarnell, L. M., Stafford, R. E., Neff, K. D., Reilly, E. D., Knox, M. C., & Mullarkey, M. (2015). Meta-analysis of gender differences in self-compassion. *Self and Identity, 14*(5), 499–520. https://doi.org/10.1080/15298868.2015.1029966

14 Caring for the Carer – Self-Compassion in the Health Professions

Alina Pavlova and Nathan S. Consedine

Introduction

Caring for the mental and physical health of others is among the most noble of human endeavours and is variously a job, a career, and a calling. As healthcare costs soar and resourcing become tighter, however, healthcare is also a profession in which stress and burnout become ordinary (Innstrand et al., 2011; Heinemann & Heinemann, 2017) and depression and anxiety are common (Shanafelt et al., 2015; Mata et al., 2015). Burnout is more common in healthcare workers than in other professions (Shanafelt et al., 2019), taking a huge toll on the physical and mental wellbeing of this critical workforce (Raab, 2014). The objective pressures of health workplaces are likely compounded by professional identities characterised by idealism and perfectionism (Enns et al., 2001; Gaufberg et al., 2010). Mistakes are routinely met with rumination, catastrophising, self-blame, and self-criticism (Bria et al., 2012; Duarte & Pinto-Gouveia, 2017) leading to issues in performance, absenteeism and turn-over not to mention mental health and quality of life.

In this light, sustainably caring for others as a part of professional responsibilities poses a very particular set of challenges; "caring for the self" risks becoming an "optional extra". Unfortunately, while there is a burgeoning literature on the factors impacting other-focused compassion as well as increasing evidence regarding the importance of self-care among healthcare professionals, research on self-compassion in this area remains in its infancy. Prima facie, however, the attributes that define self-compassion – perspective-taking, self-kindness, and recognition of common humanity during failure or difficulties – seem well-suited to the challenges that characterise health work environments (Neff & Pommier, 2013).

In this chapter, we briefly characterise the challenges facing healthcare professionals and consider why cultivating self-compassion specifically may be of benefit. We review existing empirical work linking self-compassion to outcomes relevant to the healthcare professions, then consider data regarding the efficacy and feasibility of self-compassion-based interventions in healthcare groups. Finally, we reflect on what is known (and not known) about why self-compassion could work in the very particular context of healthcare, consider whether self-compassion might be more or less beneficial for particular groups, and assess the specific outcomes that self-compassion interventions might be most useful for. We also consider the intriguing possibility of a link between self- and other-focused care. Overall, we suggest that while conceptual and practical difficulties in defining

A. Pavlova (✉) · N. S. Consedine
Department of Psychological Medicine, University of Auckland, Auckland, New Zealand
e-mail: alina.pavlova@auckland.ac.nz

A. Finlay-Jones et al. (eds.), *Handbook of Self-Compassion*, Mindfulness in Behavioral Health,
https://doi.org/10.1007/978-3-031-22348-8_14

and applying self-compassion remain, the development of self-compassion has potential and may usefully be integrated in the daily life and training of health professionals.

Can Self-Compassion Solve the Problems of Modern Healthcare?

The Challenges of Modern Healthcare Environments

Continuous exposure to illness and suffering, long working hours, and increasing workloads put healthcare professionals at high risk for stress-related problems (Innstrand et al., 2011; Heinemann & Heinemann, 2017). Burnout, secondary traumatic stress, depression, and anxiety are now widely evident (Shanafelt et al., 2015; Mata et al., 2015); rates may be increasing (Shanafelt et al., 2019), especially under COVID-19 (Alrawashdeh et al., 2021). Meta-analyses of global prevalence studies show burnout symptoms in more than one-tenth of nurses (Woo et al., 2020) and one-fifth of physicians (Zhou et al., 2020), and up to one-third of residents, especially those in surgical/urgency (SU) specialties (Rodrigues et al., 2018). Anxiety and depression are similarly high (Alharthy et al., 2017), as is secondary traumatic stress, especially in emergency and oncology staff (Dominguez-Gomez & Rutledge, 2009; Quinal et al., 2009; Roden-Foreman et al., 2017). Burnout negatively impacts personal lives, job satisfaction, and health, as well as being linked to other issues such as substance abuse and suicide (Raab, 2014; Beyond Blue, 2013). For healthcare systems, burnout and declines in well-being can increase error rates, compromise decision-making, and reduce prosocial responding, resulting in poorer patient care (Raab, 2014).

Such contextual issues may well be compounded by the personality and coping patterns that seem to typify healthcare professionals. Medical trainees are characterised by the desire to "make a difference," as well as a profound professional idealism (Gaufberg et al., 2010). Self-selection and training contingencies favour the selection and retention of perfectionistic high-achievers (Enns et al., 2001) who, when confronted with the realities and constraints of modern healthcare environments, may struggle to maintain a clear sense of purpose (Mader et al., 2014; Triffaux et al., 2019; Dyrbye et al., 2005). Maladaptive coping styles are evident even among early trainees. For example, while medical students acknowledge not being able to 'fix' everyone, they report finding this difficult and avoid asking for help because of fears of appearing incapable or weak (Tierney et al., 2018; Singh et al., 2018). Similarly, a constant need to "prove oneself" among junior doctors may increase their risk for negative psychological and physical health outcomes (Brewin & Firth-Cozens, 1997; Neumann et al., 2011; Mahoney et al., 2016; Overland et al., 2019), especially when senior physicians – their role models and educators – do not exemplify healthy self-care themselves (Dyrbye et al., 2005; Satele et al., 2014). High workloads and low service resources may compound the effects of dysfunctional coping (Mills & Chapman, 2016; Dobkin, 2011; Neff, 2011; Neff & Seppälä, 2016; Bria et al., 2012; Duarte & Pinto-Gouveia, 2017). In terms of stress-related issues then, modern healthcare is a near-perfect storm.

This problematic dynamic has strong institutional and training origins. While compassion for others is a moral and professional expectation, self-care is only beginning to become a focus in medical curricula. Healthcare professionals operate in biomedically-focused competitive environments (Jacobs et al., 2013; De Zulueta, 2015), where idealistic and perfectionistic tendencies are implicitly or explicitly valued (Haider et al., 2020), resulting in avoiding (disclosing) shame (Lindström et al., 2011). Compassion literacy is not a priority (Tierney et al., 2018; Burridge et al., 2017) and is poorly trained (Mills & Chapman, 2016; Mills et al., 2017, 2015), while prioritising self-care may be seen as selfish (Mills

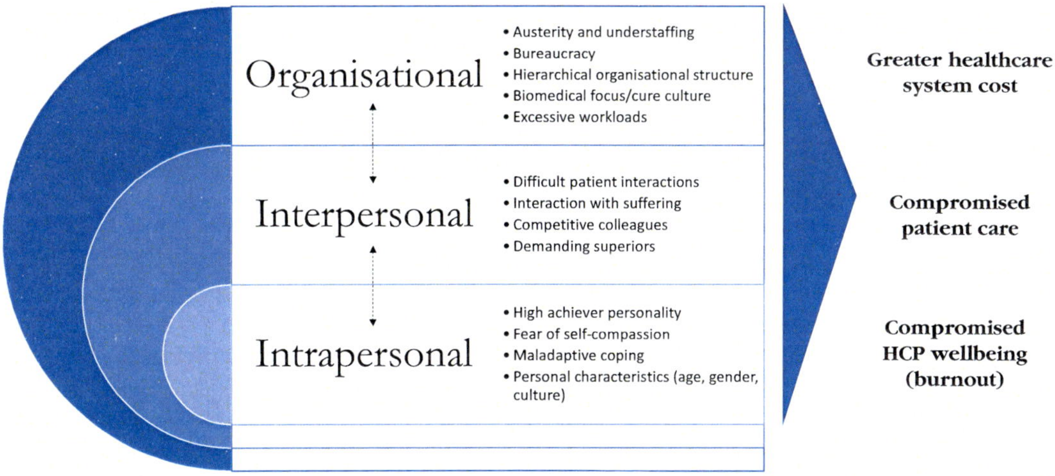

Fig. 14.1 – Challenges of modern healthcare

et al., 2015). Thus, healthcare workers often end up at a crossroads where psychologically challenging environments, high professional demands, and high achievement cultures are layered with poor training and suboptimal coping (see Fig. 14.1). The sustainability of patient care and retention of staff are almost certainly affected by this scenario.

Can Self-Compassion Be a Solution?

Given this catastrophic characterisation of modern healthcare, there are still a significant number of healthcare professionals that do not meet criteria for burnout and who are also more likely to flourish (Vetter et al., 2018). The question, of course, is how? What is it about these individuals that allows them to manage professional demands such that they sustain their commitment and care? In this chapter, we consider the possibility that self-compassion might have a role in facilitating adaptation to the very real demands of working in healthcare. Below, we (1) offer a theoretical rationale as to why self-compassion might facilitate adaptation to the challenges confronting healthcare professionals and, (2) review the empirical literature linking self-compassion to outcomes in these groups.

Why Might Self-Compassion Benefit Healthcare Professionals? Theoretical Underpinnings

There are good reasons to expect that self-compassion might have benefits for healthcare professionals. For one, the fear of failures in medicine – some of which can be minor and inevitable (i.e., adverse events, communication challenges, inability to solve all problems) and some of which can be fatal – is constant and unavoidable. The stress from continuous threat and the related negative emotions are said to compromise problem-solving and decision making (Gray, 1999), while healthcare professionals' tendency to self-blame, other-blame, catastrophise, and ruminate puts them at risk of negative outcomes (Bria et al., 2012; Duarte & Pinto-Gouveia, 2017), further compromising their ability to function under stress and making errors more probable (Prins et al., 2009; Linden et al., 2005).

As a way of relating to the self, self-compassion is the antithesis of this manner of working. Broadly defined as the ability or tendency to respond to one's own difficulties with kindness, self-compassion is thought to allow people to remain present with threatening feelings without self-judgment or criticism, reduce overidentification, and decrease feelings of isolation (Neff et al., 2018). Relating to challenges in this way may reduce the need to use finite psychological resources to protect the self from threats, such as those that healthcare professionals may encounter in their professional environments. Put simply, the challenges of this professional path seem well-suited to the particular benefits self-compassion may bring.

As a slight digression (but importantly in terms of the specific challenges and obligations in healthcare), there is also the suggestion that other-related compassion can be enhanced and sustained among those who can first relate compassionately to themselves (Neff, 2003; Barnard & Curry, 2011; Hofmann et al., 2011; Neff & Pommier, 2013). Contemplative and spiritual practices, along with self-care, are increasingly seen as facilitative of compassion towards others (Maslach & Leiter, 2005; Singh et al., 2018). Because self-compassion may help us recognise the universality of human suffering, imperfections, and (fear of making) mistakes (Neff & Pommier, 2013), it may encourage humbleness, reduced self-judgment and a reduced tendency to blame others for *their* suffering (Gilbert & Procter, 2006). Thus, in addition to contributing to self-care, self-compassion has the potential to contribute to more person-centred care by enabling healthcare workers to treat their patients as "people like me".

Empirical Studies of Self-Compassion in Healthcare

Observational Studies

Empirically, the small body of available data are consistent with the notion that self-compassion may be beneficial to healthcare workers. Healthcare professionals that report greater self-compassion also report lower stress (Finlay-Jones et al., 2015; Kemper et al., 2015, 2019, 2020), lower burnout (Abdollahi et al., 2021; Dev et al., 2020; Kemper et al., 2020; Schabram & Heng, 2021); a study of psychologists found that self-compassion predicted fewer emotion regulation difficulties (Finlay-Jones et al., 2015), the latter mediating the relationships between low self-compassion and stress. Self-compassion also weakens the negative relationships between poor sleep and mental health problems (Kotera & Sheffield, 2020), stress and burnout (Abdollahi et al., 2021), exhaustion and burnout (Schabram & Heng, 2021), and burnout and depression (McCade et al., 2021) in nurses and allied health samples. Moreover, self-compassion weakens the relationships between burnout and lower quality of care (Dyrbye et al., 2005; Satele et al., 2014) in medical students and early career physicians, and reduces the extent to which difficult patient encounters, stress, and clinical complexity interfere with care (Dev et al., 2020), suggesting that it may be protective for patients as well as professionals.

Additionally, self-compassion in healthcare professionals can be potentially an *enhancing* trait predicting greater happiness (Benzo et al., 2017), resilience (Kemper et al., 2015, 2019, 2020), self-care (Mills et al., 2018; Miller et al., 2019), and confidence in providing compassionate care (Kemper et al., 2019, 2020). Further, qualitative work suggests that self-compassion could facilitate other-focused care by facilitating the management of the challenges, including working with suffering in an environment of austerity while unable help everyone (Patsiopoulos & Buchanan, 2011; Wiklund Gustin & Wagner, 2013; McPherson et al., 2016). Testament to how complex these relationships may be, however, at least one study has found that self-compassion was weakly (but significantly) associated with *lower* other-focused compassion (Mills et al., 2018).

Importantly, in terms of the ability of research to inform healthcare policy, while there are cross-sectional works with more robust analyses methods (above) and a longitudinal evidence base is emerging (Kemper et al., 2019, 2020; Schabram & Heng, 2021), the large part of available work remains near-exclusively correlational in design. The correlational studies point at associations between self-compassion and fewer mental health problems ($r = -0.68$) (Kotera & Sheffield, 2020), and less secondary traumatic stress ($r = -62$) (Delaney, 2018), fears of compassion ($r = -0.56$) (Scarlet et al., 2017), depression

(r = −0.40) (McCade et al., 2021), and shame (Kotera & Sheffield, 2020), as well as with greater mindfulness (r = 0.62), resilience (r = 0.37) (Olson et al., 2015), engagement ($0.24<r<0.33$) (Babenko & Guo, 2019; Babenko et al., 2019), autonomy (r = 0.24), competence (r = 0.24), relatedness (r = 0.30), and achievement (vs. avoidance) goals (r = −0.31) (Babenko & Oswald, 2019). However, these correlational studies do not shed light on the causal direction of the relationships between self-compassion and these outcomes; these effects may therefore be in the other direction (e.g., those with lower depression may find it easier to be compassionate towards themselves), bidirectional, or complicated by additional variables (Dodson & Heng, 2021). Hence, randomized controlled trial (RCT) data in healthcare samples that could help us to understand causality are urgently needed.

Self-Compassion Interventions

While most of the literature related to self-compassion in healthcare professionals remains observational, interventional studies are starting to emerge. Yet, as is common in early literatures, the studies that are available are small and involve weaker designs, while studies directly testing a possible effect on patient care (see examples of similar research - (Burns et al., 2003; Halbesleben & Rathert, 2008)) are completely absent at this point. Yet, even with these limitations in mind, it is worth noting that clinicians see the potential merits in self-compassion (Neff et al., 2020; Tierney et al., 2018), that self-compassion interventions are well-received by practitioners (Maslach & Leiter, 2005; Neff et al., 2020; Singh et al., 2018; Tierney et al., 2018), and that self-compassion can be taught (Scarlet et al., 2017). Below we describe six interventional studies conducted to date.

A first study that piloted a self-compassion intervention among postgraduate psychology trainees (n = 20, 89% female) used a pre-post design based on a 6-week *online* self-compassion intervention (Finlay-Jones et al., 2017). Analyses indicated improvements in self-compassion, emotional regulation, and happiness, and reductions in depression, anxiety, and stress (Cohen's d CI 95% [0.5–0.7]). However, attrition was high (46%) and non-completers were more stressed than completers. Another study tested the effects of an 8-week Mindful Self-compassion (MSC) programme – a face-to-face programme that combines the skills in mindfulness and self-compassion to build emotional resilience – in 13 female nurses (Delaney, 2018). Although conventional significance was not reported and attrition was 28%, analyses suggested the programme increased self-compassion and mindfulness, reduced secondary traumatic stress and burnout, and contributed to resilience and compassion satisfaction (d>0.82). In a more robust design, albeit one nonetheless relying on self-reported outcomes, Eriksson et al. (2018) conducted an RCT for a shorter face-to-face 6-week programme, contrasting a "mindfulness and compassion with self and others" training against waitlist control in a sample of 101 psychologists (97% female, 80.2% retention). Mixed linear models showed clear time by group effects, with greater improvements in self-compassion scores (d=0.94), mindfulness (d=0.60), and greater reductions in self-coldness (d=0.73), perceived stress (d=0.59) and burnout (d=0.44) in the experimental group.

Subsequently, the 6-week Self-Compassion for Healthcare Communities (SCHC) programme has been developed (Neff et al., 2020). This framework incorporates "compassion with equanimity" practices designed to help carers release feelings of control over outcomes when trying to alleviate suffering. Early data are promising. Pre-/post-intervention results in a mixed sample of physicians, nurses, and other health professionals (n = 58, 86% female), showed improvements in self-compassion, mindfulness, compassion, and reductions in depression and stress in the first intervention group (n = 25), and, additionally, burnout, secondary traumatic stress, exhaustion, and feelings of accomplishment in an additional intervention group (n = 23) with medium-to-large effect sizes (d CI 95% [0.52-0.80]).

More recently, a pre-/post- design evaluated the results of both 8-week MSC (n = 10) and

6-weeks SCHC programmes in a sample ($n = 20$) of certified nurse assistants working in nursing homes (97% female) (Bluth et al., 2021). While the interventions did not differ from one another, pooled analyses showed improvements in self-compassion post intervention as well as at 3- and 6-month follow ups, and reduction in stress and depression scores post intervention and at 3-months follow up. Interestingly, there was also evidence for a "maturational" type effect in which "recognition of personhood" was greater at 6-months but not at earlier measurement points. There were no effects on job satisfaction, likelihood of leaving one's job in a year time, or burnout (except for a decrease in depersonalisation post intervention that was not maintained). A follow up qualitative study with the participants revealed that self-compassion training was feasible and beneficial for the stressors experienced by certified nurse assistants (Lathren et al., 2021).

Finally, the most recent study evaluated SCHC as a one-day face-to-face intervention in paediatric nurses ($n = 22$, 95% females, 86% White) with 100% completion rate (Franco & Christie, 2021). As in other studies, analyses indicated positive changes in self-compassion, mindfulness, and compassion satisfaction, and decreases in burnout, anxiety, and stress that were maintained at 3-month follow up; additionally, increases in compassion and resilience were reported at the 3-month follow-up. Other analyses contrasted this group with a non-randomised waitlist control ($n = 26$). Time by group interactions were evident for self-compassion, mindfulness, compassion to others, compassion satisfaction, resiliency, burnout, and stress, with moderate to high effect sizes (Cohen's d CI 95% [0.59–0.87]). No interactions were evidenced for anxiety, depression, secondary traumatic stress, resiliency-activation, or job engagement.

Methodologically, while designs are improving over time (and with the notable exception of Eriksson et al., 2018), caution is needed in interpreting these data. To date, most self-compassion intervention studies lack active controls, employ non-randomized designs, only sporadically employ appropriate control over confounds, are limited by issues in retention and self-selection, and rely heavily on self-report. Small, self-selected samples increase the odds of Type 1 error obscuring clear estimates of interventional effectiveness (Landers & Behrend, 2015), particularly when changes are well within one standard deviation from the baseline. If studies are to impact policy and interventions of this kind are to get traction in healthcare workplaces, more robust evidence is needed, as is a focus on the sorts of outcomes that healthcare administrators are concerned about.

These caveats noted, early data are nonetheless broadly consistent with theoretical suggestions and evidence from other populations, and offer greater confidence in the directionality implied in cross-sectional work; self-compassion-based interventions appear to be of benefit in groups of healthcare workers. Studies are generally acceptable and pose limited risks. Across studies, evidence suggests improvements in self-compassion, mindfulness, stress, and burnout, with other effects in mental health (i.e., anxiety, depression, secondary traumatic stress), wellbeing (i.e., personal accomplishment, happiness), and in other focused care (i.e., compassion, compassion satisfaction, recognition of personhood) that are more mixed and evident in fewer studies. While mechanisms remain unknown, that these interventions increase self-compassion among healthcare professionals is increasingly clear, as is the likelihood that other outcomes are also positively influenced.

Commentary and Key Issues

In addition to continuing to improve design robustness and outcomes measurement, there are several questions regarding self-compassion in healthcare that remain unclear. Below, we outline the queries we feel will be of particular benefit. In developing this agenda, this summative section will first question and theorise *why* we should focus on self-compassion specifically (rather than focus on other constructs such as mindfulness, compassion, or self-care). Second, bearing in mind that healthcare workers are a professionally diverse group, we offer some commentary regarding the subgroups for whom self-compassion might be of greatest benefit. Three, we consider the outcomes of greatest relevance to healthcare professionals (and healthcare environments) for which self-compassion is most likely to be of benefit, given the proposed mechanisms. Finally, we look at evidence regarding one of the most poorly understood and contentious areas of compassion research in healthcare, the question of whether or not the development of self-compassion is able to positively influence other-related compassion.

Why Target Self-Compassion (Rather than Something Else) in Healthcare? One initial question that arises when considering the possible role (or benefits) of self-compassion for healthcare professionals regards the focus on self-compassion itself. Although it may seem counterintuitive in the context of such a volume, we start this discussion by considering the key question of why we should focus efforts to improve functioning in this key workforce via self-compassion rather than some other outcome. In addition to ongoing debates regarding construct validity and measurement (Muris, 2016; Muris & Otgaar, 2020; Neff, 2020, Sinclair et al., 2017), empirically separating self-compassion from other useful, analytically solid constructs, notably compassion, mindfulness and self-care is challenging and whether self-compassion is inherently positive is also unclear (Dodson & Heng, 2021). Further, there are fears that self-compassion might promote selfishness or the normalisation of undesirable behaviours (Morton & Postmes, 2011; Haslam & Bain, 2007), although no empirical studies have yet linked self-compassion to these negative outcomes.

In considering these complex issues somewhat further, it is worth recalling that self-compassion, compassion, mindfulness, and self-care often exhibit similar relationships with outcomes (e.g., burnout, depression, empathy) (Fernando et al., 2017; Barbosa et al., 2013; Schneider et al., 2014). Compounded by lack of head-to-head trials and limited insight into what works best for whom, such a pattern makes it unclear whether self-compassion *specifically* is where we should target our efforts. For example, self-compassion and mindfulness share many conceptual and operational features (Duarte et al., 2016; Duarte & Pinto-Gouveia, 2017; Sinclair et al., 2017), with both traits implying the capacity to keep perspective when experiencing difficulties and the replacement of reactivity with acceptance. Similarly, self-compassion and other-focused compassion converge in recognising common humanity in relation to suffering (Neff, 2003; Ling et al., 2021). Experimentally, prosocial behavior has shown to result in greater increases in psychological flourishing than does kindness related to self (Nelson et al., 2016). Likewise, self- and other-compassion have both shown to alleviate burnout, albeit in a different ways (Schabram & Heng, 2021).

In addition to similarities within the theoretical constructs, mindfulness and compassion-based interventions seem to arrive at similar results (Kılıç et al., 2020; Kirby et al., 2017; Ferrari et al., 2019; Wasson et al., 2020; Conversano et al., 2020). In healthcare professionals, a study by Sansó et al. (2019) evaluated mindfulness-based stress reduction (MBSR) and compassion cultivation training (CCT) interventions among 50 Spanish primary care physicians; both programmes increased self-compassion and professional quality of life (reduced compassion fatigue, increased compassion satisfaction, and lowered burnout). An interventional CCT study by Scarlet et al. (2017) among 62 healthcare workers showed improvements in both mindfulness and self-compassion as well as reduced fears of self-compassion and giving compassion to others. In a study by Orosa-Duarte et al. (2021) among medical students (n = 84), both a mindfulness-based mobile app intervention and an in-person mindfulness-based program (IMBP) increased self-compassion and mindfulness as compared to waitlist controls. Comparable effects for diverse interventions highlight the fact that interventions purported to increase self-compassion have such a broad effect on general affective functioning that specificity of effects is obscured; mediation analyses are urgently needed to understand specific effects of self-compassion as compared to other constructs (Sinclair et al., 2017; Kotera & Van Gordon, 2021).

Another related dilemma in seeking to understanding self-compassion's effect relates to the self-compassion scale itself. Most commonly thought to comprise six dimensions (self-kindness, common humanity, mindfulness, self-judgement, isolation, and over-identification), this complexity makes it difficult to tell which elements of self-compassion are impacting which outcomes. To give an example, a study by Duarte et al. (2016) showed that while lower self-judgement, isolation, and higher mindfulness all predicted lower burnout, only lower self-judgement predicted lower compassion fatigue and only higher mindfulness predicted compassion satisfaction. Other studies are similarly challenging to interpret. For instance, Montero-Marin et al. (2016) showed that only dimensions of the non-compassionate self predicted burnout, while Gracia-Gracia & Oliván-Blázquez (2017) showed that the burnout subscales of emotional exhaustion and depersonalization were affected by self-kindness/self-judgement and common humanity/isolation dimensions only. Although it has been argued that the different components of self-compassion work together synergistically (Neff, 2022), the question remains whether self-compassion as an undivided phenomenon is as effective as its separate parts.

Difficulties related to the self-compassion construct noted, the clinical effectiveness of self-compassion training seems clear (Neff, 2020). As such, healthcare professionals who are often self-critical, prone to taking unwarranted responsibility and, being the caregivers, may neglect taking care of themselves (Beydoun et al., 2019), should find self-compassion beneficial regardless of exactly how it is conceptualised or measured. Moreover, we posit that self-compassion may be uniquely suited to the challenges and personalities in healthcare and contribute to worker well-being beyond general self-care. Although self-care is often prominent among recommended burnout prevention strategies for healthcare professionals (Kravits et al., 2010; Skovholt & Trotter-Mathison, 2014) and self-compassion may predict self-care (Mills et al., 2018, 2019) or encourage self-care *as* self-compassionate action (Dodson & Heng, 2021; Gilbert et al., 2017; Gu et al., 2020; Schabram & Heng, 2021), self-compassion is a deeper internal practice. Self-care typically revolves around behaviours and practices that occur outside of the work situation (e.g., exercise, healthy diet, or socializing) (Kravits et al., 2010; Cook-Cottone, 2015; Hernandez, 2009), is often prescriptive, and is an additional activity – potentially yet another extra thing on one's to-do list. Conversely, self-compassion can be practiced "in the moment" (K. D. Neff et al., 2020), does not require a commitment, and arises from how one relates to oneself, which is ultimately soothing and supportive. Moreover, self-compassion entails full presence, while self-care does not. Thus, while healthcare professionals who engage in self-care may be better able to "tune-out" and be generally less prone to stress (and may thus be better resourced), self-compassion costs no time and can help to "tune-in", being deployed *at the time* of the emotionally difficult situations that are so frequent.

Finally, self-compassion might have a *special* role in helping to mitigate the effects of compulsive caregiving (Hermanto & Zuroff, 2016), "omnipotent saviour" dynamics (Breithaupt, 2019), and pathological altruism (Oakley et al., 2011). In some views, individuals have independent care-seeking and care-giving tendencies (Gilbert, 1989, 2000). While this might mean that self-compassion might not directly predict compassion towards others, a "balance" between these systems often appears lacking in the healthcare professions and workers are routinely more compassionate towards others than they are towards themselves (López et al., 2018). Empirically, while the marginalisation of one's own needs or interests can serve as a means of obtaining a sense of control or gaining rewards (Hermanto & Zuroff, 2016), it has been associated with emotional and interpersonal difficulties (Gerber et al., 2015; Hermanto & Zuroff, 2016) and may contribute to caring in ways that are not necessarily beneficial for the other (Oakley et al., 2011). Combined with the possibility that self-compassion can be stigmatised in the helping professions (Mills et al., 2015), self-care may thus be compromised (Hermanto & Zuroff, 2016). On the other hand, enhancing self-compassionate traits may facilitate both physician wellbeing, and a more balanced and sustainable patient-focused practice.

However, we should also be mindful of how little we know about how self-compassion operates in contexts where other-focused care is a key professional responsibility. Because self-compassion does not specifically call for selflessness (MacBeth & Gumley, 2012; Sinclair et al., 2016, 2017), concerns that self-compassion might "bleed" into professional styles that justify self-centredness or professional negligence have been expressed. In some views, seeing one's flaws as 'only human' has the potential to be used as a justificatory tool to mitigate individual responsibility (Haslam & Bain, 2007; Morton & Postmes, 2011; Zessin et al., 2015). Empirically, the evidence for such a "risk" remains weak (Mills et al., 2018) and positive changes in self-compassion resulting from interventions do not seem to lower one's ability to care. More broadly, this way of thinking about self-compassion is inconsistent with the way in which the construct is generally discussed. In most views, the emphasis in self-compassion training rests on seeing the commonalities between our own experiences and those of others (i.e., of *not* seeing the self or one's suffering as unique). It is this tendency that leads to accepting suffering with kindness and reduction in self-criticism (Neff & Seppälä, 2016). In the healthcare professions, many of which are highly prestigious, staffed by high achievers, and in which the risk of failure is high, finding humility by accepting one's own flaws, taking responsibility for one's mistakes rather than avoiding them or trying to cover them up may help to maintain self-worth and offset some of the risks posed by styles of functioning that continuously call for perfection. Thus, although some issues remain, self-compassion may offer a unique set of benefits for health professionals in helping to address challenges particular to this workforce.

Who is Self-Compassion Good For? A second key question in this area involves systematically identifying the subgroups of healthcare professionals for whom self-compassion may be of benefit. There are three reasons to suspect such variation may be uncovered. First, levels of self-compassion vary both within (Mills et al., 2018) and between healthcare occupations (Dev et al., 2020). Such data, coupled with evidence that self-compassion may moderate the associations between predictors and outcomes (Dev et al., 2018, 2020; K. D. Neff et al., 2020), creates a prima facie case for the suggestion that self-compassion may have different effects in different groups. Second, work in non-healthcare samples suggests that the links between self-compassion and outcomes vary as a function of gender, age, and personality (Baker & McNulty, 2011; Kelly et al., 2010). Finally, and most broadly, the specific challenges, systems, training, and personalities that predominate in different healthcare specialties create fertile grounds to expect that the effects of self-compassion might vary across professional groups (Mills et al., 2018; Gleichgerrcht & Decety, 2013; Dev et al., 2019; Fernando & Consedine, 2017; Claxton-Oldfield & Banzen, 2010).

To exemplify how such personal and contextual influences may affect how self-compassion relates to outcomes in the healthcare workforce, let us first consider the potential relevance of *gender*. As was noted above, a large proportion of the participants in studies of self-compassion in healthcare workforces are female, presumably partly as a result of gender imbalances in the workforce (e.g., nursing, allied health) (Murphy, 2019; Snyder & Green, 2008). However, males generally report greater self-compassion (Yarnell et al., 2015). Future research might thus profitably consider whether self-compassion interventions are comparably effective in occupations that are predominantly female (or male). Similarly, greater *age* (Zessin et al., 2015) and levels of *previous exposure to crisis* (Tierney et al., 2018) tend to predict greater self-compassion, possibly explaining Dev et al.'s (2020) finding that medical students have lower self-compassion than already established physicians and nurses. Such a pattern might imply that self-compassion interventions may be more effective in younger groups, possibly because there is more "room to move" for self-compassion development, because the personality system is more flexible, or, hypothetically, because resistance to self-compassion less well-developed.

More broadly, investigating the possibility that persons who are lower in self-compassion might derive greater benefit from the development of this way of relating to the self seems likely to pay dividends in healthcare groups known to vary (and lack) in self-compassion. For example, Neff et al. (2020) found that people who had lower self-compassion at baseline saw larger increases in self-compassion (and possibly associated benefits) than people who were already highly self-compassionate. Equally, analyses from a sample of 799 nurses showed that while self-compassion predicted fewer barriers to care (Dev et al., 2018), the link between burnout and burnout-related barriers to care was stronger among those with greater self-compassion. Similarly, while self-compassion predicted better outcomes in a sample of 1700 healthcare professionals and trainees (Dev et al., 2020), the link between stress and burnout was stronger in *more* self-compassionate nurses (but not doctors or medical students). While numerous explanatory possibilities exist, including a tradition of other-focused care in nursing (Boyle, 2011), lower power (Daiski, 2004), or dynamics in which trait self-compassion changes how burnout is *reported* (Dev et al., 2018), findings of this kind suggest self-compassion may not be comparably beneficial for all groups; other factors (e.g., self-compassion at baseline, gender, professional cultures) might moderate self-compassion's effects.

Such analyses are, however, uncommon thus far. Yet, moderation analyses are useful both in terms of (a) identifying groups that may differentially benefit and/or (b) identifying groups in whom other associations are differentially altered (e.g., Dev et al., 2020). For example, it may be that factors such as self-esteem, self-criticism, perceptions of agency, and self-stigma – all of which relate to professional status – are relevant. Studies in other samples are consistent with these possibilities. For example, a meta-analysis showed that higher self-esteem weakens the correlation between self-compassion and well-being (Zessin et al., 2015). Other studies show that health professionals with high trait self-criticism or control/agency perceptions may see self-compassion as more selfish (Robinson et al., 2016) or see it as stigmatising (Mills et al., 2015). Thus, while it is typically of benefit, there is also grounds for caution. Well-accepted interventions such as mindfulness have shown negative effects in some groups (Reynolds et al., 2017) and considering the possibility that self-compassion interventions may have different effects (including both null or "negative" effects) in some samples is important.

Contextual Factors that Facilitate or Hinder the Development and Practice of Self-Compassion Comparably, just as with personal factors, contextual factors could also be influential in moderating the effects of self-compassion within the healthcare domain. Professions, institutions, and even individual wards have their own particularities, training requirements, and unspoken rules (Boyle, 2011; Daiski, 2004; Reyes, 2012; Mills et al., 2015; Robinson et al., 2016) that structure and reinforce ways of professional conduct and intrapersonal functioning. Contexts exert a huge, albeit poorly understood influence on compassion in healthcare (Fernando et al., 2016) and there is reason to suspect that organisational and professional norms may also influence whether self-compassion training is of benefit and whether any benefits can be sustained. Different healthcare professions and organisations are quite distinct, varying in the extent to which self-reflection and self-care are normed and encouraged, how heavy workloads are managed and viewed, how organisational hierarchies and team communication dynamics operate. Such factors clearly have the capacity to interfere with self-compassion and its cultivation (Dodson & Heng, 2021; Egan et al., 2019).

Research on compassion in healthcare makes it clear that organisational values regarding efficiency and commerce (over compassion and humanity) are thought to interfere with patient care (Pavlova et al., 2021). Ideally, co-operative and caring colleagues and superiors serve as the positive role models, improving other-focused care (Dyrbye et al., 2005; Pavlova et al., 2021). Although evidence is absent, it seems likely that both explicit and implicit organisational values regarding the place of compassion in health will "echo" through the values regarding self-care and self-compassion.

Professional healthcare staff do not operate in cultural vacuums and, even supposing a healthcare practitioner could soothe their own inner critic following a mistake, error, or failure, external criticism might override or undermine any benefits. Killian (2008) has previously suggested that self-compassion was more likely to be engaged when clinicians were surrounded by compassionate co-workers and supervisors who were willing to offer emotional support and advice. Conversely, in organisational cultures manifesting limited compassion towards others (e.g., patients, colleagues) or where organisations are experienced as machines that "treat individuals as automated cogs carrying out rigidly prescribed activities" (de Zulueta, 2016, p. 6), both workers and patients risk being dehumanised. Operating in environments characterised by belittling, blame, or abuse almost certainly undermines self-compassion, as will environments in which self-compassion is seen as weak, lazy, or self-indulgent (Gilbert & Procter, 2006; Miron et al., 2014; Robinson et al., 2016).

An emphasis on the likely importance of organisational characteristics is consistent with the conclusions of a recent realist review by Sinclair et al. (2021) suggesting that other-focused compassion training is more likely to be successful when it is reflected in the infrastructure and values of organisations. The review suggests that organisational contexts can either hinder or facilitate compassion training and our sense is that similar perils and promise will confront self-compassion in healthcare. Put simply, it seems likely that self-compassion training will be more successful where self-compassion is reflected throughout organisations' broader vision.

Although almost nothing is known empirically, it seems reasonable to suspect that employers and line managers may fear that self-compassion interventions have negative effects on productivity. According to Dodson and Heng (2021) traditional organisational perspective sees an ideal employee as exclusively dedicated to their work (Dumas & Sanchez-Burks, 2015), effectively ignoring individual needs (Liedtka, 1989). Alternate approaches such as the human sustainability perspective (Spreitzer et al., 2012) are available. This view might suggest that incentivising self-compassion in employees might contribute to a more sustainable workforce, although whether self-compassion is seen as capable of improving work performance-related outcomes remains unclear (Dodson & Heng, 2021). What is clear is that empirical evidence regarding the effects of self-compassion on the sorts of outcomes that organisations value (e.g., absenteeism, turnover) is sorely needed here.

Perhaps most importantly, it must be noted that the deployment of interventions and the development of self-compassion in a professional workforce is inherently a systemic challenge (Sinclair et al., 2021). Attempting to change individual behaviour and process in the face of obstructive systemic dynamics is unlikely to be effective and multi-level research is clearly needed. Considering that self- and other focused compassion share the value of common humanity, it is possible that intervening at both organisational and personal level could provide greater benefits. Ultimately, however, the question of whether self-compassion is of equal benefit to healthcare professionals as a function of gender, specialty, and professional environments remains an empirical one.

To Which Outcomes is Self-compassion Primarily Relevant? Related to the issue of which groups of healthcare professionals and healthcare contexts are likely to see the greatest benefits from self-compassion are questions regarding which outcomes are most likely to be affected. To date, the outcomes being assessed in healthcare samples are similar to those studied in other samples (e.g., anxiety and depression, burnout, and well-being). Objectively, however, if we are to suggest that self-compassion is the "right tool for the job," evidence showing that it impacts the specific outcomes that are strategically important to patients, workers, and healthcare management are clearly needed.

One possibility proposed in recent research among healthcare workers is that self-compassion will likely be of greater benefit regarding outcomes impacted by the more *internal* elements of the stress response process. Having found that self-compassion was linked to lower burnout and patient/family barriers to care (but not clinical or environmental barriers exemplified by clinical complexity, interruptions, hectic environments, bureaucracy), Dev et al. (2018) suggested that "the capacity to be kind to the self is perhaps more relevant to barriers reflecting one's own conduct rather than factors in the workplace that are outside of the individual's control" (p.86). Although this report is concentrated on the barriers to other-focused care, the possibility that self-compassion might be less relevant in mitigating the impact of external factors is worth further investigation.

More broadly, the question of a possible association between self- and other-related compassion remains unclear. Neff and Pommier (2013) found a weak correlation between self- and empathic concern in community ($r = 0.15$) and practicing meditators samples ($r = 0.28$), but no correlation in students ($r = 0.01$). López et al. (2018) and Oslon and Kemper (2015) also found no correlation among community and residents' samples ($0.10<r<0.17$), while the study by Mills et al. (2018) showed a weak negative correlation ($r = -0.12$) in a sample of healthcare professionals. Thus, while self- and other-related compassion stem from the same evolved caregiving systems (Gilbert, 2000), they do not seem to be closely related at a cross-sectional level. For healthcare employers, the possibility of a negative link between self- and other-related compassion would likely be concerning. However, self-compassion (Boellinghaus et al., 2014; Lamothe et al., 2016; López et al., 2018; Raab, 2014; Sinclair et al., 2017; Lim & DeSteno, 2019) and mindfulness-based interventions with self-compassion components tend to improve prosocial behaviours (Bazarko et al., 2013) and promote compassion or motivation to help in healthcare (Barbosa et al., 2013; Fernando et al., 2017; Schneider et al., 2014). To this point then, the evidence for a conceptually important link between self- and other-focused compassion is tentatively favourable, but as yet, poorly understood.

Equally, it is important when discussing outcomes in relation to the place of self-compassion in healthcare that we do not neglect to consider the fact that self-compassionate behaviour occurs within systems; patient, practitioner, and system outcomes are all simultaneously important. One intriguing possibility, for example, is that self-compassion enhances patient care *indirectly* through a process in which adaptive ways of responding to the self are modelled by carers. Prior work suggests that patients' perceptions of physician well-being and demeanour influences how advice is received (Fraser et al., 2013). Similarly, because senior physicians' ways of coping with stress are likely transmitted to future cohorts of professionals via role modelling, the development of self-compassion may have "spill over" effects in subsequent generations (Dyrbye et al., 2005; Satele et al., 2014). Other studies show that individual self-compassionate practice improves collegiality within organisations, with conflicts appearing less frequently and compromises more common (Yarnell & Neff, 2013), as well as resulting in lower odds of emotional exhaustion after negative interactions (Anjum et al., 2020). Hence, unless its actively frowned upon (Dodson & Heng, 2021), self-compassion can contribute to outcomes indirectly by improving self-management and, as a consequence, organisational working environments. More broadly, it seems clear that "compassion in health is a systemic problem that requires systemic solutions" (Dev et al., 2019, p. 2), and we must be equally systemic in our thinking about the outcomes self-compassion may impact in this context.

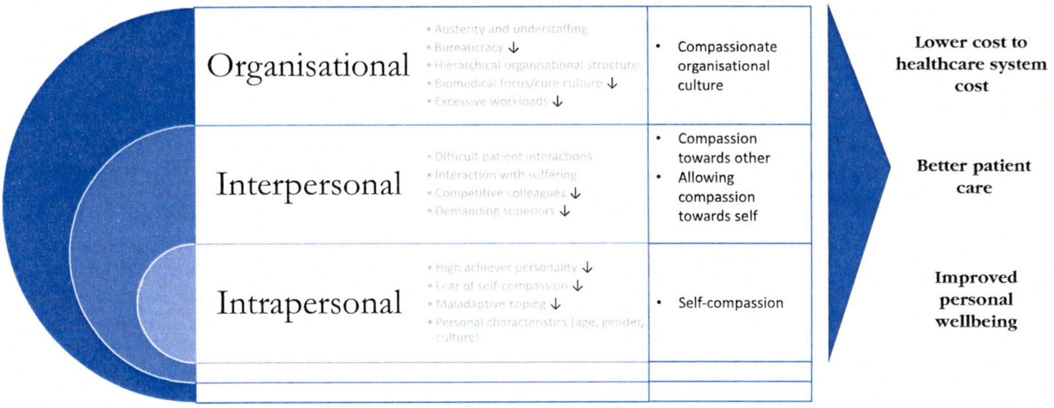

Fig. 14.2 Systemic framework to organisation pathways linking self-compassion to outcomes in healthcare

Finally, there is a clear need for research that considers the sorts of healthcare-specific outcomes that influence the values and policy of governments and healthcare organisations. Demonstrating that the development of self-compassion in the professional healthcare workforce impacts the sorts of outcomes that drive policy (including, most obviously, cost) is critical to ensure "buy in" within bureaucracies and policy makers (Fig. 14.2). Securing data demonstrating that self-compassion does not pose a threat to motivation or performance (Robinson et al., 2016) as well as assessing links to patient perceptions of service quality, absenteeism, complaints, and staff turnover are key agendas for self-compassion research in this applied context. Put simply, we need data to assess whether self-compassion benefits the outcomes this key stakeholder group values because it is these concerns that determine how finite healthcare resources are prioritised and allocated.

Concluding Remarks

Although compassion towards patients is expected, morally mandated, and required in professional codes of practice, healthcare professionals routinely neglect treating themselves with compassion and care. While this accomplished professional group typically appear competent and robust, nearly one-third of some groups of healthcare professional suffer from burnout and similar issues are seen in mental and physical health. Characteristic perfectionism and high-achievement normatively lead to self-blame and self-criticism, compounding the effects of environments in which resources are stretched and disappointments common; instead of recognising the common humanity of negative experiences, doctors proceed to treat burnout as a badge of honour (Rowe & Kidd, 2009). While our review suggests self-compassion research among healthcare workers remains in its infancy, self-compassion appears acceptable to this group, well suited to many of the challenges facing healthcare professionals, and evidence that it may have a host of benefits for practitioners, patients, and healthcare systems is growing.

The promise of self-compassion in healthcare noted, yet serious challenges remain. Most obviously, its potential is limited by the absence of high-quality data demonstrating the efficacy of this specific class of intervention in managing the challenges healthcare professionals face. Similarly, besides a small possibility that self-compassion will differentially benefit healthcare practitioners earlier in their career, we know little regarding for whom self-compassion is of greatest assistance and/or the sorts of outcomes that are affected. These are key agendas for future work. Perhaps most broadly, however, we must be careful in assuming that self-compassion can offer anything approaching a universal panacea for the ills of modern healthcare. While developing the capacity to be kind to the self during times of difficulty will likely benefit practitioners (although not all equally) and may have positive spill over effects for patients and systems, it can only ever be a part solution. There remain serious financial, logistical, and ideological barriers to compassionate care, with respect to both patients and carers. Such barriers are yet to be investigated; the challenge of self-compassion and care in health is a systemic problem requiring systemic solutions. Our hope is that this chapter contribute in some way towards this agenda.

References

Abdollahi, A., Taheri, A., & Allen, K. A. (2021). Perceived stress, self-compassion and job burnout in nurses: The moderating role of self-compassion. *Journal of Research in Nursing, 26*(3), 182–191. https://doi.org/10.1177/1744987120970612

Alharthy, N., Alrajeh, O. A., Almutairi, M., & Alhajri, A. (2017). Assessment of anxiety level of emergency health-care workers by generalized anxiety disorder-7 tool. *International Journal of Applied and Basic Medical Research, 7*(3), 150. https://doi.org/10.4103/2229-516X.212963

Alrawashdeh, H. M., Ala'a, B., Alzawahreh, M. K., Al-Tamimi, A., Elkholy, M., Al Sarireh, F., et al. (2021). Occupational burnout and job satisfaction among physicians in times of COVID-19 crisis: A convergent parallel mixed-method study. *BMC Public Health, 21*(1), 1–18. https://doi.org/10.1186/s12889-021-10897-4

Anjum, M. A., Liang, D., Durrani, D. K., & Parvez, A. (2020). Workplace mistreatment and emotional exhaustion: The interaction effects of self-compassion. *Current Psychology*, 1–12. https://doi.org/10.1007/s12144-020-00673-9

Babenko, O., & Guo, Q. (2019). Measuring self-compassion in medical students: Factorial validation of the self-compassion scale–short form (SCS-SF). *Academic Psychiatry, 43*(6), 590–594. https://doi.org/10.1007/s40596-019-01095-x

Babenko, O., & Oswald, A. (2019). The roles of basic psychological needs, self-compassion, and self-efficacy in the development of mastery goals among medical students. *Medical Teacher, 41*(4), 478–481. https://doi.org/10.1080/0142159X.2018.1442564

Babenko, O., Mosewich, A. D., Lee, A., & Koppula, S. (2019). Association of physicians' self-compassion with work engagement, exhaustion, and professional life satisfaction. *Medical Sciences, 7*(2). https://doi.org/10.3390/medsci7020029

Baker, L. R., & McNulty, J. K. (2011). Self-compassion and relationship maintenance: The moderating roles of conscientiousness and gender. *Journal of Personality and Social Psychology, 100*(5), 853–873. https://doi.org/10.1037/a0021884

Barbosa, P., Raymond, G., Zlotnick, C., Wilk, J., Toomey, R., 3rd, & Mitchell, J., 3rd. (2013). Mindfulness-based stress reduction training is associated with greater empathy and reduced anxiety for graduate healthcare students. *Education for Health, 26*(1), 9–14. https://doi.org/10.4103/1357-6283.112794

Barnard, L. K., & Curry, J. F. (2011). Self-compassion: Conceptualizations, correlates, & interventions. *Review of General Psychology, 15*(4), 289–303. https://doi.org/10.1037/a0025754

Bazarko, D., Cate, R. A., Azocar, F., & Kreitzer, M. J. (2013). The impact of an innovative mindfulness-based stress reduction program on the health and well-being of nurses employed in a corporate setting. *Journal of Workplace Behavioral Health, 28*(2), 107–133. https://doi.org/10.1080/15555240.2013.779518

Benzo, R. P., Kirsch, J. L., & Nelson, C. (2017). Compassion, mindfulness, and the happiness of healthcare workers. *Explore, 13*(3), 201–206. https://doi.org/10.1016/j.explore.2017.02.001

Beydoun, J., Nasrallah, L., Sabrah, T., & Caboral-Stevens, M. (2019). Towards a definition of caregiver fatigue: A concept analysis. *Advances in Nursing Science, 42*(4), 297–306. https://doi.org/10.1097/ANS.0000000000000262

Beyond Blue. (2013). *National mental health survey of doctors and medical students: Beyond Blue*. Beyond Blue. https://www.beyondblue.org.au/docs/default-source/research-project-files/bl1132-report%2D%2D-nmhdmss-full-report_web

Bluth, K., Lathren, C., Silbersack Hickey, J. V., Zimmerman, S., Wretman, C. J., & Sloane, P. D. (2021). Self-compassion training for certified nurse assistants in nursing homes. *Journal of the American Geriatrics Society, 69*(7), 1896–1905. https://doi.org/10.1111/jgs.17155

Boellinghaus, I., Jones, F. W., & Hutton, J. (2014). The role of mindfulness and loving-kindness meditation in cultivating self-compassion and other-focused concern in health care professionals. *Mindfulness, 5*(2), 129–138. https://doi.org/10.1007/s12671-012-0158-6

Boyle, D. A. (2011). Countering compassion fatigue: A requisite nursing agenda. *The Online Journal of Issues in Nursing, 16*(1). https://doi.org/10.3912/OJIN.Vol16No01Man02

Breithaupt, F. (2019). False empathy, filtered empathy. In F. Breithaupt (Ed.), *The dark sides of empathy* (pp. 131–160). Cornell University Press.

Brewin, C. B., & Firth-Cozens, J. (1997). Dependency and self-criticism as predictors of depression in young doctors. *Journal of Occupational Health Psychology, 2*(3), 242–246. https://doi.org/10.1037/1076-8998.2.3.242

Bria, M., Baban, A., & Dumitrascu, D. L. (2012). Systematic review of burnout risk factors among European healthcare professionals. *Cognition, Brain, Behavior: An Interdisciplinary Journal, 16*(3), 423–452.

Burns, R., Nichols, L. O., Martindale-Adams, J., Graney, M. J., & Lummus, A. (2003). Primary care interventions for dementia caregivers: 2-year outcomes from the REACH study. *The Gerontologist, 43*(4), 547–555. https://doi.org/10.1093/geront/43.4.547

Burridge, L. H., Winch, S., Kay, M., & Henderson, A. (2017). Building compassion literacy: Enabling care in primary health care nursing. *Collegian, 24*(1), 85–91. https://doi.org/10.1016/j.colegn.2015.09.004

Claxton-Oldfield, S. & Banzen, Y. (2010). Personality characteristics of hospice palliative care volunteers: The "Big Five" and empathy. *American Journal of Hospice and Palliative Medicine®, 27*(6), 407–412. https://doi.org/10.1177/1049909110364017.

Conversano, C., Ciacchini, R., Orrù, G., Di Giuseppe, M., Gemignani, A., & Poli, A. (2020). Mindfulness, compassion, and self-compassion among health care professionals: What's new? *Frontiers in Psychology, 11*, 1683. https://doi.org/10.3389/fpsyg.2020.01683

Cook-Cottone, C. P. (2015). Incorporating positive body image into the treatment of eating disorders: A model for attunement and mindful self-care. *Body Image, 14*, 158–167. https://doi.org/10.1016/j.bodyim.2015.03.004

Daiski, I. (2004). Changing nurses' dis-empowering relationship patterns. *Journal of Advanced Nursing, 48*(1), 43–50. https://doi.org/10.1111/j.1365-2648.2004.03167.x

De Zulueta, P. C. (2015). Suffering, compassion and 'doing good medical ethics'. *Journal of Medical Ethics, 41*(1), 87–90. https://doi.org/10.1136/medethics-2014-102355

de Zulueta, P. C. (2016). Developing compassionate leadership in health care: An integrative review. *Journal of Healthcare Leadership, 8*, 1–10. https://doi.org/10.2147/JHL.S93724

Delaney, M. C. (2018). Caring for the caregivers: Evaluation of the effect of an eight-week pilot mindful self-compassion (MSC) training program on nurses' compassion fatigue and resilience. *PloS one, 13*(11), e0207261. https://doi.org/10.1371/journal.pone.0207261

Dev, V., Fernando, A. T. R., Lim, A. G., & Consedine, N. S. (2018). Does self-compassion mitigate the relationship between burnout and barriers to compassion? A cross-sectional quantitative study of 799 nurses. *International Journal of Nursing Studies, 81*, 81–88. https://doi.org/10.1016/j.ijnurstu.2018.02.003

Dev, V., Fernando, A. T., Kirby, J. N., & Consedine, N. S. (2019). Variation in the barriers to compassion across healthcare training and disciplines: A cross-sectional study of doctors, nurses, and medical students. *International Journal of Nursing Studies, 90*, 1–10. https://doi.org/10.1016/j.ijnurstu.2018.09.015

Dev, V., Fernando, A. T., & Consedine, N. S. (2020). Self-compassion as a stress moderator: A cross-sectional study of 1700 doctors, nurses, and medical students. *Mindfulness, 11*(5), 1170–1181. https://doi.org/10.1007/s12671-020-01325-6

Dobkin, P. L. (2011). Mindfulness and whole person care. In T. A. Hutchinson (Ed.), *Whole person care: A new paradigm for the 21st century* (pp. 69–82). Springer. https://doi.org/10.1007/978-1-4419-9440-0_7

Dodson, S. J., & Heng, Y. T. (2021). Self-compassion in organizations: A review and future research agenda. *Journal of Organizational Behavior, 43*(2), 168–196. https://doi.org/10.1002/job.2556

Dominguez-Gomez, E., & Rutledge, D. N. (2009). Prevalence of secondary traumatic stress among emergency nurses. *Journal of Emergency Nursing, 35*(3), 199–204. https://doi.org/10.1016/j.jen.2008.05.003

Duarte, J., & Pinto-Gouveia, J. (2017). Empathy and feelings of guilt experienced by nurses: A cross-sectional study of their role in burnout and compassion fatigue symptoms. *Applied Nursing Research, 35*, 42–47. https://doi.org/10.1016/j.apnr.2017.02.006

Duarte, J., Pinto-Gouveia, J., & Cruz, B. (2016). Relationships between nurses' empathy, self-compassion and dimensions of professional quality of life: A cross-sectional study. *International Journal of Nursing Studies, 60*, 1–11. https://doi.org/10.1016/j.ijnurstu.2016.02.015

Dumas, T. L., & Sanchez-Burks, J. (2015). The professional, the personal, and the ideal worker: Pressures and objectives shaping the boundary between life domains. *The Academy of Management Annals, 9*(1), 803–843. https://doi.org/10.1080/19416520.2015.1028810

Dyrbye, L. N., Thomas, M. R., & Shanafelt, T. D. (2005). Medical student distress: Causes, consequences, and proposed solutions. *Mayo Clinic Proceedings, 80*(12), 1613–1622. https://doi.org/10.4065/80.12.1613

Egan, H., Keyte, R., McGowan, K., Peters, L., Lemon, N., Parsons, S., et al. (2019). 'You before me': A qualitative study of health care professionals' and students' understanding and experiences of compassion in the workplace, self-compassion, self-care and health behaviours. *Health Professions Education, 5*(3), 225–236. https://doi.org/10.1016/j.hpe.2018.07.002

Enns, M. W., Cox, B. J., Sareen, J., & Freeman, P. (2001). Adaptive and maladaptive perfectionism in medical students: A longitudinal investigation. *Medical Education, 35*(11), 1034–1042. https://doi.org/10.1111/j.1365-2923.2001.01044.x

Eriksson, T., Germundsjö, L., Åström, E., & Rönnlund, M. (2018). Mindful self-compassion training reduces stress and burnout symptoms among practicing psychologists: A randomized controlled trial of a brief web-based intervention. *Frontiers in Psychology, 9*, 2340. https://doi.org/10.3389/fpsyg.2018.02340

Fernando, A. T., & Consedine, N. S. (2017). Barriers to medical compassion as a function of experience and specialization: Psychiatry, pediatrics, internal medicine, surgery, and general practice. *Journal of Pain and Symptom Management, 53*(6), 979–987. https://doi.org/10.1016/j.jpainsymman.2016.12.324

Fernando, A. T., Arroll, B., & Consedine, N. S. (2016). Enhancing compassion in general practice: It's not all about the doctor. *British Journal of General*

Practice, 66(648), 340–341. https://doi.org/10.3399/bjgp16X68574
Fernando, A. T., Skinner, K., & Consedine, N. S. (2017). Increasing compassion in medical decision-making: Can a brief mindfulness intervention help? *Mindfulness, 8*(2), 276–285. https://doi.org/10.1007/s12671-016-0598-5
Ferrari, M., Hunt, C., Harrysunker, A., Abbott, M. J., Beath, A. P., & Einstein, D. A. (2019). Self-compassion interventions and psychosocial outcomes: A meta-analysis of RCTs. *Mindfulness, 10*(8), 1455–1473. https://doi.org/10.1007/s12671-019-01134-6
Finlay-Jones, A. L., Rees, C. S., & Kane, R. T. (2015). Self-compassion, emotion regulation and stress among Australian psychologists: Testing an emotion regulation model of self-compassion using structural equation modeling. *PloS one, 10*(7), e0133481. https://doi.org/10.1371/journal.pone.0133481
Finlay-Jones, A., Kane, R., & Rees, C. (2017). Self-compassion online: A pilot study of an internet-based self-compassion cultivation program for psychology trainees. *Journal of Clinical Psychology, 73*(7), 797–816. https://doi.org/10.1002/jclp.22375
Franco, P. L., & Christie, L. M. (2021). Effectiveness of a one day self-compassion training for pediatric nurses' resilience. *Journal of Pediatric Nursing, 61*, 109–114. https://doi.org/10.1016/j.pedn.2021.03.020
Fraser, S., Leveritt, M., & Ball, L. (2013). Patients' perceptions of their general practitioner's health and weight influences their perceptions of nutrition and exercise advice received. *Journal of Primary Health Care, 5*(4), 301–307.
Gaufberg, E. H., Batalden, M., Sands, R., & Bell, S. K. (2010). The hidden curriculum: What can we learn from third-year medical student narrative reflections? *Academic Medicine, 85*(11), 1709–1716. https://doi.org/10.1097/ACM.0b013e3181f57899
Gerber, Z., Tolmacz, R., & Doron, Y. (2015). Self-compassion and forms of concern for others. *Personality and Individual Differences, 86*, 394–400. https://doi.org/10.1016/j.paid.2015.06.052
Gilbert, P. (1989). *Human nature and suffering*. Psychology Press.
Gilbert, P. (2000). Social mentalities: internal social conflict and the role of inner warmth and compassion in cognitive therapy. In P. Gilbert & K. G. Bailey (Eds.), *Genes on the couch: explorations in evolutionary psychotherapy* (pp. 118–150). Brunner-Routledge.
Gilbert, P., & Procter, S. (2006). Compassionate mind training for people with high shame and self-criticism: Overview and pilot study of a group therapy approach. *Clinical Psychology & Psychotherapy, 13*(6), 353–379. https://doi.org/10.1002/cpp.507
Gilbert, P., Catarino, F., Duarte, C., Matos, M., Kolts, R., Stubbs, J., et al. (2017). The development of compassionate engagement and action scales for self and others. *Journal of Compassionate Health Care, 4*(1), 1–24. https://doi.org/10.1186/s40639-017-0033-3
Gleichgerrcht, E., & Decety, J. (2013). Empathy in clinical practice: How individual dispositions, gender, and experience moderate empathic concern, burnout, and emotional distress in physicians. *PLoS One, 8*(4). https://doi.org/10.1371/journal.pone.0061526
Gracia-Gracia, P., & Oliván-Blázquez, B. (2017). Burnout and mindfulness self-compassion in nurses of intensive care units: Cross-sectional study. *Holistic Nursing Practice, 31*(4), 225–233. https://doi.org/10.1097/hnp.0000000000000215
Gray, J. R. (1999). A bias toward short-term thinking in threat-related negative emotional states. *Personality and Social Psychology Bulletin, 25*(1), 65–75. https://doi.org/10.1177/0146167299025001006
Gu, J., Baer, R., Cavanagh, K., Kuyken, W., & Strauss, C. (2020). Development and psychometric properties of the Sussex-Oxford compassion scales (SOCS). *Assessment, 27*(1), 3–20. https://doi.org/10.1177/1073191119860911
Haider, S. I., Riaz, Q., & Gill, R. C. (2020). Empathy in clinical practice: A qualitative study of early medical practitioners and educators. *JPMA. The Journal of the Pakistan Medical Association, 70*(1), 116. https://doi.org/10.5455/jpma.14408
Halbesleben, J. R., & Rathert, C. (2008). Linking physician burnout and patient outcomes: Exploring the dyadic relationship between physicians and patients. *Health Care Management Review, 33*(1), 29–39. https://doi.org/10.1097/01.HMR.0000304493.87898.72
Haslam, N., & Bain, P. (2007). Humanizing the self: Moderators of the attribution of lesser humanness to others. *Personality and Social Psychology Bulletin, 33*(1), 57–68. https://doi.org/10.1177/0146167206293191
Heinemann, L. V., & Heinemann, T. (2017). Burnout research: Emergence and scientific investigation of a contested diagnosis. *SAGE Open, 7*(1), 2158244017697154. https://doi.org/10.1177/2158244017697154
Hermanto, N., & Zuroff, D. C. (2016). The social mentality theory of self-compassion and self-reassurance: The interactive effect of care-seeking and caregiving. *The Journal of Social Psychology, 156*(5), 523–535. https://doi.org/10.1080/00224545.2015.1135779
Hernandez, G. (2009). The heART of self-CARING: A journey to becoming an optimal healing presence to ourselves and our patients. *Creative Nursing, 15*(3), 129–133. https://doi.org/10.1891/1078-4535.15.3.129
Hofmann, S. G., Grossman, P., & Hinton, D. E. (2011). Loving-kindness and compassion meditation: Potential for psychological interventions. *Clinical Psychology Review, 31*(7), 1126–1132. https://doi.org/10.1016/j.cpr.2011.07.003
Innstrand, S. T., Langballe, E. M., Falkum, E., & Aasland, O. G. (2011). Exploring within-and between-gender differences in burnout: 8 different occupational groups. *International Archives of Occupational and Environmental Health, 84*(7), 813–824. https://doi.org/10.1007/s00420-011-0667-y
Jacobs, R., Mannion, R., Davies, H. T., Harrison, S., Konteh, F., & Walshe, K. (2013). The relationship between organizational culture and performance in

acute hospitals. *Social Science & Medicine, 76*, 115–125. https://doi.org/10.1016/j.socscimed.2012.10.014

Kelly, A. C., Zuroff, D. C., Foa, C. L., & Gilbert, P. (2010). Who benefits from training in self-compassionate self-regulation? A study of smoking reduction. *Journal of Social and Clinical Psychology, 29*(7), 727–755. https://doi.org/10.1521/jscp.2010.29.7.727

Kemper, K. J., Mo, X., & Khayat, R. (2015). Are mindfulness and self-compassion associated with sleep and resilience in health professionals? *Journal of Alternative and Complementary Medicine, 21*(8), 496–503. https://doi.org/10.1089/acm.2014.0281

Kemper, K. J., McClafferty, H., Wilson, P. M., Serwint, J. R., Batra, M., Mahan, J. D., et al. (2019). Do mindfulness and self-compassion predict burnout in pediatric residents? *Academic Medicine, 94*(6), 876–884. https://doi.org/10.1097/acm.0000000000002546

Kemper, K. J., Schwartz, A., Wilson, P. M., Mahan, J. D., Schubert, C. J., Staples, B. B., et al. (2020). Burnout in pediatric residents: Three years of national survey datare mindfulness and self-compassion associated with sleep and resilience in health professionals? *Pediatrics, 145*(1), e20191030. https://doi.org/10.1542/peds.2019-1030

Kılıç, A., Hudson, J., McCracken, L. M., Ruparelia, R., Fawson, S., & Hughes, L. D. (2020). A systematic review of the effectiveness of self-compassion related interventions for individuals with chronic physical health conditions. *Behavior Therapy, 52*(3), 602–635. https://doi.org/10.1016/j.beth.2020.08.001

Killian, K. D. (2008). Helping till it hurts? A multimethod study of compassion fatigue, burnout, and self-care in clinicians working with trauma survivors. *Traumatology, 14*(2), 32–44. https://doi.org/10.1177/1534765608319083

Kirby, J. N., Tellegen, C. L., & Steindl, S. R. (2017). A meta-analysis of compassion-based interventions: Current state of knowledge and future directions. *Behavior Therapy, 48*(6), 778–792. https://doi.org/10.1016/j.beth.2017.06.003

Kotera, Y., & Sheffield, D. (2020). Revisiting the self-compassion scale-short form: Stronger associations with self-inadequacy and resilience. *SN Comprehensive Clinical Medicine, 2*, 761–769. https://doi.org/10.1007/s42399-020-00309-w

Kotera, Y., & Van Gordon, W. (2021). Effects of self-compassion training on work-related well-being: A systematic review. *Frontiers in Psychology, 12*, 1142. https://doi.org/10.3389/fpsyg.2021.630798

Kravits, K., McAllister-Black, R., Grant, M., & Kirk, C. (2010). Self-care strategies for nurses: A psycho-educational intervention for stress reduction and the prevention of burnout. *Applied Nursing Research, 23*(3), 130–138. https://doi.org/10.1016/j.apnr.2008.08.002

Lamothe, M., Rondeau, É., Malboeuf-Hurtubise, C., Duval, M., & Sultan, S. (2016). Outcomes of MBSR or MBSR-based interventions in health care providers: A systematic review with a focus on empathy and emotional competencies. *Complementary Therapies in Medicine, 24*, 19–28. https://doi.org/10.1016/j.ctim.2015.11.001

Landers, R. N., & Behrend, T. S. (2015). An inconvenient truth: Arbitrary distinctions between organizational, Mechanical Turk, and other convenience samples. *Industrial and Organizational Psychology, 8*(2), 142–164. https://doi.org/10.1017/iop.2015.13

Lathren, C., Sheffield-Abdullah, K., Sloane, P. D., Bluth, K., Hickey, J. V., Wretman, C. J., et al. (2021). Certified nursing assistants' experiences with self-compassion training in the nursing home setting. *Geriatric Nursing, 42*(6), 1341–1348. https://doi.org/10.1016/j.gerinurse.2021.08.013

Liedtka, J. M. (1989). Value congruence: The interplay of individual and organizational value systems. *Journal of Business Ethics, 8*(10), 805–815. https://doi.org/10.1007/BF00383780

Lim, D., & DeSteno, D. (2019). Past adversity protects against the numeracy bias in compassion. *Emotion, 20*(8), 1344–1356. https://doi.org/10.1037/emo0000655

Linden, D. V. D., Keijsers, G. P., Eling, P., & Schaijk, R. V. (2005). Work stress and attentional difficulties: An initial study on burnout and cognitive failures. *Work & Stress, 19*(1), 23–36. https://doi.org/10.1080/02678370500065275

Lindström, U. H., Hamberg, K., & Johansson, E. E. (2011). Medical students' experiences of shame in professional enculturation. *Medical Education, 45*(10), 1016–1024. https://doi.org/10.1111/j.1365-2923.2011.04022.x

Ling, D., Petrakis, M., & Olver, J. H. (2021). The use of common humanity scenarios to promote compassion in healthcare workers. *Australian Social Work, 74*(1), 110–121. https://doi.org/10.1080/0312407X.2020.1808031

López, A., Sanderman, R., Ranchor, A. V., & Schroevers, M. J. (2018). Compassion for others and self-compassion: Levels, correlates, and relationship with psychological well-being. *Mindfulness, 9*(1), 325–331. https://doi.org/10.1007/s12671-017-0777-z

MacBeth, A., & Gumley, A. (2012). Exploring compassion: A meta-analysis of the association between self-compassion and psychopathology. *Clinical Psychology Review, 32*(6), 545–552. https://doi.org/10.1016/j.cpr.2012.06.003

Mader, E. M., Roseamelia, C., & Morley, C. P. (2014). The temporal decline of idealism in two cohorts of medical students at one institution. *BMC Medical Education, 14*(1), 58. https://doi.org/10.1186/1472-6920-14-58

Mahoney, S., Sladek, R. M., & Neild, T. (2016). A longitudinal study of empathy in pre-clinical and clinical medical students and clinical supervisors. *BMC Medical Education, 16*(1), 270. https://doi.org/10.1186/s12909-016-0777-z

Maslach, C., & Leiter, M. P. (2005). Reversing burnout. *Stanford Social Innovation Review*, 43–49.

Mata, D. A., Ramos, M. A., Bansal, N., Khan, R., Guille, C., Di Angelantonio, E., et al. (2015). Prevalence of depression and depressive symptoms among resident physicians: A systematic review and meta-analysis.

JAMA, 314(22), 2373–2383. https://doi.org/10.1001/jama.2015.15845

McCade, D., Frewen, A., & Fassnacht, D. B. (2021). Burnout and depression in Australian psychologists: The moderating role of self-compassion. *Australian Psychologist, 56*(2), 111–122. https://doi.org/10.1080/00050067.2021.1890979

McPherson, S., Hiskey, S., & Alderson, Z. (2016). Distress in working on dementia wards–a threat to compassionate care: A grounded theory study. *International Journal of Nursing Studies, 53*, 95–104. https://doi.org/10.1016/j.ijnurstu.2015.08.013

Miller, J., Lee, J., Niu, C., Grise-Owens, E., & Bode, M. (2019). Self-compassion as a predictor of self-care: A study of social work clinicians. *Clinical Social Work Journal, 47*(4), 321–331. https://doi.org/10.1007/s10615-019-00710-6

Mills, J., & Chapman, M. (2016). Compassion and self-compassion in medicine. *Australasian Medical Journal, 9*(5), 87–91.

Mills, J., Wand, T., & Fraser, J. (2015). On self-compassion and self-care in nursing: Selfish or essential for compassionate care? *International Journal of Nursing Studies, 52*(4), 791–793. https://doi.org/10.1016/j.ijnurstu.2014.10.009

Mills, J., Wand, T., & Fraser, J. A. (2017). Palliative care professionals' care and compassion for self and others: A narrative review. *International Journal of Palliative Nursing, 23*(5), 219–229. https://doi.org/10.12968/ijpn.2017.23.5.219

Mills, J., Wand, T., & Fraser, J. A. (2018). Examining self-care, self-compassion and compassion for others: A cross-sectional survey of palliative care nurses and doctors. *International Journal of Palliative Nursing, 24*(1), 4–11. https://doi.org/10.12968/ijpn.2018.24.1.4

Miron, L. R., Orcutt, H. K., Hannan, S. M., & Thompson, K. L. (2014). Childhood abuse and problematic alcohol use in college females: The role of self-compassion. *Self and Identity, 13*(3), 364–379. https://doi.org/10.1080/15298868.2013.836131

Montero-Marin, J., Zubiaga, F., Cereceda, M., Piva Demarzo, M. M., Trenc, P., & Garcia-Campayo, J. (2016). Burnout subtypes and absence of self-compassion in primary healthcare professionals: A cross-sectional study. *PLoS One, 11*(6), e0157499. https://doi.org/10.1371/journal.pone.0157499

Morton, T. A., & Postmes, T. (2011). Moral duty or moral defence? The effects of perceiving shared humanity with the victims of ingroup perpetrated harm. *European Journal of Social Psychology, 41*(1), 127–134. https://doi.org/10.1002/ejsp.751

Muris, P. (2016). A protective factor against mental health problems in youths? A critical note on the assessment of self-compassion. *Journal of Child and Family Studies, 25*, 1461–1465. https://doi.org/10.1007/s10826-015-0315-3

Muris, P., & Otgaar, H. (2020). The process of science: A critical evaluation of more than 15 years of research on self-compassion with the self-compassion scale. *Mindfulness, 11*(6), 1469–1482. https://doi.org/10.1007/s12671-020-01363-0

Murphy, B. (2019). *These medical specialties have the biggest gender imbalances*. American Medical Association. https://www.ama-assn.org/residents--students/specialty-profiles/these-medical-specialties--have-biggest-gender-imbalances.

Neff, K. D. (2003). The development and validation of a scale to measure self-compassion. *Self and Identity, 2*(3), 223–250. https://doi.org/10.1080/15298860309027

Neff, K. D. (2011). *Self-compassion*. Hachette.

Neff, K. D. (2020). Commentary on Muris and Otgaar (2020): Let the empirical evidence speak on the self-compassion scale. *Mindfulness, 11*, 1900–1909. https://doi.org/10.1007/s12671-020-01411-9

Neff, K. D. (2022). The differential effects fallacy in the study of self-compassion: Misunderstanding the nature of bipolar continuums. *Mindfulness*. https://doi.org/10.1007/s12671-022-01832-8

Neff, K. D., & Pommier, E. (2013). The relationship between self-compassion and other-focused concern among college undergraduates, community adults, and practicing meditators. *Self and Identity, 12*(2), 160–176. https://doi.org/10.1080/15298868.2011.649546

Neff, K. D., & Seppälä, E. (2016). Compassion, Well-being, and the hypo-egoic self. In K. W. Brown & M. R. Leary (Eds.), *Oxford handbook of hypo-egoic phenomena: Theory and research on the quiet ego* (pp. 189–202). Oxford Press.

Neff, K. D., Tóth-Király, I., & Colosimo, K. (2018). Self-compassion is best measured as a global construct and is overlapping with but distinct from neuroticism: A response to Pfattheicher, Geiger, Hartung, Weiss, and Schindler (2017). *European Journal of Personality, 32*(4), 371–392. https://doi.org/10.1002/per.2148

Neff, K. D., Knox, M. C., Long, P., & Gregory, K. (2020). Caring for others without losing yourself: An adaptation of the mindful self-compassion program for health-care communities. *Journal of Clinical Psychology, 76*(9), 1543–1562. https://doi.org/10.1002/jclp.23007

Nelson, S. K., Layous, K., Cole, S. W., & Lyubomirsky, S. (2016). Do unto others or treat yourself? The effects of prosocial and self-focused behavior on psychological flourishing. *Emotion, 16*(6), 850. https://doi.org/10.1037/emo0000178

Neumann, M., Edelhäuser, F., Tauschel, D., Fischer, M. R., Wirtz, M., Woopen, C., et al. (2011). Empathy decline and its reasons: A systematic review of studies with medical students and residents. *Academic Medicine, 86*(8), 996–1009. https://doi.org/10.1097/ACM.0b013e318221e615

Oakley, B., Knafo, A., Madhavan, G., & Wilson, D. S. (2011). *Pathological altruism*. Oxford University Press.

Olson, K., Kemper, K. J., & Mahan, J. D. (2015). What factors promote resilience and protect against burnout in first-year pediatric and medicine-pediatric residents? *Journal of Evidence-Based Complementary*

& Alternative Medicine, 20(3), 192–198. https://doi.org/10.1177/2156587214568894

Orosa-Duarte, Á., Mediavilla, R., Muñoz-Sanjose, A., Palao, Á., Garde, J., López-Herrero, V., et al. (2021). Mindfulness-based mobile app reduces anxiety and increases self-compassion in healthcare students: A randomised controlled trial. *Medical Teacher, 43*(6), 686–693. https://doi.org/10.1080/0142159X.2021.1887835

Overland, M. K., Zumsteg, J. M., Lindo, E. G., Sholas, M. G., Montenegro, R. E., Campelia, G. D., et al. (2019). Microaggressions in clinical training and practice. *PM&R, 11*(9), 1004–1012. https://doi.org/10.1002/pmrj.12229

Patsiopoulos, A. T., & Buchanan, M. J. (2011). The practice of self-compassion in counseling: A narrative inquiry. *Professional Psychology: Research and Practice, 42*(4), 301–307. https://doi.org/10.1037/a0024482

Pavlova, A., Wang, C. X. Y., Boggiss, A. L., O'Callaghan, A., & Consedine, N. S. (2021). Predictors of physician compassion, empathy, and related constructs: A systematic review. *Journal of General Internal Medicine, 37*(4), 900–911. https://doi.org/10.1007/s11606-021-07055-2

Prins, J. T., Van Der Heijden, F., Hoekstra-Weebers, J., Bakker, A., van de Wiel, H. B., Jacobs, B., et al. (2009). Burnout, engagement and resident physicians' self-reported errors. *Psychology, Health & Medicine, 14*(6), 654–666. https://doi.org/10.1080/13548500903311554

Quinal, L., Harford, S., & Rutledge, D. N. (2009). Secondary traumatic stress in oncology staff. *Cancer Nursing, 32*(4), E1–E7. https://doi.org/10.1097/NCC.0b013e31819ca65a

Raab, K. (2014). Mindfulness, self-compassion, and empathy among health care professionals: A review of the literature. *Journal of Health Care Chaplaincy, 20*(3), 95–108. https://doi.org/10.1080/08854726.2014.913876

Reyes, D. (2012). Self-compassion: A concept analysis. *Journal of Holistic Nursing, 30*(2), 81–89. https://doi.org/10.1177/0898010111423421

Reynolds, L. M., Bissett, I. P., Porter, D., & Consedine, N. S. (2017). A brief mindfulness intervention is associated with negative outcomes in a randomised controlled trial among chemotherapy patients. *Mindfulness, 8*(5), 1291–1303. https://doi.org/10.1007/s12671-017-0705-2

Robinson, K. J., Mayer, S., Allen, A. B., Terry, M., Chilton, A., & Leary, M. R. (2016). Resisting self-compassion: Why are some people opposed to being kind to themselves? *Self and Identity, 15*(5), 505–524. https://doi.org/10.1080/15298868.2016.1160952

Roden-Foreman, J. W., Bennett, M. M., Rainey, E. E., Garrett, J. S., Powers, M. B., & Warren, A. M. (2017). Secondary traumatic stress in emergency medicine clinicians. *Cognitive Behaviour Therapy, 46*(6), 522–532. https://doi.org/10.1080/16506073.2017.1315612

Rodrigues, H., Cobucci, R., Oliveira, A., Cabral, J. V., Medeiros, L., Gurgel, K., et al. (2018). Burnout syndrome among medical residents: A systematic review and meta-analysis. *PLoS One, 13*(11), e0206840. https://doi.org/10.1371/journal.pone.0206840

Rowe, L., & Kidd, M. (2009). *First do no harm: Being a resilient doctor in the 21st century*. McGraw-Hill.

Sansó, N., Galiana, L., González, B., Sarmentero, J., Reynes, M., Oliver, A., et al. (2019). Differential effects of two contemplative practice-based programs for health care professionals. *Psychosocial Intervention, 28*(3), 131–138. https://doi.org/10.5093/pi2019a12

Satele, D., Dyrbye, L. N., West, C. P., Boone, S., Tan, L., Sloan, J., et al. (2014). Burnout among US medical students, residents, and early career physicians relative to the general US population. *Academic Medicine, 89*(3), 443–451. https://doi.org/10.1097/ACM.0000000000000134

Scarlet, J., Altmeyer, N., Knier, S., & Harpin, R. E. (2017). The effects of Compassion Cultivation Training (CCT) on health-care workers. *Clinical Psychologist, 21*(2), 116–124. https://doi.org/10.1111/cp.12130

Schabram, K., & Heng, Y. T. (2021). How other-and self-compassion reduce burnout through resource replenishment. *Academy of Management Journal, 65*(2). https://doi.org/10.5465/amj.2019.0493

Schneider, S., Kingsolver, K., & Rosdahl, J. (2014). Physician coaching to enhance well-being: A qualitative analysis of a pilot intervention. *Explore, 10*(6), 372–379. https://doi.org/10.1016/j.explore.2014.08.007

Shanafelt, T. D., Hasan, O., Dyrbye, L. N., Sinsky, C., Satele, D., Sloan, J., et al. (2015). Changes in burnout and satisfaction with work-life balance in physicians and the general us working population between 2011 and 2014. *Mayo Clinic Proceedings, 90*(12), 1600–1613. https://doi.org/10.1016/j.mayocp.2015.08.023

Shanafelt, T. D., West, C. P., Sinsky, C., Trockel, M., Tutty, M., Satele, D. V., et al. (2019). Changes in burnout and satisfaction with work-life integration in physicians and the general us working population between 2011 and 2017. *Mayo Clinic Proceedings, 94*(9), 1681–1694. https://doi.org/10.1016/j.mayocp.2018.10.023

Sinclair, S., Norris, J. M., McConnell, S. J., Chochinov, H. M., Hack, T. F., Hagen, N. A., et al. (2016). Compassion: A scoping review of the healthcare literature. *BMC Palliative Care, 15*(1), 6. https://doi.org/10.1186/s12904-016-0080-0

Sinclair, S., Kondejewski, J., Raffin-Bouchal, S., King-Shier, K. M., & Singh, P. (2017). Can self-compassion promote healthcare provider well-being and compassionate care to others? Results of a systematic review. *Applied Psychology: Health and Well-Being, 9*(2), 168–206. https://doi.org/10.1111/aphw.12086

Sinclair, S., Kondejewski, J., Jaggi, P., Roze des Ordons, A. L., Kassam, A., Hayden, K. A., et al. (2021). What works for whom in compassion training programs offered to practicing healthcare providers: A realist review. *BMC Medical Education, 21*(1), 1–17. https://doi.org/10.1186/s12909-021-02863-w

Singh, P., Raffin-Bouchal, S., McClement, S., Hack, T. F., Stajduhar, K., Hagen, N. A., et al. (2018). Healthcare providers' perspectives on perceived barriers and facilitators of compassion: Results from a grounded theory study. *Journal of Clinical Nursing, 27*(9–10), 2083–2097. https://doi.org/10.1111/jocn.14357

Skovholt, T. M., & Trotter-Mathison, M. (2014). *The resilient practitioner: Burnout prevention and self-care strategies for counselors, therapists, teachers, and health professionals*. Routledge.

Snyder, K. A., & Green, A. I. (2008). Revisiting the glass escalator: The case of gender segregation in a female dominated occupation. *Social Problems, 55*(2), 271–299. https://doi.org/10.1525/sp.2008.55.2.271

Spreitzer, G., Porath, C. L., & Gibson, C. B. (2012). Toward human sustainability: How to enable more thriving at work. *Organizational Dynamics, 41*(2), 155–162. https://doi.org/10.1016/j.orgdyn.2012.01.009

Tierney, S., Ozer, C.-T., & Perry, S. (2018). Having the "headspace" for compassion toward self and others: A qualitative study of medical students' views and experiences. *Teaching and Learning in Medicine, 30*(3), 274–283. https://doi.org/10.1080/10401334.2018.1423973

Triffaux, J. M., Tisseron, S., & Nasello, J. A. (2019). Decline of empathy among medical students: Dehumanization or useful coping process? *L' Encéphale, 45*(1), 3–8. https://doi.org/10.1016/j.encep.2018.05.003

Vetter, M. H., Vetter, M. K., & Fowler, J. (2018). Resilience, hope and flourishing are inversely associated with burnout among members of the Society for Gynecologic Oncology. *Gynecologic Oncology Reports, 25*, 52–55. https://doi.org/10.1016/j.gore.2018.06.002

Wasson, R. S., Barratt, C., & O'Brien, W. H. (2020). Effects of mindfulness-based interventions on self-compassion in health care professionals: A meta-analysis. *Mindfulness, 11*(8), 1914–1934. https://doi.org/10.1007/s12671-020-01342-5

Wiklund Gustin, L., & Wagner, L. (2013). The butterfly effect of caring – Clinical nursing teachers' understanding of self-compassion as a source to compassionate care. *Scandinavian Journal of Caring Sciences, 27*(1), 175–183. https://doi.org/10.1111/j.1471-6712.2012.01033.x

Woo, T., Ho, R., Tang, A., & Tam, W. (2020). Global prevalence of burnout symptoms among nurses: A systematic review and meta-analysis. *Journal of Psychiatric Research, 123*, 9–20. https://doi.org/10.1016/j.jpsychires.2019.12.015

Yarnell, L. M., & Neff, K. D. (2013). Self-compassion, interpersonal conflict resolutions, and well-being. *Self and Identity, 12*(2), 146–159. https://doi.org/10.1080/15298868.2011.649545

Yarnell, L. M., Stafford, R. E., Neff, K. D., Reilly, E. D., Knox, M. C., & Mullarkey, M. (2015). Meta-analysis of gender differences in self-compassion. *Self and Identity, 14*(5), 499–520. https://doi.org/10.1080/15298868.2015.1029966

Zessin, U., Dickhäuser, O., & Garbade, S. (2015). The relationship between self-compassion and well-being: A meta-analysis. *Applied Psychology: Health and Well-Being, 7*(3), 340–364. https://doi.org/10.1111/aphw.12051

Zhou, A. Y., Panagioti, M., Esmail, A., Agius, R., Van Tongeren, M., & Bower, P. (2020). Factors associated with burnout and stress in trainee physicians: A systematic review and meta-analysis. *JAMA Network Open, 3*(8), e2013761–e2013761. https://doi.org/10.1001/jamanetworkopen.2020.13761

15 Self-Compassion in Relationships and Caregiving Contexts

Christine Lathren

Self-compassion, a personal resource entailing a supportive attitude toward oneself during challenges, contributes to many aspects of *individual* well-being. A burgeoning area of research extends these findings by examining self-compassion's role in *relational* well-being. In fact, our capacity for self-compassion may be rooted in our earliest childhood relationships and closely tied to attachment, or the enduring emotional bond a child develops with primary caregivers (Gilbert & Proctor, 2006). Experiences with warm, supportive parents and primary caregivers (herein referred to as parents) promote secure attachment while simultaneously providing a foundational "template" for warm, supportive *self*-directed coping strategies later in life (Shaver et al., 2016). To this point, adolescents and young adults who rate their childhood experiences with parents as high in warmth and nurturance are more likely to have high self-compassion (Kelly & Dupasquier, 2016; Neff & McGehee, 2010; Temel & Atalay, 2018). Meanwhile, those who report receiving inconsistent, harsh, or unsupportive responses to their needs as children are more likely to have low self-compassion in later years (Pepping et al., 2015; Tanaka et al., 2011). Without a template for care under distress, we may have a compromised ability to self-soothe, and may be more likely to respond to challenges with harshness or self-criticism. Thus, the capacity for self-compassion may be facilitated through parents or caregivers who have served as models of compassion.

However, the connection between self-compassion and relationship health does not end in childhood; it is likely complex, bidirectional, and relevant to numerous close relationships in adolescence and adulthood. The tendency to relate self-compassionately to one's own distress may be associated with adaptive thoughts, emotions, and behaviors that nourish healthy families, friendships, romantic partnerships, and more. To explore this possibility, this chapter is organized into two main sections. The first section theorizes why self-compassion may be beneficial to relationship health and describes the current evidence linking self-compassion to various interpersonal measures and factors, including relationship quality and satisfaction, prosocial behaviors and attitudes, conflict and transgression behaviors, communication, and autonomy. The second part of the chapter is dedicated to exploring self-compassion within the special circumstance of caregiving, including parenting, formal (i.e., professional) caregiving, and informal (i.e., family) caregiving.

First, why might self-compassion be associated with high relationship quality and satisfaction? Potential factors can be framed within the context of the three components of self-compassion: self-kindness, common humanity, and mindfulness. First, self-compassionate people are *kind* to themselves when they make a mis-

C. Lathren (✉)
UNC-Chapel Hill School of Medicine, University of North Carolina at Chapel Hill, Chapel Hill, NC, USA
e-mail: lathren@email.unc.edu

A. Finlay-Jones et al. (eds.), *Handbook of Self-Compassion*, Mindfulness in Behavioral Health,
https://doi.org/10.1007/978-3-031-22348-8_15

take or are experiencing difficulty. This self-kindness may steady them and help them to feel less threatened in negative interpersonal encounters. Indeed, evidence demonstrates that individuals with higher self-compassion have adaptive responses to threatening stimuli; they are more likely to disengage their attention from situations that evoke feelings of inadequacy or shame (Yip & Tong, 2021) and are quicker to recover from social evaluative threats (Arch et al., 2014). Importantly, however, this does not negate taking responsibility for one's actions: in fact, people with higher levels of self-compassion have been found to be more likely to take responsibility for their mistakes than those who are less self-compassionate (Leary et al., 2007). The self-compassionate person's ability to self-soothe and avoid harsh self-judgement may mean they require less reassurance from relationship partners, bounce back more easily after relationship ruptures, and have the courage to admit wrongdoing and address problems constructively.

The *common humanity* component of self-compassion may help people to accept that, like themselves, relationship partners will make mistakes, have weaknesses, and feel unpleasant emotions; this ability to view things from a shared perspective may help people to see the best in others, forgive, and remain steadily satisfied despite conflicts or upsets. The common humanity element of self-compassion may also serve to attenuate feelings of shame and support people to communicate in ways that are likely to promote trust and intimacy in relationships. For example, research has found that practicing self-compassion supports people to disclose experiences that threaten their self-esteem (Dupasquier et al., 2020) and increase the likelihood of disclosing experiences of distress even when they fear receiving compassion from others (Dupasquier et al., 2017). Likewise, given self-compassion includes holding difficult feelings with *mindful awareness* without avoiding or overidentifying with them, it may assist one to be authentic and communicative about their feelings, while enabling them to advocate for themselves in a constructive manner. The three components work dynamically, facilitating well-being (Zessin et al., 2015), adaptive stress coping (Allen & Leary, 2010), low psychopathology (MacBeth & Gumley, 2012), and a stable sense of self-worth (Neff & Vonk, 2009) while linking to a host of socially adaptive traits such as agreeableness, conscientiousness, and extraversion (Neff et al., 2007) – arguably all factors that should be beneficial to relational health. In sum, it is likely that compassion toward oneself is not selfish, but rather socially advantageous and connection-building.

Indeed, current evidence shows that people high in self-compassion generally report both high-quality close relationships and high satisfaction in those relationships. For example, in adolescents and young adults, self-compassion has been linked to various indices of positive family functioning such as family cohesion, supportiveness, and flexibility (Berryhill et al., 2018; Hayes et al., 2016; Hood et al., 2020; Jiang et al., 2016; Neff & McGehee, 2010). College students higher in self-compassion report higher levels of relational well-being – conceptualized as one's self-esteem and emotional well-being within the relationship (Yarnell & Neff, 2013) – and relationship quality, including measures of support, depth, and conflict (Huang & Berenbaum, 2017), with a variety of close others in their lives, including friends. In the context of romantic partnerships, self-compassion is positively linked to relationship satisfaction (Fahimdanesh et al., 2020; Maleki et al., 2019; Neff & Beretvas, 2013; Shahabi et al., 2019), relational well-being (Neff & Beretvas, 2013), and to more stable marital satisfaction over time (Baker & McNulty, 2011). These studies point to overall positive connection between self-compassion and healthy, satisfying interpersonal relationships across contexts.

Other evidence describes how self-compassion is linked to a variety of prosocial behaviors that may facilitate positive emotions between relationship partners. For example, Neff and Beretvas (2013) studied heterosexual romantic partners who had been in a relationship for a year or more and found that individuals who were high in self-compassion were perceived as behaving more favorably (i.e., more caring, accepting, and autonomy-granting, while also less controlling, aggressive, and detached) and had partners who were more satisfied in the relationship, compared to those lower in self-compassion. Other studies have supported these findings, suggesting that self-compassion may facilitate having a more compassionate, accepting view of others' flaws and shortcomings (Zhang et al., 2019), having higher levels of compassionate goals within friendships (Crocker & Canevello, 2008), as well as the ability to be empathetic and see things from another's perspective (Fuochi et al., 2018; Neff & Pommier, 2013). Notably, not all studies have found significant positive correlations between self-compassion and compassion for others (e.g., López et al., 2018); this may be because many people who are low in self-compassion are compassionate toward others. However, in many instances, self-compassionate tendencies may promote healthy relationship functioning and thus contribute to relationship thriving.

Given self-compassion provides anchoring and self-soothing during times of hardship, high levels of self-compassion are likely to be associated with adaptive responses to inevitable relationship difficulties – including conflict and transgression scenarios. In college students, high self-compassion is associated with increased likelihood to compromise as opposed to either self-subordinate or self-prioritize when resolving conflicts with people close to them; self-compassionate respondents also rated their resolution choice as more authentic and causing lower levels of emotional distress (Yarnell & Neff, 2013). In this case, self-compassion may promote respect for one's own needs and well-being, while simultaneously recognizing the perspectives and "common humanity" of others, leading to solutions that are sensitive to both parties. In challenging romantic contexts, self-compassion is related to decreased jealousy, increased willingness to forgive (Tandler & Petersen, 2018), adjustment after divorce (Sbarra et al., 2012), and in certain circumstances, motivation to repair after a mistake (Baker & McNulty, 2011) or to self-improve after a break-up (Zhang & Chen, 2017). Perhaps related to these findings, several studies have shown that self-compassionate people have lower levels of negative emotional responses to unpleasant interpersonal events (Leary et al., 2007; Purdie & Morley, 2015) and are less sensitive to interpersonal rejection (Gerber et al., 2015). Thus, in the context of various types of relationship breaches or unpleasantries, self-compassion may allow one to put one's own weaknesses and the weaknesses of others in perspective. It may promote the emotional stability needed to move forward with kindness and understanding toward *both* oneself and relationship partners, leading to more effective problem-solving, self-improving, and rupture-repairing behaviors.

Similarly, self-compassion may be linked to improved communication in relationship partners, particularly if the discussion topic brings up difficult emotions. Given self-compassionate individuals are more likely to accept their difficult emotions in a supportive and non-avoidant manner, individuals who are self-compassionate may have more emotional clarity and have an easier time discussing difficulties or admitting shortcomings with others without becoming controlling or detached. At the same time, self-compassion may facilitate improved listening skills via emotion regulation and decreased threat perception. Women high in self-compassion who were facing infertility, for example, reported greater ease in talking about infertility with their romantic partner (Raque-Bogdan & Hoffman, 2015). Likewise, in couples facing lung cancer, patients' self-compassion was significantly associated with better self-reported communication with their partner about the cancer (Schellekens et al., 2017). Recent examination of communication styles in organizational contexts also supports the link between self-compassion and more effective, person-centered listening styles (Salazar, 2017). Meanwhile, in parenting contexts (described in more detail later in this chapter), parent self-compassion is associated with higher self-reported levels of mindful parenting behaviors; these behaviors include listening to the child with full attention and being aware of and sensitive to the emotional cues from the child (Moreira et al., 2016). Thus, in a variety of contexts, self-compassion is associated with the ability to express oneself and likewise, be mindfully attuned to others.

Healthy relationships also thrive when there is a balance between one's own feelings and needs and those of relationship partners. Attending to the needs of others to the detriment of one's own needs or becoming enmeshed with or distressed by the needs of others can negatively impact relational health. Self-compassion has been linked to lower levels of overdependency (Denckla et al., 2017) and greater levels of autonomy, self-awareness, and competence, which may protect against caregiving burnout and pathological investment in other needs (Gerber et al., 2015; Gerber & Anaki, 2021). Likewise, in families where dysfunctional relationships cause overinvolvement of family members in one another's lives, self-compassion may be useful in promoting improved mental health outcomes (Berryhill et al., 2018).

Balance may be particularly salient in parent–child and other caregiving relationships, which are the topics of the rest of this chapter. Caregiving relationships are unique given they are often high intensity, with parents and caregivers influenced by societal, cultural, and self-imposed expectations to provide compassionate, selfless care for extended periods of time to the detriment of their own health. While there are many positive and fulfilling aspects of caregiving, stress, dysfunctional coping patterns, caregiving fatigue, burden, and burnout are also common – bringing with them negative consequences for the relationship between caregiver and care recipient(s) and the quality of care provided.

Meanwhile, theoretically, the three components of self-compassion may assuage the difficult aspects of the caregiving experience. Self-kindness may temper caregivers' perfectionism and self-judgement, allowing caregivers to let go of unrealistic expectations, accept their human limitations, and include themselves in the circle of care. Moreover, when caregivers can meet their own emotional needs, fewer demands may be placed on the care recipient, who may be unable (e.g., family members with dementia, young children) to offer reciprocal emotional care. These healthy boundaries and rebalancing of the flow of care may result in improved relationship functioning. The mindfulness component of self-compassion may help caregivers to maintain perspective regarding their role. By being aware of emotions without becoming overwhelmed or ruminating on them, caregivers may also be more skillful in challenging interpersonal encounters and may deescalate emotionally charged caregiving situations. Finally, common humanity reminds caregivers that their difficult feelings make sense and are to be expected under challenging circumstances, keeping them connected to rather than isolated from others. It may also help caregivers to remember the humanity and imperfection of those they care for. This may engender patience, forgiveness, and compassionate caregiving behaviors, which ultimately strengthens the relationship. In sum, self-compassion may be an exceptionally well-suited resource for overextended parents and caregivers, with interwoven benefits for caregiver, care recipient, and the relationship between them.

As expected, the literature supports self-compassion as a beneficial resource for struggling parents, as it promotes healthy stress coping and adaptive attitudes toward parenting challenges. Self-compassion is associated with lower levels of parental stress in general community samples (Gouveia et al., 2016; Moreira et al., 2015), as well as lower distress and improved well-being for parents of children who are on the autism spectrum (Neff & Faso, 2015), have intellectual and developmental disabilities (Robinson et al., 2017), and have mental health problems (Shenaar-Golan et al., 2021). In contexts where children often display challenging emotions and behaviors, parents can be faced with feelings of inadequacy, frustration, self-criticism and self-blame, and isolation. Self-compassion allows parents to meet these difficulties with kindness and understanding, normalizing them as part of life's ups and downs rather than abnormal or shameful. Parents of children on the autism spectrum report feeling less stigmatized (Wong et al., 2016), more hopeful, satisfied with life, self-efficacious in their parental role (Neff & Faso, 2015), and more forgiving of their perceived weaknesses as a parent (Bohadana et al., 2020).

Even in response to everyday parenting challenges, self-compassion appears beneficial in tempering feelings of guilt and shame with respect to not showing up in an ideal way during a difficult parenting event (Sirois et al., 2019). Meanwhile, for parents of children with various mental health problems, self-compassion was a stronger predictor of higher levels of positive feelings and lower levels of negative feelings in the parental role than the severity of child mental health symptoms (Shenaar-Golan et al., 2021). These findings suggest that self-compassion may help parents cope with the demands of parenthood from a balanced, connected perspective – a view that may engender more harmony in the parent–child relationship.

Likewise, parents who are high in self-compassion may extend a kind and understanding orientation toward their children, resulting in a less judgmental and more accepting view of their child's struggles. For example, Neff and Faso (2015) showed parents of children with autism who had high self-compassion were less likely to rate their child's behavior as difficult or problematic, and more likely to feel satisfied in their relationship with their child. Similarly, in a sample of parents with a history of depression, parents high in self-compassion were gentler in their assessment of their child's difficult behavior, attributing the behavior to transient or situational factors as opposed to static character traits (Psychogiou et al., 2016). Thus, by helping parents remain open and forgiving, allowing them to see the best in both themselves and their child, self-compassion may promote better functioning under challenging parenting conditions.

Closely connected to adaptive coping and attitudes, parents with high self-compassion may display more sensitive behaviors, as they may be better attuned to the emotional cues of their child. Parents with high self-compassion are more likely to be authoritative in their parenting, a style considered both warm and supportive (Gouveia et al., 2016). Numerous studies have also linked high self-compassion in parents to mindful parenting (Duncan et al., 2009), described as a five-dimensional set of practices or skills that relate to high-quality parenting behaviors: (1) listening to child with full attention, (2) emotional awareness of self and child, (3) self-regulation in the parenting relationship, (4) nonjudgmental acceptance of self and child, and (5) compassion for self and child. Parents who have high self-compassion are more likely to interact with their child using mindful parenting practices (Kim et al., 2019; Moreira et al., 2016; Moreira & Canavarro, 2017). Mindful parenting, in turn, is linked to adolescents' secure attachment to parents (Moreira et al., 2018).

Other related research posits that parents who have high self-compassion may be generally attuned to and comfortable with negative emotions in both themselves and their child (Lathren et al., 2020), linking to child development literature on emotion socialization processes (Eisenberg et al., 1998) and emotion coaching behaviors (Gottman et al., 1996). Parents with high self-compassion are mindfully aware of their own negative emotions and related needs, attending to these emotions supportively rather than avoiding or becoming overwhelmed. It makes sense, then, that these parents would approach negative emotions similarly when displayed by their child. Parents may play the role of an "emotion coach" who recognizes, validates, labels, comforts, and welcomes negative emotions in their child as an opportunity for intimacy and problem-solving support. This approach contrasts with parents who themselves are uncomfortable with difficult feelings, and who may tend to minimize or deny feelings like sadness or anger in their children, or even punish children for having these emotions (Gottman et al., 1996). Thus, parents with high self-compassion may develop stronger bonds with their children through behaviors which are attentive, attuned, compassionate, and "coach" the child to navigate emotional challenges in a healthy, validating manner.

Professional caregivers are another group for whom self-compassion is highly relevant. Evidence to date supports the link between self-compassion and decreased caregiving stress in a wide variety of professional caregivers, including nursing home nursing assistants, hospital and community nurses, physicians, and health-care students. These studies have shown self-compassion is associated with lower levels of secondary trauma symptoms (i.e., experiencing trauma due to exposure to the trauma and suffering of others), compassion fatigue (i.e., exhaustion and dysfunction due to prolonged exposure to the suffering and stress of others), and various measures of burnout, while associated with higher levels of compassion satisfaction (e.g., fulfillment in one's caregiving role) and quality of life (Bluth et al., 2021; Delaney, 2018; Duarte et al., 2016; Durkin et al., 2016; Upton, 2018). In these settings, self-compassion may improve caregiving capacities by strengthening *internal* support mechanisms in order to revitalize and fuel *outward* support.

Similarly, researchers have developed interventions aimed to raise self-compassion levels for hospital nurses (Delaney, 2018), mixed health-care communities (Neff et al., 2020), and nursing home nursing assistants (Bluth et al., 2021). These studies show self-compassion interventions are feasible and acceptable for professional caregivers and are linked to improvements in well-being, burnout, and compassion-related measures. However, studies to date have lacked a direct measurement of potential impact on dyadic relationships between caregiver and care recipient, nor related outcomes such as care quality and satisfaction with care. These issues will be important to explore in future work.

Family caregivers, particularly those caring for loved ones with dementia or another debilitating illness, similarly experience high levels of stress and burden. Evidence suggests self-compassion may help relieve stress and burden in family caregivers via connection with lower levels of dysfunctional coping strategies (Lloyd et al., 2019), lower caregiver burden (Xu et al., 2020), and improved quality of life (Hlabangana & Hearn, 2020). Other evidence shows that self-compassion moderates the relationship between caregiving stress and depression (Hsieh et al., 2021). However, despite high promise, interventions explicitly focused on raising family caregivers' self-compassion and improving family relationships have not yet been explored (Murfield et al., 2020).

Other studies provide a more nuanced view of how self-compassion may be useful in relieving caregiver stress by examining impact on unhelpful or unrealistic thought patterns or behaviors related to the caregiving role. For example, some family caregivers experience unhealthy guilt – feeling guilty for feeling sad or hopeless – related to societal or cultural norms regarding familism and the importance of honoring and caring for loved ones (Sayegh & Knight, 2011). Similarly, family caregiver accounts commonly reveal self-sacrificing behaviors, a strong commitment to caregiver identity and reluctance to prioritize oneself due to fear of detracting from one's responsibilities or becoming selfish (Berardini et al., 2021; Diggory & Reeves, 2021). The culture of selflessness and denial of self-needs also pervades health-care settings, where providers have high workloads and lack opportunities for self-care; the expectation is to set one's needs aside, "people-please" and prioritize caregiving tasks to the detriment of provider well-being (Andrews et al., 2020; Lathren et al., 2021).

Self-compassion appears to soften this view, promoting a more balanced understanding of caregiving situations and acceptance of caregiver's often limited role in fully ameliorating the challenges faced by the care recipient. This reframing is particularly useful for combatting caregivers' feelings of shame or guilt for being an imperfect human who has needs of their own; these findings are similar to themes within parenting literature discussed earlier. For example, for mothers caring for a child on the autism spectrum, self-compassion is described as allowing caregivers to challenge unrealistic internal and societal expectations of themselves, helping them to forgive their shortcomings, and promoting emotional self-care without sinking into self-judgment and despair (Bohadana et al., 2020). Similarly, for nursing home nursing assistants in a professional caregiving context, self-compassion intervention promoted self-appreciation, validation of one's needs as a human, and a more realistic view of their capacities and limitations as caregivers (Lathren et al., 2021). Caregivers described how practicing self-compassion alleviated burden and self-imposed stress and resulted in perceived increases in kindness, patience, and understanding toward care recipients. Thus, whether personal or professional, a "rebalancing" of compassion and care *toward the caregiver* may ultimately provide the internal resources to increase compassion and care *toward those they provide for*.

Surprisingly, while mindfulness studies are relatively common, to date there have been few self-compassion-focused intervention studies that target parents and caregivers – despite numerous potential benefits for both formal caregivers and family caregivers. Given the well-established benefits of self-compassion, a pragmatic approach may be appropriate. For health-care professionals, these interventions could be woven into educational curricula, professional development, and in services. To address critical shortages of health-care workers and burnout, policies must create a cultural shift away from self-sacrifice and denial to one of support and replenishment. Ultimately, a self-compassionate clinical workplace culture would benefit *all* relationships – including relationships between administrators, staff (office staff, physicians, nurses, aides, etc.), and with patients and their families. Meanwhile, family caregivers could be offered access to self-compassion programming through physicians' offices and community organizations that offer support and educational programming. For example, organizations that provide resources to caregivers for persons with dementia, cancer, or autism support, organizations could offer self-compassion training – in various formats – to supplement traditional caregiving support services like peer support groups.

Self-compassion training should also be offered more widely to both new parents and to parents and their young children early in family life, particularly for families facing adversity or dysfunction. These offerings could originate through pediatric offices, community outreach organizations, schools, and churches. Offerings that provide skills to parents and children simultaneously would not only benefit parent and child well-being but also develop a broader family culture that values emotional expression, acceptance, and self-care. In this way, family self-compassion training promotes a *powerful "dual model" of support* by strengthening both the relational support family members provide to one another and each person's individual capacity for self-directed support.

Moreover, evidence described in this chapter suggests that interpersonal, longitudinal effects may be seen. For example, youth who learn self-compassion skills early may be more likely to develop healthy, fulfilling relationships in adolescence and adulthood, eventually influencing *their own* capacity to be a compassionate, nurturing parent and caregiver. Thus, self-compassion may have implications across generations, making it a particularly valuable resource for healing cycles of dysfunction passed on via strained family relationships and attachment insecurity. For these reasons, research that applies self-compassion interventions to relational contexts and examines longitudinal, cross-generational effects hold significant promise.

To summarize, self-compassion and healthy, satisfying, supportive human relationships appear to go hand in hand. As human relationships go through inevitable periods of hardship – conflict, transgressions, stress, trauma, illness – self-compassion serves to stabilize, comfort, remind us that we are good enough just as we are, and that we are all worthy of giving and receiving love and forgiveness. In a nutshell, self-compassion allows us to show up for ourselves, so that we can similarly show up for those we care about.

References

Allen, A. B., & Leary, M. R. (2010). Self-compassion, stress, and coping. *Social and Personality Psychology Compass, 4*(2), 107–118. https://doi.org/10.1111/j.1751-9004.2009.00246.x

Andrews, H., Tierney, S., & Seers, K. (2020). Needing permission: The experience of self-care and self-compassion in nursing: A constructivist grounded theory study. *International Journal of Nursing Studies, 101*, 103436. https://doi.org/10.1016/J.IJNURSTU.2019.103436

Arch, J. J., Warren Brown, K., Dean, D. J., Landy, L. N., Brown, K., & Laudenslager, M. L. (2014).

Self-compassion training modulates alpha-amylase, heart rate variability, and subjective responses to social evaluative threat in women. *Psychoneuroendocrinology, 42*, 49–58. https://doi.org/10.1016/j.psyneuen.2013.12.018
Baker, L. R., & McNulty, J. K. (2011). Self-compassion and relationship maintenance: The moderating roles of conscientiousness and gender. *Journal of Personality and Social Psychology, 100*(5), 853–873. https://doi.org/10.1037/a0021884
Berardini, Y., Chalmers, H., & Ramey, H. (2021). Unfolding what self-compassion means in young carers' lives. *Child and Adolescent Social Work Journal, 38*, 533–545. https://doi.org/10.1007/s10560-021-00791-8
Berryhill, M. B., Hayes, A., & Lloyd, K. (2018). Chaotic-enmeshment and anxiety: The mediating role of psychological flexibility and self-compassion. *Contemporary Family Therapy: An International Journal, 40*, 326–337. https://doi.org/10.1007/s10591-018-9461-2
Bluth, K., Lathren, C., Silbersack Hickey, J. V. T., Zimmerman, S., Wretman, C. J., & Sloane, P. D. (2021). Self-compassion training for certified nurse assistants in nursing homes. *Journal of the American Geriatrics Society, 69*(7), 1896–1905. https://doi.org/10.1111/jgs.17155
Bohadana, G., Morrissey, S., & Paynter, J. (2020). Self-compassion in mothers of children with autism spectrum disorder: A qualitative analysis. *Journal of Autism and Developmental Disorders, 51*(4), 1290–1303. https://doi.org/10.1007/S10803-020-04612-2
Crocker, J., & Canevello, A. (2008). Creating and undermining social support in communal relationships: The role of compassionate and self-image goals. *Journal of Personality and Social Psychology, 95*(3), 555–575. https://doi.org/10.1037/0022-3514.95.3.555
Delaney, M. C. (2018). Caring for the caregivers: Evaluation of the effect of an eight-week pilot mindful self-compassion (MSC) training program on nurses' compassion fatigue and resilience. *PLoS One, 13*(11), e0207261. https://doi.org/10.1371/journal.pone.0207261
Denckla, C. A., Consedine, N. S., & Bornstein, R. F. (2017). Self-compassion mediates the link between dependency and depressive symptomatology in college students. *Self and Identity, 16*(4), 373–383. https://doi.org/10.1080/15298868.2016.1264464
Diggory, K., & Reeves, A. (2021). 'Permission to be kind to myself'. The experiences of informal carers of those with a life-limiting or terminal illness of a brief self-compassion-based self-care intervention. *Progress in Palliative Care, 30*(3), 149–157. https://doi.org/10.1080/09699260.2021.1972722
Duarte, J., Pinto-Gouveia, J., & Cruz, B. (2016). Relationships between nurses' empathy, self-compassion and dimensions of professional quality of life: A cross-sectional study. *International Journal of Nursing Studies, 60*, 1–11. https://doi.org/10.1016/j.ijnurstu.2016.02.015
Duncan, L. G., Coatsworth, J. D., & Greenberg, M. T. (2009). A model of mindful parenting: Implications for parent-child relationships and prevention research. *Clinical Child and Family Psychology Review, 12*(3), 255–270. https://doi.org/10.1007/s10567-009-0046-3
Dupasquier, J. R., Kelly, A. C., Moscovitch, D. A., & Vidovic, V. (2017). Practicing self-compassion weakens the relationship between fear of receiving compassion and the desire to conceal negative experiences from others. *Mindfulness, 9*, 500–511. https://doi.org/10.1007/s12671-017-0792-0
Dupasquier, J. R., Kelly, A. C., Moscovitch, D. A., & Vidovic, V. (2020). Cultivating self-compassion promotes disclosure of experiences that threaten self-esteem. *Cognitive Therapy and Research, 44*, 108–119. https://doi.org/10.1007/s10608-019-10050-x
Durkin, M., Beaumont, E., Hollins Martin, C. J., & Carson, J. (2016). A pilot study exploring the relationship between self-compassion, self-judgement, self-kindness, compassion, professional quality of life and wellbeing among UK community nurses. *Nurse Education Today, 46*, 109–114. https://doi.org/10.1016/j.nedt.2016.08.030
Eisenberg, N., Cumberland, A., & Spinrad, T. (1998). Parental socialization of emotion. *Psycholgical Inquiry, 9*(4), 241–273. https://doi.org/10.1207/s15327965pli0904
Fahimdanesh, F., Noferesti, A., & Tavakol, K. (2020). Self-compassion and forgiveness: Major predictors of marital satisfaction in young couples. *American Journal of Family Therapy, 48*(3), 221–234. https://doi.org/10.1080/01926187.2019.1708832
Fuochi, G., Veneziani, C. A., & Voci, A. (2018). Exploring the social side of self-compassion: Relations with empathy and outgroup attitudes. *European Journal of Social Psychology, 48*(6), 769–783. https://doi.org/10.1002/ejsp.2378
Gerber, Z., & Anaki, D. (2021). The role of self-compassion, concern for others, and basic psychological needs in the reduction of caregiving burnout. *Mindfulness, 12*(3), 741–750. https://doi.org/10.1007/s12671-020-01540-1
Gerber, Z., Tolmacz, R., & Doron, Y. (2015). Self-compassion and forms of concern for others. *Personality and Individual Differences, 86*, 394–400. https://doi.org/10.1016/j.paid.2015.06.052
Gilbert, P., & Procter, S. (2006). Compassionate mind training for people with high shame and self-criticism: Overview and pilot study of a group therapy approach. *Clinical Psychology & Psychotherapy, 13*, 353–379. https://doi.org/10.1002/cpp.507
Gottman, J. M., Katz, L. F., & Hooven, C. (1996). Parental meta-emotion philosophy and the emotional life of families: Theoretical models and preliminary data. *Journal of Family Psychology, 10*(3), 243–268. https://doi.org/10.1037/0893-3200.10.3.243
Gouveia, M. J., Carona, C., Canavarro, M. C., & Moreira, H. (2016). Self-compassion and dispositional mindfulness are associated with parenting styles and parenting stress: The mediating role of mindful parent-

ing. *Mindfulness, 7*, 700–712. https://doi.org/10.1007/s12671-016-0507-y

Hayes, J. A., Lockard, A. J., Janis, R. A., & Locke, B. D. (2016). Construct validity of the self-compassion scale-short form among psychotherapy clients. *Counselling Psychology Quarterly, 29*(4), 405–422. https://doi.org/10.1080/09515070.2016.1138397

Hlabangana, V., & Hearn, J. H. (2020). Depression in partner caregivers of people with neurological conditions; associations with self-compassion and quality of life. *Journal of Mental Health, 29*(2), 176–181. https://doi.org/10.1080/09638237.2019.1630724

Hood, C. O., Thomson Ross, L., & Wills, N. (2020). Family factors and depressive symptoms among college students: Understanding the role of self-compassion. *Journal of American College Health, 68*(7), 683–687. https://doi.org/10.1080/07448481.2019.1596920

Hsieh, C.-C., Lin, Z.-Z., Ho, C.-C., Yu, C.-J., Chen, H.-J., Chen, Y.-W., & Hsiao, F.-H. (2021). The short- and long-term causal relationships between self-compassion, trait mindfulness, caregiver stress, and depressive symptoms in family caregivers of patients with lung cancer. *Mindfulness, 12*(7), 1812–1821. https://doi.org/10.1007/S12671-021-01642-4

Huang, A. B., & Berenbaum, H. (2017). Accepting our weaknesses and enjoying better relationships: An initial examination of self-security. *Personality and Individual Differences, 106*, 64–70. https://doi.org/10.1016/j.paid.2016.10.031

Jiang, Y., You, J., Hou, Y., Du, C., Lin, M.-P., Zheng, X., & Ma, C. (2016). Buffering the effects of peer victimization on adolescent non-suicidal self-injury: The role of self-compassion and family cohesion. *Journal of Adolescence, 53*, 107–115. https://doi.org/10.1016/j.adolescence.2016.09.005

Kelly, A. C., & Dupasquier, J. (2016). Social safeness mediates the relationship between recalled parental warmth and the capacity for self-compassion and receiving compassion. *Personality and Individual Differences, 89*, 157–161. https://doi.org/10.1016/j.paid.2015.10.017

Kim, E., Krägeloh, C. U., Medvedev, O. N., Duncan, L. G., & Singh, N. N. (2019). Interpersonal mindfulness in parenting scale: Testing the psychometric properties of a Korean version. *Mindfulness, 10*, 516–528. https://doi.org/10.1007/s12671-018-0993-1

Lathren, C., Bluth, K., & Zvara, B. (2020). Parent self-compassion and supportive responses to child difficult emotion: An intergenerational theoretical model rooted in attachment. *Journal of Family Theory and Review, 12*(3), 368–381. https://doi.org/10.1111/jftr.12388

Lathren, C., Sheffield-Abdullah, K., Sloane, P. D., Bluth, K., Hickey, J. V. T. S., Wretman, C. J., Phillips, L. P., & Zimmerman, S. (2021). Certified nursing assistants' experiences with self-compassion training in the nursing home setting. *Geriatric Nursing, 42*, 1341–1348. https://doi.org/10.1016/j.gerinurse.2021.08.013

Leary, M. R., Tate, E. B., Adams, C. E., Allen, A. B., & Hancock, J. (2007). Self-compassion and reactions to unpleasant self-relevant events: The implications of treating oneself kindly. *Journal of Personality and Social Psychology, 92*(5), 887–904. https://doi.org/10.1037/0022-3514.92.5.887

Lloyd, J., Muers, J., Patterson, T. G., & Marczak, M. (2019). Self-compassion, coping strategies, and caregiver burden in caregivers of people with dementia. *Clinical Gerontologist, 42*(1), 47–59. https://doi.org/10.1080/07317115.2018.1461162

López, A., Sanderman, R., Ranchor, A. V., & Schroevers, M. J. (2018). Compassion for others and self-compassion: Levels, correlates, and relationship with psychological well-being. *Mindfulness, 9*(1), 325–331. https://doi.org/10.1007/s12671-017-0777-z

MacBeth, A., & Gumley, A. (2012). Exploring compassion: A meta-analysis of the association between self-compassion and psychopathology. *Clinical Psychology Review, 32*(6), 545–552. https://doi.org/10.1016/j.cpr.2012.06.003

Maleki, A., Veisani, Y., Aibod, S., Azizifar, A., Alirahmi, M., & Mohamadian, F. (2019). Investigating the relationship between conscientiousness and self-compassion with marital satisfaction among Iranian married employees. *Journal of Education and Health Promotion, 8*(76), 1–7. https://doi.org/10.4103/jehp.jehp_105_18

Moreira, H., & Canavarro, M. C. (2017). Psychometric properties of the interpersonal mindfulness in parenting scale in a sample of Portuguese mothers. *Mindfulness, 8*, 691–706. https://doi.org/10.1007/s12671-016-0647-0

Moreira, H., Gouveia, M. J., Carona, C., Silva, N., & Canavarro, M. C. (2015). Maternal attachment and children's quality of life: The mediating role of self-compassion and parenting stress. *Journal of Child and Family Studies, 24*(8), 2332–2344. https://doi.org/10.1007/s10826-014-0036-z

Moreira, H., Carona, C., Silva, N., Nunes, J., & Canavarro, M. C. (2016). Exploring the link between maternal attachment-related anxiety and avoidance and mindful parenting: The mediating role of self-compassion. *Psychology and Psychotherapy: Theory, Research and Practice, 89*(4), 369–384. https://doi.org/10.1111/papt.12082

Moreira, H., Gouveia, M. J., & Canavarro, M. C. (2018). Is mindful parenting associated with adolescents' Well-being in early and middle/late adolescence? The mediating role of adolescents' attachment representations, self-compassion and mindfulness. *Journal of Youth and Adolescence, 47*(8), 1771–1788. https://doi.org/10.1007/s10964-018-0808-7

Murfield, J., Moyle, W., & O'Donovan, A. (2020). Self-compassion as an applicable intervention target for family carers of older adults: A conceptual commentary. *International Journal of Geriatric Psychiatry, 35*(4), 376–383. https://doi.org/10.1002/gps.5257

Neff, K. D., & Beretvas, S. N. (2013). The role of self-compassion in romantic relationships. *Self and*

Identity, 12(1), 78–98. https://doi.org/10.1080/15298868.2011.639548

Neff, K. D., & Faso, D. J. (2015). Self-compassion and well-being in parents of children with autism. *Mindfulness, 6*(4), 938–947. https://doi.org/10.1007/s12671-014-0359-2

Neff, K. D., & McGehee, P. (2010). Self-compassion and psychological resilience among adolescents and young adults. *Self and Identity, 9*(3), 225–240. https://doi.org/10.1080/15298860902979307

Neff, K., & Pommier, E. (2013). The relationship between self-compassion and other-focused concern among college undergraduates, community adults, and practicing meditators. *Self and Identity, 12*, 160–176. https://doi.org/10.1080/15298868.2011.649546

Neff, K. D., & Vonk, R. (2009). Self-compassion versus global self-esteem: Two different ways of relating to oneself. *Journal of Personality, 77*(1), 23–50. https://doi.org/10.1111/j.1467-6494.2008.00537.x

Neff, K. D., Rude, S. S., & Kirkpatrick, K. L. (2007). An examination of self-compassion in relation to positive psychological functioning and personality traits. *Journal of Research in Personality, 41*(4), 908–916. https://doi.org/10.1016/j.jrp.2006.08.002

Neff, K. D., Knox, M. C., Long, P., & Gregory, K. (2020). Caring for others without losing yourself: An adaptation of the mindful self-compassion program for healthcare communities. *Journal of Clinical Psychology, 76*, 1543–1562. https://doi.org/10.1002/jclp.23007

Pepping, C. A., Davis, P. J., O'Donovan, A., & Pal, J. (2015). Individual differences in self-compassion: The role of attachment and experiences of parenting in childhood. *Self and Identity, 14*(1), 104–117. https://doi.org/10.1080/15298868.2014.955050

Psychogiou, L., Legge, K., Parry, E., Mann, J., Nath, S., Ford, T., & Kuyken, W. (2016). Self-compassion and parenting in mothers and fathers with depression. *Mindfulness, 7*, 896–908. https://doi.org/10.1007/s12671-016-0528-6

Purdie, F., & Morley, S. (2015). Self-compassion, pain, and breaking a social contract. *Pain, 156*(11), 2354–2363. https://doi.org/10.1097/j.pain.0000000000000287

Raque-Bogdan, T. L., & Hoffman, M. A. (2015). The relationship among infertility, self-compassion, and well-being for women with primary or secondary infertility. *Psychology of Women Quarterly, 39*(4), 484–496. https://doi.org/10.1177/0361684315576208

Robinson, S., Hastings, R. P., Weiss, J. A., Pagavathsing, J., & Lunsky, Y. (2017). Self-compassion and psychological distress in parents of young people and adults with intellectual and developmental disabilities. *Journal of Applied Research in Intellectual Disabilities, 31*(3), 454–458. https://doi.org/10.1111/jar.12423

Salazar, L. R. (2017). The influence of business students' listening styles on their compassion and self-compassion. *Business and Professional Communication Quarterly, 80*(4), 226–244.

Sayegh, P., & Knight, B. G. (2011). The effects of familism and cultural justification on the mental and physical health of family caregivers. *The Journals of Gerontology: Series B, 66B*(1), 3–14. https://doi.org/10.1093/GERONB/GBQ061

Sbarra, D. A., Smith, H. L., & Mehl, M. R. (2012). When leaving your ex, love yourself: Observational ratings of self-compassion predict the course of emotional recovery following marital separation. *Psychological Science, 23*(3), 261–269. https://doi.org/10.1177/0956797611429466

Schellekens, M. P. J., Karremans, J. C., van der Drift, M. A., Molema, J., van den Hurk, D. G. M., Prins, J. B., & Speckens, A. E. M. (2017). Are mindfulness and self-compassion related to psychological distress and communication in couples facing lung cancer? A dyadic approach. *Mindfulness, 8*(2), 325–336. https://doi.org/10.1007/s12671-016-0602-0

Shahabi, B., Shahabi, R., & Foroozandeh, E. (2019). Analysis of the self-compassion and cognitive flexibility with marital compatibility in parents of children with autism spectrum disorder. *International Journal of Developmental Disabilities, 66*(4), 282–288. https://doi.org/10.1080/20473869.2019.1573000

Shaver, P., Mikulincer, M., Sahdra, B., & Gross, J. T. (2016). Attachment security as a foundation for kindness toward self and others. In K. W. Brown & M. R. Leary (Eds.), *The Oxford handbook of hypo-egoic phenomena* (pp. 223–242). Oxford University Press. https://doi.org/10.1093/oxfordhb/9780199328079.013.15

Shenaar-Golan, V., Wald, N., & Yatzkar, U. (2021). Parenting a child with mental health problems: The role of self-compassion. *Mindfulness, 12*, 2810–2819. https://doi.org/10.1007/s12671-021-01744-z

Sirois, F. M., Bögels, S., & Emerson, L.-M. (2019). Self-compassion improves parental well-being in response to challenging parenting events. *The Journal of Psychology, 153*(3), 327–341. https://doi.org/10.1080/00223980.2018.1523123

Tanaka, M., Wekerle, C., Schmuck, M. L., Paglia-Boak, A., & MAP Research Team. (2011). The linkages among childhood maltreatment, adolescent mental health, and self-compassion in child welfare adolescents. *Child Abuse & Neglect, 35*(10), 887–898. https://doi.org/10.1016/j.chiabu.2011.07.003

Tandler, N., & Petersen, L.-E. (2018). Are self-compassionate partners less jealous? Exploring the mediation effects of anger rumination and willingness to forgive on the association between self-compassion and romantic jealousy. *Current Psychology: A Journal for Diverse Perspectives on Diverse Psychological Issues, 39*, 750–760. https://doi.org/10.1007/s12144-018-9797-7

Temel, M., & Atalay, A. A. (2018). The relationship between perceived maternal parenting and psychological distress: Mediator role of self-compassion. *Current Psychology, 39*, 2203–2210. https://doi.org/10.1007/s12144-018-9904-9

Upton, K. V. (2018). An investigation into compassion fatigue and self-compassion in acute medical care hospital nurses: A mixed methods study. *Journal of Compassionate Health Care, 5*(7). https://doi.org/10.1186/s40639-018-0050-x

Wong, C. C. Y., Mak, W. W. S., & Liao, K. Y.-H. (2016). Self-compassion: A potential buffer against affiliate stigma experienced by parents of children with autism spectrum disorders. *Mindfulness, 7*(6), 1385–1395. https://doi.org/10.1007/s12671-016-0580-2

Xu, S., Zhang, H., & Wang, J. (2020). Caregiver burden and depression among Chinese family caregivers: The role of self-compassion. *Mindfulness, 11*(7), 1647–1654. https://doi.org/10.1007/S12671-020-01378-7

Yarnell, L. M., & Neff, K. D. (2013). Self-compassion, interpersonal conflict resolutions, and well-being. *Self and Identity, 12*(2), 146–159. https://doi.org/10.1080/15298868.2011.649545

Yip, V. T., & Tong, E. (2021). Self-compassion and attention: Self-compassion facilitates disengagement from negative stimuli. *The Journal of Positive Psychology, 16*(5), 593–609. https://doi.org/10.1080/17439760.2020.1778060

Zessin, U., Dickhäuser, O., & Garbade, S. (2015). The relationship between self-compassion and well-being: A meta-analysis. *Applied Psychology: Health and Well-Being, 7*(3), 340–364. https://doi.org/10.1111/aphw.12051

Zhang, J. W., & Chen, S. (2017). Self-compassion promotes positive adjustment for people who attribute responsibility of a romantic breakup to themselves. *Self and Identity, 16*(6), 732–759. https://doi.org/10.1080/15298868.2017.1305985

Zhang, J., Chen, S., & Tomova Shakur, T. (2019). From me to you: Self-compassion predicts acceptance of own and others' imperfections. *Personality and Social Psychology Bulletin*, 1–15. https://doi.org/10.1177/0146167219853846

Self-Compassion in Parenting

16

Helena Moreira

Introduction

Being a parent is one of the most significant and gratifying experiences in an individual's life, but it is also one of the most challenging and demanding tasks. Raising a child takes courage and commitment, time and dedication, and it is often exhausting physically and emotionally. In everyday life, parenting is often experienced more as a set of tasks to be accomplished than as pleasurable interaction with the child. Most parents make a concerted effort to reconcile work and family-related responsibilities and face numerous daily stressors associated with parenting (e.g., taking children to different activities on time, preparing meals, soothing a baby that will not stop crying, dealing with a toddler's tantrum or with a teenager's defiant behaviors). Thus, although in the relationship with children there are many moments of joy and deep connection, there are also many moments of stress that often lead parents to experience strong negative emotions such as anxiety, frustration, sadness, guilt, and even anger and resentment directed at both themselves and the child.

Experiencing high levels of parenting stress is becoming increasingly prevalent in modern society (Barroso et al., 2018; Roskam et al., 2021), and all parents, regardless of their sociocultural background and mental health status, experience some degree of parenting stress at some point in their lives (Crnic & Greenberg., 1990; Deater-Deckard, 2004). Parenting stress is one of the strongest risk factors for a range of adverse parenting outcomes (e.g., Anthony et al., 2005; Niu et al., 2018), and consequently, it can seriously affect child development and increase the likelihood of psychopathology and adjustment problems (Deater-Deckard & Panneton, 2017; Deater-Deckard, 1998). For instance, among clinically referred and community samples, parenting stress was shown to be associated with children's externalizing problems (Schleider et al., 2015; Stone et al., 2016), internalizing problems (Stone et al., 2016; Rodriguez, 2011), lower quality of life (Moreira et al., 2015), aggressive behavior (Krahé et al., 2015), and decreased coping competence (Moreland et al., 2016).

Therefore, it is essential to identify modifiable resources that can help parents experience lower levels of parenting stress. Self-compassion and mindfulness are modifiable internal resources that can help parents feel less stressed and that can be developed through mindfulness- and compassion-based programs, such as the Mindful Self-Compassion program (MSC; Neff & Germer, 2013), Compassion-Focused Therapy (CFT; Gilbert, 2009b), the Mindfulness-Based Stress Reduction program (MBSR; Kabat-Zinn,

H. Moreira (✉)
Center for Research in Neuropsychology and Cognitive-Behavioural Intervention, University of Coimbra, Coimbra, Portugal

A. Finlay-Jones et al. (eds.), *Handbook of Self-Compassion*, Mindfulness in Behavioral Health,
https://doi.org/10.1007/978-3-031-22348-8_16

1990), or Mindfulness-Based Cognitive Therapy (MBCT; Segal et al., 2002). In recent years, these programs and their adaptations have been offered to parents who experience high stress or specific challenges in parenting (e.g., Bazzano et al., 2015; Neece et al., 2019; Weitlauf et al., 2020), showing promising results in reducing parents' stress levels and in improving their well-being. The inclusion of mindfulness and compassion components in parenting interventions has also been found to increase the effectiveness of these interventions in reducing parental stress, anxiety, and depression and in promoting mindfulness and self-compassion skills (Jefferson et al., 2020; Coatsworth et al., 2014). At the same time, mindful parenting programs aimed at promoting a mindful approach to parenting (e.g., Mindful Parenting Training, Bögels et al., 2014; Mindful With Your Baby, Potharst et al., 2017, 2022, 2019) have also been shown to be effective in reducing parental stress and in improving several child outcomes (e.g., internalizing and externalizing problems; Burgdorf et al., 2019). Thus, while parenting stress can be a risk factor for negative parenting, several intervention studies have shown that it is possible to help parents better cope with stress by developing skills such as self-compassion and mindfulness.

In this chapter, we will review the empirical evidence that documents the role of parents' self-compassion in parenting. Then, we propose a conceptual model that describes how incorporating self-compassion and mindfulness into the parent–child relationship can help parents adopt more positive parenting behaviors and experience less parenting stress. In this model, we also propose that this parental approach contributes to a child's secure attachment and the development of important internal resources (e.g., self-compassion, emotional regulation) and, consequently, to a better child's mental health. The empirical evidence that supports the associations established in the model is thoroughly discussed.

Self-Compassion as an Inner Resource Associated with Better Parenting Outcomes

Self-compassion is an adaptive type of self-to-self relating that involves the recognition of and the desire to alleviate one's own suffering with kindness and care and in a nonjudgmental way. It is grounded in the recognition that we all share a common human condition (i.e., that all human beings are imperfect; Neff, 2003b). In recent years, research into the role of self-compassion in a variety of individual and interpersonal outcomes has grown exponentially. Self-compassion has been consistently associated with healthy psychological functioning (Neff et al., 2007, 2018) and, although with less empirical support, with adaptive interpersonal functioning (Neff & Pommier, 2013; Yang et al., 2019).

With regard to psychological functioning, there is already solid evidence from meta-analysis studies that self-compassion is strongly associated with lower levels of psychological distress in adolescents ($r = -.55$; Marsh et al., 2018) and with lower levels of psychopathology ($r = .54$; MacBeth & Gumley, 2012) and higher levels of well-being in adults ($r = .47$; Zessin et al., 2015). Increasing evidence suggests that self-compassion is not only associated with greater well-being (e.g., Neff et al., 2018) and healthy psychological functioning (e.g., Bluth et al., 2017; Krieger et al., 2013; Neff, 2003a) in non-clinical populations but it also seems to help individuals better cope with challenging life circumstances, including infertility (Galhardo et al., 2013), cancer (Pinto-Gouveia et al., 2014), trauma (Thompson & Waltz, 2008), pain (Wren et al., 2012), divorce (Sbarra et al., 2012), and HIV (Kemppainen et al., 2013).

With regard to interpersonal functioning, some studies have shown that self-compassionate individuals tend to exhibit more prosocial behaviors (Yang et al., 2019) and to be more compassionate toward other people (e.g., Crocker & Canevello, 2008; Gillath et al., 2005; Mikulincer et al., 2005; see Chap. 15 for a detailed discussion). In addition, it has been demonstrated that compassion toward the self and others stimulate identical parts of the brain (Longe et al., 2010), which suggests that a common process may underlie both self-compassion and compassion toward other people, including toward one's own children. According to this view, Neff and Pommier (2013) found that among adults from the general population and practicing meditators, higher levels of self-compassion were associated with greater compassion for humanity, empathetic concern for others, and altruism. Among adults from the general population, practicing meditators, and undergraduate students, higher levels of self-compassion were also associated with higher levels of perspective-taking and forgiveness and lower levels of personal distress.

The results of studies that explored the contribution of self-compassion to romantic and other interpersonal relationships (Neff & Beretvas, 2013; Yarnell & Neff, 2013) also support the view that self-compassionate individuals are more likely to focus on others' concerns. For instance, Neff and Beretvas (2013) found that self-compassionate individuals were described by their partners as more caring, accepting, and supportive of their partner's autonomy. Despite the empirical evidence supporting the view that self-compassion and compassion or concern for others are positively related, it is important to note that in some studies this association was shown to be weak (Mills et al., 2018) or even nonsignificant (Gerber et al., 2015). This is because often people are very compassionate toward others but not toward themselves. A key finding, however, is that self-compassion training increases compassion for others (Neff & Germer, 2013). Furthermore, it is important to bear in mind that the measurement of compassion is still problematic, as different authors operationalize and assess compassion for others in different ways (Strauss et al., 2016). In addition, people might differ with regard to their fears of expressing and receiving compassion (for others, from others, and for self; Gilbert et al., 2011), which can make it easier for some people to feel self-compassion and more difficult to feel compassion for others and vice versa.

Although the parent–child relationship is a particular type of interpersonal relationship in which self-compassion may be important, the role of self-compassion in parenting has only recently begun to be explored. Among the extant literature, several studies suggests that parents' self-compassion can play a key role in how parents perceive and experience parenting and their relationships with their children (Jefferson et al., 2020). For instance, studies conducted among parents of school-aged children and adolescents from the general population have shown that higher levels of self-compassion are positively associated with lower parenting stress (Gouveia et al., 2016; Moreira et al., 2015) and with more authoritative and less permissive and authoritarian parenting styles (Gouveia et al., 2016). In a study that included parents with a history of depression, greater self-compassion was found to be associated with lower levels of mothers' child-directed criticism and with lower levels of fathers' distressed reactions to their children's negative emotions (Psychogiou et al., 2016). In addition, Psychogiou et al. (2016) have found that parents who reported higher levels of self-compassion tended to attribute the cause of their children's behaviors to external factors (such as situational demands and environmental constraints) rather than to internal factors (i.e., global and stable personal traits and attributes), which may have important implications for parenting as this external locus of control can promote a less critical and a more open stance toward the child's behavior.

Gilbert's tripartite model of affect regulation (Gilbert, 2005a, 2009a) can offer insight into why self-compassionate parents may feel less stressed by the daily challenges of rearing a child and be better able to be more compassionate toward their children and other people in general. This model postulates three evolved systems oriented toward threats (the threat and self-protection system, which is responsible for detecting threats and quickly selecting a response, such as fight, flight, or freeze), resources (the drive system, which guides resources seeking and goal attainment), and affiliation (the affiliative and soothing system, which guides attachment and caregiving behaviors and is responsible for generating feelings of contentment, soothing, safeness, and connectedness). According to this model, capacities for compassion for oneself and others have the same physiological underpinnings; in other words, they are rooted in the affiliative and soothing system of affect regulation (Depue & Morrone-Strupinsky, 2005; Gilbert, 2005a, 2009a, b). Accordingly, those who are more able to access their affiliative/soothing system are more likely to experience higher levels of compassion for self and others. In addition, because (self-)compassion facilitates access to this soothing system and deactivates the threat system, self-compassionate individuals can more easily regulate their emotions and soothe themselves in times of stress and suffering (Gilbert, 2009b). In contrast, individuals with low levels of self-compassion usually find it difficult to access the soothing system, and the threat and/or drive systems can be easily and frequently activated. Therefore, in stressful moments with a child, self-compassionate parents may be better able to deactivate the threat system (which would most likely lead to a fight, flight, or freeze response, such as screaming, criticizing, or punishing their child or themselves) and more effectively counteract the negative emotions that may arise by soothing themselves and, consequently, responding in a more compassionate, calm, and regulated way.

Self-Compassion and Raising a Child in Challenging Circumstances

Many parents face major challenges in raising their children. For instance, caring for a child with a developmental problem (e.g., autism, intellectual disability) or with a chronic health condition is a demanding and potentially stressful task for parents that may result in substantial levels of parenting stress and adjustment difficulties (e.g., Kiami & Goodgold, 2017; Moreira & Canavarro, 2016; Moreira et al., 2013a, b). The results from a few studies suggest that self-compassion can play an important role in how parents cope with the unique challenges of raising a child with a developmental or chronic health problem (Cousineau et al., 2019). For instance, in a sample of parents and primary caregivers of children with a burn injury, self-compassion was found to predict fewer symptoms of depression and of posttraumatic stress syndrome (Hawkins et al., 2019), while in a sample of parents of children with a range of chronic illness, self-compassion was inversely associated with parental burnout (Gerber et al., 2021). Together, these findings suggest that experimental studies examining the impact of self-compassion intervention on parental mental health and burnout are warranted, particularly among parents who have a child with additional needs.

There is also some evidence that self-compassion can help parents of children diagnosed with autism spectrum disorder be more emotionally resilient and better cope with the stress associated with raising an autistic child. For instance, Neff and Faso (2015) found that parents of an autistic child with greater self-compassion exhibited less parental stress and depression, were more satisfied with life and more hopeful about the future, experienced greater goal reengagement (i.e., were more willing to pursue new goals when prior ones are unattainable), and had a more functional relationship with their children. In two recent studies also conducted among parents of autistic children, self-compassion was associated with lower parenting stress, higher levels of subjective well-being, and better quality of life (Bohadana et al., 2019; Torbet et al., 2019). Similar results were found in a study that included parents of adults with an intellectual or developmental disability, in which self-compassion was shown to be associated with lower levels of depressive symptoms and stress (Robinson et al., 2018).

Self-compassion also seems to play an important role in how parents of autistic children experience different forms of stigma. Specifically, self-compassion seems to protect individuals against feelings of affiliate stigma (i.e., internalized stigma experienced by family members or primary caregivers of stigmatized individuals; Torbet et al., 2019; Wong et al., 2016), public stigma (i.e., negative perceptions directed toward individuals with a disability or with a mental or physical disease), and courtesy stigma (i.e., negative perceptions directed toward individuals because of their association with stigmatized individuals; Torbet et al., 2019). It also seems to protect parents against psychological distress linked to affiliate stigma. For instance, Wong et al. (2016) found that affiliate stigma was significantly associated with greater psychological distress only when parents had lower levels of self-compassion; among those who exhibited high levels of self-compassion, affiliate stigma was not associated with psychological distress.

Taken together, the results of these studies suggest that self-compassion can serve as an adaptive coping strategy for parents of children with additional challenges, including developmental and/or health difficulties. These parents devote a large part of their time and energy to caring for their children and can often feel more exhausted and stressed than parents of typically developing children (Lindström et al., 2010; Pinquart, 2018). Arguably, self-compassion can help these parents create a healthier balance between the demands of caring for their children and caring for themselves. In addition, by cultivating self-compassion, parents may feel less isolated in their suffering because they recognize that suffering is part of the human experience and that many parents deal with similar difficulties and circumstances. They may also feel less guilty or responsible for their child's problems and distress and be better able to accept their mistakes and limitations as parents as well as their children's limitations and difficult behaviors.

A Mindful and Compassionate Approach to Parenting

Self-compassion is not only an internal resource that can help parents better deal with the challenges of parenting but it can also be a defining element of parenting itself. In recent years, there has been growing interest among clinicians, researchers, and the population at large in so-called "mindful parenting," a parental approach that applies mindfulness- and compassion-based principles to the parent–child relationship (Bögels & Restifo, 2014; Kabat-Zinn & Kabat-Zinn, 1997). Mindful parenting was first described by Kabat-Zinn and Kabat-Zinn (1997) as a type of parenting characterized by being present and paying nonjudgmental attention to the child. Based on the work of Kabat-Zinn and Kabat-Zinn (1997) and subsequent research (e.g., Duncan et al., 2009; Bögels & Restifo, 2014), mindful parenting has been described as a parenting approach that involves not only bringing mindful awareness to the parent–child relationship but also adopting a compassionate stance toward oneself as a parent and toward the child. Although the designation *mindful parenting* highlights only the mindfulness dimension, it is important to note that this approach also implies greater acceptance and compassion for oneself and for the child. Accordingly, we consider that it can be better described as a *mindful and compassionate parenting approach*, a designation that we will adopt in the current work. Another important defining characteristic of this parenting approach is the parents' greater ability to regulate their emotions and behaviors in interactions with the child, as opposed to reacting automatically. Therefore, mindful and compassionate parenting can be described as a parenting approach that encompasses three essential components: (1) mindfulness (which allows parents to be truly present when they interact with their children, listening to them with full attention, and identifying their emotional changes even if they are subtle), (2) compassion (for oneself and the child), and (3) self-regulation (which allows parents to act calmly and in line with their values and goals and hinders them from reacting automatically).

Grounded in theoretical and empirical research on mindfulness and mindfulness-based interventions (Baer, 2003; Kabat-Zinn, 2003), Duncan et al. (2009) proposed a theoretical model to explain how mindful parenting can have positive effects on the psychological functioning of parents and children. According to this model, mindful parenting can influence child psychosocial adjustment and functioning (e.g., self-regulation, psychological functioning) through its effect on several aspects of the parent–child relationship, including parenting (e.g., communication, parental self-efficacy), child management practices, parental well-being, and parent–child affection.

According to the authors, mindful parenting encompasses five interrelated dimensions: (1) listening with full attention to the child (i.e., directing complete attention to the child and being fully present during parent–child interactions); (2) nonjudgmental acceptance of the self and the child (i.e., accepting the characteristics and behaviors of the child, the self as a parent, and the challenges of parenting); (3) emotional awareness of the self and the child (i.e., noticing and correctly identifying one's emotions when interacting with the child as well as the child's emotions); (4) self-regulation in the parenting relationship (i.e., being able to regulate one's own emotions and behaviors in parent–child interactions and being able to pause before reacting); and (5) compassion toward the self as a parent and toward the child (i.e., being kind to and supportive of the child, sensitive and responsive to the child's needs, and compassionate toward oneself as a parent).

A Conceptual Model of a Mindful and Compassionate Approach to Parenting

To extend the previous work on mindful parenting, we propose an integrative conceptual model of a mindful and compassionate parenting approach that aims to highlight the main factors that may account for the individual differences in mindful and compassionate parenting and explain how this parenting approach can influence parenting behaviors and the socioemotional functioning of children and adolescents (see Fig. 16.1). In developing this model, we drew from the Duncan et al.'s (2009) model of mindful parenting, the process model of the determinants of parenting (according to which parenting is multidimensional and can be influenced by intersecting parent, child, and social variables; Belsky, 1984), and the existing empirical evidence on the determinants and effects of mindful and compassionate parenting. Starting from the mindful parenting model (Duncan et al., 2009) and taking into account subsequent studies that showed a clear separation between child-focused and parent- or parenting-focused mindful parenting dimensions (e.g., de Bruin et al., 2014; Moreira

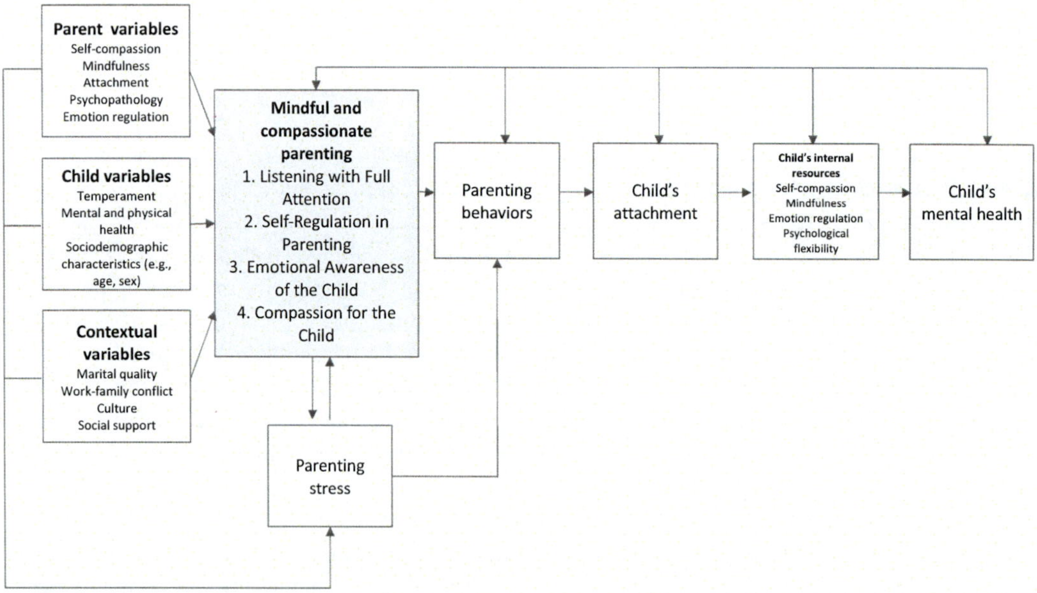

Fig. 16.1 Conceptual model of a mindful and compassionate approach to parenting

& Canavarro, 2017), we conceptualize mindful and compassionate parenting as a multidimensional construct that encompasses the following interrelated dimensions: (1) listening with full attention, which pertains to the ability to listen to the child with full attention; (2) self-regulation in parenting, which pertains to the ability to self-regulate in parent–child interactions; (3) emotional awareness of the child, which pertains to the ability to be aware of the child's emotions; (4) compassion for the child, which pertains to an attitude of compassion toward the child (i.e., an attitude of kindness, sensitivity, and responsiveness to the child's needs); and (5) nonjudgmental acceptance of parental functioning, which pertains to an attitude of nonjudgmental acceptance of parental functioning (i.e., accepting without judging the self as a parent and the challenges of parenting).

In the following sections, we will analyze in detail each component of the model as well as the empirical evidence supporting the proposed associations, as illustrated in Fig. 16.1.

Determinants of Mindful and Compassionate Parenting

According to the model, mindful and compassionate parenting, like other parenting styles and practices, is multidimensional and can be influenced by parents' characteristics (e.g., personality and mental health), children's characteristics (e.g., temperament), and contextual sources of stress and support (e.g., parents' work context, marital relationship, and social networks/support; Belsky, 1984). Knowing which factors may account for the variability in this parenting approach and understanding why some parents are more likely to be mindful and compassionate in the parent–child relationship than others may contribute to a better understanding of this construct and may have important clinical implications, such as the identification of modifiable factors that can be targeted in mindfulness- and compassion-based parenting programs.

Parent Variables All parents have psychological, physiological, and experiential differences that interact with child characteristics to influence the parent–child relationship. Certain parental characteristics, such as self-compassion, mindfulness, attachment, emotion regulation, and mental health, may be particularly important determinants of the quality of mindful and compassionate parenting.

Self-Compassion and Mindfulness There is some evidence that parents who are more mindful with regard to their own experiences (Gouveia et al., 2016; Parent et al., 2016b) and have high levels of self-compassion (Gouveia et al., 2016; Moreira et al., 2016) are more able to enact mindfulness and compassion in their parenting relationships. While these associations are not surprising, since mindfulness and self-compassion are essential and defining elements of a mindful and compassionate parenting approach, they have important clinical implications, as they suggest that the cultivation of these individual resources or skills can spill over into the relational domain and influence the relationship between parents and children.

Each of the three positive components of self-compassion (i.e., self-kindness, common humanity, and mindfulness; Neff, 2003b) can help explain why self-compassion and mindfulness can contribute to mindful and compassionate parenting. First, higher levels of self-kindness (i.e., facing difficulties, failures, and suffering with a caring and understanding attitude rather than being self-critical and judgmental) may help parents accept without judgment, the shortcomings and limitations they perceive in themselves as parents and those they perceive in their children. Rather than criticizing themselves and criticizing the child, self-compassionate parents may more easily adopt an attitude of kindness and acceptance, particularly in times of parenting stress. For example, after yelling at her child who was having a tantrum, a mother with higher levels of self-compassion may more easily view the situation as a moment of suffering during which she also deserves comfort and acceptance instead of criticizing herself for her behavior ("I am a terrible mother for screaming at my son").

Second, a greater sense of common humanity (i.e., acknowledging that one's painful experiences are part of a common human condition rather than feeling isolated, ashamed, or different) can help parents understand that difficulties and suffering are part of parenting, rather than interpreting their struggles as unique personal failures. By understanding that *all parents fail, suffer, and make mistakes*, self-compassionate parents may be able to more easily accept their limitations and imperfections as parents without criticizing themselves and feeling isolated. In addition, they may be less likely to react to child misbehavior with anger or punishment as they are less likely to interpret these behaviors as a parenting "failure."

Third, the greater capacity of self-compassionate individuals to be mindful of their painful emotions and thoughts instead of denying, avoiding, or ruminating about them can help them to have a greater ability to decenter and not overidentify with negative thoughts about their children or their parental role. For example, such parents might think of a time when they were unable to validate their child's emotions without engaging in ruminative thoughts about that specific behavior or other negative aspects of their parenting and without automatically believing that moment reflects poor parenting skills. In addition, this greater capacity for mindfulness may help parents become more aware of their own emotional states and of the emotional states of their children, allowing them to more easily regulate their behaviors and emotions in interactions with the child. As will be discussed later, this greater capacity for self-regulation is essential in enabling parents to be less reactive in interactions with the child and to act in accordance with their true parenting values by adopting more sensitive and responsive parenting behaviors.

Parents' Attachment Experiences Parents' insecure attachment with their own parents has been shown to predict increased difficulties in practicing mindful and compassionate parenting (Moreira et al., 2016; Moreira & Canavarro, 2015). These results are consistent with many studies that demonstrate that a secure attachment is an important determinant of a caring and supportive relationship with one's children (for a review, see Jones et al., 2015a, b). In general, secure attachment has been consistently related to a wide range of positive parenting characteristics and outcomes, such as higher parental sensitivity, responsiveness, and supportiveness; less parenting stress; and stronger feelings of closeness to one's children. Conversely, parents' insecure attachment (i.e., high levels of attachment-related avoidance and/or anxiety) has been associated with more negative parental caregiving behaviors, emotions, and cognitions (Jones et al., 2015a, b). Specifically, it has been suggested that because avoidant individuals have more difficulty in assuming caregiving roles (Gillath et al., 2005; Mikulincer et al., 2005), they tend to experience more stress in their parenting roles (e.g., Moreira et al., 2015; Nordahl et al., 2020; Rholes et al., 2006), perceive parenthood as less meaningful and satisfying (Cohen & Finzi-Dottan, 2005; Vieira et al., 2012), and behave in a less mindful (Moreira & Canavarro, 2015; Moreira et al., 2015, 2016), warm, close, and supportive manner toward their children (Edelstein et al., 2004; Rholes et al., 1995) compared with securely attached parents. Avoidant parents may also feel more uncertain about or less interested in having children (Rholes et al., 1997, 2006) and they may lack confidence in their parenting abilities (Rholes et al., 1995). In comparison, anxiously attached individuals tend to be self-centered, worry about their own attachment needs, and feel distressed when other people, including their children, need their assistance (Gillath et al., 2005; Mikulincer et al., 2005). Like avoidant parents, they may also report more negative attitudes toward parenthood than secure parents do (Rholes et al., 1997), experience increased parenting stress (Moreira et al., 2015; Rholes et al., 2006), and behave in a less mindful and compassionate manner toward their children (Moreira & Canavarro, 2015; Moreira et al., 2015, 2016). In contrast to avoidant individuals, they are less able to provide a secure base for their children's exploration behaviors (Adam et al., 2004) and they tend to have more idealized and perfectionist conceptions of themselves as future parents (Mikulincer & Shaver, 2016).

Emotion Regulation Relatedly, some studies point to emotion regulation as a key variable influencing one's tendency to enact mindful and compassionate parenting practices. Specifically, emotion regulation difficulties such as rumination can significantly interfere with parents' ability to practice mindful and compassionate parenting. One study has shown that higher levels of self-critical rumination, a specific type of rumination focused on the content of self-critical thoughts (Smart et al., 2016), predicted higher levels of parenting stress and that this association was mediated by two dimensions of mindful and compassionate parenting: emotional awareness of the child and nonjudgmental acceptance of parental functioning (Moreira & Canavarro, 2018a).

The hypothesis that rumination interferes with this parenting approach is also supported by studies showing that parents who tend to experience negative affect more frequently and persistently (i.e., who present higher levels of neuroticism; Moreira et al., 2020) and who experience more anxious and depressive symptoms are less able to be mindful and compassionate in their relationships with their children (Fernandes et al., 2021; Moreira & Canavarro, 2018b). In fact, individuals high in neuroticism or who experience depressive and/or anxious symptomatology tend to ruminate and worry about negative events and to get caught in downward spirals of negative mood and thinking (Muris et al., 2005). Rumination and worry may occupy the attention and consume cognitive resources, thus making it difficult for parents to redirect their attention to the environment and, consequently, to be fully aware of the present moment when interacting with the child (Beebe et al., 2007; DeJong et al., 2016; Moreira & Canavarro, 2018a; Stein et al., 2012). Hence, rumination may affect parents' ability to process children's cues (Beebe et al., 2007; Stein et al., 2012), leading them to experience greater difficulty in perceiving their children's emotional expressions and to be less responsive and sensitive to their children's needs.

In addition, rumination is associated with greater attentional bias toward negative information (Donaldson et al., 2007). This attentional bias may, on the one hand, lead parents to focus more on their perceived limitations and shortcomings as parents, contributing to lower levels of nonjudgmental acceptance of parental functioning, and, on the other hand, lead parents to focus on the negative aspects of their child's behaviors, contributing to lower levels of compassion for the child. Based on Gilbert's (2010) tripartite model of affect regulation, one might think that this greater focus on the negative aspects of the self or the child may continually stimulate the threat system, as self-critical dialogue and rumination are often perceived as (internal) threats. This constant stimulation creates a difficult-to-break internal feedback loop between the content and focus of thoughts and the threat system, which maintains the sense of threat (Gilbert, 2010), making it very difficult for parents to implement a mindful and compassionate style of parenting that calls for the activation of the soothing system.

Parent Mental Health Some studies have suggested that experiencing more anxiety and depressive symptoms (Moreira et al., 2019; Fernandes et al., 2021) as well as more stress in the parental role (Cheung et al., 2019; Fernandes et al., 2021; Moreira et al., 2019) can predispose parents to more reactive parenting, which is consistent with previous research that has consistently shown that parents with psychopathology struggle to engage in optimal parenting (Lovejoy et al., 2000). Distressed parents might be more ruminative (Dar & Iqbal, 2015; Nolen-Hoeksema et al., 2008), and, as explained above, this may lead them to be less able to bring mindful attention and awareness to interactions with their children. In addition, anxiety and depression increase parents' focus on their own needs and promote self-focused attention (Dix & Meunier, 2009; Ingram, 1990). By focusing their attention on themselves and on their own needs, distressed parents are less likely to direct their attention toward their children and to be sensitive and responsive to their needs. Distressed parents may also be more self-critical and, consequently, they may be more likely to endorse a negative view of themselves as parents (Goodman & Gotlib, 1999). They may therefore struggle to accept their perceived shortcomings and limitations as parents and feel that they fail to meet their self-defined high standards in their relationships with their children.

Child Variables Children's characteristics, particularly their temperament, can also influence parenting behaviors. For instance, research has shown that children with a difficult temperament or who are highly reactive tend to elicit less optimal parenting behaviors, including harsh control (Kyrios & Prior, 1990; Porter et al., 2005), punitive and distressed reactions to children's negative emotions (Eisenberg et al., 1999), or inconsistent discipline (Lengua & Kovacs, 2005; Lengua, 2006). Conversely, fearful or behaviorally inhibited children tend to elicit more acceptance and protective responses from their parents (Kiel & Buss, 2012; Shamir-Essakow et al., 2005). There is also preliminary evidence that child-related factors can play a role in parents' capacity to enact mindful and compassionate parenting. For instance, one study found that mothers who have a more negative perception of their infant's temperament presented lower levels of mindful parenting (Fernandes et al., 2021). Consistent with our model, this relationship was shown to be both direct and mediated by parenting stress. Similar results were found among parents of school-aged children, with children's negative reactivity predicting lower levels of mindful parenting (Moreira et al., 2020). These findings are consistent with the previous research showing that a child's negative affectivity or reactivity may evoke more negative parenting, such as more controlling or hostile parental behaviors (Rothbart & Bates, 2007). Parents of children who frequently express strong and negative emotions (e.g., irritation, frustration, anger, sadness) may excessively focus on the child's negative behavior and thus find it difficult to adopt an attentive and compassionate stance in their parenting interactions or to regulate their emotions and behaviors when the child has, for example, an outburst of anger or a tantrum. In these situations, it can be particularly difficult for parents to regulate their emotions and to not act automatically in an attempt to regulate their child's behavior. Moreover, one may think that parents of children who display high levels of challenging behavior may be more critical of their own parental functioning and blame themselves for their child's behavior. Future research should investigate the role of other important child variables on mindful and compassionate parenting, including their mental and physical health and sociodemographic characteristics such as age and sex.

Contextual Variables The context in which parents are embedded also shapes their parenting behaviors (Belsky, 1984). For example, the support that parents receive from family and friends (Angley et al., 2015), the marital relationship (Gao et al., 2019), and the work context (Perry-Jenkins et al., 2017) have a considerable impact on parenting behaviors and on the overall quality of family life. Research on the role of contextual variables in mindful and compassionate parenting is still scarce; however, it stands to reason that parents who experience higher levels of environmental stress may experience more challenges to mindful and compassionate parenting. Accordingly, a study that explored work-related variables shown that parents with flexible work schedules presented significantly higher levels of self-regulation in parenting and nonjudgmental acceptance of parental functioning than parents with a shift work schedules (Moreira et al., 2019). This study also found that higher levels of work–family conflict (i.e., a conflict between the competing responsibilities and demands of work and family contexts, which leads to participation in the family role being hampered by participation in the work role; Weer & Greenhaus, 2014) were associated with lower levels of all mindful parenting dimensions. These relationships between work–family conflict and mindful parenting were mediated by anxiety/depression symptoms and parenting stress, which is congruent with the conceptual model we propose. This study shows that experiencing work-related stress can spill over into the parenting context, leading parents to experience higher levels of parenting stress and, in turn, engage in less mindful and compassionate parenting (Moreira et al., 2019). Less is understood about the role of macro-level stressors (such as socioeconomic stress) or protective factors (such as social support or healthy marital relationships) in adopting this parental approach.

Mindful and Compassionate Parenting and Parenting Stress

As presented in Fig. 16.1, and consistent with previous studies (Beer et al., 2013; Bögels et al., 2014; Fernandes et al., 2021; Gouveia et al., 2016; Moreira & Canavarro, 2018a), we propose that mindful and compassionate parenting creates favorable conditions for parents to experience lower levels of parenting stress. However, we also consider that the association between parenting stress and mindful and compassionate parenting is bidirectional and that parenting stress can impair the ability of parents to adopt a mindful and compassionate stance in parenting. Experiencing high levels of parenting stress may lead parents to experience parenting as more stressful and taxing than rewarding, which is not conducive to them being mindful and compassionate in their relationships with their children. Therefore, it can be particularly helpful for these parents to learn mindfulness and self-compassion skills. For instance, mindful parenting training (Bögels & Restifo, 2014), an 8-week program for parents with high levels of parenting stress, based on the MBCT (Segal et al., 2002) and on the MBSR (Kabat-Zinn, 1990), has been shown to be effective in reducing parenting stress in parents of infants (Potharst et al., 2017), toddlers (Potharst et al., 2021), and children/adolescents (Bögels et al., 2010, 2014).

Mindful and Compassionate Parenting and Parenting Behaviors

Based on prior research (e.g., Duncan et al., 2015; Parent et al., 2016a; Wang et al., 2018), we propose that mindful and compassionate parenting can influence other parenting practices and behaviors. We consider that bringing the qualities of mindfulness and compassion to parenting can help parents establish a relationship with their children guided by greater sensitivity, responsiveness, acceptance, and connection. For instance, *listening with full attention* can help parents to be more attuned to their child's verbal and nonverbal communication, which may help them to be better able to notice and respond more accurately to their children's needs. Developing greater *awareness of the child's emotional states* can assist parents in correctly identifying their child's emotional states and reducing automatic patterns of response that may negatively affect parenting practices. By practicing greater *self-regulation in interactions with the child*, parents might be better able to avoid negative cycles of reactivity and maladaptive parenting interactions in stressful situations and to choose parenting behaviors that are consistent with their parenting values (Duncan et al., 2009). Mindful parenting is the opposite of automatic or reactive parenting, as mindfulness prevents automatic reactions to parenting stress and creates a "space" between the stressful event and the response in which parents can intentionally choose how to respond to the stressful situation (Bögels, 2020; Bögels & Restifo, 2014). Greater *compassion for themselves as parents* helps parents to accept their mistakes and limitations more easily and to realize that all parents struggle and make mistakes in their relationships with their children. Finally, greater *compassion for the child* may allow parents to accept their behaviors, thoughts, and emotions without judging them and automatically reacting to them. This greater acceptance helps parents to truly *see* and understand their children and, consequently, to act in a more sensitive and attuned way with the child (Bluth & Wahler, 2011). However, it is important to note that although being a mindful and compassionate parent entails adopting an attitude of compassion and kindness in the relationship with the child, it also entails imposing limits and boundaries that teach children social rules (Bögels, 2020; Bögels & Restifo, 2014).

A growing evidence base suggests that mindful and compassionate parenting is associated with more positive parenting styles and practices, such as more authoritative and less authoritarian styles (Gouveia et al., 2016; Williams & Wahler, 2010), greater warmth and affection (Duncan et al., 2015; Parent et al., 2016a; Wang et al., 2018), and positive reinforcement (Parent et al., 2016a). It has also been shown to be associated with less dysfunctional disciplinary practices, such as laxness, overreactivity, and verbosity (de Bruin et al., 2014) or harsh and inconsistent discipline (Duncan et al., 2015; Parent et al., 2016a). Similarly, intervention studies suggest that mindful parenting training can help parents adopt parenting styles characterized by less rejection and greater autonomy encouragement (Bögels et al., 2014), endorse a less overreactive parenting style (Emerson et al., 2021; Potharst et al., 2017, 2019), and be more responsive and affectionate and less hostile toward the child (Potharst et al., 2017). Mindful and compassionate parents seem also to be more likely to employ supportive emotion socialization practices (e.g., encouraging emotional expression, comforting the child, and assisting the child in problem-solving) when the child expresses a negative emotion and less likely to employ nonsupportive emotion socialization practices (e.g., being distressed, exhibiting punitive reactions, and minimizing the child's distress; McKee et al., 2017).

Other studies suggest that higher levels of mindful and compassionate parenting may help parents adopt a less controlling parenting approach. Lippold et al. (2015) observed that adolescents whose mothers reported higher levels of mindful parenting viewed their mothers as less overcontrolling, which, according to the authors, may suggest that these parents may be better able to accept children's need for autonomy and privacy and feel more comfortable giving their children more independence. Likewise, in a study with mothers of school-aged children, Moreira et al. (2020) found that higher scores in the mindful and compassionate parenting dimension pertaining to the nonjudgmental acceptance of parental functioning were negatively associated with overprotection behaviors. These results suggest that parents who feel more compassion for themselves as parents and are better able to accept their limitations and imperfections in the parenting role may feel more confident in their parenting skills and, thus, more secure about granting their children more autonomy. Interestingly, the results of this study also showed that being better able to notice the child's emotions and to adopt a compassionate attitude toward the child can contribute to higher levels of overprotection behaviors. Mindful and compassionate parents are usually more attuned to the child; consequently, they may be better able to detect distress in their child and to connect to the child's suffering and thus feel more compelled to protect the child from situations perceived as threatening or that provoke distressing emotions in the child. Although overprotection is considered a nondesirable and intrusive parental behavior, it has a component of warmth that derives from parents' intention to protect their child from potential threats (Holmbeck et al., 2002). Hence, overprotection often reflects sensitive parenting (Buss & Kiel, 2011), which may explain why more compassionate parents can also be more overprotective.

By promoting higher quality parenting, the adoption of this present-centered, compassionate, and nonreactive parenting approach is a vehicle for promoting positive interactions between parents and children and higher quality parent–child communication. A study has shown that parents with higher levels of mindful parenting were more likely to ask their adolescent child for information (e.g., about school) and that children were more likely to disclose routine information about their activities with them (Lippold et al., 2015). In a more recent study, Lippold et al. (2019) corroborated that mindful parenting may foster better parent–child communication by showing longitudinally that mindful parenting predicted increased levels of adolescent disclosure and parental solicitation.

Mindful and Compassionate Parenting and Children's Outcomes

Attachment Considering the effect that mindful parenting may have on parental behaviors and, consequently, on the quality of the parent–child relationship, it has been suggested that mindful parenting is an important pathway for the development of a secure attachment relationship between the child and his or her parents (Duncan et al., 2009). There is some preliminary evidence from cross-sectional studies that mindful and compassionate parenting, and particularly the dimensions of listening with full attention and compassion for the child, can contribute to more positive and secure representations of their relationships with their parents in preschool and school-aged children and adolescents (Moreira et al., 2018; Medeiros et al., 2016; Zhang et al., 2019). Although most studies are cross-sectional, the effect of this parental approach on child outcomes seems to be both direct and indirect, occurring through parenting behaviors (Parent et al., 2016b) and child attachment (Moreira et al., 2018).

As will be explained below, secure attachment lays the foundation for the development of important psychological resources for the child's mental health, such as dispositional mindfulness, self-compassion, emotion regulation, and psychological flexibility. This hypothesis, and our model in general, is consistent with a recent intergenerational model that explains how self-compassion develops in the context of the parent–child relationship (Lathren et al., 2020). This model explains that parents with a secure attachment orientation have higher levels of self-compassion and, consequently, tend to exhibit more supportive responses when their child experiences difficult emotions (e.g., anger, sadness, shame). These responses, in turn, promote the development of a child's secure attachment, resulting in higher levels of self-compassion and more positive socioemotional and behavioral outcomes across several domains.

Children's Internal Resources In our model of mindful and compassionate parenting, we consider that this parenting approach can provide the necessary and fundamental foundations for the development of important internal resources (i.e., characteristics that, when present, help the child better cope with difficult emotions and suffering), such as emotion regulation, mindfulness, self-compassion, and psychological flexibility. We highlight these variables because they are well-known psychological resources that have a strong protective effect on children's and adolescents' psychological functioning and mental health (e.g., Bluth & Blanton, 2015; Bluth et al., 2017; Kashdan & Rottenberg, 2010; Livheim et al., 2016; Muris et al., 2016, 2017; Neff & McGehee, 2010) and because their relationships with this parenting approach has already received empirical support. Nevertheless, we recognize that other variables may be influenced by parenting and the perception of security that the child develops in the parent–child relationship and in turn influence the child's mental health.

Emotion Regulation Emotion dysregulation or difficulty-regulating emotions has been consistently identified as a transdiagnostic vulnerability factor that is linked to numerous difficulties and forms of psychopathology across the lifespan (Berking & Wupperman, 2012; Riediger & Klipker, 2014; Sheppes et al., 2015). Among children, maladaptive strategies of emotion regulation (e.g., rumination, catastrophizing) were shown to be associated with children's emotional problems and psychopathology, whereas adaptive strategies (e.g., positive reappraisal) seem to have a protective function with regard to children's mental health (e.g., Chan et al., 2016; Garnefski et al., 2007; Legerstee et al., 2010; Liu et al., 2016; Orgiles et al., 2018).

Research indicates that although the transmission of emotion regulation strategies from parents to children may have a genetic basis (Goldsmith et al., 2008), the family context, particularly parenting practices and behaviors, plays a critical role in the development of adaptive emotion regulation ability during childhood and adolescence (Eisenberg et al., 1998; Jaffe et al., 2010; McEwen & Flouri, 2009; Melnick & Hinshaw, 2000; Morris et al., 2007; Rutherford et al., 2015; Zeman et al., 2006). A few studies have shown that a mindful and compassionate approach to parenting may play an important role in shaping this capacity. For instance, in a study with parents of children aged between 3 and 7 years, mindful parenting and the quality of parent–child attachment mediated the association between parents' dispositional mindfulness and children's emotional lability/negativity and emotion regulation (Zhang et al., 2019). According to the authors, mindful parents tend to accept children's emotions and fulfill their emotional needs, which are factors known to facilitate secure attachment and the development of adaptive emotion regulation skills (Gottman et al., 1996; Morris et al., 2007). Similarly, in a study that included mother–adolescent dyads, Moreira and Canavarro (2019) found that higher levels of mindful parenting, particularly in the dimensions of compassion for the child, listening with full attention, and nonjudgmental acceptance of parental functioning, were associated with lower levels of emotion regulation difficulty in adolescents.

The processes through which mindful and compassionate parenting may foster adaptive emotion regulation in children and adolescents can be diverse. Morris et al.'s (2007) tripartite model of the impact of the family on children's emotion regulation and adjustment postulates that the family context may affect the development of children's emotion regulation via three processes: (1) parenting practices, (2) the emotional climate of the family, and (3) observation. First, parenting practices, particularly emotion-related practices (i.e., the way parents socialize their children with emotions; Eisenberg et al., 2003), may have a profound impact on children's regulatory ability. The existing research suggests that parents who employ supportive parental emotion socialization practices (i.e., those who allow children to express their emotions and who comfort, encourage, and help the child solve problems) usually have children who are better able to regulate their emotions (e.g., Jin et al., 2017), whereas nonsupportive reactions to children's negative emotions (e.g., punishing or minimizing children's emotional expression) are usually associated with children's emotional dysregulation and internalizing symptoms (e.g., Sanders et al., 2015). As already mentioned in a previous section of this chapter, a study has shown that mindful parenting is associated with more supportive emotion socialization practices (e.g., greater encouragement of children's emotional expression; more emotion-focused responses, such as comforting the child; and more problem-focused responses, such as helping the child solve a problem) and fewer nonsupportive practices (i.e., minimization of child distress, punitive reactions, and experience of distress; McKee et al., 2017). Thus, it can be hypothesized that mindful and compassionate parenting, as a parenting approach that facilitates appropriate emotion socialization, can also contribute to the development of more adaptive emotional regulation throughout development.

Second, the emotional climate of the family, which is determined by several processes and dynamics, including the parent–child attachment relationship, the parenting style, the marital relationship, and the emotional expressivity of family members (Morris et al., 2007), also plays an important role in how parents influence their children's emotion regulation abilities. In a positive family climate, children feel secure in feeling and expressing their emotions because they expect to be accepted and understood and they know that their emotional needs will be validated and satisfied. In contrast, when the emotional climate of the family is negative (e.g., due to negative parenting practices such as psychological control or negative affective interactions), children feel less secure and may experience greater difficulty adaptively regulating their emotions (Sim et al., 2009). Mindful parents seem to create a positive family climate, adopting parenting styles characterized by emotional validation, warmth, and affection (Duncan et al., 2015; Parent et al., 2016a; Wang et al., 2018), which in turn facilitates the development of a secure relationship with their children (Medeiros et al., 2016; Zhang et al., 2019) and creates the ideal foundation for the development of adaptive emotion regulation.

Third, children can learn emotion regulation strategies through observation or modeling. By repeatedly observing how their parents respond verbally and behaviorally to situations that elicit emotions (i.e., how they usually regulate their emotions), children can internalize these regulatory strategies and begin to use them in the same emotion-eliciting situations (Rutherford et al., 2015). Research has shown that mindful and compassionate parents are more likely to employ adaptive emotion regulation skills (Gouveia et al., 2019). In addition, they tend to present higher levels of self-compassion (Gouveia et al., 2016; Moreira et al., 2016) and dispositional mindfulness (Gouveia et al., 2016; Han et al., 2021), characteristics that likely help them to regulate their negative emotions more adaptively in times of stress or suffering (e.g., Chiesa et al., 2013; Fogarty et al., 2015; Roemer et al., 2015), particularly those that arise in interactions with their children. In fact, mindfulness is associated with a greater ability to attend to internal and external events, which in turn promotes the individual's ability to detect the need to implement emotion regulation strategies (Roemer et al., 2015). By promoting greater sensitivity to affective cues and an early awareness of subtle changes in emotional states that signal the need for control, mindfulness can enhance executive control and emotion regulation (Teper et al., 2013). In addition, as mindfulness involves nonjudgmental acceptance of emotions and thoughts and psychological flexibility rather than avoidance, it may also decrease the intensity of emotional responses and increase the tolerance of negative affect, which are important aspects of emotion regulation (Roemer et al., 2015) that children can learn through observation. Previous studies have also found a consistent link between self-compassion and self-report measures and biological indices of emotion regulation (Diedrich et al., 2014; Svendsen et al., 2016). Being self-compassionate entails not avoiding or repressing negative emotions and, instead, acknowledging them, viewing them as part of the human condition, and experiencing a true desire to alleviate one's own suffering (Neff, 2003b). Therefore, self-compassion allows individuals to have a more balanced perception of negative emotions and to cope with negative emotions without avoiding them, amplifying them, or overidentifying with them. In addition, self-compassionate individuals are less likely to criticize, blame, or judge themselves in difficult situations and are more likely to view painful situations as a normal part of life. Therefore, they usually cope better with stress and negative emotions and may consequently be better able to effectively regulate negative emotional states when they arise (Allen & Leary, 2010).

Therefore, we propose that because of their self-compassion and mindfulness skills, mindful and compassionate parents may be better able to regulate their emotions and behaviors in stressful moments, particularly those that arise in interactions with their children. They may also be less likely to criticize or blame themselves during difficult times with the child. As they observe their parents coping with stressful situations in an accepting and nonjudgmental manner and without acting impulsively, children may learn and internalize this adaptive way of regulating negative emotions. In contrast, if children observe their parents being self-critical or unable to regulate their emotions and behaviors in an adaptive manner, they may learn and internalize the maladaptive strategies their parents use to regulate their negative emotions (Moreira et al., 2018).

Dispositional Mindfulness As depicted in Fig. 16.1, mindfulness is another psychological resource that can be developed in the context of a mindful and compassionate parenting, both directly and indirectly through parenting behaviors and, particularly, through the development of a secure attachment in children. According to this hypothesis, some cross-sectional studies with parent–adolescent dyads have already shown that mindful and compassionate parenting is associated with higher levels of adolescent dispositional mindfulness (Moreira et al., 2018; Moreira & Canavarro, 2018b) and that this association can be mediated by the degree of security the adolescent perceives in the parent–child relationship (Moreira et al., 2018). These results are in line with previous studies that suggested that the capacity for mindful awareness is related to the quality of one's early relationships with attachment figures (Caldwell & Shaver, 2013; Pepping & Duvenage, 2016; Ryan et al., 2007). Individuals who were raised by sensitive and responsive caregivers (i.e., who were attuned to, mirrored, and empathized with the child's experiences) and who developed a secure attachment seem to be more likely to develop the reflective, regulatory, and self-observing capacities that characterize dispositional mindfulness (Caldwell & Shaver, 2013, 2015; Melen et al., 2017; Pepping et al., 2013, 2015; Ryan et al., 2007; Shaver et al., 2007). For instance, Pepping and Duvenage (2016) found that adolescents' recollection or current experiences of parental warmth and rejection were associated with their current levels of dispositional mindfulness through their attachment orientations. Although prior research has suggested that mindfulness skills have their roots in family dynamics and parent–child interactions, further studies are needed to establish that mindful parenting can lead to the development of mindfulness skills in children/adolescents.

Self-Compassion The way parents interact with their children may also have a strong impact on the development of children's self-compassionate or self-critical inner dialogues (Gilbert & Procter, 2006; Lathren et al., 2020; Neff, 2011). Although limited, existing research found that children of mindful and compassionate parents also tend to report higher levels of self-compassion (Moreira et al., 2018; Moreira & Canavarro, 2019). In fact, it has been argued that self-compassion develops in the context of positive and security-boosting interactions with attachment figures (Gilbert, 2005a; Gilbert & Procter, 2006; Neff & McGehee, 2010; Neff & Beretvas, 2013; Shaver et al., 2017). In an optimal caring environment with a consistently supportive and caring attachment figure, the child may develop internal working models of the self and others as being reliable and worthy of care and love (Collins et al., 2004), and the soothing system of affect regulation may develop adequately, allowing the child to cultivate compassionate self-to-self relationships (Gilbert, 2005b; Gilbert & Procter, 2006). These positive experiences promote the development of emotional memories of being soothed, protected, and cared for (Lee, 2012), allowing the individual to regulate his or her emotions with affection, warmth, and care when needed. In contrast, when parents are unresponsive or inconsistently responsive or are even neglectful or abusive, they do not provide the fundamental conditions for the development of secure attachment (Mikulincer & Shaver, 2016) and, consequently, for the development of self-compassion (Shaver et al., 2017). In such a relational context, the soothing system is underdeveloped and understimulated, and the child/adolescent is likely to become more self-critical and less self-compassionate (Gilbert, 2005a; Gilbert & Procter, 2006). Individuals raised in these relational contexts may have fewer available emotional memories of being loved and soothed when they need to regulate their emotions in difficult moments (Lee, 2012).

There is empirical evidence of the associations among early experiences with caregivers, attachment styles, and levels of self-compassion. For instance, some studies have shown that an individual's level of self-compassion is associated with his or her attachment orientation (Moreira et al., 2015, 2016; Neff & Beretvas, 2013). Neff and McGehee (2010) also found that adolescents with low levels of self-compassion are more likely to have critical mothers, dysfunctional family environments, and insecure attachment styles. Based on these results, the authors concluded that self-compassion can be viewed as an "internal reflection of the parent–child relationship" (Neff & McGehee, 2010, p. 236), which means that children with cold and critical parents may internalize a cold and critical internal dialogue, whereas those with warm and caring parents are more likely to internalize a self-compassionate way of relating to themselves.

Parents who adopt a compassionate and mindful approach in parenting tend to be more caring, available, and responsive to their child's needs, thereby creating the necessary foundation for the development of a secure attachment (Duncan et al., 2009; Medeiros et al., 2016) and, consequently, for the development of self-compassion (Gilbert, 2005a; Gilbert & Procter, 2006). In addition, as already mentioned, as these parents tend to have higher levels of self-compassion (Gouveia et al., 2016; Moreira et al., 2016) and of dispositional mindfulness (Gouveia et al., 2016; Parent et al., 2016b; Zhang et al., 2019), they may also model an adaptive way of coping with stressful life events and with difficult thoughts and emotions. According to Neff (2011), being self-compassionate in front of a child is one of the most powerful ways to help children develop self-compassion. If children see their parents coping with difficult situations in a mindful and compassionate manner, they may learn this adaptive self-to-self relating through observation. In contrast, children whose parents usually cope with difficult situations with self-criticism do not have a compassionate and mindful model to follow and, instead, may learn that difficult situations and moments of suffering should be handled with negative emotions and harsh self-judgment.

Psychological Flexibility Finally, our model also predicts that mindful and compassionate parenting may foster children's psychological flexibility, which is another psychological resource know to be associated with several indicators of emotional well-being (Kashdan & Rottenberg, 2010; Livheim et al., 2016; Muris et al., 2017). Psychological flexibility has been described as one's ability to stay in contact with the present moment and to change or persist in a certain behavior based on personal values (Hayes et al., 2006). The inability to remain in contact with the present moment and to engage in patterns of effective action that are linked to one's life values (i.e., psychological inflexibility) results from several processes, including experiential avoidance (unwillingness to experience certain unwanted private events, including thoughts, memories, emotions, and bodily sensations, and efforts to avoid, suppress, or eliminate those unwanted private experiences or otherwise control their frequency, form, or situational sensitivity; Hayes et al., 2012) and cognitive fusion (a process that refers to an attachment or entanglement with the content of private events and responding to the content of private events as if they were accurate representations of reality; Greco et al., 2008).

The limited research conducted on the relationship between mindful and compassionate parenting and children's psychological flexibility suggests that this parenting approach, particularly the dimension of listening with full attention, can indeed foster adolescents' psychological flexibility (Moreira & Canavarro, 2019). Parents who are mindfully aware in interactions with their child may model an attitude of being present, that is, an attitude of nonjudgmental awareness of psychological and environmental events as they occur, which contrasts with experiential avoidance. In addition, as already mentioned, mindful and compassionate parents seem to help their children develop mindfulness skills (Moreira et al., 2018), which may enable their children to better notice and observe their thoughts and feelings without becoming entangled in the content of those private events and behaving as if they were literally true.

Mental Health The ultimate outcome in our conceptual model is the child's mental health. From our perspective and based on the findings of various empirical studies, mindful and compassionate parenting can have a beneficial effect on a child's mental health by enabling the child to develop secure attachment and internal resources such as adaptive emotion regulation strategies, mindfulness, self-compassion, or psychological flexibility, which are known to promote adaptive psychological functioning and child mental health (Bluth & Blanton, 2015; de Bruin et al., 2011; Muris et al., 2016, 2017; Neff & McGehee, 2010). In fact, mindful parenting has been associated with diverse positive adjustment outcomes, including lower levels of internalizing (e.g., depression, anxiety) and externalizing (e.g., behavior problems) problems in children (Calvete et al., 2020; Han et al., 2021; Parent et al., 2016b), better generic or disease-specific quality of life

(Medeiros et al., 2016; Moreira et al., 2018; Serkel-Schrama et al., 2016), and a lower likelihood of adolescent substance use (Turpyn & Chaplin, 2016). Studies assessing the effects of mindfulness-based parenting programs on children's psychological functioning also support the role of this parenting approach on children's adjustment (Bögels et al., 2010, 2014; Coatsworth et al., 2010; van der Oord et al., 2012). For instance, Bögels et al. (2014) found that after completing a mindful parenting program, parents of children with a diagnosed psychiatric disorder reported that their children less often presented internalizing and externalizing difficulties.

Despite the increasing evidence demonstrating a positive association between this parenting approach and children's positive psychosocial adjustment, only a few studies have attempted to understand which mechanisms might be responsible for this relationship. As highlighted in the previous sections of this chapter, and consistent with our model, different variables (parental behaviors, child attachment, and child internal resources) may explain why mindful parenting can have such an important effect on children's mental health. Understanding these mechanisms can not only contribute to a significant advance of the scientific knowledge in this area, but also guide the development of parenting interventions. Therefore, future studies should continue to explore the processes linking this parental approach to the child's mental health, and particularly the processes that explain the therapeutic change observed in the psychological adjustment of children whose parents participated in a mindfulness- and compassion-based parenting intervention.

Conclusions

In this chapter, we have sought to demonstrate how parents' self-compassion can influence their parenting. Although research into the role of self-compassion in parenting is still in its infancy, there is preliminary empirical evidence that self-compassion can help parents experience less parenting stress and adopt more positive parenting styles and practices. Based on existing research and previous theoretical contributions, we proposed a conceptual model that seeks to explain how a parenting approach characterized by high levels of (self-)compassion and mindfulness (i.e., a mindful and compassionate parenting approach) can promote better mental health in children through a set of processes. This model stresses the importance of perceiving parenting as a process resulting from the influence of several interrelated factors, namely, parenting factors (e.g., parents' self-compassion), child factors (e.g., temperament), and contextual factors (e.g., work context). According to this model, this parenting approach fosters lower levels of parenting stress and parental behaviors that are more sensitive and responsive to the child's needs and that promote the development and strengthening of a secure attachment. In turn, a secure attachment establishes the necessary and essential basis for the development of critical internal resources, such as mindfulness, self-compassion, psychological flexibility, and emotional regulation, that protect the child's mental health throughout development.

The proposed model is an intergenerational model that highlights the importance of parenting practices as a privileged vehicle for conveying important skills such as self-compassion and mindfulness. In fact, theory and research have shown that we tend to treat our children and ourselves in the same way that we were treated as a child. Parents who had sensitive caregivers and developed a secure attachment learned to regulate their emotions more adaptively and were able to develop a compassionate intrapersonal relationship. These individual characteristics decisively influence the way parents later relate to their children. Secure and compassionate parents are more likely to adopt a mindful and compassionate stance in parenting, thereby providing the ideal conditions for the development of a secure attachment and for the flourishing of compassion and other protective factors of their child's mental health. Therefore, learning to be more mindful and compassionate, both in the relationship with oneself and children, can help parents break the maladaptive intergenerational cycles that perpetuate attachment insecurity, negative parenting styles, and poorer psychological functioning in children, who may later become parents and maintain or break these cycles.

The early stages of life represent an extraordinarily important opportunity to promote lifelong mental health. Considering the significant impact that the parent–child relationship has on children's development and mental health, helping parents to reduce parental stress and adopt more positive parenting behaviors through the development of greater self-compassion and of mindful and compassionate parenting could be an extraordinarily effective way to promote more adaptive developmental trajectories for future generations. Thus, future research should continue to focus on the study of the role of self-compassion in parenting as well as on the study of the different associations presented in the conceptual model described in the present work. It is also essential to continue to develop and study the effectiveness of parenting interventions designed to promote a mindful and compassionate parenting approach.

References

Adam, E. K., Gunnar, M. R., & Tanaka, A. (2004). Adult attachment, parent emotion, and observed parenting behavior: Mediator and moderator models. *Child Development, 75*(1), 110–122. https://doi.org/10.1111/j.1467-8624.2004.00657.x

Allen, A. B., & Leary, M. R. (2010). Self-compassion, stress, and coping. *Social and Personality Psychology Compass, 4*(2), 107–118. https://doi.org/10.1111/j.1751-9004.2009.00246.x

Angley, M., Divney, A., Magriples, U., & Kershaw, T. (2015). Social support, family functioning and parenting competence in adolescent parents. *Maternal and Child Health Journal, 19*(1), 67–73. https://doi.org/10.1007/s10995-014-1496-x

Anthony, L. G., Anthony, B. J., Glanville, D. N., Naiman, D. Q., Waanders, C., & Shaffer, S. (2005). The relationships between parenting stress, parenting behaviour and preschoolers' social competence and behaviour problems in the classroom. *Infant and Child Development, 14*(2), 133–154. https://doi.org/10.1002/icd.385

Baer, R. (2003). Mindfulness training as a clinical intervention: A conceptual and empirical review. *Clinical Psychology: Science and Practice, 10*(2), 125–143. https://doi.org/10.1093/clipsy.bpg015

Barroso, N. E., Mendez, L., Graziano, P. A., & Bagner, D. M. (2018). Parenting stress through the lens of different clinical groups: A systematic review & meta-analysis. *Journal of Abnormal Child Psychology, 46*(3), 449–461. https://doi.org/10.1007/s10802-017-0313-6

Bazzano, A., Wolfe, C., Zylowska, L., Wang, S., Schuster, E., Barrett, C., & Lehrer, D. (2015). Mindfulness Based Stress Reduction (MBSR) for parents and caregivers of individuals with developmental disabilities: A community-based approach. *Journal of Child and Family Studies, 24*(2), 298–308. https://doi.org/10.1007/s10826-013-9836-9

Beebe, B., Jaffe, J., Buck, K., Chen, H., Cohen, P., Blatt, S., Kaminer, T., Feldstein, S., & Andrews, H. (2007). Six-week postpartum maternal self-criticism and dependency and 4-month mother-infant self- and interactive contingencies. *Developmental Psychology, 43*(6), 1360–1376. https://doi.org/10.1037/0012-1649.43.6.1360

Beer, M., Ward, L., & Moar, K. (2013). The relationship between mindful parenting and distress in parents of children with an autism spectrum disorder. *Mindfulness, 4*(2), 102–112. https://doi.org/10.1007/s12671-012-0192-4

Belsky, J. (1984). The determinants of parenting: A process model. *Child Development, 55*(1), 83–96. https://doi.org/10.2307/1129836

Berking, M., & Wupperman, P. (2012). Emotion regulation and mental health: Recent findings, current challenges, and future directions. *Current Opinion in Psychiatry, 25*(2), 128–134. https://doi.org/10.1097/YCO.0b013e3283503669

Bluth, K., & Blanton, P. W. (2015). The influence of self-compassion on emotional well-being among early and older adolescent males and females. *The Journal of Positive Psychology, 10*(3), 219–230. https://doi.org/10.1080/17439760.2014.936967

Bluth, K., & Wahler, R. G. (2011). Does effort matter in mindful parenting? *Mindfulness, 2*(3), 175–178. https://doi.org/10.1007/s12671-011-0056-3

Bluth, K., Campo, R. A., Futch, W. S., & Gaylord, S. A. (2017). Age and gender differences in the associations of self-compassion and emotional well-being in a large adolescent sample. *Journal of Youth and Adolescence, 46*(4), 840–853. https://doi.org/10.1007/s10964-016-0567-2

Bögels, S. (2020). *Mindful parenting: Finding space to be – In a world of to do*. Pavilion Publishing & Media Ltd.

Bögels, S., & Restifo, K. (2014). *Mindful parenting: A guide for mental health practitioners*. Springer.

Bögels, S., Lehtonen, A., & Restifo, K. (2010). Mindful parenting in mental health care. *Mindfulness, 1*(2), 107–120. https://doi.org/10.1007/s12671-010-0014-5

Bögels, S., Hellemans, J., van Deursen, S., Römer, M., & van der Meulen, R. (2014). Mindful parenting in mental health care: Effects on parental and child psychopathology, parental stress, parenting, coparenting, and marital functioning. *Mindfulness, 5*(5), 536–551. https://doi.org/10.1007/s12671-013-0209-7

Bohadana, G., Morrissey, S., & Paynter, J. (2019). Self-compassion: A novel predictor of stress and quality of life in parents of children with autism spectrum disorder. *Journal of Autism and Developmental Disorders, 49*(10), 4039–4052. https://doi.org/10.1007/s10803-019-04121-x

Burgdorf, V., Szabó, M., & Abbott, M. J. (2019). The effect of mindfulness interventions for parents on parenting stress and youth psychological outcomes: A systematic review and meta-analysis. *Frontiers in Psychology, 10*, Article 1336. https://doi.org/10.3389/fpsyg.2019.01336

Buss, K. A., & Kiel, E. J. (2011). Do maternal protective behaviors alleviate toddlers' fearful distress? *International Journal of Behavioral Development, 35*(2), 136–143. https://doi.org/10.1177/0165025410375922

Caldwell, J. G., & Shaver, P. R. (2013). Mediators of the link between adult attachment and mindfulness. *Interpersona: An International Journal on Personal Relationships, 7*(2), 299–310. https://doi.org/10.5964/ijpr.v7i2.133

Caldwell, J. G., & Shaver, P. R. (2015). Promoting attachment-related mindfulness and compassion: A wait-list-controlled study of women who were mistreated during childhood. *Mindfulness, 6*(3), 624–636. https://doi.org/10.1007/s12671-014-0298-y

Calvete, E., Gómez-Odriozola, J., & Orue, I. (2020). Differential susceptibility to the benefits of mindful parenting depending on child dispositional mindfulness. *Mindfulness, 12*, 405–418. https://doi.org/10.1007/s12671-020-01467-7

Chan, S. M., Poon, S. F. O., & Hang Tang, E. M. (2016). Daily hassles, cognitive emotion regulation and anxiety in children. *Vulnerable Children and Youth Studies, 11*(3), 238–250. https://doi.org/10.1080/17450128.2016.1214887

Cheung, R. Y. M., Leung, S. S. W., & Mak, W. W. S. (2019). Role of mindful parenting, affiliate stigma, and parents' well-being in the behavioral adjustment of children with autism spectrum disorder: Testing parenting stress as a mediator. *Mindfulness, 10*(11), 2352–2362. https://doi.org/10.1007/s12671-019-01208-5

Chiesa, A., Serretti, A., & Jakobsen, J. C. (2013). Mindfulness: Top-down or bottom-up emotion regulation strategy? *Clinical Psychology Review, 33*(1), 82–96. https://doi.org/10.1016/j.cpr.2012.10.006

Coatsworth, J. D., Duncan, L. G., Greenberg, M. T., & Nix, R. L. (2010). Changing parent's mindfulness, child management skills and relationship quality with their youth: Results from a randomized pilot intervention trial. *Journal of Child and Family Studies, 19*(2), 203–217. https://doi.org/10.1007/s10826-009-9304-8

Coatsworth, J. D., Duncan, L., Berrena, E., Bamberger, K. T., Loeschinger, D., Greenberg, M. T., & Nix, R. L. (2014). The Mindfulness-Enhanced Strengthening Families Program: Integrating brief mindfulness activities and parent training within an evidence-based prevention program. *New Directions for Youth Development, 2014*(142), 45–58. https://doi.org/10.1002/yd.20096

Cohen, O., & Finzi-Dottan, R. (2005). Parent–child relationships during the divorce process; from attachment theory and intergenerational perspective. *Contemporary Family Therapy, 27*(1), 81–99. https://doi.org/10.1007/s10591-004-1972-3

Collins, N. L., Guichard, A. C., Ford, M. B., & Feeney, B. C. (2004). Working models of attachment: New developments and emerging themes. In W. S. Rholes & J. A. Simpson (Eds.), *Adult attachment: Theory, research, and clinical implications* (pp. 196–239). Guilford Publications.

Cousineau, T. M., Hobbs, L. M., & Arthur, K. C. (2019). The role of compassion and mindfulness in building parental resilience when caring for children with chronic conditions: A conceptual model. *Frontiers in Psychology, 10*, Article 1602. https://doi.org/10.3389/fpsyg.2019.01602

Crnic, K. A., & Greenberg, M. T. (1990). Minor parenting stresses with young children. *Child Development, 61*(5), 1628–1637. https://doi.org/10.1111/j.1467-8624.1990.tb02889.x

Crocker, J., & Canevello, A. (2008). Creating and undermining social support in communal relationships: The role of compassionate and self-image goals. *Journal of Personality and Social Psychology, 95*(3), 555–575. https://doi.org/10.1037/0022-3514.95.3.555

Dar, K. A., & Iqbal, N. (2015). Worry and rumination in generalized anxiety disorder and obsessive compulsive disorder. *The Journal of Psychology, 149*(8), 866–880. https://doi.org/10.1080/00223980.2014.986430

de Bruin, E. I., Zijlstra, B. J. H., van de Weijer-Bergsma, E., & Bögels, S. M. (2011). The mindful attention awareness scale for adolescents (MAAS-A): Psychometric properties in a Dutch sample. *Mindfulness, 2*(3), 201–211. https://doi.org/10.1007/s12671-011-0061-6

de Bruin, E. I., Zijlstra, B. J. H., Geurtzen, N., van Zundert, R. M. P., van de Weijer-Bergsma, E., Hartman, E. E., Nieuwesteeg, A. M., Duncan, L. G., & Bögels, S. M. (2014). Mindful parenting assessed further: Psychometric properties of the Dutch version of the Interpersonal Mindfulness in Parenting scale (IM-P). *Mindfulness, 5*(2), 200–212. https://doi.org/10.1007/s12671-012-0168-4

Deater-Deckard, K. (1998). Parenting stress and child adjustment: Some old hypotheses and new questions. *Clinical Psychology: Science and Practice, 5*(3), 314–332. https://doi.org/10.1111/j.1468-2850.1998.tb00152.x

Deater-Deckard, K. (2004). *Parenting stress*. Yale University Press.

Deater-Deckard, K., & Panneton, R. (Eds.). (2017). *Parental stress and early child development: Adaptive and maladaptive outcomes*. Springer.

DeJong, H., Fox, E., & Stein, A. (2016). Rumination and postnatal depression: A systematic review and a cognitive model. *Behaviour Research and Therapy, 82*, 38–49. https://doi.org/10.1016/j.brat.2016.05.003

Depue, R. A., & Morrone-Strupinsky, J. V. (2005). A neurobehavioral model of affiliative bonding. *Behavioral and Brain Sciences, 28*, 313–395. https://doi.org/10.1016/j.biopsycho.2007.09.001

Diedrich, A., Grant, M., Hofmann, S. G., Hiller, W., & Berking, M. (2014). Self-compassion as an emotion regulation strategy in major depressive disorder. *Behaviour Research and Therapy, 58*, 43–51. https://doi.org/10.1016/j.brat.2014.05.006

Dix, T., & Meunier, L. N. (2009). Depressive symptoms and parenting competence: An analysis of 13 regulatory processes. *Developmental Review, 29*(1), 45–68. https://doi.org/10.1016/j.dr.2008.11.002

Donaldson, C., Lam, D., & Mathews, A. (2007). Rumination and attention in major depression. *Behaviour Research and Therapy, 45*(11), 2664–2678. https://doi.org/10.1016/j.brat.2007.07.002

Duncan, L. G., Coatsworth, J. D., & Greenberg, M. T. (2009). A model of mindful parenting: Implications for parent–child relationships and prevention research. *Clinical Child and Family Psychology Review, 12*(3), 255–270. https://doi.org/10.1007/s10567-009-0046-3

Duncan, L. G., Coatsworth, J. D., Gayles, J. G., Geier, M. H., & Greenberg, M. T. (2015). Can mindful parenting be observed? Relations between observational ratings of mother-youth interactions and mothers' self-report mindful parenting. *Journal of Family Psychology, 29*(2), 276–282. https://doi.org/10.1037/a0038857

Edelstein, R. S., Alexander, K. W., Shaver, P. R., Schaaf, J. M., Quas, J. A., Lovas, G. S., & Goodman, G. S. (2004). Adult attachment style and parental responsiveness during a stressful event. *Attachment & Human Development, 6*(1), 31–52. https://doi.org/10.1080/14616730310001659584

Eisenberg, N., Cumberland, A., & Spinrad, T. L. (1998). Parental socialization of emotion. *Psychological Inquiry, 9*(4), 241–273. https://doi.org/10.1207/s15327965pli0904_1

Eisenberg, N., Fabes, R. A., Shepard, S. A., Guthrie, I. K., Murphy, B. C., & Reiser, M. (1999). Parental reactions to children's negative emotions: Longitudinal relations to quality of children's social functioning. *Child Development, 70*, 513–534. https://doi.org/10.1111/1467-8624.00037

Eisenberg, N., Valiente, C., Morris, A. S., Fabes, R. A., Cumberland, A., Reiser, M., Gershoff, E. T., Shepard, S. A., & Losoya, S. (2003). Longitudinal relations among parental emotional expressivity, children's regulation, and quality of socioemotional functioning. *Developmental Psychology, 39*(1), 3–19. https://doi.org/10.1037/0012-1649.39.1.3

Emerson, L. M., Aktar, E., de Bruin, E. I., Potharst, E., & Bögels, S. (2021). Mindful parenting in secondary child mental health: Key parenting predictors of treatment effects. *Mindfulness, 12*, 532–542. https://doi.org/10.1007/s12671-019-01176-w

Fernandes, D. V., Canavarro, M. C., & Moreira, H. (2021). The mediating role of parenting stress in the relationship between anxious and depressive symptomatology, mothers' perception of infant temperament, and mindful parenting during the postpartum period. *Mindfulness, 12*(2), 275–290. https://doi.org/10.1007/s12671-020-01327-4

Fogarty, F. A., Lu, L. M., Sollers, J. J., Krivoschekov, S. G., Booth, R. J., & Consedine, N. S. (2015). Why it pays to be mindful: Trait mindfulness predicts physiological recovery from emotional stress and greater differentiation among negative emotions. *Mindfulness, 6*(2), 175–185. https://doi.org/10.1007/s12671-013-0242-6

Galhardo, A., Cunha, M., Pinto-Gouveia, J., & Matos, M. (2013). The mediator role of emotion regulation processes on infertility-related stress. *Journal of Clinical Psychology in Medical Settings, 20*(4), 497–507. https://doi.org/10.1007/s10880-013-9370-3

Gao, M., Du, H., Davies, P. T., & Cummings, E. M. (2019). Marital conflict behaviors and parenting: Dyadic links over time. *Family Relations, 68*(1), 135–149. https://doi.org/10.1111/fare.12322

Garnefski, N., Rieffe, C., Jellesma, F., Terwogt, M. M., & Kraaij, V. (2007). Cognitive emotion regulation strategies and emotional problems in 9 - 11-year-old children: The development of an instrument. *European Child & Adolescent Psychiatry, 16*(1), 1–9. https://doi.org/10.1007/s00787-006-0562-3

Gerber, Z., Tolmacz, R., & Doron, Y. (2015). Self-compassion and forms of concern for others. *Personality and Individual Differences, 86*, 394–400. https://doi.org/10.1016/j.paid.2015.06.052

Gerber, Z., Davidovics, Z., & Anaki, D. (2021). The relationship between self-compassion, concern for others, and parental burnout in child's chronic care management. *Mindfulness, 12*, 2920–2928. https://doi.org/10.1007/s12671-021-01752-z

Gilbert, P. (2005a). Compassion and cruelty: A biopsychosocial approach. In P. Gilbert (Ed.), *Compassion: Conceptualisations, research and use in psychotherapy* (pp. 3–74). Routledge.

Gilbert, P. (Ed.). (2005b). *Compassion: Conceptualisations, research and use in psychotherapy*. Routledge.

Gilbert, P. (2009a). *The compassionate mind*. Constable & Robinson.

Gilbert, P. (2009b). Introducing compassion-focused therapy. *Advances in Psychiatric Treatment, 15*, 199–208. https://doi.org/10.1192/apt.bp.107.005264

Gilbert, P. (2010). *Compassion focused therapy*. Routledge.

Gilbert, P., & Procter, S. (2006). Compassionate mind training for people with high shame and self-criticism: Overview and pilot study of a group therapy approach. *Clinical Psychology & Psychotherapy, 13*(6), 353–379. https://doi.org/10.1002/cpp.507

Gilbert, P., McEwan, K., Matos, M., & Rivis, A. (2011). Fears of compassion: Development of three self-report measures. *Psychology and Psychotherapy: Theory, Research and Practice, 84*(3), 239–255. https://doi.org/10.1348/147608310X526511

Gillath, O., Shaver, P. R., & Mikulincer, M. (2005). An attachment-theoretical approach to compassion and altruism. In P. Gilbert (Ed.), *Compassion: Conceptualizations, research and use in psychotherapy* (pp. 121–147). Routledge.

Goldsmith, H. H., Pollak, S. D., & Davidson, R. J. (2008). Developmental neuroscience perspectives on emotion regulation. *Child Development Perspectives, 2*(3), 132–140. https://doi.org/10.1111/j.1750-8606.2008.00055.x

Goodman, S. H., & Gotlib, I. H. (1999). Risk for psychopathology in the children of depressed mothers: A developmental model for understanding mechanisms of transmission. *Psychological Review, 106*(3), 458–490. https://doi.org/10.1037/0033-295X.106.3.458

Gottman, J. M., Katz, L. F., & Hooven, C. (1996). Parental meta-emotion philosophy and the emotional life of families: Theoretical models and preliminary data. *Journal of Family Psychology, 10*(3), 243–268. https://doi.org/10.1037/0893-3200.10.3.243

Gouveia, M. J., Carona, C., Canavarro, M. C., & Moreira, H. (2016). Self-compassion and dispositional mindfulness are associated with parenting styles and parenting stress: The mediating role of mindful parenting. *Mindfulness, 7*(3), 700–712. https://doi.org/10.1007/s12671-016-0507-y

Gouveia, M. J., Canavarro, M. C., & Moreira, H. (2019). Linking mothers' difficulties in emotion regulation to children/adolescents' emotional eating in pediatric obesity: The mediating role of mindful parenting and children/adolescents' depressive symptoms. *Mindfulness, 10*, 877–893. https://doi.org/10.1007/s12671-018-1055-4

Greco, L. A., Lambert, W., & Baer, R. A. (2008). Psychological inflexibility in childhood and adolescence: Development and evaluation of the Avoidance and Fusion Questionnaire for Youth. *Psychological Assessment, 20*(2), 93–102. https://doi.org/10.1037/1040-3590.20.2.93

Han, Z. R., Ahemaitijiang, N., Yan, J., Hu, X., Parent, J., Dale, C., DiMarzio, K., & Singh, N. N. (2021). Parent mindfulness, parenting, and child psychopathology in China. *Mindfulness, 12*, 334–343. https://doi.org/10.1007/s12671-019-01111-z

Hawkins, L., Centifanti, L. C. M., Holman, N., & Taylor, P. (2019). Parental adjustment following pediatric burn injury: The role of guilt, shame, and self-compassion. *Journal of Pediatric Psychology, 44*(2), 229–237. https://doi.org/10.1093/jpepsy/jsy079

Hayes, S. C., Luoma, J. B., Bond, F. W., Masuda, A., & Lillis, J. (2006). Acceptance and commitment therapy: Model, processes and outcomes. *Behaviour Research and Therapy, 44*(1), 1–25. https://doi.org/10.1016/j.brat.2005.06.006

Hayes, S. C., Pistorello, J., & Levin, M. E. (2012). Acceptance and commitment therapy as a unified model of behavior change. *The Counseling Psychologist, 40*(7), 976–1002. https://doi.org/10.1177/0011000012460836

Holmbeck, G. N., Johnson, S. Z., Wills, K. E., McKernon, W., Rose, B., Erklin, S., & Kemper, T. (2002). Observed and perceived parental overprotection in relation to psychosocial adjustment in preadolescents with a physical disability: The mediational role of behavioral autonomy. *Journal of Consulting and Clinical Psychology, 70*(1), 96–110. https://doi.org/10.1037//0022-006x.70.1.96

Ingram, R. E. (1990). Self-focused attention in clinical disorders: Review and a conceptual model. *Psychological Bulletin, 107*(2), 156–176. https://doi.org/10.1037/0033-2909.107.2.156

Jaffe, M., Gullone, E., & Hughes, E. K. (2010). The roles of temperamental dispositions and perceived parenting behaviours in the use of two emotion regulation strategies in late childhood. *Journal of Applied Developmental Psychology, 31*(1), 47–59. https://doi.org/10.1016/j.appdev.2009.07.008

Jefferson, F. A., Shires, A., & McAloon, J. (2020). Parenting self-compassion: A systematic review and meta-analysis. *Mindfulness, 11*(9), 2067–2088. https://doi.org/10.1007/s12671-020-01401-x

Jin, Z., Zhang, X., & Han, Z. R. (2017). Parental emotion socialization and child psychological adjustment among Chinese urban families: Mediation through child emotion regulation and moderation through dyadic collaboration. *Frontiers in Psychology, 8*, Article 2198. https://doi.org/10.3389/fpsyg.2017.02198

Jones, J. D., Cassidy, J., & Shaver, P. R. (2015a). Adult attachment style and parenting. In J. A. Simpson & W. S. Rholes (Eds.), *Attachment theory and research: New directions and emerging themes*. The Guilford Press.

Jones, J. D., Cassidy, J., & Shaver, P. R. (2015b). Parents' self-reported attachment styles: A review of links with parenting behaviors, emotions, and cognitions. *Personality and Social Psychology Review, 19*(1), 44–76. https://doi.org/10.1177/1088868314541858

Kabat-Zinn, J. (1990). *Full catastrophe living: Using the wisdom of your body and mind to face stress, pain, and illness*. Dell Publishing.

Kabat-Zinn, J. (2003). Mindfulness-based interventions in context: Past, present, and future. *Clinical Psychology: Science and Practice, 10*(2), 144–156. https://doi.org/10.1093/clipsy.bpg016

Kabat-Zinn, J., & Kabat-Zinn, M. (1997). *Everyday blessings: The inner work of mindful parenting*. Hyperion.

Kashdan, T. B., & Rottenberg, J. (2010). Psychological flexibility as a fundamental aspect of health. *Clinical Psychology Review, 30*(7), 865–878. https://doi.org/10.1016/j.cpr.2010.03.001

Kemppainen, J., Johnson, M. O., Phillips, J. C., Sullivan, K. M., Corless, I. B., Reid, P., Lipinge, S., Chaiphibalsarisdi, M. P., Sefcik, E., Chen, W. T., Kirksey, K., Voss, J., Rivero-Méndez, M., Tyer-Viola, L., Rose, C. D., Webel, A., Nokes, K., Portillo, C., Holzemer, W. L., et al. (2013). A multinational study of self-compassion and human immunodeficiency virus-related anxiety. *International Nursing Review, 60*(4), 477–486. https://doi.org/10.1111/inr.12056

Kiami, S. R., & Goodgold, S. (2017). Support needs and coping strategies as predictors of stress level among mothers of children with autism spectrum disorder. *Autism Research and Treatment, 2017*, 8685950–8685950. https://doi.org/10.1155/2017/8685950

Kiel, E. J., & Buss, K. A. (2012). Associations among context-specific maternal protective behavior, toddlers' fearful temperament, and maternal accuracy and goals. *Social Development, 21*(4), 742–760. https://doi.org/10.1111/j.1467-9507.2011.00645.x

Krahé, B., Bondü, R., Höse, A., & Esser, G. (2015). Child aggression as a source and a consequence of parenting stress: A three-wave longitudinal study. *Journal of Research on Adolescence, 25*(2), 328–339. https://doi.org/10.1111/jora.12115

Krieger, T., Altenstein, D., Baettig, I., Doerig, N., & Holtforth, M. G. (2013). Self-compassion in depression: Associations with depressive symptoms, rumination, and avoidance in depressed outpatients. *Behavior Therapy, 44*(3), 501–513. https://doi.org/10.1016/j.beth.2013.04.004

Kyrios, M., & Prior, M. (1990). Temperament, stress, and family factors in behavioral adjustment of 3–5-year-old children. *International Journal of Behavioral Development, 13*, 67–93. https://doi.org/10.1177/016502549001300105

Lathren, C., Bluth, K., & Zvara, B. (2020). Parent self-compassion and supportive responses to child difficult emotion: An intergenerational theoretical model rooted in attachment. *Journal of Family Theory & Review, 12*(3), 368–381. https://doi.org/10.1111/jftr.12388

Lee, D. (2012). *Recovering from trauma using compassion focused therapy*. Robinson.

Legerstee, J. S., Garnefski, N., Jellesma, F. C., Verhulst, F. C., & Utens, E. M. (2010). Cognitive coping and childhood anxiety disorders. *European Child & Adolescent Psychiatry, 19*(2), 143–150. https://doi.org/10.1007/s00787-009-0051-6

Lengua, L. J. (2006). Growth in temperament and parenting as predictors of adjustment during children's transition to adolescence. *Developmental Psychology, 42*(5), 819–832. https://doi.org/10.1037/0012-1649.42.5.819

Lengua, L. J., & Kovacs, E. A. (2005). Bidirectional associations between temperament and parenting and the prediction of adjustment problems in middle childhood. *Journal of Applied Developmental Psychology, 26*(1), 21–38. https://doi.org/10.1016/j.appdev.2004.10.001

Lindström, C., Aman, J., & Norberg, A. L. (2010). Increased prevalence of burnout symptoms in parents of chronically ill children. *Acta Paediatrica, 99*(3), 427–432. https://doi.org/10.1111/j.1651-2227.2009.01586.x

Lippold, M. A., Duncan, L. G., Coatsworth, J. D., Nix, R. L., & Greenberg, M. T. (2015). Understanding how mindful parenting may be linked to mother–adolescent communication. *Journal of Youth and Adolescence, 44*(9), 1663–1673. https://doi.org/10.1007/s10964-015-0325-x

Lippold, M. A., Jensen, T. M., Duncan, L. G., Nix, R. L., Coatsworth, J. D., & Greenberg, M. T. (2019). Mindful parenting, parenting cognitions, and parent-youth communication: Bidirectional linkages and mediational processes. *Mindfulness, 12*, 381–391. https://doi.org/10.1007/s12671-019-01119-5

Liu, W., Chen, L., & Blue, P. R. (2016). Chinese adaptation and psychometric properties of the child version of the Cognitive Emotion Regulation Questionnaire. *PLoS One, 11*(2), e0150206. https://doi.org/10.1371/journal.pone.0150206

Livheim, F., Tengström, A., Bond, F. W., Andersson, G., Dahl, J., & Rosendahl, I. (2016). Psychometric properties of the Avoidance and Fusion Questionnaire for Youth: A psychological measure of psychological inflexibility in youth. *Journal of Contextual Behavioral Science, 5*(2), 103–110. https://doi.org/10.1016/j.jcbs.2016.04.001

Longe, O., Maratos, F. A., Gilbert, P., Evans, G., Volker, F., Rockliffe, H., & Rippon, G. (2010). Having a word with yourself: Neural correlates of self-criticism and self-reassurance. *NeuroImage, 49*, 1849–1856. https://doi.org/10.1016/j.neuroimage.2009.09.019

Lovejoy, M. C., Graczyk, P. A., O'Hare, E., & Neuman, G. (2000). Maternal depression and parenting behavior: A meta-analytic review. *Clinical Psychology Review, 20*(5), 561–592. https://doi.org/10.1016/S0272-7358(98)00100-7

MacBeth, A., & Gumley, A. (2012). Exploring compassion: A meta-analysis of the association between self-compassion and psychopathology. *Clinical Psychology Review, 32*(6), 545–552. https://doi.org/10.1016/j.cpr.2012.06.003

Marsh, I. C., Chan, S. W. Y., & MacBeth, A. (2018). Self-compassion and psychological distress in adolescents: A meta-analysis. *Mindfulness, 9*(4), 1011–1027. https://doi.org/10.1007/s12671-017-0850-7

McEwen, C., & Flouri, E. (2009). Fathers' parenting, adverse life events, and adolescents' emotional and eating disorder symptoms: The role of emotion regulation. *European Child & Adolescent Psychiatry, 18*(4), 206–216. https://doi.org/10.1007/s00787-008-0719-3

McKee, L. G., Parent, J., Zachary, C. R., & Forehand, R. (2017). Mindful parenting and emotion socialization practices: Concurrent and longitudinal associations. *Family Process, 57*(3), 752–766. https://doi.org/10.1111/famp.12329

Medeiros, C., Gouveia, M. J., Canavarro, M. C., & Moreira, H. (2016). The indirect effect of the mindful parenting of mothers and fathers on the child's perceived well-being through the child's attachment to parents. *Mindfulness, 7*(4), 916–927. https://doi.org/10.1007/s12671-016-0530-z

Melen, S., Pepping, C. A., & O'Donovan, A. (2017). Social foundations of mindfulness: Priming attachment anxiety reduces emotion regulation and mindful attention. *Mindfulness, 8*(1), 136–143. https://doi.org/10.1007/s12671-016-0587-8

Melnick, S. M., & Hinshaw, S. P. (2000). Emotion regulation and parenting in AD/HD and comparison boys: Linkages with social behaviors and peer preference. *Journal of Abnormal Child Psychology, 28*(1), 73–86. https://doi.org/10.1023/a:1005174102794

Mikulincer, M., & Shaver, P. R. (2016). *Attachment in adulthood: Structure, dynamics, and change*. Guilford Press.

Mikulincer, M., Shaver, P. R., Gillath, O., & Nitzberg, R. A. (2005). Attachment, caregiving, and altruism: Boosting attachment security increases compassion and helping. *Journal of Personality and Social Psychology, 89*(5), 817–839. https://doi.org/10.1037/0022-3514.89.5.817

Mills, J., Wand, T., & Fraser, J. A. (2018). Examining self-care, self-compassion and compassion for others: A cross-sectional survey of palliative care nurses and doctors. *International Journal of Palliative Nursing, 24*(1), 4–11. https://doi.org/10.12968/ijpn.2018.24.1.4

Moreira, H., & Canavarro, M. C. (2015). Individual and gender differences in mindful parenting: The role of attachment and caregiving representations. *Personality and Individual Differences, 87*, 13–19. https://doi.org/10.1016/j.paid.2015.07.021

Moreira, H., & Canavarro, M. C. (2016). Parental attachment insecurity and parenting stress: The mediating role of parents' perceived impact of children's diabetes on the family. *Family, Systems, and Health, 34*(3), 240–209. https://doi.org/10.1037/fsh0000211

Moreira, H., & Canavarro, M. C. (2017). Psychometric properties of the Interpersonal Mindfulness in Parenting Scale in a sample of Portuguese mothers. *Mindfulness, 8*(3), 691–706. https://doi.org/10.1007/s12671-016-0647-0

Moreira, H., & Canavarro, M. C. (2018a). The association between self-critical rumination and parenting stress: The mediating role of mindful parenting. *Journal of Child and Family Studies, 27*, 2265–2275. https://doi.org/10.1007/s10826-018-1072-x

Moreira, H., & Canavarro, M. C. (2018b). Does the association between mindful parenting and adolescents' dispositional mindfulness depend on the levels of anxiety and depression symptomatology in mothers? *Journal of Adolescence, 68*, 22–31. https://doi.org/10.1016/j.adolescence.2018.07.003

Moreira, H., & Canavarro, M. C. (2019). Mindful parenting is associated with adolescents' difficulties in emotion regulation through adolescents' psychological inflexibility and self-compassion. *Journal of Youth and Adolescence, 49*(1), 192–211. https://doi.org/10.1007/s10964-019-01133-9

Moreira, H., Carona, C., Silva, N., Frontini, R., Bullinger, M., & Canavarro, M. C. (2013a). Psychological and quality of life outcomes in pediatric populations: A parent-child perspective. *Journal of Pediatrics, 163*(5), 1471–1478. https://doi.org/10.1016/j.jpeds.2013.06.028

Moreira, H., Frontini, R., Bullinger, M., & Canavarro, M. C. (2013b). Caring for a child with type 1 diabetes: Links between family cohesion, perceived impact, and parental adjustment. *Journal of Family Psychology, 27*(5), 731–742. https://doi.org/10.1037/a0034198

Moreira, H., Gouveia, M. J., Carona, C., Silva, N., & Canavarro, M. C. (2015). Maternal attachment and children's quality of life: The mediating role of self-compassion and parenting stress. *Journal of Child and Family Studies, 24*(8), 2332–2344. https://doi.org/10.1007/s10826-014-0036-z

Moreira, H., Carona, C., Silva, N., Nunes, J., & Canavarro, M. C. (2016). Exploring the link between maternal attachment-related anxiety and avoidance and mindful parenting: The mediating role of self-compassion. *Psychology and Psychotherapy: Theory, Research and Practice, 89*(4), 369–384. https://doi.org/10.1111/papt.12082

Moreira, H., Gouveia, M. J., & Canavarro, M. C. (2018). Is mindful parenting associated with adolescents' well-being in early and middle/late adolescence? The mediating role of adolescents' attachment representations, self-compassion and mindfulness. *Journal of Youth and Adolescence, 47*(8), 1771–1788. https://doi.org/10.1007/s10964-018-0808-7

Moreira, H., Fonseca, A., Caiado, B., & Canavarro, M. C. (2019). Work-family conflict and mindful parenting: The mediating role of parental psychopathology symptoms and parenting stress in a sample of Portuguese employed parents. *Frontiers in Psychology, 10*, Article 635. https://doi.org/10.3389/fpsyg.2019.00635

Moreira, H., Caiado, B., & Canavarro, M. C. (2020). Is mindful parenting a mechanism that links parents' and children's tendency to experience negative affect to overprotective and supportive behaviors? *Mindfulness, 12*, 319–333. https://doi.org/10.1007/s12671-020-01468-6

Moreland, A. D., Felton, J. W., Hanson, R. F., Jackson, C., & Dumas, H. E. (2016). The relation between parenting stress, locus of control and child outcomes: Predictors of change in a parenting intervention. *Journal of Child and Family Studies, 25*, 2046–2054. https://doi.org/10.1007/s10826-016-0370-4

Morris, A. S., Silk, J. S., Steinberg, L., Myers, S. S., & Robinson, L. R. (2007). The role of the family context in the development of emotion regulation. *Social Development, 16*(2), 361–388. https://doi.org/10.1111/j.1467-9507.2007.00389.x

Muris, P., Roelofs, J., Rassin, E., Franken, I., & Mayer, B. (2005). Mediating effects of rumination and worry on the links between neuroticism, anxiety and depression. *Personality and Individual Differences, 39*(6), 1105–1111. https://doi.org/10.1016/j.paid.2005.04.005

Muris, P., Meesters, C., Pierik, A., & de Kock, B. (2016). Good for the self: Self-compassion and other self-related constructs in relation to symptoms of anxiety and depression in non-clinically youths. *Journal of Child and Family Studies, 25*(2), 607–617. https://doi.org/10.1007/s10826-015-0235-2

Muris, P., Meesters, C., Herings, A., Jansen, M., Vossen, C., & Kersten, P. (2017). Inflexible youngsters: Psychological and psychopathological correlates of the Avoidance and Fusion Questionnaire for youths in nonclinical Dutch adolescents. *Mindfulness, 8*(5), 1381–1392. https://doi.org/10.1007/s12671-017-0714-1

Neece, C. L., Chan, N., Klein, K., Roberts, L., & Fenning, R. M. (2019). Mindfulness-based stress reduction for parents of children with developmental delays: Understanding the experiences of Latino families. *Mindfulness, 10*(6), 1017–1030. https://doi.org/10.1007/s12671-018-1011-3

Neff, K. (2003a). The development and validation of a scale to measure self-compassion. *Self and Identity, 2*(3), 223–250. https://doi.org/10.1080/15298860309027

Neff, K. (2003b). Self-compassion: An alternative conceptualization of a healthy attitude toward oneself. *Self and Identity, 2*(2), 85–101. https://doi.org/10.1080/15298860309032

Neff, K. (2011). *Self-compassion*. Morrow.

Neff, K., & Beretvas, S. (2013). The role of self-compassion in romantic relationships. *Self and Identity, 12*(1), 78–98. https://doi.org/10.1080/15298868.2011.639548

Neff, K., & Faso, D. (2015). Self-compassion and well-being in parents of children with autism. *Mindfulness, 6*(4), 938–947. https://doi.org/10.1007/s12671-014-0359-2

Neff, K., & Germer, C. (2013). A pilot study and randomized controlled trial of the mindful self-compassion program. *Journal of Clinical Psychology, 69*(1), 28–44. https://doi.org/10.1002/jclp.21923

Neff, K., & McGehee, P. (2010). Self-compassion and psychological resilience among adolescents and young adults. *Self and Identity, 9*(3), 225–240. https://doi.org/10.1080/15298860902979307

Neff, K., & Pommier, E. (2013). The relationship between self-compassion and other-focused concern among college undergraduates, community adults, and practicing meditators. *Self and Identity, 12*(2), 160–176. https://doi.org/10.1080/15298868.2011.649546

Neff, K., Kirkpatrick, K., & Rude, S. (2007). Self-compassion and its link to adaptive psychological functioning. *Journal of Research in Personality, 41*, 139–154. https://doi.org/10.1016/j.jrp.2006.03.004

Neff, K., Long, P., Knox, M. C., Davidson, O., Kuchar, A., Costigan, A., Williamson, Z., Rohleder, N., Tóth-Király, I., & Breines, J. G. (2018). The forest and the trees: Examining the association of self-compassion and its positive and negative components with psychological functioning. *Self and Identity, 17*(6), 627–645. https://doi.org/10.1080/15298868.2018.1436587

Niu, H., Liu, L., & Wang, M. (2018). Intergenerational transmission of harsh discipline: The moderating role of parenting stress and parent gender. *Child Abuse & Neglect, 79*, 1–10. https://doi.org/10.1016/j.chiabu.2018.01.017

Nolen-Hoeksema, S., Wisco, B. E., & Lyubomirsky, S. (2008). Rethinking rumination. *Perspectives on Psychological Science, 3*(5), 400–424. https://doi.org/10.1111/j.1745-6924.2008.00088.x

Nordahl, D., Rognmo, K., Bohne, A., Landsem, I. P., Moe, V., Wang, C. E. A., & Høifødt, R. S. (2020). Adult attachment style and maternal-infant bonding: The indirect path of parenting stress. *BMC Psychology, 8*(1), Article 58. https://doi.org/10.1186/s40359-020-00424-2

Orgiles, M., Morales, A., Fernandez-Martinez, I., Ortigosa-Quiles, J. M., & Espada, J. P. (2018). Spanish adaptation and psychometric properties of the child version of the Cognitive Emotion Regulation Questionnaire. *PLoS One, 13*(8), e0201656. https://doi.org/10.1371/journal.pone.0201656

Parent, J., McKee, L. G., Anton, M., Gonzalez, M., Jones, D. J., & Forehand, R. (2016a). Mindfulness in parenting and coparenting. *Mindfulness, 7*(2), 504–513. https://doi.org/10.1007/s12671-015-0485-5

Parent, J., McKee, L. G., Rough, J. N., & Forehand, R. (2016b). The association of parent mindfulness with parenting and youth psychopathology across three developmental stages. *Journal of Abnormal Child Psychology, 44*(1), 191–202. https://doi.org/10.1007/s10802-015-9978-x

Pepping, C. A., & Duvenage, M. (2016). The origins of individual differences in dispositional mindfulness. *Personality and Individual Differences, 93*, 130–136. https://doi.org/10.1016/j.paid.2015.05.027

Pepping, C. A., Davis, P. J., & O'Donovan, A. (2013). Individual differences in attachment and dispositional mindfulness: The mediating role of emotion regulation. *Personality and Individual Differences, 54*, 453–456. https://doi.org/10.1016/j.paid.2012.10.006

Pepping, C. A., Davis, P. J., & O'Donovan, A. (2015). The association between state attachment security and state mindfulness. *PLoS One, 10*(3), e0116779. https://doi.org/10.1371/journal.pone.0116779

Perry-Jenkins, M., Smith, J. Z., Wadsworth, L. P., & Halpern, H. P. (2017). Workplace policies and mental health among working-class, new parents. *Community, Work & Family, 20*(2), 226–249. https://doi.org/10.1080/13668803.2016.1252721

Pinquart, M. (2018). Parenting stress in caregivers of children with chronic physical condition-A meta-analysis. *Stress and Health, 34*(2), 197–207. https://doi.org/10.1002/smi.2780

Pinto-Gouveia, J., Duarte, C., Matos, M., & Fráguas, S. (2014). The protective role of self-compassion in relation to psychopathology symptoms and quality of life in chronic and in cancer patients. *Clinical Psychology & Psychotherapy, 21*, 311–323. https://doi.org/10.1002/cpp.1838

Porter, C. L., Hart, C. H., Yang, C., Robinson, C. C., Olsen, S. F., Zeng, Q., & Olsen, J. A. (2005). A comparative study of child temperament and parenting in Beijing, China and the western United States. *International Journal of Behavioral Development, 29*, 541–551. https://doi.org/10.1080/01650250500147402

Potharst, E. S., Aktar, E., Rexwinkel, M., Rigterink, M., & Bögels, S. M. (2017). Mindful with your baby: Feasibility, acceptability, and effects of a mindful parenting group training for mothers and their babies in a mental health context. *Mindfulness, 8*(5), 1236–1250. https://doi.org/10.1007/s12671-017-0699-9

Potharst, E. S., Boekhorst, M., Cuijlits, I., van Broekhoven, K., Jacobs, A., Spek, V., Nyklíček, I., Bögels, S., & Pop, V. J. M. (2019). A randomized control trial evaluating an online mindful parenting training for mothers with elevated parental stress. *Frontiers in*

Psychology, 10, Article 1550. https://doi.org/10.3389/fpsyg.2019.01550

Potharst, E. S., Zeegers, M., & Bögels, S. M. (2021). Mindful with your toddler group training: Feasibility, acceptability, and effects on subjective and objective measures. *Mindfulness, 12*, 489–503. https://doi.org/10.1007/s12671-018-1073-2

Potharst, E. S., Veringa-Skiba, I., van Broekhuizen, E., & Bögels, S. M. (2022). Mindful with your baby for mothers of infants with (parental) stress in a non-clinical setting: A wait-list controlled pilot trial. *BMC Pregnancy and Childbirth, 22*(1), 1–11. https://doi.org/10.1186/s12884-022-04640-z

Psychogiou, L., Legge, K., Parry, E., Mann, J., Nath, S., Ford, T., & Kuyken, W. (2016). Self-compassion and parenting in mothers and fathers with depression. *Mindfulness, 7*(4), 896–908. https://doi.org/10.1007/s12671-016-0528-6

Rholes, W. S., Simpson, J. A., & Blakely, B. S. (1995). Adult attachment styles and mothers' relationships with their young children. *Personal Relationships, 2*(1), 35–54. https://doi.org/10.1111/j.1475-6811.1995.tb00076.x

Rholes, W. S., Simpson, J. A., Blakely, B. S., Lanigan, L., & Allen, E. A. (1997). Adult attachment styles, the desire to have children, and working models of parenthood. *Journal of Personality, 65*, 357–385. https://doi.org/10.1111/j.1467-6494.1997.tb00958.x

Rholes, W. S., Simpson, J. A., & Friedman, M. (2006). Avoidant attachment and the experience of parenting. *Personality and Social Psychology Bulletin, 32*, 275–285. https://doi.org/10.1177/0146167205280910

Riediger, M., & Klipker, K. (2014). Emotion regulation in adolescence. In J. J. Gross (Ed.), *Handbook of emotion regulation* (2nd ed., pp. 187–202). Guilford Press.

Robinson, S., Hastings, R. P., Weiss, J. A., Pagavathsing, J., & Lunsky, Y. (2018). Self-compassion and psychological distress in parents of young people and adults with intellectual and developmental disabilities. *Journal of Applied Research in Intellectual Disabilities, 31*(3), 454–458. https://doi.org/10.1111/jar.12423

Rodriguez, C. M. (2011). Association between independent reports of maternal parenting stress and children's internalizing symptomatology. *Journal of Child and Family Studies, 20*, 631–639. https://doi.org/10.1007/s10826-010-9438-8

Roemer, L., Williston, S. K., & Rollins, L. G. (2015). Mindfulness and emotion regulation. *Current Opinion in Psychology, 3*, 52–57. https://doi.org/10.1016/j.copsyc.2015.02.006

Roskam, I., Aguiar, J., Akgun, E., Arikan, G., Artavia, M., Avalosse, H., Aunola, K., Bader, M., Bahati, C., Barham, E. J., Besson, E., Beyers, W., Boujut, E., Brianda, M. E., Brytek-Matera, A., Carbonneau, N., César, F., Chen, B., Dorard, G., et al. (2021). Parental burnout around the globe: A 42-country study. *Affective Science, 2*(1), 58–79. https://doi.org/10.1007/s42761-020-00028-4

Rothbart, M. K., & Bates, J. E. (2007). Temperament. In W. Damon & R. M. Lerner (Eds.), *Social, emotional and personality development* (6th ed.). Wiley.

Rutherford, H., Wallace, N. S., Laurent, H. K., & Mayes, L. C. (2015). Emotion regulation in parenthood. *Developmental Review, 36*, 1–14. https://doi.org/10.1016/j.dr.2014.12.008

Ryan, R. M., Brown, K. W., & Creswell, J. D. (2007). How integrative is attachment theory? Unpacking the meaning and significance of felt security. *Psychological Inquiry, 18*(3), 177–182. https://doi.org/10.1080/10478400701512778

Sanders, W., Zeman, J., Poon, J., & Miller, R. (2015). Child regulation of negative emotions and depressive symptoms: The moderating role of parental emotion socialization. *Journal of Child and Family Studies, 24*(2), 402–415. https://doi.org/10.1007/s10826-013-9850-y

Sbarra, D. A., Smith, H. L., & Mehl, M. R. (2012). When leaving your ex, love yourself: Observational ratings of self-compassion predict the course of emotional recovery following marital separation. *Psychological Science, 23*(3), 261–269. https://doi.org/10.1177/0956797611429466

Schleider, J. L., Patel, A., Krumholz, L., Chorpita, B. F., & Weisz, J. R. (2015). Relation between parent symptomatology and youth problems: Multiple mediation through family income and parent–youth stress. *Child Psychiatry & Human Development, 46*(1), 1–9. https://doi.org/10.1007/s10578-014-0446-6

Segal, Z. V., Williams, M. J., & Teasdale, J. D. (2002). *Mindfulness-based cognitive therapy for depression.* Guilford Press.

Serkel-Schrama, I. J. P., de Vries, J., Nieuwesteeg, A. M., Pouwer, F., Nyklíček, I., Speight, J., de Bruin, E. I., Bögels, S. M., & Hartman, E. E. (2016). The association of mindful parenting with glycemic control and quality of life in adolescents with type 1 diabetes: Results from diabetes MILES—The Netherlands. *Mindfulness, 7*(5), 1227–1237. https://doi.org/10.1007/s12671-016-0565-1

Shamir-Essakow, G., Ungerer, J. A., & Rapee, R. M. (2005). Attachment, behavioral inhibition, and anxiety in preschool children. *Journal of Abnormal Child Psychology, 33*(2), 131–143. https://doi.org/10.1007/s10802-005-1822-2

Shaver, P. R., Lavy, S., Saron, C. D., & Mikulincer, M. (2007). Social foundations of the capacity for mindfulness: An attachment perspective. *Psychological Inquiry, 18*(4), 264–271. https://doi.org/10.1080/10478400701598389

Shaver, P. R., Mikulincer, M., Sahdra, B., & Gross, J. (2017). Attachment security as a foundation for kindness toward self and others. In K. W. Brown & M. R. Leary (Eds.), *The Oxford handbook of hypo-egoic phenomena*. Oxford University Press.

Sheppes, G., Suri, G., & Gross, J. J. (2015). Emotion regulation and psychopathology. *Annual Review of Clinical Psychology, 11*, 379–405. https://doi.org/10.1146/annurev-clinpsy-032814-112739

Sim, L., Adrian, M., Zeman, J., Cassano, M., & Friedrich, W. N. (2009). Adolescent deliberate self-harm: Linkages to emotion regulation and family emotional climate. *Journal of Research on Adolescence, 19*(1), 75–91. https://doi.org/10.1111/j.1532-7795.2009.00582.x

Smart, L. M., Peters, J. R., & Baer, R. A. (2016). Development and validation of a measure of self-critical rumination. *Assessment, 23*(3), 321–332. https://doi.org/10.1177/1073191115573300

Stein, A., Craske, M. G., Lehtonen, A., Harvey, A., Savage-McGlynn, E., Davies, B., Goodwin, J., Murray, L., Cortina-Borja, M., & Counsell, N. (2012). Maternal cognitions and mother–infant interaction in postnatal depression and generalized anxiety disorder. *Journal of Abnormal Psychology, 121*, 795–809. https://doi.org/10.1037/a0026847

Stone, L., Mares, S., Otten, R., Engels, R., & Janssens, J. (2016). The co-development of parenting stress and childhood internalizing and externalizing problems. *Journal of Psychopathology and Behavioral Assessment, 38*(1), 76–86. https://doi.org/10.1007/s10862-015-9500-3

Strauss, C., Lever Taylor, B., Gu, J., Kuyken, W., Baer, R., Jones, F., & Cavanagh, K. (2016). What is compassion and how can we measure it? A review of definitions and measures. *Clinical Psychology Review, 47*, 15–27. https://doi.org/10.1016/j.cpr.2016.05.004

Svendsen, J. L., Osnes, B., Binder, P.-E., Dundas, I., Visted, E., Nordby, H., Schanche, E., & Sørensen, L. (2016). Trait self-compassion reflects emotional flexibility through an association with high vagally mediated heart rate variability. *Mindfulness, 7*(5), 1103–1113. https://doi.org/10.1007/s12671-016-0549-1

Teper, R., Segal, Z. V., & Inzlicht, M. (2013). Inside the mindful mind: How mindfulness enhances emotion regulation through improvements in executive control. *Current Directions in Psychological Science, 22*(6), 449–454. https://doi.org/10.1177/0963721413495869

Thompson, B. L., & Waltz, J. (2008). Self-compassion and PTSD symptom severity. *Journal of Traumatic Stress, 21*(6), 556–558. https://doi.org/10.1002/jts.20374

Torbet, S., Proeve, M., & Roberts, R. M. (2019). Self-compassion: A protective factor for parents of children with autism spectrum disorder. *Mindfulness, 10*(12), 2492–2506. https://doi.org/10.1007/s12671-019-01224-5

Turpyn, C. C., & Chaplin, T. M. (2016). Mindful parenting and parents' emotion expression: Effects on adolescent risk behaviors. *Mindfulness, 7*(1), 246–254. https://doi.org/10.1007/s12671-015-0440-5

van der Oord, S., Bögels, S., & Peijnenburg, D. (2012). The effectiveness of mindfulness training for children with ADHD and mindful parenting for their parents. *Journal of Child and Family Studies, 21*(1), 139–147. https://doi.org/10.1007/s10826-011-9457-0

Vieira, J. M., Ávila, M., & Matos, P. M. (2012). Attachment and parenting: The mediating role of work-family balance in Portuguese parents of preschool children. *Family Relations, 61*(1), 31–50. https://doi.org/10.1111/j.1741-3729.2011.00680.x

Wang, Y., Liang, Y., Fan, L., Lin, K., Xie, X., Pan, J., & Zhou, H. (2018). The indirect path from mindful parenting to emotional problems in adolescents: The role of maternal warmth and adolescents' mindfulness. *Frontiers in Psychology, 9*, Article 546. https://doi.org/10.3389/fpsyg.2018.00546

Weer, C., & Greenhaus, J. H. (2014). Work-to-family conflict. In A. C. Michalos (Ed.), *Encyclopedia of quality of life and well-being research* (pp. 7244–7245). Springer.

Weitlauf, A. S., Broderick, N., Stainbrook, J. A., Taylor, J. L., Herrington, C. G., Nicholson, A. G., Santulli, M., Dykens, E. M., Juárez, A. P., & Warren, Z. E. (2020). Mindfulness-based stress reduction for parents implementing early intervention for autism: An RCT. *Pediatrics, 145*(Suppl 1), S81–s92. https://doi.org/10.1542/peds.2019-1895K

Williams, K. L., & Wahler, R. G. (2010). Are mindful parents more authoritative and less authoritarian? An analysis of clinic-referred mothers. *Journal of Child and Family Studies, 19*(2), 230–235. https://doi.org/10.1007/s10826-009-9309-3

Wong, C., Mak, W., & Liao, K. (2016). Self-compassion: A potential buffer against affiliate stigma experienced by parents of children with autism spectrum disorders. *Mindfulness, 7*(6), 1385–1395. https://doi.org/10.1007/s12671-016-0580-2

Wren, A. A., Somers, T. J., Wright, M. A., Goetz, M. C., Leary, M. R., Fras, A. M., Huh, B. K., Rogers, L. L., & Keefe, F. J. (2012). Self-compassion in patients with persistent musculoskeletal pain: Relationship of self-compassion to adjustment to persistent pain. *Journal of Pain and Symptom Management, 43*(4), 759–770. https://doi.org/10.1016/j.jpainsymman.2011.04.014

Yang, Y., Guo, Z., Kou, Y., & Liu, B. (2019). Linking self-compassion and prosocial behavior in adolescents: The mediating roles of relatedness and trust. *Child Indicators Research, 12*(6), 2035–2049. https://doi.org/10.1007/s12187-019-9623-2

Yarnell, L. M., & Neff, K. (2013). Self-compassion, interpersonal conflict resolutions, and well-being. *Self and Identity, 12*(2), 146–159. https://doi.org/10.1080/15298868.2011.649545

Zeman, J., Cassano, M., Perry-Parrish, C., & Stegall, S. (2006). Emotion regulation in children and adolescents. *Journal of Developmental and Behavioral Pediatrics, 27*(2), 155–168. https://doi.org/10.1097/00004703-200604000-00014

Zessin, U., Dickhäuser, O., & Garbade, S. (2015). The relationship between self-compassion and well-being: A meta-analysis. *Applied Psychology: Health and Well-Being, 7*(3), 340–364. https://doi.org/10.1111/aphw.12051

Zhang, W., Wang, M., & Ying, L. (2019). Parental mindfulness and preschool children's emotion regulation: The role of mindful parenting and secure parent-child attachment. *Mindfulness, 10*(12), 2481–2491. https://doi.org/10.1007/s12671-019-01120-y

The Psychophysiology of Self-Compassion

17

Elizabeth T. Slivjak, Alex Kirk, and Joanna J. Arch

Introduction

Self-compassion has become an established area of scientific inquiry. In this chapter, we focus on the peripheral physiological effects of self-compassion, including physiological measures linked to the autonomic nervous system and to immune functioning. We review the evidence for whether self-compassion affects peripheral stress and immune response systems in a manner that aligns with greater stress resilience and with more adaptive autonomic and immune functioning. Studies encompass various forms of self-compassion, including dispositional or trait self-compassion, defined herein as the self-reported tendency to generally embody a compassionate state or perspective regarding one's own experience; induced state self-compassion, which reflects brief inductions or brief trainings in self-compassion intended to produce an immediate self-compassionate state; and regularly trained self-compassion, which reflects more ongoing, multi-week, formal interventions that aim to increase self-compassion in a more enduring manner. This chapter reviews the findings on self-compassion with regard to peripheral physiology, including psychophysiology and peripheral immune and inflammatory markers.

The stress system in humans is located both within the central nervous system and the periphery. The autonomic nervous system, a division of the periphery, is consisted of three branches: the sympathetic, parasympathetic, and enteric nervous systems. When homeostasis is disrupted by detection of an internal or external stressor or threat (e.g., infection, emotional distress), the sympathetic nervous system and hypothalamic–pituitary–adrenal (HPA) axis are activated (Miller & O'Callaghan, 2002). Activation of these two systems results in the body's stress or fight-or-flight response, wherein the brain stimulates changes in behavior and in periphery physiology to enhance survival, including increased heart rate, blood pressure, respiration rate, peripheral vasoconstriction, and reduced appetite. These changes result in heightened arousal and alertness and increased vigilance and attention, facilitating detecting and responding to threat. Threat detection also results in decreased parasympathetic nervous system activation, which is largely responsible for bodily functions that occur at rest, such as digestion. Whereas time-limited activation of the autonomic stress response systems provides benefits for survival, its prolonged or chronic activation is associated with adverse consequences and poor physical health, including suppressing key functions such as immune functioning. Thus, researchers have been interested in

E. T. Slivjak · A. Kirk · J. J. Arch (✉)
Department of Psychology and Neuroscience, University of Colorado Boulder, Boulder, CO, USA
e-mail: Elizabeth.Slivjak@colorado.edu; Joanna.Arch@colorado.edu

A. Finlay-Jones et al. (eds.), *Handbook of Self-Compassion*, Mindfulness in Behavioral Health,
https://doi.org/10.1007/978-3-031-22348-8_17

understanding the effects of interventions that dampen or enhance recovery of biological responses to acute and chronic stress.

The studies discussed in this chapter evaluate a range of peripheral physiological effects of self-compassion. The most common physiological measure used in these studies is high frequency heart rate variability (HF-HRV) and related HRV measures that assess the parasympathetic nervous system influence on the time intervals between heartbeats via the vagus nerve (the tenth cranial nerve). HF-HRV has been conceptualized as "a transdiagnostic biomarker of self-regulation and cognitive control" (Beauchaine & Thayer, 2015, p. 338), with higher HF-HRV (in general) reflecting greater levels of self-regulation and cognitive control. Heart rate (HR), reported in several relevant studies, reflects both sympathetic and parasympathetic autonomic nervous system influences (Berntson et al., 2007). Skin conductance, the activity of sweat glands, reflects sympathetic nervous system activity (Dawson et al., 2007), and thus is considered a more direct measure of sympathetic activation. Salivary alpha-amylase (sAA), an enzyme found in saliva, also tracks predominantly with the sympathetic nervous system, with higher levels reflecting higher levels of sympathetic activation (Rohleder et al., 2004). Cortisol is produced by the body's other major stress response system, the HPA axis, which in humans tends to be slower acting than the sympathetic nervous system and to respond most robustly to stressors characterized as uncontrollable and socially threatening (Dickerson & Kemeny, 2004).

In addition, significant work has established a robust relationship between higher levels of psychological stress and increased immune activation (Marsland et al., 2017). In regard of this connection, research to date has begun to examine the extent to which self-compassion training can improve immune outcomes, including assessing two markers as correlates of immune system activation: interleukin 6 (IL-6), measured from blood plasma, and salivary immunoglobulin A (sIgA). IL-6 is a commonly studied cytokine, a class of immune system protein broadly involved in cellular signaling and messaging, and has received attention as a proinflammatory cytokine associated with higher levels of psychological stress (Slavich & Irwin, 2014). Though IL-6 is a complicated protein involved in both pro- and anti-inflammatory processes (Del Giudice & Gangestad, 2018), elevations in IL-6 can be interpreted as a metric of proinflammatory activity when they occur in conjunction with the activation of other physiological stress systems, particularly the sympathetic nervous system (Michopoulos et al., 2017). Salivary IgA is a secreted antibody involved in neutralizing and blocking pathogens from accessing various tissues. As it relates to stress research, sIgA is under regulatory control from neuroendocrine circuits activated by psychological stress (Bosch et al., 2002). Lower levels of sIgA (Phillips et al., 2006) are associated with higher rates of illness and infection (Pilette et al., 2001). Thus, sIgA is a metric of immune activity influenced by neural regions associated with psychological stress.

Each of these physiological measures, as well as the biological systems in which they are embedded, is highly complex, responsive to shifts in environmental demands, and interacts with a range of other biological systems, often including one another. However, a full description of their complexity lies beyond the scope of this chapter.

Most of the work on self-compassion and peripheral physiology relies on Neff's (2003a, b) definition of self-compassion, which encompasses three dimensions: "being kind and understanding toward oneself in instances of pain or failure rather than being harshly self-critical; perceiving one's experiences as part of the larger human experience rather than seeing them as isolating; and holding painful thoughts and feelings in mindful awareness, rather than overidentifying with them" (p. 223). Based on this three-part definition of self-compassion, Neff (2003a) developed and validated a corresponding 26-item Self-Compassion Scale (SCS), which serves as the primary self-report measure for trait or dispositional self-compassion used in the studies in this chapter.

While writing this chapter, we identified 23 studies documenting the relationship between self-compassion and physiology. Most of these were published recently, demonstrating an emerging interest in the relationship between self-compassion and physiological outcomes. The included articles reported on a range of biomarkers, and they mostly included healthy adult samples and stress (as opposed to mental health related) outcomes. Ultimately, we categorized the included articles into three sections: (1) observational studies of trait self-compassion, (2) brief laboratory studies and experiments designed to train self-compassion skills, and (3) lengthier clinical interventions. We begin with reviewing the evidence for links between self-compassion and peripheral physiology by focusing on observational studies of trait self-compassion and stress.

Observational Studies of Trait Self-Compassion and Stress

Observational studies provide insight into the relationship between individual differences in self-compassion and physiological responses to laboratory-induced stressors. Biomarkers that have been studied in relation to self-compassion in observational studies include IL-6, sAA, heart rate (HR), HF-HRV, blood pressure (BP), and cortisol. Most of the studies in this area have assessed self-reported self-compassion and biomarkers at baseline, followed by exposure to the Trier Social Stress Test (TSST), a standardized laboratory social performance stressor that includes anticipation, stress, and recovery phases (Kirschbaum et al., 2008); during the TSST in the identified studies, tasks typically included a 5-min speech and a 5-min mental math task, both performed in front of two or three study judges.

Despite limitations including small, homogenous samples and lack of direct manipulation of self-compassion, the available observational studies provide evidence that is consistent with a protective role of trait self-compassion in buffering biological stress responses. For example, in a study of 41 healthy young adults (ages 18–35 years), researchers found that trait self-compassion led to lower proinflammatory markers (IL-6) in response to an initial social performance stressor and lower anticipatory proinflammatory response to a repeated social performance stressor, suggesting greater physiological resilience to social performance stress (Breines et al., 2014). Participants took part in two TSSTs over 2 consecutive days to evaluate responses to both a novel and repeated stressor. On both days, blood draws to assess for IL-6 were taken at baseline as well as 30 and 120 min following the TSST. As expected, IL-6 levels increased following the social stressor during both laboratory sessions, without evidence of a habituated IL-6 response to the repeated stressor. Additionally, as predicted, baseline trait self-compassion negatively predicted day 1 IL-6 response to the TSST, such that higher trait self-compassion was associated with lower levels of a proinflammatory response to the novel stressor. Surprisingly, however, baseline trait self-compassion did not predict Day 2 IL-6 response to the TSST. In additional analyses conducted to better understand this lack of association, findings showed an elevated IL-6 response prior to the TSST on Day 2 for individuals with lower trait levels of self-compassion. The authors hypothesized that this elevated anticipatory IL-6 response could have been due to anticipatory anxiety of the upcoming TSST or rumination regarding the previous day's TSST and prevented a more elevated IL-6 response to the TSST.

Having demonstrated a relationship between trait self-compassion and proinflammatory responding, Breines et al. (2015) next sought to examine whether individuals with greater trait self-compassion display less sympathetic nervous system activation in response to novel and repeated stressors. Given that the proinflammatory response to stress is largely driven by increased sympathetic nervous system activity, lower IL-6 in the previous study would suggest a lower sympathetic nervous system response, as measured here by lower sAA. Thirty-three healthy young adults (ages 18–34 years) completed baseline self-report measures including the SCS (Neff, 2003a). Participants then completed initial saliva collection, followed by the TSST, with additional saliva collection at multiple timepoints over a 60-min period. Participants completed the TSST again the following day. As predicted, for each individual day of the TSST, higher levels of trait self-compassion were associated with significantly lower sAA responses to the TSST with medium-to-large effect sizes. Additionally, unlike the robust IL-6 response observed across 2 days of social stress tasks (Breines et al., 2014), sAA responses decreased significantly across the repeated stressors, although the overall habituation in sAA from Day 1 to day 2 was not associated with baseline self-compassion, potentially due to an already low sAA response to the novel stressor among self-compassionate individuals.

Importantly, while these studies have contributed to our understanding of the relationship between trait self-compassion and physiological stress, not all researchers view self-compassion as a unidimensional construct, but instead as consisting of two distinct factors composed of positive and negative characteristics (Costa et al., 2015; López et al., 2015). Neff's SCS (Neff, 2003a) assesses a total self-compassion score made up of six subscales – three positive (e.g., self-kindness, common humanity, and mindfulness) and three negative subscales (e.g., self-judgment, self-isolation, and overidentification). Neff et al. (2018) argue that both positive and negative subscales are central to self-compassion. Others posit that the SCS's negative subscales are significantly more related to psychopathology than the positive subscales, thus disproportionately increasing the link between a total SCS score and psychopathology (Muris & Petrocchi, 2017). Moreover, researchers have argued that differential physiological systems might be involved for positive versus negative aspects of compassion (Gilbert et al., 2011).

Given this controversy, Neff et al. (2018) thoughtfully reanalyzed data from the two aforementioned studies (Breines et al., 2014, 2015) to evaluate the relationship between physiological stress response and various subcomponents of self-compassion. A reanalysis was conducted in which data for sAA and IL-6 were available for 33 and 41 healthy young adults, respectively (Neff et al., 2018). In their reanalysis, Neff et al. (2018) examined correlations between these physiological markers and baseline self-compassion scores, including total SCS, positive subscale, negative subscale, and the individual six subscale scores. For the TSST completed on the first day, total SCS and both positive and negative subscale scores all moderately correlated with sAA and IL-6, with no significant difference in the correlations between the physiological markers and each of the positive and negative subscale scores. In terms of individual subscales, two positive subscales (self-kindness and mindfulness) and two negative subscales (isolation and overidentification) significantly correlated with sAA and IL-6 in the expected direction (i.e., negative correlations for the positive subscales and positive correlations for the negative subscales), whereas common humanity and self-judgment did not. On the second day of the TSST, only sAA significantly correlated with SCS scores; again, total SCS and both positive and negative subscale scores moderately correlated with sAA. Of the individual subscales, only isolation was significantly associated with sAA, again in the expected direction. Importantly, the authors noted a smaller sample size as well as smaller intercorrelations for the second day of the TSST. Given these overall findings, the authors concluded that the positive and negative subscales do not each uniquely predict distinct underlying physiology, as both correlate moderately with sympathetic nervous system and pro-inflammatory responding.

In addition to sAA and IL-6, researchers have examined the relationship between dispositional self-compassion and vagally mediated heart rate variability (vmHRV), a biological marker of the parasympathetic nervous system control over heart rate variability and a proposed indicator of the ability to adaptively regulate emotions and stress. In a study of 53 healthy university students, participants completed baseline measurements of trait self-compassion as assessed by the SCS as well as resting vmHRV (assessed with the root mean square of successive differences [RMSSD], e.g., between R-R intervals in the

ECG), measured during a 5-min interval in which participants were instructed to breath slowly and relax (Svendsen et al., 2016). Additionally, a subsample of participants ($n = 26$) wore heart monitors to collect physiological data over 24 h. As predicted, young adults with higher trait self-compassion demonstrated higher resting vmHRV during the baseline period ($r = .52$, $p < .01$). This positive correlation was also found with a 24-h measure of vmHRV assessed outside of the laboratory ($r = .50$, $p < .02$), demonstrating the continued association between higher self-compassion and greater vmHRV in a more naturalistic setting.

Luo et al. (2018) extended the work conducted by Svendsen et al. (2016) by examining vmHRV in the context of the TSST, as opposed to resting and naturalistic vmHRV. In their study, 85 male university students in China completed the SCS, and students with self-compassion scores in the upper and lower 27% were asked to participate in a laboratory session as part of high- and low-self-compassion groups, respectively ($n = 17$ for both groups). Results showed that HR did not differ significantly between groups but vmHRV did differ. Similar to results from Svendsen et al. (2016), baseline vmHRV (assessed via RMSSD) was higher among self-compassionate students. Additionally, self-compassionate students showed significantly higher vmHRV during the social stressor and recovery phases on the TSST, although no difference was found during the anticipation stage of the stressor.

Whereas most studies examined the relationship between self-compassion and physiological functioning among healthy young adults, Bluth et al. (2016) hypothesized a protective function for self-compassion during adolescence, a life stage marked by transitions that can increase stress. Twenty-eight adolescents (ages 13–18 years) completed a laboratory session consisting of physiological (e.g., BP, HR, HF-HRV, and salivary cortisol) and psychological measures as well as the TSST. Analyses were conducted using a median split approach that categorized adolescents into those with higher ($n = 16$) and lower ($n = 12$) baseline trait self-compassion. Systolic BP response to stress was the only physiological measure significantly buffered by trait self-compassion, with an inverse relationship between higher self-compassion and lower systolic BP. No significant differences across the groups emerged in change of HR during stressor tasks or in HF-HRV. Similarly, despite lower overall cortisol output among the more highly self-compassionate adolescents, a group difference did not emerge (Hedge's $g = 0.12$). The authors highlighted ceiling effects in HR on stressor tasks as well as a lack of gender balance between groups (including no male participants in the low self-compassion group) as plausible explanations for a failure to find physiological differences across the two groups.

Like adolescence, older adulthood can represent a life stage often characterized by unique age-related stressors – in this case, more persistent and less controllable stressors like declining physical health. To examine whether self-compassion serves a protective function during older adulthood, a pioneering study evaluated the relationship between trait self-compassion and diurnal cortisol secretion among 233 community-dwelling older adults ages 59–93 years (Herriot et al., 2018). Five salivary cortisol samples were collected per day over 3 nonconsecutive days in a given week; researchers used area-under-the-curve (AUC) to calculate daily cortisol levels, and then computed average cortisol secretion using daily AUC cortisol levels. Self-compassion was assessed using the 12-item Self-Compassion Scale (Raes et al., 2011), a short-form of Neff's SCS (Neff, 2003a). The authors found that among older adults who reported high physical health problems, functional disabilities, and life regrets, only those who were low in self-compassion showed higher levels of cortisol (AUC). Thus, the detrimental effects of physical health problems and functional disability in influencing cortisol levels may be buffered by higher levels of self-compassion. Importantly, there was no evidence for a main effect of stressors on cortisol levels, indicating that individual differences in interpretation of stressors in self-compassionate (or non-self-compassionate) terms rather than the stressors themselves may most significantly impact physiological responses to stress.

In addition to research among adolescents and older adults, one study examined self-compassion and physiological health in a group that may be prone to frequent failure and thus self-criticism – athletes (Ceccarelli et al., 2019). Following baseline measures of HF-HRV and trait self-compassion among 91 adult university or national level athletes (ages 18–40 years) across a variety of sports, participants took part in a standardized laboratory stressor where a researcher read aloud from a guided imagery script, asking participants to recall a recent sport failure. The researcher prompted athletes to recall a mistake or setback in as much detail as possible, focusing attention on emotions and physical sensations experienced at the time of the failure as well as reexperienced in the present moment. Results showed that baseline trait self-compassion was significantly associated with HF-HRV during the brief (2-min) recollection of sport-related failure, such that more self-compassionate athletes displayed increased parasympathetic activity (i.e., higher HF-HRV), and thus increased self-regulation during the stress induction. However, this relationship was not found during a brief (2-min) recovery phase immediately following the stressor, suggesting that trait self-compassion increased athletes' physiological regulation primarily during the stressor itself.

Summary of Observational Studies on Self-Compassion and Stress Physiology

In summary, although research on the association between self-compassion and physiology is in its early stages, a few trends across studies emerged. Across the eight reviewed studies, emerging evidence suggests that dispositional self-compassion can buffer physiological responses to laboratory stressors. Specifically, in response to standardized laboratory stressors, trait self-compassion may positively correlate with HF-HRV and negatively correlate with systolic BP, cortisol, IL-6, and sAA, though several findings were mixed. First, examining indices of heart functioning, studies found some evidence for higher HRV among more trait self-compassionate individuals both at rest (Luo et al., 2018; Svendsen et al., 2016) and during a laboratory stressor (Ceccarelli et al., 2019; Luo et al., 2018) and during stressor recovery in one study (Luo et al., 2018) but not the other (Ceccarelli et al., 2019). In addition, no group differences were found in HRV among high- and low-self-compassionate adolescents completing a social stressor (Bluth et al., 2016); only systolic BP differed. Two studies similarly found no evidence of a significant association between self-compassion and HR (Bluth et al., 2016; Luo et al., 2018), though both were small and thus likely underpowered. Second, two studies evaluated cortisol as an index of HPA axis activity. Herriot et al. (2018) reported a significant moderation effect in which higher self-compassion correlated with lower diurnal cortisol levels in older adults who reported high physical health problems, functional disabilities, and life regrets. Finally, markers of inflammation (IL-6) and sympathetic nervous system activation (sAA) were each examined in only one observational study (Breines et al., 2014, 2015). Breines et al. (2014) found evidence for lower IL-6 response to a novel stressor among self-compassionate individuals, though this association was not significant for a repeated stressor. Similarly, Breines et al. (2015) demonstrated an inverse relationship between trait self-compassion and sAA response to a social stressor, suggesting a protective role of self-compassion.

Given the largely small, homogeneous samples predominantly focused on healthy adults and few studies per biomarker, more research is needed. Future directions might include assessing the link between trait self-compassion and longer term health outcomes, examining trait self-compassion as a continuous variable (rather than use a median split, for example), and assessing the impact of trait self-compassion in response to more naturalistic stressors. A more direct method of examining the influence of self-compassion to stress responding is to manipulate self-compassion directly; we turn next to these studies.

Experimental Studies Related to Self-Compassion and Physiological Stress

Five laboratory experiments have explored the relationship between self-compassion and physiological stress responding. These experiments are largely characterized by brief manipulations of self-compassion in healthy adults to assess both physiological and self-reported stress responses. Biomarkers from these studies included cortisol as a metric of the HPA axis stress response; sAA, SCL, and HR as metrics of the sympathetic stress response; and HF-HRV as a metric of the parasympathetic stress response. Although these studies are largely limited in their use of brief manipulation strategies applied to healthy adults within laboratory settings, the overall results lend greater support to a causal relationship between increased self-compassion and decreased physiological stress responding.

In a study of 105 healthy undergraduate women, researchers examined the impact of a brief self-compassion training on stress-related biomarkers including sAA, HF-HRV, and cortisol (Arch et al., 2014). Participants were randomized to either a *metta* (loving-kindness) meditation condition, an attention placebo condition focused on cognition and problem-solving, or a control condition receiving no intervention. The *metta* meditation condition used both traditional *metta* phrases ("May I be happy. May I be healthy and strong…") and study-specific phrases ("May I know that others struggle along with me. May I love and accept myself completely, just as I am…") that focused on cultivating a sense of common humanity, well-being, and acceptance toward oneself and, to a lesser extent, toward others. All participants attended two laboratory sessions. The first session involved brief condition-specific training (in the *metta* meditation or attention placebo control). The second session involved completion of the TSST, a standardized social stress paradigm that requires participants to prepare and deliver a speech without notes and complete a challenging math task, both in front of judges (Kirschbaum et al., 2008). Between the two sessions, participants in both the self-compassion and attention placebo conditions completed three 10-min recorded training sessions in their assigned intervention. Finally, they completed a briefer training session (in *metta* or placebo control) immediately prior to being introduced to the TSST. Biomarkers assessed at the second TSST-focused laboratory session included sAA and salivary cortisol (collected at 5 points during baseline through 35 min post-TSST) and HF-HRV (collected continuously during baseline through 10 min post-TSST). Results showed that the self-compassion condition endorsed greater trait and state self-compassion from the first to second laboratory session relative to both control conditions. The self-compassion condition also showed a steeper decrease in state anxiety from the TSST speech preparation through the post-TSST recovery phases but not at other study points, relative to both control conditions. Regarding physiological outcomes, the self-compassion condition showed significantly lower sAA (in terms of area under the curve with respect to increase) relative to both control groups in response to the TSST, indicative of a reduced sympathetic stress response (Thayer et al., 2012; Thayer & Lane, 2000). In addition, the self-compassion condition had a more engaged or stable HF-HRV during both the speech preparation phase in anticipation of the TSST and in the recovery phase following the TSST relative to both control groups indicative of increased or maintained parasympathetic control and associated with improved emotion regulation in prior studies (Porges, 2007). No group differences emerged in cortisol responding, perhaps because the TSST represents an uncontrollable performance stressor – characteristics that elicit a robust cortisol response (Dickerson & Kemeny, 2004).

In summary, these results suggest that brief self-compassion training can improve both sympathetic (sAA) and parasympathetic (HF-HRV) markers of stress in the context of a social stress task, though in this study, did not influence HPA axis activation (in the form of cortisol). Together with the self-report findings, the study thus suggests that brief self-compassion meditation training led to shifts in how participants responded to the stressor – that is, with greater self-compassion and self-regulation – rather than in their experience of the stressor's controllability.

Following this experimental study, the authors subsequently examined predictors and moderators of the beneficial physiological outcomes associated with self-compassion meditation (Arch et al., 2016). Specifically, they sought to determine what role two stress vulnerability traits (social anxiety, rumination) and two resiliency traits (self-compassion, nonattachment, i.e., the Buddhist notion of release from mental fixations) might have in moderating the effects of brief self-compassion meditation training on sAA and HF-HRV (vs. control conditions). Results showed that relative to the control groups, higher reported baseline nonattachment (i.e., lower attachment) within the self-compassion group predicted lower TSST-related increase in sAA and lower self-reported anxiety than did lower baseline nonattachment (e.g., higher attachment). In contrast, levels of nonattachment did not influence sAA outcomes for the two control groups compared to the self-compassion group. However, trait rumination was significantly more influential in predicting sAA increases within the control groups compared to the self-compassion group. In summary, the physiological benefits of brief self-compassion training (relative to the two control conditions) were robust across various levels of baseline social anxiety and trait self-compassion but were moderated by baseline levels of nonattachment and rumination, such that participants with higher levels of attachment or rumination benefitted less from self-compassion training than those with lower levels of attachment or rumination. As participants with these baseline characteristics did not benefit equally, the authors concluded that individuals with higher levels of attachment or rumination might require a more extensive or tailored self-compassion training.

Whereas the aforementioned studies examined self-compassion training in non-clinical samples, Ascone et al. (2017) evaluated the physiological impact of a single session of self-compassion-focused guided imagery among 51 psychiatric patients (including 36 psychiatry inpatients and 15 outpatients) receiving treatment for paranoid ideation. Skin conductance levels (SCL) were measured throughout the experiment by a recording bracelet. Following baseline assessment, participants underwent a 3–5-min negative mood induction in which they recalled a recent social stressor that induced either fear or shame. Participants were then randomly assigned to either self-compassion imagery (n = 26) or control imagery (n = 25). In the intervention group, experimenters read from a standard script aloud for 10 min, instructing participants to create a mental image of a person or object that elicits compassion and warmth for the participants. In the control group, experimenters described a chair in the experiment room, matching the self-compassion imagery script in style and length. Compared to the control group, those receiving self-compassion imagery reported significant increases in happiness and self-reassurance, but there were no differences between conditions in SCL or symptoms of paranoia. Given that this experiment was conducted in a single session, the authors suggest that participants may not have experienced sufficient practice creating a self-compassionate mental image, resulting in increased effort during the intervention that impeded their ability to relax physiologically.

While these studies support the role of self-compassion in improving physiological markers of stress, more recent work has begun to elucidate the intervention processes that account for these physiological benefits. Specifically, one study sought to examine the precise intervention mechanisms that drive the beneficial physiological effects of self-compassion by randomizing 135 adults in equal numbers (n = 27 per condition) to five conditions including a loving-kindness meditation condition where compassion is directly instructed (direct compassion), a compassionate body scan condition where participants attend to bodily sensations with a sense of calm acceptance (indirect compassion), a positive-excitement condition, a self-critical rumination condition, and a neutral control condition (Kirschner et al., 2019). Thus, the study aimed to test whether self-compassion uniquely affects physiological responding, or whether it confers these benefits only to the extent that it increases positive affect.

This study consisted of a single laboratory session where participants listened to one 11.5-min tape, which guided participants through an induction specific to their condition. For example, participants in the direct compassion loving-kindness condition were instructed to direct loving thoughts toward someone close to them and then to direct those same thoughts toward themselves (in the spirit of *metta* meditation), whereas those in the indirect compassion body scan condition were guided to direct kind attention to body sensations. Researchers tracked HF-HRV, SCL, and HR for the duration of the taped experimental inductions. Results from this brief intervention showed decreased HR throughout the entire audio exercise in the direct loving-kindness group and for the first 8 min in the compassionate body scan group, relative to the three comparison groups. Further, HF-HRV was elevated in both self-compassion groups throughout the exercise (except the final minute of the compassionate body scan) relative to the three comparison groups. Fewer robust differences were seen in SCL, though the loving-kindness group did show a significant decrease relative to the neutral control group for the first 7 min of the exercise.

Regarding mechanisms, this study found that HR fully mediated the relationship between the loving-kindness condition and self-reported self-criticism and partially mediated the relationship between the compassionate body scan condition and self-reported levels of both self-criticism and positive affiliative affect. Thus, inducing self-compassion may be beneficial in improving self-criticism and affiliative affect to the extent that it calms HR-based physiological arousal. In sum, the authors report that brief self-compassion inductions may confer physiological benefits in two ways: first, through activation of the parasympathetic positive affect system (HF-HRV), which corresponds to reduced stress and enhanced emotion regulation capabilities; and second, by increasing a positive sense of self while decreasing a negative sense of self. Importantly, these improvements cannot be attributed simply to increasing positive affect, suggesting that inducing self-compassion confers unique benefits in improving physiological markers of stress.

Beyond the use of physiological measures associated with stress, other work has extended our understanding of self-compassion into the study of physical pain. Luo et al. (2019) randomized 29 healthy adults to undergo both compassionate and neutral self-talk protocols in a randomized order to measure their impact on HF-HRV and pain ratings during a cold pain exposure. To accomplish this, these researchers first had participants generate four self-compassion statements, followed by the start of the experimental manipulation trials. For each trial, participants were presented with either a self-compassion (e.g., "I understand your pain") or neutral control (e.g., "The store is selling fruits") statement for 10 s that they were asked to read aloud. After 10 s, participants underwent a 3-min cold pain exposure by holding a bottle filled with iced water while recording pain ratings every 30 s. All four self-compassion statements along with four control statements were presented three times each, resulting in a total of 2 min of exposure to self-compassion statements and 2 min of exposure to control statements. Results showed that reported pain was significantly lower during the self-compassion statement trials compared to the control statement trials. In addition, HF-HRV during cold pain exposure was higher in the self-compassion trials as compared to the control trials. Further, higher levels of HF-HRV were more significantly associated with lower pain ratings during the self-compassion trials than in the control trials. This study thus extends past findings by showing that relative to a neutral control condition, very brief self-compassion in the context of thermal pain can improve both perceived physical pain and physiological markers of self-regulation.

The studies described above indicate that brief self-compassion interventions can improve physiological markers of stress in response to both physical pain and psychological stressors. Additional work has sought to examine the extent to which small adjustments in self-compassion exercises bolster these outcomes. One such study by Petrocchi et al. (2016) tested whether repeating self-generated, self-compassionate phrases while looking at a mirror would improve outcomes as compared to not using a mirror, as is typically done. They recruited 86 adults from the general population and had each of them generate four phrases they would use to soothe or encourage a friend going through a difficult time. Participants then wrote about a recent situation during which they harshly criticized themselves out of shame or disappointment. The self-criticism writing exercise successfully increased HR and negative affect, while decreasing positive affect and HRV, suggesting that the researchers successfully induced a state of self-criticism. Participants then were randomized to one of three conditions for a 5-min experimental manipulation, including repeating their four encouraging phrases toward themselves while looking at themselves in a mirror ($n = 30$), repeating their four encouraging phrases toward themselves without looking into a mirror ($n = 28$), or looking at themselves in a mirror without repeating the phrases ($n = 28$). Self-report and ECG measures of HR and HRV were incorporated throughout all procedures. The group that recited self-compassionate statements while looking in the mirror showed significantly larger increases in HRV and reported more soothing positive affect relative to the two comparison groups. The relationship between condition and increase in soothing positive affect was partially mediated by an increased sense of common humanity, a facet of self-compassion. Overall, while the precise mechanisms are not fully understood, the authors purport that reciting self-compassionate phrases while looking into a mirror may improve outcomes by increasing the number of positive social signals being communicated, facilitating a sense of self-compassion that extends beyond verbal language, or seeing ourselves from an external, more objective point of view in a manner that reduces self-critical biases.

Summary of Experimental Studies on Self-Compassion and Stress Physiology

Overall, the limited number of laboratory studies that manipulate self-compassion show that such manipulations often result in improved physiological markers of stress, particularly those associated with sympathetic (e.g., sAA, HR) and parasympathetic (i.e., HF-HRV) nervous system responding. These improvements in physiological stress were observed in the context of both psychological and physical stress tasks. Thus, briefly trained or induced self-compassion appears to cultivate resilience in the face of diverse acute stressors.

Research has begun to explore mediators and moderators of the physiological benefits of induced self-compassion and to elucidate the specific approaches to inducing self-compassion that have the greatest effect on physiological self-regulation and stress response. Indeed, these beneficial effects can be bolstered with minor manipulations, such as reciting self-compassion phrases into a mirror as compared to not using a mirror (Petrocchi et al., 2016). Importantly, one study found that the beneficial effects of self-compassion training were not attributable solely to increased positive affect but rather were unique to self-compassion processes (Kirschner et al., 2019).

While these initial findings are promising, more work remains to be done in order to better understand the physiological benefits of briefly induced or trained self-compassion. For example, additional laboratory-based studies are needed to replicate and extend current findings, and particularly to clarify the potential effects of self-compassion training on skin conductance and cortisol, which are key indices of sympathetic and HPA axis activation, respectively. In addition, future studies can attempt to extend these findings into more vulnerable populations, including in clinical groups that are more likely to suffer from lower levels of self-compassion (Macbeth & Gumley, 2012).

Clinical Interventions Related to Stress and Mental Health

Expanding on brief manipulations in laboratory settings, eight studies examined the impact of lengthier clinical interventions in more naturalistic settings. Similar to most of the aforementioned work, these studies were predominantly conducted with healthy adult populations, although three focused on populations facing potentially heightened stress, recruiting individuals who had experienced early life adversity, parenting stress, or cancer biopsy (Pace et al., 2013; Poehlmann-Tynan et al., 2020; Wren et al., 2019). Interventions included at-home meditation practice using audio recordings as well as group-based self-compassion training, ranging in length from a relatively brief manipulation (2 weeks) to a 15-week course. Self-compassion was cultivated using several formal interventions, including Cognitively-Based Compassion Therapy (CBCT; Pace et al., 2009), Emotion Focused Training for Self-Compassion and Self-Protection (EFT-SC; Halamová et al., 2019), and Compassionate Mind Training (CMT; Gilbert, 2000, 2009, 2010), among others. In addition to self-reported outcomes, the following biomarkers were assessed: sAA and HR to assess sympathetic nervous system response, cortisol as a marker of the HPA axis, and HRV as an indicator of parasympathetic nervous system response. Additionally, immune system functioning was explored through IL-6, salivary C-reactive protein (CRP), and sIgA.

In one such early clinical intervention study, researchers examined the impact of CBCT training on IL-6 and cortisol in response to a TSST among 61 healthy university students ages 17–19 years (Pace et al., 2009). Recruited from a health education class, participants were randomized to take part in either CBCT (n = 33), compassion meditation training based on Tibetan Buddhist mind-training or *lojong* practices, or a health discussion group (n = 28) over the course of 6 weeks, totaling 12 intervention hours for both conditions. In contrast to mindfulness practices that encourage nonjudgmental awareness of thought processes and emotions, *lojong* practices first challenge individuals' preconceived thoughts and emotions toward others and then guide them to develop spontaneous feelings of empathy, love, and compassion toward all people, beginning with the self and expanding to include others whom individuals dislike or with whom they have conflicts. CBCT was designed to first teach attention and mindfulness techniques as a foundation, and then gradually shifted toward focusing on *lojong* meditations. In addition to 12 h of group training, CBCT participants were encouraged to practice exercises daily at home using audio recordings. The health discussion groups reviewed standardized topics related to mental and physical health among university students, and participants wrote weekly papers applying material to their lives to control for time and practice outside of class. To evaluate the impact of meditation training on physiological response to a novel stressor, participants completed the TSST between 8 and 10 weeks from the initiation of the study. Surprisingly, researchers did not find evidence for a main effect of compassion meditation on IL-6, cortisol, or self-reported subjective distress following the TSST. However, after examining variability in weekly meditation practice among participants, researchers conducted a median split analysis for high and low practice. Using this approach, significant correlations were found, demonstrating an inverse relationship between practice, IL-6, and distress. Specifically, a greater number of practice sessions per week was associated with lower immune response and subjective distress, a finding which highlights the importance of committed practice of compassion exercises in improving psychological and physical outcomes.

In a subsequent study, Pace et al. (2013) examined the impact of CBCT on salivary C-reactive protein (CRP) among 55 adolescents in the foster care system, seeking to understand the impact of compassion training in a population impacted by early life adversity. Adolescents in this trial were randomly assigned to either a 6-week CBCT intervention (n = 29) or a waitlist control (n = 26). CBCT participants attended class twice a week, completing a total of 12 intervention hours; additionally, they were encouraged to practice daily exercises at home using an audio recording. To examine CRP, saliva samples were collected immediately upon awakening prior to and following the intervention. As with findings from Pace et al. (2009), no group differences were found for salivary CRP. However, when researchers examined a subset of adolescents with complete data on weekly practice time, an inverse relationship was found between number of practice sessions and morning salivary CRP levels across the 6-week study, such that more practice was associated with lower CRP levels. Consistent with previous findings (Pace et al., 2009), the authors concluded that participation in a CBCT class might not be enough to reduce markers of inflammation; instead, active at-home practice may be particularly important.

Like Pace et al. (2009, 2013), Poehlmann-Tynan et al. (2020) also employed CBCT to manipulate self-compassion, although the authors examined a different population potentially prone to higher stress levels – parents of young children. In a randomized preliminary efficacy study, 39 parents of children ages 9 months to 5 years 4 months were randomized to either CBCT ($n = 25$) or a waitlist control ($n = 14$). Parents in the CBCT condition completed eight weekly group sessions in addition to one full day retreat, totaling 20 intervention hours. Hair cortisol concentration (HCC) was analyzed for both parents and children at pre- and post-intervention assessments along with psychosocial measures of stress and parenting stress. Among parents in both CBCT and the waitlist control group, average HCC increased over time, with no significant difference between groups. Whereas no main effect of the intervention was found for parent HCC, there was a significant group difference in children's HCC. Specifically, among children of parents in the CBCT condition, average HCC decreased from pre- to post-intervention. Furthermore, the average HCC for children of parents on the waitlist control increased across the intervention, which the authors found surprising given that children's HCC tends to decrease with age after age 1 year.

Paralleling CBCT, another study tested the effects of group meditation training focused first on mindfulness training followed by self-compassion training. In this small, uncontrolled study, 31 healthy adults (90% female) participated in an 8-week group-based program (Bellosta-Batalla et al., 2018) derived from Mindfulness-Based Stress Reduction (MBSR; Kabat-Zinn, 1990) and Compassion-Focused Therapy (CFT; Gilbert, 2010). Participants completed a 2-h group session each week for 8 weeks and were assessed at pre- and post-intervention on self-reported variables, sIgA, and salivary cortisol levels. Neither sIgA nor salivary cortisol levels changed significantly over the 8 weeks of the intervention. However, cortisol decreased significantly *within* the first and last intervention sessions, whereas sIgA levels increased within the last intervention session. Though at-home practice was not assessed, this small study suggests that compassion practice can immediately benefit stress- and immune-related biomarkers, similar to previous research (Pace et al., 2009, 2013). Given that this study was small and did not include a control group, results should be interpreted with caution.

Another more recently developed intervention, Emotion-Focused Training for Self-Compassion and Self-Protection (EFT-SCP; Halamová et al., 2019), aims not only to bolster self-compassion but also to explicitly reduce self-criticism by increasing protective anger, or an ability to stand up and assert oneself in response to self-criticism. In a quasi-experimental study evaluating EFT-SCP, 73 healthy university students were recruited and assigned to the following conditions by clustering based on year in school: (1) EFT-SCP ($n = 19$), (2) an active control consisting of expressive writing ($n = 20$), (3) or a no-treatment control condition ($n = 34$) (Halamová et al., 2019). During a baseline laboratory session, participants completed psychosocial measures followed by a 10-min audio-taped guided imagery task that began with 1-min of relaxation and then three, 3-min imagery tasks that intentionally cultivated sequential feelings of self-criticism, self-protection, and self-compassion. In the first imagery task, participants remembered a situation in which they felt self-criticism and were asked to describe their inner self-critic. Second, participants imagined the part of them that would want to defend or protect them against their inner critic. Third, participants imagined the part of them that is self-compassionate and loving at times of difficulty and thought about how they might respond to their self-critic. HF-HRV (via root mean square of successive R-R intervals) was assessed during this imagery exercise. Following this session, students in the EFT-SCP condition completed a 12-week group course, totaling 1.5 intervention hours per week and consisting of experiential exercises designed to elicit self-compassion (e.g., self-compassionate body scan and breathing, imagining a safe place) and reduce self-criticism (e.g., imagining one's self-critical part to build awareness). In addition, EFT-SCP participants completed daily expressive writing tasks at home associated with that week's EFT-SCP content to cultivate self-compassion (e.g., writing a letter to yourself as a child expressing compassion, practicing self-compassion in front of the mirror) or lessen self-criticism (e.g., changing your self-critical dialogue). In contrast, students participating in the active control completed a weekly at-home expressive writing task. After 12 weeks, all participants returned to the laboratory to complete the audio-taped imagery exercise for a second time, during which HF-HRV was again assessed. Findings showed that compared to both control groups, HF-HRV for students in the EFT-SCP intervention condition increased significantly during the imagery exercise from pre- to post-test, specifically during the guided self-critical imagery task where participants described their inner self-critic and the guided self-compassion imagery task, in which participants took a self-compassionate perspective. This suggests that the EFT-SCP self-compassion training led to greater self-regulation in the context of both induced self-compassion and induced self-criticism.

Ko et al. (2018) also examined a group-based university course of self-compassion, randomly assigning 41 university students (ages 18–22 years) to either a 15-week academic Seminar on Compassion (n = 21) or a waitlist control group (n = 20). Content in this Seminar on Compassion included a range of self-compassion, compassion, and mindfulness meditation and related practices, reading and discussing biographical information on world leaders in compassion, and compassion teachings from diverse world religions. In this study, students completed a baseline laboratory session at the end of their fall semester and prior to beginning the compassion course, consisting of pre-intervention self-report measures as well as saliva collection to assess sAA. Students in the intervention group were then asked to attend the compassion course twice a week for 80 min per class throughout the semester, totaling 40 intervention hours. During the last week of the spring semester, post-intervention measures and saliva collection were again completed. Findings showed a significant group difference in sAA, such that sAA decreased among students taking the compassion course, whereas sAA increased among students in the waitlist control group.

Whereas the aforementioned interventions utilized group training, a few studies relied largely or solely on individual at-home practice of self-compassion exercises. One of these studies examined compassionate mind training (CMT), a component of Compassion-Focused Therapy (CFT; Gilbert, 2009, 2010), that emphasizes soothing breathing, attention and awareness, and guided imagery to develop compassionate images of others and self (Matos et al., 2017). Ninety-three healthy adults and university students (ages 18–43 years) were randomly assigned to participate in either a 2-week CMT training program (n = 56) or a waitlist control (n = 37). All participants completed a laboratory session at pre- and post-intervention. During these laboratory sessions, participants completed self-report measures, and resting HRV in the form of RMSSD was assessed over a 5-min relaxation period. In between the two laboratory sessions, participants in the CMT training program completed an initial 2-h group session where they learned CMT exercises and were provided with a manual outlining CMT theory as well as audio recordings of CMT exercises to facilitate at-home practice. Participants were invited to practice the exercises at home over

2 weeks and could independently decide whether to use the audio recordings to guide this practice. Following 2 weeks, all participants returned to the laboratory to complete the post-intervention assessment, again consisting of self-report measures and an assessment of resting HRV. As hypothesized, participants in the CMT condition demonstrated a significant increase in HRV from pre- to post-intervention, whereas there was no significant change in HRV over time for participants in the control condition. This study suggests that self-compassion can be practiced largely independently over a relatively short period of time, and still provide benefits for physiological self-regulation (in the form of increases in resting HRV).

Wren et al. (2019) also relied on at-home meditation practice, reporting on the physiological response to a loving-kindness meditation intervention among breast cancer patients. In this study, 56 participants underwent four assessments of diastolic and systolic BP and HR: (1) pre-biopsy and prior to randomization, (2) post-biopsy, (3) after learning of biopsy results if abnormal, and (4) 1-week post-surgery. All assessments of BP and HR were completed during laboratory sessions, except for the final two assessments, which were taken from patients' electronic medical records. Following the pre-biopsy assessment, patients were randomly assigned to receive one of the following conditions during a biopsy: (1) a loving-kindness meditation administered by audio recording ($n = 23$), (2) music from a genre that the patients selected ($n = 16$), or (3) usual care consisting of support from the biopsy team ($n = 17$). Following biopsy, patients in the loving-kindness and music conditions received condition-specific instructions for condition-specific at-home practice. Specifically, loving-kindness participants received an audio recording of loving-kindness meditation exercises, a rationale for daily practice, and instructions for at-home 20-min daily practice, whereas the music control group received an audio recording with their selected music genre, a rationale for the soothing effect of music, and instructions to listen every day at home for 20 min. Both the loving-kindness intervention and music groups participated in a scripted booster phone call to assess and solve practice barriers. Results indicated a significant group difference in HR for those in the loving-kindness meditation group compared to the two control groups, where HR remained stable over time for the intervention group but increased for both the music and usual care groups. Furthermore, the authors found that *within* the loving-kindness group, more time spent practicing loving-kindness meditation predicted a lower HR over time, reflecting findings from previous studies regarding the physiological benefits of more daily self-compassion practice (Pace et al., 2009, 2013). However, the authors did not find evidence for group differences in either diastolic or systolic BP.

In summary, more extended self-compassion interventions appear to improve some measures of physiological response to stress, particularly parasympathetic (e.g., sAA and HRV) and sympathetic (e.g., HR, though HR also reflects parasympathetic influence) system responses (Halamová et al., 2019; Ko et al., 2018; Matos et al., 2017; Wren et al., 2019). Whereas no group differences emerged from pre- to post-intervention on biomarkers assessing immune functioning (e.g., IL-6, salivary CRP, sIgA) or HPA axis stress response (e.g., cortisol; Bellosta-Batalla et al., 2018; Pace et al., 2009, 2013), two studies demonstrated the benefits on immune system functioning (e.g., IL-6, CRP; Pace et al., 2009, 2013) and sympathetic responding (e.g., HR; Wren et al., 2019) for individuals within self-compassion interventions who committed more to daily self-compassion practice. Additionally, one study demonstrated benefits of self-compassion training for children of parents undergoing self-compassion training, finding improved cortisol response among children with parents participating in a group-based self-compassion intervention (Poehlmann-Tynan et al., 2020). Thus, pending replication, self-compassion intervention studies point toward greater physiological benefits for those who practice more in their daily lives, an important but unsurprising finding, and extend to those who are being parented by the practitioner, suggesting the broader benefits of parenting training.

Ultimately, extant findings suggest that training in self-compassion can improve biomarkers associated with stress and immune response systems, likely reflecting greater physiological resilience. Caveats include the fact that research in this area is still emerging, and that most studies did not use acute stress or criticism paradigms to evaluate whether benefits extend to the contexts in which they are most needed. Future self-compassion intervention studies should extend the work of Pace et al. (2009) and Halamová et al. (2019) by examining more systematically whether more adaptive physiological responses following self-compassion training extend to a range of stressful contexts.

Conclusions

This chapter characterizes the growing literature on observational, laboratory, and intervention studies that evaluate the benefits of trait, briefly induced state, or more extensively trained self-compassion on physiological outcomes related to stress and immune functioning. Evidence derived from diverse methodologies show that self-compassion in multiple forms (trait, brief state, and enduringly trained self-compassion) can benefit autonomic and immune biomarkers associated with stress resilience and self-regulation. Most extant studies, however, have been conducted in healthy samples – an important focus for improving general population well-being. However, research has identified populations who are particularly low in self-compassion, including women (Neff, 2003a) and those with various symptoms of psychopathology (Macbeth & Gumley, 2012). Individuals undergoing acute or chronic stressors also represent appropriate targets for self-compassion training. In addition, adults with anxiety disorders, posttraumatic stress disorder, major depression, and other forms of psychopathology, as well as adults and children undergoing acute and chronic stressors, often evidence poorer autonomic regulation (Beauchaine & Thayer, 2015) and immune functioning (e.g., Danese & Baldwin, 2017), thus recommending them to interventions that help to improve such functioning. Thus, future research can build on the small body of physiologically focused self-compassion studies that target these more vulnerable populations (e.g., Arch et al., 2014; Wren et al., 2019) with the intention of including those who stand to benefit the most from training in self-compassion.

References

Arch, J. J., Brown, K. W., Dean, D. J., Landy, L. N., Brown, K. D., & Laudenslager, M. L. (2014). Self-compassion training modulates alpha-amylase, heart rate variability, and subjective responses to social evaluative threat in women. *Psychoneuroendocrinology, 42*, 49–58. https://doi.org/10.1016/j.psyneuen.2013.12.018

Arch, J. J., Landy, L. N., & Brown, K. W. (2016). Predictors and moderators of biopsychological social stress responses following brief self-compassion meditation training. *Psychoneuroendocrinology, 69*, 35–40. https://doi.org/10.1016/j.psyneuen.2016.03.009

Ascone, L., Sundag, J., Schlier, B., & Lincoln, T. M. (2017). Feasibility and effects of a brief compassion-focused imagery intervention in psychotic patients with paranoid ideation: A randomized experimental pilot study. *Clinical Psychology & Psychotherapy, 24*, 348–358.

Beauchaine, T. P., & Thayer, J. F. (2015). Heart rate variability as a transdiagnostic biomarker of psychopathology. *International Journal of Psychophysiology, 98*(2), 338–350. https://doi.org/10.1016/j.ijpsycho.2015.08.004

Bellosta-Batalla, M., Ruiz-Robledillo, N., Sariñana-González, P., Capella-Solano, T., Vitoria-Estruch, S., Hidalgo-Moreno, G., Pérez-Blasco, J., Romero-Martínez, A., & Moya-Albiol, L. (2018). Increased salivary IgA response as an indicator of immunocompetence after a mindfulness and self-compassion-based intervention. *Mindfulness, 9*, 905–913. https://doi.org/10.1007/s12671-017-0830-y

Berntson, G. G., Quiqley, D. S., & Lozano, D. (2007). Cardiovascular psychophysiology. In J. T. Cacioppo, L. G. Tassinary, & G. G. Berntson (Eds.), *Handbook of psychophysiology* (3rd ed., pp. 182–210). Cambridge University Press.

Bluth, K., Roberson, P. N. E., Gaylord, S. A., Faurot, K. R., Grewen, K. M., Arzon, S., & Girdler, S. S. (2016). Does self-compassion protect adolescents from stress? *Journal of Child and Family Studies, 25*(4), 1098–1109. https://doi.org/10.1007/s10826-015-0307-3

Bosch, J. A., Ring, C., de Geus, E. J. C., Veerman, E. C. I., & Nieuw Amerongen, A. V. (2002). Stress and secretory immunity. *International Review of Neurobiology, 52*, 213–253. https://doi.org/10.1016/S0074-7742(02)52011-0

Breines, J. G., Thoma, M. V., Gianferante, D., Hanlin, L., Chen, X., & Rohleder, N. (2014). Self-compassion as a predictor of interleukin-6 response to acute psychosocial stress. *Brain, Behavior, and Immunity, 37*, 109–114. https://doi.org/10.1016/j.bbi.2013.11.006

Breines, J. G., McInnis, C. M., Kuras, Y. I., Thoma, M. V., Gianferante, D., Hanlin, L., Chen, X., & Rohleder, N. (2015). Self-compassionate young adults show lower salivary alpha-amylase responses to repeated psychosocial stress. *Self and Identity, 14*(4), 390–402. https://doi.org/10.1080/15298868.2015.1005659

Ceccarelli, L. A., Giuliano, R. J., Glazebrook, C. M., & Strachan, S. M. (2019). Self-compassion and psychophysiological recovery from recalled sport failure. *Frontiers in Psychology, 10*, Article 1564. https://doi.org/10.3389/fpsyg.2019.01564

Costa, J., Marôco, J., Pinto-Gouveia, J., Ferreira, C., & Castilho, P. (2015). Validation of the psychometric properties of the Self-Compassion Scale. *Clinical Psychology & Psychotherapy, 23*, 460–468. https://doi.org/10.1002/cpp.1974

Danese, A., & Baldwin, J. R. (2017). Hidden wounds? Inflammatory links between childhood trauma and psychopathology. *Annual Review of Psychology, 68*, 517–544. https://doi.org/10.1146/annurev-psych-010416-044208

Dawson, M., Schell, A. M., & Filion, D. L. (2007). The electrodermal system. In J. T. Cacioppo, L. G. Tassinary, & G. Berntson (Eds.), *Handbook of psychophysiology*. Cambridge University Press.

Del Giudice, M., & Gangestad, S. W. (2018). Rethinking IL-6 and CRP: Why they are more than inflammatory biomarkers, and why it matters. *Brain, Behavior, and Immunity, 70*, 61–75. https://doi.org/10.1016/j.bbi.2018.02.013

Dickerson, S. S., & Kemeny, M. E. (2004). Acute stressors and cortisol responses: A theoretical integration and synthesis of laboratory research. *Psychological Bulletin, 130*(3), 355–391. https://doi.org/10.1037/0033-2909.130.3.355

Gilbert, P. (2000). The relationship of shame, social anxiety and depression: The role of the evaluation of social rank. *Clinical Psychology & Psychotherapy, 7*(3), 174–189. https://doi.org/10.1002/1099--0897(200007)7:3<174::AID-CPP236>3.0.CO;2-U

Gilbert, P. (2009). *The compassionate mind: A new approach to life's challenges*. Constable and Robinson.

Gilbert, P. (2010). *Compassion-focused therapy: Distinctive features*. Routledge.

Gilbert, P., McEwan, K., Matos, M., & Rivis, A. (2011). Fears of compassion: Development of three self-report measures. *Psychology and Psychotherapy, 84*(3), 239–255. https://doi.org/10.1348/147608310X526511

Halamová, J., Koróniová, J., Kanovský, M., Kénesy Túniyová, M., & Kupeli, N. (2019). Psychological and physiological effects of emotion focused training for self-compassion and self-protection. *Research in Psychotherapy: Psychopathology, Process and Outcome, 22*(2), 265–280. https://doi.org/10.4081/ripppo.2019.358

Herriot, H., Wrosch, C., & Gouin, J.-P. (2018). Self-compassion, chronic age-related stressors, and diurnal cortisol secretion in older adulthood. *Journal of Behavioral Medicine, 41*(6), 850–862. https://doi.org/10.1007/s10865-018-9943-6

Kabat-Zinn, J. (1990). *Full catastrophe living: Using the wisdom of your body and mind to face stress, pain, and illness*. Delacorte Press.

Kirschbaum, C., Pirke, K. M., & Hellhammer, D. H. (2008). The 'Trier Social Stress Test' – A tool for investigating psychobiological stress responses in a laboratory setting. *Neuropsychobiology, 28*(1-2), 76–81. https://doi.org/10.1159/000119004

Kirschner, H., Kuyken, W., Wright, K., Roberts, H., Brejcha, C., & Karl, A. (2019). Soothing your heart and feeling connected: A new experimental paradigm to study the benefits of self-compassion. *Clinical Psychological Science, 7*(3), 545–565. https://doi.org/10.1177/2167702618812438

Ko, C. M., Grace, F., Chavez, G. N., Grimley, S. J., Dalrymple, E. R., & Olson, L. E. (2018). Effect of seminar on compassion on student self-compassion, mindfulness, and well-being: A randomized controlled trial. *Journal of American College Health, 66*(7), 537–545. https://doi.org/10.1080/07448481.2018.1431913

López, A., Sanderman, R., Smink, A., Zhang, Y., van Sonderen, E., Ranchor, A., & Schroevers, M. J. (2015). A reconsideration of the Self-Compassion Scale's total score: Self-compassion versus self-criticism. *PLoS One, 10*(7), e0132940. https://doi.org/10.1371/journal.pone.0132940

Luo, X., Qiao, L., & Che, X. (2018). Self-compassion modulates heart rate variability and negative affect to experimentally induced stress. *Mindfulness, 9*, 1522–1528. https://doi.org/10.1007/s12671-018-0900-9

Luo, X., Liu, J., & Che, X. (2019). Investigating the influence and a potential mechanism of self-compassion on experimental pain: Evidence from a compassionate self-talk protocol and heart rate variability. *Journal of Pain, 21*(7-8), 790–797. https://doi.org/10.1016/j.jpain.2019.11.006

Macbeth, A., & Gumley, A. (2012). Exploring compassion: A meta-analysis of the association between self-compassion and psychopathology. *Clinical Psychology Review, 32*(6), 545–552. https://doi.org/10.1016/j.cpr.2012.06.003

Marsland, A. L., Walsh, C., Lockwood, K., & John-Henderson, N. A. (2017). The effects of acute psychological stress on circulating and stimulated inflammatory markers: A systematic review and meta-analysis. *Brain, Behavior, and Immunity, 64*, 208–219. https://doi.org/10.1016/j.bbi.2017.01.011

Matos, M., Duarte, C., Duarte, J., Pinto-Gouveia, J., Petrocchi, N., Basran, J., & Gilbert, P. (2017). Psychological and physiological effects of Compassionate Mind Training: A pilot randomised controlled study. *Mindfulness, 8*, 1699–1712. https://doi.org/10.1007/s12671-017-0745-7

Michopoulos, V., Powers, A., Gillespie, C. F., Ressler, K. J., & Jovanovic, T. (2017). Inflammation in fear-

and anxiety-based disorders: PTSD, GAD, and beyond. *Neuropsychopharmacology, 42*(1), 254–270. https://doi.org/10.1038/npp.2016.146

Miller, D. B., & O'Callaghan, J. P. (2002). Neuroendocrine aspects of the response to stress. *Metabolism, 51*(6): 5–10. http://dx.doi.org/10.1053/meta.2002.33184

Muris, P., & Petrocchi, N. (2017). Protection or vulnerability? A meta-analysis of the relations between the positive and negative components of self-compassion and psychopathology. *Clinical Psychology & Psychotherapy, 24*(2), 373–383. https://doi.org/10.1002/cpp.2005

Neff, K. D. (2003a). The development and validation of a scale to measure self-compassion. *Self and Identity, 2*(3), 223–250. https://doi.org/10.1080/15298860309027

Neff, K. D. (2003b). Self-Compassion: An alternative conceptualization of a healthy attitude toward oneself. *Self and Identity, 2*(2), 85–101. https://doi.org/10.1080/15298860309032

Neff, K. D., Long, P., Knox, M. C., Davidson, O., Kuchar, A., Costigan, A., Williamson, Z., Rohleder, N., Tóth-Király, I., & Breines, J. G. (2018). The forest and the trees: Examining the association of self-compassion and its positive and negative components with psychological functioning. *Self and Identity, 17*(6), 627–645. https://doi.org/10.1080/15298868.2018.1436587

Pace, T. W. W., Negi, L. T., Adame, D. D., Cole, S. P., Sivilli, T. I., Brown, T. D., Issa, M. J., & Raison, C. L. (2009). Effect of compassion meditation on neuroendocrine, innate immune and behavioral responses to psychosocial stress. *Psychoneuroendocrinology, 34*(1), 87–98. https://doi.org/10.1016/j.psyneuen.2008.08.011

Pace, T. W. W., Negi, L. T., Dodson-Lavelle, B., Ozawa-de Silva, B., Reddy, S. D., Cole, S. P., Danese, A., Craighead, L. W., & Raison, C. L. (2013). Engagement with cognitively-based compassion training is associated with reduced salivary C-reactive protein from before to after training in foster care program adolescents. *Psychoneuroendocrinology, 38*(2), 294–299. https://doi.org/10.1016/j.psyneuen.2012.05.019

Petrocchi, N., Ottaviani, C., & Couyoumdjian, A. (2016). Compassion at the mirror: Exposure to a mirror increases the efficacy of a self-compassion manipulation in enhancing soothing positive affect and heart rate variability. *The Journal of Positive Psychology, 12*(6), 525–536. https://doi.org/10.1080/17439760.2016.1209544

Phillips, A. C., Carroll, D., Evans, P., Bosch, J. A., Clow, A., Hucklebridge, F., & Der, G. (2006). Stressful life events are associated with low secretion rates of immunoglobulin A in saliva in the middle aged and elderly. *Brain, Behavior, and Immunity, 20*(2), 191–197. https://doi.org/10.1016/j.bbi.2005.06.006

Pilette, C., Ouadrhiri, Y., Godding, V., Vaerman, J. P., & Sibille, Y. (2001). Lung mucosal immunity: Immunoglobulin-A revisited. *European Respiratory Journal, 18*(3), 571–588. https://doi.org/10.1183/09031936.01.00228801

Poehlmann-Tynan, J., Engbretson, A., Vigna, A. B., Weymouth, L. A., Burnson, C., Zahn-Waxler, C., Kapoor, A., Gerstein, E. D., Fanning, K. A., & Raison, C. L. (2020). Cognitively-based compassion training for parents reduces cortisol in infants and young children. *Infant Mental Health Journal, 41*(1), 126–144. https://doi.org/10.1002/imhj.21831

Porges, S. W. (2007). The polyvagal perspective. *Biological Psychology, 74*(2), 116–143. https://doi.org/10.1016/j.biopsycho.2006.06.009

Raes, F., Pommier, E., Neff, K. D., & Van Gucht, D. (2011). Construction and factorial validation of a short form of the Self-Compassion Scale. *Clinical Psychology & Psychotherapy, 18*(3), 250–255. https://doi.org/10.1002/cpp.702

Rohleder, N., Natar, U. M., Wolf, J. M., Ehlert, U., & Kirschbaum, C. (2004). Psychosocial stress-induced activation of salivary alpha-amylase: An indicator of sympathetic activity? In R. Yehuda & B. McEwen (Eds.), *Biobehavioral stress response: Protective and damaging effects* (pp. 258–263). New York Academy of Sciences.

Slavich, G. M., & Irwin, M. R. (2014). From stress to inflammation and major depressive disorder: A social signal transduction theory of depression. *Psychological Bulletin, 140*(3), 774–815. https://doi.org/10.1037/a0035302

Svendsen, J. L., Osnes, B., Binder, P., Dundas, I., Visted, E., Nordby, H., Schanche, E., & Sørensen, L. (2016). Trait self-compassion reflects emotional flexibility through an association with high vagally mediated heart rate variability. *Mindfulness, 7*(5), 1103–1113. https://doi.org/10.1007/s12671-016-0549-1

Thayer, J. F., & Lane, R. D. (2000). A model of neurovisceral integration in emotion regulation and dysregulation. *Journal of Affective Disorders, 61*(3), 201–216. https://doi.org/10.1016/s0165-0327(00)00338-4

Thayer, J. F., Åhs, F., Fredrikson, M., Sollers, J. J., & Wager, T. D. (2012). A meta-analysis of heart rate variability and neuroimaging studies: Implications for heart rate variability as a marker of stress and health. *Neuroscience and Biobehavioral Reviews, 36*(2), 747–756. https://doi.org/10.1016/j.neubiorev.2011.11.009

Wren, A. A., Shelby, R. A., Soo, M. S., Huysmans, Z., Jarosz, J. A., & Keefe, F. J. (2019). Preliminary efficacy of a lovingkindness meditation intervention for patients undergoing biopsy and breast cancer surgery: A randomized controlled pilot study. *Supportive Care in Cancer, 27*(9), 3583–3592. https://doi.org/10.1007/s00520-019-4657-z

Check for updates

A Triadic Pathway Model of Self-Compassion and Health

18

Fuschia M. Sirois

Introduction

In recent years, self-compassion has emerged as an important quality for promoting mental health and well-being, with a growing evidence base providing robust support for its links with higher levels of well-being, less psychopathology, and greater life satisfaction (MacBeth & Gumley, 2012; Zessin et al., 2015). Parallel to this research is a growing interest in and examination of the potential benefits of self-compassion for promoting physical health. Research has demonstrated that self-compassion can have positive consequences for a range of inter-related physical health trajectories and outcomes including subjective health, stress, sleep, and health behaviors (Brown et al., 2021; Phillips & Hine, 2019; Sirois, 2020). Although the pathways explaining why self-compassion may promote physical health are continuing to be explored, current theories on the role of self-regulation in health, and specifically linking individual differences to health, offer insights for understanding the possible ways in which self-compassion can contribute to better overall physical health.

Although there are different perspectives on what comprises self-compassion (e.g., Gilbert, 2019), one conceptualization that is most commonly used in research investigating the links with health is that of Neff (2003). Self-compassion is defined as responding to difficulties, suffering and perceived personal shortcomings in a kind, connected, and mindful manner. As a positive self-evaluation, self-compassion is further posited to be comprised of three bipolar components that operate synergistically to improve well-being and motivation to persist despite difficulties: self-kindness versus self-judgment, common humanity versus isolation, and mindfulness versus over-identification (Neff, 2011). Self-compassion can also be viewed as an enduring trait-like quality that nonetheless can be developed through practice and intervention (e.g., Wilson et al., 2019). It is this latter property of self-compassion that makes it particularly valuable as a target for maximizing health-promoting trajectories and minimizing health-risk trajectories.

This chapter will discuss current evidence and theory supporting a role for self-compassion in promoting physical health. Drawing on insights from relevant theories on self-regulation and behavior change, this chapter will outline and review the key pathways and processes that may explain how and why responding to one's failures and shortcomings with kindness, connectedness, and mindfulness may promote physical health. In

F. M. Sirois (✉)
Department of Psychology, Durham University, Durham, UK
e-mail: fuschia.sirois@durham.ac.uk

A. Finlay-Jones et al. (eds.), *Handbook of Self-Compassion*, Mindfulness in Behavioral Health,
https://doi.org/10.1007/978-3-031-22348-8_18

doing so, it will present a new model of self-compassion and health that incorporates these insights to provide a provisional framework to guide further research in this rapidly growing research area.

Current Evidence on the Role of Self-Compassion for Physical Health

The idea that self-compassion may be beneficial for physical health has gained considerable attention from researchers in recent years. This interest has generated a growing body of research that largely supports this notion and provides evidence of the links between self-compassion and array of outcomes relevant for physical health states. In general, this research has examined self-compassion with respect to two key indicators of physical health states: physical symptoms and subjective health status. Whereas physical symptoms refer to the perception of physical discomfort in specific areas of the body (e.g., difficulty breathing) which are perceived as a sign of an underlying illness, subjective health status is an overall perception of one's general state of health (e.g., feeling generally unwell; Jylhä, 2009). Although each of these markers of physical health are subjective in nature, a number of studies have also demonstrated that they have reliable links to more objective measures of physical health (see Jylhä, 2009 for a review), and therefore can be considered reasonable proxies of people's overall state of physical health.

Self-Compassion and Physical Symptoms

Physical symptoms can vary widely as a function of the underlying illness or physical health issue that they reflect and are also influenced by individual differences in the way that people experience, interpret, and respond to physiological stimuli. Research on self-compassion and physical symptoms indicates that people's experience of physical symptoms may vary as a function of the degree to which they are able to treat themselves compassionately. Pain is one of the more common physical symptoms that people report when they are ill (Taylor et al., 2020). Several studies have found that self-compassion is associated with less pain (Allen et al., 2011; Carvalho et al., 2018), and less pain intensity (Gregory et al., 2017; Wren et al., 2012), in both general and clinical samples (See Chap. 19 for a more complete review).

There is also evidence that self-compassion may be beneficial for physical health when measured with composite measures of physical symptoms that include fatigue, nausea, headaches, and general discomfort. For example, in a study that includes matched samples of psychiatric patients and adults from the general population, self-compassion was associated with fewer physical health symptoms in both samples (Dewsaran-van der Ven et al., 2018). Similar results have been found with undergraduate student samples (Dunne et al., 2016). Whether particular components of self-compassion might account for the link with fewer health symptoms has yet to be fully explored. However, at least one study has examined this issue. In a cross-sectional study of college students, the self-kindness/low self-judgment and common humanity/low isolation components of the self-compassion scale were associated with fewer self-reported physical symptoms, whereas the mindfulness/low over-identification component was not significantly related to symptoms (Hall et al., 2013).

Self-Compassion and Overall Physical Health

The potential benefits of self-compassion for physical health have also been investigated in relation to measures of overall health. Overall physical health is often assessed using either a multi-item composite score, such as the physical health subscale of the SF-12 (Ware et al., 1996), or the SF-36 from the Medical Outcomes Survey (Ware & Sherbourne, 1992), or a single-item rating of global or self-rated health taken from these same measures. The multi-item, composite measures are well-known to predict consequential health outcomes such as hospitalization and mortality (e.g., Thombs et al., 2008). However, the self-rated health single item is also a robust measure of physical health. Despite the apparent simplicity of this approach, decades of research have demonstrated that assessing health in this manner provides a highly reliable index of health (Jylhä, 2009). For example, single-item self-rated health is known to predict a number of objective measures of physical health including cortisol responses to stress, health behaviors, morbidity, mortality, and even future health (Benyamini, 2011; Kristenson et al., 2005; Mora et al., 2013; Tamayo-Fonseca et al., 2013).

Several studies have examined how self-compassion relates to composite measures of physical health across a range of populations. For example, cross-sectional studies have demonstrated that self-compassion is associated with better overall physical health in older adults (Allen et al., 2011; Smith, 2015), middle-aged women (Brown et al., 2016), and among community adults (Homan & Sirois, 2017). Importantly, there is also evidence that self-compassion predicts better overall physical health over time. In one 7-year longitudinal study of adults ranging in age from 27 to 101 years, both baseline self-compassion and changes in self-compassion over the course of the study were associated with improved physical health at the 7-year follow-up, but only for those under 60 (Lee et al., 2021).

Research on the health benefits of self-compassion for self-rated health has yielded similar findings. For example, self-compassion was associated with better self-rated health in health-care social workers (Lianekhammy et al., 2018), and men living with HIV (Skinta et al., 2019). The former study also examined the subscales of the self-compassion scale in relation to self-rated health. Interestingly, only the over-identification and self-judgment subscales were associated with self-rated health, with low scores on these two components linking to better overall self-rated health. In one of the most comprehensive investigations of self-compassion and self-rated health, associations were meta-analyzed across 26 samples of students, community adults, and people with chronic illness, including over 6000 people (Sirois, 2020). Self-compassion was significantly associated with better self-rated health across the samples, and this average association was not influenced by gender, age, or the sample population. Because self-reports of health can be influenced by positive and negative states (Howren & Suls, 2011), further analyses were conducted to control for positive and negative affect in the associations of self-compassion with self-rated health. These analyses yielded the same significant associations, suggesting that self-compassion is linked to self-rated health in meaningful ways that go beyond possible reporting biases.

When taken at face value, this evidence might suggest that self-compassion is associated with better overall physical health, insomuch that self-compassionate people tend to not report many general physical symptoms. However, because the evidence presented above relies on cross-sectional data collected at a single time point, these "snapshots" of the links between self-compassion and health tell us little about whether self-compassion contributes to better physical health in the form of fewer symptoms. One could potentially argue that not experiencing physical symptoms may make it easier to have a mind-set that is conducive to being self-compassionate. For example, someone struggling with pain, fatigue, and other uncomfortable physical symptoms that are known to contribute to negative mood states may be more likely to engage in thoughts focused on self-blame and feeling isolated and have difficulty disengaging from the negative states brought on by their physical symptoms. Such negative states can in turn potentially amplify the intensity of these symptoms, increase vigilance to them, and affect the extent to which they are reported (Howren & Suls, 2011). While hypothetically it is in these very instances of suffering that self-compassion is proposed to be most beneficial, evidence based on cross-sectional data is limited in terms of addressing the potential directionality of effects in the link between self-compassion and physical symptoms.

To address this important issue, at least two studies have investigated the effects of self-compassion interventions on physical health symptoms over time. Bellosta-Batalla et al. (2018) tested the effects of a one-week mindfulness and self-compassion-based intervention on several health outcomes, including a composite measure of physical health symptoms, in a small sample of adults. Compared to the baseline symptom reporting, the levels of postintervention symptoms were significantly reduced. It is worth noting though that the study did not include a control group as a comparison for the changes in symptoms, and that the effectiveness of the intervention for increasing self-compassion was not directly tested. This latter point raises the issue of whether it was in fact self-compassion that was contributing to the improvement in physical health, or mindfulness, a synergy between the two, or some other factor. Although the compassion-focused therapy integrated into the intervention is effective for cultivating self-compassion (Wilson et al., 2019), its use alongside mindfulness-based stress reduction muddies the conceptual waters in terms of understanding the unique contribution of self-compassion to the improvements in physical health observed. Reductions in negative affect and anxiety as a result of the intervention were tested as possible pathways and did explain in part the improvements in health; however, it is unclear what contribution self-compassion alone had on these effects.

Some of the strongest evidence of the effects of self-compassion on physical health symptoms were demonstrated in a longitudinal intervention study conducted with university students (Wong & Mak, 2016). This study had a number of methodological strengths, including randomization to a control versus intervention group, a manipulation check of the effectiveness of the intervention, and 1-month and 3-month follow-ups to assess whether any effects were maintained over time. The intervention involved a 3-day self-compassion writing intervention that asks participants to reflect on a difficult situation and reframe it using a self-compassionate attitude. The researchers found that the intervention was effective in reducing physical symptoms compared to the control group at both the 1-month and the 3-month follow-up. Although there was evidence of self-compassion increasing during the intervention, there was no significant change in self-compassion at either of the follow-ups, leaving open the question of the mechanisms through which the improvements in physical health occurred.

Thus, current research provides compelling evidence of the benefits of self-compassion for physical health, whether health is assessed via symptom reporting, or composite or summary measures of overall health status. However, physical health can be assessed using a variety of methods, including measures of psychophysiological reactivity to stress. Indeed, a recent meta-analysis of self-compassion and health took a very inclusive approach to assessing health across a number of domains, that included such physiological markers (Phillips & Hine, 2019). This yielded an overall small, but significant association (average $r = 0.18$). In the current chapter, these psychophysiological markers will be covered in a later section on self-compassion and stress, as arguably, measures such as cortisol levels and immunological responses are considered key indicators of the presence and extent to which stress is perceived and experienced. Moving from the descriptive to the explanatory, the next sections will examine theoretical perspectives to answer the important question of why self-compassion may promote physical health.

Why Is Self-Compassion Associated with Physical Health?

Despite the surge of research interest into the role of self-compassion in health in recent years, investigations into *why* self-compassion leads to important health outcomes are limited (e.g., Dunne et al., 2016; Homan & Sirois, 2017; Sirois et al., 2015a). The first wave of research into a new domain generally seeks to establish linkages with the outcomes of interest, in this case whether self-compassion is associated with physical health. Investigating such associations across a variety of contexts, and using different methodological approaches are also essential to build a strong evidence base and lay the foundation for research into whether self-compassion interventions can be harnessed to improve health, especially among those most vulnerable for poor health outcomes.

The second wave of research into a new domain then focuses on addressing the important question of *which* factors and processes contribute to and explain the association established in the first wave. It is this second wave of research on self-compassion and health that is still evolving and that requires further theoretical development to galvanize its continued growth both within the field of self-compassion research, and the field of health psychology more generally. Understanding the processes and contributing factors that underpin the effects of self-compassion on health is crucial for theoretical development. But these insights are also valuable from an applied perspective. For example, in addition to physical health status, self-compassion interventions may also have immediate effects on intermediate or process-related outcomes, such as stress and health behaviors (e.g., Biber & Ellis, 2019), which contribute to better physical health.

One way to understand how self-compassion can contribute to better physical health is to consider this question from the lens of models that link personality and individual differences to health. Although self-compassion can be cultivated, it is also commonly measured as an enduring tendency or mind-set that shapes people's responses to failures, difficulties, and personal shortcomings across a variety of contexts. Generally, these models suggest that enduring characteristics and tendencies influence health through two key pathways that are well-known to have consequences for the promotion and maintenance of health, namely stress and health behaviors (Friedman, 2000; Smith, 2006; Suls & Rittenhouse, 1990).

The psychophysiological arousal associated with stress is well-known to have both immediate and downstream effects on health. The activation of the hypothalamic-pituitary-adrenal-cortical (HPAC) axis as a result of acute stress can have several adverse health impacts, including suppressing immune functioning and increasing vulnerability for infections in the short-term. When stress becomes chronic, the repeated activation of the HPAC can contribute to dysregulation of cortisol responses to stress, and inflammatory responses in the long-term (Cohen et al., 2012; McEwen, 2007). These can be particularly damaging to health, as both are known precursors to the development of a number of chronic diseases, including cardiovascular disease, diabetes, and even cancer (McEwen, 2007).

Health behaviors are another important pathway for understanding health trajectories. Often referred to as modifiable risk factors for the development of disease, health behaviors play a central role in promoting life-long health and preventing illness (World Health Organization, 2015). The practice of health-promoting behaviors such as healthy eating and engaging in regular physical activity contributes to better health outcomes, whereas health-risk behaviors such as excessive alcohol use, smoking, and substance use create vulnerability for poor health and increase risk for disease. Unlike stress which can have both immediate and long-term effects on health, the benefits or risks of health behaviors tend to accumulate over time after repeated instances, often culminating in their full effects on health at midlife or later.

Triadic Pathways to Health

In addition to these two key pathways, there is a third route to health outcomes that has relevance for understanding the potential health benefits of self-compassion: sleep. Sleep quality is an important and widely recognized factor for understanding health trajectories. As a key factor that contributes to health trajectories, sleep quality includes not only the duration of sleep but also whether sleep is broken, and if there are difficulties falling and staying asleep (St-Onge et al., 2016). Poor sleep quality not only compromises health (Reidy et al., 2016), but can also significantly impact daily functioning in ways that create further risks for poor outcomes (Daley et al., 2009). For example, poor sleep quality and insufficiency are linked to greater risk for health problems (Zuraikat et al., 2020). However, the reasons why sleep increases vulnerability for poor health are complex and involve multiple psychological and physiological pathways that are often reciprocally related to both stress and health behaviors. Sleep disturbances can interfere with the self-regulation capacities necessary to maintain exercise routines and resist unhealthy behaviors, such as smoking and heavy alcohol use (Strine & Chapman, 2005), which in turn can further contribute to poor sleep (Haario et al., 2013).

Current evidence supports the idea that self-compassion contributes to better sleep quality. For example, across two studies, people with major depressive disorder and students who received a brief or daily self-compassion intervention reported increases in sleep quality (Butz & Stahlberg, 2018). The links between self-compassion and better sleep quality were summarized in a recent meta-analysis which found that across 17 studies, self-compassion had a small to moderate average association with self-reported sleep quality (Brown et al., 2021). The researchers also conducted further tests to examine whether it was the positive or negatively worded items in the self-compassion scale that were linked to better sleep quality. Although both were significantly linked to sleep quality in the expected directions, the associations with negatively worded items were stronger, suggesting that a lack of self-coldness may better account for why self-compassion can promote quality sleep.

Stress is known to disrupt the quality of sleep (Kashani et al., 2012). Indicators of poor sleep quality, such as insomnia and sleep deprivation, can lower the stress tolerance threshold, leading to stressors being perceived as more stressful and intense (Minkel et al., 2012; Morin et al., 2003). Importantly, stress is implicated in the mechanisms through which poor sleep contributes to metabolic issues (Hirotsu et al., 2015). If we also consider that stress is linked to poor health behaviors in general (Homan & Sirois, 2017), less frequent exercise (Ng & Jeffery, 2003; Stults-Kolehmainen & Sinha, 2014), and unhealthy eating habits (Ng & Jeffery, 2003; Steptoe et al., 1998), then it becomes clear that this triad of pathways to physical health holds significant promise for gaining a more complete understanding of how self-compassion may contribute to health outcomes.

The Role of Self-Regulation in Linking Self-Compassion to Health

How might self-compassion reduce stress, improve sleep quality, and promote engaging in health behaviors whilst minimizing health-risk behaviors? One important capacity that has relevance for self-compassion and that is known to play a central role in each of these health pathways is self-regulation. Often defined as the processes involved in regulating thoughts, emotions, and behaviors while working toward one's goals (Carver & Scheier, 1982), self-regulation is an important capacity for understanding health trajectories. Self-regulation encompasses several key processes including setting goals, engaging in action to reach these goals, monitoring and evaluating progress toward these goals, regulating emotional responses to the inevitable ups and downs of goal pursuit. Not surprisingly, previous research has found that self-compassion is positively associated with a number of processes involved in self-regulation (Terry et al., 2013).

It is worth mentioning that the goals set do not necessarily need to be health-focused for the processes involved in self-regulation to impact health. Difficulties and challenges experienced with *any* goals can be stressful and therefore potentially impact health outcomes such as sleep and health behaviors. From this perspective, self-compassion may provide benefits for health through several key processes tied to self-regulatory abilities. Figure 18.1 provides an overview of three of these processes - coping strategies, emotion regulation, and efficacy beliefs – and how they link self-compassion to health through the triadic pathways. As will be discussed in the following sections, theory and research suggest that self-compassion has robust links to each of these key self-regulation factors, as well as the triadic pathways that contribute to physical health.

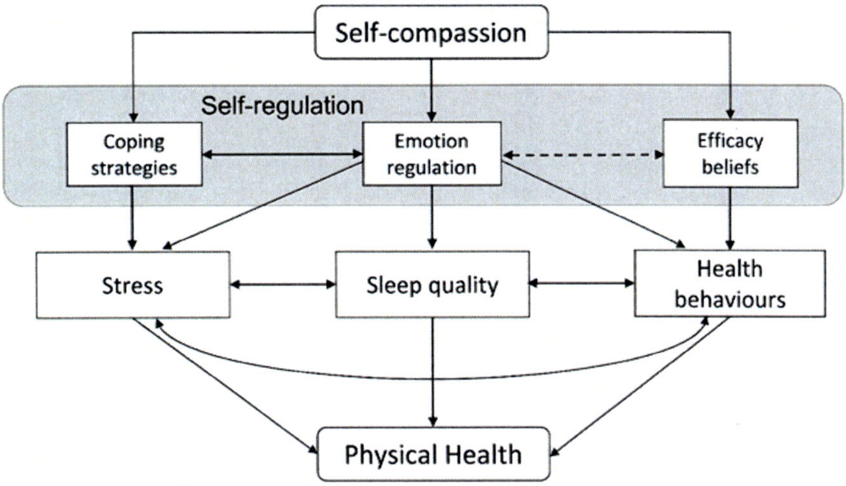

Fig. 18.1 A triadic model of self-compassion and physical health. Dashed lines indicate provisional linkages

Self-Compassion and Stress

A growing evidence base supports the role of self-compassion for reducing stress across multiple context and populations. This research has demonstrated that self-compassion is associated with lower self-reported perceived stress in university students (Sirois, 2014; Stutts et al., 2018; Zhang et al., 2016), community adults (Homan & Sirois, 2017), athletes (Mosewich et al., 2019), and those with chronic health conditions (Pinto-Gouveia et al., 2014; Sirois & Hirsch, 2019; Sirois et al., 2015b; Trindade & Sirois, 2021). Although much of this research has been cross-sectional and examined dispositional self-compassion, there is also some evidence that interventions designed to increase self-compassion can be effective for reducing perceived stress (Ferrari et al., 2019). For example, in a randomized controlled trial of the Mindful Self-Compassion intervention, those in the treatment group reported higher self-compassion and less stress following the intervention, and these improvements were maintained at the 6-month and 1-year follow-up (Neff & Germer, 2013). Similar results in terms of reducing stress have been found for Mindful Self-Compassion Training among practicing psychologists at risk for burnout (Eriksson et al., 2018).

Researchers have also investigated whether and how self-compassion might play a role in reducing stress by examining physiological measures of stress. This research has the advantage of measuring stress using more objective means that are free from the reporting biases that can plague self-report measures. It also provides key insights into how self-compassion is implicated in the physiological processes involved in stress that have relevance for physical health (for a more complete review, see Chap. 17). For example, one study found that adolescents who were high in self-compassion demonstrated lower physiological reactivity in response to the Trier Social Stress test, a standard procedure for inducing stress in a lab setting (Bluth et al., 2016). Although the differences in changes in salivary cortisol, heart rate, blood pressure, and heart rate variability in comparison to a control group were small, the findings supported an overall profile of less reactivity to stress for those with higher levels of self-compassion. Similar results were found with young adults exposed to stress-inducing lab tasks, with those scoring higher on self-compassion having lower levels of stress-induced inflammation (as measured by interleukin-6) compared to a control following the stress induction (Breines et al., 2014).

However, it is unclear whether self-compassion may be more beneficial for attenuating activation of the sympathetic nervous system or the HPAC axis or engaging parasympathetic responses that quiesce physiological reactivity. One study found that brief self-compassion training reduced sympathetic (salivary alpha-amylase), and cardiac parasympathetic responses to a stress-inducing lab task, but had no effect on HPAC axis activity as measured by salivary cortisol (Arch et al., 2014). Contrasting this, other researchers have found that the positive and negative poles of self-compassion were associated with lower reactivity as measured with markers of both sympathetic (salivary alpha amylase) and HPAC axis (interleukin-6 levels) responses following a stressful situation (Neff et al., 2018). Although it is possible that responding to stressors with self-compassion engages all three systems, more work is needed to verify this proposition and examine which aspects of self-compassion might benefit which systems.

Classic psychological models define stress as a transaction between the individual and their environment. When the individual appraises the demands of the environment as being harmful to their well-being, and taxing or overwhelming the internal and external resources that they have available, they experience stress (Lazarus & Folkman, 1984). Appraisal is also viewed as an ongoing process that can lead to different evaluations of the demands of the situation and how they can be met depending on the resources that are available and mobilized to respond to the demands. When viewed from this perspective, self-compassion may help to reduce stress through several key processes tied to the interpretation of the stressor and one's resources, as well as the response to the challenging event. Theory and evidence suggest that coping strategies and emotion regulation are two such processes.

Coping as a Process Linking Self-Compassion to Lower Stress

Consistent with appraisal-based models of stress, coping is viewed as a process involving cognitive and behavioral responses that aim to manage or control the stressor or reduce the associated stress (Lazarus & Folkman, 1984). Effective coping is achieved when the source of stress is eliminated or diminished, or one interacts or views the stressor in a way that is no longer harmful or threatening. Self-compassion may promote both healthy cognitive responses and behavioral responses to stress in several ways. Appraising a stressor as less threatening or harmful, by not over-identifying with stressful negative emotions either initially or after assessing one's resources, can reduce or diminish stress reactivity and allow a faster return to prestress levels of functioning. Additionally, acknowledging one's struggle and difficulties with a stressor as being a common human experience can reduce any feelings of shame that might prevent seeking support, information, or advice for addressing or even removing the stressor. The cognitive and emotional resources that are freed up from not ruminating about the stressor and reactivating its harmful effects (e.g., Smyth et al., 2013), can therefore be mobilized to taking constructive action to problem solve and eliminate or reduce the stressor.

Prior research indicates that overall, self-compassionate people tend to use a healthier repertoire of coping strategies to deal with stressors. This evidence was summarized in a recent meta-analysis of 136 studies which found that self-compassion was associated with greater use of adaptive coping strategies, and less use of maladaptive coping (Ewert et al., 2021). Specifically, self-compassion was associated with greater use of emotional approach coping strategies, such as seeking emotional social support, positive reframing, and acceptance of stressors, as well as greater use of problem-focused coping strategies, including active coping, planning, and seeking instrumental social support. Such strategies are well-known to be effective for reducing stress (Carver & Connor-Smith, 2010; Carver et al., 1989). In contrast, coping strategies such as self-blame, denial, rumination, disengaging from and avoiding the stressor, and substance use to withdraw from the stressor provide immediate relief from stress, but at the cost of prolonging or worsening stress in the long run as the stressor remains. Not surprisingly, self-compassion was moderately and negatively associated with the use of these maladaptive strategies in the studies included in the meta-analysis (Ewert et al., 2021).

In addition to the reasons noted earlier, self-compassion may promote adaptive coping through lower perceptions of threat, and increased perceptions of control over the stressor. For example, in one longitudinal study of Japanese undergraduate students, baseline self-compassion predicted greater use of adaptive coping strategies 1 month later, with reduced perceptions of threat and greater perceptions of controllability accounting for this association (Chishima et al., 2018). The links between self-compassion and reduced threat appraisals and enhanced control appraisals have also been noted in a study of female athletes (Mosewich et al., 2019). Viewing stressors as common human experiences, rather than something that is unique to the individual or a sign of failure can make the stressor seem less threatening. However, it is likely that this relationship is bidirectional. Reappraisal of a stressor to be less harmful and threatening may also enhance feelings of control that are essential for taking constructive action toward dealing with the stressor and using problem-focused and approach-oriented strategies.

Self-Compassion, Emotion Regulation, and Stress

Several researchers have suggested that emotion regulation is a core process within self-compassion (Barnard & Curry, 2011; Finlay-Jones et al., 2015; Inwood & Ferrari, 2018). This is not surprising given the capacity of self-compassion for enhancing positive mood states and reducing negative mood states aligns with classic definitions of emotion regulation. For example, the process model of emotion regulation defines emotion regulation as "the processes by which individuals influence which emotions they have, when they have them, and how they experience and express these emotions" (Gross, 1998, p. 275). Adaptive emotion regulation further involves upregulating positive emotions and downregulating negative emotions such as distress, often through the use of specific strategies such as cognitive reappraisal (reframing situations to change their emotional impact) and suppression (decreasing the experiences and behaviors linked to an emotion) (Gross & John, 2003). Like adaptive coping, emotion regulation is considered successful when the changes experienced are lasting rather than short-term. Emotion regulation can also be viewed from a more holistic perspective, as part of global self-regulatory system that includes physiological regulation of heart rate, for example, that can have a direct impact on physical health (Butler, 2011). From this perspective, success in regulating emotions should also translate into positive physiological changes such as healthy responsiveness of the parasympathetic nervous system to quiesce stress.

Current evidence supports the linkages of self-compassion to healthy emotion regulation as a process for understanding stress responses. Cross-sectional studies have noted strong linkages between self-compassion and cognitive reappraisal but not suppression (Sirois et al., 2019), and between low self-compassion and difficulties in emotion regulation (Finlay-Jones et al., 2015). There is also some evidence that self-compassion is linked to physiological markers of healthy emotion regulation, such as vagally mediated heart rate variability (Svendsen et al., 2016), even after controlling for related relevant constructs such as mindfulness (Svendsen et al., 2020). These findings make sense if we consider how the synergistic actions of the three bimodal components of self-compassion may operate. Responding with self-kindness and common humanity provides opportunities to reappraise one's difficulties from a less critical and more connected perspective that can reduce negative emotions and enhance positive emotional states. Mindfulness rather than over-identifying can at the same time foster stepping back and viewing one's emotional experiences during difficulties from a more balanced perspective.

Although there are few direct tests of the role of emotion regulation for explaining why self-compassion may reduce stress, at least one study provides clear evidence supporting this proposition. In a study of practicing and trainee psychologists, fewer emotion regulation difficulties such as nonacceptance of emotional responses and lack of emotional clarity, mediated the link between self-compassion and stress (Finlay-Jones et al., 2015). Notably, these results were found after controlling for neuroticism, a personality trait well known to predict stress and negative emotional states.

It is also worth noting that emotion regulation has some conceptual overlap with coping and there are likely complex pathways between self-compassion, emotion regulation, and coping. For example, certain forms of emotion-focused coping may help regulate emotions, and regulating negative emotions by being self-compassionate may make it easier to implement more rational, problem-focused coping. However, emotion regulation also includes processes such as enhancing and maintaining positive emotions that are not part of the coping process (Gross, 1998). Accordingly, self-compassion may help to reduce stress and promote more adaptive functioning via emotion regulation, in ways that are distinct from coping (see Chap. 24 for a more detailed discussion).

Self-Compassion and Health Behaviors

On the surface, it may appear that self-compassion would promote engaging in healthier behaviors simply because doing so reflects self-kindness or self-care. However, this simplistic view neglects a full consideration of the psychological processes involved in being self-compassionate, as well as the complexities involved in understanding when, why, and whether people choose to engage in certain health behaviors. Most people are aware that certain behaviors are healthier than others, and that uptake of such behaviors has the potential to improve health and quality of life. But this knowledge alone is not enough to motivate people to stop unhealthy habits such as smoking or excessive alcohol use, or to adopt better eating habits and stay physically active. Indeed, the tripling of worldwide obesity rates since 1975 (World Health Organization, 2021) attests to the idea that simply knowing what is and is not healthy is not enough to motivate behavior change.

It is also important to highlight that self-compassion is not equivalent to self-care. Self-compassion involves a set of compassionate responses to suffering that arises from failure, difficulties, or perceived personal shortcomings that help to reduce the suffering, re-establish balance, and motivate steps toward self-improvement (Neff, 2003; Neff et al., 2007). Self-care, in contrast, involves engaging in behaviors aimed at caring for the self, either more generally, such as having a relaxing day at the spa, or specifically in the context of health circumstances which require that certain self-management behaviors are practiced to ensure that health is maintained. Sticking to a low sugar diet in the context of diabetes or managing stress in the context of depression or anxiety, are common examples of self-care practices. Although engaging in self-care behaviors such as these are associated with higher levels of self-compassion (e.g., Sirois & Hirsch, 2019), their linkage alone tells us little about the reasons *why* self-compassion may promote their practice.

A further consideration that is needed before discussing why self-compassion may promote better physical health through health behaviors is the dimensionality of health behaviors. Theory and evidence have demonstrated that health behaviors form two separate and replicable dimensions: (1) health-promoting and maintaining behaviors (e.g., physical activity, healthy eating, and medical care seeking); and (2) health-risk behaviors (e.g., smoking, substance use, and excessive alcohol use) (Lippke et al., 2012; Vickers et al., 1990). Given the outcomes of these behaviors can lead to better or worse health, respectively, we might expect that self-compassion would be differentially related to health-promoting and health-risk behaviors. Overall, current evidence supports this assumption. For example, a meta-analysis of self-compassion and health behaviors found an overall positive association with a range of health-promoting behaviors, as well as with health-risk behaviors that had been recoded to reflect less engagement in healthy behaviors (Phillips & Hine, 2019). Consistent with these findings, another meta-analysis found that self-compassion (both dispositional and induced) was associated with greater engagement in physical activity (Wong et al., 2021). Other reviews have examined the effectiveness of self-compassion interventions for promoting health behavior changes. One review found that self-compassion interventions were effective for eating-related behaviors and weight loss (Rahimi-Ardabili et al., 2018), whereas another review found that they were just as effective as more traditional approaches for fostering the self-regulation of health behavior (Biber & Ellis, 2019).

Traditionally, models for understanding when and why people do or do not engage in health behaviors focus on social cognitive factors, such as attitudes, social norms, and perceptions of control. These social cognitive models, such as the Theory of Planned Behavior (Ajzen, 1991), highlight the central role of cognitions in behavior, often overlooking or downplaying the influence of emotions and mood states in people's health behavior choices. Increasingly though, researchers are acknowledging the importance of also considering these *hot* or *automatic* processes that govern behavior choices, which can provide additional insights and predictive power to the more *cold* or *reflective* processes involved in decision-making (e.g., Gerrard et al., 2008). Emotion-regulation processes fall under these more automatic processes and have relevance for understanding why self-compassion can help promote engagement with health-promoting behaviors, whilst also minimizing temptations or motivations to engage in health-risky behaviors. The following sections discuss possible reasons why self-compassion is linked to better health behaviors, drawing on both traditional and contemporary theories of self-regulation of health behaviors, and the evidence that supports these views.

Self-Compassion, Self-Efficacy, and Health Behaviors

A core concept within several social cognitive theories of health behavior change is that people will engage and persist with their behavior change efforts to the extent that feel they have some control over the behavior. For example, perceptions of control figure prominently in explaining intentions and actual behaviors in the Theory of Planned Behavior (TPB; Ajzen, 1991), one of the most widely used theories for understanding the self-regulation of health behaviors. Self-efficacy is another control-related concept that is often used as a proxy for perceived behavioral control from the TPB to understand the regulation of health behaviors. Self-efficacy reflects the extent to which an individual believes they have the capacity and ability to execute the actions required to achieve their goals (Bandura, 1977).

The influence of self-compassion on perceptions of and reactions to failures is one reason why self-compassion may foster feelings of self-efficacy about reaching health goals. Self-compassionate people may notice and remember not just their failures when trying to make changes to their health behaviors, but also their successes. The mindfulness component of self-compassion can also promote taking a balanced view of one's health behavior efforts and progress so that failures do not overshadow successes. Taking a kind and accepting stance to difficulties experienced when trying to change health behaviors can promote viewing these lapses not as failures, but as a natural part of any behavior change effort that anyone may experience. In short, self-compassion can reduce self-defeating thinking in response to the inevitable lapses that occur when people try to change their health behaviors.

From the perspective of Social Cognitive Theory, these shifts in perspective are crucial for building self-efficacy (Bandura, 1977). When people view their challenging experiences with a particular behavior as failures, it diminishes their confidence in their ability to be successful. But the reverse is also true. Focusing on the successes experienced with a behavior can help to build confidence about one's capacity to continue to be successful with similar behaviors. This increased self-efficacy can then help fuel persistence in the face of any minor lapses whilst trying to change behavior, something that is all too common in the context of health behaviors.

Current evidence supports these explanations of why self-compassion may promote self-efficacy. For example, a meta-analysis of 60 studies found that self-compassion overall was associated with higher self-efficacy, and that each of the three positive subscales were positively linked to self-efficacy, and each of the three negative subscales were negatively linked to self-efficacy (Liao et al., 2021).

Although there has been less research testing whether self-efficacy explains the link between self-compassion and health behaviors, two studies provide supportive evidence for this proposition. In one study of emerging adults, higher levels of self-compassion were associated with stronger intentions to engage in health-promoting behaviors such as staying physically active, eating healthy, and managing stress within the next 6 months (Sirois, 2015). Consistent with the TPB, higher levels of health-specific self-efficacy explained this link. In another study, overweight and obese adults trying to lose weight by restricting their diet were followed for 2 weeks using Ecological Momentary Assessment with twice daily sampling (Thøgersen-Ntoumani et al., 2021). Self-compassionate responses to dietary lapses and temptations were associated both with stronger intentions to continue dieting, and greater self-efficacy for dieting, with lower feelings of guilt explaining these associations. These findings support the idea that the healthy emotion regulation linked to being self-compassionate can contribute to feelings of self-efficacy, and in turn health behaviors, as proposed by the triadic model of self-compassion and health (Fig. 18.1).

Emotional Regulation as a Self-Regulation Resource for Health Behaviors

Making healthy behavior changes can be challenging for most people. The path to successfully achieving health goals typically follows a trajectory of peaks, when progress is made, and troughs, when lapses and temptations may temporarily derail progress. It is at these challenging points on the path to changing health behaviors that people are at their greatest risk for putting off or even abandoning their health goals, in part because of their emotional reactions to these lapses (e.g., Sirois & Giguère, 2018). Rather than motivating persistence, feelings of guilt, self-criticism, and other negative emotions arising from lapses can impair self-regulation by shifting attention away from longer-term health goals and placing it on the immediate goal of regulating current negative mood states (Sirois & Pychyl, 2013; Wagner & Heatherton, 2015). As a result, people may use avoidant coping strategies and other self-defeating, but immediately gratifying, behaviors, such as overindulging in food, or binge-watching their favorite shows rather than exercising, as a way to regulate these negative emotions (Baumeister et al., 2007).

Successful navigation of these lapses and temptations requires healthy emotion regulation skills to reduce the negative emotional responses to such challenges and enhance positive states that can fuel motivation and persistence to continue (e.g., Sin et al., 2015). This is where theory and evidence suggest that self-compassion may be particularly beneficial for promoting the practice of healthy behaviors. Accepting lapses as a normal part of health behavior change that anyone can experience can reduce negative self-defeating thoughts and emotions and taking a balanced view of the lapse in the context progress made rather than ruminating about it can further reduce negative emotional responses and increase positive feelings about overall progress. Research on how people respond to exercise setbacks supports these ideas. In one study, self-compassion was associated with lower levels of negative affect, and greater goal re-engagement following the recall an exercise lapse, even after controlling for self-esteem (Semenchuk et al., 2018).

From a self-regulation perspective, the healthy balance between high levels of positive emotions and low levels of negative emotions that characterize self-compassionate responses to setbacks can be viewed as resources that promote effective self-regulation of health behaviors. According the Temporal Self-Regulation Resource Model (TASRR; Sirois, 2015, 2016; Sirois & Hirsch, 2015), the extent to which individual differences are characterized by high levels of positive affect and future orientation, and low levels of negative affective states, will predict the effectiveness of self-regulation. In essence, positive affect and future orientation (i.e., a tendency to consider future time frames when making decisions) serve as resources to draw upon when facing self-regulation threats (Baird et al., 2021; Fredrickson, 1998), whereas negative affect is a liability for effective self-regulation (see Wagner & Heatherton, 2015 for a review). Health behaviors are often considered the prototypical self-regulation task (Baumeister et al., 1994), because they require regulating and monitoring emotions to maintain a focus on long-term goals rather than allowing emotions to drive choices for immediate gratification (Tice & Bratslavsky, 2000). Individual differences that are characterized by healthy emotion regulation skills and affective balance are therefore likely to promote the practice of health-promoting behaviors. Self-compassion can be viewed as one such individual difference.

Current research provides some support for this emotion regulation account of how self-compassion promotes engaging in healthy behaviors. In one study of mothers who had young children, those with higher levels of self-compassion experienced lower levels of guilt about taking time for themselves to engage in health-promoting behaviors, and this lower guilt explained their more frequent practice of health behaviors (Miller & Strachan, 2020). Similar results with respect to negative emotions were found in a study of university students. Lower levels of negative affect, but not higher levels of positive affect, explained the association between self-compassion and stronger intentions to engage in health-promoting behaviors after accounting for current body mass index (BMI) and current practice of health behaviors (Sirois, 2015).

More compelling evidence for the idea that a healthy affective balance links self-compassion to health behaviors comes from a meta-analysis of 15 samples that included over 3200 participants (Sirois et al., 2015a). In all 15 samples, dispositional self-compassion was significantly associated with more frequent practice of a set of health behaviors such as eating healthy, staying physically active, and managing stress, with a meta-analysis of these effects showing a small but significant association. Additional analyses with eight samples testing the indirect effects through positive and negative affect revealed results consistent with the TASRR model. Across each sample, the association between self-compassion and more frequent health-promoting behaviors was explained in part by higher positive affect and lower negative affect. Although the above findings rely upon cross-sectional data which preclude drawing conclusions about causality, they are consistent with theory, which assumes the temporal precedence of dispositional self-compassion as contributing to health behaviors rather than the reverse. Nonetheless, longitudinal research testing these links over time would provide more conclusive evidence.

Self-Compassion, Emotion Regulation, and Sleep Quality

In addition to ameliorating stress and supporting effective self-regulation of health behaviors, emotion regulation figures prominently as a process to explain why self-compassion can contribute to better sleep quality. Emotion regulation processes that are supported by taking a self-compassionate response to daily difficulties may be particularly useful for helping individuals to wind down after a difficult day rather than ruminating about things that did not go as well as expected. Positive presleep cognitions are known to impact sleep quality (Wood et al., 2009), and may be fostered by making adaptive cognitive reappraisals of difficult events. Consistent with this idea, a study of university students found that self-compassion was associated with better sleep quality, and this link was explained by less use of harmful cognitive emotional regulation strategies, such as self-blame (Semenchuk et al., 2021). Other research has directly tested the role of adaptive cognitive reappraisals for reducing negative affect and improving sleep-related outcomes. In one study, self-compassion was associated with engaging in less bedtime procrastination, a sleep behavior known to contribute to poor sleep quality (Kroese et al., 2014), and this association was explained by greater use of emotion regulation in the form of cognitive reappraisal, which in turn contributed to lower levels of negative emotions (Sirois et al., 2019).

Conclusions and Future Directions

From the research reviewed in this chapter, it is clear that self-compassion has important implications for physical health. Both dispositional self-compassion and self-compassion interventions are linked to reduced physical symptoms and better overall physical health.

The current chapter introduced a Triadic Pathway model of self-compassion and physical health as a provisional framework to conceptually organize the current research on the possible routes and self-processes linking self-compassion to health. It is hoped that this model can guide future "second wave" research into the reasons why self-compassion may be beneficial for physical health and provide a foundation for generating new theoretical insights. Lower stress, better sleep quality, and engaging in a repertoire of healthier behaviors, such as staying physically active and eating a healthy diet, are three key-related pathways through which self-compassion is proposed to exert its effects on health. One key message from the theory and evidence reviewed is that the value of self-compassion for promoting the practice of healthy behaviors may be due to its effectiveness in helping people bridge the gap between the initiation of health behaviors and the maintenance of those behaviors over time, both through increasing perceptions of control and through adaptive emotional responses to setbacks and lapses. Indeed, several processes and factors that are central for effective self-regulation appear to hold the most promise for understanding why responding to difficulties and failures in a kind connected, and mindful manner can maximize health-promoting trajectories and minimize health-risk trajectories, not only with respect to health behaviors, but also for stress and sleep quality.

Despite the promise of this research, there remains several gaps that future research on self-compassion would be wise to address to elucidate the links between self-compassion and health. The Triadic Pathway model highlights the synergistic nature of the three health pathways *and* the processes linking self-compassion to each. Yet research to date has not considered these three pathways simultaneously to understand their possible dynamic and reciprocally reinforcing linkages or how the different components of self-compassion may be involved. Often stress and/or health behaviors are examined as pathways linking self-compassion to health (e.g., Dunne et al., 2016; Homan & Sirois, 2017), but rarely is sleep considered as a key pathway. Research examining one or more of the proposed processes and how they contribute to the three health pathways is rare, despite the fact that emotion regulation is often considered a central feature of self-compassion and has clear relevance for understanding stress, sleep quality, and health behaviors. As the field of self-compassion and health continues to advance, it will be important that investigations consider the simultaneous contributions of the self-regulation processes proposed by the Triadic Pathway model and use robust experimental and longitudinal research designs to provide further clarity regarding how and why self-compassion can benefit physical health.

References

Ajzen, I. (1991). The theory of planned behavior. *Organizational Behavior and Human Decision Processes, 50*, 179–211. https://doi.org/10.1016/0749-5978(91)90020-T

Allen, A. B., Goldwasser, E. R., & Leary, M. R. (2011). Self-compassion and well-being among older adults. *Self and Identity, 11*(4), 428–453. https://doi.org/10.1080/15298868.2011.595082

Arch, J. J., Brown, K. W., Dean, D. J., Landy, L. N., Brown, K. D., & Laudenslager, M. L. (2014). Self-compassion training modulates alpha-amylase, heart rate variability, and subjective responses to social evaluative threat in women. *Psychoneuroendocrinology, 42*(0), 49–58. https://doi.org/10.1016/j.psyneuen.2013.12.018

Baird, H. M., Webb, T. L., Sirois, F. M., & Gibson-Miller, J. (2021). Understanding the effects of time perspective: A meta-analysis testing a self-regulatory framework. *Psychological Bulletin, 147*(3), 233–267. https://doi.org/10.1037/bul0000313

Bandura, A. (1977). Self-efficacy: Toward a unifying theory of behavioral change. *Psychological Review, 84*(2), 191–215.

Barnard, L. K., & Curry, J. F. (2011). Self-compassion: Conceptualizations, correlates, & interventions. *Review of General Psychology, 15*(4), 289–303. https://doi.org/10.1037/a0025754

Baumeister, R. F., Heatherton, T. F., & Tice, D. M. (1994). *Losing control: How and why people fail at self-regulation*. Academic Press.

Baumeister, R. F., Zell, A. L., & Tice, D. M. (2007). How emotions facilitate and impair self-regulation. In A. L. Zell & D. M. Tice (Eds.), *Handbook of emotion regulation* (pp. 408–426). Guilford Press.

Bellosta-Batalla, M., Ruiz-Robledillo, N., Sariñana-González, P., Capella-Solano, T., Vitoria-Estruch, S., Hidalgo-Moreno, G., et al. (2018). Increased salivary IgA response as an indicator of immunocompetence after a mindfulness and self-compassion-based intervention. *Mindfulness, 9*(3), 905–913. https://doi.org/10.1007/s12671-017-0830-y

Benyamini, Y. (2011). Why does self-rated health predict mortality? An update on current knowledge and a research agenda for psychologists. *Psychology & Health, 26*(11), 1407–1413. https://doi.org/10.1080/08870446.2011.621703

Biber, D. D., & Ellis, R. (2019). The effect of self-compassion on the self-regulation of health behaviors: A systematic review. *Journal of Health Psychology, 24*(14), 2060–2071. https://doi.org/10.1177/1359105317713361

Bluth, K., Roberson, P. N. E., Gaylord, S. A., Faurot, K. R., Grewen, K. M., Arzon, S., & Girdler, S. S. (2016). Does self-compassion protect adolescents from stress? *Journal of Child and Family Studies, 25*(4), 1098–1109. https://doi.org/10.1007/s10826-015-0307-3

Breines, J. G., Thoma, M. V., Gianferante, D., Hanlin, L., Chen, X., & Rohleder, N. (2014). Self-compassion as a predictor of interleukin-6 response to acute psychosocial stress. *Brain, Behavior, and Immunity, 37*(0), 109–114. https://doi.org/10.1016/j.bbi.2013.11.006

Brown, L., Bryant, C., Brown, V., Bei, B., & Judd, F. (2016). Self-compassion, attitudes to ageing and indicators of health and well-being among midlife women. *Aging & Mental Health, 20*(10), 1035–1043. https://doi.org/10.1080/13607863.2015.1060946

Brown, L., Houston, E. E., Amonoo, H. L., & Bryant, C. (2021). Is self-compassion associated with sleep quality? A Meta-analysis. *Mindfulness, 12*(1), 82–91. https://doi.org/10.1007/s12671-020-01498-0

Butler, E. A. (2011). Three views of emotion regulation and health. *Social and Personality Psychology Compass, 5*(8), 563–577. https://doi.org/10.1111/j.1751-9004.2011.00372.x

Butz, S., & Stahlberg, D. (2018). Can self-compassion improve sleep quality via reduced rumination? *Self and Identity, 17*(6), 666–686. https://doi.org/10.1080/15298868.2018.1456482

Carvalho, S. A., Gillanders, D., Palmeira, L., Pinto-Gouveia, J., & Castilho, P. (2018). Mindfulness, selfcompassion, and depressive symptoms in chronic pain: The role of pain acceptance. *Journal of Clinical Psychology, 74*(12), 2094–2106. https://doi.org/10.1002/jclp.22689

Carver, C. S., & Connor-Smith, J. (2010). Personality and coping. *Annual Review of Psychology, 61*, 679–704. https://doi.org/10.1146/annurev.psych.093008.100352

Carver, C. S., & Scheier, M. F. (1982). Control theory: A useful conceptual framework for personality–social, clinical, and health psychology. *Psychological Bulletin, 92*(1), 111–135.

Carver, C. S., Scheier, M. F., & Weintraub, J. K. (1989). Assessing coping strategies: A theoretically based approach. *Journal of Personality and Social Psychology, 56*(2), 267–283.

Chishima, Y., Mizuno, M., Sugawara, D., & Miyagawa, Y. (2018). The influence of self-compassion on cognitive appraisals and coping with stressful events. *Mindfulness, 9*(6), 1907–1915. https://doi.org/10.1007/s12671-018-0933-0

Cohen, S., Janicki-Deverts, D., Doyle, W. J., Miller, G. E., Frank, E., Rabin, B. S., & Turner, R. B. (2012). Chronic stress, glucocorticoid receptor resistance, inflammation, and disease risk. *Proceedings of the National Academy of Sciences, 109*(16), 5995–5999. https://doi.org/10.1073/pnas.111835510

Daley, M., Morin, C. M., LeBlanc, M., Grégoire, J. P., Savard, J., & Baillargeon, L. (2009). Insomnia and its relationship to health-care utilization, work absenteeism, productivity and accidents. *Sleep Medicine, 10*(4), 427–438. https://doi.org/10.1016/j.sleep.2008.04.005

Dewsaran-van der Ven, C., van Broeckhuysen-Kloth, S., Thorsell, S., Scholten, R., De Gucht, V., & Geenen, R. (2018). Self-compassion in somatoform disorder. *Psychiatry Research, 262*, 34–39. https://doi.org/10.1016/j.psychres.2017.12.013

Dunne, S., Sheffield, D., & Chilcot, J. (2016). Brief report: Self-compassion, physical health and the mediating role of health-promoting behaviors. *Journal of Health Psychology, 23*(7), 993–999. https://doi.org/10.1177/1359105316643377

Eriksson, T., Germundsjö, L., Åström, E., & Rönnlund, M. (2018). Mindful self-compassion training reduces stress and burnout symptoms among practicing psychologists: A randomized controlled trial of a brief web-based intervention. *Frontiers in Psychology, 9*(2340). https://doi.org/10.3389/fpsyg.2018.02340

Ewert, C., Vater, A., & Schröder-Abé, M. (2021). *Self-compassion and coping: A meta-analysis. Mindfulness, 12*(5), 1063–1077. https://doi.org/10.1007/s12671-020-01563-8

Ferrari, M., Hunt, C., Harrysunker, A., Abbott, M. J., Beath, A. P., & Einstein, D. A. (2019). Self-compassion interventions and psychosocial outcomes: A meta-analysis of RCTs. *Mindfulness, 10*(8), 1455–1473. https://doi.org/10.1007/s12671-019-01134-6

Finlay-Jones, A. L., Rees, C. S., & Kane, R. T. (2015). Self-compassion, emotion regulation and stress among Australian psychologists: Testing an emotion regulation model of self-compassion using structural equation modeling. *PLoS One, 10*(7), e0133481. https://doi.org/10.1371/journal.pone.0133481

Fredrickson, B. L. (1998). What good are positive emotions? *Review of General Psychology, 2*(3), 300–319.

Friedman, H. S. (2000). Long-term relations of personality and health: Dynamisms, mechanisms, tropisms. *Journal of Personality, 68*(6), 1089–1107. https://doi.org/10.1111/1467-6494.00127

Gerrard, M., Gibbons, F. X., Houlihan, A. E., Stock, M. L., & Pomery, E. A. (2008). A dual-process approach to health risk decision making: The prototype willingness model. *Developmental Review, 28*(1), 29–61. https://doi.org/10.1016/j.dr.2007.10.001

Gilbert, P. (2019). Explorations into the nature and function of compassion. *Current Opinion in Psychology, 28*, 108–114. https://doi.org/10.1016/j.copsyc.2018.12.002

Gregory, W. E., Glazer, J. V., & Berenson, K. R. (2017). Self-compassion, self-injury, and pain. *Cognitive Therapy and Research, 41*(5), 777–786. https://doi.org/10.1007/s10608-017-9846-9

Gross, J. J. (1998). The emerging field of emotion regulation: An integrative review. *Review of General Psychology, 2*(3), 271–299. https://doi.org/10.1037/1089-2680.2.3.271

Gross, J. J., & John, O. P. (2003). Individual differences in two emotion regulation processes: Implications for affect, relationships, and well-being. *Journal of Personality and Social Psychology, 85*(2), 348–362. https://doi.org/10.1037/0022-3514.85.2.348

Haario, P., Rahkonen, O., Laaksonen, M., Lahelma, E., & Lallukka, T. E. A. (2013). Bidirectional associations between insomnia symptoms and unhealthy behaviors. *Journal of Sleep Research, 22*(1), 89–95. https://doi.org/10.1111/j.1365-2869.2012.01043.x

Hall, C. W., Row, K. A., Wuensch, K. L., & Godley, K. R. (2013). The role of self-compassion in physical and psychological well-being. *The Journal of Psychology, 147*(4), 311–323. https://doi.org/10.1080/00223980.2012.693138

Hirotsu, C., Tufik, S., & Andersen, M. L. (2015). Interactions between sleep, stress, and metabolism: From physiological to pathological conditions. *Sleep Science, 8*(3), 143–152. https://doi.org/10.1016/j.slsci.2015.09.002

Homan, K. J., & Sirois, F. M. (2017). Self-compassion and physical health: Exploring the roles of perceived stress and health-promoting behaviors. *Health Psychology Open, 4*(2). https://doi.org/10.1177/2055102917729542

Howren, M. B., & Suls, J. (2011). The symptom perception hypothesis revised: Depression and anxiety play different roles in concurrent and retrospective physical symptom reporting. *Journal of Personality and Social Psychology, 100*, 182–195. https://doi.org/10.1037/a0021715

Inwood, E., & Ferrari, M. (2018). Mechanisms of change in the relationship between self-compassion, emotion regulation, and mental health: A systematic review. *Applied Psychology: Health and Well-Being, 10*(2), 215–235. https://doi.org/10.1111/aphw.12127

Jylhä, M. (2009). What is self-rated health and why does it predict mortality? Towards a unified conceptual model. *Social Science & Medicine, 69*(3), 307–316. https://doi.org/10.1016/j.socscimed.2009.05.013

Kashani, M., Eliasson, A., & Vernalis, M. (2012). Perceived stress correlates with disturbed sleep: A link connecting stress and cardiovascular disease. *Stress, 15*(1), 45–51. https://doi.org/10.3109/10253890.2011.578266

Kristenson, M., Olsson, A., & Kucinskiene, Z. (2005). Good self-rated health is related to psychosocial resources and a strong cortisol response to acute stress: The LiVicordia study of middle-aged men. *International Journal of Behavioral Medicine, 12*(3), 153–160. https://doi.org/10.1207/s15327558ijbm1203_4

Kroese, F. M., Evers, C., Adriaanse, M. A., & de Ridder, D. T. D. (2014). Bedtime procrastination: A self-regulation perspective on sleep insufficiency in the general population. *Journal of Health Psychology*. https://doi.org/10.1177/1359105314540014

Lazarus, R. S., & Folkman, S. (1984). *Stress, appraisal, and coping*. Springer.

Lee, E. E., Govind, T., Ramsey, M., Wu, T. C., Daly, R., Liu, J., et al. (2021). Compassion toward others and self-compassion predict mental and physical well-being: A 5-year longitudinal study of 1090 community-dwelling adults across the lifespan. *Translational Psychiatry, 11*(1), 397. https://doi.org/10.1038/s41398-021-01491-8

Lianekhammy, J., Miller, J. J., Lee, J., Pope, N., Barnhart, S., & Grise-Owens, E. (2018). Exploring the self-compassion of health-care social workers: How do they fare? *Social Work in Health Care, 57*(7), 563–580. https://doi.org/10.1080/00981389.2018.1471017

Liao, K. Y.-H., Stead, G. B., & Liao, C.-Y. (2021). A meta-analysis of the relation between self-compassion and self-efficacy. *Mindfulness, 12*(8), 1878–1891. https://doi.org/10.1007/s12671-021-01626-4

Lippke, S., Nigg, C. R., & Maddock, J. E. (2012). Health-promoting and health-risk behaviors: Theory-driven analyses of multiple health behavior change in three international samples. *International Journal of Behavioral Medicine, 19*(1), 1–13. https://doi.org/10.1007/s12529-010-9135-4

MacBeth, A., & Gumley, A. (2012). Exploring compassion: A meta-analysis of the association between self-compassion and psychopathology. *Clinical*

Psychology Review, 32(6), 545–552. https://doi.org/10.1016/j.cpr.2012.06.003

McEwen, B. S. (2007). Physiology and neurobiology of stress and adaptation: Central role of the brain. *Physiological Reviews, 87*(3), 873–904. https://doi.org/10.1152/physrev.00041.2006

Miller, C. L., & Strachan, S. M. (2020). Understanding the role of mother guilt and self-compassion in health behaviors in mothers with young children. *Women & Health, 60*(7), 763–775. https://doi.org/10.1080/03630242.2020.1713966

Minkel, J. D., Banks, S., Htaik, O. M., Marisa, C., Jones, C. W., McGlinchey, E. L., Simpson, N. S., & Dinges, D. F. (2012). Sleep deprivation and stressors: Evidence for elevated negative affect in response to mild stressors when sleep deprived. *Emotion, 12*, 1015–1020. https://doi.org/10.1037/a0026871

Mora, P. A., Orsak, G., DiBonaventura, M. D., & Leventhal, E. A. (2013). Why do comparative assessments predict health? The role of self-assessed health in the formation of comparative health judgments. *Health Psychology, 32*(11), 1175–1178. https://doi.org/10.1037/a0032044

Morin, C. M., Rodrigue, S., & Ivers, H. (2003). Role of stress, arousal, and coping skills in primary insomnia. *Psychosomatic Medicine, 65*(2), 259–267. https://doi.org/10.1097/01.PSY.0000030391.09558.A3

Mosewich, A. D., Sabiston, C. M., Kowalski, K. C., Gaudreau, P., & Crocker, P. R. E. (2019). Self-compassion in the stress process in women athletes. *The Sport Psychologist, 33*(1), 23–34. https://doi.org/10.1123/tsp.2017-0094

Neff, K. D. (2003). Self-compassion: An alternative conceptualization of a healthy attitude toward oneself. *Self and Identity, 2*(2), 85–101. https://doi.org/10.1080/15298860390129863

Neff, K. D. (2011). Self-compassion, self-esteem, and well-being. *Social and Personality Psychology Compass, 5*(1), 1–12. https://doi.org/10.1111/j.1751-9004.2010.00330.x

Neff, K. D., & Germer, C. K. (2013). A pilot study and randomized controlled trial of the mindful self-compassion program. *Journal of Clinical Psychology, 69*(1), 28–44. https://doi.org/10.1002/jclp.21923

Neff, K. D., Kirkpatrick, K. L., & Rude, S. S. (2007). Self-compassion and adaptive psychological functioning. *Journal of Research in Personality, 41*(1), 139–154. https://doi.org/10.1016/j.jrp.2006.03.004

Neff, K. D., Long, P., Knox, M. C., Davidson, O., Kuchar, A., Costigan, A., et al. (2018). The forest and the trees: Examining the association of self-compassion and its positive and negative components with psychological functioning. *Self and Identity, 17*(6), 627–645. https://doi.org/10.1080/15298868.2018.1436587

Ng, D. M., & Jeffery, R. W. (2003). Relationships between perceived stress and health behaviors in a sample of working adults. *Health Psychology, 22*(6), 638–642. https://doi.org/10.1037/0278-6133.22.6.638

Phillips, W. J., & Hine, D. W. (2019). Self-compassion, physical health, and health behavior: A meta-analysis. *Health Psychology Review*, 1–27. https://doi.org/10.1080/17437199.2019.1705872

Pinto-Gouveia, J., Duarte, C., Matos, M., & Fráguas, S. (2014). The protective role of self-compassion in relation to psychopathology symptoms and quality of life in chronic illness and in cancer patients. *Clinical Psychology & Psychotherapy, 21*(4), 311–323. https://doi.org/10.1002/cpp.1838

Rahimi-Ardabili, H., Reynolds, R., Vartanian, L. R., McLeod, L. V. D., & Zwar, N. (2018). A systematic review of the efficacy of interventions that aim to increase self-compassion on nutrition habits, eating behaviors, body weight and body image. *Mindfulness, 9*(2), 388–400. https://doi.org/10.1007/s12671-017-0804-0

Reidy, B. L., Raposa, E. B., Brennan, P. A., Hammen, C. L., Najman, J. M., & Johnson, K. C. (2016). Prospective associations between chronic youth sleep problems and young adult health. *Sleep Health, 2*(1), 69–74. https://doi.org/10.1016/j.sleh.2015.11.005

Semenchuk, B. N., Strachan, S. M., & Fortier, M. (2018). Self-compassion and the self-regulation of exercise: Reactions to recalled exercise setbacks. *Journal of Sport & Exercise Psychology, 40*(1), 31–39. https://doi.org/10.1123/jsep.2017-0242

Semenchuk, B. N., Onchulenko, S., & Strachan, S. M. (2021). Self-compassion and sleep quality: Examining the mediating role of taking a proactive health focus and cognitive emotional regulation strategies. *Journal of Health Psychology*, 13591053211047148. https://doi.org/10.1177/13591053211047148

Sin, N. L., Moskowitz, J. T., & Whooley, M. A. (2015). Positive affect and health behaviors across 5 years in patients with coronary heart disease: The heart and soul study. *Psychosomatic Medicine, 77*(9), 1058–1066. https://doi.org/10.1097/PSY.0000000000000238

Sirois, F. M. (2014). Procrastination and stress: Exploring the role of self-compassion. *Self and Identity, 13*(2), 128–145. https://doi.org/10.1080/15298868.2013.763404

Sirois, F. M. (2015). A self-regulation resource model of self-compassion and health behavior intentions in emerging adults. *Preventive Medicine Reports, 2*, 218–222. https://doi.org/10.1016/j.pmedr.2015.03.006

Sirois, F. M. (2016). Perfectionism and health behaviors: A self-regulation resource perspective. In F. M. Sirois & D. S. Molnar (Eds.), *Perfectionism, health and well-being* (pp. 45–68). Springer.

Sirois, F. M. (2020). The association between self-compassion and self-rated health in 26 samples. *BMC Public Health, 20*(1), 74. https://doi.org/10.1186/s12889-020-8183-1

Sirois, F. M., & Giguère, B. (2018). Giving in when feeling less good: Procrastination, action control, and social temptations. *British Journal of Social Psychology, 57*(2), 404–427. https://doi.org/10.1111/bjso.12243

Sirois, F. M., & Hirsch, J. K. (2015). Big five traits, affect balance and health behaviors: A self-regulation resource perspective. *Personality and Individual*

Differences, 87, 59–64. https://doi.org/10.1016/j.paid.2015.07.031

Sirois, F. M., & Hirsch, J. K. (2019). Self-compassion and adherence in five medical samples: The role of stress. *Mindfulness, 10*(1), 46–54. https://doi.org/10.1007/s12671-018-0945-9

Sirois, F. M., & Pychyl, T. (2013). Procrastination and the priority of short-term mood regulation: Consequences for future self. *Social and Personality Psychology Compass, 7*(2), 115–127. https://doi.org/10.1111/spc3.12011

Sirois, F. M., Kitner, R., & Hirsch, J. K. (2015a). Self-compassion, affect, and health-promoting behaviors. *Health Psychology, 34*(6), 661–669. https://doi.org/10.1037/hea0000158

Sirois, F. M., Molnar, D. S., & Hirsch, J. K. (2015b). Self-compassion, stress, and coping in the context of chronic illness. *Self and Identity*, 1–14. https://doi.org/10.1080/15298868.2014.996249

Sirois, F. M., Nauts, S., & Molnar, D. S. (2019). Self-compassion and bedtime procrastination: An emotion regulation perspective. *Mindfulness, 10*(3), 434–445. https://doi.org/10.1007/s12671-018-0983-3

Skinta, M. D., Fekete, E. M., & Williams, S. L. (2019). HIV-stigma, self-compassion, and psychological well-being among gay men living with HIV. *Stigma and Health, 4*(2), 179–187. https://doi.org/10.1037/sah0000133

Smith, T. W. (2006). Personality as risk and resilience in physical health. *Current Directions in Psychological Science, 15*, 227–231. https://doi.org/10.1111/j.1467-8721.2006.00441.x

Smith, J. L. (2015). Self-compassion and resilience in senior living residents. *Seniors Housing & Care Journal, 23*, 17–31.

Smyth, J., Zawadzki, M., & Gerinm, W. (2013). Stress and disease: A structural and functional analysis. *Social and Personality Psychology Compass, 7*, 217–227. https://doi.org/10.1111/spc3.12020

Steptoe, A., Lipsey, Z., & Wardle, J. (1998). Stress, hassles and variations in alcohol consumption, food choice and physical exercise: A diary study. *British Journal of Health Psychology, 3*(1), 51–63. https://doi.org/10.1111/j.2044-8287.1998.tb00555.x

St-Onge, M.-P., Grandner, M. A., Brown, D., Conroy, M. B., Jean-Louis, G., Coons, M., & Bhatt, D. L. (2016). Sleep duration and quality: Impact on lifestyle behaviors and cardiometabolic health: A scientific statement from the American Heart Association. *Circulation, 134*(18), e367–e386. https://doi.org/10.1161/CIR.0000000000000444

Strine, T. W., & Chapman, D. P. (2005). Associations of frequent sleep insufficiency with health-related quality of life and health behaviors. *Sleep Medicine, 6*(1), 23–27. https://doi.org/10.1016/j.sleep.2004.06.003

Stults-Kolehmainen, M. A., & Sinha, R. (2014). The effects of stress on physical activity and exercise. *Sports Medicine (Auckland, N.Z.), 44*(1), 81–121. https://doi.org/10.1007/s40279-013-0090-5

Stutts, L. A., Leary, M. R., Zeveney, A. S., & Hufnagle, A. S. (2018). A longitudinal analysis of the relationship between self-compassion and the psychological effects of perceived stress. *Self and Identity, 17*(6), 609–626. https://doi.org/10.1080/15298868.2017.1422537

Suls, J., & Rittenhouse, J. D. (1990). Models of linkages between personality and disease. In H. S. Friedman (Ed.), *Personality and disease* (pp. 38–63). Wiley.

Svendsen, J. L., Osnes, B., Binder, P.-E., Dundas, I., Visted, E., Nordby, H., et al. (2016). Trait self-compassion reflects emotional flexibility through an association with high vagally mediated heart rate variability. *Mindfulness, 7*(5), 1103–1113. https://doi.org/10.1007/s12671-016-0549-1

Svendsen, J. L., Schanche, E., Osnes, B., Vøllestad, J., Visted, E., Dundas, I., et al. (2020). Is dispositional self-compassion associated with psychophysiological flexibility beyond mindfulness? An exploratory pilot study. *Frontiers in Psychology, 11*, 614. https://doi.org/10.3389/fpsyg.2020.00614

Tamayo-Fonseca, N., Quesada, J. A., Nolasco, A., Melchor, I., Moncho, J., Pereyra-Zamora, P., et al. (2013). Self-rated health and mortality: A follow-up study of a Spanish population. *Public Health, 127*(12), 1097–1104. https://doi.org/10.1016/j.puhe.2013.09.003

Taylor, S. E., Sirois, F. M., & Molnar, D. S. (2020). *Health psychology* (Fifth Canadian edition ed.). McGraw-Hill Ryerson.

Terry, M. L., Leary, M. R., Mehta, S., & Henderson, K. (2013). Self-compassionate reactions to health threats. *Personality and Social Psychology Bulletin, 39*(7), 911–926. https://doi.org/10.1177/0146167213488213

Thøgersen-Ntoumani, C., Dodos, L. A., Stenling, A., & Ntoumanis, N. (2021). Does self-compassion help to deal with dietary lapses among overweight and obese adults who pursue weight-loss goals? *British Journal of Health Psychology, 26*(3), 767–788. https://doi.org/10.1111/bjhp.12499

Thombs, B. D., Ziegelstein, R. C., Stewart, D. E., Abbey, S. E., Parakh, K., & Grace, S. L. (2008). Physical health status assessed during hospitalization for acute coronary syndrome predicts mortality 12 months later. *Journal of Psychosomatic Research, 65*(6), 587–593. https://doi.org/10.1016/j.jpsychores.2008.06.004

Tice, D. M., & Bratslavsky, E. (2000). Giving in to feel good: The place of emotion regulation in the context of general self-control. *Psychological Inquiry, 11*, 149–159. https://doi.org/10.1207/S15327965PLI1103_03

Trindade, I. A., & Sirois, F. M. (2021). The prospective effects of self-compassion on depressive symptoms, anxiety, and stress: A study in inflammatory bowel disease. *Journal of Psychosomatic Research, 146*, 110429. https://doi.org/10.1016/j.jpsychores.2021.110429

Vickers, R. R., Conway, T. L., & Hervig, L. K. (1990). Demonstration of replicable dimensions of health behaviors. *Preventive Medicine, 19*, 377–401. https://doi.org/10.1016/0091-7435(90)90037-K

Wagner, D. D., & Heatherton, T. F. (2015). Self-regulation and its failure: The seven deadly threats to self-regulation. In *APA handbook of personality and social psychology, Volume 1: Attitudes and social cognition* (pp. 805–842). American Psychological Association.

Ware, J., Jr., Kosinski, M., & Keller, S. D. (1996). A 12-item short-form health survey: Construction of scales and preliminary tests of reliability and validity. *Medical Care, 34*(3), 220–233. https://doi.org/10.1097/00005650-199603000-00003

Ware, J. E. J., & Sherbourne, C. D. (1992). The MOS 36-item short-form health survey (SF-36): I. Conceptual framework and item selection. *Medical Care, 30*(6), 473–483. https://www.jstor.org/stable/3765916

Wilson, A. C., Mackintosh, K., Power, K., & Chan, S. W. Y. (2019). Effectiveness of self-compassion-related therapies: A systematic review and meta-analysis. *Mindfulness, 10*(6), 979–995. https://doi.org/10.1007/s12671-018-1037-6

Wong, C. C. Y., & Mak, W. W. S. (2016). Writing can heal: Effects of self-compassion writing among Hong Kong Chinese college students. *Asian American Journal of Psychology, 7*(1), 74–82. https://doi.org/10.1037/aap0000041

Wong, M. Y. C., Chung, P.-K., & Leung, K.-M. (2021). The relationship between physical activity and self-compassion: A systematic review and meta-analysis. *Mindfulness, 12*(3), 547–563. https://doi.org/10.1007/s12671-020-01513-4

Wood, A. M., Joseph, S., Lloyd, J., & Atkins, S. (2009). Gratitude influences sleep through the mechanism of pre-sleep cognitions. *Journal of Psychosomatic Research, 66*(1), 43–48. https://doi.org/10.1016/j.jpsychores.2008.09.002

World Health Organization. (2015). *Global status report on noncommunicable diseases 2014*. Retrieved from https://www.who.int/publications/i/item/9789241564854

World Health Organization. (2021). *Obesity and overweight Fact sheet.* https://www.who.int/news-room/fact-sheets/detail/obesity-and-overweight

Wren, A. A., Somers, T. J., Wright, M. A., Goetz, M. C., Leary, M. R., Fras, A. M., et al. (2012). Self-compassion in patients with persistent musculoskeletal pain: Relationship of self-compassion to adjustment to persistent pain. *Journal of Pain and Symptom Management, 43*(4), 759–770. https://doi.org/10.1016/j.jpainsymman.2011.04.014

Zessin, U., Dickhäuser, O., & Garbade, S. (2015). The relationship between self-compassion and well-being: A meta-analysis. *Applied Psychology: Health and Well-Being, 7*(3), 340–364. https://doi.org/10.1111/aphw.12051

Zhang, Y., Luo, X., Che, X., & Duan, W. (2016). Protective effect of self-compassion to emotional response among students with chronic academic stress. *Frontiers in Psychology, 7*(1802). https://doi.org/10.3389/fpsyg.2016.01802

Zuraikat, F. M., Makarem, N., Redline, S., Aggarwal, B., Jelic, S., & St-Onge, M.-P. (2020). Sleep regularity and cardiometabolic heath: Is variability in sleep patterns a risk factor for excess adiposity and glycemic dysregulation? *Current Diabetes Reports, 20*(8), 38. https://doi.org/10.1007/s11892-020-01324-w

19 Self-Compassion and Chronic Medical Conditions

Amy Finlay-Jones, Anna Boggiss, and Anna Serlachius

Introduction

Chronic medical conditions (CMCs) are health problems that are likely to require treatment, including hospital stays, medication, or outpatient supports, for longer than 12 months (Hardelid et al., 2014). Examples of CMCs include cancer, diabetes, autoimmune disorders, asthma, and myalgic encephalomyelitis/chronic fatigue syndrome; collectively, these conditions affect between 40 and 65% of adults (Chapel et al., 2017; Harrison et al., 2017; Hvidberg et al., 2020), with a substantial proportion experiencing multimorbidity (Newman et al., 2020). The impact of CMCs is pervasive and multidimensional. Individuals living with CMCs often experience limitations in their physical functioning and report disruptions to family and peer relationships, school and employment, and activities of daily living. In addition, they frequently experience ongoing burden as a result of treatment and physical symptoms such as pain and fatigue (Lambert & Keogh, 2015). Unsurprisingly, individuals with CMCs may experience difficulties with adjustment and vulnerability to mental health difficulties (Daré et al., 2019; Katon et al., 2007; Tegethoff et al., 2015). For example, a recent study conducted in the United States found a 51% greater risk of mental health conditions among young people with CMCs compared to those without (Adams et al., 2019). These relationships may be bidirectional, and mental and physical health symptoms can compound one another, leading to a vicious cycle of chronic comorbidity (Katon et al., 2007; Moussavi et al., 2007). Accordingly, there is a need to understand modifiable factors that are associated with mental well-being and resilience among individuals living with CMCs. The current chapter explores *self-compassion* as a modifiable, transdiagnostic resilience factor that is associated with better physical and mental health outcomes among individuals living with CMCs. We provide specific examples from the literature on chronic pain, diabetes, and cancer, and discuss future directions for self-compassion and CMC research.

Self-Compassion as a Psychological Resource in CMCs

There are several theoretical frameworks that describe the process of adjustment for individuals with CMCs, and these models vary in the extent to which they emphasize the role of biological, psychological, and social factors in determining an individual's well-being. While

A. Finlay-Jones (✉)
Telethon Kids Institute, Nedlands, WA, Australia
e-mail: amy.finlay-jones@telethonkids.org.au

A. Boggiss · A. Serlachius
Department of Psychological Medicine, University of Auckland, Auckland, New Zealand

A. Finlay-Jones et al. (eds.), *Handbook of Self-Compassion*, Mindfulness in Behavioral Health,
https://doi.org/10.1007/978-3-031-22348-8_19

the biomedical model highlights disease-related processes as the primary determinant of adjustment, psychological and biopsychosocial models emphasize the key role of psychological functioning and coping resources in influencing adjustment outcomes (Walker et al., 2004). Examples of these latter models include Lansing and Berg's (2014) self-regulation model, and Dall'Oglio et al's (2021) integrated self-care model, which highlight how psychological resources interact with biological and environmental factors to influence self-regulation and self-care, which in turn promote better adjustment and well-being. Through the lens of psychological and biopsychosocial models, self-compassion can be viewed as an important psychological resource that can support individuals to cope with the various challenges associated with living with a CMC. Self-compassion involves responding to oneself in a supportive manner during times of difficulty, by taking a mindful perspective on the difficult experience, remembering that one is not alone, and treating oneself with kindness and understanding (Neff, 2003). Prior work has underscored the role of self-compassion in promoting adaptive coping and emotion regulation (Finlay-Jones, 2017; Inwood & Ferrari, 2018) in the context of mental health difficulties, indicating that self-compassion may act in a similar way among individuals with CMCs (Prentice et al., 2021).

Empirical evidence indicates that self-compassion is consistently related to better adjustment and psychological health in individuals with a range of different CMCs. For example, self-compassion is associated with more resilience, greater quality of life and coping efficacy, and less depression and anxiety among individuals with epilepsy (Baker et al., 2019; Clegg et al., 2019) and multiple sclerosis (Gedik and Idiman, 2020; Nery-Hurwit et al., 2018). A recent systematic review found that across CMCs, self-compassion shares moderate-large inverse associations with depression and anxiety (Hughes et al., 2021). While most research has been conducted with adults, Prentice et al. (2021) found self-compassion was associated with both well-being and distress in a sample of adolescents and young adults with a range of CMCs. Together, these findings suggest that self-compassion is a valuable transdiagnostic intervention target for promoting better adjustment and mental health among individuals with CMCs (Finlay-Jones et al., 2020).

One of the ways in which self-compassion may serve a protective function for individuals with CMCs is by reducing the experience of shame and self-stigma. Individuals living with CMCs often report high levels of shame, self-criticism, self-blame, and stigma (Casati et al., 2000). This may be heightened when an individual's physical appearance is altered because of their CMC, such as in the case of chronic skin conditions (Clarke et al., 2020) or when stigma exists around other aspects of the disease. For example, individuals with inflammatory bowel disease report perceived stigma around their condition as a "dirty disease" that is socially unacceptable (Dibley et al., 2018), while people with diabetes have described stigma regarding injecting insulin due to the association between needle use and illicit drug use (Schabert et al., 2013). Shame and self-stigma may also be exacerbated when individuals experience decreases in functioning or increased dependence on others, which may lead to feelings of uselessness or being a burden on others (Ambridge et al., 2020). In turn, experiences of stigma, shame, and self-blame are associated with poorer psychological outcomes among people living with CMCs (Bennett et al., 2005; Clarke et al., 2020; Phelan et al., 2013; Trindade et al., 2018, 2019), and may also impact on help-seeking, treatment engagement, and health behavior change (Casati et al., 2000; Joachim & Acorn, 2000; Valdiserri, 2002).

Hughes et al.'s (2021) systematic review found that across studies of people with CMCs, the association between self-compassion and poor mental health (i.e., anxiety and depression) was mediated by shame, as well as depressive brooding and worry. Further, studies have revealed that among individuals with HIV, self-compassion buffers the adverse impact of self-stigma on life satisfaction (Yang & Mak, 2017) and is inversely associated with shame (Brion et al., 2014), while self-compassion also attenuates the impact of self-directed disgust on symptoms of depression in people living with chronic skin conditions (Clarke et al., 2020). Connected to, but distinct from self-criticism and shame, a second pathway by which self-compassion may promote more adaptive outcomes for people living with CMC is by promoting more positive health behaviors, such as treatment adherence and reducing health-risk behaviors, such as smoking, illicit drug use, and unprotected sex (Dawson Rose et al., 2014; Sirois & Hirsch, 2019; Sirois & Rowse, 2016). For example, self-compassion was found to moderate the link between shame and health-risk behaviors in people with HIV, such that those who were higher in self-compassion were more likely to disclose their HIV status to others and adhere to healthy self-management practices (Brion et al., 2014). Chapter 18 in this Handbook describes the associations between self-compassion and various health-promoting behaviors in greater detail.

Self-Compassion and Chronic Pain

Chronic pain provides one example of how self-compassion might help to promote better outcomes for individuals living with CMCs. Chronic pain is one of the most common CMCs and one of the leading causes of disability worldwide (Vos et al., 2012). In addition to functional impairment, individuals living with chronic pain often experience difficult emotional and affective experiences, such as fear of pain, and of activities that may cause pain (Bunzli et al., 2015) and shame regarding their condition (J. A. Smith & Osborn, 2007), which may exacerbate their pain symptoms and compound disability (Crombez et al., 2012). In the fear-avoidance model of chronic pain, cognitive factors and experiential avoidance act as key mechanisms in the onset and maintenance of chronic pain. Interpretation of pain is a core component of the model. When individuals respond to pain by catastrophizing about it, they are more likely to engage in hypervigilance or experiential avoidance. In turn, hypervigilance and avoidance lead to reduced engagement in daily activities, over time increasing the risk of physical deterioration and vulnerability to mental health problems.

Self-compassion appears to play a meaningful role in disrupting the fear-avoidance model of chronic pain, via reductions in pain-related fear (Edwards et al., 2019), pain catastrophizing (Wren et al., 2012), avoidance (Costa & Pinto-Gouveia, 2013), and other maladaptive cognitive emotion regulation strategies (Purdie & Morley, 2015). Other research demonstrates that self-compassion is meaningfully associated with important pain-related clinical and functional outcomes. For example, one study found that people with obesity and chronic pain who were more self-compassionate reported lower levels of pain-related disability (Wren et al., 2012) while other studies found self-compassion to be associated with greater engagement in daily activities (Costa & Pinto-Gouveia, 2011). Conversely, individuals with low self-compassion – particularly those who feel alone in their pain - report higher levels of pain severity (Chang et al., 2019).

In addition to reducing avoidance, self-compassion is associated with factors that are recognized to protect individuals from the deleterious effects of chronic pain. For example, self-compassion is associated with greater pain acceptance (Costa & Pinto-Gouveia, 2011, 2013) and use of adaptive strategies to cope with pain (Barnes et al., 2018). Unsurprisingly, across samples of individuals with chronic pain, self-compassion is associated with better mental health outcomes, including symptoms of depression, anxiety, and stress (Carvalho et al., 2018; Costa & Pinto-Gouveia, 2011, 2013). Additionally, self-compassion may alleviate the impact of *cognitive fusion,* or the tendency to believe the literal content of one's thoughts. Cognitive fusion is a known risk factor for pain-related disability and bears similarities with the overidentification facet of Neff's (2003) model of self-compassion. Conversely, the mindfulness facet of Neff's model provides an adaptive alternative mode of relating to one's difficult thoughts. For example, with cognitive fusion or overidentification, the thought *"this pain is never going to go away"* may provoke a strong fear reaction, whereas, with mindfulness, the stance that *"I am having the thought that this pain is never going to go away (but that is not necessarily true)"* may help to decouple the thought from the affective response. Indeed, it has been found that while cognitive fusion mediates the relationship between pain severity and depressive symptoms, this relationship is moderated by self-compassion, such that those with high self-compassion are less likely to experience adverse outcomes (Carvalho et al., 2018). Interestingly, while there are substantial numbers of studies attesting to the role of mindfulness in protecting against pain-related fear and avoidance, a longitudinal study found that baseline self-compassion, but not mindfulness, predicted depressive symptoms over time, above and beyond baseline impairment, and depressive symptoms (Carvalho et al., 2019). This study indicates that self-compassion may play an important role in preventing depression among individuals experiencing pain-related functional impairment.

Psychological interventions for those who experience chronic pain aim to support patients to change their cognitive, affective, and behavioral responses to the physical sensation of pain. To date, two studies have trialed self-compassion-based approaches as a means of promoting more adaptive responding in the face of pain, with demonstrated benefit. Montero-Marin et al. (2017) compared Attachment-Based Compassion Therapy (ABCT) with a relaxation training condition to determine the effect on functional impairment and mental health in people with fibromyalgia. The ABCT intervention was delivered over 8 weeks (one two-hour session per week), with three booster sessions. The focus of ABCT is to support participants to be kinder to themselves and others, through exercises focused on mindfulness, self-compassion, and early attachment experiences. Functional outcomes, depression, and anxiety in the ABCT group were significantly better than those in the relaxation group following the intervention, and these differences were maintained over three months with a large effect size. Follow-up studies found that the ABCT training had greater cost utility than the relaxation comparator (D'Amico et al., 2020), and also reported treatment effects on mindfulness and self-compassion for the ABCT group, with moderately large to large effects (Montero-Marin et al., 2020). While changes in functional impairment were not mediated by either mindfulness or self-compassion, analyses of self-compassion subscales showed that changes in the common humanity facet of self-compassion mediated improvements in anxiety and depression. These findings indicate that by being able to connect their individual struggles with that of the shared human experience, individuals with fibromyalgia participating in the ABCT program were less likely to experience symptoms of anxiety and depression.

Self-Compassion and Diabetes

Further examples of how self-compassion benefits those living with CMCs can be found in the diabetes literature. Diabetes mellitus is a cluster of metabolic disorders of which the most common types are type 1 and type 2 diabetes, with type 2 diabetes accounting for approximately 90% of cases worldwide (Chatterjee et al., 2017). More than 451 million adults are estimated to live with diabetes worldwide, which is expected to increase to 693 million by 2045 (Cho et al., 2018). The economic costs of diabetes are considerable, with type 2 diabetes estimated to cost the global economy US$825 billion (Seuring et al., 2015). As well as the economic and public health impact of diabetes, the psychological impact of living with diabetes is substantial and is estimated to double the odds of developing mental health problems such as depression (Anderson et al., 2001; Young-Hyman et al., 2016).

Diabetes requires daily, onerous self-care behaviors. Achieving optimal glycemic control in diabetes and preventing hyperglycemia (high blood sugar levels) is paramount, as tighter glycemic control in diabetes is associated with slower progression of diabetes-related complications and reduced mortality (Diabetes Control and Complications Trial Research Group, 1993; Holman et al., 2008; Stratton et al., 2000; Zoungas et al., 2012). Self-management of type 2 diabetes is multifaceted and includes modifying lifestyle behaviors, such as diet and physical activity, as well as adherence to medication, regular monitoring of blood glucose levels, and foot care. Arguably type 1 diabetes is characterized by an even more invasive treatment regimen, consisting of administering insulin (via multiple daily injections or pump therapy), 4–8 daily blood tests (via finger pricks or using continuous glucose monitoring devices), and the ongoing need for close monitoring of food intake and the dose of insulin required to cover food that is consumed. People living with type 1 and type 2 diabetes must also attend regular GP or diabetes outpatients' appointments to assess their glycemic control (HbA1c) and screen for diabetes-related complications.

The self-discipline needed to optimize glycemic control and prevent hyperglycemia or hypoglycemia (low blood sugar levels) can be immensely challenging and stressful for people living with diabetes. Due to the vast challenges associated with managing diabetes, it is not surprising that diabetes is associated with adverse mental health outcomes including depression, anxiety, and impaired quality of life (Ducat et al., 2014; Semenkovich et al., 2015; Smith et al., 2013). The prevalance of depression and other psychiatric disorders are significantly higher in people living with diabetes, with adults living with diabetes being three to four times more likely to be diagnosed with depression or anxiety (Ali et al., 2006; Barnard et al., 2006; Nouwen et al., 2010). In youth with type 1 diabetes, the prevalence of psychiatric disorders including depression, anxiety, and eating disorders are estimated to be as high as 30–50% (Hislop et al., 2008; Jones et al., 2000; Kovacs et al., 1997). The high rates of psychiatric disorders are especially alarming due to the association between mental illness and sub-optimal glycemic control and increased risk of life-threatening diabetes-related complications (de Groot et al., 2001; van Dooren et al., 2013; Van Tilburg et al., 2001).

Self-Compassion and Physical and Mental Health Outcomes in Diabetes Due to the challenges associated with the treatment regimen, the excessive focus on numbers and achieving "good" glycemic control, as well as the stigmatizing nature of diabetes (Liu et al., 2017), self-compassion offers an empathetic approach to dealing with the never-ending burden of striving to achieve optimal diabetes management. Self-compassion seems an especially good fit with type 1 diabetes, which is often diagnosed in childhood with peak onset in adolescence (Tuomilehto, 2013). Adolescence is in itself a period of significant physical, cognitive, and emotional change and upheaval, with an increased risk for depression, poor self-esteem, and body image (Hazen et al., 2008; Thapar et al., 2012). Teenagers living with diabetes must navigate this challenging developmental period while also living with a demanding chronic illness.

Despite the good conceptual fit and potential utility in managing diabetes, the evidence base for self-compassion is limited. A handful of observational studies have examined the association between self-compassion and mental and health outcomes in diabetes, using either the Self-Compassion Scale (SCS) (Neff, 2003) or the Self-Compassion Scale–Short Form (Raes et al., 2011). One of the first studies to examine the association between self-compassion and diabetes outcomes was a cross-sectional study conducted in New Zealand with sample of 110 adults with type 1 or type 2 diabetes (Friis et al., 2015). Friis et al. observed that higher self-compassion was correlated with lower levels of diabetes distress and depression and that self-compassion moderated the relationship between diabetes distress and glycemic control. In another cross-sectional study of 310 adults with type 1 or type 2 diabetes conducted in Australia, self-compassion was associated with improved self-care behaviors, lower glycemic control, and higher well-being (Ferrari et al., 2017). Similar findings were observed in a cross-sectional study of 176 adults in the UK with type 2 diabetes, where higher self-compassion was associated with more optimal glycemic control (Morrison et al., 2019). These findings were further supported by a larger (n = 1907) cross-sectional study of adults in Australia with type 1 or type 2 diabetes (Ventura et al., 2019), where higher self-compassion was positively associated with more optimal glycemic control, self-management behaviors, and negatively associated with diabetes distress and depression. To our knowledge, only one study has explored self-compassion and physical health outcomes in youth with diabetes. A recent cross-sectional study conducted in New Zealand with 113 young adults aged 17–25 years with type 1 diabetes (Loseby et al., 2022) explored associations between self-compassion, self-care behaviors, and glycemic control. Higher self-compassion was associated with more optimal self-care behaviors and demonstrated a trend toward more optimal glycemic control.

In comparison to the studies noted above which all used the SCS as a unitary scale, a longitudinal study of 120 adults in the US with type 2 diabetes (Kane et al., 2018) examined the predictive ability of the positive and negative domains of the SCS and their association with diabetes distress three months later. It was only the negative domain of the SCS (items assessing negative self-responding) that was independently associated with diabetes distress at the three-month follow-up. These findings add to the ongoing debate (see Chap. 1) of the role of positive versus negative self-responding and the validity of assessing the scale as a whole as well as which aspects are more important to target in psychological interventions (Kane et al., 2018).

When considering the observational research examining self-compassion and its relationship to diabetes outcomes, there are several potential mechanisms that may play a role. Studies in healthy participants demonstrate that higher self-compassion may decrease stress and reduce cortisol levels (Rockliff et al., 2008), may improve health behaviors and reduce inflammatory responses (Homan & Sirois, 2017), and in adults and youth with diabetes may buffer against diabetes distress (Friis et al., 2016) and improve self-care behaviors and lower stress leading to more optimal HbA1c (Loseby et al., 2022). However, further longitudinal and experimental studies are required to explore potential mechanisms between self-compassion and diabetes outcomes.

Diabetes-Specific Measures of Self-Compassion

General measures of self-compassion such as the SCS may not adequately measure or capture the numerous, daily diabetes-specific challenges and how people living with diabetes cope with feelings of self-criticism when these challenges arise. To try to capture diabetes-specific self-compassion, Tanenbaum et al. (2018) developed and validated the Diabetes-Specific Self-Compassion Scale (SCS-D) for adults living with diabetes and a version for parents of children with diabetes (SCS-Dp) (Tanenbaum et al., 2020), who often shoulder the burden of diabetes management. The original 26-item SCS was adapted and items which could be reworded were modified to capture diabetes-specific situations and experiences. The new 19-item SCS-D demonstrated acceptable reliability and validity in a sample of 542 adults with type 1 diabetes, with higher diabetes-specific self-compassion associated with lower diabetes distress, higher diabetes empowerment, and lower glycemic control. The subsequent parent version (SCS-Dp) was further reworded to capture how parents respond and treat themselves in the context of their child's diabetes-specific challenges. In a sample of 198 parents (Tanenbaum et al., 2020), higher parental self-compassion was negatively associated with youth diabetes distress and diabetes empowerment, but not with youth glycemic control.

Self-Compassion Interventions in Diabetes

Despite the limited but promising observational research, self-compassion interventions developed for diabetes patients are even more scarce, possibly a reflection of the intensive training required to administer standardized interventions such as Mindful Self- Compassion (MSC; Neff & Germer, 2013). To our knowledge, there are only three published interventions, two conducted in New Zealand (Boggiss et al., 2020a, b; Friis et al., 2016) and a pre-post intervention study conducted in Iran (Karami et al., 2018). The first intervention was conducted by Friis et al. (2016), who trialed MSC in a cohort of adults with type 1 and type 2 diabetes. In this wait-list controlled trial, 63 participants completed the standardized 8-week MSC intervention and were followed up for three months. Findings demonstrated significant and clinically meaningful within-group improvements in depression, diabetes distress, and glycemic control in the MSC group from baseline to three-months follow-up with no changes observed in the control group. A more recent feasibility study (Boggiss et al., 2020b) trialed an adapted 2-session version of the 8-session teen-specific version of MSC (Mindful Self-compassion for Teens) for adolescents with type 1 diabetes and disordered eating behavior to assess feasibility and acceptability of the 2-session intervention. Nineteen adolescents aged between 12 and 16 years attended the program (with ten participants completing both sessions) and reported an increased sense of common humanity, mindfulness, and coping strategies. Despite qualitative data reporting high acceptability, feasibility was limited by the reported difficulties of attending the face-to-face sessions. In a follow-up qualitative study, the majority of adolescents with type 1 diabetes described that they would prefer to learn self-compassion digitally, using a chatbot, as opposed to face-to-face or other digital options, such as a website or app (Boggiss et al., 2021). As such, a digital adaptation of this trial is currently underway aiming to deliver MSC for teens with type 1 diabetes using a chatbot, adapting the content to be delivered in 5-to-10-minute conversational lessons. Lastly, a pre-post intervention conducted by Karami et al. (2018) (n = 20) assessed self-compassion training delivered in eight sessions for adults with type 2 diabetes in Iran compared to a control group (type of control not specified). However, the small sample size and limited data reporting mean that conclusions about efficacy cannot be drawn.

Self-Compassion and Cancer

A final example of self-compassion applied in the context of CMCs can be derived from the oncology literature. Cancer is one of the most significant health problems in the world. In 2018, it was estimated 43.8 million people were living with cancer, 18.1 million cases diagnosed, and 9.6 million cancer deaths worldwide (Bray et al., 2018). The most commonly diagnosed cancers worldwide are lung (1.61 million), breast (1.38 million), and colorectal cancers (1.23 million), with lung, stomach, and liver cancers accounting for the most deaths. As the second leading cause of death globally, cancer brings an increasing and rising public health and economic burden, with the continuing growth and aging of the world's population (Bray et al., 2018).

Unsurprisingly, cancer is associated with significant physical, financial, and emotional strains to the individuals and their families and communities. A cancer diagnosis brings significant distress due to uncertainty of diagnosis, survival odds, demanding exams and treatments, confronting complex decisions, changing relationship dynamics, and financial burden (Gorman, 1998). In addition, treatments often include surgery, chemotherapy, hormonal therapy, and radiation treatments, alone or in combination, and have significant and complex side effects, such as fatigue, insomnia, nausea/ vomiting, limited physical functioning, problems with memory and attention, and often appearance altering side effects, such as scars, hair loss, skin discoloration, and loss or removal of body parts (Coates et al., 1983; Mustian et al., 2012). As a result, clinical distress is reported by approximately 35 to 45% of people diagnosed with cancer (Carlson et al., 2004; Mitchell et al., 2011; Zabora et al., 2001) and a significant number of cancer patients develop clinically significant levels of anxiety and depression (Pirl & Roth, 1999; Stark et al., 2002). Further, the comorbidities of anxiety and depression can negatively impact cancer progression, symptomology, and pain (Spiegel & Giese-Davis, 2003), negatively affect adherence to medical treatment and screening (Lerman et al., 1994; Pirl & Roth, 1999), quality of life (Spiegel & Giese-Davis, 2003), and ultimately survival rates (Pinquart & Duberstein, 2010).

While many cancer survivors return to normal functioning, some physical and psychological problems last even after treatment is finished. Long-term consequences frequently impact the quality of life and everyday functioning of many cancer survivors and can include weakness, numbness, or pain, persisting fatigue, cognitive or sexual difficulties, continuing anxiety and depression, and functional difficulties with returning to work and physical and social activities (Stein et al., 2008). In addition, fear of cancer recurrence has been shown to persist five or more years after initial diagnosis (Koch et al., 2013), with 42–70% of survivors reporting clinically significant levels of fear of recurrence (Thewes et al., 2012).

Several individual factors have been shown to predict short- and long-term adjustment and distress following a cancer diagnosis, such as coping strategies (Carver et al., 1999), social support (Carver et al., 2005; Morris & Shakespeare-Finch, 2011), illness cognitions (Morris & Shakespeare-Finch, 2011), personality traits, such as optimism (Carver et al., 2005), psychological resilience (Harms et al., 2019; Mohlin et al., 2020), perceived self-efficacy (Boehmer et al., 2007), attachment styles (Arambasic et al., 2019; Porter et al., 2012; Rodin et al., 2007), and medical factors, such as the stage of disease or choice of treatment (Glanz & Lerman, 1992; Moyer & Salovey, 1996). Significantly, social support has been shown to be a strong predictor of adjustment, with evidence of associations with higher post-traumatic growth (Morris & Shakespeare-Finch, 2011) and better subjective well-being (McDonough et al., 2014). For example, in a study of women with breast cancer, women who were socially isolated before diagnosis had a 66% increased risk of mortality (Kroenke et al., 2006). In addition, coping strategies such as acceptance and the use of humor have shown to predict lower levels of distress (Carver et al., 1999), whereas coping strategies of denial, disengagement (Carver et al., 1999), and avoidance have shown to predict poorer adjustment (Stanton et al., 2013). Regarding cognition, threatening appraisals of cancer and cognitive fusion have been shown to be strong predictors of distress and lowered quality of life after cancer (Gillanders et al., 2015). Together, these studies illustrate how individual differences in protective factors can influence not only psychological adjustment but also cancer progression.

Self-compassion has recently shown to be an important protective factor in promoting better adjustment in cancer patients generally. Studies of mixed samples of cancer patients have shown higher levels self-compassion to associate with better psychological function, shown in fewer symptoms of anxiety, depression, stress (Gillanders et al., 2015; Pinto-Gouveia et al., 2014), fatigue (Zhu et al., 2019), and higher quality of life (Gillanders et al., 2015; Pinto-Gouveia et al., 2014). In a longitudinal study of cancer patients, the level of positive self-compassion at diagnosis was the most consistent predictor of symptoms of depression, anxiety, and fatigue, over the period of their cancer treatment (Zhu et al., 2019). Similarly, higher levels of self-compassion have been shown to strongly relate to depressive symptoms and affect, with the individual facets of higher levels of isolation and reduced levels of mindfulness most significantly associated (van der Donk et al., 2020). Together, these findings suggest that for cancer patients generally, being kind and understanding toward themselves, reminding themselves of common humanity, and holding an accepting and mindful awareness of their challenges appear to be more beneficial than being self-critical of changes in physical appearance, psychosocial difficulties or life limitations imposed by cancer, feeling different or isolated, or focusing on negative experiences in the long run.

When investigating these effects across different forms of cancer, results are consistent. In lung cancer patients, higher levels of self-compassion have been demonstrated as an important predictor of mental adjustment to cancer diagnosis (Batista et al., 2015), with higher levels of self-compassion significantly related to higher social support and fewer symptoms of depression, anxiety (Batista et al., 2015), distress (Arambasic et al., 2019), shame (Siwik et al., 2019), and fatigue over time (Hsieh et al., 2019). Similarly, in breast cancer, survivor's lower levels of self-compassion were significantly related to higher levels of distress (Arambasic et al., 2019; Boyle et al., 2017; Przezdziecki & Sherman, 2016) and higher perceived negative impact of cancer (Arambasic et al., 2019). Self-compassion has also been shown to reduce negative cognitions such as threat-related rumination and worry in breast cancer survivors (Brown et al., 2020). In prostate cancer patients, results indicated that higher levels of self-compassion moderate the distress related to the emasculating aspects of their prostate cancer experience (Lennon et al., 2018).

A limited amount of evidence also suggests self-compassion can improve interpersonal functioning (Schellekens et al., 2017). In a study conducted with couples facing lung cancer, higher levels of self-compassion were associated with better communication with their partner about their emotional experience of their cancer diagnosis. Secondly, the association between self-compassion and psychological distress was shown to depend on their partner's level of self-compassion. Thus, suggesting that if one partner displays less self-compassion, the other may compensate by showing more self-compassion, alleviating the distress on both people. Together, showing the associations with self-compassion and better psychological functioning in cancer patients is not only limited to individual coping but may also be able to impact functioning in romantic relationships.

In addition, a limited number of studies have shown higher levels of self-compassion to associate with better clinical outcomes and self-care behaviors in cancer patients. In a study of breast cancer patients, self-compassion was found to moderate the relationship between stress and self-care behaviors, such as engaging in physical exercise (Abdollahi et al., 2020). Self-compassion has also been shown to associate with better sleep quality and fewer cancer-related psychological and physical symptoms in a mixed sample of cancer patients (Wan et al., 2016). Also of note, ongoing research is investigating the relationship between inflammation, postcancer symptoms, and self-compassion (Cohen et al., 2017), following evidence suggesting self-compassion may be a protective factor against inflammatory disease and stress-induced inflammation in healthy participants (Breines et al., 2014).

Furthermore, self-compassion also offers a promising approach to the body image distress that frequently arises following a cancer diagnosis (Fingeret et al., 2014) and impacts quality of life, identity, sense of self, sexuality, and further depression and anxiety (Przezdziecki & Sherman, 2016). For example, in studies in breast cancer samples, higher levels of self-compassion have shown associations with lower levels of depression, anxiety, stress, body image distress (Przezdziecki & Sherman, 2016; Todorov et al., 2019), and higher levels of body appreciation (Przezdziecki & Sherman, 2016), and hope (Todorov et al., 2019). More specifically, in a study of breast cancer survivors who underwent surgery, self-compassion was found to mediate the relationship between body image disturbance and distress, suggesting self-compassion may serve as a protective factor (Przezdziecki et al., 2013). Similarly, self-compassion has also been shown to moderate the relationship between body image disturbance and psychological distress, in those survivors who underwent mastectomy and breast reconstruction (Sherman et al., 2018).

Self-compassion may also have the protective ability on psychological distress and clinical outcomes, through the mechanisms of threat-related rumination and worry. When people face a cancer diagnosis or any type of adversity, people commonly respond by ruminating about the past or worrying about the future (S. L. Brown et al., 2020), activating our threat system and increasing psychological distress (Gilbert, 2017). In a study conducted with breast cancer patients (Brown et al., 2020), findings demonstrated worry and depressive brooding (a form of rumination) as meditators between the relationship of self-compassion subscales of mindfulness and self-kindness, and anxiety. In addition, depressive rumination was shown to mediate the relationship between mindfulness and self-kindness, and depression. These findings are supported by the results of a qualitative study that explored changes in self-compassion in cancer patients after completing a mindfulness intervention. Along with reports of improved emotion regulation, a key finding was that participants reported less engagement in thoughts and indicted decreases in rumination (L'Estrange et al., 2016). However, more research is needed in determining these causal mechanisms and in diverse cancer samples.

Self-Compassion Interventions in Cancer Recently, self-compassion interventions have been conducted in cancer, both in young adult and adult samples. In young adult cancer survivors, a combination of the 8-week MSC and "Making Friends with Yourself" (MFY) programs has been trialed using videoconferencing. The study found the program to be feasible and acceptable, and revealed significant improvements, with moderate to large effect sizes, for mindfulness, self-compassion, anxiety, depression, body image, social isolation, and posttraumatic growth, from pre- to postintervention (Campo et al., 2017). A qualitative analysis was then conducted analyzing the transcripts from the 8-week program, revealing the program addressed their key concerns of peer isolation, body image concerns, and health-related anxiety through increasing self-reliance for emotional support, common humanity, gratitude, self-kindness, acceptance, and self-care behaviors (Lathren et al., 2018).

In adults, a limited number of randomized controlled trials (RCTs) and feasibility and acceptability studies have been conducted. An RCT in a sample of colorectal cancer patients investigated the effectiveness of the 8-week MSC program. Participants showed improvements in self-compassion; however, no significant effect on anxiety, depression, or cancer-related symptomatology was seen, when compared to a wait-list control group (Ho et al., 2018). In a more recent, feasibility and acceptability study, the MSC program was shown to be acceptable and feasible among 32 cancer patients. Sixty-three percent of participants perceived an improvement in their mental well-being throughout the program and although under-powered, small-to-medium effects were observed in fear of cancer recurrence, depressive symptoms, stress, loneliness, body image satisfaction, mindfulness, and self-compassion (Brooker et al., 2020). In addition to the MSC program, a brief online self-compassion writing intervention "My Changed Body" (MyCB) has also been shown to be acceptable (Przezdziecki et al., 2016) and effective for breast cancer patients (Mifsud et al., 2021; Przezdziecki & Sherman, 2016; Sherman et al., 2018). In comparison with control conditions, improvements have been demonstrated in negative affect, self-compassion (Przezdziecki & Sherman, 2016), body image-related distress, and body appreciation (Sherman et al., 2018). Altogether, showing promise for the use of self-compassion interventions to improve well-being and body-image distress for cancer patients and survivors.

Summary and Conclusion

Taken as a whole, the existing research on self-compassion in CMCs suggests self-compassion supports adjustment and well-being in the face of the challenges inherent in living with a CMC. There are several mechanisms through which self-compassion may exert positive effects, including through reducing shame and self-stigma, improving self-care behaviors, and potentially, through modulating physiological markers of disease. It follows that self-compassion interventions may be beneficial for individuals with chronic conditions, although a recent meta-analysis found that overall, the quality of the available literature in this area was low (Mistretta & Davis, 2022). Further, it has been acknowledged that transdiagnostic approaches are useful as they target common processes across CMCs and therefore have the potential to be used by a larger proportion of the population, as well as those with multimorbidity (Finlay-Jones et al., 2020; Prentice et al., 2021). Despite this, there are limited studies of the effectiveness of self-compassion-based interventions among samples with a range of different chronic conditions. Available literature documenting study protocols (Finlay-Jones et al., 2020) and pilot studies (Brown et al., 2019) suggests that such evidence may be available in the future. Given the range of different self-compassion training approaches available, we also echo Kılıç et al.'s (2021) recommendation that future work examines which methods are most effective for promoting self-compassion among people with CMCs.

In addition to understanding the role of self-compassion across different chronic illness experiences, further insight into the benefits of self-compassion for specific groups is needed. However, the literature exploring the associations between self-compassion and adaptive outcomes in individual conditions is small, and additional research exploring the associations between self-compassion and physiological disease indices is warranted to understand the psychophysiological profile of self-compassion in people with different types of CMC. Integration of such markers into outcome measurement in experimental studies of self-compassion-based interventions is also an important direction for future research. Furthermore, given preliminary evidence of the association between self-compassion and putative mechanisms such as health behaviors, shame, and self-stigma, future work should determine whether such pathways mediate intervention effects.

Given that individuals with CMCs face several barriers to intervention access, consideration must be given to how such interventions can be optimized for individuals with accessibility requirements. Efforts to provide web-based (Finlay-Jones et al., 2020) and videoconferencing (Lathren et al., 2018) interventions are an important contribution; however, such approaches may not be suitable for people with certain types of functional impairment. Involving those living with different types of CMCs in the co-design of self-compassion intervention approaches is critical to developing accessible and engaging supports for this population. Insights into the meaning of self-compassion for individuals living with CMCs would complement this work and could be gathered through qualitative studies exploring how people with different levels of self-compassion navigate the chronic illness experience. Further, consideration should be given to how self-compassion cultivation can be supported through the existing health, education, and community support systems in which those living with CMCs already engage. For example, there is the opportunity to examine whether supporting caregivers of individuals with CMCs to practice self-compassion may simultaneously support the cultivation of self-compassion in those for whom they provide care. This potentially has the dual benefit of potentially supporting well-being in caregivers, who are an at-risk group for psychological distress (Cousineau et al., 2019), as well as supporting more sustained intervention effects in people living with CMCs.

References

Abdollahi, A., Taheri, A., & Allen, K. A. (2020). Self-compassion moderates the perceived stress and self-care behaviors link in women with breast cancer. *Psycho-Oncology, 29*(5), 927–933. https://doi.org/10.1002/pon.5369

Adams, J. S., Chien, A. T., & Wisk, L. E. (2019). Mental illness among youth with chronic physical conditions. *Pediatrics, 144*(1). https://doi.org/10.1542/peds.2018-1819

Ali, S., Stone, M., Peters, J., Davies, M., & Khunti, K. (2006). The prevalence of co-morbid depression in adults with Type 2 diabetes: A systematic review and meta-analysis. *Diabetic Medicine, 23*(11), 1165–1173. https://doi.org/10.1111/j.1464-5491.2006.01943.x

Ambridge, J., Fleming, P., & Henshall, L. (2020). The influence of self-compassion on perceived responsibility and shame following acquired brain injury. *Brain Injury, 34*(7), 945–957. https://doi.org/10.1080/02699052.2020.1763466

Anderson, R. J., Freedland, K. E., Clouse, R. E., & Lustman, P. J. (2001). The prevalence of comorbid depression in adults with diabetes: A meta-analysis. *Diabetes Care, 24*(6), 1069–1078. https://doi.org/10.2337/diacare.24.6.1069

Arambasic, J., Sherman, K. A., Elder, E., & Australia, B. C. N. (2019). Attachment styles, self-compassion, and psychological adjustment in long-term breast cancer survivors. *Psycho-Oncology, 28*(5), 1134–1141. https://doi.org/10.1002/pon.5068

Baker, D. A., Caswell, H. L., & Eccles, F. J. R. (2019). Self-compassion and depression, anxiety, and resilience in adults with epilepsy. *Epilepsy & Behavior, 90*, 154–161. https://doi.org/10.1016/j.yebeh.2018.11.025

Barnard, K., Skinner, T., & Peveler, R. (2006). The prevalence of co-morbid depression in adults with Type 1 diabetes: Systematic literature review. *Diabetic Medicine, 23*(4), 445–448. https://doi.org/10.1111/j.1464-5491.2006.01814.x

Barnes, A., Adam, M. E., Eke, A. O., & Ferguson, L. J. (2018). Exploring the emotional experiences of young women with chronic pain: The potential role of self-compassion. *Journal of Health Psychology, 26*(3), 367–377. https://doi.org/10.1177/1359105318816509

Batista, R., Cunha, M., Galhardo, A., Couto, M., & Massano-Cardoso, I. (2015). Psychological adjustment to lung cancer: The role of self-compassion and social support. *European Psychiatry, 30*, 1511. https://doi.org/10.1016/S0924-9338(15)31168-8

Bennett, K. K., Compas, B. E., Beckjord, E., & Glinder, J. G. (2005). Self-blame and distress among women with newly diagnosed breast cancer. *Journal of Behavioral Medicine, 28*(4), 313–323. https://doi.org/10.1007/s10865-005-9000-0

Boehmer, S., Luszczynska, A., & Schwarzer, R. (2007). Coping and quality of life after tumor surgery: Personal and social resources promote different domains of quality of life. *Anxiety, Stress, & Coping, 20*(1), 61–75. https://doi.org/10.1080/10615800701195439

Boggiss, A. L., Consedine, N., Jefferies, C., Bluth, K., Hofman, P., & Serlachius, A. (2020a). Protocol for a feasibility study: A brief self-compassion intervention for adolescents with type 1 diabetes and disordered eating. *BMJ Open, 10*(2). https://doi.org/10.1136/bmjopen-2019-034452

Boggiss, A. L., Consedine, N., Schache, K., Jefferies, C., Bluth, K., Hofman, P., et al. (2020b). A brief self-compassion intervention for adolescents with type 1 diabetes and disordered eating: A feasibility study. *Diabetic Medicine : A Journal of the British Diabetic Association, 37*(11), 1854–1860. https://doi.org/10.1111/dme.14352

Boggiss, A. L., Consedine, N. S., Schache, K. R., Wallace-Boyd, K., Cao, N., Hofman, P. L., & Serlachius, A. S. (2021). Exploring the views of adolescents with type 1 diabetes on digital mental health interventions: What functionality and content do they want? *Diabetic Medicine, 38*(11), e14591. https://doi.org/10.1111/dme.14591

Boyle, C. C., Stanton, A. L., Ganz, P. A., Crespi, C. M., & Bower, J. E. (2017). Improvements in emotion regulation following mindfulness meditation: Effects on depressive symptoms and perceived stress in younger breast cancer survivors. *Journal of Consulting and Clinical Psychology, 85*(4), 397–402. https://doi.org/10.1037/ccp0000186

Bray, F., Ferlay, J., Soerjomataram, I., Siegel, R. L., Torre, L. A., & Jemal, A. (2018). Global cancer statistics 2018: GLOBOCAN estimates of incidence and mortality worldwide for 36 cancers in 185 countries. *CA: A Cancer Journal for Clinicians, 68*(6), 394–424. https://doi.org/10.3322/caac.21492

Breines, J. G., Thoma, M. V., Gianferante, D., Hanlin, L., Chen, X., & Rohleder, N. (2014). Self-compassion as a predictor of interleukin-6 response to acute psychosocial stress. *Brain, Behavior, and Immunity, 37*, 109–114. https://doi.org/10.1016/j.bbi.2013.11.006

Brion, J. M., Leary, M. R., & Drabkin, A. S. (2014). Self-compassion and reactions to serious illness: The case of HIV. *Journal of Health Psychology, 19*(2), 218–229. https://doi.org/10.1177/1359105312467391

Brooker, J., Julian, J., Millar, J., Prince, H. M., Kenealy, M., Herbert, K., et al. (2020). A feasibility and acceptability study of an adaptation of the Mindful Self-Compassion program for adult cancer patients. *Palliative & Supportive Care, 18*(2), 130–140. https://doi.org/10.1017/S1478951519000737

Brown, L., Karmakar, C., Flynn, M., Motin, M. A., Palaniswami, M., Celano, C. M., et al. (2019). A self-compassion group intervention for patients living with chronic medical illness: Treatment development and feasibility study. *The Primary Care Companion for CNS Disorders, 21*(5). https://doi.org/10.4088/PCC.19m02470

Brown, S. L., Hughes, M., Campbell, S., & Cherry, M. G. (2020). Could worry and rumination mediate relationships between self-compassion and psychological distress in breast cancer survivors? *Clinical Psychology & Psychotherapy, 27*(1), 1–10. https://doi.org/10.1002/cpp.2399

Bunzli, S., Smith, A., Schütze, R., & O'Sullivan, P. (2015). Beliefs underlying pain-related fear and how they evolve: A qualitative investigation in people with chronic back pain and high pain-related fear. *BMJ Open, 5*(10). https://doi.org/10.1136/bmjopen-2015-008847

Campo, R. A., Bluth, K., Santacroce, S. J., Knapik, S., Tan, J., Gold, S., et al. (2017). A mindful self-compassion videoconference intervention for nationally recruited post-treatment young adult cancer survivors: Feasibility, acceptability, and psychosocial outcomes. *Supportive Care in Cancer, 25*(6), 1759–1768. https://doi.org/10.1007/s00520-017-3586-y

Carlson, L., Angen, M., Cullum, J., Goodey, E., Koopmans, J., Lamont, L., et al. (2004). High levels of untreated distress and fatigue in cancer patients. *British Journal of Cancer, 90*(12), 2297–2304. https://doi.org/10.1038/sj.bjc.6601887

Carvalho, S. A., Pinto-Gouveia, J., Gillanders, D., & Castilho, P. (2018). Pain and depressive symptoms: Exploring cognitive fusion and self-compassion in a moderated mediation model (General Psychology 2100). *The Journal of Psychology: Interdisciplinary and Applied, 153*(2), 173–186. https://doi.org/10.1080/00223980.2018.1507990

Carvalho, S. A., Trindade, I. A., Gillanders, D., Pinto-Gouveia, J., & Castilho, P. (2019). Self-compassion and depressive symptoms in chronic pain (CP): A 1-year longitudinal study. *Mindfulness, 11*(3), 709–719. https://doi.org/10.1007/s12671-019-01292-7

Carver, C. S., Pozo, C., Harris, S. D., Noriega, V., Scheier, M. F., Robinson, D. S., Ketcham, A. S., Moffat, F. L., & Clark, K. C. (1999). How coping mediates the effect of optimism on distress: A study of women with early stage breast cancer. *Journal of Personality and Social Psychology, 65*(2), 375–390. https://doi.org/10.1037/10338-005

Carver, C. S., Smith, R. G., Antoni, M. H., Petronis, V. M., Weiss, S., & Derhagopian, R. P. (2005). Optimistic personality and psychosocial well-being during treatment predict psychosocial well-being among long-term survivors of breast cancer. *Health Psychology, 24*(5), 508. https://doi.org/10.1037/0278-6133.24.5.508

Casati, J., Toner, B. B., De Rooy, E. C., Drossman, D. A., & Maunder, R. G. (2000). Concerns of patients with inflammatory bowel disease : A review of emerging themes. *Digestive Diseases and Sciences, 45*(1), 26–31. https://doi.org/10.1023/A:1005492806777

Chang, E. C., Lucas, A. G., Chang, O. D., Angoff, H. D., Li, M., Duong, A. H., et al. (2019). Relationship between future orientation and pain severity in fibromyalgia patients: Self-compassion as a coping mechanism. *Social Work, 64*(3), 253–258. https://doi.org/10.1093/sw/swz013

Chapel, J. M., Ritchey, M. D., Zhang, D., & Wang, G. (2017). Prevalence and medical costs of chronic diseases among adult Medicaid beneficiaries. *American Journal of Preventive Medicine, 53*(6), S143–S154. https://doi.org/10.1016/j.amepre.2017.07.019

Chatterjee, S., Khunti, K., & Davies, M. J. (2017). Type 2 diabetes. *The Lancet, 389*(10085), 2239–2251. https://doi.org/10.1016/S0140-6736(17)30058-2

Cho, N., Shaw, J., Karuranga, S., Huang, Y., da Rocha Fernandes, J., Ohlrogge, A., & Malanda, B. (2018). IDF Diabetes Atlas: Global estimates of diabetes prevalence for 2017 and projections for 2045. *Diabetes Research and Clinical Practice, 138*, 271–281. https://doi.org/10.1016/j.diabres.2018.02.023

Clarke, E. N., Thompson, A. R., & Norman, P. (2020). Depression in people with skin conditions: The effects of disgust and self-compassion. *British Journal of Health Psychology, 25*(3), 540–557. https://doi.org/10.1111/bjhp.12421

Clegg, S., Sirois, F. M., & Reuber, M. (2019). Self-compassion and adjustment in epilepsy and psychogenic nonepileptic seizures. *Epilepsy & Behavior, 100*, 106490. https://doi.org/10.1016/j.yebeh.2019.106490

Coates, A., Abraham, S., Kaye, S. B., Sowerbutts, T., Frewin, C., Fox, R., & Tattersall, M. (1983). On the receiving end—patient perception of the side-effects of cancer chemotherapy. *European Journal of Cancer and Clinical Oncology, 19*(2), 203–208. https://doi.org/10.1016/0277-5379(83)90418-2

Cohen, M., Levkovich, I., Fried, G., & Pollack, S. (2017). Post-treatment symptoms and cytokine levels in breast cancer patients: The moderating role of self-compassion and optimism. *Psycho-Oncology, 26*, 75.

Costa, J., & Pinto-Gouveia, J. (2011). Acceptance of pain, self-compassion and psychopathology: Using the Chronic Pain Acceptance Questionnaire to identify patients' subgroups. *Clinical Psychology & Psychotherapy, 18*(4), 292–302. https://doi.org/10.1002/cpp.718

Costa, J., & Pinto-Gouveia, J. (2013). Experiential avoidance and self-compassion in chronic pain. *Journal of Applied Social Psychology, 43*(8), 1578–1591. https://doi.org/10.1111/jasp.12107

Cousineau, T. M., Hobbs, L. M., & Arthur, K. C. (2019). The role of compassion and mindfulness in building parental resilience when caring for children with chronic conditions: A conceptual model. *Frontiers in Psychology, 10*, 1602. https://doi.org/10.3389/fpsyg.2019.01602

Crombez, W. S. G., Eccleston, W. S. C., Van Damme, W. S. S., Vlaeyen, W. S. J., & Karoly, W. S. P. (2012). Fear-avoidance model of chronic pain: The next generation. *The Clinical Journal of Pain, 28*(6), 475–483. https://doi.org/10.1097/AJP.0b013e3182385392

D'Amico, F., Feliu-Soler, A., Montero-Marin, J., Penarrubia-Maria, M. T., Navarro-Gil, M., Van Gordon, W., et al. (2020). Cost-utility of attachment-based compassion therapy (ABCT) for fibromyalgia compared to relaxation: A pilot randomized controlled trial. *Journal of Clinical Medicine, 9*(3), 726. https://doi.org/10.3390/jcm9030726

Dall'Oglio, I., Gasperini, G., Carlin, C., Biagioli, V., Gawronski, O., Spitaletta, G., Grimaldi Capitello, T., Salata, M., Vanzi, V., Rocco, G., Tiozzo, E., Vellone, E., & Raponi, M. (2021). Self-care in pediatric patients with chronic conditions: A systematic review of theoretical models. *International Journal of Environmental Research and Public Health, 18*(7). https://doi.org/10.3390/ijerph18073513

Daré, L. O., Bruand, P.-E., Gérard, D., Marin, B., Lameyre, V., Boumédiène, F., et al. (2019). Co-morbidities of mental disorders and chronic physical diseases in developing and emerging countries: A meta-analysis. *BMC Public Health, 19*(1), 304. https://doi.org/10.1186/s12889-019-6623-6

Dawson Rose, C., Webel, A., Sullivan, K. M., Cuca, Y. P., Wantland, D., Johnson, M. O., Brion, J., Portillo, C. J., Corless, I. B., Voss, J., Chen, W.-T., Phillips, J. C., Tyer-Viola, L., Marta, R.-M., Nicholas, P. K., Nokes, K., Kemppainen, J., Sefcik, E., Eller, L. S., et al. (2014). Self-compassion and risk behavior among people living with HIV/AIDS. *Research in Nursing & Health, 37*(2), 98–106. https://doi.org/10.1002/nur.21587

de Groot, M., Anderson, R., Freedland, K. E., Clouse, R. E., & Lustman, P. J. (2001). Association of depression and diabetes complications: A meta-analysis. *Psychosomatic Medicine, 63*(4), 619–630.

Dibley, L., Norton, C., & Whitehead, E. (2018). The experience of stigma in inflammatory bowel disease: An interpretive (hermeneutic) phenomenological study. *Journal of Advanced Nursing, 74*(4), 838–851. https://doi.org/10.1111/jan.13492

Diabetes Control and Complications Trial Research Group (1993). The effect of inten-sive treatment of diabetes on the development and progression of long-term compli-cations in insulin-dependent diabetesmellitus. *The New England Journal of Medicine, 329*, 977–986.

Ducat, L., Philipson, L. H., & Anderson, B. J. (2014). The mental health comorbidities of diabetes. *JAMA, 312*(7), 691–692. https://doi.org/10.1001/jama.2014.8040

Edwards, K. A., Pielech, M., Hickman, J., Ashworth, J., Sowden, G., & Vowles, K. E. (2019). The relation of self-compassion to functioning among adults with chronic pain. *European Journal of Pain (United Kingdom), 23*(8), 1538–1547. https://doi.org/10.1002/ejp.1429

Ferrari, M., Dal Cin, M., & Steele, M. (2017). Self-compassion is associated with optimum self-care behavior, medical outcomes and psychological well-being in a cross-sectional sample of adults with diabetes. *Diabetic Medicine, 34*(11), 1546–1553. https://doi.org/10.1111/dme.13451

Fingeret, M. C., Teo, I., & Epner, D. E. (2014). Managing body image difficulties of adult cancer patients: Lessons from available research. *Cancer, 120*(5), 633–641. https://doi.org/10.1002/cncr.28469

Finlay-Jones, A. (2017). The relevance of self-compassion as an intervention target in mood and anxiety disorders: A narrative review based on an emotion regulation framework. *Clinical Psychologist, 21*, 90–103. https://doi.org/10.1111/cp.12131

Finlay-Jones, A., Boyes, M., Perry, Y., Sirois, F. M., Lee, R., & Rees, C. S. (2020). Online self-compassion training to improve the wellbeing of youth with chronic medical conditions: Protocol for a randomised control trial. *BMC Public Health, 20*(1), 106. https://doi.org/10.1186/s12889-020-8226-7

Friis, A., Johnson, M., Cutfield, R., & Consedine, N. (2015). Does kindness matter? Self-compassion buffers the negative impact of diabetes-distress on HbA1c. *Diabetic Medicine, 32*(12), 1634–1640. https://doi.org/10.1111/dme.12774

Friis, A., Johnson, M., Cutfield, R., & Consedine, N. (2016). Kindness matters: A randomized controlled trial of a mindful self-compassion intervention improves depression, distress, and HbA1c among patients with diabetes. *Diabetes Care, 39*(11), 1963–1971. https://doi.org/10.2337/dc16-0416

Gedik, Z., & Idiman, E. (2020). Health-related quality of life in multiple sclerosis: Links to mental health, self-esteem, and self-compassion. *Dusunen Adam: The Journal of Psychiatry and Neurological Sciences, 33*, 59–70. https://doi.org/10.14744/DAJPNS.2019.00061

Gilbert, P. (2017). A brief outline of the evolutionary approach for compassion focused therapy. *EC Psychology and Psychiatry, 3*(6), 218–227.

Gillanders, D. T., Sinclair, A. K., MacLean, M., & Jardine, K. (2015). Illness cognitions, cognitive fusion, avoidance and self-compassion as predictors of distress and quality of life in a heterogeneous sample of adults, after cancer. *Journal of Contextual Behavioral Science, 4*(4), 300–311. https://doi.org/10.1016/j.jcbs.2015.07.003

Glanz, K., & Lerman, C. (1992). Psychosocial impact of breast cancer: A critical review. *Annals of Behavioral Medicine, 14*(3), 204–212. https://doi.org/10.1093/abm/14.3.204

Gorman, L. (1998). The psychosocial impact of cancer on the individual, family, and society. *Psychosocial Nursing Care Along the Cancer Continuum*, 3–25.

Hardelid, P., Dattani, N., & Gilbert, R. (2014). Estimating the prevalence of chronic conditions in children who die in England, Scotland and Wales: A data linkage cohort study. *BMJ Open, 4*, e005331. https://doi.org/10.1136/bmjopen-2014-005331

Harms, C. A., Cohen, L., Pooley, J. A., Chambers, S. K., Galvão, D. A., & Newton, R. U. (2019). Quality of life and psychological distress in cancer survivors: The role of psycho-social resources for resilience. *Psycho-Oncology, 28*(2), 271–277. https://doi.org/10.1002/pon.4934

Harrison, C., Henderson, J., Miller, G., & Britt, H. (2017). The prevalence of diagnosed chronic conditions and multimorbidity in Australia: A method for estimating population prevalence from general practice patient encounter data. *PLoS One, 12*(3), e0172935. https://doi.org/10.1371/journal.pone.0172935

Hazen, E., Schlozman, S., & Beresin, E. (2008). Adolescent psychological development: A review. *Pediatrics in Review, 29*, 161–168. https://doi.org/10.1542/pir.29-5-161

Hislop, A. L., Fegan, P. G., Schlaeppi, M. J., Duck, M., & Yeap, B. B. (2008). Prevalence and associations of psychological distress in young adults with Type 1 diabetes. *Diabetic Medicine, 25*(1), 91–96. https://doi.org/10.1111/j.1464-5491.2007.02310.x

Ho, R. T., Wan, A. H., Yau, J. C., Chan, C. L., Chan, S. M., Chung, K. F., & Ng, S. N. (2018). A Mindful self-compassion pathway to cancer survivorship: Preliminary results from a randomized-controlled trial of colorectal cancer patients. *Psycho-Oncology, 27*, 159–160.

Holman, R. R., Paul, S. K., Bethel, M. A., Matthews, D. R., & Neil, H. A. (2008). 10-year follow-up of intensive glucose control in type 2 diabetes. *The New England Journal of Medicine, 359*(15), 1577–1589. https://doi.org/10.1056/NEJMoa0806470. Epub 2008 Sep 10. PMID: 18784090.

Homan, K. J., & Sirois, F. M. (2017). Self-compassion and physical health: Exploring the roles of perceived stress and health-promoting behaviors. *Health Psychology Open, 4*(2), 2055102917729542. https://doi.org/10.1177/2055102917729542

Hsieh, C. C., Yu, C. J., Chen, H. J., Chen, Y. W., Chang, N. T., & Hsiao, F. H. (2019). Dispositional mindfulness, self-compassion, and compassion from others as moderators between stress and depression in caregivers of patients with lung cancer. *Psycho-Oncology, 28*(7), 1498–1505. https://doi.org/10.1002/pon.5106

Hughes, M., Brown, S. L., Campbell, S., Dandy, S., & Cherry, M. G. (2021). Self-compassion and anxiety and depression in chronic physical illness populations: A systematic review. *Mindfulness, 12*(7), 1597–1610. https://doi.org/10.1007/s12671-021-01602-y

Hvidberg, M. F., Johnsen, S. P., Davidsen, M., & Ehlers, L. (2020). A nationwide study of prevalence rates and characteristics of 199 chronic conditions in Denmark. *PharmacoEconomics – Open, 4*(2), 361–380. https://doi.org/10.1007/s41669-019-0167-7

Inwood, E., & Ferrari, M. (2018). Mechanisms of change in the relationship between self-compassion, emotion regulation, and mental health: A systematic review. *Applied Psychology: Health and Well-Being, 10*(2), 215–235. https://doi.org/10.1111/aphw.12127

Joachim, G., & Acorn, S. (2000). Living with chronic illness: The interface of stigma and normalization. *Canadian Journal of Nursing Research, 32*(3), 37–48.

Jones, J. M., Lawson, M. L., Daneman, D., Olmsted, M. P., & Rodin, G. (2000). Eating disorders in adolescent females with and without type 1 diabetes: Cross sectional study. *British Medical Journal, 320*(7249), 1563–1566. https://doi.org/10.1136/bmj.320.7249.1563

Kane, N., Hoogendoorn, C., Tanenbaum, M., & Gonzalez, J. (2018). Physical symptom complaints, cognitive emotion regulation strategies, self-compassion and diabetes distress among adults with Type 2 diabetes. *Diabetic Medicine, 35*(12), 1671–1677. https://doi.org/10.1111/dme.13830

Karami, J., Rezaei, M., Karimi, P., & Rafiee, Z. (2018). Effectiveness of self-compassion intervention training on glycemic control in patients with diabetes. *Journal of Kermanshah University of Medical Sciences, 22*(2). https://doi.org/10.5812/jkums.83282

Katon, W., Lin, E. H., & Kroenke, K. (2007). The association of depression and anxiety with medical symptom burden in patients with chronic medical illness. *General Hospital Psychiatry, 29*(2), 147–155. https://doi.org/10.1016/j.genhosppsych.2006.11.005

Kılıç, A., Hudson, J., McCracken, L. M., Ruparelia, R., Fawson, S., & Hughes, L. D. (2021). A systematic review of the effectiveness of self-compassion-related interventions for individuals with chronic physical health conditions. *Behavior Therapy, 52*(3), 607–625. https://doi.org/10.1016/j.beth.2020.08.001

Koch, L., Jansen, L., Brenner, H., & Arndt, V. (2013). Fear of recurrence and disease progression in long-term (≥ 5 years) cancer survivors – A systematic review of quantitative studies. *Psycho-Oncology, 22*(1), 1–11. https://doi.org/10.1002/pon.3022

Kovacs, M., Goldston, D., Obrosky, D. S., & Bonar, L. K. (1997). Psychiatric disorders in youths with IDDM: Rates and risk factors. *Diabetes Care, 20*(1), 36–44. https://doi.org/10.2337/diacare.20.1.36

Kroenke, C. H., Kubzansky, L. D., Schernhammer, E. S., Holmes, M. D., & Kawachi, I. (2006). Social networks, social support, and survival after breast cancer diagnosis. *Journal of Clinical Oncology, 24*(7), 1105–1111. https://doi.org/10.1200/JCO.2005.04.2846

L'Estrange, K., Timulak, L., Kinsella, L., & D'Alton, P. (2016). Experiences of changes in self-compassion following mindfulness-based intervention with a cancer population. *Mindfulness, 7*(3), 734–744. https://doi.org/10.1007/s12671-016-0513-0

Lambert, V., & Keogh, D. (2015). Striving to live a normal life: A review of children and young people's experience of feeling different when living with a long-term condition. *Journal of Pediatric Nursing, 30*(1), 63–77. https://doi.org/10.1177/1367493512473852

Lansing, A. H., & Berg, C. A. (2014). Topical review: Adolescent self-regulation as a foundation for chronic illness self-management. *Journal of Pediatric Psychology, 39*(10), 1091–1096. https://doi.org/10.1093/jpepsy/jsu067

Lathren, C., Bluth, K., Campo, R., Tan, W., & Futch, W. (2018). Young adult cancer survivors' experiences with a mindful self-compassion (MSC) video-chat intervention: A qualitative analysis. *Self and Identity, 17*(6), 646–665. https://doi.org/10.1080/15298868.2018.1451363

Lennon, J., Hevey, D., & Kinsella, L. (2018). Gender role conflict, emotional approach coping, self-compassion, and distress in prostate cancer patients: A model of direct and moderating effects. *Psycho-Oncology, 27*(8), 2009–2015. https://doi.org/10.1002/pon.4762

Lerman, C., Kash, K., & Stefanek, M. E. (1994). Younger women at increased risk for breast cancer: Perceived risk, psychological well-being, and surveillance behavior. *Journal of the National Cancer Institute. Monographs*, (16), 171–176.

Liu, N. F., Brown, A. S., Close, K. L., Younge, M. F., Wood, R., & Guzman, S. J. (2017). Stigma in people with type 1 or type 2 diabetes. *Clinical Diabetes, 35*(1), 27–34. https://doi.org/10.2337/cd16-0020

Loseby, P., Schache, K., Cavadino, A., Young, S., Hofman, P. L., & Serlachius, A. (2022). The role of protective psychological factors, self-care behaviors, and HbA1c in young adults with type 1 diabetes. *Pediatric Diabetes, 23*(3), 380–389. https://doi.org/10.1111/pedi.13306. Epub 2022 Jan 7. PMID: 34967089.

McDonough, M. H., Sabiston, C. M., & Wrosch, C. (2014). Predicting changes in post-traumatic growth and subjective well-being among breast cancer survivors: The role of social support and stress. *Psycho-Oncology, 23*(1), 114–120. https://doi.org/10.1002/pon.3380

Mifsud, A., Pehlivan, M. J., Fam, P., O'Grady, M., van Steensel, A., Elder, E., Gilchrist, J., & Sherman, K. A. (2021). Feasibility and pilot study of a brief self-compassion intervention addressing body image distress in breast cancer survivors. *Health Psychology and Behavioral Medicine, 9*(1), 498–526. https://doi.org/10.1080/21642850.2021.1929236

Mistretta, E. G., & Davis, M. C. (2022). Meta-analysis of self-compassion interventions for pain and psychological symptoms among adults with chronic illness. *Mindfulness, 13*(2), 267–284. https://doi.org/10.1007/s12671-021-01766-7

Mitchell, A. J., Hussain, N., Grainger, L., & Symonds, P. (2011). Identification of patient-reported distress by clinical nurse specialists in routine oncology practice: A multicentre UK study. *Psycho-Oncology, 20*(10), 1076–1083. https://doi.org/10.1002/pon.1815

Mohlin, Å., Axelsson, U., Bendahl, P.-O., Borrebaeck, C., Hegardt, C., Johnsson, P., Rahm Hallberg, I., & Rydén, L. (2020). Psychological resilience and health-related quality of life in swedish women with newly diagnosed breast cancer. *Cancer Management and Research, 12*, 12041–12051. https://doi.org/10.2147/CMAR.S268774

Montero-Marín, J., Navarro-Gil, M., Puebla-Guedea, M., Luciano, J. V., Van Gordon, W., Shonin, E., & García-Campayo, J. (2017). Efficacy of "attachment-based compassion therapy" in the treatment of fibromyalgia: A randomized controlled trial. *Frontiers in Psychiatry, 16*(8), 307. https://doi.org/10.3389/fpsyt.2017.00307

Montero-Marin, J., Van Gordon, W., Shonin, E., Navarro-Gil, M., Virginia, G., Yolanda, L.-d.-H., Luciano, J. V., & Garcia-Campayo, J. (2020). Attachment-based compassion therapy for ameliorating fibromyalgia: Mediating role of mindfulness and self-compassion. *Mindfulness, 11*(3), 816–828. https://doi.org/10.1007/s12671-019-01302-8

Morris, B. A., & Shakespeare-Finch, J. (2011). Rumination, post-traumatic growth, and distress: Structural equation modelling with cancer survivors. *Psycho-Oncology, 20*(11), 1176–1183. https://doi.org/10.1002/pon.1827

Morrison, A. E., Zaccardi, F., Chatterjee, S., Brady, E., Doherty, Y., Robertson, N., Hadjiconstantinou, M., Daniels, L., Hall, A., Khunti, K., & Davies, M. J. (2019). Self-compassion, metabolic control and health status in individuals with type 2 diabetes: A UK observational study. *Experimental and Clinical Endocrinology & Diabetes, 129*(06), 413–419. https://doi.org/10.1055/a-0897-3772

Moussavi, S., Chatterji, S., Verdes, E., Tandon, A., Patel, V., & Ustun, B. (2007). Depression, chronic diseases, and decrements in health: Results from the World Health Surveys. *Lancet, 370*(9590), 851–858. https://doi.org/10.1016/S0140-6736(07)61415-9

Moyer, A., & Salovey, P. (1996). Psychosocial sequelae of breast cancer and its treatment. *Annals of Behavioral Medicine, 18*(2), 110–125. https://doi.org/10.1007/BF02909583

Mustian, K. M., Sprod, L. K., Janelsins, M., Peppone, L. J., & Mohile, S. (2012). Exercise recommendations for cancer-related fatigue, cognitive impairment, sleep problems, depression, pain, anxiety, and physical dysfunction: A review. *Oncology & Hematology Review, 8*(2), 81. https://doi.org/10.17925/ohr.2012.08.2.81

Neff, K. D. (2003). The development and validation of a scale to measure self-compassion. *Self and Identity, 2*(3), 223–250. https://doi.org/10.1080/15298860390129863

Neff, K. D., & Germer, C. K. (2013). A pilot study and randomized controlled trial of the mindful self-compassion program. *Journal of Clinical Psychology, 69*(1), 28–44. https://doi.org/10.1002/jclp.21923

Nery-Hurwit, M., Yun, J., & Ebbeck, V. (2018). Examining the roles of self-compassion and resilience on health-related quality of life for individuals with Multiple Sclerosis. *Disability and Health Journal, 11*(2), 256–261. https://doi.org/10.1016/j.dhjo.2017.10.010

Newman, D., Tong, M., Levine, E., & Kishore, S. (2020). Prevalence of multiple chronic conditions by US state and territory, 2017. *PLoS One, 15*(5), e0232346. https://doi.org/10.1371/journal.pone.0232346

Nouwen, A., Winkley, K., Twisk, J., Lloyd, C. E., Peyrot, M., Ismail, K., & Pouwer, F. (2010). Type 2 diabetes mellitus as a risk factor for the onset of depression: A systematic review and meta-analysis. *Diabetologia, 53*(12), 2480–2486. https://doi.org/10.1007/s00125-010-1874-x

Phelan, S. M., Griffin, J. M., Jackson, G. L., Zafar, S. Y., Hellerstedt, W., Stahre, M., Nelson, D., Zullig, L. L., Burgess, D. J., & van Ryn, M. (2013). Stigma, perceived blame, self-blame, and depressive symptoms in men with colorectal cancer. *Psycho-Oncology, 22*(1), 65–73. https://doi.org/10.1002/pon.2048

Pinquart, M., & Duberstein, P. (2010). Depression and cancer mortality: A meta-analysis. *Psychological Medicine, 40*(11), 1797–1810. https://doi.org/10.1017/S0033291709992285

Pinto-Gouveia, J., Duarte, C., Matos, M., & Fráguas, S. (2014). The protective role of self-compassion in relation to psychopathology symptoms and quality of life in chronic and in cancer patients. *Clinical Psychology & Psychotherapy, 21*(4), 311–323. https://doi.org/10.1002/cpp.1838

Pirl, W. F., & Roth, A. J. (1999). Diagnosis and treatment of depression in cancer patients. *Depression, 13*(9).

Porter, L. S., Keefe, F. J., Davis, D., Rumble, M., Scipio, C., & Garst, J. (2012). Attachment styles in patients with lung cancer and their spouses: Associations with patient and spouse adjustment. *Supportive Care in Cancer, 20*(10), 2459–2466. https://doi.org/10.1007/s00520-011-1367-6

Prentice, K., Rees, C., & Finlay-Jones, A. (2021). Self-compassion, wellbeing, and distress in adolescents and young adults with chronic medical conditions: The mediating role of emotion regulation difficulties. *Mindfulness, 12*(9), 2241–2252. https://doi.org/10.1007/s12671-021-01685-7

Przezdziecki, A., & Sherman, K. A. (2016). Modifying affective and cognitive responses regarding body image difficulties in breast cancer survivors using a self-compassion-based writing intervention. *Mindfulness, 7*(5), 1142–1155. https://doi.org/10.1007/s12671-016-0557-1

Przezdziecki, A., Sherman, K. A., Baillie, A., Taylor, A., Foley, E., & Stalgis-Bilinski, K. (2013). My changed body: Breast cancer, body image, distress and self-compassion. *Psycho-Oncology, 22*(8), 1872–1879. https://doi.org/10.1002/pon.3230

Przezdziecki, A., Alcorso, J., & Sherman, K. A. (2016). My changed body: Background, development and acceptability of a self-compassion based writing activity for female survivors of breast cancer. *Patient Education and Counseling, 99*(5), 870–874. https://doi.org/10.1016/j.pec.2015.12.011

Purdie, F., & Morley, S. (2015). Self-compassion, pain, and breaking a social contract. *Pain, 156*(11), 2354–2363. https://doi.org/10.1097/j.pain.0000000000000287

Raes, F., Pommier, E., Neff, K. D., & Van Gucht, D. (2011). Construction and factorial validation of a short form of the self-compassion scale. *Clinical Psychology & Psychotherapy, 218*(3), 250–255. https://doi.org/10.1002/cpp.702

Rockliff, H., Gilbert, P., McEwan, K., Lightman, S., & Glover, D. (2008). A pilot exploration of heart rate variability and salivary cortisol responses to compassion-focused imagery. *Journal of Clinical Neuropsychiatry, 5*(3), 132–139.

Rodin, G., Walsh, A., Zimmermann, C., Gagliese, L., Jones, J., Shepherd, F. A., Moore, M., Braun, M.,

Donner, A., & Mikulincer, M. (2007). The contribution of attachment security and social support to depressive symptoms in patients with metastatic cancer. *Psycho-Oncology: Journal of the Psychological, Social and Behavioral Dimensions of Cancer, 16*(12), 1080–1091. https://doi.org/10.1002/pon.1186
Schabert, J., Browne, J. L., Mosely, K., & Speight, J. (2013). Social stigma in diabetes: A framework to understand a growing problem for an increasing epidemic. *The Patient – Patient-Centered Outcomes Research, 6*(1), 1–10. https://doi.org/10.1007/s40271-012-0001-0
Schellekens, M. P., Karremans, J. C., van der Drift, M. A., Molema, J., van den Hurk, D. G., Prins, J. B., et al. (2017). Are mindfulness and self-compassion related to psychological distress and communication in couples facing lung cancer? A dyadic approach. *Mindfulness, 8*(2), 325–336. https://doi.org/10.1007/s12671-016-0602-0
Semenkovich, K., Brown, M. E., Svrakic, D. M., & Lustman, P. J. (2015). Depression in type 2 diabetes mellitus: Prevalence, impact, and treatment. *Drugs, 75*(6), 577–587. https://doi.org/10.1007/s40265-015-0347-4
Seuring, T., Archangelidi, O., & Suhrcke, M. (2015). The economic costs of type 2 diabetes: A global systematic review. *PharmacoEconomics, 33*(8), 811–831. https://doi.org/10.1007/s40273-015-0268-9
Sherman, K. A., Przezdziecki, A., Alcorso, J., Kilby, C. J., Elder, E., Boyages, J., Koelmeyer, L., & Mackie, H. (2018). Reducing body image–related distress in women with breast cancer using a structured online writing exercise: Results from the My Changed Body randomized controlled trial. *Journal of Clinical Oncology, 36*(19), 1930–1940. https://doi.org/10.1200/JCO.2017.76.3318
Sirois, F. M., & Hirsch, J. K. (2019). Self-compassion and adherence in five medical samples: The role of stress. *Mindfulness, 10*(1), 46–54. https://doi.org/10.1007/s12671-018-0945-9
Sirois, F. M., & Rowse, G. (2016). The role of self-compassion in chronic illness care. *Journal of Clinical Outcomes Management, 23*(11), 521–527.
Siwik, C. J., Phillips, K., Zimmaro, L., Cash, L., Hicks, A., & Sephton, S. E. (2019). The relationship between shame, guilt, and depressive symptoms and the ameliorative role of self-compassion among lung cancer patients. *Psychosomatic Medicine, 81*(4), A100–A101.
Smith, J. A., & Osborn, M. (2007). Pain as an assault on the self: An interpretative phenomenological analysis of the psychological impact of chronic benign low back pain. *Psychology & Health, 22*(5), 517–534. https://doi.org/10.1080/14768320600941756
Smith, K. J., Béland, M., Clyde, M., Gariépy, G., Pagé, V., Badawi, G., Rémi, R.-L., & Schmitz, N. (2013). Association of diabetes with anxiety: A systematic review and meta-analysis. *Journal of Psychosomatic Research, 74*(2), 89–99. https://doi.org/10.1016/j.jpsychores.2012.11.013
Spiegel, D., & Giese-Davis, J. (2003). Depression and cancer: Mechanisms and disease progression. *Biological Psychiatry, 54*(3), 269–282. https://doi.org/10.1016/S0006-3223(03)00566-3
Stanton, A. L., Luecken, L. J., MacKinnon, D. P., & Thompson, E. H. (2013). Mechanisms in psychosocial interventions for adults living with cancer: Opportunity for integration of theory, research, and practice. *Journal of Consulting and Clinical Psychology, 81*(2), 318. https://doi.org/10.1037/a0028833
Stratton, I. M., Adler, A. I., Neil, H. A., Matthews, D. R., Manley, S. E., Cull, C. A., et al. (2000). Association of glycaemia with macrovascular and microvascular complications of type 2 diabetes (UKPDS 35): prospective observational study. *BMJ, 321*(7258), 405–412. https://doi.org/10.1136/bmj.321.7258.405. PMID: 10938048; PMCID: PMC27454.
Stark, D., Kiely, M., Smith, A., Velikova, G., House, A., & Selby, P. (2002). Anxiety disorders in cancer patients: Their nature, associations, and relation to quality of life. *Journal of Clinical Oncology, 20*(14), 3137–3148. https://doi.org/10.1200/JCO.2002.08.549
Stein, K. D., Syrjala, K. L., & Andrykowski, M. A. (2008). Physical and psychological long-term and late effects of cancer. *Cancer, 112*(S11), 2577–2592. https://doi.org/10.1002/cncr.23448
Tanenbaum, M., Adams, R., Gonzalez, J., Hanes, S., & Hood, K. (2018). Adapting and validating a measure of diabetes-specific self-compassion. *Journal of Diabetes and its Complications, 32*(2), 196–202. https://doi.org/10.1016/j.jdiacomp.2017.10.009
Tanenbaum, M. L., Adams, R. N., Wong, J. J., & Hood, K. K. (2020). Diabetes-specific self-compassion: A new measure for parents of youth with type 1 diabetes. *Journal of Pediatric Psychology, 45*(5), 488–497. https://doi.org/10.1093/jpepsy/jsaa011
Tegethoff, M., Belardi, A., Stalujanis, E., & Meinlschmidt, G. (2015). Association between mental disorders and physical diseases in adolescents from a nationally representative cohort. *Psychosomatic Medicine, 77*(3), 319–332. https://doi.org/10.1097/PSY.0000000000000151
Thapar, A., Collishaw, S., Pine, D. S., & Thapar, A. K. (2012). Depression in adolescence. *The Lancet, 379*(9820), 1056–1067. https://doi.org/10.1016/S0140-6736(11)60871-4
Thewes, B., Butow, P., Zachariae, R., Christensen, S., Simard, S., & Gotay, C. (2012). Fear of cancer recurrence: A systematic literature review of self-report measures. *Psycho-Oncology, 21*(6), 571–587. https://doi.org/10.1002/pon.2070
Todorov, N., Sherman, K. A., Kilby, C. J., & Breast Cancer Network Australia. (2019). Self-compassion and hope in the context of body image disturbance and distress in breast cancer survivors. *Psycho-Oncology, 28*(10), 2025–2032. https://doi.org/10.1002/pon.5187
Trindade, I. A., Duarte, J., Ferreira, C., Coutinho, M., & Pinto-Gouveia, J. (2018). The impact of illness-related shame on psychological health and social relationships: Testing a mediational model in students with chronic

illness. *Clinical Psychology and Psychotherapy, 25*(3), 408–414. https://doi.org/10.1002/cpp.2175

Trindade, I. A., Irons, C., Ferreira, C., Portela, F., & Pinto-Gouveia, J. (2019). The influence of self-criticism on depression symptoms among ambulatory patients with inflammatory bowel disease. *Clinical Psychology and Psychotherapy, 26*(6), 743–750. https://doi.org/10.1002/cpp.2398

Tuomilehto, J. (2013). The Emerging Global Epidemic of Type 1 Diabetes. *Current Diabetes Reports, 13*(6), 795–804. https://doi.org/10.1007/s11892-013-0433-5

Valdiserri, R. O. (2002). HIV/AIDS stigma: An impediment to public health. *American Journal of Public Health, 92*(3), 341–342. https://doi.org/10.2105/AJPH.92.3.341

van der Donk, L. J., Fleer, J., Tovote, A., Ranchor, A. V., Smink, A., Mul, V. E. M., Sanderman, R., & Schroevers, M. J. (2020). The role of mindfulness and self-compassion in depressive symptoms and affect: A comparison between cancer patients and healthy controls. *Mindfulness, 11*(4), 883–894. https://doi.org/10.1007/s12671-019-01298-1

van Dooren, F. E. P., Nefs, G., Schram, M. T., Verhey, F. R. J., Denollet, J., & Pouwer, F. (2013). Depression and risk of mortality in people with diabetes mellitus: A systematic review and meta-analysis. *PLoS One, 8*(3), e57058. https://doi.org/10.1371/journal.pone.0057058

Van Tilburg, M. A. L., McCaskill, C. C., Lane, J. D., Edwards, C. L., Bethel, A., Feinglos, M. N., & Surwit, R. S. (2001). Depressed mood is a factor in glycemic control in type 1 diabetes. *Psychosomatic Medicine, 63*(4), 551–555.

Ventura, A. D., Nefs, G., Browne, J. L., Friis, A. M., Pouwer, F., & Speight, J. (2019). Is self-compassion related to behavioral, clinical and emotional outcomes in adults with diabetes? Results from the second Diabetes MILES—Australia (MILES-2) Study. *Mindfulness, 10*(7), 1222–1231. https://doi.org/10.1007/s12671-018-1067-0

Vos, T., Naghavi, M., Michaud, C., Ali, M. K., Anderson, L., Atkinson, C., et al. (2012). Years lived with disability (YLDs) for 1160 sequelae of 289 diseases and injuries 1990–2010: A systematic analysis for the Global Burden of Disease Study 2010. *The Lancet (British edition), 380*(9859), 2163–2196. https://doi.org/10.1016/S0140-6736(12)61729-2

Walker, J. G., Jackson, H. J., & Littlejohn, G. O. (2004). Models of adjustment to chronic illness: Using the example of rheumatoid arthritis. *Clinical Psychology Review, 24*(4), 461–488. https://doi.org/10.1016/j.cpr.2004.03.001

Wan, H., Ho, R. T., & Hon, T. (2016). What good is self-compassion in cancer coping? A cross-sectional study of bi-psychosocial-spiritual well-being of people with cancer. In *The 18th World Congress of Psycho Oncology Society (IPOS 2018)*. Wiley.

Wren, A. A., Somers, T. J., Wright, M. A., Goetz, M. C., Leary, M. R., Fras, A. M., Huh, B. K., Rogers, L. L., & Keefe, F. J. (2012). Self-compassion in patients with persistent musculoskeletal pain: Relationship of self-compassion to adjustment to persistent pain. *Journal of Pain and Symptom Management, 43*(4), 759–770. https://doi.org/10.1016/j.jpainsymman.2011.04.014

Yang, X., & Mak, W. W. S. (2017). The differential moderating roles of self-compassion and mindfulness in self-stigma and well-being among people living with mental illness or HIV. *Mindfulness, 8*(3), 595–602. https://doi.org/10.1007/s12671-016-0635-4

Young-Hyman, D., De Groot, M., Hill-Briggs, F., Gonzalez, J. S., Hood, K., & Peyrot, M. (2016). Psychosocial care for people with diabetes: A position statement of the American Diabetes Association. *Diabetes Care, 39*(12), 2126–2140. https://doi.org/10.2337/dc16-2053

Zabora, J., BrintzenhofeSzoc, K., Curbow, B., Hooker, C., & Piantadosi, S. (2001). The prevalence of psychological distress by cancer site. *Psycho-Oncology, 10*(1), 19–28.

Zhu, L., Yao, J., Wang, J., Wu, L., Gao, Y., Xie, J., Liu, A., Ranchor, A. V., & Schroevers, M. J. (2019). The predictive role of self-compassion in cancer patients' symptoms of depression, anxiety, and fatigue: A longitudinal study. *Psycho-Oncology, 28*(9), 1918–1925. https://doi.org/10.1002/pon.5174

Zoungas, S., Chalmers, J., Ninomiya, T., Li, Q., Cooper, M. E., Colagiuri, S., et al. (2012). ADVANCE Collaborative Group. Association of HbA1c levels with vascular complications and death in patients with type 2 diabetes: evidence of glycaemic thresholds. *Diabetologia, 55*(3), 636–643. https://doi.org/10.1007/s00125-011-2404-1. Epub 2011 Dec 21. PMID: 22186981.

Check for updates

20 Self-Compassion in Trauma Treatment

Christine Brähler

Introduction

Self-compassion refers to the capacity to respond mindfully to difficult experiences, be kind to ourselves in the midst of suffering, and see ourselves as connected to humankind when we experience distress (Neff, 2003a, b). Shame, on the other hand, can be defined as seeing oneself as unworthy of love and attention and as uniquely isolated from the rest of humankind. Self-compassion thus appears to represent an obvious antidote to shame. Shame is a much-neglected social emotion, whose role in psychopathology is being increasingly recognized. Shame has been found to contribute to, trigger, maintain, and exacerbate a wide range of psychiatric conditions such as depression (Kim et al., 2011), eating disorders (Kelly & Tasca, 2016), psychosis (Carden et al., 2020), obsessive-compulsive disorder (Weingarden et al., 2016), complex Post-Traumatic Stress Disorder (PTSD; see this chapter), and borderline personality disorder (Brown et al., 2009). The lack of self-compassion combined with the high prevalence of shame in most individuals with psychiatric conditions has led clinicians and researchers to hypothesize that interventions promoting self-compassion hold great promise for clinical populations. Whilst the evidence suggests that interventions are safe and acceptable, the evidence for clear-cut benefits in clinical populations remains limited, as meta-analyses usually combine clinical and nonclinical populations (Ferrari et al., 2019; Kirby et al., 2017b).

In this chapter, I aim to expand the early rationale for integrating self-compassion into psychological treatment for psychiatric conditions by drawing on the latest clinical research and clinical experience and by highlighting the common pitfalls and how to navigate around them. Specifically, I will focus on the relevance of self-compassion for the psychological treatment of complex PTSD. Frequently, those who meet diagnostic criteria for complex PTSD lack in self-compassion, have high fears of compassion, high shame, and a history of attachment traumas and other traumas (Maercker et al., 2013a, b). I will address the relevance of self-compassion for treating complex PTSD within a phase-based approach (Herman, 2015). Herman, a pioneer of trauma research and therapy, summarizes the work of trauma recovery as involving "overcoming barriers to shame and secrecy, making intolerable feelings bearable through connection with others, grieving the past, and coming to a new perspective with a more compassionate view of oneself in the present." (Herman, 2015, p. 276), thus placing self-compassion at the heart of trauma recovery.

C. Brähler (✉)
University of Glasgow, Glasgow, Scotland

A. Finlay-Jones et al. (eds.), *Handbook of Self-Compassion*, Mindfulness in Behavioral Health,
https://doi.org/10.1007/978-3-031-22348-8_20

Trauma Exposure and Complex PTSD

Trauma exposure is common. In a global survey of approximately 51,000 people across low-, middle-, and high-income countries, seventy percent of adults were found to have experienced at least one traumatic event over their lifetime (Kessler et al., 2017). For trauma exposure to lead to PTSD or to other mental disorders depends on a variety of inner and outer risk and resiliency factors. Exposure to a traumatic event per se hence does not say anything about the mental health of an individual. However, childhood and adult trauma exposure is known to be a nonspecific etiological factor for most mental health problems. Prior work has found that mental health problems increase in a dose-response relationship with exposure to adverse childhood experiences (see Zarse et al., 2019, for a review).

In the eleventh revision of the International Classification of Diseases (ICD-11), the symptom profile of PTSD comprises six symptoms distributed across three symptom clusters: (1) re-experiencing the traumatic event in the here-and-now; (2) avoidance of stimuli that remind one of the traumatic event; and (3) a sense of threat (First et al., 2015). Unlike simple PTSD which focuses on the impact of a single traumatic event, complex PTSD refers to the impact of cumulative traumatic events occurring across childhood and adulthood (van der Kolk et al., 1993). These traumatic events are often interpersonal as opposed to impersonal (car accident, natural disaster), often occur in relationships to attachment figures, and thus often occur repeatedly (Hyland et al., 2017). Complex PTSD is defined as consisting of the existing PTSD criteria plus a set of symptoms referred to as *disturbances in self-organization*, which summarize the pervasive dysregulating effects of chronic victimization on affect regulation involving shame, fear, guilt, anger, the sense of self, and relational functioning (Maercker et al., 2013a). Dissociation is another common adaptation to chronic victimization, and the fifth edition of the Diagnostic and Statistical Manual for Mental Disorders (DSM-V), includes a dissociative subtype encompassing additional criteria of depersonalization (i.e., feeling as though one is detached from oneself or observing oneself from the outside) and/or derealization (i.e., the feeling that one's experiences are surreal or distorted) (American Psychiatric Association, 2013).

Self-Compassion and PTSD

Authors of a systematic review of 35 studies on PTSD and self-compassion confirmed the results from earlier studies, which found self-compassion to be negatively related to PTSD symptomatology in general and to the avoidance cluster of symptoms in particular (Winders et al., 2020). The studies, however, were largely based on student or general population samples, thus limiting their transferability to clinical populations. Karatzias et al. (2017) examined the link between self-compassion and complex PTSD in a largely female sample referred to trauma therapy services. Interestingly, self-compassion was found to be negatively associated with the *disturbance in self-organization* factor, but not with a general PTSD factor (Karatzias et al., 2017). More specifically, low self-compassion was linked to negative self-concept, relationship difficulties, and affect dysregulation—particularly hypoactivation strategies. Hypoactivation of distress involves attempts to downregulate, numb, and turn the distress inwards, which are in keeping with shame, defeat, dissociation, and feelings of depression, whereas hyperactivating strategies involve increasing arousal, expressing distress, and potentially becoming aggressive. Such an internalization of abusive treatment by others is a common consequence of chronic interpersonal trauma, especially when it occurs early in life and in caregiving contexts (van der Kolk et al., 2005). In contrast, increasing self-compassion among individuals with PTSD is hypothesized to help them to improve their affect regulation, self-concept, and relational functioning by reducing their feelings of shame, guilt, failure, and defeat (Lee et al., 2001).

Self-Compassion, Shame, and PTSD

Self-compassion may be an effective antidote to the shame experienced by those with PTSD, where feelings of kindness, common humanity, and mindfulness can come to replace feelings of self-blame, isolation, and emotional avoidance. Although shame is known to exacerbate many types of psychological disorders, including PTSD, it is still frequently overlooked in treatment, which has historically prioritized anxiety, fear, and anger (Taylor, 2015). Shame is a universal emotion (Sznycer et al., 2018) characterized by a state of hypoarousal and submission that evolved to avert attack by members from one's own group (Keltner & Harker, 1998). To escape the attack, shame activates submissive or aggressive defenses. If those are ineffective and the individual cannot find safety or support with another person or group, then primary consciousness shuts down leading to dissociation (Schore, 2015). Shame can have a paralyzing effect and thus being shamed can be considered traumatic if it happens early in development and if done by a caregiver (Matos & Pinto-Gouveia, 2010). Submissive strategies involve pleasing powerful others by adapting and giving up one's own will, internalizing their opinions about oneself, correcting, or punishing oneself, suppressing anger and self-protective impulses in the body, losing any boundaries with powerful others, and becoming complacent. Aggressive strategies involve some form of counterattack such as blaming others overtly or covertly by talking negatively about people behind their backs or by using physical aggression.

Shame may maintain and exacerbate PTSD by increasing the severity and duration of illness (Brewin & Holmes, 2003), over and above the impact of exposure to trauma (DePrince et al., 2011a, b). For example, intrusive memories of traumatic experiences were found to be accompanied more often by feelings of shame than by feelings of fear, horror, or helplessness (Holmes et al., 2005). Other recent research indicates that shame may mediate the relationship between PTSD symptom severity and suicidal ideation among veterans with PTSD (Cunningham et al., 2018). In contrast, self-compassion might help veterans to better adjust once back home by reducing their shame, as higher self-compassion has been linked to lower PTSD symptoms, general psychopathology, and better functioning among returning veterans, irrespective of trauma exposure (Dahm et al., 2015; Hiraoka et al., 2015).

Shame is often a result of childhood trauma, which is a risk factor for the later development of PTSD and other pathologies. Childhood maltreatment has been linked to higher shame and lower self-compassion and worse PTSD symptoms (Andrews et al., 2000), via emotion dysregulation (Barlow et al., 2017; Scoglio et al., 2018; Vettese et al., 2011). Children may internalize their caregivers' contemptuous and hostile intentions by developing deeply shaming core beliefs such as "I am bad/evil/disgusting/unworthy," which resolve some of the cognitive dissonance (i.e., the discomfort associated with holding opposing beliefs, ideas or values; Briere, 1992). Not surprisingly, higher doses of early trauma are associated with lower levels of self-compassion (Játiva & Cerezo, 2014; Tanaka et al., 2011). For example, highly critical or otherwise dysfunctional family environments and emotional abuse have been linked to low self-compassion and insecure attachment (Neff & McGehee, 2010; Tanaka et al., 2011). In turn, low levels of self-compassion have been associated with depression and anxiety (Joeng et al., 2017), self-harming behavior (Jiang et al., 2016), anxiety (Berryhill et al., 2018), and PTSD symptoms (Bistricky et al., 2017).

Betrayal Traumas and Shame

In cases of interpersonal violence at the hands of a close one, distress among survivors is often fueled by a sense of betrayal (Freyd, 1996). The closer and the more trusting the relationship, the worse the sense of betrayal. High betrayal traumas have been associated with greater shame and dissociation (DePrince et al., 2011b; Platt & Freyd, 2015) and with more severe PTSD symptoms than low betrayal traumas such as natural disasters or abuse by someone not close to the victim (DePrince & Freyd, 2002). Victims of interpersonal violence may blame themselves for not defending themselves, for trusting the person in the first place, for not detecting the threat sooner, or, in the case of repeated violence, for not being able to separate or distance themselves from the perpetrator. Partner violence in adulthood, incest, or other abuse by caregivers in childhood are amongst the high betrayal traumas and more frequently suffered by women than by men (Freyd, 1996).

When a person is subjected to abuse by a caregiver or close one during childhood, the dependency on that person creates such a high level of threat in the child that the experience might be partly or completely dissociated (Liotti, 2004). Many survivors of such high betrayal trauma who begin to remember the trauma more clearly (traumatic memories are often fragmented or due to dissociative amnesia cannot be recalled) struggle to break the silence of this shameful secret as they fear not being believed, not being supported, or on the contrary to be betrayed and punished again and to lose any important attachment relationships. They may avoid seeking treatment for fear of being misunderstood, blamed, invalidated, let down, and retraumatized by mental health providers. The lack of inner safety, trust in others and in solidarity as well as the lack of self-protective anger or fierce self-compassion, maintains PTSD and keeps them suffering in silence and isolation.

Taken together, trauma survivors who experience shame during or after the traumatic event have worse outcomes as evidenced by higher symptom load, longer duration of illness, greater suicidality in war veterans, higher retraumatization rates in betrayal trauma, and less engagement with mental health services (Saraiya & Lopez-Castro, 2016). PTSD sufferers who experienced childhood maltreatment by caregivers are more prone to feel shame and to have low self-compassion (Matos et al., 2017). Shame in its different guises clearly constitutes a valid target for treatment for many PTSD clients. Self-compassion has the potential to buffer against the impact of the trauma via cognitive appraisal and emotion regulation routes (Barlow et al., 2017; Zeller et al., 2015) and via relational routes such as by helping the person to use social supports (Maheux & Price, 2016).

Self-Compassion in the Treatment of Complex PTSD

Although the correlational research to date clearly suggests that self-compassion may protect against the toxic effects of shame in PTSD, a systematic review concluded that there is only tentative evidence that "interventions based, in part or in whole, on a self-compassion model potentially reduce PTSD symptoms" (Winders et al., 2020, p. 300) and they conclude that "self-compassion may be agent of change in PTSD symptom reduction and also in trauma-related guilt and shame" (Winders et al., 2020, p. 321). Whilst some of these interventions integrated self-compassion into a standard evidence-based treatment for simple PTSD such as cognitive behavior therapy (CBT), other interventions included ultra-brief bibliotherapy interventions, which were unintegrated into any other evidence-based treatment for PTSD. These findings might be explained by: (1) self-compassion being offered as a stand-alone intervention instead of being integrated into an evidence-based treatment for complex PTSD; (2) by interventions being too brief (Galili-Weinstock et al., 2019); (3) by self-compassion being introduced without providing the person with a relational template for compassion, e.g., receiving compassion from peers in groups therapy or from the therapist in individual therapist; (4) by self-compassion not being offered within a safe compassionate therapeutic relationship where working through common fears and obstacles to compassion is possible; or (5) by research focusing on symptomatic outcomes instead of capturing the mechanism of change (i.e., self-compassion). Regarding the latter point, increasing self-compassion would be expected to bring about a change in the *relationship* to the symptoms and to the person experiencing it instead of reducing the symptoms per se. This in turn may help to reduce the functional impact of PTSD symptoms.

Measures of life satisfaction, quality of life or other indicators of psychological resilience could help to determine the broader impacts of such interventions on adaptive functioning, while longer follow-up periods may be helpful in determining whether changes in self-compassion drive reductions in PTSD symptoms over time.

Self-Compassion for Complex Mental Health Problems

There are two common misconceptions about self-compassion in clinical practice. Firstly, it is commonly viewed solely as a technique that can be successfully trained by the client with guided meditations. Whilst this is the case for healthy populations (McEwan & Gilbert, 2016), it promotes an overly mechanistic view of clinical psychotherapy with the therapist being a mere mental health technician prescribing automated interventions to fix the problem. Secondly, it is represented as a stand-alone psychotherapy for individuals with psychiatric disorders. This way of overstating the scope of self-compassion in the clinical domain often results from the powerlessness clinicians feel when faced with complex clinical issues and the wish for a simple and reliable solution. Clinicians therefore need to learn to hold the uncertainty and the complexity lightly without being overwhelmed and without running the risk to resort to a one-method-cures-all attitude that oversimplifies complex mental health issues and runs the risk of harming clients.

Self-compassion is a radical new way of meeting suffering. It is therefore primarily an attitude of care and goodwill toward difficult emotions and experiences. Techniques and interventions have been developed to help cultivate this attitude. In a therapeutic setting, our goal is to help the client develop a compassionate way of relating to herself and to the inner experiences that she finds painful or unlikable. Several studies have shown that improved self-compassion in the client is associated with reduced symptoms following a whole range of different diagnoses and therapies (Galili-Weinstock et al., 2017; Hoffart et al., 2015; Schanche et al., 2011), none of which explicitly cultivated self-compassion. These findings tell us that a kinder relationship to oneself is a common mechanism of change that can have a significant positive impact on symptoms without including explicit interventions. Instead, self-compassion seems to be at play implicitly through other therapist, client, and relational variables. These findings also support the idea that compassion and self-compassion can be integrated into any acceptance-based approach to psychotherapy. This has been done with complex mental health issues such as psychosis (Braehler et al., 2013a,b) and eating disorders (Kelly et al., 2017) through Compassion-Focused Therapy (CFT; Gilbert, 2010; see Chap. 23). CFT is the only clinical compassion-based approach which has integrated explicit compassion practices into CBT. Core CBT interventions are given a compassionate mind focus and modules are added to standard psychological treatment, which consists of standard disorder-specific CBT protocols. Being familiar with standard CBT interventions is thus necessary to implement CFT. Evidence-based phase-based treatments for complex PTSD are Skills Training in Affective and Interpersonal Regulation (STAIR; Cloitre et al., 2002) and Dialectical Behavior Therapy-Post-Traumatic Stress Disorder (DBT-PTSD; Bohus et al., 2019). A compassion-based module is being integrated into DBT-PTSD to target shame more explicitly.

Whilst healthy groups can benefit from practicing with guided exercises on their own, clinical groups encounter many more fears and barriers and shame (Kirby et al., 2019). They therefore need modeling from the therapist as well as guidance to overcome obstacles in a way that is attuned and suited to their needs and limitations before they can give it to themselves. For this to be genuine and authentic, a therapist needs to have integrated these various qualities to embody them. A therapist also needs to know ways to return to a place of good will and open-heartedness and clarity toward the client whenever he or she has lost it, got triggered or distracted and to transform empathy distress into compassion – not just to prevent burnout but also malpractice. A therapist also requires an existing model of an acceptance-based psychotherapy suited for the client's issues in which to integrate self-compassion and compassion across three relational levels: (1) for therapist from therapist; (2) from therapist to client; and (3) for client from client.

Understanding Fears and Barriers to Compassion

Asking a person who is more familiar with being in abusive than nonabusive relationships to "treat herself like a dear friend," as is done in some self-compassion practices, may simply be impossible to do, as she has no template to draw upon. It is like asking someone to imagine the taste of chocolate who has never tasted chocolate. Being asked to hold yourself in a loving embrace might bring up confusing and distressing feelings if you have never been held unconditionally, or were held but also abused or neglected by the same person, or you lost them suddenly. The repair of the attachment system of this person must start by drawing up a new template for healthy caring relationships with oneself and with others that the client would like to work toward. Slowly developing a secure attachment to the therapist or to other trusted sentient beings in her life may be a first longer-term goal.

Our Attachment Past Determines Our Compassion Present

Why are some people mistrusting or afraid of kindness and care? Evidence suggests that barriers and facilitators for self-compassion are closely linked to early attachment experiences (Bowlby, 1982; see Chap. 5). When early experiences of needing or receiving care are associated with emotions such as shame, anger, loneliness, fear, panic, despair, vulnerability, powerlessness, or disgust, an individual is much more likely to feel an aversion or fear toward these care experiences, even as adults. In turn, the capacity to experience affiliative emotions such as love, and compassion in a safe and nurturing way can thus be severely compromised. In effect, experiences of love and connection as an adult can activate experiences of shame and disconnection one was exposed to as a child. Germer and Neff (2019; Neff & Germer, 2018) encapsulate this experience by saying: "When we give ourselves unconditional love, we uncover the conditions under which we have not been loved." We can extend this by adding "or under which we were maltreated or otherwise traumatized."

Such emotional conditioning is largely unconscious and only manifests in an attachment context such as during a close relationship or eventually in the therapeutic relationship. Once these patterns become manifest, they typically reveal meta-cognitive beliefs about negative consequences or feelings that may occur if the person receives or gives kindness or compassion to self or others (Gilbert et al., 2014c). These so called "fears of compassion," tap into people's fears about what reactions they may have to giving or receiving compassion for others, from others, and from themselves (Gilbert et al., 2011). "People will take advantage of me if I am too compassionate or forgiving" suggests the need to stay harsh to protect oneself. "I worry that people are only kind toward me when they want something," refers to experiences of earlier experiences of manipulation. Others involve a fear of dependency on others, a sense of unworthiness of kindness, a fear of being overwhelmed by difficult emotions, and a preference for self-punishment over self-forgiveness (Gilbert et al., 2011). Among people with clinical depression, fears of compassion are associated with greater attachment insecurity (Gilbert et al., 2014a, c). Fears and resistance to compassion are associated with greater anxiety and depression in the general population (Gilbert et al., 2012) and in clinical cohorts (Gilbert et al., 2014c). People who have early memories of shame and who lack early experiences of warmth and safeness may be more liable to fear receiving compassion (from self or others) and also more likely to experience anxiety, depression, and paranoia (Matos et al., 2017). One study found that together, fears of compassion accounted for 53% of variance in depressive symptoms and were strongly associated with self-criticism, which in turn is a risk factor for depression (Gilbert et al., 2014a). Unsurprisingly, greater fear of compassion is associated with increased PTSD symptomatology (Winders et al., 2020). A meta-analysis found fears of self-compassion and of receiving compassion to be linked to shame, self-criticism, and depression. These associations were strongest in clinical versus healthy populations (Kirby et al., 2019). This highlights the need to take into account these fears in clinical populations instead of generalizing findings from healthy populations.

Trust First, Compassion Second

The anecdotal evidence of the importance of attachment and safeness for self-compassion was first backed up by somewhat accidental research findings by Rockliff and colleagues (Rockliff et al., 2011). In a double-blind placebo-controlled trial, participants from the general public received either oxytocin nasal spray or placebo before being guided in a compassion-focused imagery. Oxytocin is a neuropeptide involved in affiliative behavior (Bartz et al., 2011). The authors detected significant individual differences in the response to the compassion-focused imagery, which were the most pronounced under oxytocin. As expected, about half the participants experienced the compassionate friend imagery as leading to a safe/content/relaxing feeling in both oxytocin and placebo groups. Those with higher self-criticism, lower social safeness (trust), and with attachment insecurity however experienced more resistance to receiving compassionate feelings when they received the oxytocin spray and felt less safe/content than before starting the exercise. They reported feeling angry, frustrated, sad, scared, and lonely. In keeping with other oxytocin research (Bartz et al., 2010), oxytocin spray is thought to directly activate socially salient memories of affiliation, which explains the negative emotions and aversion to care in those with attachment insecurity. This points to the importance of titrating the dose of any compassion intervention in individuals with high self-criticism, low social safeness, and attachment insecurity as not to "overdose" and trigger an adverse "allergic" reaction to the "medicine." Clinicians would not ordinarily give oxytocin spray, so the adverse effects may be less strong, yet we need to take into account that this study was conducted with healthy participants. What about individuals who suffer from psychological distress and whose attachment systems are thus on high alert?

Ebert et al. (2018) measured plasma oxytocin, fears of compassion, and parental warmth in 57 clients with Borderline Personality Disorder (BPD) and compared those to healthy controls. BPD patients had lower levels of plasma oxytocin compared to healthy controls. Within the BPD group, greater fears of compassion from self, from others, and for others were correlated with lower oxytocin. In keeping with the hypothesis that oxytocin is related to salient memories of affiliation, patients with BDP who recalled less emotional warmth from their parents had lower levels of oxytocin. No such relationships were found amongst the healthy group. BPD is associated with traumatic invalidation and shaming from attachment figures in childhood. This study supports the hypothesis that early adverse attachment experiences change the endocrine substrate of the attachment system, thus likely leading to different responses to training in self-compassion designed for healthy populations and thus pointing to the need for adaptations.

Heart rate variability (HRV) is thought to be a physiological measure of the functioning of the parasympathetic nervous system and in particular of the ability to self-soothe (Kirby et al., 2017a). Low HRV is considered indicative of poorer self-soothing ability whereas high HRV is indicative of a better soothing capacity. Rockliff et al. (2008) explored both HRV and cortisol in response to compassionate friend imagery relative to control imagery. One group showed an activation of the soothing system indicated by an increase in HRV and decrease in cortisol. Another group, however, showed a decrease in HRV and similar cortisol levels following the compassion imagery, indicating an increase in the physiological threat response. The main distinguishing variable between the two groups was the degree of social safeness (or sense of trust in others). Nonsignificant trends showed that the threat group was more self-critical, colder toward themselves, more anxiously attached, and had more psychiatric symptoms.

Trust in others appears to be a key prerequisite for being able to benefit from compassion imagery. What if you prime people to think of someone they trust? Baldwin et al. (2020) found that this experimental manipulation was associated with an increase in HRV following compassion imagery in 68 students with attachment insecurity thus suggesting that helping to build trust first made it easier for the person to receive compassion from a being in their imagination. The researchers found a decrease in HRV (threat response) in those same participants when they asked them to do the compassion imagery without the prime (Baldwin et al., 2020; Galili-Weinstock et al., 2017; Kelly & Tasca, 2016) Taken together, the neurophysiological and endocrinological studies show that clients with attachment insecurity or disorganization may not only have a different HPA axis functioning but also a different oxytocinergic system. Beliefs such as "Others cannot be trusted. Kindness from others is not genuine. Giving kindness to others means I would let them off the hook." (Gilbert et al., 2011) might easily be triggered by compassion exercises and create adverse emotional reactions. To prepare clients emotionally and physiologically, therapists first need to establish trust in the therapeutic relationship and within the client. External safety is of course necessary before any therapeutic work can begin.

Consider the Capacity to Mentalize

Another obstacle to receiving compassion from others can be related to mentalizing difficulties. Mentalizing difficulties include a compromised ability to infer and reflect on the mental states of oneself and others and to take different perspectives (Fonagy et al., 2002), similar to psychological inflexibility (Miron et al., 2016). Gilbert found that individuals with greater fears of compassion were more self-critical, but also struggled to label their emotions and to talk about their feelings, making it more difficult to notice and feel their pain (Gilbert et al., 2012). Fear of self-compassion and high psychological inflexibility have been shown to interact to predict PTSD symptom severity in students with trauma exposure (Boykin et al., 2018; Hiraoka et al., 2015). Clients with the highest levels of shame and self-criticism experience the lowest level of self-compassion and the greatest fears of receiving compassion from others including from oneself, which correlates with insecure attachment (Gilbert et al., 2014c), low mentalizing ability (Boykin et al., 2018), and worse outcomes (Kelly & Carter, 2013; Miron et al., 2016; Vettese et al., 2011). Therefore, therapists need to assess and consider not just fears of compassion but also their clients' capacity to mentalize and to differentiate their emotions when planning any compassion-based interventions. Directive interventions based on CBT and DBT that helped clients differentiate negative emotions associated with self-criticism were associated with a greater increase in self-compassion over the course of therapy compared to nondirective interventions (Galili-Weinstock et al., 2019). In the later section on differentiation, I give some guidance on how to adjust interventions for individuals with low mentalizing ability.

An Attachment-Based Reformulation of Self-Compassion in Trauma Treatment

Self-compassion has the potential to buffer against the impact of traumatic stress through the process of cognitive appraisal and emotion regulation (Barlow et al., 2017; Játiva & Cerezo, 2014; Zeller et al., 2015) and by helping individuals to make better use of social support (Maheux & Price, 2016). Much like attachment security, self-compassion offers an inner safe haven, where individuals can seek refuge and recover when distressed, and a secure base, from which they can explore the world and connect to others to feel energized again (Bowlby, 1988). Capacities for emotion regulation and mentalizing/psychological flexibility develop in early attachment contexts and are critical for later emotional and interpersonal functioning (Fonagy et al., 2002). Self-compassion involves both emotional regulation (self-kindness) and mentalizing (mindfulness and common humanity). Research demonstrates that attachment styles are, however, fluid across the lifespan, and that a secure attachment style can be developed later in life through corrective experiences with another attachment figure, such as a teacher, therapist, a romantic partner, or a spiritual being—an "earned secure" style (Roisman et al., 2002). It could be argued that self-compassion is the result of such corrective experiences, which have been shown to buffer against the impact of dysfunctional family experiences (Berryhill et al., 2018; Homan, 2016; Jiang et al., 2017).

Safely Navigating Unchartered Attachment Trauma Territory

If clients fear the care they long for, then how can we safely integrate self-compassion into therapy? Ask yourself: "How can I, as a therapist, safely access the care system in this person without unnecessarily activating the threat system?" Self-report questionnaires might help to assess conscious fears (Gilbert et al., 2011), although most are unconscious unless activated in an attachment context. Knowing a client's attachment style and mentalizing capacity can also provide helpful guidance for how to attune your role and the type of interventions you use.

Adapt to the Attachment Style

Clients with *anxious-preoccupied* attachment tendencies might present as needing the therapist or others for support, guidance, reassurance, and comfort. This is often because the client does not trust their ability to self-soothe. Thinking about their own needs may bring up feelings of confusion, despair, or fear. In this scenario, the therapist can play a role in supporting the client to recognize and express her own needs, while acknowledging and reinforcing experiences where the client has effectively responded to her own needs. In some cases, it may be easier for clients to infer their needs based on their expectation of what another person might need in a similar situation. An important part of this exploration may also be workshopping potential ways that the client can meet her own needs in a situation without always relying on others.

In contrast, *dismissive-avoidant* attachment tendencies tend to manifest as exclusive self-reliance, as caregivers were unreliable in times of need. In this case, the prospect of seeking or receiving help from others may trigger fears of abandonment or being let down or disappointed. Given that the client's self-reliance has long felt critical to their survival, relying on others can also evoke feelings of powerlessness or shame. Therapists can start to engage such clients by acknowledging the strength that the client has shown through their ability to survive without others' help, while also validating that this cannot have been an easy task. It is important, however, that validation of these experiences comes from a place of curiosity and compassion, rather than sympathy or pity. Respecting and honoring the client's need for autonomy is critical. As described in Braehler and Neff (2020), focused self-disclosures can also be helpful in de-shaming the client with dismissive-avoidant attachment tendencies. For example, in acknowledging and validating the steps a client has taken to seek help by coming to therapy, the therapist may wish to disclose a time when they also struggled to confide or trust in others for support.

Fearful-avoidant attachment tendencies in adults result from a sense of being unable to trust neither themselves nor others to care for them at times of distress, creating confusion, dissociation, and intense despair. Such expectations are strongly associated with sexual, physical, or emotional abuse and neglect by caregivers during childhood (Van Ijzendoorn et al., 1999). The attachment figure was a source of fear, accounting for the intensely ambivalent and confusing behavior in relationships. Seeking help is likely to be a struggle for clients with this attachment conditioning, as they touch on the fear of being abused again. Significant courage and motivation to help oneself is needed to overcome this fear. Expressions of care from the therapist are likely to be met with intense fear and reactivation of early relational trauma, such as by flashbacks, dissociation, or numbing. Rage or anger might be another response to create distance again from the therapist who is seeking to come closer. Clients with such distressing early experiences require the utmost stability, attunement, flexibility, patience, long-term commitment, and good will from the therapist. Any rigid expectations with regard to how the client should respond, behave, or how therapy should unfold may be met with increased dysregulation and could potentially result in the client dropping out of therapy prematurely.

Principles for Selecting Interventions

Following an assessment and formulation, including experiences of safeness, warmth and care, the therapist might want to formulate some hypotheses about where roadblocks may occur. Despite the best analysis and preparation, many attachment traumas remain hidden until they are activated during therapy. The following three principles based on my clinical experience may help to prevent unnecessary activation of the threat system and help overcome fears of compassion safely.

Desensitization: Warming Up Slowly Accessing the care and attachment system (for example, by receiving care and compassion with the therapist) can be extremely threatening for the client with attachment trauma. Accordingly, the therapist needs to "warm up slowly" by working with the client to find a safe way of bringing the care system "online". This includes first working with the protective parts and exploring their fears instead of pushing them to the side and forcing your way to the vulnerable parts without their consent thus destabilizing the client. Therapists might want to acknowledge and normalize the paradox that while the care and attachment system is typically associated with "pleasant" feelings such as connectedness and warmth, when it has been "offline" for some time, reactivation can feel quite unpleasant. This can be likened to running one's hands under warm water after being out in the snow. Although the warmth is needed (and will become pleasant over time), the initial experience of warmth can be a painful one. However, this explanation should not be viewed as carte blanche to trigger negative emotions. Instead, the therapist should always titrate the dose of their own compassionate presence with the client - as well as that of any practices – to avoid such adverse effects as far as possible. Warmth, kindness, and wisdom are qualities we would hope to find in all therapists. What is important is that any warmth does not emerge from a script of how one should be as a therapist but emerges from resonance with the client and good will, which helps therapists to provide sensitive, attuned, and thus flexible care. Flexible means also toning down one's expression of kindness if it is "too much of a good thing" for the client. By titrating exposure to the care system in a sensitive and flexible manner, therapists can help their clients stay within the window of tolerance and keep this window open, rather than causing the client to shut down.

Direction: Accessing Compassion Where It Flows Freely Just as fears of compassion can manifest as fears of receiving from others, giving to others, and receiving from self, so too can the flow of compassion be directed from self to other, from other to self, or from self to self. It is likely that not all directions flow as freely for each client. Accordingly, therapists can support the clients to become accustomed to accessing the care and attachment system where it flows most freely. Allowing the client to choose how far away or how close to them they want to experience care can help to establish feelings of safety and trust in the process. Ask yourself: "In which direction can the client feel trust, kindness, compassion, or happiness most easily without feeling threatened?"

Differentiation: Who Needs What from Whom and How? Self-compassion practices frequently involve differentiating various "parts" or selves, such as the inner self-critic and the inner compassionate voice. For clients who are easily able to identify and differentiate these parts, such exercises can provide a useful demonstration of how one can utilize one's internal resources to respond to internal difficulties. However, clients with complex PTSD may be more likely to overidentify with one part of themselves at the expense of being able to acknowledge or hold other parts or selves. For example, they may feel overidentified with a younger, more vulnerable self at the expense of honoring and validating their stronger, adult self.

Over the course of therapy, clients may become more aware of, and able to integrate these different parts. For example, they may identify both protector parts, and vulnerable, exiled parts that are connected to specific autobiographical memories. In doing so, the client may experience the capacity to offer care from their compassionate self to another part and to integrate traumatic memories with present-day realities. However, this parts-based approach is only recommended where clients are reasonably stable, have good mentalizing ability, and have capacity to access their compassionate self. When working with individuals with low mentalizing abilities, it is recommended that you make any intervention concrete, visible (interpersonal), and focused on the here-and-now instead of making it abstract, invisible (intrapersonal), and focusing it on distress in the there-and-then (Bateman & Fonagy, 2006). Undifferentiated practices keep the level of differentiation of the quality of care and of the object of care low as well as concrete, visible, and based in the here-and-now. Touch or practical behaviors are examples of undifferentiated care, and nurturing the body through diet, exercise, touch, massage or pets, plants, nature or simply breathing provides opportunities for here-and-now, concrete, and visible care experiences.

Differentiation may also be applied to the quality of care. What is needed? Providing the client with examples of some common needs in relation to specific experiences may help to expand their vocabulary of needs-based language if this is an unfamiliar concept. For example, in the context of insecurity, encouragement is a common need; in the context of abuse, protection; when one has experienced neglect, they may need a sense of providing; if they experience grief, they may need comfort; in the context of fear, they may need calming; and in the context of shame, they may need validation and a sense of belonging.

Self-Compassion Within Phase-Based Therapy for Complex PTSD

I now highlight a few processes during any phase-based treatment of complex PTSD (Herman, 1992) where self-compassion comes into play in its different guises. Self-compassion for the therapist strengthens compassionate presence, which, in turn, strengthens the therapeutic relationship, and lastly allows for sensitive direct interventions with the client.

Troubleshooting for Triggered Parts in the Therapist It is normal to experience empathic distress when sitting with clients with significant trauma. We may feel a whole range of distressing emotions in response to the wide range of distressing images, feelings, and thoughts the client is carrying with her. The first and most significant step is to recognize the activation of your own threat system, to stay grounded in your body, and to offer love and understanding to your parts that have been triggered in order to return to a sufficiently open, relaxed, and caring state. If you have taken self-compassion training, you can insert informal practices in your clinical day to practice silently whilst sitting with a client or in your break in response to any of those exploratory questions:

- How am I feeling right now? Is there distress in me? How am I relating to my (empathic) distress? What do I need to alleviate this distress somewhat for now? What gets in the way of giving that to myself?
- Why am I doing what I am doing right now? What part is in the driver's seat? Powerless,

overwhelmed, traumatized, despairing, numbing, intellectualizing, angry, afraid of emotions, shaming, punitive, or striving to rescue, succeed, control, lecture, or please? Offer love and understanding to the part that is trying to protect you or that feels vulnerable and has been activated.

Therapists thereby take full responsibility for regulating any of their triggered parts. Therapists can be transparent about their emotions such as feeling compassion for a part of the client if it serves the therapeutic process, but they need to be careful not to burden the client with their own unresolved issues. For example, co-regulating yourself by guiding a practice with the client who only serves to calm the therapist down would be unethical as the therapist would be putting part of the responsibility for him or herself onto the client. This would be violating the therapist's duty of care. We need to ensure that practices serve the client and that we are here to support and guide, not the other way around. If you notice being triggered regularly, then personal therapy, mindfulness, and self-compassion training are strongly advised for the therapist as well as considering referring the client to a more experienced therapist. A study found that therapists who were more self-critical were more likely to be critical and hostile to their clients leading to worse outcomes (Henry et al., 1990).

Empowerment Through Fierce Self-Compassion An important step in recovery from high betrayal traumas is when the person begins to experience healthy anger (fierce self-compassion). Righteous outrage about the injustices helps the survivor to move out of dissociation, shame, and self-attacking into groundedness, self-confidence, calm strength, and readiness to protect herself, which is profoundly empowering. Feeling appropriate anger about the physical or emotional cruelties inflicted on one's body and mind also allows the person to begin to grieve. A common misunderstanding amongst therapists is that compassion means only to be nurturing, calming, and soothing (yin qualities) instead of also including assertiveness, protection, and encouragement (yang qualities). Compassion means being sensitive to suffering and alleviating it with a quality of care that is appropriate to the suffering.

For those who have experienced betrayal traumas, it is fundamental to validate the anger and the pain connected to the betrayal. Validating the injustice of the experience should continue for as long as the person needs validation. Many survivors may be accustomed to having their experiences invalidated and may even engage in self-invalidation. Accordingly, the role of the therapist is to model healthy self-protective anger so that the client is able to recognize the full weight of their experience. Over time, as clients start to believe that what was done to them is wrong, unjust, and in most cases, illegal, they develop the capacity to validate their own anger. This is the first step of Yang self-compassion as described in the Yang Self-compassion break (Braehler & Neff, 2020). However, this is not necessarily a linear process. Often, shame will re-emerge, leading the survivor to feel guilt regarding their anger. This in turn can activate a submissive defeat state, triggering depressive mood (Catarino et al., 2014). Accordingly, it is vital that the therapist can remain patient and present with the client's experience. Creating space to gently explore and acknowledge the fear of anger that survivors may carry provides a firm foundation for clients to gradually experience, express, and validate their own anger.

The therapist's own relationship with anger plays an important role in this process. If therapists have not yet integrated healthy anger in their own personality, their own habitual reactions to anger in the client may trigger their own fear or shame, which might drive the wish to stop the anger for fear of aggression or the wish to tell the person off for being angry. If a therapist cannot tolerate a patient's anger well, then the therapist is likely to shame the client for being angry or to react with anger or even go into a freeze response themselves. Consequently, the patient is likely to feel invalidated or shamed, thus potentially dropping out of therapy or returning to submissive relational patterns. Personal inner work is necessary on the side of the therapists to empathize with the anger or rage the client might be feeling and to tolerate it easily to avoid harming the client with pseudo yin compassion offered in the service of suppression of anger, shaming and ultimately retraumatization.

Once clients have stopped dissociating as much from their suffering, they begin to see the measure of cruelty that has been inflicted on them. Whilst this is an important step to integrating traumatic memories, it understandably can give rise to rage and revenge fantasies against the perpetrators and any bystanders who allowed this to happen. Rage and revenge fantasies are an attempt to reduce the feeling of powerlessness experienced at the time of the traumatic events and experience now at not being able to change the past. Through validation and staying present with these feelings and fantasies, they can be transformed into healthy self-protective anger and assertiveness and eventually give way for grief to naturally arise once there is enough of a secure base within the client.

Therapists working with complex PTSD clients are likely to suffer empathy fatigue as they absorb the many intense emotions the client experiences. Self-compassion training outside the office can help therapists build the resilience necessary to regulate their own empathic distress and shift to a compassionate stance that allows them to be exactly where the client is, not where the therapist wants them to be. If therapists are not conscious of their own aversion to anger or revenge fantasies, then they may tend to want to bypass anger in the client instead of working directly with it and helping the client transform rage into assertiveness. The therapist may push the client to consider the forgiveness prematurely by inviting the client to empathize with the perpetrators instead of remaining the client's loyal and wise advocate who validates the victim's perspective fully first. Clients are usually deeply afraid of their own anger and rage, as expressing these feelings at the time of the trauma would have meant more cruelty and even death. The therapist needs to empathize with the victim and these fears yet see that this is a healthy and important step on the healing journey.

Given the abundance of "forgiveness practices" available, therapist may think that forgiveness is a skill to be practiced and a task to be accomplished instead of a process that unfolds naturally within the client as a by-product of the other therapeutic work over a long period of time. Once dissociation stops and all feelings have been felt and integrated, a bigger perspective of oneself and of life and on one's painful past and hopefully positive future naturally emerges. This increased wisdom and equanimity usually gives rise to a more balanced outlook, including at the perpetrators. Actively guiding the client to get into the heads of the perpetrator from a place of goodwill when their intentions were ill will, could be considered manipulative and retraumatizing. Furthermore, clients may bypass the outrage about the atrocities they have experienced by escaping into fantasies of being able to forgive and absolve the perpetrator (Herman, 1992, p. 189). Therapists who are unaware of the importance of integrating outrage and anger may breathe a sigh of relief at hearing this and may inadvertently feed this fantasy of grand forgiveness by introducing forgiveness practices. All extreme fantasies of the client either annihilating or of loving the perpetrator can be viewed as attempts to deal with the underlying powerlessness and hurt and prevent the person from mourning the losses. Therapists thus need to let go of any agendas and meet the clients where they are at. Therapists have a duty of care, thus they need to be open to all emotions and take personal responsibility for their own well-being by regulating the vicarious distress they might be feeling themselves instead of by closing down or manipulating the experience of the client.

Offering Solidarity We can apply the three components of yang self-compassion to the therapeutic relationship. Firstly, the client should feel safe to vent her rage and anger with the therapist and be met with radical understanding and acceptance (mindfulness). Secondly, to help move from fantasies of rage and revenge to a more grounded and ultimately powerful feeling of "righteous indignation" (Herman, 1992, p. 189), the trauma victim needs to feel actual support from others, i.e., to directly experience solidarity instead of standing alone and having to resort to grandiose ideas of power though love or hatred (common humanity). This involves being believed by others and having authorities and others join forces to bring the perpetrator to justice. The feeling of safety in numbers allows the victim to voice her outrage, too. The therapist needs to convey clearly that they believe the client and offer themselves as advocates who share the outrage about the injustice and support the client in accessing a wider sense of solidarity and strength in numbers. Group therapy with other trauma survivors or other peer groups might help at this stage, if the person feels ready. Thirdly, the question arises what the wisest action would be in this moment that would promote the well-being in the client in the longer-term. The therapist should trust the wisdom of the client to know and if the client struggles to work toward finding functional ways to protect and assert themselves. These may vary depending on the situation. It is important to guide the client to connect with her inner wisdom instead of staying stuck in rage or submission. All these three steps of yang compassion require courage, strength, and a degree of fearlessness of the therapists.

Informal Practice for Trauma Survivors: Yang Self-Compassion Break for Protection As clients learn to manage feelings of shame and guilt and develop the capacity to practice fierce self-compassion with regard to past experiences, they may feel more confident asserting themselves. Being reminded of the "steps" to follow when injustice is experienced (or re-experienced) can be a supportive informal practice for when these moments arise. The Yang Self-Compassion Break (Box 20.1), taken from Braehler and Neff (2020) was developed to validate the truth of trauma survivors' experience, to nurture self-protection, and to evoke courage and wisdom. While the practice is infused with fierce self-compassion, through a strong, calm, and assertive stance, it is not aggressive.

Grief and Mourning

After shame and guilt have abated and righteous anger has been transformed into assertiveness and self-protection, it is common for trauma survivors to move into an experience of grief and mourning. Grief may relate to both the atrocities endured and feelings of loss connected to the impact of these experiences in quality of life. Mourning is therefore a central part of trauma recovery. By providing a yin compassionate presence to bear witness to the client's pain, therapists can support clients to gradually grieve whilst feeling held. As clients become more accustomed to receiving compassion from the therapist, fears of compassion may gradually diminish.

Over time, the client may also start to internalize both the yin and yang aspects of compassion, so that they are able to hold the assertive and the vulnerable parts of themselves simultaneously. A typical barrier to mourning and crying is the belief that it is a weakness, which makes clients vulnerable to further attacks. Again, the therapist never pursues an agenda to make the client feel anything as this would be trying to control or manipulate when in fact the trauma client needs to be empowered. Instead, the therapist holds the client in their hearts and thus creates a loving space in which the client can go at his or her own pace. Once enough trust and compassion is present within the client, grief naturally arises.

Box 20.1: Yang Self-Compassion Break for Protection (from Braehler & Neff, 2020)

1. **Mindfulness of Suffering**
   (a) Validating the pain or hurt or injustice you are experiencing.
   "This is my truth. I believe what I experienced. I trust myself despite what others might be saying to invalidate me!"
   (b) Courage.
   "I dare speak my truth starting with myself."
2. **Common Humanity and Solidarity**
   (a) Empowering yourself to reach out to and to trust in others to share your truth.
   (b) Feeling connected in suffering to others who experience similar suffering and to feel solidarity in protecting yourself against future hurts.
3. **Fierce Self-Kindness**
   (a) What do I truly need to protect myself or to support myself or to stand up for myself?
   (b) What is the wisest thing for me to do in this situation in the short-term and in the longer-term? To say no? To draw my boundaries? To speak my truth? To walk away? To gather support to stand our ground and keep making our voices being heard?

Informal Practice for Trauma Survivors: Yin Self-Compassion Break for Grief When clients experience grief and are in mourning, *The Yin Self-Compassion Break for Grief* (Braehler & Neff, 2020; see Box 20.2) can be a supportive practice. This practice can support clients to soothe themselves and feel more confident maintaining what they have learnt in therapy. While the practice can be applied between sessions, it is important that the client knows how to connect with their bodily experiences without dissociating and to access their soothing system in a safe way.

Box 20.2: Yin Self-Compassion Break for Grief

1. **Mindfulness: To Notice, Name, and Validate the Distress**
   Feel the grief in your body for now. Let yourself know that you get why you feel this now. For example: *"It is tragic that this happened to me. It is so painful to have missed out on happiness, joy and ease. It is so understandable that I would feel like this at this point."*
2. **Common Humanity: To Help You Feel Connected Instead of Isolated**
   Think of at least one person you know who has experienced similar pain. You do not even have to know them personally. For example: *"Even though it feels like I am all alone, I also feel some connection to my two friends from group therapy. They are people who really understand what it is like to have experienced these unspeakable things."*
3. **Soothing Touch: To Bring the Care Physiology Online**
   If it is comfortable, place the hand on a part of the body where touch feels comforting or touches a warm blanket or warm mug of tea or stroke a pet.
4. **Self-Kindness: To Offer Yourself Inner Guidance and Comfort**
   Ask yourself: what comforting words or gestures do you need to receive right now? From whom would they feel safest to receive? For example: *"I am here for you, my dear. Luckily the hurt is over. May you be extra gentle and kind with yourself as you go through this grief remembering that it will pass, too. I will take good care of you and allow you to feel whatever you are feeling."*

The Final Phase: Learning to Thrive

The final phase of working with trauma moves beyond merely surviving and toward learning to thrive. This involves becoming acquainted – and comfortable with – feeling states and experiences that may be very foreign to the client, such as happiness and play. Because they have spent most of their life in survival mode, most clients with complex PTSD experience chronic tension and hyperarousal, sleep and digestive problems, restlessness, and hypervigilance. Just as early experiences can give rise to fears of compassion, these experiences may also make the client vulnerable to fears about experiencing joy and happiness (Şar et al., 2019). For instance, a client who had early experiences of punishment or shame when they were happy or at play will learn quickly to suppress happiness and may be actively avoidant or dismissive of experiences akin to play or relaxation. Similar to fears of compassion, fears of happiness are related to both attachment insecurity and alexithymia in clinically depressed groups (Joshanloo, 2018; Gilbert et al. 2014b). By modeling and supporting the safe expression of joy and happiness and finding opportunities for playfulness once trust is established, the therapist can work with the client to develop a more accepting and less fearful response to these emotions.

Conclusion

Evidence suggests that self-compassion can be a powerful antidote to shame and other difficult interpersonal experiences that characterize complex PTSD. However, the mechanistic perspective of adding self-compassion "tools" to the toolbox of treatment of complex of PTSD without developing a more comprehensive interpersonal and attachment-based formulation including early caregiving experiences, attachment patterns, mentalizing capacity, shame, guilt, betrayal, fears of compassion, and happiness is unlikely to accommodate the many variables influencing treatment outcome. Compassion-based work is attachment-based work and taps directly into early experiences of attention, care, love, and appreciation, or lack thereof. As with all trauma-based works, it requires clinicians to be sensitive, flexible, diligent, and willing to become a safe haven and secure base for their clients until they learn to become a safe haven and secure base for themselves. This chapter provides comprehensive guidance on how therapists can work with their own and their clients' emotional and relational experiences in a compassionate way in the treatment of complex PTSD. I trust that the processes and principles presented in this chapter can easily be transferred to the integration of self-compassion into the psychological conceptualization and treatment of other complex mental health disorders.

References

American Psychiatric Association. (2013). *Diagnostic and statistical manual of mental disorders* (5th ed.). American Psychiatric Publishing.

Andrews, B., Brewin, C. R., Rose, S., & Kirk, M. (2000). Predicting PTSD symptoms in victims of violent crime: The role of shame, anger, and childhood abuse. *Journal of Abnormal Psychology, 109*(1), 69. https://doi.org/10.1037//0021-843x.109.1.69

Baldwin, S., Bandarian-Balooch, S., & Adams, R. (2020). Attachment and compassion-threat: Influence of a secure attachment-prime. *Psychology and Psychotherapy: Theory, Research and Practice, 93*, 520–536. https://doi.org/10.1111/papt.12244

Barlow, M. R., Goldsmith Turow, R. E., & Gerhart, J. (2017). Trauma appraisals, emotion regulation difficulties, and self-compassion predict post-traumatic stress symptoms following childhood abuse. *Child Abuse & Neglect, 65*, 37–47. https://doi.org/10.1016/j.chiabu.2017.01.006

Bartz, J. A., Zaki, J., Bolger, N., & Ochsner, K. N. (2011). Social effects of oxytocin in humans: context and person matter. *Trends in Cognitive Sciences, 15*(7), 301–309. https://doi.org/10.1016/j.tics.2011.05.002

Bartz, J. A., Zaki, J., Ochsner, K. N., Bolger, N., Kolevzon, A., Ludwig, N., & Lydon, J. E. (2010). Effects of oxytocin on recollections of maternal care and closeness. *Proceedings of the National Academy of Sciences,*

107(50), 21371–21375. https://doi.org/10.1073/pnas.1012669107

Bateman, A., & Fonagy, P. (2006). *Mentalization-based treatment for borderline personality disorder: A practical guide*. OUP Oxford.

Berryhill, M. B., Hayes, A., & Lloyd, K. (2018). Chaotic-enmeshment and anxiety: the mediating role of psychological flexibility and self-compassion. *Contemporary Family Therapy, 40*, 326–337. https://doi.org/10.1007/s10591-018-9461-2

Bistricky, S. L., Gallagher, M. W., Roberts, C. M., Ferris, L., Gonzalez, A. J., & Wetterneck, C. T. (2017). Frequency of interpersonal trauma types, avoidant attachment, self-compassion, and interpersonal competence: A model of persisting post-traumatic symptoms. *Journal of Aggression, Maltreatment & Trauma, 26*(6), 608–625. https://doi.org/10.1080/10926771.2017.1322657

Bohus, M., Schmahl, C., Fydrich, T., Steil, R., Müller-Engelmann, M., Herzog, J., Ludäscher, P., Kleindienst, N., & Priebe, K. (2019). A research programme to evaluate DBT-PTSD, a modular treatment approach for Complex PTSD after childhood abuse. *Borderline Personality Disorder and Emotion Dysregulation, 6*(1), 7. https://doi.org/10.1186/s40479-019-0099-y

Bowlby, J. (1982). *Attachment*. Hogarth Press and the Institute of Psycho-Analysis.

Bowlby, J. (1988). *A secure base: Clinical applications of attachment theory*. Routledge.

Boykin, D. M., Himmerich, S. J., Pinciotti, C. M., Miller, L. M., Miron, L. R., & Orcutt, H. K. (2018). Barriers to self-compassion for female survivors of childhood maltreatment: The roles of fear of self-compassion and psychological inflexibility. *Child Abuse & Neglect, 76*(Supplement C), 216–224. https://doi.org/10.1016/j.chiabu.2017.11.003

Braehler, C., Gumley, A., Harper, J., Wallace, S., Norrie, J., & Gilbert, P. (2013a). Exploring change processes in compassion-focused therapy in psychosis: Results of a feasibility randomized controlled trial. *British Journal of Clinical Psychology, 52*(2), 199–214. https://doi.org/10.1111/bjc.12009

Braehler, C., Harper, J., & Gilbert, P. (2013b). Compassion-focused group therapy for recovery after psychosis. In C. Steel (Ed.), *Cognitive behaviour therapy for schizophrenia: Evidence-based interventions and future directions* (pp. 236–266). Wiley. https://doi.org/10.13140/2.1.3956.2886.

Braehler, C., & Neff, K. D. (2020). Self-compassion in PTSD. In M. T. Tull & N. A. Kimbrel (Eds.), *Emotion in post-traumatic stress disorder (pp. 567–596)*. Elsevier Academic Press. https://doi.org/10.1016/B978-0-12-816022-0.00020-X

Brewin, C. R., & Holmes, E. A. (2003). Psychological theories of post-traumatic stress disorder. *Clinical Psychology Review, 23*(3), 339–376. https://doi.org/10.1016/S0272-7358(03)00033-3

Briere, J. N. (1992). *Child abuse trauma: Theory and treatment of the lasting effects*. Sage Publications.

Brown, M. Z., Linehan, M. M., Comtois, K. A., Murray, A., & Chapman, A. L. (2009). Shame as a prospective predictor of self-inflicted injury in borderline personality disorder: A multi-modal analysis. *Behaviour Research and Therapy, 47*(10), 815–822. https://doi.org/10.1016/j.brat.2009.06.008

Carden, L. J., Saini, P., Seddon, C., Watkins, M., & Taylor, P. J. (2020). Shame and the psychosis continuum: A systematic review of the literature. *Psychology and Psychotherapy: Theory, Research and Practice, 93*(1), 160–186. https://doi.org/10.1111/papt.12204

Catarino, F., Gilbert, P., McEwan, K., & Baião, R. (2014). Compassion motivations: Distinguishing submissive compassion from genuine compassion and its association with shame, submissive behavior, depression, anxiety and stress. *Journal of Social and Clinical Psychology, 33*(5), 399–412. https://doi.org/10.1521/jscp.2014.33.5.399

Cloitre, M., Koenen, K. C., Cohen, L. R., & Han, H. (2002). Skills training in affective and interpersonal regulation followed by exposure: a phase-based treatment for PTSD related to childhood abuse. *Journal of Consulting and Clinical Psychology, 70*(5), 1067–1074. https://doi.org/10.1037/0022-006X.70.5.1067

Cunningham, K. C., LoSavio, S. T., Dennis, P. A., Farmer, C., Clancy, C. P., Hertzberg, M. A., Kimbrel, N. A., Calhoun, P. S., & Beckham, J. C. (2018). Shame as a mediator between post-traumatic stress disorder symptoms and suicidal ideation among veterans. *Journal of Affective Disorders, 243*, 216–219. https://doi.org/10.1016/j.jad.2018.09.040

Dahm, K. A., Meyer, E. C., Neff, K. D., Kimbrel, N. A., Gulliver, S. B., & Morissette, S. B. (2015). Mindfulness, self-compassion, post-traumatic stress disorder symptoms, and functional disability in US Iraq and Afghanistan war veterans. *Journal of Traumatic Stress, 28*(5), 460–464. https://doi.org/10.1002/jts.22045

DePrince, A. P., Chu, A. T., & Pineda, A. S. (2011a). Links between specific post-trauma appraisals and three forms of trauma-related distress. *Psychological Trauma: Theory, Research, Practice, and Policy, 3*(4), 430. https://doi.org/10.1037/a0021576.

DePrince, A. P., Chu, A. T., & Pineda, A. S. (2011b). Links between specific post-trauma appraisals and three forms of trauma-related distress. *Psychological Trauma: Theory, Research, Practice, and Policy, 3*(4), 430–441. https://doi.org/10.1037/a0021576

DePrince, A. P., & Freyd, J. J. (2002). The intersection of gender and betrayal in trauma. In R. Kimerling,

P. C. Ouimette, & J. Wolfe (Eds.), *Gender and PTSD* (pp. 98–113). Guilford Press.

Ebert, A., Edel, M. A., Gilbert, P., & Brüne, M. (2018). Endogenous oxytocin is associated with the experience of compassion and recalled upbringing in Borderline Personality Disorder. *Depression and Anxiety, 35*(1), 50–57. https://doi.org/10.1002/da.22683

Ferrari, M., Hunt, C., Harrysunker, A., Abbott, M. J., Beath, A. P., & Einstein, D. A. (2019). Self-compassion interventions and psychosocial outcomes: a meta-analysis of RCTs. *Mindfulness, 10*(8), 1455–1473. https://doi.org/10.1007/s12671-019-01134-6

First, M. B., Reed, G. M., Hyman, S. E., & Saxena, S. (2015). The development of the ICD-11 clinical descriptions and diagnostic guidelines for mental and behavioural disorders. *World Psychiatry, 14*(1), 82–90. https://doi.org/10.1002/wps.20189

Fonagy, P., Gergely, G., Jurist, E. L., & Target, M. (2002). *Affect regulation, mentalization, and the development of the self.* Other Press.

Freyd, J. J. (1996). *Betrayal trauma: The logic of forgetting childhood abuse*. Harvard University Press.

Galili-Weinstock, L., Chen, R., Atzil-Slonim, D., Bar-Kalifa, E., Peri, T., & Rafaeli, E. (2017). The association between self-compassion and treatment outcomes: Session-level and treatment-level effects. *Journal of Clinical Psychology, 74*(6), 849–866. https://doi.org/10.1002/jclp.22569

Galili-Weinstock, L., Chen, R., Atzil-Slonim, D., Rafaeli, E., & Peri, T. (2019). Enhancement of self compassion in psychotherapy: The role of therapists' interventions. *Psychotherapy Research, 30*(6), 1–14. https://doi.org/10.1080/10503307.2019.1650979

Gilbert, P. (2010). *Compassion-focused therapy: Distinctive features.* Routledge.

Gilbert, P., McEwan, K., Catarino, F., & Baião, R. (2014a). Fears of compassion in a depressed population implication for psychotherapy. *Journal of Depression and Anxiety, S2*(003), 1–8. https://doi.org/10.4172/2167-1044.S2-003

Gilbert, P., McEwan, K., Catarino, F., & Baião, R. (2014b). Fears of negative emotions in relation to fears of happiness, compassion, alexithymia and psychopathology in a depressed population: A preliminary study. *Journal of Depression and Anxiety, S2*(004), 1–7. https://doi.org/10.4172/2167-1044.S2-004

Gilbert, P., McEwan, K., Catarino, F., Baião, R., & Palmeira, L. (2014c). Fears of happiness and compassion in relationship with depression, alexithymia, and attachment security in a depressed sample. *British Journal of Clinical Psychology, 53*(2), 228–244. https://doi.org/10.1111/bjc.12037

Gilbert, P., McEwan, K., Gibbons, L., Chotai, S., Duarte, J., & Matos, M. (2012). Fears of compassion and happiness in relation to alexithymia, mindfulness, and self-criticism. *Psychology and Psychotherapy: Theory, Research and Practice, 85*(4), 374–390. https://doi.org/10.1111/j.2044-8341.2011.02046.x

Gilbert, P., McEwan, K., Matos, M., & Rivis, A. (2011). Fears of compassion: Development of three self-report measures. *Psychology and Psychotherapy: Theory, Research and Practice, 84*(3), 239–255. https://doi.org/10.1348/147608310X526511

Henry, W. P., Schacht, T. E., & Strupp, H. H. (1990). Patient and therapist introject, interpersonal process, and differential psychotherapy outcome. *Journal of Consulting and Clinical Psychology, 58*(6), 768–774. https://doi.org/10.1037/0022-006X.58.6.768

Herman, J. L. (1992). Complex PTSD: A syndrome in survivors of prolonged and repeated trauma. *Journal of Traumatic Stress, 5*(3), 377–391. https://doi.org/10.1002/jts.2490050305

Herman, J. L. (2015). *Trauma and recovery: The aftermath of violence - From domestic abuse to political terror.* Hachette UK.

Hiraoka, R., Meyer, E. C., Kimbrel, N. A., DeBeer, B. B., Gulliver, S. B., & Morissette, S. B. (2015). Self-compassion as a prospective predictor of PTSD symptom severity among trauma-exposed U.S. Iraq and Afghanistan war veterans. *Journal of Traumatic Stress, 28*(2), 127–133. https://doi.org/10.1002/jts.21995

Hoffart, A., Øktedalen, T., & Langkaas, T. F. (2015). Self-compassion influences PTSD symptoms in the process of change in trauma-focused cognitive-behavioral therapies: a study of within-person processes. *Frontiers in Psychology, 6*, 1273. https://doi.org/10.3389/fpsyg.2015.01273

Holmes, E. A., Grey, N., & Young, K. A. D. (2005). Intrusive images and "hotspots" of trauma memories in Post-traumatic Stress Disorder: An exploratory investigation of emotions and cognitive themes. *Journal of Behavior Therapy and Experimental Psychiatry, 36*(1), 3–17. https://doi.org/10.1016/j.jbtep.2004.11.002

Homan, K. J. (2016). Self-compassion and psychological well-being in older adults. *Journal of Adult Development, 23*(2), 111–119. https://doi.org/10.1007/s10804-016-9227-8

Hyland, P., Murphy, J., Shevlin, M., Vallières, F., McElroy, E., Elklit, A., Christofferson, M., & Cloitre, M. (2017). Variation in post-traumatic response: the role of trauma type in predicting ICD-11 PTSD and CPTSD symptoms. *Social Psychiatry and Psychiatric Epidemiology, 52*(6), 727–736. https://doi.org/10.1007/s00127-017-1350-8

Játiva, R., & Cerezo, M. A. (2014). The mediating role of self-compassion in the relationship between victimization and psychological maladjustment in a sample of adolescents. *Child Abuse & Neglect, 38*(7), 1180–1190. https://doi.org/10.1016/j.chiabu.2014.04.005

Jiang, Y., You, J., Hou, Y., Du, C., Lin, M.-P., Zheng, X., & Ma, C. (2016). Buffering the effects of peer victimization on adolescent non-suicidal self-injury: The role of self-compassion and family cohesion. *Journal of*

Adolescence, 53, 107–115. https://doi.org/10.1016/j.adolescence.2016.09.005

Jiang, Y., You, J., Zheng, X., & Lin, M.-P. (2017). The qualities of attachment with significant others and self-compassion protect adolescents from non-suicidal self-injury. *School Psychology Quarterly, 32*(2), 143–155. https://doi.org/10.1037/spq0000187

Joeng, J. R., Turner, S. L., Kim, E. Y., Choi, S. A., Lee, Y. J., & Kim, J. K. (2017). Insecure attachment and emotional distress: Fear of self-compassion and self-compassion as mediators. *Personality and Individual Differences, 112*, 6–11. https://doi.org/10.1016/j.paid.2017.02.048

Joshanloo, M. (2018). Fear and fragility of happiness as mediators of the relationship between insecure attachment and subjective well-being. *Personality and Individual Differences, 123*, 115–118. https://doi.org/10.1016/j.paid.2017.11.016

Karatzias, T., Shevlin, M., Fyvie, C., Hyland, P., Efthymiadou, E., Wilson, D., Roberts, N., Bisson, J. I., Brewin, C. R., & Cloitre, M. (2017). Evidence of distinct profiles of post-traumatic stress disorder (PTSD) and complex post-traumatic stress disorder (CPTSD) based on the new ICD-11 Trauma Questionnaire (ICD-TQ). *Journal of Affective Disorders, 207*, 181–187. https://doi.org/10.1016/j.jad.2016.09.032

Kelly, A. C., & Carter, J. C. (2013). Why self-critical patients present with more severe eating disorder pathology: The mediating role of shame. *British Journal of Clinical Psychology, 52*(2), 148–161. https://doi.org/10.1111/bjc.12006

Kelly, A. C., & Tasca, G. A. (2016). Within-persons predictors of change during eating disorders treatment: An examination of self-compassion, self-criticism, shame, and eating disorder symptoms: Within-person predictors. *The International Journal of Eating Disorders, 49*(7), 716–722. https://doi.org/10.1002/eat.22527

Kelly, A. C., Wisniewski, L., Martin-Wagar, C., & Hoffman, E. (2017). Group-based compassion-focused therapy as an adjunct to outpatient treatment for eating disorders: A pilot randomized controlled trial. *Clinical Psychology & Psychotherapy, 24*(2), 475–487. https://doi.org/10.1002/cpp.2018

Keltner, D., & Harker, L. (1998). The forms and functions of the nonverbal signal of shame.

Kessler, R. C., Aguilar-Gaxiola, S., Alonso, J., Benjet, C., Bromet, E. J., Cardoso, G., Degenhardt, L., de Girolamo, G., Dinolova, R. V., Ferry, F., Florescu, S., Gureje, O., Haro, J. M., Huang, Y., Karam, E. G., Kawakami, N., Lee, S., Lepine, J., Levinson, D., et al. (2017). Trauma and PTSD in the WHO World Mental Health Surveys. *European Journal of Psychotraumatology, 8*(sup5), 1353383–1353383. https://doi.org/10.1080/20008198.2017.1353383

Kim, S., Thibodeau, R., & Jorgensen, R. S. (2011). Shame, guilt, and depressive symptoms: a meta-analytic review. *Psychological Bulletin, 137*(1), 68–96. https://doi.org/10.1037/a0021466

Kirby, J. N., Day, J., & Sagar, V. (2019). The 'Flow' of compassion: A meta-analysis of the fears of compassion scales and psychological functioning. *Clinical Psychology Review, 70*, 26–39. https://doi.org/10.1016/j.cpr.2019.03.001

Kirby, J. N., Doty, J. R., Petrocchi, N., & Gilbert, P. (2017a). The current and future role of heart rate variability for assessing and training compassion. *Frontiers in Public Health, 5*, Article 40. https://doi.org/10.3389/fpubh.2017.00040.

Kirby, J. N., Tellegen, C. L., & Steindl, S. R. (2017b). A meta-analysis of compassion-based interventions: Current state of knowledge and future directions. *Behavior Therapy, 48*(6), 778–792. https://doi.org/10.1016/j.beth.2017.06.003

Lee, D. A., Scragg, P., & Turner, S. (2001). The role of shame and guilt in traumatic events: A clinical model of shame-based and guilt-based PTSD. *British Journal of Medical Psychology, 74*(4), 451–466. https://doi.org/10.1348/000711201161109

Liotti, G. (2004). Trauma, dissociation, and disorganized attachment: Three strands of a single braid. *Psychotherapy: Theory, Research, Practice, Training, 41*(4), 472–486. https://doi.org/10.1037/0033-3204.41.4.472

Maercker, A., Brewin, C. R., Bryant, R. A., Cloitre, M., Reed, G. M., van Ommeren, M., Humayun, A., Jones, L. M., Kagee, A., Llosa, A. E., Rousseau, C., Somasundaram, D. J., Souza, R., Suzuki, Y., Weissbecker, I., Wessely, S. C., First, M. B., & Saxena, S. (2013a). Proposals for mental disorders specifically associated with stress in the International Classification of Diseases-11. *The Lancet, 381*(9878), 1683–1685. https://doi.org/10.1016/S0140-6736(12)62191-6

Maercker, A., Brewin, C. R., Bryant, R. A., Cloitre, M., van Ommeren, M., Jones, L. M., Humayan, A., Kagee, A., Llosa, A. E., Rousseau, C., Somasundaram, D. J., Souza, R., Suzuki, Y., Weissbecker, I., Wessely, S. C., First, M. B., & Reed, G. M. (2013b). Diagnosis and classification of disorders specifically associated with stress: proposals for ICD-11. *World Psychiatry, 12*(3), 198–206. https://doi.org/10.1002/wps.20057

Maheux, A., & Price, M. (2016). The indirect effect of social support on post-trauma psychopathology via self-compassion. *Personality and Individual Differences, 88*, 102–107. https://doi.org/10.1016/j.paid.2015.08.051

Matos, M., Duarte, J., & Pinto-Gouveia, J. (2017). The origins of fears of compassion: Shame and lack of safeness memories, fears of compassion and psychopathology. *The Journal of Psychology, 151*(8), 804–819. https://doi.org/10.1080/00223980.2017.1393380

Matos, M., & Pinto-Gouveia, J. (2010). Shame as a traumatic memory. *Clinical Psychology & Psychotherapy, 17*(4), 299–312. https://doi.org/10.1002/ccp.659

McEwan, K., & Gilbert, P. (2016). A pilot feasibility study exploring the practising of compassionate imagery exercises in a nonclinical population. *Psychology and Psychotherapy: Theory, Research and Practice, 89*(2), 239–243. https://doi.org/10.1111/papt.12078

Miron, L. R., Seligowski, A. V., Boykin, D. M., & Orcutt, H. K. (2016). The potential indirect effect of childhood abuse on post-trauma pathology through self-compassion and fear of self-compassion. *Mindfulness, 7*(3), 596–605. https://doi.org/10.1007/s12671-016-0493-0

Neff, K. D. (2003a). The development and validation of a scale to measure self-compassion. *Self and Identity, 2*(3), 223–250. https://doi.org/10.1080/15298860309027

Neff, K. D. (2003b). Self-compassion: An alternative conceptualization of a healthy attitude toward oneself. *Self and Identity, 2*(2), 85–101. https://doi.org/10.1080/15298860309032

Neff, K. D., & Germer, C. (2018). *The mindful self-compassion workbook: A proven way to accept yourself, build inner strength, and thrive*. Guilford Press.

Neff, K. D., & McGehee, P. (2010). Self-compassion and psychological resilience among adolescents and young adults. *Self and Identity, 9*(3), 225–240. https://doi.org/10.1080/15298860902979307

Platt, M. G., & Freyd, J. J. (2015). Betray my trust, shame on me: Shame, dissociation, fear, and betrayal trauma. *Psychological Trauma: Theory, Research, Practice, and Policy, 7*(4), 398–404. https://doi.org/10.1037/tra0000022

Rockliff, H., Gilbert, P., McEwan, K., Lightman, S., & Glover, D. (2008). A pilot exploration of heart rate variability and salivary cortisol responses to compassion-focused imagery. *Clinical Neuropsychiatry, 5*(3), 132–139.

Rockliff, H., Karl, A., McEwan, K., Gilbert, J., Matos, M., & Gilbert, P. (2011). Effects of intranasal oxytocin on compassion-focused imagery'. Emotion, 11(6), 1388–1396. https://doi.org/10.1037/a0023861.

Roisman, G. I., Padrón, E., Sroufe, L. A., & Egeland, B. (2002). Earned–secure attachment status in retrospect and prospect. *Child Development, 73*(4), 1204–1219. https://doi.org/10.1111/1467-8624.00467

Şar, V., Türk, T., & Öztürk, E. (2019). Fear of happiness among college students: The role of gender, childhood psychological trauma, and dissociation. *Indian Journal of Psychiatry, 61*(4), 389–394. https://doi.org/10.4103/psychiatry.IndianJPsychiatry_52_17

Saraiya, T., & Lopez-Castro, T. (2016). Ashamed and afraid: A scoping review of the role of shame in post-traumatic stress disorder (PTSD). *Journal of Clinical Medicine, 5*(11), 94. https://doi.org/10.3390/jcm5110094

Schanche, E., Stiles, T. C., McCullough, L., Svartberg, M., & Nielsen, G. H. (2011). The relationship between activating affects, inhibitory affects, and self-compassion in patients with Cluster C personality disorders. *Psychotherapy, 48*(3), 293–303. https://doi.org/10.1037/a0022012

Schore, A. N. (2015). *Affect regulation and the origin of the self: The neurobiology of emotional development*. Routledge.

Scoglio, A. J., Rudat, D. A., Garvert, D., Jarmolowski, M., Jackson, C., & Herman, J. L. (2018). Self-compassion and responses to trauma: The role of emotion regulation. *Journal of Interpersonal Violence, 33*(13), 2016–2036. https://doi.org/10.1177/0886260515622296

Sznycer, D., Xygalatas, D., Agey, E., Alami, S., An, X. F., Ananyeva, K. I., Atkinson, Q. D., Broitman, B. R., Conte, T. J., Flores, C., Fukushima, S., Hitokoto, H., Kharitonov, A. N., Onyishi, C. N., Romero, P. P., Schrock, J. M., Snodgradd, J. J., Sugiyama, L. S., Takemura, K., et al. (2018). Cross-cultural invariances in the architecture of shame. *Proceedings of the National Academy of Sciences, 115*(39), 9702–9707. https://doi.org/10.1073/pnas.1805016115

Tanaka, M., Wekerle, C., Schmuck, M. L., & Paglia-Boak, A. (2011). The linkages among childhood maltreatment, adolescent mental health, and self-compassion in child welfare adolescents. *Child Abuse & Neglect, 35*(10), 887–898. https://doi.org/10.1016/j.chiabu.2011.07.003

Taylor, T. F. (2015). The influence of shame on post-trauma disorders: Have we failed to see the obvious? *European Journal of Psychotraumatology, 6*(1), Article 28847.. https://doi.org/10.3402/ejpt.v6.28847.

van der Kolk, B. A., Roth, S., Pelcovitz, D., & Mandel, F. (1993). *Complex PTSD: Results of the PTSD field trial for DSM-IV*. American Psychiatric Association.

van der Kolk, B. A., Roth, S., Pelcovitz, D., Sunday, S., & Spinazzola, J. (2005). Disorders of extreme stress: The empirical foundation of a complex adaptation to trauma. *Journal of Traumatic Stress, 18*(5), 389–399. https://doi.org/10.1002/jts.20047

Van Ijzendoorn, M. H., Schuengel, C., & Bakermans–Kranenburg, M. J. (1999). Disorganized attachment in early childhood: Meta-analysis of precursors, concomitants, and sequelae. *Development and Psychopathology, 11*(2), 225–250. https://doi.org/10.1017/s0954579499002035

Vettese, L. C., Dyer, C. E., Li, W. L., & Wekerle, C. (2011). Does self-compassion mitigate the association between childhood maltreatment and later emotion regulation difficulties? A preliminary investigation. *International Journal of Mental Health and Addiction, 9*(5), 480. https://doi.org/10.1007/s11469-011-9340-7

Weingarden, H., Renshaw, K. D., Wilhelm, S., Tangney, J. P., & DiMauro, J. (2016). Anxiety and shame as risk factors for depression, suicidality, and functional impairment in body dysmorphic disorder and obsessive compulsive disorder. *The Journal of Nervous and Mental Disease, 204*(11), 832–839. https://doi.org/10.1097/NMD.0000000000000498

Winders, S.-J., Murphy, O., Looney, K., & O'Reilly, G. (2020). Self-compassion, trauma, and post-traumatic stress disorder: A systematic review. *Clinical Psychology & Psychotherapy, 27*(3), 300–329. https://doi.org/10.1002/cpp.2429

Zarse, E. M., Neff, M. R., Yoder, R., Hulvershorn, L., Chambers, J. E., & Chambers, R. A. (2019). The adverse childhood experiences questionnaire: Two decades of research on childhood trauma as a primary cause of adult mental illness, addiction, and medical diseases. *Cogent Medicine, 6*(1), Article 1581447. https://doi.org/10.1080/2331205X.2019.1581447.

Zeller, M., Yuval, K., Nitzan-Assayag, Y., & Bernstein, A. (2015). Self-compassion in recovery following potentially traumatic stress: Longitudinal study of at-risk youth. *Journal of Abnormal Child Psychology, 43*(4), 645–653. https://doi.org/10.1007/s10802-014-9937-y

21 Self-Compassion and Non-suicidal Self-Injury

Penelope Hasking

What Is NSSI?

Non-suicidal self-injury (NSSI) is a growing health concern for many clinicians, parents, teachers, and the general public. Defined as the deliberate damage to body tissue without suicidal intent, and for purposes not socially or culturally sanctioned, NSSI often includes behaviours such as cutting or burning the skin, self-battery, and severe scratching (International Society for the Study of Self-Injury [ISSS], 2020). Typically first emerging among young people aged 12–14 years (Plener et al., 2015), onset of NSSI in early adulthood is also common (Kiekens et al., 2018). In community samples, approximately 18% of adolescents, 13% of young adults, and 5% of adults over 25 years of age report a history of self-injury (Swannell et al., 2014). However, university students appear more likely to engage in self-injury than their same-aged peers (20%), and 15% of students first engage in NSSI during the first 2 years of university (Kiekens et al., 2019). In clinical samples, rates of NSSI are higher, with up to 50% reporting a history of self-injury (Glenn & Klonsky, 2013; Groschwitz et al., 2015; Kaess et al., 2013).

For some time, it was believed that girls and women self-injure at higher rates than boys and men. However, meta-analytic studies suggest that there is little difference in rates of self-injury across sexes in community samples (Bresin & Schoenleber, 2015). Early studies in the area tended to include clinical samples, where females are over-represented, studies of people with borderline personality disorder, for which NSSI is one diagnostic criterion, or limited forms of self-injury (e.g. cutting) which may be more commonly reported by females. When these methodological factors are taken into account, we see that girls and women may be more likely to try self-injury, but there is little sex difference in who continues to engage in the behaviour (Swannell et al., 2014). In clinical samples, girls and women are more likely to report a history of self-injury than boys and men, but this could be due to sex differences in help-seeking, over-representation of individuals diagnosed with borderline personality disorder (more often diagnosed among women), or a tendency for girls/women to be more comfortable reporting NSSI.

Individuals can engage in NSSI for a multitude of reasons; however, the most commonly reported reason to self-injure is to reduce or avoid intense or unwanted emotion (Taylor et al., 2018). Not surprisingly then, NSSI is associated with a number of psychological concerns (e.g. low self-esteem, stress), and a range of emotional disorders, including depressive and anxiety disorders, panic disorder, and post-traumatic stress disorder

P. Hasking (✉)
School of Population Health, Curtin University, Kent St, Bentley, WA, Australia
e-mail: Penelope.Hasking@curtin.edu.au

A. Finlay-Jones et al. (eds.), *Handbook of Self-Compassion*, Mindfulness in Behavioral Health,
https://doi.org/10.1007/978-3-031-22348-8_21

(Bentley et al., 2015). Although explicitly not engaged with conscious suicidal intent, NSSI is the single most reliable predictor of subsequent suicidal thoughts and behaviours (Franklin et al., 2017; Ribeiro et al., 2016). Even among individuals with no diagnosis of mental illness, individuals reporting a history of NSSI are 5.3 times more likely to report a subsequent suicide attempt (Kiekens et al., 2018). Of note, NSSI first appears approximately 3 years before a suicide attempt, providing a critical window for intervention.

Theoretical Accounts of NSSI

Given the central role of emotion regulation in the initiation and maintenance of NSSI, it is not surprising that most theoretical accounts of NSSI focus on the experience and regulation of emotion. One of the earliest accounts, the Experiential Avoidance Model (Chapman et al., 2006) proposes that individuals vary in the extent to which they wish to avoid unpleasant or unwanted emotions. Individuals at the higher end of this spectrum are more likely to use NSSI as a means of avoiding emotions, particularly when coupled with experiencing more intense emotions (anger, shame, sadness, frustration), difficulty regulating emotion, and poor distress tolerance. NSSI provides temporary relief from the negative emotional experience and is thus negatively reinforced, and repeated when future negative emotional experiences are felt.

Nock's (2009) integrated model of NSSI incorporates distal risk factors (genetic predisposition, childhood abuse, family criticism) which are proposed to lead to intrapersonal and interpersonal vulnerabilities (e.g. poor distress tolerance, poor communication skills) that increase risk of NSSI in a stressful situation. Nock (2009) proposes that there are a number of reasons an individual may self-injure rather than engage in other coping strategies, including learning of the behaviour from others, self-punishment, or to communicate distress to others. The Emotional Cascade Model (Selby & Joiner, 2009) was one of the first to highlight the interaction between emotion and cognition in explaining why someone may engage in self-injury. Selby and Joiner (2009) propose that a repeated focus on negative affective experience, through rumination, amplifies the negative affect, which in turn encourages greater rumination. In this way, emotion and cognition interact to produce an emotional cascade, resulting in escalating negative emotional experiences. They propose that less potent coping strategies are insufficient to break this cascade, but that NSSI can effectively divert attention away from the cascade by directing attention to the self-injury.

Another model to incorporate an interaction between emotion and cognition is the Cognitive Emotional Model of self-injury (Hasking et al., 2016). In this model, Hasking et al. (2016) highlight a specific role for outcome expectancies and self-efficacy in the initiation, maintenance, and recovery from self-injury. They argue that even without having directly experienced self-injury, an individual holds beliefs about what may happen if they did engage in the behaviour (outcome expectancies). In line with operant principles, if an individual expects favourable outcomes from self-injury (e.g. emotional relief), they are more likely to self-injure than if they hold unfavourable outcomes (e.g. physical pain). These outcome expectancies interact with a person's belief in their ability to deliberately hurt themselves, or belief in their ability to resist an urge to self-injure (self-efficacy). Individuals bring these beliefs to any given situation, and when encountering an emotionally volatile situation act in accordance with these beliefs, and in line with their existing coping and emotion regulation strategies.

The most recent theoretical account of NSSI deviates away from a focus on the experience and regulation of emotion to outline the potential benefits and barriers to self-injury. In the Benefits and Barriers model, Hooley and Franklin (2017) note that self-injury serves a purpose for those who engage in the behaviour and thus has several benefits. These include emotion regulation, gratifying the desire for self-punishment, bonding with peers, and communicating distress or strength. Yet there are a number of barriers that prevent most people engaging in self-injury, including having not previously come across the behaviour and thus not being aware it exists, a reluctance to experience physical pain, a positive view of the self, and adherence to social norms that view NSSI in a negative light.

All theoretical accounts have a significant amount of empirical work to support their propositions. Of note, individuals who self-injure consistently report greater levels of experiential avoidance and use of avoidant coping strategies (Hasking et al., 2008), greater levels of negative affect and rumination (Hasking et al., 2019; Richmond et al., 2017), poor distress tolerance (Anestis et al., 2013; Slabbert et al., 2018), and a lack of alternate emotion regulation strategies (Tatnell et al., 2018; Voon et al., 2014). Individuals who self-injure report stronger expectations that NSSI will alleviate negative affect and weaker expectations of physical pain than individuals with no history of NSSI (Dawkins et al., 2019), findings which have been supported in experimental work (Dawkins et al., 2021a, b). Further, self-efficacy to resist self-injury reliably differentiates individuals with no history of NSSI, from those who have a history of NSSI, but have not engaged in the behaviour in the last 12 months, from those who continue to engage in the behaviour (Dawkins et al., 2021a, b).

Self-Criticism and NSSI

One of the most common findings in the self-injury literature is that individuals who self-injure are more self-critical than individuals who do not (Cohen et al., 2015; Itzhaky et al., 2015). Many suggest that the relationship between self-criticism and NSSI may stem from a history of childhood maltreatment or abuse. Individuals who experience a critical, hostile, or abusive childhood environment are more likely to internalise this criticism and self-injure in an effort to punish the self (Glassman et al., 2007; Hooley et al., 2018). Supporting this, perceived parental criticism is associated with self-criticism among individuals who self-injure (Baetens et al., 2015).

Of note, self-criticism appears to mediate the relationship between a number of family factors and NSSI. Among adolescents and young adults, self-criticism mediates the relationship between parent expressed emotion (i.e. the extent to which families are critical hostile, or overinvolved) and NSSI (Ammerman & Brown, 2018), as well as between parental self-criticism and NSSI (Gromtasky et al., 2017). Self-criticism also mediates the relationship between childhood emotional abuse and NSSI (Glassman et al., 2007). Further, the relationships between negative childhood experiences, the absence of positive experiences, and NSSI are mediated by self-criticism (Xavier et al., 2016c). Relatedly, Swannell and her colleagues (2012) also found the relationship between child maltreatment and NSSI was mediated by self-blame.

Several longitudinal studies support self-criticism as a temporal predictor of later NSSI (Perkins et al., 2020; Smith et al., 2020; Zelkowitz & Cole, 2019). Yet the relationship between self-criticism and NSSI may be bidirectional. Among first-year university students, NSSI at baseline predicted self-criticism 1 year later, but the reverse was not observed (Daly & Willoughby, 2019). In longitudinal work, Xavier et al. (2017) noted that NSSI at baseline predicted subsequent self-hatred, which in turn was related to NSSI 6 months later. Similarly, among a sample of adolescents from Hong Kong, You et al. (2017) observed parental control to be related to later NSSI, which in turn was associated with self-criticism, but not NSSI, 6 months later. In the same sample, You et al. (2015) observed self-criticism to interact with features of borderline personality disorder and negative emotions to predict later NSSI. Self-criticism also interacted with baseline NSSI to predict NSSI at two later time points, each 6 months apart.

While affect regulation is the most commonly reported reason for NSSI, most studies find self-punishment to be the second most common reason, and this is linked to self-criticism (Hooley et al., 2018; Nock, 2009). The self-punitive model proposes that some people are highly sensitive to failure and that real or perceived failures can impact self-worth. Consistent with this, perfectionism, parental criticism, shame, and self-criticism are strongly associated with NSSI, especially among women (Flett et al., 2012; Gong et al., 2019). Self-criticism also impacts emotional responses to pain (Fox et al., 2017). Participants who are highly self-critical report less pain during experimental pain induction, suggesting that highly critical people may use the pain of self-injury as a form of self-punishment (Fox et al., 2019). As such, self-criticism may not only remove a barrier to self-injury but may enhance the benefits. Consistent with this, cognitive interventions designed to improve self-worth are associated with both a decrease in pain tolerance and willingness to endure pain among individuals who self-injure (Hooley & St Germain, 2014). A journaling intervention, in which participants were asked to write for 5 min each day about something that made them feel good about themselves, was associated with reductions in self-criticism; there was a reduction in NSSI during the treatment period (4 weeks), but this change was not maintained over time (Hooley et al., 2018). These results might suggest that fostering a more enduring sense of self-compassion could mitigate against NSSI.

Self-Compassion and NSSI

In contrast to self-criticism, self-compassion involves relating to oneself with kindness, taking a mindful, balanced perspective on difficult experiences, and remembering that one is not alone in times of failure or struggle (Neff, 2003). Consistent with findings regarding the association between NSSI and elevated self-criticism, individuals who self-injure also report reduced self-compassion. A systematic review of studies examining the association between self-compassion and NSSI or suicidal ideation noted that all 16 included studies found a relationship between self-compassion and lower levels of self-injury, with several suggesting that self-compassion may weaken the relationship between negative life events and NSSI (Cleare et al., 2019). More recently, a meta-analysis of 18 studies reported substantial heterogeneity across studies, and a small effect size for the relationship between self-compassion and NSSI (Suh & Jeong, 2021). The authors highlighted the importance of considering both underactivation of self-compassion and overexposure to risk factors when attempting to understand the pathways that lead to NSSI.

Among adolescents, self-compassion mediates the relationship between a reduced sense of closeness to mothers, fathers, and peers and NSSI (Jiang et al., 2017b). Adolescents with a history of self-injury report greater self-judgement, isolation, and over-identification than adolescents with no such history. Of interest, adolescents who had thought about self-injury but not engaged in the behaviour reported more self-kindness and common humanity than youth who acted on these thoughts, suggesting that fostering self-kindness may be preventative and may disrupt the link between injurious thoughts and behaviour (Jiang et al., 2017a, b). Self-compassion scores are also negatively associated with using self-injury for self-punishment suggesting that fostering self-compassion may negate the need to punish oneself by self-injuring (Tuna & Gençöz, 2021). Fear or resistance associated with self-compassion also appears to play a role in moderating the link between rejection sensitivity and NSSI, such that those with high rejection sensitivity are more likely to experience NSSI when they have greater fears of self-compassion (Jiang et al., 2021).

Although self-compassion mediates the relationship between negative affect and NSSI (Hasking et al., 2019), this relationship may also be bidirectional, with lowered self-compassion associated with symptoms of depression and daily hassles, which in turn are associated with NSSI (Xavier et al., 2016a). Self-compassion may also act as a buffer in the relationship between peer victimisation and NSSI (Jiang et al., 2016) and mediate associations between gratitude, hope, and reduced NSSI (Jiang et al., 2020). Among university students, self-compassion attenuates the relationships between depressive symptoms, anxiety symptoms, and NSSI. Specifically, self-kindness and common humanity were the two salient moderators of both the depression–NSSI and anxiety–NSSI relationships (Kaniuka et al., 2020). Similar results have been observed among adolescents in Portugal, with high levels of self-compassion attenuating the relationship between symptoms of depression and NSSI (Xavier et al., 2016a, b, c). Self-kindness and mindfulness were particularly salient in reducing the strength of this relationship, while the negative scales of isolation, and over-identification increased the impact of depressive symptoms (Xavier et al., 2016a, b, c). In longitudinal work, Wu et al. (2019) observed self-compassion to moderate the relationship between behavioural impulsivity and NSSI among Chinese adolescents; a negative relationship between self-compassion and NSSI was observed for students reporting high impulsivity, suggesting self-compassion may counter the risk conferred by high levels of impulsivity.

In one of the early first-hand accounts of the role of self-compassion in NSSI, Sutherland et al. (2014) analysed comments that people made about their self-injury online (e.g. in discussion forums). The authors identified that individuals made comments that reflected self-kindness, common humanity, and mindfulness in efforts to understand and provide meaning to their self-injury. Individuals described understanding why they self-injured, acknowledging that it was a response to particularly difficult times, not a sign of psychopathology or that something was "wrong" with them. This self-kindness allowed participants to stop blaming and criticising themselves and promoted self-care through acts such as engaging in hobbies, practising mindfulness, exercising, and relaxing. This self-kindness ultimately led to individuals re-affirming a sense of self and seeing themselves as more than their self-injury.

Common humanity was evident in posts that reflected receiving compassion and understanding from others, understanding the impact of self-injury on others, and coming to terms with disclosures of self-injury (Sutherland et al., 2014). For most people, self-injury is a hidden behaviour. The significant stigma associated with self-injury means individuals are reluctant to disclose their experiences for fear of judgement from others (Staniland et al., 2020). When posting online, individuals talked about coming to accept care from others, to avoid internalising social judgement, and being more open and vulnerable with others. Finally, mindfulness was also evident in the posts. This included individuals becoming more accepting of their self-injury as well as being able to recognise distress as temporary. This appeared to foster a sense of hope for the future, even among individuals who were still engaging in self-injury. These facets of self-compassion were seen as central to NSSI recovery.

NSSI Recovery

In recent years, there has been a shift from exploring factors associated with the initiation and maintenance of NSSI to identify factors associated with recovery. In this research, "recovery" is often conceptualised as not having engaged in self-injury for over 12 months. However, when talking to people with lived experience of self-injury, the idea of recovery is much more nuanced (Lewis et al., 2019). People experience ongoing thoughts and urge to self-injure long after ceasing to engage in the behaviour, and these may never fully dissipate, although the strength will reduce over time (Kelada et al., 2018). For this reason, it is not surprising that individuals experience setbacks throughout the recovery process. A key part of NSSI recovery is finding alternate emotion regulation strategies. However, other strategies may not be as immediately effective, and it takes trial and error, and concerted effort, to practice alternate coping strategies. Along with this, an individual will develop self-efficacy to engage other coping strategies and increased self-efficacy to resist ongoing thoughts or urges to self-injure (Lewis & Hasking, 2020).

NSSI can also come with physical scars that may be visible to others. Because of this, NSSI recovery requires that the individual is able to accept any NSSI scars or make decisions about whether they choose not to conceal them (Lewis & Mehrabkhani, 2016). In line with this, NSSI recovery may involve a number of disclosures about NSSI. This could be voluntary disclosure to another individual, or it could be that another person notices NSSI scars. Either way, an individual who has self-injured needs to be able to consider the consequences of disclosure, and consider, to whom, when, and how any such disclosures may take place (Lewis & Hasking, 2020). Given NSSI is associated with significant stigma (Staniland et al., 2020), this process can be extremely difficult, and an individual may struggle with talking about or disclosing their self-injury (Rosenrot & Lewis, 2020). Given the risk of enacted stigma, disclosure may risk relationships with others, or result in a stigmatising or unhelpful response that can have a detrimental effect on the individual with lived experience of self-injury.

The relationship between self-compassion and NSSI disclosure has not been directly studied. However, prior work has demonstrated that self-compassion facilitates distress disclosure (Dupasquier et al., 2019) and help-seeking (Hermanto et al., 2017; Hermanto & Zuroff, 2016), thereby suggesting that those with greater self-compassion may be more likely to disclose their NSSI.

In summary, NSSI recovery is not simply about stopping the behaviour. Rather, it involves accepting that ongoing thoughts and urges to self-injure, and setbacks are a normal part of the recovery process, that developing new ways of coping will take time and effort, accepted or attributed meaning to scars and considering approaches to disclosure. In this way, an individual can move towards self-acceptance and self-compassion. Further, individuals can build resilience and find meaning in their history of self-injury (Lewis & Hasking, 2020).

Self-Compassion Interventions

Given the role of self-compassion in NSSI recovery, there have been surprisingly few efforts to trial self-compassion interventions for NSSI. In 2011, van Vliet and Kalnins proposed that compassion-focused therapy may have applicability in counselling clients who self-injure. They proposed that mindfulness practice may come to replace self-injury as a means of self-soothing and that acceptance of emotional experiences without judgement may counter self-criticism. They also recommend compassionate imagery (e.g. a loved one wrapping them in a blanket) to help reduce urges to self-injure. A self-report diary in which individuals record self-critical thoughts, and then construct reassuring thoughts that counter these, is offered as one strategy that may promote self-compassion and reduce self-injury. Finally, compassionate behaviours are not only alternative behaviours to self-injury but involve fostering self-efficacy and positive relationships with others.

Despite clear evidence that self-compassion may play a role in reducing NSSI, it is interesting that there are no published studies of compassion-focused therapy for NSSI. Individual writing exercises have been trialled. Similar to Hooley's (2018) writing task described above, undergraduate students tasked with writing a brief essay about their best value (5 min) reported elevated self-compassion relative to students who wrote a brief essay about their lowest ranked value and why it might be important to someone else (Gregory et al., 2017). Students with a history of NSSI who completed the values-affirmation task also rated their pain, in a cold pressor task, as more intense than students in the control condition, and could tolerate pain for a shorter period of time (Gregory et al., 2017). This is significant given theories that those who engage in self-injury believe that they "deserve" pain and are therefore willing to endure it for longer (Glenn et al., 2014, Hamza et al., 2014, Hooley et al., 2010).

The writing tasks outlined above demonstrate some promise in reducing NSSI, but to date there has not been a committed effort to exploring the potential benefits of compassion-focused therapy or other self-compassion-based interventions. Given the central role of affect regulation, and the salience of self-criticism and shame in the onset and maintenance of self-injury, this seems like a lost opportunity. The combination of acceptance, mindfulness, and self-soothing taught through compassion-focused therapy (Gilbert, 2009) and mindful self-compassion training (Neff & Germer, 2013) seems ideally matched to the holistic concept of NSSI recovery, which goes beyond cessation of the behaviour to foster self-acceptance and self-compassion.

Conclusion

Non-suicidal self-injury is a behaviour that can be confusing to understand, both for those who engage in it, as well as those who care for them. Self-criticism plays a pivotal role in the onset and maintenance of self-injury and can impede recovery efforts. Of note, repeated and ongoing thoughts or urges to self-injure can be met with self-criticism, furthering the sense of shame and failure individuals may feel if they continue to self-injure. Learning to accept these thoughts and setbacks which may be associated with them as a normal part of the recovery process and learning self-compassion can assist individuals to achieve a sense of resilience and optimism about the future. Future efforts to harness the potential benefits of compassion-focussed therapy in assisting individuals who self-injure are warranted.

References

Ammerman, B. A., & Brown, S. (2018). The mediating role of self-criticism in the relationship between parental expressed emotion and NSSI. *Current Psychology, 37*, 325–333. https://doi.org/10.1007/s12144-016-9516-1

Anestis, M. D., Pennings, S. M., Lavender, J. M., Tull, M. T., & Gratz, K. L. (2013). Low distress tolerance as an indirect risk factor for suicidal behavior: Considering the explanatory role of non-suicidal self-injury. *Comprehensive Psychiatry, 54*(7), 996–1002. https://doi.org/10.1016/j.comppsych.2013.04.005

Baetens, I., Claes, L., Hasking, P., Smits, D., Grietens, H., Onghena, P., & Martin, G. (2015). The relationship between parental expressed emotions and non-suicidal self-injury: The mediating roles of self-criticism and depression. *Journal of Child and Family Studies, 24*(2), 491–498. https://doi.org/10.1007/s10826-013-9861-8

Bentley, K. H., Cassiello-Robbins, C. F., Vitterio, L., Sauer-Zavala, S., & Barlow, D. H. (2015). The association between nonsuicidal self-injury and the emotional disorders: A meta-analytic review. *Clinical Psychology Review, 37*, 72–88. https://doi.org/10.1016/j.cpr.2015.02.006

Bresin, K., & Schoenleber, M. (2015). Gender differences in the prevalence of nonsuicidal self-injury: A meta-analysis. *Clinical Psychology Review, 38*, 55–64. https://doi.org/10.1016/j.cpr.2015.02.009

Chapman, A. L., Gratz, K. L., & Brown, M. Z. (2006). Solving the puzzle of deliberate self-harm: The experiential avoidance model. *Behaviour Research and Therapy, 44*(3), 371–394. https://doi.org/10.1016/j.brat.2005.03.005

Cleare, S., Gumley, A., & O'Connor, R. (2019). Self-compassion, self-forgiveness, suicidal ideation and self-harm: A systematic review. *Clinical Psychology and Psychotherapy, 26*(5), 511–530. https://doi.org/10.1002/cpp.2372

Cohen, J. N., Stange, J. P., Hamilton, J. L., Burke, T. A., Jenkins, A., Ong, M., Heimberg, R. G., Abramson, L. Y., & Alloy, L. B. (2015). The interaction of affective states and cognitive vulnerabilities in the prediction of non-suicidal self-injury. *Cognition and Emotion, 29*(3), 539–547. https://doi.org/10.1080/02699931.2014.918872

Daly, O., & Willoughby, T. (2019). A longitudinal study investigating bidirectionality among nonsuicidal self-injury, self-criticism, and parental criticism. *Psychiatry Research, 271*, 678–683. https://doi.org/10.1016/j.psychres.2018.12.056

Dawkins, J., Hasking, P., Boyes, M., Greene, D., & Passchier, C. (2019). Applying a cognitive-emotional model to nonsuicidal self-injury. *Stress and Health, 35*(1), 39–48. https://doi.org/10.1002/smi.2837

Dawkins, J., Hasking, P., & Boyes, M. (2021a). Thoughts and beliefs about nonsuicidal self-injury: An application of Social Cognitive Theory. *Journal of American College Health, 69*(4), 428–434. https://doi.org/10.1080/07448481.2019.1679817

Dawkins, J., Hasking, P., Luck, C., & Boyes, M. (2021b). Implicit assessment of self-injury related outcome expectancies: A comparison of three behavioural tasks. *Psychological Reports, 124*(6), 003329412096151. https://doi.org/10.1177/0033294120961512

Dupasquier, J. R., Kelly, A. C., Moscovitch, D. A., & Vidovic, V. (2019). Cultivating self-compassion promotes disclosure of experiences that threaten self-esteem. *Cognitive Therapy and Research, 44*(1), 108–119. https://doi.org/10.1007/s10608-019-10050-x

Flett, G. L., Goldstein, A. L., Hewitt, P. L., & Wekerle, C. (2012). Predictors of deliberate self-harm behaviour among emerging adolescents: An initial test of a self-punitiveness model. *Current Psychology, 31*, 49–64. https://doi.org/10.1007/s12144-012-9130-9

Fox, K. R., Toole, K. E., Franklin, J. C., & Hooley, J. M. (2017). Why does nonsuicidal self-injury improve mood? A preliminary test of three hypotheses. *Clinical Psychological Science, 5*(1), 111–121. https://doi.org/10.1177/2167702616662270

Fox, K. R., O'Sullivan, I. M., Wang, S. B., & Hooley, J. M. (2019). Self-criticism impacts emotional responses to pain. *Behaviour Therapy, 50*(2), 410–420. https://doi.org/10.1016/j.beth.2018.07.008

Franklin, J. C., Ribeiro, J. D., Fox, K. R., Bentley, K. H., Kleiman, E. M., Huang, X., Musacchio,

K. M., Jaroszewski, A. C., Chang, B. P., & Nock, M. K. (2017). Risk factors for suicidal thoughts and behaviours: A meta-analysis of 50 years of research. *Psychological Bulletin, 143*(2), 187–232. https://doi.org/10.1037/bul0000084

Gilbert, P. (2009). Introducing compassion focused therapy. *Advances in Psychiatric Treatment, 15*(3), 199–208. https://doi.org/10.1192/apt.bp.107.005264

Glassman, L. H., Weierich, M. R., Hooley, J. M., Deliberto, T. L., & Nock, M. K. (2007). Child maltreatment, non-suicidal self-injury, and the mediating role of self-criticism. *Behaviour Research and Therapy, 45*(10), 2483–2490. https://doi.org/10.1016/j.brat.2007/04.002

Glenn, C. R., & Klonsky, E. D. (2013). Nonsuicidal self-injury disorder: An empirical investigation in adolescent psychiatric patients. *Journal of Clinical Child and Adolescent Psychology, 42*(4), 496–507. https://doi.org/10.1080/15374416.2013.794699

Glenn, J. J., Michel, B. D., Franklin, J. C., Hooley, J. M., & Nock, M. K. (2014). Pain analgesia among self-injurers. *Psychiatry Research, 220*(3), 921–926. https://doi.org/10.1016/j.psychres.2014.08.016

Gong, T., Ren, Y., Wu, J., Jiang, Y., Hu, W., & You, J. (2019). The associations among self-criticism, hopelessness, rumination, and NSSI in adolescents: A moderated mediation model. *Journal of Adolescence, 72*, 1–9. https://doi.org/10.1016/j.adolescence.2019.01.007

Gregory, W. E., Glazer, J. V., & Berenson, K. R. (2017). Self-compassion, self-injury, and pain. *Cognitive Research and Therapy, 41*(5), 777–786. https://doi.org/10.1007/s10608-017-9846-9

Gromtasky, M. A., Waszczuk, M. A., Perkman, G., Salis, K. L., Klein, D. N., & Kotov, R. (2017). The role of parental psychopathology and personality in adolescent non-suicidal self-injury. *Journal of Psychiatric Research, 85*, 15–23. https://doi.org/10.1016/j.psychires.2016.10.013

Groschwitz, R. C., Plener, P. L., Schumacher, T., Stoehr, R., Boege, I., & Kaess, M. (2015). The situation of former adolescent self-injurers as young adults: A follow-up study. *BMC Psychiatry, 15*, 1. https://doi.org/10.1186/s12888-015-0555-1

Hamza, C. A., Willoughby, T., & Armiento, J. (2014). A laboratory examination of pain threshold and tolerance among nonsuicidal self-injurers with and without self-punishing motivations. *Archives of Scientific Psychology, 2*(1), 33–42. https://doi.org/10.1037/arc0000008

Hasking, P. A., Momeni, R., Swannell, S., & Chia, S. (2008). The nature and extent of deliberate self-injury in a non-clinical sample of young adults. *Archives of Suicide Research, 12*(3), 208–218. https://doi.org/10.1080/13811110802100957

Hasking, P., Whitlock, J., Voon, D., & Rose, A. (2016). A cognitive-emotional model of NSSI: Using emotion regulation and cognitive processes to explain why people self-injure. *Cognition and Emotion, 31*(8), 1543–1556. https://doi.org/10.1080/02699931.2016.1241219

Hasking, P., Boyes, M. E., Finlay-Jones, A., McEvoy, P. M., & Rees, C. S. (2019). Common pathways to NSSI and suicide ideation: The roles of rumination and self-compassion. *Archives of Suicide Research, 23*(2), 247–260. https://doi.org/10.1080/13811118.2018.1468836

Hermanto, N., & Zuroff, D. C. (2016). The social mentality theory of self-compassion and self-reassurance: The interactive effect of care-seeking and caregiving. *The Journal of Social Psychology, 156*(5), 523–535. https://doi.org/10.1080/00224545.2015.1135779

Hermanto, N., Zuroff, D. C., Kelly, A. C., & Leybman, M. J. (2017). Receiving support, giving support, and self-reassurance: A daily diary test of social mentality theory. *Personality and Individual Differences, 107*, 37–42. https://doi.org/10.1016/j.paid.2016.11.013

Hooley, J. M., & Franklin, J. C. (2017). Why do people hurt themselves? A new conceptual model of nonsuicidal self-injury. *Clinical Psychological Science, 6*(3), 428–451. https://doi.org/10.1177/2167702617745641

Hooley, J. H., & St Germain, S. A. (2014). Nonsuicidal self-injury, pain, and self-criticism: Does changing self-worth change pain endurance in people who engage in self-injury? *Clinical Psychological Science, 2*(3), 297–305. https://doi.org/10.1177/2167702613509372

Hooley, J. M., Ho, D. T., Slater, J., & Lockshin, A. (2010). Pain perception and non-suicidal self-injury: A laboratory investigation. *Personality Disorders: Theory, Research, and Treatment, 1*(3), 170–179. https://doi.org/10.1037/a0020106

Hooley, J. M., Fox, K. R., Wang, S. B., & Kwashie, A. N. D. (2018). Novel online daily diary interventions for nonsuicidal self-injury: A randomized controlled trial. *BMC Psychiatry, 18*, Article 264. https://doi.org/10.1186/s12888-018-1840-6

International Society for the Study of Self-Injury. (2020). *What is self-injury?* Retrieved from: https://itriples.org/category/about-self-injury/

Itzhaky, L., Shahar, G., Stein, D., & Fennig, S. (2015). In eating disordered inpatient adolescents self-criticism predicts nonsuicidal self-injury. *Suicide and Life-threatening Behaviour, 46*(4), 385–397. https://doi.org/10.1111/sltb.12223

Jiang, Y., You, J., Hou, Y., Du, C., Lin, M., Zheng, X., & Ma, C. (2016). Buffering the effects of peer victimization on adolescent non-suicidal self-injury: The role of self-compassion and family cohesion. *Journal of Adolescence, 53*, 107–115. https://doi.org/10.1016/j.adolescence.09.005

Jiang, Y., You, J., Ren, Y., Sun, R., Liao, S., Zhu, J., & Ma, N. (2017a). Brief report: A preliminary comparison of self-compassion between adolescents with nonsuicidal self-injury thoughts and actions. *Journal of Adolescence, 59*, 124–128. https://doi.org/10.1016/j.adolesence.2017.05.019

Jiang, Y., You, J., Zheng, X., & Lin, M.-P. (2017b). The qualities of attachment with significant others and self-compassion protect adolescents from non-suicidal self-injury. *School Psychology Quarterly, 32*(2), 143–155. https://doi.org/10.1037/spq0000187

Jiang, Y., Ren, Y., Zhu, J., & You, J. (2020). Gratitude and hope relate to adolescent nonsuicidal self-injury: Mediation through self-compassion and family and school experiences. *Current Psychology, 41*(2), 935–942. https://doi.org/10.1007/s12144-020-00624-4

Jiang, Y., Ren, Y., Liu, T., & You, J. (2021). Rejection sensitivity and adolescent non-suicidal self-injury: Mediation through depressive symptoms and moderation by fear of self-compassion. *Psychology and Psychotherapy: Theory, Research and Practice, 94*(S2), 481–496. https://doi.org/10.1111/papt.12293

Kaess, M., Parzer, P., Mattern, M., Plener, P. L., Bifulco, A., Resch, F., & Brunner, R. (2013). Adverse childhood experiences and their impact on frequency, severity, and the individual function of nonsuicidal self-injury in youth. *Psychiatry Research, 206*(2–3), 265–272. https://doi.org/10.1016/j.psychres.2012.10.012

Kaniuka, A. R., Kelliher-Rabon, J., Chang, E. C., Sirois, F. M., & Hirsch, J. K. (2020). Symptoms of anxiety and depression and suicidal behaviour in college students: Conditional indirect effects of non-suicidal self-injury and self-compassion. *Journal of College Student Psychotherapy, 34*(4), 316–338. https://doi.org/10.1080/87568225.2019.1601048

Kelada, L., Hasking, P., Melvin, G., Whitlock, J., & Baetens, I. (2018). "I do want to stop, at least I think I do": An international comparison of recovery from nonsuicidal self-injury among young people. *Journal of Adolescent Research, 33*(4), 416–441. https://doi.org/10.1177/0743558416684954

Kiekens, G., Hasking, P., Claes, L., Boyes, M., Mortier, P., Auerbach, R. P., Cuijpers, P., Demyttenaere, K., Green, J. G., Kessler, R. C., Nock, M. K., & Bruffaerts, R. (2018). The associations between non-suicidal self-injury and first onset suicidal thoughts and behaviors. *Journal of Affective Disorders, 239*, 171–179. https://doi.org/10.1016/j.jad.2018.06.033

Kiekens, G., Hasking, P., Claes, L., Boyes, M., Mortier, P., Auerbach, R. P., Cuijpers, P., Demyttenaere, K., Green, J. G., Kessler, R. C., Myin-Germeys, I., Nock, M. K., & Bruffaerts, R. (2019). Predicting the incidence of non-suicidal self-injury in college students. *European Psychiatry, 59*, 44–51. https://doi.org/10.1016/j.eurpsy.2019.04.002

Lewis, S. P., & Hasking, P. (2020). Rethinking self-injury recovery: A commentary and conceptual reframing. *BJPsych Bulletin, 44*(2), 44–46. https://doi.org/10.1192/bjb.2019.51

Lewis, S. P., & Mehrabkhani, S. (2016). Every scar tells a story: Insight into people's self-injury scar experiences. *Counselling Psychology Quarterly, 29*(3), 1–15. https://doi.org/10.1080/09515070.2015.1088431

Lewis, S. P., Kenny, T. E., Whitfield, K., & Gomez, J. (2019). Understanding self-injury recovery: Views from individuals with lived experience. *Journal of Clinical Psychology, 75*(12), 2119–2139. https://doi.org/10.1002/jclp.22834

Neff, K. D. (2003). Self-compassion: An alternative conceptualization of a healthy attitude toward oneself. *Self and Identity, 2*(2), 85–101. https://doi.org/10.1080/15298860309032

Neff, K. D., & Germer, C. K. (2013). A pilot study and randomized controlled trial of the mindful self-compassion program. *Journal of Clinical Psychology, 69*(1), 28–44. https://doi.org/10.1002/jclp.21923

Nock, M. K. (2009). Why do people hurt themselves? New insights into the nature and function of self-injury. *Current Direction in Psychological Science, 18*(2), 78–83. https://doi.org/10.1111/j.1467-8721.2009.01613.x

Perkins, N., Ortiz, S. N., & Smith, A. R. (2020). Self-criticism longitudinally predicts nonsuicidal self-injury in eating disorders. *Eating Disorders, 28*(2), 157–170. https://doi.org/10.1080/10640266.2019.1695450

Plener, P. L., Schumacher, T. S., Munz, L. M., & Groschwitz, R. C. (2015). The longitudinal course of non-suicidal self-injury and deliberate self-harm: A systematic review of the literature. *Borderline Personality Disorder and Emotion Dysregulation, 2*, Article 2. https://doi.org/10.1186/s40479-014-0024-3

Ribeiro, J. D., Franklin, J. C., Fox, K. R., Bentley, K. H., Kleiman, E. M., Chang, B. P., & Nock, M. K. (2016). Self-injurious thoughts and behaviours as risk factors for future suicide ideation, attempts, and death: A meta-analysis of longitudinal studies. *Psychological Medicine, 46*(2), 225–236. https://doi.org/10.1017/S0033291715001804

Richmond, S., Hasking, P., & Meaney-Tavares, R. (2017). Psychological distress and non-suicidal self-injury: The mediating roles of rumination, cognitive reappraisal and expressive suppression. *Archives of Suicide Research, 21*(1), 62–72. https://doi.org/10.1080/13811118.2015.1008160

Rosenrot, S. A., & Lewis, S. P. (2020). Barriers and responses to the disclosure of non-suicidal self-injury: A thematic analysis. *Counselling Psychology Quarterly, 33*(2), 121–141. https://doi.org/10.1080/09515070.2018.1489220

Selby, E. A., & Joiner, T. E. (2009). Cascades of emotion: The emergence of Borderline Personality Disorder from emotional and behavioural dysregulation. *Review of General Psychology, 13*(3), 219–229. https://doi.org/10.1037/a0015687

Slabbert, A., Hasking, P., & Boyes, M. (2018). Riding the emotional roller coaster: Testing the role of distress tolerance in non-suicidal self-injury. *Psychiatry Research, 269*, 309–315. https://doi.org/10.1016/j.psychres.2018.08.061

Smith, D. M., Wang, S. B., Carter, M. L., Fox, K. R., & Hooley, J. M. (2020). Longitudinal predictors of self-injurious thoughts and behaviours in sexual and gender minority adolescents. *Journal of Abnormal Psychology, 129*(1), 114–121. https://doi.org/10.1037/abn0000483

Staniland, L., Hasking, P., Boyes, M., & Lewis, S. (2020). Stigma and nonsuicidal self-injury: Application of a conceptual framework. *Stigma and Health, 6*(3), 312–323. https://doi.org/10.1037/sah0000257

Suh, H., & Jeong, J. (2021). Association of self-compassion with suicidal thoughts and behaviors and non-suicidal self injury: A meta-analysis. *Frontiers in Psychology, 12*, Article 633482. https://doi.org/10.3389/fpsyg.2021.633482

Sutherland, O., Dawczyk, A., de Leon, K., Cripps, J., & Lewis, S. P. (2014). Self-compassion in online accounts of nonsuicidal self-injury: An interpretive phenomenological analysis. *Counselling Psychology Quarterly, 27*(4), 409–433. https://doi.org/10.1080/09515070.2014.948809

Swannell, S., Martin, G., Page, A., Hasking, P., Hazell, P., Taylor, A., & Protani, M. (2012). Child maltreatment, subsequent non-suicidal self-injury and the mediating roles of dissociation, alexithymia and self-blame. *Child Abuse & Neglect, 36*(7–8), 572–584. https://doi.org/10.1016/j.chiabu.2012.05.005

Swannell, S. V., Martin, G. E., Page, A., Hasking, P., & St John, N. J. (2014). Prevalence of nonsuicidal self-injury in nonclinical samples: Systematic review, meta-analysis and meta-regression. *Suicide and Life-threatening Behavior, 44*(3), 273–303. https://doi.org/10.1111/sltb.12070

Tatnell, R., Hasking, P., & Newman, L. (2018). Multiple mediation modelling exploring relationships between specific aspects of attachment, emotion regulation and non-suicidal self-injury. *Australian Journal of Psychology, 70*(1), 48–56. https://doi.org/10.1111/ajpy.12166

Taylor, P. J., Jomar, K., Dhingra, K., Forrester, R., Shahmalak, U., & Dickson, J. (2018). A meta-analysis of the prevalence of different functions of non-suicidal self-injury. *Journal of Affective Disorders, 227*, 759–769. https://doi.org/10.1016/j.jad.2017.11.073

Tuna, E., & Gençöz, T. (2021). Pain perception, distress tolerance, and self-compassion in Turkish young adults with and without a history of non-suicidal self-injury. *Current Psychology, 40*, 4143–4155. https://doi.org/10.1007/s12144-020-00634-2

van Vliet, K. J., & Kalnins, G. R. C. (2011). A compassion-focussed approach to nonsuicidal self-injury. *Journal of Mental Health Counselling, 33*(4), 295–311. https://doi.org/10.17744/mehc.33.4.j7540338q223t417

Voon, D., Hasking, P., & Martin, G. (2014). Change in emotion regulation strategy use and its impact on adolescent non-suicidal self-injury: A three-year longitudinal analysis using latent growth modelling. *Journal of Abnormal Psychology, 123*(3), 487–498. https://doi.org/10.1037/a0037024

Wu, N., Hou, Y., Chen, P., & You, J. (2019). Peer acceptance and nonsuicidal self-injury among Chinese adolescents: A longitudinal moderated mediation model. *Journal of Youth and Adolescence, 48*(9), 1806–1817. https://doi.org/10.1007/s10964-019-01093-0

Xavier, A., Pinto-Gouveia, J. P., & Cunha, M. (2016a). Non-suicidal self-injury in adolescence: The role of shame, self-criticism and fear of self-compassion. *Child & Youth Care Forum, 45*(4), 571–586. https://doi.org/10.1007/s10566-016-9346-1

Xavier, A., Pinto-Gouveia, J. P., & Cunha, M. (2016b). The protective role of self-compassion on risk factors for non-suicidal self-injury in adolescence. *School Mental Health, 8*(4), 476–485. https://doi.org/10.1007/s12310-016-9197-9

Xavier, A., Pinto-Gouveia, J. P., Cunha, M., & Carvalho, S. (2016c). Self-criticism and depressive symptoms mediate the relationship between emotional experiences with family and peers and self-injury in adolescence. *The Journal of Psychology, 150*(8), 1046–1061. https://doi.org/10.1080/00223980.2016.1235538

Xavier, A., Pinto-Gouveia, J. P., Cunha, M., & Dinis, A. (2017). Longitudinal pathways for the maintenance of non-suicidal self-injury in adolescence: The pernicious blend of depressive symptoms and self-criticism. *Child and Youth Care Forum, 46*(6), 841–856. https://doi.org/10.1007/s10566-017-9406-1

You, J., Lin, M., & Leung, F. (2015). A longitudinal moderated mediation model of nonsuicidal self-injury among adolescents. *Journal of Abnormal Child Psychology, 43*(2), 381–390. https://doi.org/10.1007/s10802-014-9901-x

You, J., Jiang, Y., Zhang, M., Du, C., Lin, M., & Leung, F. (2017). Perceived parental control, self-criticism, and nonsuicidal self-injury among adolescents: Testing the reciprocal relationships by a three-wave cross-lag model. *Archives of Suicide Research, 21*(3), 379–391. https://doi.org/10.1080/13811118.2016.1199989

Zelkowitz, R. L., & Cole, D. A. (2019). Self-criticism as a transdiagnostic process in nonsuicidal self-injury and disordered eating: Systematic review and meta-analysis. *Suicide and Life-threatening Behaviour, 49*(1), 310–327. https://doi.org/10.1111/sltb.12436

Check for updates

22 Self-Compassion in Psychotherapy: Clinical Integration, Evidence Base, and Mechanisms of Change

Christopher Germer

Introduction

Imagine that you are a psychotherapist and have a client who suffers from anxiety or depression. During therapy, your client becomes deeply compassionate toward herself. That means that when things go wrong in her life, your client can recognize and validate for herself how she feels rather than getting lost in rumination (mindfulness); she feels connected to other people in the midst of her troubles rather than feeling alone (common humanity); and her internal conversation is mostly reassuring and supportive rather than self-critical (self-kindness). With these mental habits, there is a good chance that her anxiety or depression would have already subsided and she does not need as much therapy anymore, although the challenges of life will inevitably remain.

As this book amply demonstrates, self-compassion is a key ingredient in mental health and psychological well-being. The beneficial impact of self-compassion is perhaps even more evident in psychotherapy where people bring their most challenging life experiences. The purpose of this chapter is to offer a panoramic view of how to integrate self-compassion into *any* form of treatment and also to outline the evidence base and putative mechanisms of change underlying self-compassion in psychotherapy. A few empirically supported models of therapy are explicitly compassion-based, such as compassion-focused therapy (CFT; Gilbert, 2009), emotion-focused therapy (EFT; Greenberg, 2006), and internal family systems (IFS; Schwartz, 1995). This chapter is primarily informed by CFT, the most distinctly compassion-based treatment model, and by Mindful Self-Compassion program (MSC; Germer & Neff, 2019; Neff & Germer, 2018), a structured, empirically supported training for the general public designed specifically to cultivate *self*-compassion.

The chapter begins by locating self-compassion in the context of psychotherapy, past and present. Next, we outline the evidence for self-compassion as a transdiagnostic and transtheoretical mechanism of action in therapy. The majority of this chapter describes three levels by which self-compassion can be integrated into psychotherapy—*compassionate presence, compassionate relationship,* and *compassionate interventions*—along with supporting research. When all three levels are part of treatment, it can be considered fully self-compassion based. Finally, we explore emotion regulation as the basic mechanism by which self-compassion works in psychotherapy, along with underlying neurophysiological and psychological processes, especially the cultivation of secure attachment and the alleviation of shame.

C. Germer (✉)
Harvard Medical School, Cambridge Health Alliance, Cambridge, MA, USA

A. Finlay-Jones et al. (eds.), *Handbook of Self-Compassion*, Mindfulness in Behavioral Health,
https://doi.org/10.1007/978-3-031-22348-8_22

Historical Context

Self-compassion has been part of psychotherapy for over a century under the umbrella of "self-acceptance." Psychotherapy giants such as William James, Sigmund Freud, and B. F. Skinner all considered acceptance of oneself and others to be psychologically beneficial (Williams & Lynn, 2010). Carl Rogers (1951) and other humanistic therapists elevated self-acceptance to the status of a core change process in psychotherapy. Interestingly, both Freud (1957) and Rogers (1951) considered self-acceptance to be a *precursor* to positive therapeutic change and acceptance of others, and this perspective became a focus of empirical investigation well into the 1980s. In the 1990s, clinical research shifted away from acceptance of the "self" to acceptance of "moment-to-moment experience" with the introduction of Buddhist-inspired mindfulness and acceptance-based treatments such as dialectical behavior therapy (DBT; Linehan, 1993), acceptance and commitment therapy (ACT; Hayes et al., 2011), and mindfulness-based cognitive therapy (MBCT; Segal & Teasdale, 2018). Recently, the pendulum has begun to swing back to include acceptance of both moment-to-moment *experience* (mindfulness) and the *experiencer* (self-compassion)—the "self."

Empathy and Compassion

For most of the history of psychotherapy, the term "compassion" has been relatively absent from the research literature. However, we can assume that compassion has not been absent from psychotherapy, but rather implied in the definition of empathy. Empathy refers to experiencing the world of another person as one's own. Carl Rogers (1951) wrote the following about empathy in client-centered therapy:

> It is the counselor's function to assume…the internal frame of reference of the client, to perceive the world as the client sees it, to perceive the client himself as he is seen by himself, to lay aside all perceptions from the external frame of reference while doing so…(p. 29).

A modern definition of *compassion* is:

> … a multidimensional process comprised of four key components: (1) an *awareness* of suffering (cognitive/empathic awareness), (2) *sympathetic concern* related to being emotionally moved by suffering (affective component), (3) a *wish* to see the relief of that suffering (intention), and (4) a *responsiveness* or readiness to help relieve that suffering (motivational) (Jinpa (2010) in Jazaieri et al., 2013).

In this definition of compassion, *awareness* and *concern* refer to cognitive and affective empathy, and the *wish* and *readiness* to alleviate suffering are special attributes of compassion—the added component of goodwill in the face of suffering. Since the premise of psychotherapy is the alleviation of psychological distress, we can assume that the wish and readiness to alleviate suffering, or compassion, have always been implied in our clinical understanding empathy.

Empathy is a key ingredient in effective therapy relationships (Elliott et al., 2011; Norcross & Lambert, 2018). Our understanding of empathy has continued to evolve since Carl Rogers, especially with the advent of social neuroscience in the 1990s. Empathy now encompasses not only experiencing the world of another as one's own, but also having *perspective* on the emotional state of another person and the ability to *regulate* one's own empathic distress in order to maintain a compassionate state of mind (Eisenberg & Eggum, 2009). With the addition of perspective and emotion regulation to sustain a positive attitude, the modern definition of empathy moves closer to our understanding of compassion.

Discovering Self-Compassion

Most people are more compassionate toward others than themselves (Knox et al., 2016; Pommier et al., 2020). This disparity comes at a price, however, especially in the clinical arena. Twenty years ago, in 2000, Paul Gilbert had a revolutionary insight while treating depressed clients with cognitive-behavioral therapy (CBT). He realized that his clients could become adept at identifying cognitive distortions and replacing them with more balanced thoughts, but their symptoms of depression did not diminish as long as they spoke to themselves in a harsh and demanding tone. What his clients needed, Gilbert concluded, was to "warm up the conversation" (personal communication). This understanding led to the development of CFT, a therapy model based on evolutionary psychology, attachment theory, and Tibetan Buddhist compassion practice, that is currently researched and practiced by clinicians throughout the world (Gilbert, 2010a, b).

Three years later, Kristin Neff (2003), a developmental psychologist, operationalized the construct of self-compassion and published the self-report scale, the Self-Compassion Scale (SCS), that is used in most research on self-compassion. Since then, the number of studies on self-compassion has grown exponentially. Most self-compassion research is still correlational, but studies with laboratory mood manipulations and outcome studies on self-compassion training are increasing as self-compassion training and compassion-based therapy becomes widely disseminated. Self-compassion may be considered a psychological *construct* (e.g., Neff's three component model), a *trait* or *state* (Neff, 2003; Neff et al., 2020, respectively), a *practice* (e.g., meditation), or a psychological *process* (mechanism of change). It is possible, for example, to measure self-compassion as an underlying process in therapy without the treatment containing any identifiable self-compassion practices (Galili-Weinstock et al., 2018; Kelly & Tasca, 2016). However, the construct of self-compassion, especially Neff's three component model, guides our understanding of self-compassion as a trait, state, practice, or process.

A New Paradigm

Within the CBT tradition, a "third wave" of therapies (after behavioral and cognitive approaches) began emerging in the 1990s that focus on "being with" challenging moment-to-moment experiences rather than directly changing them (Dimidjian et al., 2016; UliaSzek et al., 2020). This approach has since embraced compassion-oriented treatment and coalesced into a new paradigm—*mindfulness-, acceptance-, and compassion-based psychotherapy* (Germer & Siegel, 2012; Germer et al., 2013). Within this new paradigm, some therapists focus more on mindfulness, others on acceptance, and still others on compassion as the primary mechanism of change in treatment. For example, mindfulness-based clinicians (e.g., Segal et al., 2012; Shapiro & Carlson, 2009; Siegel, 2009) tend to emphasize the role of attention and awareness in how we create, experience, and alleviate emotional suffering. Acceptance-based clinicians (e.g., Hayes et al., 2011; Roemer et al., 2008) focus more on non-avoidance and acceptance of moment-to-moment experience, along with values-based living, and they are less likely to prescribe meditation than mindfulness-based clinicians. Compassion-based therapies, such as CFT, EFT, and IFS, focus primarily on emotion regulation through care and connection. However, elements of mindfulness, acceptance, and compassion can be found in all therapies contained within this paradigm.

Compassion can be directed toward oneself or others and also received from others. However, these directions are not entirely distinct. For example, research shows that cultivating self-compassion often increases other-compassion (Neff & Germer, 2013), and increasing other-compassion enhances self-compassion (Breines & Chen, 2013). Although a therapy model like CFT enhances the flow of compassion in all directions, the primary focus of compassion-based therapy is on helping the client to develop *self*-compassion. Nonetheless, compassion is embedded in all aspects of compassion-based therapy—self-compassion by the client, compassion for the client, and compassion by and for the therapist.

Transdiagnostic and Transtheoretical Change Process

Self-compassion has been proposed as a mechanism of change in mindfulness- and acceptance-based treatment (Baer, 2010). Increasing evidence shows that self-compassion improves mental health in diverse clinical populations, suggesting that it is a *transdiagnostic* mechanism of change. Empirical evidence also indicates that self-compassion increases in clients in different kinds of therapy, suggesting that self-compassion is also a *transtheoretical* change process.

Transdiagnostic Process

There is a growing body of research demonstrating the potential of compassion-based psychotherapy across a range of clinical disorders, including anxiety (Haj Sadeghi et al., 2018), depression (Kirby, 2017), trauma (Au et al., 2017), social anxiety disorder (Gharraee et al., 2018), eating disorders (Kelly et al., 2017), (Braehler et al., 2013), dementia (Craig et al., 2018), addictions (Kelly et al., 2010), and personality disorders (Feliu-Soler et al., 2017; Lucre & Corten, 2013). CFT was the main treatment model studied and the data are encouraging, although many studies were pilot or feasibility studies and more randomized controlled trials are needed (Craig et al., 2020).

A number of meta-analyses support self-compassion as a transdiagnostic mechanism of change. In a meta-analysis of compassion-based interventions for a variety of different diagnoses, treatment significantly relieved psychological distress and increased self-compassion, even when the studies used active control groups (Kirby et al., 2017b). A meta-analysis specifically for *self*-compassion-based interventions found strong effect sizes for eating behavior and rumination and moderate effect sizes for stress, anxiety, depression, and self-criticism (Ferrari et al., 2019). Interestingly, a meta-analysis of compassion-related therapies for chronic *physical* health conditions also found that the treatments increased self-compassion and improved various outcomes (Kılıç et al., 2020).

Self-compassion training programs designed for the *general public* have also been effective in reducing various kinds of psychological distress (Finlay-Jones, 2017). MSC reduced anxiety and depression among adults in the community (Neff & Germer, 2013) as well as depressive symptoms among diabetes patients (Friis et al., 2016). Compassion cultivation training (CCT; Goldin & Jazaieri, 2017) decreased worry and emotional suppression (Jazaieri et al., 2014), cognitively based compassion training (CBCT; Ash et al., 2019), reduced cancer stress (Gonzalez-Hernandez et al., 2018), and online Mindfulness Based Compassionate Living (MBCL; Van den Brink & Koster, 2015) lowered self-criticism and raised self-compassion (Halamová et al., 2020). An online version of Compassionate Mind Training (CMT), a structured program based on CFT principles and practices, decreased self-criticism in a non-clinical sample (Halamová et al., 2020) and reduced analgesic use among people suffering from chronic pain (Dhokia et al., 2020).

In general, trait self-compassion (measured by the SCS) is associated with mental health in clinical and non-clinical populations. High-trait self-compassion is associated with decreased psychopathology in adults (MacBeth & Gumley, 2012) and adolescents (Marsh et al., 2018), and reduced suicidality (Kelliher Rabon et al., 2018; Xavier et al., 2016). Self-compassionate people tend to ruminate less (Fresnics et al., 2019); they recover from negative moods more easily (Diedrich et al., 2017) and are less vulnerable to shame (Ewert et al., 2018; Zhang et al., 2018). Self-compassion is associated with a more positive body image and fewer eating disorders (Braun et al., 2016; Webb et al., 2016). Higher levels of self-compassion are also associated with lessened symptomology among individuals with schizophrenia (Eicher et al., 2013) and obsessive-compulsive disorder (Wetterneck et al., 2013). Self-compassion levels tend to be lower among people with bipolar disorder (Døssing et al., 2015), depression (Krieger et al., 2013), generalized anxiety disorder (Hoge et al., 2013), social anxiety disorder (Werner et al., 2012), substance use disorder (Phelps et al., 2018), and persecutory delusions (Collett et al., 2016). In a systematic review of studies on self-compassion and trauma, Winders et al. (2020) found that self-compassion was consistently associated with reduced PTSD symptomatology. In sum, based on outcome research with clinical and non-clinical populations, and on correlational research on self-compassion and mental health, self-compassion appears to be an underlying change process that alleviates distress and enhances well-being.

Transtheoretical Process

Although there are currently over a thousand different kinds of psychotherapies, they may be broadly categorized into four paradigms: psychodynamic, cognitive-behavioral, humanistic, and third-wave therapies. These traditions differ theoretically, yet preliminary research evidence suggests that they all increase self-compassion, which implies that self-compassion is a *transtheoretical* mechanism of action.

Schanche et al. (2011) found that both *psychodynamic* and *cognitive* therapy for Cluster C personality disorders (avoidant, dependent, obsessive-compulsive) increased self-compassion, and self-compassion predicted decreases in psychiatric symptoms, interpersonal problems, and personality pathology. Galili-Weinstock et al. (2018) also treated clients with psychodynamic psychotherapy and found that self-compassion levels predicted session-by-session improvement as well as overall therapy outcome.

EFT is a type of *humanistic* psychotherapy and a program adapted for the general public (with elements from CFT and MSC) increased self-compassion and reduced self-criticism (Halamová & Kanovský, 2019). Neff et al. (2007) conducted a study with the two-chair technique from EFT and also found that it increased self-compassion and decreased self-criticism along with fewer experiences of depression, rumination, thought suppression, and anxiety.

Cognitive-behavioral therapy also enhances self-compassion. For example, Hoffart et al. (2015) looked at within-person change in CBT for post-traumatic stress disorder (PTSD) and found that changes in self-compassion predicted PTSD symptoms, while the opposite was not true. Wadsworth et al. (2018) treated patients with CBT and DBT and reported that improvements in anxiety and depression were related to changes in self-compassion.

Unsurprisingly, within the paradigm of *mindfulness-*, *acceptance-*, and *compassion-based* therapies, self-compassion is also related to positive outcomes. Participation in Mindfulness-based Stress Reduction (MBSR; Kabat-Zinn, 2013) and MBCT (Birnie et al., 2010; Goodman et al., 2014; Raab et al., 2015; Taylor et al., 2014) increase self-compassion even though self-compassion is usually taught implicitly in those programs. In mediation analysis—a statistical method used to determine underlying causal mechanisms between two variables (MacKinnon & Lueken, 2008)—self-compassion mediated many of the positive effects in MBSR (Keng et al., 2012; Shapiro et al., 2005) and MBCT (Greenberg et al., 2018; Kuyken et al., 2010). Van Dam et al. (2011) found that self-compassion predicts mental health even more strongly than mindfulness in a large community sample, accounting for ten times more unique variance in symptom severity (anxiety, depression) and quality of life, although mindfulness remains difficult to measure by self-report scales (Park et al., 2013). Hildebrandt et al. (2017) found that mindfulness training and affect training with an explicit emphasis on self-compassion led to a greater increase in self-compassion than mindfulness training alone.

ACT is an acceptance-based therapy model that teaches clients to be kind to themselves in various ways without necessarily mentioning self-compassion (Neff & Tirch, 2013). An ACT training that specifically targeted self-compassion increased self-compassion while reducing psychological distress and anxiety, and a key process in ACT, cognitive flexibility, mediated changes in self-compassion (Yadavaia et al., 2014).

Since self-compassion seems to be implicitly influencing outcomes in therapy, it makes sense to target self-compassion directly in treatment. In a meta-analysis of all three "third wave" therapies—mindfulness-, acceptance- and compassion-based therapy—that supposedly targeted self-compassion, Wilson et al. (2019) found that these therapies all enhanced self-compassion and reduced anxiety and depression, but change scores were not significantly different from those of active control groups. The authors concluded that targeting self-compassion may not make a treatment more powerful than an active control group. Challenging this conclusion, Kirby and Gilbert (2019) noted that the studies included in the Wilson et al. meta-analysis were actually not all compassion-based treatments. For example, Wilson's meta-analysis included manualized MBCT training that contains no explicit self-compassion training. In their own meta-analysis (Kirby et al., 2017a, b), compassion-based interventions reduced psychological distress and increased self-compassion even when compared to active control groups.

Overall, research showing that self-compassion is linked to positive treatment outcomes in different kinds of therapies indicates that self-compassion is a transtheoretical mechanism of action in psychotherapy. Interestingly, ordinary, non-clinical activities also increase self-compassion, such as practicing yoga (Crews et al., 2016), owning a dog (Bergen-Cico et al., 2018), and spending time in nature (Kotera & Fido, 2020). It appears that anything we do for ourselves that promotes a sense of well-being is likely to increase our self-compassion.

Three Levels of Integration into Psychotherapy

Since self-compassion is strongly related to mental health and appears to increase during successful psychotherapy, clinicians are beginning to ask, "Can I help my clients become more self-compassionate while continuing to practice psychotherapy in my own way?" This is possible by considering three levels of integration and their corresponding mechanisms of action. The levels are as follows: (1) *compassionate presence*—how therapists relate to their experience of themselves and the client, mostly non-verbally, (2) *compassionate alliance*—how therapists engage with their clients, verbally and non-verbally, and (3) *compassionate interventions*—how clients relate to themselves, especially during home practice. In the research literature, these categories correspond to therapeutic presence, therapeutic alliance, and therapeutic interventions, respectively. The qualifier, "therapeutic," has been substituted by "compassionate" in this discussion to make the implicit quality of compassion more explicit at each level.

Level 1: Compassionate Presence

Presence is about being with our moment-to-moment experience in a clear, open, and direct way, often without thoughts or words (Brach, 2012; Morgan et al., 2013). The term "presence" is closely related to "mindfulness" (Bourgault & Dionne, 2019; Epstein, 1999). Mindfulness refers to spacious, non-judgmental awareness of what is occurring in the present moment (Germer, 2013; Kabat-Zinn, 2003). Current evidence shows a positive relationship between a therapist's mindfulness (usually measured with trait mindfulness scales) and the therapy alliance (Leonard et al., 2018; Razzaque et al., 2015; Ryan et al., 2012), but the relationship between therapist mindfulness and treatment outcome is less reliable (Escuriex & Labbé, 2011; Grepmair et al., 2007; Perich et al., 2013; Stanley et al., 2006).

Mindfulness and self-compassion may be considered two wings of a bird—we need both to stay present. Mindfulness appears to regulate emotions primarily through attention regulation (Bishop et al., 2004; Malinowski, 2013) whereas self-compassion regulates emotions through care and connection (Finlay-Jones, 2017; Gilbert, 2009). Self-compassion is the *attitude* of mindfulness in the midst of suffering. Mindfulness is likely to diminish when we suffer, for example, when we experience fear and our perceptual field contracts (Bezdek et al., 2015; Zadra & Clore,

2011). Self-compassion—a warm and kind attitude toward the suffering "self"—can support mindfulness and the practice of mindfulness meditation (Evans et al., 2018; Rowe et al., 2016). Clinicians who wish to enhance their presence in psychotherapy are therefore encouraged to have a personal practice of both mindfulness and self-compassion. Personal practice has been emphasized when teaching mindfulness (Shohin & Van Gordon, 2015) and self-compassion (Germer & Neff, 2019) and also when practicing psychotherapy (Bennett-Levy & Finlay-Jones, 2018; Gale et al., 2017; Kolts et al., 2018).

Research has shown that MBSR, the gold standard of mindfulness training, raises levels of mindfulness which can lead to functional and structural changes in the brain (Gotink et al., 2016; Hölzel et al., 2011a, b; Young et al., 2018). Self-compassion training, such as MSC, also raises mindfulness (Neff & Germer, 2013) and mindfulness training increases self-compassion (Golden et al., 2020; Keng et al., 2012; Kuyken et al., 2010). Self-compassion can also be learned by participating in compassion-based training. Yela et al. (2020) found, for example, that MSC increased self-compassion in clinical psychology trainees (but only when the trainees were committed to the training). Furthermore, research shows that individual meditation practices can raise self-compassion, such as loving-kindness meditation (Engel et al., 2019; Weibel et al., 2017) or visualizing a compassionate image (Allen et al., 2015; Naismith et al., 2018).

Mindfulness and self-compassion meditation are usually practiced in isolation, but presence in psychotherapy is distinctly interpersonal. According to Geller (2017), *therapeutic presence* means that the therapist is:

> ...first (a) open and *receptive* to clients' experience, attuning to their verbal and nonverbal expressions. You then (b) *attune inwardly* to your resonance with clients' in-the-moment experience, which serves as a guide to (c) *extend and promote contact* (p. 19).

Openness to the experience of the client comes before attuning inwardly because our primary obligation is to the well-being of the client. However, to be open and receptive to our clients, we need to be open and receptive to our own experience, especially the *impact* our clients are having on us, mentally, physically, and emotionally. "Am I feeling afraid?" "Inspired?" "Annoyed?" If we are capable of holding the "client within" in compassionate awareness, we are more likely to relate compassionately to the client sitting in front of us. Research has shown that self-compassion usually, but not always, enhances compassion for others (Bayır-Toper et al., 2020; Neff & Germer, 2013; Neff et al., 2020).

Bourgault and Dionne (2019) found that self-compassion is clearly linked to therapeutic presence. The authors speculate that self-compassion works by enhancing presence in three ways: with *oneself* by reducing self-criticism and fostering self-kindness (Neff & Vonk, 2009), with the *client* by enhancing consideration of others (Neff & Pommier, 2013), and in the therapy *relationship* by increasing awareness of common humanity and lowering reactivity to relational difficulties (Leary et al., 2007).

There are many legitimate reasons why a clinician might *not* be present during a therapy session. For example, listening to a client's traumatic experience might trigger a similar experience in the therapist and hijack the therapist's attention. Or it might be too distressing to hear what a client is saying so the therapist stops listening. Such lapses are not a problem as long as therapists can find their way back to compassionate presence. Research suggests that loving-kindness and compassion meditation can help to enhance empathy and restore therapeutic presence (Bibeau et al., 2016; Mascaro et al., 2013).

Working with Empathic Distress Psychotherapy can be a challenging profession because clinicians listen to painful experiences of others all day long. Since human beings are hardwired to feel the emotions of others as their own (see below), therapists inevitably experience *empathic distress*. How do therapists cope with empathic distress and still manage to maintain compassionate presence?

Compassion seems to mitigate the negative impact of empathic distress. Tania Singer and colleagues (Klimecki et al., 2014; Singer & Klimecki, 2014) identified non-overlapping neural networks for empathic distress and for compassion. Empathic distress activated areas associated with empathy and negative affect (anterior insula and the anterior middle cingulate cortex, respectively) and compassion-activated brain areas associated with positive affect and affiliation (medial orbitofrontal cortex and the ventral striatum, respectively). In other words, the pain of others is less stressful when it is wrapped in compassion.

Self-compassion is a way of bringing compassion back into the therapy room. Research shows that self-compassion training protects healthcare professionals against caregiver fatigue (Neff et al., 2020) and that self-compassion buffers caregiver fatigue and burnout among student counselors (Beaumont et al., 2016). One explanation is that self-compassion activates a physiological state of safety and deactivates the threat state associated with empathic distress (Svendsen et al., 2020). Increasing self-compassion also enhances compassion for others (Neff et al., 2020). A practical method for activating self-compassion in therapy is the Giving and Receiving Compassion meditation from the MSC training (Germer & Neff, 2019). When a therapist notices she is anxious or distracted, she can bring her attention to the sensation of breathing, allowing each inhalation to be for herself and each exhalation for her client. Anecdotal evidence suggests that paying attention in this way during therapy enhances compassionate presence.

Mechanisms of Compassionate Presence What underlying mechanisms could help to explain how compassionate presence alone, without even talking, might have a positive impact on a client? To explore this question, we turn to the neurobiology of empathy, mirror neurons, and interpersonal synchrony.

Human beings are hardwired for empathy—to feel in our own bodies what others are feeling in theirs (Bernhardt & Singer, 2012; Decety, 2011; Nummenmaa et al., 2008; Singer & Lamm, 2009). We have specialized neurons dedicated to this process—mirror neurons (Kilner & Lemon, 2013; Rizzolatti et al., 1996). Mirror neurons are activated when we perform an action and when we observe another person performing the same action, thereby allowing us to directly experience in our own bodies what others are experiencing. Similar circuits in the brain are also stimulated when people observe an emotion in others or experience the same emotion themselves (Decety & Lamm, 2006; Keysers et al., 2010). For example, witnessing another person in pain activates similar brain structures in the observer (Marsh, 2018; Saarela et al., 2007). This process has been variously called "brain-to-brain coupling" (Hasson et al., 2012), "neural resonance" (Krautheim et al., 2019), and "interpersonal synchrony" (Cacioppo et al., 2014).

Interpersonal synchrony is a robust, multidisciplinary field of research which explores how two or more people synchronize their neural, perceptual, affective, physiological, verbal, and behavioral responses (Koole & Tschacher, 2016; Wheatley et al., 2012). There are many methods of measuring interpersonal synchrony, including brain imaging (Lecchi et al., 2019) heart rate (Feldman et al., 2011), breathing (McFarland, 2001), pupil size (Kret & De Dreu, 2017), hormone levels (Edelstein et al., 2017), as well as vocal pitch (Imel et al., 2014), skin conductance (Palmieri et al., 2018), and body movements (Gupta et al., 2019; Ramseyer & Tschacher, 2011). Sophisticated technologies are used to measure brain synchrony, for example, dual EEG (Lecchi et al., 2019) and functional near-infrared spectroscopy (Zhang et al., 2020).

Koole and Tschacher (2016) argue that interpersonal synchrony is important for emotion regulation. In research with children, interpersonal synchrony has been shown to enhance a child's ability to downregulate emotional distress (Feldman, 2015). Parent–child synchrony generates a sense of familiarity in the child, which enhances the child's ability to internalize the parent and use the parent's image to manage distress even when the parent is absent (Symons & Johnson, 1997). Interpersonal synchrony also increases compassion, presumably due to enhanced familiarity (Valdesolo & DeSteno, 2011).

In psychotherapy, interpersonal synchrony is likely to function in a similar manner: synchrony helps a client and therapist to feel emotionally connected, it facilitates the therapist's compassion, encourages the client to internalize the therapist, and promotes emotion regulation in the client. For example, if a therapist maintains compassionate presence while a client describes a traumatic experience, the therapist's compassion will be felt by the client through emotional attunement, compassion will co-mingle with the traumatic memory, and the memory will be recalled slightly differently going forward. Over time, exposure to a therapist with compassionate presence, and with whom the client identifies, and is likely to change how a client thinks and feels about herself. Therefore, if therapists want their clients to become more self-compassionate, the first step is for the therapist to cultivate compassionate presence.

Level 2: Compassionate Alliance

The next level of integration of self-compassion in psychotherapy is the *compassionate alliance*, variously called the therapeutic alliance, therapeutic relationship, working relationship, or therapeutic bond. Whereas compassionate presence is mostly about how therapists relate non-verbally to their experience of the client and themselves in the therapy relationship, the compassionate alliance involves another layer—verbal interaction.

Sigmund Freud (1927, in Horvath & Luborsky, 1993) was the first to recognize that a positive relationship is necessary to do the difficult work of therapy. Almost 50 years later, Bordin (1975, 1994) proposed the *therapeutic alliance* as a common factor in effective therapy. The alliance has 3 factors: agreement on the *goals* of therapy (e.g., alleviation of anxiety or shame), consensus on the *tasks* that they will engage in to reach those goals (e.g., dialogue, home practice), and a strong, positive *bond* between the client and therapist (e.g., rapport and mutual regard). In this widely accepted view of the alliance, the therapy *relationship* refers specifically to the bond between client and therapist and is not identical to the alliance. However, it is difficult to accomplish any tasks or goals, especially challenging ones, without a strong bond between the client and therapist. Research shows that the therapeutic alliance is a robust predicter of therapy outcome across many different kinds of therapy (Horvath et al., 2011).

Overall, research on psychotherapy outcome indicates that 75–80% of people who participate in psychotherapy receive some benefit (Lambert & Ogles, 2004). Averaging across thousands of outcome studies, Norcross and Lambert (2019) determined that approximately 30% of therapy outcomes are attributable to "common factors," most notably the therapy relationship and client and therapist factors. Treatment techniques only account for 15% of improvement and the placebo effect is another 15% (the largest percentage of therapy outcome, 40%, is due to factors outside therapy such as social support or getting a new job). Among the factors related to therapy itself, 30% of therapy outcome is attributable to the patient, 15% to the therapy relationship, 10% to the therapist, and the treatment method accounts for about 10% of outcome (Norcross & Lambert, 2019). However, the therapeutic alliance and treatment methods cannot be clearly delineated. For example, Barber et al. (2006) found that when there was a strong therapeutic alliance, adherence to a treatment manual did not affect outcome whereas a weak alliance required a moderate level of adherence for the best outcome. In other words, "treatment methods are relational acts" (Safran & Muran, 2000).

Meta-analyses of psychotherapy outcome highlight the importance of compassion in the therapy alliance. Bohart et al. (2002) determined in their meta-analysis that empathy accounts for more outcome variance than treatment interventions. Other researchers found that empathy and positive regard (e.g., affirmation, respect, warmth, support, validation, prizing) are important aspects of the therapy relationship (Elliott et al., 2019; Farber et al., 2019) and that therapists who score higher on interpersonal skills like empathy and warmth have better treatment outcomes (Anderson et al., 2009). Lambert and Ogles (2004) concluded in their meta-analysis that successful therapists tended to be warmer, more empathic, understanding, and supportive of their clients, and are less likely to blame, ignore, neglect, or reject the clients.

Unfortunately, although research indicates that therapists can learn empathy skills in the classroom, those skills do not necessarily generalize to the therapy office (Lambert & Ogles, 1997). Could compassion training enhance warmth and empathy in psychotherapy? Bibeau et al. (2016) reviewed the literature on meditation as a means of cultivating empathy among psychotherapists. Three decades of research on mindfulness meditation have been encouraging, but not conclusive, about whether mindfulness meditation increases empathy (Boellinghaus et al., 2013; Raab, 2014; Shapiro et al., 1998) or treatment effectiveness (Grepmair et al., 2007; Ivanovic et al., 2015; Ryan et al., 2012). In mindfulness meditation, the attitude of warmth is mostly implicit, however. Therefore, the question arises whether *explicitly* training a warm attitude through loving-kindness or compassion meditation might have a more reliable impact on empathy in therapy and treatment outcome. Boellinghaus et al. (2014) explored this question and found that mindfulness meditation tended to increase self-compassion but not necessarily compassion for others, whereas loving-kindness meditation was more likely to achieve both these outcomes. In another literature review, Bibeau et al. (2016) determined that loving-kindness and compassion meditation have a positive impact on numerous variables related to empathy, such as altruism, positive regard, prosocial behavior, affective empathy, and empathic accuracy. Later qualitative research showed that compassion meditation impacted both therapist empathy and the therapeutic relationship (Bibeau et al., 2020). At the present time, however, we still do not have enough empirical evidence to determine whether compassion training (for oneself or others) increases empathy or compassion in psychotherapy or improves treatment outcomes.

The Compassionate Therapeutic Relationship In the following discussion, we focus specifically on the therapeutic relationship, or bond, in the therapeutic alliance. There are three factors (three Rs) that constitute a compassionate therapeutic relationship—radical acceptance, resonance, and resource-building. *Radical acceptance* is the overall attitude of the treatment process, *resonance* is the primary mode of engagement, and *resource-building* is the desired outcome of the therapy relationship. The three Rs are based on the "inquiry" method used in mindfulness training (Brandsma, 2017; Wolf & Serpa, 2015) which was specifically adapted for self-compassion training in the MSC program (Germer & Neff, 2019).

R1: Radical Acceptance. Radical acceptance is the *attitude* or *intention* of a compassionate therapeutic relationship. Marsha Linehan (1993) first coined the term "radical acceptance" as the attitude that therapists need to adopt toward their clients with borderline personality disorder (BPD) in order to keep them engaged and working in therapy. Radical acceptance is also the attitude that BPD clients are encouraged to adopt toward their own emotional challenges to reduce their suffering. "Radical acceptance is the fully open experience of what is, entering into reality just as it is, at this moment. …acceptance without the haze of what one wants and does not want it to be" (Robins et al., 2004, pp. 39–40). Radical acceptance does not mean accepting harmful behavior or being complacent in the face of injustice; rather, it refers to abandoning the fight against our inner experience as it arises in the present moment.

Radical acceptance is a tall order. On a continuum on acceptance—curiosity, tolerance, allowing, and friendship (Germer, 2009, pp. 27–28)—radical acceptance refers to friendship with difficult emotions. Radical acceptance goes beyond just tolerating difficult emotions. For example, consider how much better it feels when your anger is embraced rather than just endured, when your vulnerability is welcomed rather than merely tolerated, and when your trauma is honored rather than simply acknowledged. Radical acceptance by clients means doing that for themselves. For therapists, radical acceptance is an invitation to hold the client and the client's pain in a compassionate embrace, especially when it hurts, without rushing to fix it.

There is a dialectic between radical acceptance and change (Linehan, 1993). The goal of compassion-based therapy is to alleviate suffering, but the main question is "how" therapists alleviate suffering—do we resist and avoid what is happening in the present moment, which usually makes suffering persist or amplify, or do we accept what is happening as a foundation for change? Radical acceptance takes clients just as they are, without judgment or an obligation to change. The central paradox of self-compassion is: "When we suffer, we practice self-compassion not to feel better but *because* we feel bad" (Germer & Neff, 2019, p. 109). The challenge is to temporarily suspend the change agenda and allow kindness to flow naturally in response to suffering. This is a radical step for most therapists, and a subtle distinction that can make all the difference in therapy.

R2: Resonance. Resonance is the primary *mode of engagement* in a compassionate therapeutic relationship. It refers to affective attunement between the therapist and client— a sense of "feeling felt" (Siegel, 2010, p. 136). Daniel Stern (2018) defined affective attunement as "the performance of behaviors that express the quality of feeling of a shared affect state without imitating the exact behavioral expression of the inner state" (p. 142). Attunement between a parent and child is necessary precondition for healthy emotional attachment (Schore, 1994). When attunement was lacking in childhood, it can be learned in psychotherapy. The process of attunement in psychoanalysis is known as *intersubjectivity.* Buirski et al. (2020) wrote about intersubjectivity: "We now appreciate that in addition to attuning to affect and putting words to affective experience, the analyst's care and love for the other makes a significant contribution to the transformative process" (p. 6). In compassion-based psychotherapy, resonance is enabled by the therapist's own compassionate presence, but it has the added element of being actively engaged, usually verbally, with a client.

Interpersonal synchrony is an interesting way of measuring resonance and exploring how it works in the therapeutic relationship (Altmann et al., 2020; Koole & Tschacher, 2016). For example, synchrony in vocal pitch between clients and therapists was found to be associated with therapist empathy (Imel et al., 2014) and synchrony of body movements predicted the quality of the alliance as rated by the patient at the end of therapy (Ramseyer & Tschacher, 2011). Recently, Lecchi et al. (2019) found a significant correlation between the perceived strength of the alliance and neural synchrony as measured by dual EEG in therapist–client dyads, and Zhang et al. (2018) discovered that the therapeutic alliance was particularly strong when the right temporo-parietal junction, a part of the brain associated with mentalizing and understanding, was synchronized between counselors and clients.

There appears to be a strong relationship between interpersonal synchrony and treatment *outcome* (Altmann et al., 2020; Galbusera et al., 2018; Ramseyer & Tschacher, 2011, 2014), including that low therapist–client synchrony is associated with premature termination of treatment (Schoenherr et al., 2019). Clients tend to feel seen, heard, and understood by their therapists when they are synchronized. However, Paulick et al. (2018) found that patients with the best therapy outcomes experienced only a *medium* level of interpersonal synchrony during therapy. This may be because a therapist sometimes needs to regulate a client's emotional arousal by embodying an emotional state *opposite* that of the client, such as calmness in the midst of fear. There also seems to be a trade-off between the enjoyment of interpersonal synchrony and the ability to self-regulate—the more we rely on others to regulate our emotions, the less we may regulate our own emotions (Galbusera et al., 2019).

The relationship between interpersonal synchrony and the therapeutic *alliance* is also strong, but it is less reliable than treatment outcome (Koole & Tschacher, 2016). For example, Reich and colleagues (Reich et al., 2014) found that synchrony of vocal pitch of the patient and therapist negatively correlated with ratings of the therapy relationship. The authors surmised that matching of pitch, especially if a client is depressed, could amplify the distress of the client or that it could be interpreted by the client that the therapist lacked confidence in how to move the session forward.

Currently, there does not appear to be any research on self-compassion and interpersonal synchrony. However, one study found that a sense of perceived emotional synchrony during collective dancing explained increases in kindness and a sense of common humanity on a self-report scale of compassion for others (Pizarro et al., 2020).

On a practical level, how might a therapist maintain resonance while being actively engaged with a client? A helpful practice taught in MSC teacher training that is applicable to psychotherapy is to "follow the pings" (Germer & Neff, 2019). A ping is a moment of salience that a therapist experiences in his or her body, or "what stands out" while the client is talking, usually an emotion such as fear, sadness, relief, or awe. When it is the therapist's turn to speak, the therapist can share her or his embodied experience in a validating, clinically relevant manner. For example, a therapist can say to a client, "When you spoke about how angry you were toward your son, I felt sad because I know how important it is for you to have a close relationship with your son." Carl Rogers (1951) noted that empathy in psychotherapy was not simply sharing the emotions of a client, but also "sensing meanings of which [the client] is scarcely aware" (p. 142). This is a practice of listening and speaking from embodied experience rather than getting caught up in our thoughts and having intellectual conversations in therapy.

R3: Resource-Building. The third "R" of the therapeutic bond is resource-building—the desired *outcome* of psychotherapy. In compassion-based psychotherapy, the desired outcome is enhanced emotion regulation by cultivating the resource of self-compassion. In this section, we go beyond resonance and engage our clients in explicit conversations about their experience in- and outside of therapy to help them respond to their difficulties a more compassionate manner.

Resource-building continues to rely on the inquiry method, especially listening and speaking from embodied experience. Inquiry is a self-to-other dialogue that mirrors the self-to-self relationship that mindfulness and compassion teachers wish to cultivate in their students. In MSC training, inquiry usually follows an experiential practice (e.g., meditation, class exercise), but in psychotherapy we do not necessarily have a practice to anchor the conversation so we focus on emotional challenges that the client may experience during or outside therapy. The therapist then asks the client to share precisely what was experienced and how the client responded. For example, in the clinical vignette given earlier, after the therapist shared the "ping" of feeling sad while the client was speaking, and if the client acknowledged he was actually quite sad about fighting with his son, the therapist follow up by asking, "And how do you care for yourself when you are sad?" or "Right now, what do you think you need when you feel sad like this?" or "If you had a friend in the same situation as you, what might you say to your friend, heart-to-heart?" All these questions direct the client to explore how he could respond compassionately to his emotional pain, thereby building the resource of self-compassion. The conversation also opens the door to practicing at home what was discovered in session.

In compassion-based psychotherapy, the "pings" are not always moments of pain and self-compassion is not the only resource that is cultivated. In order for clients to do the hard work of engaging with suffering, their strengths also need to be validated and reinforced, such as courage, perseverance, insight, or sense of humor. Positive regard in good times and bad is a hallmark of effective psychotherapy. In a meta-analysis of positive regard in therapy, Farber et al. (2019) found that positive regard significantly predicted treatment success. The authors suggested that positive regard strengthens a client's sense of agency and ability to succeed in therapy. They advised therapists to "allow yourself to express positive feelings to clients" but also to "monitor your positive regard and adjust it as a function of particular patients and specific situations." (p. 314). Positive regard needs to be tempered with clinical wisdom, or, as Marsha Linehan (2009) quipped, "What good is compassion if it doesn't actually help!"

In sum, the three Rs of radical acceptance, resonance, and resource-building can serve as a guide for creating a compassionate therapeutic relationship. Next, we turn our attention to the third level of integrating self-compassion into psychotherapy—compassionate interventions. Skillful application of compassionate interventions relies on the previous two levels—compassionate presence and the compassionate alliance.

Level 3: Compassionate Interventions

An intervention in psychotherapy is broadly understood as an action taken to bring about positive change in a client or patient. In the psychodynamic tradition, interventions are typically embedded in the therapeutic relationship (e.g., attunement or intersubjectivity) while the client is invited to explore hidden feelings or reflect on the experience of therapy itself. In CBT, interventions usually refer to tasks that are designed to achieve specific goals, such in vivo exposure or challenging irrational thinking to alleviate social anxiety. Interestingly, naturalistic research has shown that clinicians conducting brief psychodynamic therapy frequently included CBT interventions in their psychodynamic treatment, a phenomenon known as the "smuggling hypothesis" (Ablon & Jones, 1998; Ablon et al., 2006). CBT interventions in psychodynamic psychotherapy have been shown to enhance both the therapeutic alliance and treatment outcomes (Samstag & Norlander, 2019).

To explore the impact of therapeutic interventions on self-compassion levels and treatment outcome, Galili-Weinstock et al. (2020) compared the use of directive interventions (e.g., teaching skills, encouraging behaviors, reviewing homework) to common factors interventions (e.g., validation and empathic listening) in the context of psychodynamic psychotherapy. Positive change in therapy was predicted by directive interventions. Interestingly, clients with low self-compassion at the outset of therapy showed greater increases in self-compassion after treatment when their therapists used *less* validation and empathic listening. In light of earlier research that the alliance is less related to treatment outcome in short-term therapy (Blatt, 1995), the authors speculate that validation might be more useful in longer-term therapy. Another explanation could be that old relational wounds are more likely to be activated when a therapist is empathic and validating (see "backdraft," below).

In the current discussion, we are referring only to *directive* interventions as compassionate interventions. Less directive interventions, such as when the therapy relationship itself is the treatment intervention, were already discussed in the context of compassionate presence and the compassionate alliance. Compassionate interventions are exercises and practices that clients can practice at home between therapy sessions. Directive interventions make sense because psychotherapy is usually only 1 h per week and practicing self-compassion at home multiples the amount of time available to cultivate a new mental habit. We know from research on neuroplasticity that the practice of meditation can change the structure of the brain (Kang et al., 2013; Lazar et al., 2005; Valk et al., 2017) and that the quantity and quality of meditation practice impact the results (Goldberg et al., 2020; Hasenkamp & Barsalou, 2012).

There are still relatively few research studies that test the efficacy of individual self-compassion practices in clinical populations. Most outcome research on self-compassion training evaluates multicomponent self-compassion training (e.g., MSC, CMT). Individual meditation practices, such as loving-kindness meditation or compassion meditation, are not usually designed to cultivate *self*-compassion. There are a few exceptions in the research literature, such as brief self-compassion meditation training for body image distress (Albertson et al., 2015; Toole & Craighead, 2016) and chronic pain (Lutz et al., 2020).

Nonetheless, there are wealth of practices available to clinicians that can be customized for individual clients to cultivate self-compassion. For example, the MSC training program contains seven formal meditations and twenty informal practices (mindfulness and/or self-compassion) (Germer & Neff, 2019). CFT provides a range of compassion and self-compassion practices focused on (1) developing an inner compassionate self, (2) compassion flowing out from oneself to others, (3) compassion flowing into oneself, and (4) giving compassion for oneself (Gilbert, 2010a, b; Kolts, 2016). Workbooks are proliferating that help readers to cultivate self-compassion (e.g., Bluth, 2017; Irons & Beaumont, 2017; Neff & Germer, 2018). Learning self-compassion from a workbook alone has been shown to increase self-compassion levels (Held et al., 2018).

CBT has traditionally emphasized directive interventions and many CBT exercises can be reconfigured to cultivate self-compassion. For example, if a person wants to do exposure therapy for agoraphobia, intentional self-soothing practices can be practiced during the exposure experience. Similarly, behavioral activation for depression, such as scheduling enjoyable activities, could be accompanied by a practice to motivate oneself to do those activities with kindness and encouragement rather than self-criticism. Third-wave CBT, such as ACT, can also be enhanced with explicit self-compassion (Neff & Tirch, 2013). For example, ACT focuses on making space for cognitive distortions, rather than disputing or succumbing to them. Therapists could explicitly add self-compassion by encouraging clients to be kind and understanding toward *themselves* precisely because our minds have a tendency to distort the facts of our lives.

Interventions should be built on the foundation of therapeutic presence and the therapeutic alliance. If self-compassion practices are suggested to a client without a strong therapeutic bond and agreement about the tasks and goals of therapy, the client is less likely to practice them, especially when difficulties arise. Non-compliance with directive interventions does not have to end compassion-based therapy—it is simply an invitation to back up and collaborate more meaningfully in the process of designing home practices. Some clients entirely refuse to do home practices, which means that therapists need to focus on enhancing self-compassion through their compassionate presence and a compassionate alliance until their clients are motivated to practice on their own.

A helpful challenge for therapists who want to design home practices for their clients is to do so without ever mentioning the term "self-compassion." This is because striving to become more self-compassionate can be demoralizing, especially for clients who are low in self-compassion. Ideally, interventions should be co-created with clients based on what transpired in the therapy session. Rather than "teaching" self-compassion, clinicians can help their clients notice their emotional pain in the here-and-now, recognize ways that they are not alone, and respond with kindness as they might with a good friend. A client is more likely to follow a treatment recommendation when a home practice is a genuine relief, which makes the practice self-reinforcing. It is easy for clinicians to work in this way when they adopt an attitude of radical acceptance. Rather than striving to fix the client or remove their pain, the challenge is to meet "what is" with kindness and compassion, both in the therapy relationship and during home practices.

Self-Compassion for Emotion Regulation

Due to the proliferation of therapies, clinical scientists have turned to discovering underlying mechanisms or processes of change to make sense of what is happening in therapy (Carey et al., 2020). Emotion regulation is a key mechanism of change in psychotherapy (Gratz et al., 2015; McRae & Gross, 2020). Emotion regulation refers to the ability to attend to, appraise, and modulate the intensity and duration of emotional states (Gross & Muñoz, 1995). As described above, self-compassion also has much in common with other mechanisms of change such as the therapy alliance, empathy, positive regard, and interpersonal synchrony. However, self-compassion is most closely associated with emotion regulation in the research literature (Allen & Leary, 2010; Finlay-Jones et al., 2015; Neff et al., 2007).

The role of self-compassion as an emotion regulation process is particularly evident in the clinical arena (Inwood & Ferrari, 2018; Kraiss et al., 2020; Trompetter et al., 2017). Research shows that self-compassion helps to regulate emotions associated with depression (Bakker et al., 2019; Diedrich et al., 2017; Diedrich et al., 2014), anxiety (Bergen-Cico & Cheon, 2014; Finlay-Jones, 2017), childhood maltreatment (Vettese et al., 2011), trauma (Barlow et al., 2017; Dahm et al., 2015; Scoglio et al., 2018), substance use (Wisener & Khoury, 2020), bulimia nervosa (Hessler-Kaufmann et al., 2020), obsessive-compulsive disorder (Eichholz et al., 2020), sexual pain (Vasconcelos et al., 2020), and caregiver distress (Finlay-Jones et al., 2015; Neff et al., 2020).

Trauma treatment illustrates the role of self-compassion in emotion regulation. Most people who experience trauma do not develop PTSD. How trauma survivors regulate challenging emotions statistically predicts PTSD better than trauma exposure itself (Barlow et al., 2017). PTSD is maintained by experiential avoidance (Marx & Sloan, 2005) and self-compassion helps people to acknowledge and accept their feelings rather than avoid them (Thompson & Waltz, 2008). Research shows that self-compassion mitigates the effect of trauma among people with childhood abuse and neglect (Vettese et al., 2011) as well as women with severe and repeated interpersonal trauma (Scoglio et al., 2018). Self-compassion was also found to mediate the link between perceived parental maltreatment (abuse or indifference) and mental health symptom severity among adult psychotherapy patients (Westphal et al., 2016). In a systematic review of the literature on self-compassion, trauma, and PTSD, Winders et al. (2020) found consistent evidence that increased self-compassion was associated with reduced PTSD. Self-compassion has also been linked to greater post-traumatic growth and healing (Wong & Yeung, 2017).

An argument can be made that each of the three components of self-compassion—mindfulness, common humanity, and self-kindness—are individually effective for regulating emotion. For example, mindfulness is well-established as a change process in psychotherapy (Alsubaie et al., 2017; Hölzel et al., 2011b). Common humanity has rarely been studied as a change process in psychotherapy, but the opposite of common humanity—a sense of isolation—is known to negatively impact mental health (Leigh-Hunt et al., 2017; Ma et al., 2020; Wang et al., 2017). Research on self-kindness is also scarce, but the opposite of self-kindness—self-criticism—is prevalent in most forms of psychological distress (Kannan & Levitt, 2013; McIntyre et al., 2018). Therefore, we can conclude that each of the three components of self-compassion probably has a beneficial effect on mental health.

The impact of the three components on mental health can be measured using the SCS (Neff, 2003). For example, Van Dam et al. (2011) found that the isolation and self-criticism subscales significantly predicted anxiety symptoms and quality of life in a sample of people with mixed anxiety and depression. This is a legitimate use of the SCS, but some clinical scientists attempt to separate the positive subscales (i.e., mindfulness, common humanity, self-kindness) from the negative subscales (i.e., overidentification, isolation, self-criticism) into two categories—self-warmth and self-coldness, respectively (Brophy et al.,

2020; Muris & Otgaar, 2020). The developer of the SCS, Kristin Neff, argues that self-compassion is a dynamic system in which all six subscales change simultaneously, and there is ample evidence to support this view (Neff & Tóth-Király, 2020). Dividing the components of self-compassion into two subconstructs may be interesting but given that self-warmth and self-coldness change in tandem it is not clear how relevant it is to therapeutic interventions. Also, 95% of the reliable variance in responding to SCS scale items is explained by a general factor of self-compassion and splitting the scale into two positive and negative factors has not been shown to be psychometrically valid (Neff et al., 2019, 2020; Neff & Tóth-Király, 2020). Accordingly, self-compassion is best seen as a global psychological mindstate.

Neurophysiological Mechanisms

Increases in self-compassion are associated with changes in the sympathetic and parasympathetic branches of the autonomic nervous system. The sympathetic nervous system (SNS) directs the body's response to dangerous or stressful situations and the parasympathetic nervous system (PNS) allows the body to rest and relax following sympathetic arousal. The SNS increases heart rate and the PNS reduces heart rate. People with high self-compassion had reduced arousal of the SNS when confronted by a social stressor, as measured by both salivary alpha-amylase (Breines et al., 2015) and interleukin 6 (Breines et al., 2014). Training in self-compassion also lowered salivary alpha-amylase and subjective anxiety responses to stress (Arch et al., 2014).

People with high-trait self-compassion have more PNS activity, as measured by vagally mediated heart rate variability (vmHRV) (Kirschner et al., 2019; Petrocchi et al., 2017; Rockliff et al., 2008; Steffen et al., 2020; Svendsen et al., 2016). The PNS uses the vagus nerve to regulate heart rate, and when there is a lot of variability between heartbeats (i.e., high vmHRV), it is a sign that the PNS is actively regulating emotional arousal (Holzman & Bridgett, 2017). VmHRV may be considered a measure of self-soothing and safeness, often associated with social cues like calm voice or a gentle touch (Porges, 2007). Porges proposes that the PNS both downregulates the SNS (fight/flight) and motivates affiliative behaviors such as proximity seeking or caring responses. If a person is stuck in emotional arousal, such as during anxiety or depression (Chalmers et al., 2014; Kemp et al., 2010), then arousal is less regulated and vmHRV is reduced.

Overall, vmHRV is considered a marker for emotion regulation (Appelhans & Luecken, 2006) and compassion is positively linked to higher vmHRV (Di Bello et al., 2020). *Self*-compassion is also associated to higher vmHRV (Svendsen et al., 2020). For example, speaking to oneself in a soothing, compassionate manner while looking in a mirror has been found to increase mvHRV along with positive affect (Petrocchi et al., 2017). Self-compassionate people are also more likely to maintain higher vmHRV in response to stress (Luo et al., 2018).

The CFT model of therapy is anchored in physiology, especially by using compassion to regulate three subsystems of the autonomic nervous system—threat and protection (adrenalin and cortisol); drive and excitement (dopamine); and contentment, soothing, and safeness (oxytocin, endorphins) (Gilbert, 2010a, b; Panksepp, 1998). Most therapies help clients to manage challenges arising from overactive threat or drive systems (Gilbert, 1993). Compassion shifts our physiology into contentment, soothing, and social safeness, which is also the physiology of enhanced vmHRV (Geller & Porges, 2014; Kirby et al., 2017a, b). However, compassion is not always soothing. For example, compassion may need to be fierce and protective, such as defending a child against danger or saying "no" to an unwelcome advance. In that case, we still need to calm and steady ourselves to behave in a wise manner, perhaps like a martial artist (Clapton & Hiskey, 2020), which would engage multiple motivational systems.

Brain Research

Recent research offers insight into the neurological processes underlying self-compassion (see Stevens et al., 2018). Longe et al. (2010) explored brain activation differences between self-criticism and self-reassurance using functional magnetic resonance imaging (fMRI). They found that self-criticism was associated with activation in the dorsolateral prefrontal cortex (dlPFC; active while switching attention and response inhibition) and self-reassurance activated the anterior insula (AI; active during interoceptive awareness). However, Kim et al. (2020) found that self-criticism activated and self-reassurance de-activated the dlPFC, suggesting that the two functions operate in tandem in the same brain region.

Parrish et al. (2018) conducted a fMRI study on self-compassion during a stress test. The ventromedial prefrontal cortex (vmPFC; active during information processing and decision-making) is commonly thought to downregulate activity in the amygdala (active while feeling emotions, esp. fear) during threat-related emotion regulation. The researchers found that greater negative correlation between the vmPFC and the amygdala was associated with high-trait self-compassion and positive correlation was associated with low-trait self-compassion. Therefore, this fronto-limbic circuitry may be considered a neurological mechanism by which self-compassion protects against stress and negative emotions. In a brain imaging study (Berry et al., 2020; Lutz et al., 2020) on the impact of 2 weeks of self-compassion training on chronic back pain, where patients were exposed to self-critical thoughts while in the scanner, the dlPFC reacted more strongly to self-criticism after training, suggesting that participants were actively regulating their emotional response. Patients with high-trait self-compassion had even higher dlPFC responses, indicating that self-compassion training helped them deal better with self-criticism.

In an fMRI study exploring how self-compassion protects against depression, Liu et al. (2022) found that depressed adolescents with high-trait self-compassion had *lower* activity in the right dlPFC when looking at sad images of themselves, and they also scored lower on depression severity. The researchers speculated that self-compassionate adolescents required less cognitive effort to regulate their affect because they were more accepting and less ruminative about negative personal information. Interestingly, when viewing *another* person's sad face, self-compassionate youth had more activity in brain regions associated with empathy (insula, postcentral gyrus, and inferior parietal lobule). The combined results of this study suggest that depressed youth who are also self-compassionate are less focused on their own distress, more empathic toward others, and require less effort to regulate their emotions.

A groundbreaking study on the neurogenetic mechanisms of self-compassion found that carriers of a particular form of the gene OXTR (responsible for affiliation and associated with oxytocin) displayed high self-compassion (Wang et al., 2019). The researchers observed that activity in the empathy network of the brain (right angular gyrus, mPFC, and the anterior cingulate cortex) and also in the executive control network (right dlPFC and inferior parietal cortex) mediated the association of OXTR with self-compassion. Similar to the Liu et al. (2022) study, these findings suggest that highly self-compassionate people do not require as much effort to be empathic or to regulate their emotions. In sum, at least two areas in the prefrontal cortex that are commonly associated with emotion regulation—the vmPFC and dlPFC—seem to be associated with self-compassionate responding, but we still have much to learn about how the neurophysiology of self-compassion.

Psychological Mechanisms

Developing Secure Attachment

The psychological construct most commonly associated with self-compassion is secure attachment (Gilbert & Procter, 2006; Homan, 2018; Neff & McGehee, 2010; Shaver et al., 2017). Insecure attachment is consistently linked to psychopathology (Gazzillo, et al., 2020; Mikulincer & Shaver, 2012, 2016; Shorey & Snyder, 2006) and enhancing attachment security ("earned secure" attachment: (Levy & Johnson, 2019; Pearson et al., 1994) correlates with improvements in therapy (Kinley & Reyno, 2013; Reiner et al., 2016). Therefore, an underlying mechanism of self-compassion in psychotherapy is likely to be the development of secure attachment.

Attachment theory is based on the idea that children seek proximity with attachment figures, especially parents, in stressful situations (Bowlby, 1969, 1973, 1980). A caregiver's ability to provide comfort and security is internalized by children as an attachment style. When caregivers are sensitive and responsive, the child will develop a secure attachment style that guides the child's feelings, thoughts, and expectations about self and others. When caregivers fail to reliably comfort and soothe the child, the child develops insecure attachment, especially anxious, avoidant, or disorganized attachment (Ainsworth et al., 1978; Brennan et al., 1998). Insecure attachment is associated with emotion dysregulation (Girme et al., 2020; Mikulincer & Shaver, 2019; Moutsiana et al., 2014).

Self-compassion is shaped by a person's attachment history. Since children internalize how they are treated, and if they received comfort and support from primary caregivers when they were in distress, they will probably do the same for themselves later in life (Neff & McGehee, 2010; Ross et al., 2019). From a physiological perspective, when a child feels securely attached to caregivers, her or his soothing system has a chance to develop, and when a child is insecurely attached, the threat system will become overly developed (Gilbert & Procter, 2006). Therefore, the shift from insecure to secure attachment in compassion-based therapy is a shift from threat-dominated physiological arousal to safeness and soothing.

In correlational research, insecure attachment is associated with lower levels of self-compassion (Joeng et al., 2017; Mackintosh et al., 2018; Raque-Bogdan et al., 2011; Wei et al., 2011). Parental rejection, criticism, overprotection, and stressful family relationships are negatively related with self-compassion (Neff & McGehee, 2010, Pepping et al., 2015) and early memories of warmth and safeness are positively linked to self-compassion (Cunha et al., 2014; Homan, 2018; Steindl et al., 2018). Self-compassion appears to mediate the relationship between insecure attachment and emotional distress (Mackintosh et al., 2018), subjective well-being (Wei et al., 2011), and mental health in general (Raque-Bogdan et al., 2011). The impact of insecure attachment can also be intergenerational. Moreira et al. (2015) found that when mothers had insecure attachment to their own mothers, their children reported lower quality of life, but self-compassion mitigates the negative impact of a mother's insecure attachment on their children.

Can cultivating self-compassion *increase* attachment security? At the present time, no studies were found showing that self-compassion priming in the laboratory or self-compassion training enhanced secure attachment. However, Pepping et al. (2015) found that secure attachment priming (imagining a person with whom you felt safe) increased self-compassion. Additionally, a type of attachment-based compassion therapy increased secure attachment and self-compassion mediated that outcome (Navarro-Gil et al., 2020). In spite of the lack of direct evidence that self-compassion training increases secure attachment but considering that self-compassion mediates the link between insecure attachment and mental health, we can conclude attachment, self-compassion, and emotion regulation are important, overlapping themes in mental health (Finlay-Jones, 2017).

Working with Fears of Compassion

According to attachment theory, if a child was neglected or abused by early attachment figures, the child is likely to be dismissive, harsh, or critical as an adult and also likely to feel unworthy or unacceptable (Gilbert & Procter, 2006; Shaver & Mikulincer, 2007). Furthermore, when positive emotions arise in relationship to others, they are likely to trigger painful childhood memories of rejection or harm. These memories feel threatening just as the actual experiences were threatening in the past. When painful memories are triggered by compassion, a person is likely to develop is likely to develop "fears of compassion" (Gilbert et al., 2011; Matos et al., 2017).

Examples of fears of compassion are "If I am too compassionate with myself, others will reject me," "Being too compassionate makes people soft and easy to take advantage of," and "I try to keep my distance from others even if I know they are kind" (Gilbert et al., 2011). Research shows that fears of compassion are consistently associated with mental health difficulties (Kirby et al., 2019), including anxiety (Merritt & Purdon, 2020), depression (Hart et al., 2020), body image and disordered eating (Ferreira et al., 2019), alcohol misuse (Forkus et al., 2020), and psychosis (Martins et al., 2017). Fears of compassion can be considered patterns of avoidance that prevent painful childhood memories from being restructured in the relative safety of therapy sessions. Therefore, fears of compassion need to be explicitly addressed in therapy.

Baldwin et al. (2020) provide evidence that people with insecure attachment may feel threatened by compassion. They found that, after a compassion-focused imagery exercise, insecurely attached individuals had significantly lower vmHRV. However, after they were primed for secure attachment ("visualize someone who makes you feel safe and secure"), a subsequent imagery exercise actually increased vmHRV, suggesting that the exercise had become comforting and soothing. In other words, insecurely attached individuals may need additional support to benefit from self-compassion practices.

The distress that arises when people give compassion to themselves or receive compassion from others is known as "backdraft" (Germer, 2009, pp. 150–152; Germer & Neff, 2019). Backdraft can take the form of *thoughts* and *beliefs*, such as "I'm unlovable;" *emotions*, such as grief or shame; *body* aches and pains; and automatic *behaviors*, such as withdrawal or aggression. Backdraft is an intrinsic part of the transformation process of compassion-based therapy. Compassion activates old memories and makes them available for reprocessing—it provides an opportunity to receive the kindness and understanding that was probably lacking when the painful experiences originally occurred. This is a delicate process and therapists need to make sure that their clients remain within the "window of tolerance" (Siegel, 1999), especially when backdraft consists of traumatic memories. As the resource of self-compassion develops, however, clients feel safer within themselves and develop a "secure base" (Bowlby, 2005) from which to explore the inner and outer world. In this manner, compassion-based therapy can be understood as a process of *reparenting*. Research (cited above) showing that self-compassion mediates the link between insecure attachment and psychological well-being supports this understanding.

Alleviating Shame

The alleviation of shame is another mechanism by which self-compassion appears to work in psychotherapy. Shame is a "self-conscious" emotion characterized by negative self-evaluation. Helen Block Lewis (1987) wrote that shame is "one's own vicarious experience of the other's scorn.... the self-in-the-eyes-of-the-other" (p. 15). In a moment of shame, our attention is either directed *externally* toward what others are thinking about us, or *internally* toward our own personal characteristics and behavior, but the common denominator of shame is how we imagine ourselves to exist in the minds of others (Gilbert & Irons, 2009).

Human beings are social animals and need to be accepted and valued by others to feel safe (Baumeister & Leary, 1995). Shame seems to appear very early in life—an early precursor of shame is when an infant drops its eyes and head in response to losing facial communication with a parent (Tomkins, in Stipek, 1983). The full emotion of shame arises around the second half of the second year of life when a child becomes self-aware (Lewis, 2016). Patterns of rejection in early childhood—being blamed, attacked, ignored, or abandoned by a primary caregiver—can make a person shame-prone (Claesson &

Sohlberg, 2002). Shame is also closely linked to insecure attachment (Matos & Pinto-Gouveia, 2014; Schore, 1998). Shame is likely to arise at any stage of our lives, however, when our relationship security is compromised or our social standing is in jeopardy (Gilbert, 2007).

Shame has a long history as a predisposing factor in psychopathology, associated with a broad spectrum of disorders including depression, anxiety, eating disorders, PTSD, and substance abuse (Dearing & Tangney, 2011). Shame itself can also be traumatic. Steindl et al. (2018) argue that when a shame memory is central to an individual's personality, it is more likely to be traumatic. Shame becomes central in our lives when we relive it in words, images, smells, thoughts, behaviors, and emotions. Conversely, trauma can lead to shame when people blame themselves for their traumatic experiences, perhaps concluding that they are "bad," "defective," or "powerless" (Scoglio et al., 2018). Traumatized people may also behave in ways that cause more shame, such as engaging substance abuse, aggression, or withdrawal (Briere, 2019).

Self-compassion is the opposite of shame. The three qualities that oppose self-compassion in Neff's (2003) definition of self-compassion—self-criticism, isolation, and overidentification/rumination—are qualities of shame. Research shows that shame and self-criticism occur in tandem, especially in psychopathology (Gilbert & Irons, 2005; Werner et al., 2019). Shame is also associated with social isolation (Hartling et al., 2004; Thoresen et al., 2018) and rumination (Cândea & Szentágotai-Tătar, 2017; Orth et al., 2006). All three factors can be pathogenic. Self-criticism, in particular, is considered a transdiagnostic risk factor in depression, social anxiety, eating disorders, personality disorders, and psychotic symptoms (Löw et al., 2020; Werner et al., 2019).

Many studies have shown that self-compassion and shame are inversely associated. In the clinical arena, self-compassion was associated with lower levels of shame and psychological distress among people struggling with depression (Sick et al., 2020; Steindl et al., 2018), trauma and depression (Ross et al., 2019), eating disorders (Ferreira et al., 2014; Serpell et al., 2020), body image problems (Ferreira et al., 2013; Huellemann & Calogero, 2020), shame proneness (Woods & Proeve, 2014), and stress related to fertility (Galhardo et al., 2013), HIV (Skelton et al., 2020), imposter feelings (Wei et al., 2020), sexual minority stigma (Chan et al., 2020), parenting (Sirois et al., 2019), and caregiving (Biddle et al., 2020).

In a particularly interesting study, Zhang et al. (2018) found that self-compassion, but not contingent self-worth, mediated the link between shame and depression. Contingent self-worth refers to feeling good about ourselves based on the support we receive from external sources (e.g., family, friends) whereas self-compassion enhances self-worth by how we treat ourselves. Self-compassion is a more stable source of self-worth because it is not dependent on others. In the Zhang study, self-compassion dampened the impact of shame, and therefore the symptoms of depression, presumably by enhancing self-worth.

Shame and self-compassion were also inversely related in *therapeutic interventions* for depression (Johnson et al., 2018), PTSD (Au et al., 2017), narcissistic personality disorder (Kramer et al., 2018), eating disorders (Kelly et al., 2017), body weight shame (Carter et al., 2020), social anxiety disorder (Khoramnia et al., 2020); and stress due to intellectual disability (Goad & Parker, 2020), and body image (Albertson et al., 2015; Amy et al., 2020).

Shame is becoming a target for psychological interventions, especially with the emergence of compassion-based treatments such as CFT, EFT, and IFS. CFT was originally developed by Paul Gilbert to treat depressed people who suffered from high levels of shame and self-criticism (Gilbert, 2010a, b). He observed that people who were shamed in childhood found it difficult to access feelings of warmth, kindness, and compassion. These clients were also afraid of compassionate connections that might heal those old wounds. Therefore, the primary objectives of CFT treatment are to address fears of compassion and to help clients activate compassion for themselves. Preliminary outcome studies on CFT have shown reductions in shame along with clinical improvement in eating disorders (Carter et al., 2020; Kelly et al., 2017), depression (Gilbert & Procter, 2006), social anxiety (Boersma et al., 2015), PTSD (Au et al., 2017), perfectionism (Matos & Steindl, 2020), chronic pain (Håkansson et al., 2015), and personality disorder (Lucre & Corten, 2013) although well-controlled studies are still needed to confirm that CFT reduces shame across diagnoses.

Shame is also a focus of treatment in the EFT model and is considered a central feature of psychopathology (Greenberg, 2010; Greenberg & Iwakabe, 2011). The purpose of EFT is to help clients regulate their emotions. EFT assumes that maladaptive emotions such as toxic shame are not changed through reasoning or new skills, but rather by activating more adaptive emotions such as anger, grief (Shahar, 2020). EFT for shame begins by helping clients acknowledge shame, usually by a therapist empathizing with painful emotions underlying shame, such as betrayal or humiliation. When the client is ready to address shame, a new experience may be created to transform the painful emotion, for example, by using the two-chair dialogue technique in which clients enact two parts of themselves—the self-critic and the object of criticism (Shahar et al., 2012). The two-chair technique evokes self-compassion, which is considered in EFT to a transformative emotional experience (Shahar, 2020).

EFT and CFT are both based in attachment theory. The role of the therapist is to facilitate secure attachment by creating corrective emotional experiences that are internalized by the client over time (Karris & Caldwell, 2015). CFT therapists are more likely than EFT therapists to explicitly teach self-compassion through home practices. Karris and Caldwell (2015) recommended blending of EFT with CFT for couples struggling with trauma and shame.

IFS is another empirically supported, self-compassion-based therapy that works directly with shame. IFS was developed by Richard Schwartz (Schwartz, 1995; Schwartz & Sweezy, 2019) and works entirely with parts, or subpersonalities, of ourselves. This approach is particularly helpful for treating shame because shame can lead to dissociation and splitting of consciousness into parts (Dorahy et al., 2017; Platt et al., 2017). In IFS, the emotion of shame is a burden carried by an "exile"—a child part, tucked away and stuck in time. Self-criticism is the work of a "manager" who is trying to suppress the emotion. Unfortunately, self-criticism is likely to cause more shame, which then evokes a "firefighter" who may engage in risky behaviors or be aggressive toward others (Sweezy, 2013). Each part is trying to help us in some way and needs to be treated with deep respect and compassion, and when that happens, the part feels safer and is willing to let go of its burdens or responsibilities. The physiological basis of change in IFS is similar to CFT and EFT, namely, the reduction of threat and the cultivation of inner safety and security.

Conclusion

In summary, self-compassion is a key factor in mental health with wide-ranging implications for psychotherapy. Self-compassion has been present throughout the history of psychotherapy but is now emerging as a unique construct and specific target of treatment within the mindfulness-, acceptance-, and compassion-based psychotherapy paradigm. Compassion-based therapy is mostly oriented toward cultivating *self*-compassion in the client. Self-compassion appears to be a transdiagnostic and transtheoretical change process underlying most, if not all, forms of therapy. Self-compassion can also be integrated into therapy by cultivating compassionate presence, a compassionate therapeutic alliance, or by customizing compassionate interventions for clients. Ideally, all three levels are blended together in self-compassion-based treatment.

Understanding the mechanisms of change underlying self-compassion in psychotherapy can assist in the process of integrating self-compassion into therapy. For example, self-compassion in therapy has much in common with positive regard, empathy, and mindfulness as common factors in effective treatment. Interpersonal synchrony is also linked to self-compassion, albeit indirectly through therapeutic presence and the treatment alliance. The mechanism of change most commonly associated with self-compassion is emotion regulation. Under the umbrella of emotion regulation, self-compassion appears to calm the sympathetic nervous system and activate the parasympathetic system, along with enhancing executive functioning in the brain. Self-compassion also appears to be a resource that can help clients move from insecure to secure attachment and reduce the harmful effects of shame in their lives.

The field of self-compassion in therapy is currently in its adolescence. The empirical research base is expanding rapidly and is quite promising, but there still are relatively few, well-controlled therapy outcome studies that specifically target self-compassion. Indirect evidence for the importance of self-compassion in therapy is robust, however, such as numerous studies on self-compassion as a mediator of positive therapy outcomes and correlational studies consistently showing a close association between self-compassion and secure attachment and an inverse relationship between self-compassion and shame. Ongoing efforts by clinical scientists and practitioners will inevitably deepen our understanding of how self-compassion works in therapy and enhance the effectiveness of treatments for a broad range of psychological conditions and diverse populations.

References

Ablon, J., & Jones, E. (1998). How expert clinicians' prototypes of an ideal treatment correlate with outcome in psychodynamic and cognitive-behavioral therapy. *Psychotherapy Research, 8*(1), 71–83. https://doi.org/10.1093/ptr/8.1.71

Ablon, J. S., Levy, R. A., & Katzenstein, T. (2006). Beyond brand names of psychotherapy: Identifying empirically supported change processes. *Psychotherapy: Theory, Research, Practice, Training, 43*(2), 216. https://doi.org/10.1037/0033-3204.43.2.216

Ainsworth, M. D., Blehar, M., Waters, E., & Wall, S. (1978). *Patterns of attachment*. Erlbaum.

Albertson, E. R., Neff, K. D., & Dill-Shackleford, K. E. (2015). Self-compassion and body dissatisfaction in women: A randomized controlled trial of a brief meditation intervention. *Mindfulness, 6*(3), 444–454. https://doi.org/10.1007/s12671-014-0277-3

Allen, A. B., & Leary, M. R. (2010). Self-compassion, stress, and coping. *Social and Personality Psychology Compass, 4*(2), 107–118. https://doi.org/10.1111/j.1751-9004.2009.00246.x

Allen, A. B., Barton, J., & Stevenson, O. (2015). Presenting a self-compassionate image after an interpersonal transgression. *Self and Identity, 14*(1), 33–50. https://doi.org/10.1080/15298868.2014.946958

Alsubaie, M., Abbott, R., Dunn, B., Dickens, C., Keil, T. F., Henley, W., & Kuyken, W. (2017). Mechanisms of action in mindfulness-based cognitive therapy (MBCT) and mindfulness-based stress reduction (MBSR) in people with physical and/or psychological conditions: A systematic review. *Clinical Psychology Review, 55*, 74–91. https://doi.org/10.1016/j.cpr.2017.04.008

Altmann, U., Schoenherr, D., Paulick, J., Deisenhofer, A. K., Schwartz, B., Rubel, J. A., Stangier, U., Lutz, W., & Strauss, B. (2020). Associations between movement synchrony and outcome in patients with social anxiety disorder: Evidence for treatment specific effects. *Psychotherapy Research, 30*(5), 574–590. https://doi.org/10.1080/10503307.2019.1630779

Amy, J., Lane, B. R., & Mulgrew, K. E. (2020). A randomised controlled trial examining the effects of self-compassion meditations on women's body image. *Body Image, 35*, 22–29. https://doi.org/10.1016/j.bodyim.2020.07.009

Anderson, T., Ogles, B. M., Patterson, C. L., Lambert, M. J., & Vermeersch, D. A. (2009). Therapist effects: Facilitative interpersonal skills as a predictor of therapist success. *Journal of Clinical Psychology, 65*(7), 755–768. https://doi.org/10.1002/jclp.20583

Appelhans, B. M., & Luecken, L. J. (2006). Heart rate variability as an index of regulated emotional responding. *Review of General Psychology, 10*(3), 229–240. https://doi.org/10.1037/1089-2680.10.3.229

Arch, J. J., Brown, K. W., Dean, D. J., Landy, L. N., Brown, K. D., & Laudenslager, M. L. (2014). Self-compassion training modulates alpha-amylase, heart rate variability, and subjective responses to social evaluative threat in women. *Psychoneuroendocrinology, 42*, 49–58. https://doi.org/10.1016/j.psyneuen.2013.12.018

Ash, M., Harrison, T., Pinto, M., DiClemente, R., & Negi, L. T. (2019). A model for cognitively-based compassion training: Theoretical underpinnings and proposed mechanisms. *Social Theory & Health, 19*, 43–67. https://doi.org/10.1057/s41285-019-00124-x

Au, T. M., Sauer-Zavala, S., King, M. W., Petrocchi, N., Barlow, D. H., & Litz, B. T. (2017). Compassion-based therapy for trauma-related shame and post-traumatic stress: Initial evaluation using a multiple baseline design. *Behavior Therapy, 48*(2), 207–221. https://doi.org/10.1016/j.beth.2016.11.012

Baer, R. A. (2010). Self-compassion as a mechanism of change in mindfulness-and acceptance-based treatments. In R. Baer (Ed.), *Assessing mindfulness and acceptance processes in clients: Illuminating the theory and practice of change* (pp. 135–153). Guilford Press.

Bakker, A. M., Cox, D. W., Hubley, A. M., & Owens, R. L. (2019). Emotion regulation as a mediator of self-compassion and depressive symptoms in recurrent depression. *Mindfulness, 10*(6), 1169–1180. https://doi.org/10.1007/s12671-018-1072-3

Baldwin, S., Bandarian-Balooch, S., & Adams, R. (2020). Attachment and compassion-threat: Influence of a secure attachment-prime. *Psychology and Psychotherapy: Theory, Research and Practice, 93*(3), 520–536. https://doi.org/10.1111/papt.12244

Barber, J. P., Gallop, R., Crits-Christoph, P., Frank, A., Thase, M. E., Weiss, R. D., & Gibbons, M. B. C. (2006). The role of therapist adherence, therapist competence, and alliance in predicting outcome of individual drug counseling: Results from the National Institute Drug Abuse Collaborative Cocaine Treatment Study. *Psychotherapy Research, 16*(02), 229–240. https://doi.org/10.1080/10503300500288951

Barlow, M. R., Turow, R. E. G., & Gerhart, J. (2017). Trauma appraisals, emotion regulation difficulties, and self-compassion predict posttraumatic stress symptoms following childhood abuse. *Child Abuse & Neglect, 65*, 37–47. https://doi.org/10.1016/j.chiabu.2017.01.006

Baumeister, R. F., & Leary, M. R. (1995). The need to belong: Desire for interpersonal attachments as a fundamental human motivation. *Psychological Bulletin, 117*(3), 497. https://doi.org/10.1037/0033-2909.117.3.497

Bayır-Toper, A., Sellman, E., & Joseph, S. (2020). Being yourself for the 'greater good': An empirical investigation of the moderation effect of authenticity between self-compassion and compassion for others. *Current Psychology, 41*, 4871–4884. https://doi.org/10.1007/s12144-020-00989-6

Beaumont, E., Durkin, M., Hollins Martin, C. J., & Carson, J. (2016). Measuring relationships between self-compassion, compassion fatigue, burnout and well-being in student counsellors and student cognitive behavioural psychotherapists: A quantitative survey. *Counselling and Psychotherapy Research, 16*(1), 15–23. https://doi.org/10.1002/capr.12054

Bennett-Levy, J., & Finlay-Jones, A. (2018). The role of personal practice in therapist skill development: A model to guide therapists, educators, supervisors and researchers. *Cognitive Behaviour Therapy, 47*(3), 185–205. https://doi.org/10.1080/16506073.2018.1434678

Bergen-Cico, D., & Cheon, S. (2014). The mediating effects of mindfulness and self-compassion on trait anxiety. *Mindfulness, 5*(5), 505–519. https://doi.org/10.1007/s12-013-0205-y

Bergen-Cico, D., Smith, Y., Wolford, K., Gooley, C., Hannon, K., Woodruff, R., Spicer, M., & Gump, B. (2018). Dog ownership and training reduces post-traumatic stress symptoms and increases self-compassion among veterans: Results of a longitudinal control study. *The Journal of Alternative and Complementary Medicine, 24*(12), 1166–1175. https://doi.org/10.1089/acm.2018.0179

Bernhardt, B. C., & Singer, T. (2012). The neural basis of empathy. *Annual Review of Neuroscience, 35*, 1–23. https://doi.org/10.1146/annurev-neuro-062111-150536

Berry, M. P., Lutz, J., Schuman-Olivier, Z., Germer, C., Pollak, S., Edwards, R. R., ... & Napadow, V. (2020). Brief self-compassion training alters neural responses to evoked pain for chronic low back pain: a pilot study. *Pain Medicine, 21*(10), 2172–2185.

Bezdek, M. A., Gerrig, R. J., Wenzel, W. G., Shin, J., Revill, K. P., & Schumacher, E. H. (2015). Neural evidence that suspense narrows attentional focus. *Neuroscience, 303*, 338–345.

Bibeau, M., Dionne, F., & Leblanc, J. (2016). Can compassion contribute to the development of psychotherapists' empathy? A review. *Mindfulness, 7*(1), 255–263. https://doi.org/10.1007/s12671-015-0439-y

Bibeau, M., Dionne, F., Riera, A., & Leblanc, J. (2020). The influence of compassion meditation on the psychotherapist's empathy and clinical practice: A phenomenological analysis. *Journal of Humanistic Psychology*. https://doi.org/10.1177/0022167820953258

Biddle, Z., O'Callaghan, F. V., Finlay-Jones, A. L., & Reid, N. E. (2020). Caregivers of children with fetal alcohol spectrum disorder: Psychosocial factors and evidence for self-compassion as a potential interv target. *Mindfulness, 11*(9), 2189–2198. https://doi.org/10.1007/s12671-020-01443-1

Birnie, K., Speca, M., & Carlson, L. E. (2010). Exploring self-compassion and empathy in the context of mindfulness-based stress reduction (MBSR). *Stress and Health, 26*(5), 359–371. https://doi.org/10.1002/smi.1305

Bishop, S. R., Lau, M., Shapiro, S., Carlson, L., Anderson, N. D., Carmody, J., Segal, Z. V., Abbey, S., Speca, M., Velting, D., & Devins, G. (2004). Mindfulness: A proposed operational definition. *Clinical Psychology: Science and Practice, 11*(3), 230–241. https://doi.org/10.1093/clipsy.bph077

Blatt, S. J. (1995). The destructiveness of perfectionism: Implications for the treatment of depression. *American Psychologist, 50*(12), 1003. https://doi.org/10.1037//0003-066x.50.12.1003

Bluth, K. (2017). *The self-compassion workbook for teens: Mindfulness and compassion skills to over self-criticism and embrace who you are*. New Harbinger Publications.

Boellinghaus, I., Jones, F. W., & Hutton, J. (2013). Cultivating self-care and compassion in psychological therapists in training: The experience of practicing loving-kindness meditation. *Training and Education in Professional Psychology, 7*(4), 267. https://doi.org/10.1037/a0033092

Boellinghaus, I., Jones, F. W., & Hutton, J. (2014). The role of mindfulness and loving-kindness meditation in cultivating self-compassion and other-focused concern in health care professionals. *Mindfulness, 5*(2), 129–138. https://doi.org/10.1007/s12671-012-0158-6

Boersma, K., Håkanson, A., Salomonsson, E., & Johansson, I. (2015). Compassion focused therapy to counteract shame, self-criticism and isolation. A replicated single case experimental study for individuals with social anxiety. *Journal of Contemporary Psychotherapy, 45*(2), 89–98. https://doi.org/10.1007/s10879-014-9286-8

Bohart, A. C., Elliot, R., Greenberg, L. S., & Watson, J. C. (2002). Empathy. In J. C. Norcross (Ed.), *Psychotherapy relationships that work* (pp. 89–108). Oxford University Press.

Bordin, E. S. (1975). *The working alliance: Basis for a general theory of psychotherapy.* Paper presented at the annual meeting of the American Psychological Association, Washington, DC.

Bordin, E. S. (1994). Theory and research on the therapeutic working alliance: New directions. In A. O. Horvath & L. S. Greenberg (Eds.), *The working alliance: Theory, research, and practice* (pp. 13–37). Wiley.

Bourgault, M., & Dionne, F. (2019). Therapeutic presence and mindfulness: Mediating role of self-compassion and psychological distress among psychologists. *Mindfulness, 10*(4), 650–656. https://doi.org/10.1007/s12671-018-1015-z

Bowlby, J. (1969). *Attachment and loss: Vol. 1. Attachment*. Hogarth Press.

Bowlby, J. (1973). *Attachment and loss: Vol. 2. Separation - anxiety and anger*. Hogarth Press.

Bowlby, J. (1980). *Attachment and loss: Vol. 3. Loss – Sadness and depression*. Hogarth Press.

Bowlby, J. (2005). *A secure base: Clinical applications of attachment theory* (Vol. 393). Taylor & Francis.

Brach, T. (2012). Mindful presence: A foundation for compassion and wisdom. In C. Germer & R. Siegel (Eds.), *Wisdom and compassion in psychotherapy*. Guilford Press.

Braehler, C., Gumley, A., Harper, J., Wallace, S., Norrie, J., & Gilbert, P. (2013). Exploring change processes in compassion focused therapy in psychosis: Results of a feasibility randomized controlled trial. *British Journal of Clinical Psychology, 52*(2), 199–214. https://doi.org/10.1111/bjc.12009

Brandsma, R. (2017). *The mindfulness teaching guide: Essential skills and competencies for teaching mindfulness-based interventions*. New Harbinger Publications.

Braun, T. D., Park, C. L., & Gorin, A. (2016). Self-compassion, body image, and disordered eating: A review of the literature. *Body Image, 17*, 117–131. https://doi.org/10.1016/j.bodyim.2016.03.003

Breines, J. G., & Chen, S. (2013). Activating the inner caregiver: The role of support-giving schemas in increasing state self-compassion. *Journal of Experimental Social Psychology, 49*(1), 58–64. https://doi.org/10.1016/j.jesp.2012.07.015

Breines, J. G., Thoma, M. V., Gianferante, D., Hanlin, L., Chen, X., & Rohleder, N. (2014). Self-compassion as a predictor of interleukin-6 response to acute psychosocial stress. *Brain, Behavior, and Immunity, 37*, 109–114. https://doi.org/10.1016/j.bbi.2013.11.006

Breines, J. G., McInnis, C. M., Kuras, Y. I., Thoma, M. V., Gianferante, D., Hanlin, L., Chen, X., & Rohleder, N. (2015). Self-compassionate young adults show lower salivary alpha-amylase responses to repeated psychosocial stress. *Self and Identity, 14*(4), 390–402. https://doi.org/10.1080/15298868.2015.1005659

Brennan, K. A., Clark, C. L., & Shaver, P. R. (1998). Self-report measurement of adult attachment: An integrative overview. In J. A. Simpson & W. S. Rholes (Eds.), *Attachment theory and close relationships* (pp. 46–76). The Guilford Press.

Briere, J. (2019). *Treating risky and compulsive behavior in trauma survivors*. Guilford Publications.

Brophy, K., Brähler, E., Hinz, A., Schmidt, S., & Körner, A. (2020). The role of self-compassion in the relationship between attachment, depression, and quality of life. *Journal of Affective Disorders, 260*, 45–52. https://doi.org/10.1016/j.jad.2019.08.066

Buirski, P., Haglund, P., & Markley, E. (2020). *Making sense together: The intersubjective approach to psychotherapy*. Rowman & Littlefield Publishers.

Cacioppo, S., Zhou, H., Monteleone, G., Majka, E. A., Quinn, K. A., Ball, A. B., Norman, G. J., Semin, G. R., & Cacioppo, J. T. (2014). You are in sync with me: Neural correlates of interpersonal synchrony with a partner. *Neuroscience, 277*, 842–858. https://doi.org/10.1016/j.neuroscience.2014.07.051

Cândea, D.-M., & Szentágotai-Tătar, A. (2017). Shame as a predictor of post-event rumination in social anxiety. *Cognition and Emotion, 31*(8), 1684–1691. https://doi.org/10.1080/02699931.2016.1243518

Carey, T. A., Griffiths, R., Dixon, J. E., & Hines, S. (2020). Identifying functional mechanisms in psychotherapy: A scoping systematic review. *Frontiers in Psychiatry, 11*, 291. https://doi.org/10.3389/fpsyt.2020.00291

Carter, A., Gilbert, P., & Kirby, J. N. (2020). Compassion focused therapy for body weight shame: A mixed methods pilot trial. *Clinical Psychology & Psychotherapy, 28*(1), 93–108. https://doi.org/10.1002/cpp.2488

Chalmers, J. A., Quintana, D. S., Abbott, M. J., & Kemp, A. H. (2014). Anxiety disorders are associated with reduced heart rate variability: A meta-analysis. *Frontiers in Psychiatry, 5*, 80. https://doi.org/10.3389/fpsyt.2014.00080

Chan, K. K. S., Yung, C. S. W., & Nie, G. M. (2020). Self-compassion buffers the negative psychological impact of stigma stress on sexual minorities. *Mindfulness, 11*(10), 2338–2348. https://doi.org/10.1007/s12671-020-01451-1

Claesson, K., & Sohlberg, S. (2002). Shame and interpersonal scripts: Internalized shame and other scripts characterized by indifference, abandonment and rejection: Replicated findings. *Clinical Psychology and Psychotherapy, 9*(4), 277–284. https://doi.org/10.1002/cpp.331

Clapton, N., & Hiskey, S. (2020). Radically embodied compassion: The potential role of traditional martial arts in compassion cultivation. *Frontiers in Psychology, 11*, 555156. https://doi.org/10.3389/fpsyg.2020.555156

Collett, N., Pugh, K., Waite, F., & Freeman, D. (2016). Negative cognitions about the self in patients with persecutory delusions: An empirical study of self-compassion, self-stigma, schematic beliefs, self-esteem, fear of madness, and suicidal ideation. *Psychiatry Research, 239*, 79–84. https://doi.org/10.1016/j.psychres.2016.02.043

Craig, C., Hiskey, S., Royan, L., Poz, R., & Spector, A. (2018). Compassion focused therapy for people with dementia: A feasibility study. *International Journal of Geriatric Psychiatry, 33*(12), 1727–1735. https://doi.org/10.1002/gps.4977

Craig, C., Hiskey, S., & Spector, A. (2020). Compassion focused therapy: A systematic review of its effectiveness and acceptability in clinical populations. *Expert Review of Neurotherapeutics, 20*(4), 385–400. https://doi.org/10.1080/14737175.2020.1746184

Crews, D. A., Stolz-Newton, M., & Grant, N. S. (2016). The use of yoga to build self-compassion as a healing method for survivors of sexual violence. *Journal of Religion & Spirituality in Social Work: Social Thought, 35*(3), 139–156. https://doi.org/10.1080/15426432.2015.1067583

Cunha, M., Martinho, M. I., Xavier, A. M., & Espirito-Santo, H. (2014). Early memories of positive emotions and its relationships to attachment styles, self-compassion and psychopathology in adolescence. *European Psychiatry, 28*(Suppl. 1), 1. https://doi.org/10.1016/S0924-9338(13)76444-7

Dahm, K. A., Meyer, E. C., Neff, K. D., Kimbrel, N. A., Gulliver, S. B., & Morissette, S. B. (2015). Mindfulness, self-compassion, posttraumatic stress disorder symptoms, and functional disability in US Iraq and Afghanistan war veterans. *Journal of Traumatic Stress, 28*(5), 460–464. https://doi.org/10.1002/jts.22045

Dearing, R. L., & Tangney, J. P. E. (2011). *Shame in the therapy hour*. American Psychological Association.

Decety, J. (2011). Dissecting the neural mechanisms mediating empathy. *Emotion Review, 3*(1), 92–108. https://doi.org/10.1177/1754073910374662

Decety, J., & Lamm, C. (2006). Human empathy through the lens of social neuroscience. *The Scientific World Journal, 6*, 1146–1163. https://doi.org/10.1100/tsw.2006.221

Dhokia, M., Elander, J., Clements, K., & Gilbert, P. (2020). A randomized-controlled pilot trial of an online compassionate mind training intervention to help people with chronic pain avoid analgesic misuse. *Psychology of Addictive Behaviors, 34*(7), 726–733. https://doi.org/10.1037/adb0000579

Di Bello, M., Carnevali, L., Petrocchi, N., Thayer, J. F., Gilbert, P., & Ottaviani, C. (2020). The compassionate vagus: A meta-analysis on the connection between compassion and heart rate variability. *Neuroscience & Biobehavioral Reviews, 116*, 21–30. https://doi.org/10.1016/j.neubiorev.2020.06.016

Diedrich, A., Grant, M., Hofmann, S. G., Hiller, W., & Berking, M. (2014). Self-compassion as an emotion regulation strategy in major depressive disorder. *Behaviour Research and Therapy, 58*, 43–51. https://doi.org/10.1016/j.brat.2014.05.006

Diedrich, A., Burger, J., Kirchner, M., & Berking, M. (2017). Adaptive emotion regulation mediates the relationship between self-compassion and depression in individuals with unipolar depression. *Psychology and Psychotherapy: Theory, Research and Practice, 90*(3), 247–263. https://doi.org/10.1111/papt.12107

Dimidjian, S., Arch, J. J., Schneider, R. L., Desormeau, P., Felder, J. N., & Segal, Z. V. (2016). Considering meta-analysis, meaning, and metaphor: A systematic review and critical examination of "third wave" cognitive and behavioral therapies. *Behavior Therapy, 47*(6), 886–905. https://doi.org/10.1016/j.beth.2016.07.002

Dorahy, M. J., McKendry, H., Scott, A., Yogeeswaran, K., Martens, A., & Hanna, D. (2017). Reactive dissociative experiences in response to acute increases in shame feelings. *Behaviour Research and Therapy, 89*, 75–85. https://doi.org/10.1016/j.brat.2016.11.007

Døssing, M., Nilsson, K. K., Svejstrup, S. R., Sørensen, V. V., Straarup, K. N., & Hansen, T. B. (2015). Low self-compassion in patients with bipolar disorder. *Comprehensive Psychiatry, 60*, 53–58. https://doi.org/10.1016/j.comppsych.2015.03.010

Edelstein, R. S., Chopik, W. J., Saxbe, D. E., Wardecker, B. M., Moors, A. C., & LaBelle, O. P. (2017). Prospective and dyadic associations between expectant parents' prenatal hormone changes and postpartum parenting outcomes. *Developmental Psychobiology, 59*(1), 77–90. https://doi.org/10.1002/dev.21469

Eicher, A. C., Davis, L. W., & Lysaker, P. H. (2013). Self-compassion: A novel link with symptoms in schizophrenia? *The Journal of Nervous and Mental Disease, 201*(5), 389–393. https://doi.org/10.1097/NMD.0b013e31828e10fa

Eichholz, A., Schwartz, C., Meule, A., Heese, J., Neumüller, J., & Voderholzer, U. (2020). Self-compassion and emotion regulation difficulties in obsessive–compulsive disorder. *Clinical Psychology & Psychotherapy, 27*(5), 630–639. https://doi.org/10.1002/cpp.2451

Eisenberg, N., & Eggum, N. D. (2009). Empathic responding: Sympathy and personal distress. In J. Decety & W. Ickes (Eds.), *The social neuroscience of empathy* (pp. 71–83). MIT Press.

Elliott, R., Bohart, A. C., Watson, J. C., & Greenberg, L. S. (2011). Empathy. *Psychotherapy, 48*(1), 43. https://doi.org/10.1037/a0022187

Elliott, R., Bohart, A. C., Watson, J. C., & Murphy, D. (2019). Empathy. In J. C. Norcross & M. J. Lambert (Eds.), *Psychotherapy relationship that work* (pp. 245–664). Oxford University Press.

Engel, Y., Noordijk, S., Spoelder, A., & van Gelderen, M. (2019). Self-compassion when coping with venture obstacles: Loving-kindness meditation and entrepreneurial fear of failure. *Entrepreneurship Theory and Practice, 45*(2), 263–290. https://doi.org/10.1177/1042258719890991

Epstein, R. M. (1999). Mindful practice. *The Journal of the American Medical Association, 282*(9), 833–839. https://doi.org/10.1001/jama.282.9.833

Escuriex, B. F., & Labbé, E. E. (2011). Health care providers' mindfulness and treatment outcomes: A critical review of the research literature. *Mindfulness, 2*(4), 242–253. https://doi.org/10.1007/s12671-011-0068-z

Evans, S., Wyka, K., Blaha, K. T., & Allen, E. S. (2018). Self-compassion mediates improvement in well-being in a mindfulness-based stress reduction program in a community-based sample. *Mindfulness, 9*(4), 1280–1287. https://doi.org/10.1007/s12671-017-0872-1

Ewert, C., Gaube, B., & Geisler, F. C. M. (2018). Dispositional self-compassion impacts immediate and delayed reactions to social evaluation. *Personality and Individual Differences, 125*, 91–96. https://doi.org/10.1016/j.paid.2017.12.037

Farber, B. A., Suzuki, J. Y., & Lynch, D. A. (2019). Positive regard and affirmation. In J. C. Norcross & M. J. Lambert (Eds.), *Psychotherapy relationships that work: Evidence-based therapist contributions* (pp. 288–322). Oxford University Press. https://doi.org/10.1093/med-psych/9780190843953.003.0008

Feldman, R. (2015). Mutual influences between child emotion regulation and parent–child reciprocity support development across the first 10 years of life: Implications for developmental psychopathology. *Development and Psychopathology, 27*(4pt1), 1007–1023. https://doi.org/10.1017/S0954579415000656

Feldman, R., Magori-Cohen, R., Galili, G., Singer, M., & Louzoun, Y. (2011). Mother and infant coordinate heart rhythms through episodes of interaction synchrony. *Infant Behavior and Development, 34*(4), 569–577. https://doi.org/10.1016/j.infbeh.2011.06.008

Feliu-Soler, A., Pascual, J. C., Elices, M., Martín-Blanco, A., Carmona, C., Cebolla, A., Simón, V., & Soler, J. (2017). Fostering self-compassion and loving-kindness in patients with borderline personality disorder: A randomized pilot study. *Clinical Psychology & Psychotherapy, 24*(1), 278–286. https://doi.org/10.1002/cpp.2000

Ferrari, M., Hunt, C., Harrysunker, A., Abbott, M. J., Beath, A. P., & Einstein, D. A. (2019). Self-compassion interventions and psychosocial outcomes: A meta-analysis of RCTs. *Mindfulness, 10*(8), 1455–1473. https://doi.org/10.1007/s12671-019-01134-6

Ferreira, C., Pinto-Gouveia, J., & Duarte, C. (2013). Self-compassion in the face of shame and body image dissatisfaction: Implications for eating disorders. *Eating Behaviors, 14*(2), 207–210. https://doi.org/10.1016/j.eatbeh.2013.01.005

Ferreira, C., Matos, M., Duarte, C., & Pinto-Gouveia, J. (2014). Shame memories and eating psychopathology: The buffering effect of self-compassion. *European Eating Disorders Review, 22*(6), 487–494. https://doi.org/10.1002/erv.2322

Ferreira, C., Dias, B., & Oliveira, S. (2019). Behind women's body image-focused shame: Exploring the role of fears of compassion and self-criticism. *Eating Behaviors, 32*, 12–17. https://doi.org/10.1016/j.eatbeh.2018.11.002

Finlay-Jones, A. L. (2017). The relevance of self-compassion as an intervention target in mood and anxiety disorders: A narrative review based on an emotion regulation framework. *Clinical Psychologist, 21*(2), 90–103. https://doi.org/10.1111/cp.12131

Finlay-Jones, A. L., Rees, C. S., & Kane, R. T. (2015). Self-compassion, emotion regulation and stress among Australian psychologists: Testing an emotion regulation model of self-compassion using structural equation modeling. *PLoS One, 10*(7), e0133481. https://doi.org/10.1371/journal.pone.0133481

Forkus, S. R., Breines, J. G., & Weiss, N. H. (2020). PTSD and alcohol misuse: Examining the mediating role of fear of self-compassion among military veterans. *Psychological Trauma: Theory, Research, Practice, and Policy, 12*(4), 364. https://doi.org/10.1037/tra0000481

Fresnics, A. A., Wang, S. B., & Borders, A. (2019). The unique associations between self-compassion and eating disorder psychopathology and the mediating role of rumination. *Psychiatry Research, 274*, 91–97. https://doi.org/10.1016/j.psychres.2019.02.019

Freud, S. (1957). *Totem and taboo (1913)* (Vol. XIV, Standard ed.). Hogarth.

Friis, A. M., Johnson, M. H., Cutfield, R. G., & Consedine, N. S. (2016). Kindness matters: A randomized controlled trial of a mindful self-compassion intervention improves depression, distress, and HbA1c among patients with diabetes. *Diabetes Care, 39*(11), 1963–1971. https://doi.org/10.2337/dc16-0416

Galbusera, L., Finn, M. T., & Fuchs, T. (2018). Interactional synchrony and negative symptoms: An outcome study of body-oriented psychotherapy for schizophrenia. *Psychotherapy Research, 28*(3), 457–469. https://doi.org/10.1080/10503307.2016.1216624

Galbusera, L., Finn, M. T., Tschacher, W., & Kyselo, M. (2019). Interpersonal synchrony feels good but impedes self-regulation of affect. *Scientific Reports, 9*(1), 1–12. https://doi.org/10.1038/s41598-019-50960-0

Gale, C., Schröder, T., & Gilbert, P. (2017). 'Do you practice what you preach?' A qualitative exploration of therapists' personal practice of compassion focused therapy. *Clinical Psychology & Psychotherapy, 24*(1), 171–185. https://doi.org/10.1002/cpp.1993

Galhardo, A., Cunha, M., Pinto-Gouveia, J., & Matos, M. (2013). The mediator role of emotion regulation processes on infertility-related stress. *Journal of Clinical Psychology in Medical Settings, 20*(4), 497–507. https://doi.org/10.1007/s10880-013-9370-3

Galili-Weinstock, L., Chen, R., Atzil-Slonim, D., Bar-Kalifa, E., Peri, T., & Rafaeli, E. (2018). The association between self-compassion and treatment outcomes: Session-level and treatment-level effects. *Journal of Clinical Psychology, 74*(6), 849–866. https://doi.org/10.1002/jclp.22569

Galili-Weinstock, L., Chen, R., Atzil-Slonim, D., Rafaeli, E., & Peri, T. (2020). Enhancement of self compassion in psychotherapy: The role of therapists' interventions. *Psychotherapy Research, 30*(6), 815–828. https://doi.org/10.1080/10503307.2019.1650979

Gazzillo, F., Dazzi, N., De Luca, E., Rodomonti, M., & Silberschatz, G. (2020). Attachment disorganization and severe psychopathology: A possible dialogue between attachment theory and control-mastery theory. *Psychoanalytic Psychology, 37*(3), 173. https://doi.org/10.1037/pap0000260

Geller, S. M. (2017). *A practical guide to cultivating therapeutic presence*. American Psychological Association.

Geller, S. M., & Porges, S. W. (2014). Therapeutic presence: Neurophysiological mechanisms mediating feeling safe in therapeutic relationships. *Journal of Psychotherapy Integration, 24*(3), 178. https://doi.org/10.1037/a0037511

Germer, C. (2009). *The mindful path to self-compassion: Freeing yourself from destructive thoughts and emotions*. Guilford Press.

Germer, C. K. (2013). Mindfulness: What is it? What does it matter? In C. Germer, R. Siegel, & P. Fulton (Eds.), *Mindfulness and psychotherapy*. Guilford Press.

Germer, C., & Neff, K. (2019). *Teaching the mindful self-compassion program: A guide for professionals*. Guilford Press.

Germer, C. K., & Siegel, R. D. (2012). *Wisdom and compassion in psychotherapy: Deepening mindfulness in clinical practice*. Guilford Press.

Germer, C. K., Siegel, R. D., & Fulton, P. R. (2013). *Mindfulness and psychotherapy*. Guilford Press.

Gharraee, B., Tajrishi, K. Z., Farani, A. R., Bolhari, J., & Farahani, H. (2018). A randomized controlled trial of compassion focused therapy for social anxiety disorder. *Iranian Journal of Psychiatry and Behavioral Sciences, 12*(4). https://doi.org/10.5812/ijpbs.80945

Gilbert, P. (1993). Defence and safety: Their function in social behaviour and psychopathology. *British Journal of Clinical Psychology, 32*(2), 131–153.

Gilbert, P. (2007). The evolution of shame as a marker for relationship security: A biopsychosocial approach. In J. L. Tracy, R. W. Robins, & J. P. Tangney (Eds.), *The self-conscious emotions: Theory and research* (pp. 283–309). Guilford Press.

Gilbert, P. (2009). Introducing compassion-focused therapy. *Advances in Psychiatric Treatment, 15*(3), 199. https://doi.org/10.1192/apt.bp.107.005264

Gilbert, P. (2010a). An introduction to compassion focused therapy in cognitive behavior therapy. *International Journal of Cognitive Therapy, 3*(2), 97–112. https://doi.org/10.1521/ijct.2010.3.2.97

Gilbert, P. (2010b). *Compassion focused therapy: Distinctive features*. Routledge.

Gilbert, P., & Irons, C. (2005). Focused therapies and compassionate mind training for shame and self-attacking. In *Compassion: Conceptualisations, research and use in psychotherapy* (pp. 263–325). Routledge.

Gilbert, P., & Irons, C. (2009). Shame, self-criticism, and self-compassion in adolescence. In N. Allen & L. Sheeer (Eds.), *Adolescent emotional development and the emergence of depressive disorders, 1* (pp. 195–214). Cambridge University Press.

Gilbert, P., & Procter, S. (2006). Compassionate mind training for people with high shame and self-criticism: Overview and pilot study of a group therapy approach. *Clinical Psychology & Psychotherapy: An International Journal of Theory & Practice, 13*(6), 353–379. https://doi.org/10.1002/cpp.507

Gilbert, P., McEwan, K., Matos, M., & Rivis, A. (2011). Fears of compassion: Development of three self-report measures. *Psychology and Psychotherapy: Theory, Research and Practice, 84*(3), 239–255. https://doi.org/10.1348/147608310X526511

Girme, Y. U., Jones, R. E., Fleck, C., Simpson, J. A., & Overall, N. C. (2020). Infants' attachment insecurity predicts attachment-relevant emotion regulation strategies in adulthood. *Emotion, 1*(2), 260–272. https://doi.org/10.1037/emo0000721

Goad, E. J., & Parker, K. (2020). Compassion-focused therapy groups for people with intellectual disabilities: An extended pilot study. *Journal of Intellectual Disabilities, 25*(4), 661–679. https://doi.org/10.1177/1744629520925953

Goldberg, S. B., Knoeppel, C., Davidson, R. J., & Flook, L. (2020). Does practice quality mediate the relationship between practice time and outcome in mindfulness-based stress reduction? *Journal of Counseling Psychology, 67*(1), 115. https://doi.org/10.1037/cou0000369

Golden, H. L., Vosper, J., Kingston, J., & Ellett, L. (2020). The impact of mindfulness-based programmes on self-compassion in nonclinical populations: A systematic review and meta-analysis. *Mindfulness*, 1–24. https://doi.org/10.1007/s12671-020-01501-8

Goldin, P. R., & Jazaieri, H. (2017). The compassion cultivation training (CCT) program. In *The Oxford handbook of compassion science* (pp. 237–246). Oxford University Press.

Gonzalez-Hernandez, E., Romero, R., Campos, D., Burychka, D., Diego-Pedro, R., Baños, R., Negi, L. T., & Cebolla, A. (2018). Cognitively-based compassion training (CBCT®) in breast cancer survivors: A randomized clinical trial study. *Integrative Cancer Therapies, 17*(3), 684–696. https://doi.org/10.1177/1534735418772095

Goodman, J. H., Guarino, A., Chenausky, K., Klein, L., Prager, J., Petersen, R., Forget, A., & Freeman, M. (2014). CALM Pregnancy: Results of a pilot study of mindfulness-based cognitive therapy for perinatal anxiety. *Archives of Women's Mental Health, 17*(5), 373–387. https://doi.org/10.1007/s00737-013-0402-7

Gotink, R. A., Meijboom, R., Vernooij, M. W., Smits, M., & Hunink, M. M. (2016). 8-week mindfulness based stress reduction induces brain changes similar to traditional long-term meditation practice–a systematic review. *Brain and Cognition, 108*, 32–41. https://doi.org/10.1016/j.bandc.2016.07.001

Gratz, K. L., Weiss, N. H., & Tull, M. T. (2015). Examining emotion regulation as an outcome, mechanism, or target of psychological treatments. *Current Opinion in Psychology, 3*, 85–90. https://doi.org/10.1016/j.copsyc.2015.02.010

Greenberg, L. (2006). Emotion-focused therapy: A synopsis. *Journal of Contemporary Psychotherapy, 36*(2), 87–93. https://doi.org/10.1007/s10879-006-9011-3

Greenberg, L. S. (2010). Emotion-focused therapy: A clinical synthesis. *Focus, 8*(1), 32–42. https://doi.org/10.1176/foc.8.1.foc32

Greenberg, L. S., & Iwakabe, S. (2011). Emotion-focused therapy and shame. In R. L. Dearing & J. P. Tangney (Eds.), *Shame in the therapy hour* (pp. 69–90). American Psychological Association. https://doi.org/10.1037/12326-003

Greenberg, J., Datta, T., Shapero, B. G., Sevinc, G., Mischoulon, D., & Lazar, S. W. (2018). Compassionate hearts protect against wandering minds: Self-compassion moderates the effect of mind-wandering on depression. *Spirituality in Clinical Practice, 5*(3), 155–169. https://doi.org/10.1037/scp0000168

Grepmair, L., Mitterlehner, F., Loew, T., Bachler, E., Rother, W., & Nickel, M. (2007). Promoting mindfulness in psychotherapists in training influences the treatment results of their patients: A randomized, double-blind, controlled study. *Psychotherapy and Psychosomatics, 76*(6), 332–338. https://doi.org/10.1159/000107560

Gross, J. J., & Muñoz, R. F. (1995). Emotion regulation and mental health. *Clinical Psychology: Science and Practice, 2*(2), 151–164. https://doi.org/10.1111/j.1468-2850.1995.tb00036.x

Gupta, A., Strivens, F. L., Tag, B., Kunze, K., & Ward, J. A. (2019). *Blink as you sync: uncovering eye and nod synchrony in conversation using wearable sensing.* Paper presented at the Proceedings of the 23rd International Symposium on Wearable Computers.

Haj Sadeghi, Z., Yazdi-Ravandi, S., & Pirnia, B. (2018). Compassion-focused therapy on levels of anxiety and depression among women with breast cancer: A randomized pilot trial. *International Journal of Cancer Management, 11*(11), e67019. https://doi.org/10.5812/ijcm.67019

Håkansson, A., Friberg, M., Lidén, A., Svanberg, M., Rothman, M., & Boersma, K. (2015). *Compassion focused therapy: A transdiagnostic approach to counteract shame and self-criticism in the treatment of individuals with chronic pain and comorbid emotional problems.* Paper presented at the 45th European Association for Behavioral and Cognitive Therapies Annual Congress (EABCT 2015), Jerusalem, Israel, August 31–September 3, 2015.

Halamová, J., & Kanovský, M. (2019). Emotion-focused training for emotion coaching–an intervention to reduce self-criticism. *Human Affairs, 29*(1), 20–31. https://doi.org/10.1515/humaff-2019-0003

Halamová, J., Kanovský, M., Pačutová, A., & Kupeli, N. (2020). Randomised controlled trial of an online version of compassion mind training in a nonclinical sample. *Europe's Journal of Psychology, 16*(2), 262–279. https://doi.org/10.5964/ejop.v16i2.1683

Hart, J. S., Kirby, J. N., Steindl, S. R., Kane, R. T., & Mazzucchelli, T. G. (2020). Insecure striving, self-criticism, and depression: The prospective moderating role of fear of compassion from others. *Mindfulness, 11*, 1699–1709. https://doi.org/10.1007/s12671-020-01385-8

Hartling, L. M., Rosen, W. B., Walker, M., & Jordan, J. V. (2004). Shame and humiliation: From isolation to relational transformation. In J. V. Jordan, M. Walker, & L. M. Hartling (Eds.), *The complexity of connection: Writings from the Stone Center's Jean Baker Miller Training Institute* (pp. 103–128). Guilford Press.

Hasenkamp, W., & Barsalou, L. W. (2012). Effects of meditation experience on functional connectivity of distributed brain networks. *Frontiers in Human Neuroscience, 6*, 38. https://doi.org/10.3389/fnhum.2012.00038

Hasson, U., Ghazanfar, A. A., Galantucci, B., Garrod, S., & Keysers, C. (2012). Brain-to-brain coupling: A mechanism for creating and sharing a social world. *Trends in Cognitive Sciences, 16*(2), 114–121. https://doi.org/10.1016/j.tics.2011.12.007

Hayes, S. C., Strosahl, K. D., & Wilson, K. G. (2011). *Acceptance and commitment therapy: The process and practice of mindful change*. Guilford Press.

Held, P., Owens, G. P., Thomas, E. A., White, B. A., & Anderson, S. E. (2018). A pilot study of brief self-compassion training with individuals in substance use disorder treatment. *Traumatology, 24*(3), 219–227. https://doi.org/10.1037/trm0000146

Hessler-Kaufmann, J. B., Heese, J., Berking, M., Voderholzer, U., & Diedrich, A. (2020). Emotion regulation strategies in bulimia nervosa: An experimental investigation of mindfulness, self-compassion, and cognitive restructuring. *Borderline Personality Disorder and Emotion Dysregulation, 7*(1), 1–9. https://doi.org/10.1186/s40479-020-00129-3

Hildebrandt, L. K., McCall, C., & Singer, T. (2017). Differential effects of attention-, compassion-, and socio-cognitively based mental practices on self-reports of mindfulness and compassion. *Mindfulness, 8*(6), 1488–1512. https://doi.org/10.1007/s12671-017-0716-z

Hoffart, A., Øktedalen, T., & Langkaas, T. F. (2015). Self-compassion influences PTSD symptoms in the process of change in trauma-focused cognitive-behavioral therapies: A study of within-person processes. *Frontiers in Psychology, 6*, 1273. https://doi.org/10.3389/fpsyg.2015.01273

Hoge, E. A., Hölzel, B. K., Marques, L., Metcalf, C. A., Brach, N., Lazar, S. W., & Simon, N. M. (2013). Mindfulness and self-compassion in generalized anxiety disorder: Examining predictors of disability. *Evidence-based Complementary and Alternative Medicine, 2013*, 576258. https://doi.org/10.1155/2013/576258

Hölzel, B. K., Carmody, J., Vangel, M., Congleton, C., Yerramsetti, S. M., Gard, T., & Lazar, S. W. (2011a). Mindfulness practice leads to increases in regional brain gray matter density. *Psychiatry Research: Neuroimaging, 191*(1), 36–43. https://doi.org/10.1016/j.pscychresns.2010.08.006

Hölzel, B. K., Lazar, S. W., Gard, T., Schuman-Olivier, Z., Vago, D. R., & Ott, U. (2011b). How does mindfulness meditation work? Proposing mechanisms of action from a conceptual and neural perspective. *Perspectives on Psychological Science, 6*(6), 537–559. https://doi.org/10.1177/1745691611419671

Holzman, J. B., & Bridgett, D. J. (2017). Heart rate variability indices as bio-markers of top-down self-regulatory mechanisms: A meta-analytic review. *Neuroscience & Biobehavioral Reviews, 74*, 233–255. https://doi.org/10.1016/j.neubiorev.2016.12.032

Homan, K. J. (2018). Secure attachment and eudaimonic well-being in late adulthood: The mediating role of self-compassion. *Aging & Mental Health, 22*(3), 363–370. https://doi.org/10.1080/13607863.2016.1254597

Horvath, A. O., & Luborsky, L. (1993). The role of the therapeutic alliance in psychotherapy. *Journal of Consulting and Clinical Psychology, 61*(4), 561–573. https://doi.org/10.1037/0022-006X.61.4.561

Horvath, A. O., Del Re, A., Flückiger, C., & Symonds, D. (2011). Alliance in individual psychotherapy. *Psychotherapy, 48*(1), 9–16. https://doi.org/10.1037/a0022186

Huellemann, K. L., & Calogero, R. M. (2020). Self-compassion and body checking among women: The mediating role of stigmatizing self-perceptions. *Mindfulness, 11*, 2121–2130. https://doi.org/10.1007/s12671-020-01420-8

Imel, Z. E., Barco, J. S., Brown, H. J., Baucom, B. R., Baer, J. S., Kircher, J. C., & Atkins, D. C. (2014). The association of therapist empathy and synchrony in vocally encoded arousal. *Journal of Counseling Psychology, 61*(1), 146. https://doi.org/10.1037/a0034943

Inwood, E., & Ferrari, M. (2018). Mechanisms of change in the relationship between self-compassion, emotion regulation, and mental health: A systematic review. *Applied Psychology: Health and Well-Being, 10*(2), 215–235. https://doi.org/10.1111/aphw.12127

Irons, C., & Beaumont, E. (2017). *The compassionate mind workbook: A step-by-step guide to developing your compassionate self.* Robinson.

Ivanovic, M., Swift, J. K., Callahan, J. L., & Dunn, R. (2015). A multisite pre/post study of mindfulness training for therapists: The impact on session presence and effectiveness. *Journal of Cognitive Psychotherapy, 29*(4), 331–342. https://doi.org/10.1891/0889-8391.29.4.331

Jazaieri, H., Jinpa, G. T., McGonigal, K., Rosenberg, E. L., Finkelstein, J., Simon-Thomas, E., Cullen, M., Doty, J., Gross, J. J., & Goldin, P. R. (2013). Enhancing compassion: A randomized controlled trial of a compassion cultivation training program. *Journal of Happiness Studies, 14*(4), 1113–1126. https://doi.org/10.1007/s10902-012-9373-z

Jazaieri, H., McGonigal, K., Jinpa, T., Doty, J. R., Gross, J. J., & Goldin, P. R. (2014). A randomized controlled trial of compassion cultivation training: Effects on mindfulness, affect, and emotion regulation. *Motivation and Emotion, 38*(1), 23–35. https://doi.org/10.1007/s11031-013-9368-z

Jinpa, T. (2010). *Compassion cultivation training (CCT): Instructor's manual*. Unpublished.

Joeng, J. R., Turner, S. L., Kim, E. Y., Choi, S. A., Lee, Y. J., & Kim, J. K. (2017). Insecure attachment and emotional distress: Fear of self-compassion and self-compassion as mediators. *Personality and Individual Differences, 112*, 6–11. https://doi.org/10.1016/j.paid.2017.02.048

Johnson, S. B., Goodnight, B. L., Zhang, H., Daboin, I., Patterson, B., & Kaslow, N. J. (2018). Compassion-based meditation in African Americans: Self-criticism mediates changes in depression. *Suicide and Life-threatening Behavior, 48*(2), 160–168. https://doi.org/10.1111/sltb.12347

Kabat-Zinn, J. (2003). Mindfulness-based interventions in context: Past, present, and future. *Clinical Psychology: Science and Practice, 10*(2), 144–156. https://doi.org/10.1093/clipsy.bpg016

Kabat-Zinn, J. (2013). *Full catastrophe living, revised edition: How to cope with stress, pain and illness using mindfulness meditation*. Hachette.

Kang, D. H., Jo, H. J., Jung, W. H., Kim, S. H., Jung, Y. H., Choi, C. H., Lee, U. S., An, S. C., Jang, J. H., & Kwon, J. S. (2013). The effect of meditation on brain structure: Cortical thickness mapping and diffusion tensor imaging. *Social Cognitive and Affective Neuroscience, 8*(1), 27–33. https://doi.org/10.1093/scan/nss056

Kannan, D., & Levitt, H. M. (2013). A review of client self-criticism in psychotherapy. *Journal of Psychotherapy Integration, 23*(2), 166–178. https://doi.org/10.1037/a0032355

Karris, M., & Caldwell, B. E. (2015). Integrating emotionally focused therapy, self-compassion, and compassion-focused therapy to assist shame-prone couples who have experienced trauma. *The Family Journal, 23*(4), 346–357. https://doi.org/10.1177/1066480715601676

Kelliher Rabon, J., Sirois, F. M., & Hirsch, J. K. (2018). Self-compassion and suicidal behavior in college students: Serial indirect effects via depression and wellness behaviors. *Journal of American College Health, 66*(2), 114–122. https://doi.org/10.1080/07448481.2017.1382498

Kelly, A. C., & Tasca, G. A. (2016). Within-persons predictors of change during eating disorders treatment: An examination of self-compassion, self-criticism, shame, and eating disorder symptoms. *International Journal of Eating Disorders, 49*(7), 716–722. https://doi.org/10.1002/eat.22527

Kelly, A. C., Zuroff, D. C., Foa, C. L., & Gilbert, P. (2010). Who benefits from training in self-compassionate self-regulation? A study of smoking reduction. *Journal of Social and Clinical Psychology, 29*(7), 727–755.

Kelly, A. C., Wisniewski, L., Martin-Wagar, C., & Hoffman, E. (2017). Group-based compassion-focused therapy as an adjunct to outpatient treatment for eating disorders: A pilot randomized controlled trial. *Clinical Psychology & Psychotherapy, 24*(2), 475–487. https://doi.org/10.1002/cpp.2018

Kemp, A. H., Quintana, D. S., Gray, M. A., Felmingham, K. L., Brown, K., & Gatt, J. M. (2010). Impact of depression and antidepressant treatment on heart rate variability: A review and meta-analysis. *Biological Psychiatry, 67*(11), 1067–1074. https://doi.org/10.1016/j.biopsych.2009.12.012

Keng, S.-L., Smoski, M. J., Robins, C. J., Ekblad, A. G., & Brantley, J. G. (2012). Mechanisms of change in mindfulness-based stress reduction: Self-compassion and mindfulness as mediators of intervention outcomes. *Journal of Cognitive Psychotherapy, 26*(3), 270–280. https://doi.org/10.1891/0889-8391.26.3.270

Keysers, C., Kaas, J. H., & Gazzola, V. (2010). Somatosensation in social perception. *Nature Reviews Neuroscience, 11*(6), 417–428. https://doi.org/10.1038/nrn2833

Khoramnia, S., Bavafa, A., Jaberghaderi, N., Parvizifard, A., Foroughi, A., Ahmadi, M., & Amiri, S. (2020). The effectiveness of acceptance and commitment therapy for social anxiety disorder: A randomized clinical trial. *Trends in Psychiatry and Psychotherapy, 42*(1), 30–38. https://doi.org/10.1590/2237-6089-2019-0003

Kılıç, A., Hudson, J., McCracken, L. M., Ruparelia, R., Fawson, S., & Hughes, L. D. (2020). A systematic review of the effectiveness of self-compassion-related interventions for individuals with chronic physical health conditions. *Behavior Therapy, 52*(3), 607–625. https://doi.org/10.1016/j.beth.2020.08.001

Kilner, J. M., & Lemon, R. N. (2013). What we know currently about mirror neurons. *Current Biology, 23*(23), R1057–R1062. https://doi.org/10.1016/j.cub.2013.10.051

Kim, J. J., Parker, S. L., Doty, J. R., Cunnington, R., Gilbert, P., & Kirby, J. N. (2020). Neurophysiological and behavioural markers of compassion. *Scientific Reports, 10*(1), 1–9. https://doi.org/10.1038/s41598-020-63846-3

Kinley, J. L., & Reyno, S. M. (2013). Attachment style changes following intensive short-term group psychotherapy. *International Journal of Group Psychotherapy, 63*(1), 53–75. https://doi.org/10.1521/ijgp.2013.63.1.53

Kirby, J. N. (2017). Compassion interventions: The programmes, the evidence, and implications for research and practice. *Psychology and Psychotherapy: Theory, Research and Practice, 90*(3), 432–455. https://doi.org/10.1111/papt.12104

Kirby, J. N., & Gilbert, P. (2019). Commentary regarding Wilson et al. (2018) "Effectiveness of 'self-compassion' related therapies: A systematic review and meta-analysis." All is not as it seems. *Mindfulness, 10*(6), 1006–1016. https://doi.org/10.1007/s12671-018-1088-8

Kirby, J. N., Doty, J. R., Petrocchi, N., & Gilbert, P. (2017a). The current and future role of heart rate variability for assessing and training compassion. *Frontiers in Public Health, 5*, 40. https://doi.org/10.3389/fpubh.2017.00040

Kirby, J. N., Tellegen, C. L., & Steindl, S. R. (2017b). A meta-analysis of compassion-based interventions: Current state of knowledge and future directions. *Behavior Therapy, 48*(6), 778–792. https://doi.org/10.1016/j.beth.2017.06.003

Kirby, J. N., Day, J., & Sagar, V. (2019). The 'Flow' of compassion: A meta-analysis of the fears of compassion scales and psychological functioning. *Clinical Psychology Review, 70*, 26–39. https://doi.org/10.1016/j.cpr.2019.03.001

Kirschner, H., Kuyken, W., Wright, K., Roberts, H., Brejcha, C., & Karl, A. (2019). Soothing your heart and feeling connected: A new experimental paradigm to study the benefits of self-compassion. *Clinical Psychological Science, 7*(3), 545–565. https://doi.org/10.1177/2167702618812438

Klimecki, O. M., Leiberg, S., Ricard, M., & Singer, T. (2014). Differential pattern of functional brain plasticity after compassion and empathy training. *Social Cognitive and Affective Neuroscience, 9*(6), 873–879. https://doi.org/10.1093/scan/nst060

Knox, M., Neff, K., & Davidson, O. (2016, June). *Comparing compassion for self and others: Impacts on personal and interpersonal well-being*. Paper presented at the 14th annual Association for Contextual Behavioral Science World Conference, Seattle, WA.

Kolts, R. L. (2016). *CFT made simple: A clinician's guide to practicing compassion-focused therapy*. New Harbinger Publications.

Kolts, R. L., Bell, T., Bennett-Levy, J., & Irons, C. (2018). *Experiencing compassion-focused therapy from the inside out: A self-practice/self-reflection workbook for therapists*. Guilford Publications.

Koole, S. L., & Tschacher, W. (2016). Synchrony in psychotherapy: A review and an integrative framework for the therapeutic alliance. *Frontiers in Psychology, 7*, 862. https://doi.org/10.3389/fpsyg.2016.00862

Kotera, Y., & Fido, D. (2020). Effects of Shinrin-Yoku retreat on mental health: A pilot study in Fukushima, Japan. *International Journal of Mental Health and Addiction*. https://doi.org/10.31234/osf.io/2f7cg

Kraiss, J. T., Peter, M., Moskowitz, J. T., & Bohlmeijer, E. T. (2020). The relationship between emotion regulation and well-being in patients with mental disorders: A meta-analysis. *Comprehensive Psychiatry, 102*, 152189. https://doi.org/10.1016/j.comppsych.2020.152189

Kramer, U., Pascual-Leone, A., Rohde, K. B., & Sachse, R. (2018). The role of shame and self-compassion in psychotherapy for narcissistic personality disorder: An exploratory study. *Clinical Psychology & Psychotherapy, 25*(2), 272–282. https://doi.org/10.1002/cpp.2160

Krautheim, J. T., Dannlowski, U., Steines, M., Neziroğlu, G., Acosta, H., Sommer, J., Straube, B., & Kircher, T. (2019). Intergroup empathy: Enhanced neural resonance for ingroup facial emotion in a shared neural production-perception network. *NeuroImage, 194*, 182–190. https://doi.org/10.1016/j.neuroimage.2019.03.048

Kret, M. E., & De Dreu, C. K. (2017). Pupil-mimicry conditions trust in partners: Moderation by oxytocin and group membership. *Proceedings of the Royal Society B: Biological Sciences, 284*(1850), 20162554. https://doi.org/10.1098/rspb.2016.2554

Krieger, T., Altenstein, D., Baettig, I., Doerig, N., & Holtforth, M. G. (2013). Self-compassion in depression: Associations with depressive symptoms, rumination, and avoidance in depressed outpatients. *Behavior Therapy, 44*(3), 501–513. https://doi.org/10.1016/j.beth.2013.04.004

Kuyken, W., Watkins, E., Holden, E., White, K., Taylor, R. S., Byford, S., Evans, A., Radford, S., Teasdale, J. D., & Dalgleish, T. (2010). How does mindfulness-based cognitive therapy work? *Behaviour Research and Therapy, 48*(11), 1105–1112. https://doi.org/10.1016/j.brat.2010.08.003

Lambert, M. J., & Ogles, B. M. (1997). The effectiveness of psychotherapy supervision. In C. E. Watkins Jr. (Ed.), *Handbook of psychotherapy supervision* (pp. 421–446). Wiley.

Lambert, M., & Ogles, B. (2004). The efficacy and effectiveness of psychotherapy. In M. Lambert (Ed.), *Bergin and Garfield's handbook of psychotherapy and behavior change* (pp. 139–193). Wiley.

Lazar, S. W., Kerr, C. E., Wasserman, R. H., Gray, J. R., Greve, D. N., Treadway, M. T., McGarvey, M., Quinn, B. T., Dusek, J. A., Benson, H., Rauch, S. L., Moore, C. I., & Fischl, B. (2005). Meditation experience is asso with increased cortical thickness. *Neuroreport, 16*(17), 1893–1897. https://doi.org/10.1097/01.wnr.0000186598.66243.19

Leary, M. R., Tate, E. B., Adams, C. E., Batts Allen, A., & Hancock, J. (2007). Self-compassion and reactions to unpleasant self-relevant events: The implications of treating oneself kindly. *Journal of Personality and Social Psychology, 92*(5), 887–904. https://doi.org/10.1037/0022-3514.92.5.887

Lecchi, T., da Silva, K., Giommi, F., & Leong, V. (2019, December 28). Using dual-EEG to explore therapist client interpersonal neural synchrony. https://doi.org/10.31234/osf.io/ebkpv

Leigh-Hunt, N., Bagguley, D., Bash, K., Turner, V., Turnbull, S., Valtorta, N., & Caan, W. (2017). An overview of systematic reviews on the public health consequences of social isolation and loneliness. *Public Health, 152*, 157–171. https://doi.org/10.1016/j.puhe.2017.07.035

Leonard, H. D., Campbell, K., & Gonzalez, V. M. (2018). The relationships among clinician self-report of empathy, mindfulness, and therapeutic alliance. *Mindfulness, 9*(6), 1837–1844. https://doi.org/10.1007/s12671-018-0926-z

Levy, K. N., & Johnson, B. N. (2019). Attachment and psychotherapy: Implications from empirical research. *Canadian Psychology/Psychologie Canadienne, 60*(3), 178–193. https://doi.org/10.1037/cap0000162

Lewis, H. B. (1987). Shame and the narcissistic personality. In N. Nathanson (Ed.), *The many faces of shame* (pp. 93–132). Guilford Press.

Lewis, M. (2016). The emergence of human emotions. In L. Barrett, M. Lewis & . Haviland-Jones (Eds.). Handbook of emotions (p. 272–292). : Guilford Press.

Linehan, M. M. (1993). *Skills training manual for treating borderline personality disorder*. Guilford Press.

Linehan, M. (2009, May 1). *Discovering wisdom.* Meditation and Ps conference, Harvard Medical School, Cambridge Health Alliance.

Liu, G., Zhang, N., Teoh, J. Y., Egan, C., Zeffiro, T. A., Davidson, R. J., & Quevedo, K. (2022). Self-compassion and dorsolateral prefrontal cortex activity during sad self-face recognition in depressed adolescents. *Psychological Medicine, 52*(5), 864–873.

Longe, O., Maratos, F. A., Gilbert, P., Evans, G., Volker, F., Rockliff, H., & Rippon, G. (2010). Having a word with yourself: Neural correlates of self-criticism and self-reassurance. *NeuroImage, 49*(2), 1849–1856. https://doi.org/10.1016/j.neuroimage.2009.09.019

Löw, C. A., Schauenburg, H., & Dinger, U. (2020). Self-criticism and psychotherapy outcome: A systematic review and meta-analysis. *Clinical Psychology Review, 75*, 101808. https://doi.org/10.1016/j.cpr.2019.101808

Lucre, K., & Corten, N. (2013). An exploration of group compassion-focused therapy for personality disorder. *Psychology and Psychotherapy: Theory, Research and Practice, 86*(4), 387–400. https://doi.org/10.1111/j.2044-8341.2012.02068.x

Luo, X., Qiao, L., & Che, X. (2018). Self-compassion modulates heart rate variability and negative affect to experimentally induced stress. *Mindfulness, 9*(5), 1522–1528. https://doi.org/10.1007/s12671-018-0900-9

Lutz, J., Berry, M. P., Napadow, V., Germer, C., Pollak, S., Gardiner, P., Edwards, R. R., Desbordes, G., & Schuman-Olivier, Z. (2020). Neural activations during self-related pro in patients with chronic pain and effects of a brief self-compassion training–A pilot study. *Psychiatry Research: Neuroimaging, 304*, 111155. https://doi.org/10.1016/j.pscychresns.2020.111155

Ma, R., Mann, F., Wang, J., Lloyd-Evans, B., Terhune, J., Al-Shihabi, A., & Johnson, S. (2020). The effectiveness of interventions for reducing subjective and objective social isolation among people with mental health problems: A systematic review. *Social Psychiatry and Psychiatric Epidemiology, 55*(7), 839–876. https://doi.org/10.1007/s00127-019-01800-z

MacBeth, A., & Gumley, A. (2012). Exploring compassion: A meta-analysis of the association between self-compassion and psychopathology. *Clinical Psychology Review, 32*(6), 545–552. https://doi.org/10.1016/j.cpr.2012.06.003

MacKinnon, D. P., & Luecken, L. J. (2008). How and for whom? Mediation and moderation in health psychology. *Health Psychology, 27*(2, Suppl), S99S100. https://doi.org/10.1037/0278-6133.27.2(Suppl.).S99

Mackintosh, K., Power, K., Schwannauer, M., & Chan, S. W. (2018). The relationships between self-compassion, attachment and interpersonal problems in clinical patients with mixed anxiety and depression and emotional distress. *Mindfulness, 9*(3), 961–971. https://doi.org/10.1007/s12671-017-0835-6

Malinowski, P. (2013). Neural mechanisms of attentional control in mindfulness meditation. *Frontiers in Neuroscience, 7*, 8. https://doi.org/10.3389/fnins.2013.00008

Marsh, A. A. (2018). The neuroscience of empathy. *Current Opinion in Behavioral, 19*, 110–115. https://doi.org/10.1016/j.cobeha.2017.12.016

Marsh, I. C., Chan, S. W., & MacBeth, A. (2018). Self-compassion and psychological distress in adolescents—A meta-analysis. *Mindfulness, 9*(4), 1011–1027. https://doi.org/10.1007/s12671-017-0850-7

Martins, M. J., Castilho, P., Carvalho, C. B., Pereira, A. T., Carvalho, D., Bajouco, M., Madeira, N., Santos, V., & Macedo, A. (2017). Pathways from paranoid conviction to distress: Exploring the mediator role of Fears of Compassion in a sample of people with psychosis. *Psychosis, 9*(4), 330–337. https://doi.org/10.1080/17522439.2017.1349830

Marx, B. P., & Sloan, D. M. (2005). Peritraumatic dissociation and experiential avoidance as predictors of posttraumatic stress symptomatology. *Behaviour Research and Therapy, 43*(5), 569–583. https://doi.org/10.1016/j.brat.2004.04.004

Mascaro, J. S., Rilling, J. K., Tenzin Negi, L., & Raison, C. L. (2013). Compassion meditation enhances empathic accuracy and related neural activity. *Social Cognitive and Affective Neuroscience, 8*(1), 48–55. https://doi.org/10.1093/scan/nss095

Matos, M., & Pinto-Gouveia, J. (2014). Shamed by a parent or by others: The role of attachment in shame memories relation to depression. *International Journal of Psychology and Psychological Therapy, 14*, 217–244.

Matos, M., & Steindl, S. R. (2020). "You are already all you need to be": A case illustration of compassion-focused therapy for shame and perfectionism. *Journal of Clinical Psychology, 76*(11), 2079–2096. https://doi.org/10.1002/jclp.23055

Matos, M., Duarte, J., & Pinto-Gouveia, J. (2017). The origins of fears of compassion: Shame and lack of safeness memories, fears of compassion and psychopathology. *The Journal of Psychology, 151*(8), 804–819. https://doi.org/10.1080/00223980.2017.1393380

McFarland, D. H. (2001). Respiratory markers of conversational interaction. *Journal of Speech, Language, and Hearing Research, 44*(1), 128–143. https://doi.org/10.1044/1092-4388(2001/012

McIntyre, R., Smith, P., & Rimes, K. A. (2018). The role of self-criticism in common mental health difficulties in students: A systematic review of prospective studies. *Mental Health & Prevention, 10*, 13–27. https://doi.org/10.1016/j.mhp.2018.02.003

McRae, K., & Gross, J. J. (2020). Emotion regulation. *Emotion, 20*(1), 1–9. https://doi.org/10.1037/emo0000703

Merritt, O. A., & Purdon, C. L. (2020). Scared of compassion: Fear of compassion in anxiety, mood, and non-clinical groups. *British Journal of Clinical Psychology, 59*(3), 354–368. https://doi.org/10.1111/bjc.12250

Mikulincer, M., & Shaver, P. R. (2012). An attachment perspective on psychopathology. *World Psychiatry, 11*(1), 11–15. https://doi.org/10.1016/j.wpsyc.2012.01.003

Mikulincer, M., & Shaver, P. R. (2016). Adult attachment and emotion regulation. In J. Cassidy & P. R. Shaver (Eds.), *Handbook of attachment: Theory, research, and clinical applications* (Vol. 3, pp. 507–533). Guilford Press.

Mikulincer, M., & Shaver, P. R. (2019). Attachment orientations and emotion regulation. *Current Opinion in Psychology, 25*, 6–10. https://doi.org/10.1016/j.copsyc.2018.02.006

Moreira, H., Gouveia, M. J., Carona, C., Silva, N., & Canavarro, M. C. (2015). Maternal attachment and children's quality of life: The mediating role of self-compassion and parenting stress. *Journal of Child and Family Studies, 24*(8), 2332–2344. https://doi.org/10.1007/s10826-014-0036-z

Morgan, W., Morgan, S., & Germer, C. (2013). Cultivating attention and compassion. In C. Germer, R. Siegel, & P. Fulton (Eds.), *Mindfulness and psychotherapy* (pp. 76–93). Guilford Press.

Moutsiana, C., Fearon, P., Murray, L., Cooper, P., Goodyer, I., Johnstone, T., & Halligan, S. (2014). Making an effort to feel positive: Insecure attachment in infancy predicts the neural underpinnings of emotion regulation in adulthood. *Journal of Child Psychology and Psychiatry, 55*(9), 999–1008. https://doi.org/10.1111/jcpp.12198

Muris, P., & Otgaar, H. (2020). The process of science: A critical evaluation of more than 15 years of research on self-compassion with the self-compassion scale. *Mindfulness, 11*(6), 1469–1482. https://doi.org/10.1007/s12671-020-01363-0

Naismith, I., Mwale, A., & Feigenbaum, J. (2018). Inhibitors and facilitators of compassion-focused imagery in personality disorder. *Clinical Psychology & Psychotherapy, 25*(2), 283–291. https://doi.org/10.1002/cpp.2161

Navarro-Gil, M., Lopez-del-Hoyo, Y., Modrego-Alarcón, M., Montero-Marin, J., Van Gordon, W., Shonin, E., & Garcia-Campayo, J. (2020). Effects of attachment-based compassion therapy (ABCT) on self-compassion and attachment style in healthy people. *Mindfulness, 11*(1), 51–62. https://doi.org/10.1007/s12671-018-0896-1

Neff, K. D. (2003). The development and validation of a scale to measure self-compassion. *Self and Identity, 2*(3), 223–250. https://doi.org/10.1080/15298860309027

Neff, K. D., & Germer, C. K. (2013). A pilot study and randomized controlled trial of the mindful self-compassion program. *Journal of Clinical Psychology, 69*(1), 28–44. https://doi.org/10.1002/jclp.21923

Neff, K., & Germer, C. (2018). *The Mindful Self-Compassion workbook: A proven way to accept*

yourself, build inner strength, and thrive. Guilford Publications.
Neff, K. D., & McGehee, P. (2010). Self-compassion and psychological resilience among adolescents and young adults. *Self and Identity, 9*(3), 225–240. https://doi.org/10.1080/15298860902979307
Neff, K. D., & Pommier, E. (2013). The relationship between self-compassion and other-focused concern among college undergraduates, community adults, and practicing meditators. *Self and Identity, 12*(2), 160–176. https://doi.org/10.1080/15298868.2011.649546
Neff, K., & Tirch, D. (2013). Self-compassion and ACT. In T. Kashdan & J. Ciarrochi (Eds.), *Mindfulness, acceptance, and positive psychology: The seven foundations of well-being* (pp. 78–106). New Harbinger.
Neff, K. D., & Tóth-Király, I. (2020). Self-compassion scale. In N. Oleg, O. N. Medvedev, C. U. Krägeloh, R. J. Siegert, & N. N. Singh (Eds.), *Handbook of assessment in mindfulness*. Springer.
Neff, K. D., & Vonk, R. (2009). Self-compassion versus global self-esteem: Two different ways of relating to oneself. *Journal of Personality, 77*(1), 23–50. https://doi.org/10.1111/j.1467-6494.2008.00537.x
Neff, K. D., Kirkpatrick, K. L., & Rude, S. S. (2007). Self-compassion and adaptive psychological functioning. *Journal of Research in Personality, 41*(1), 139–154. https://doi.org/10.1016/j.jrp.2006.03.004
Neff, K. D., Tóth-Király, I., Yarnell, L. M., Arimitsu, K., Castilho, P., Ghorbani, N., Guo, H. X., Hirsch, J. K., Hupfeld, J., Hutz, C. S., Kotsou, I., Lee, W. K., Montero-Marin, J., Sirois, F. M., de Souza, L. K., Svendsen, J. L., Wilkinson, R. B., & Mantzios, M. (2019). Examining the factor structure of the Self-Compassion Scale in 20 diverse samples: Support for use of a total score and six subscale scores. *Psychological Assessment, 31*(1), 27–45. https://doi.org/10.1037/pas0000629
Neff, K. D., Knox, M. C., Long, P., & Gregory, K. (2020). Caring for others without losing yourself: An adaptation of the Mindful Self-Compassion Program for healthcare communities. *Journal of Clinical Psychology, 76*(9), 1543–1562. https://doi.org/10.1002/jclp.23007
Norcross, J. C., & Lambert, M. J. (2018). Psychotherapy relationships that work III. *Psychotherapy, 55*(4), 303–315. https://doi.org/10.1037/pst0000193
Norcross, J. C., & Lambert, M. J. (2019). *Psychotherapy relationships that work: Volume 1: Evidence-based therapist contributions*. Oxford University Press.
Nummenmaa, L., Hirvonen, J., Parkkola, R., & Hietanen, J. K. (2008). Is emotional contagion special? An fMRI study on neural systems for affective and cognitive empathy. *NeuroImage, 43*(3), 571–580. https://doi.org/10.1016/j.neuroimage.2008.08.014
Orth, U., Berking, M., & Burkhardt, S. (2006). Self-conscious emotions and depression: Rumination explains why shame but not guilt is maladaptive. *Personality and Social Psychology Bulletin, 32*(12), 1608–1619. https://doi.org/10.1177/0146167206292958
Palmieri, A., Kleinbub, J. R., Calvo, V., Benelli, E., Messina, I., Sambin, M., & Voci, A. (2018). Attachment-security prime effect on skin-conductance synchronization in psychotherapists: An empirical study. *Journal of Counseling Psychology, 65*(4), 490. https://doi.org/10.1037/cou0000273
Panksepp, J. (1998). *Affective neuroscience*. Oxford University Press.
Park, T., Reilly-Spong, M., & Gross, C. R. (2013). Mindfulness: A systematic review of instruments to measure an emergent patient-reported outcome (PRO). *Quality of Life Research, 22*(10), 2639–2659. https://doi.org/10.1007/s11136-013-0395-8
Parrish, M. H., Inagaki, T. K., Muscatell, K. A., Haltom, K. E., Leary, M. R., & Eisenberger, N. I. (2018). Self-compassion and responses to negative social feedback: The role of fronto-amygdala circuit connectivity. *Self and Identity, 17*(6), 723–738.
Paulick, J., Deisenhofer, A.-K., Ramseyer, F., Tschacher, W., Boyle, K., Rubel, J., & Lutz, W. (2018). Nonverbal synchrony: A new approach to better understand psychotherapeutic processes and drop-out. *Journal of Psychotherapy Integration, 28*(3), 367–384. https://doi.org/10.1037/int0000099
Pearson, J. L., Cohn, D. A., Cowan, P. A., & Cowan, C. P. (1994). Earned-and continuous-security in adult attachment: Relation to depressive symptomatology and parenting style. *Development and Psychopathology, 6*(2), 359–373. https://doi.org/10.1017/S0954579400004636
Pepping, C. A., Davis, P. J., O'Donovan, A., & Pal, J. (2015). Individual differences in self-compassion: The role of attachment and experiences of parenting in childhood. *Self and Identity, 14*(1), 104–117. https://doi.org/10.1080/15298868.2014.955050
Perich, T., Manicavasagar, V., Mitchell, P. B., & Ball, J. R. (2013). The association between meditation practice and treatment outcome in mindfulness-based cognitive therapy for bipolar disorder. *Behaviour Research and Therapy, 51*(7), 338–343. https://doi.org/10.1016/j.brat.2013.03.006
Petrocchi, N., Ottaviani, C., & Couyoumdjian, A. (2017). Compassion at the mirror: Exposure to a mirror increases the efficacy of a self-compassion manipulation in enhancing soothing positive affect and heart rate variability. *The Journal of Positive Psychology, 12*(6), 525–536. https://doi.org/10.1080/17439760.2016.1209544
Phelps, C. L., Paniagua, S. M., Willcockson, I. U., & Potter, J. S. (2018). The relationship between self-compassion and the risk for substance use disorder. *Drug and Alcohol Dependence, 183*, 78–81. https://doi.org/10.1016/j.drugalcdep.2017.10.026
Pizarro, J. J., Basabe, N., Amutio, A., Telletxea, S., Harizmendi, M., & Van Gordon, W. (2020). The mediating role of shared flow and perceived emotional synchrony on compassion for others in a mindful-dancing program. *Mindfulness, 11*(1), 125–139. https://doi.org/10.1007/s12671-019-01200-z

Platt, M. G., Luoma, J. B., & Freyd, J. J. (2017). Shame and dissociation in survivors of high and low betrayal trauma. *Journal of Aggression, Maltreatment & Trauma, 26*(1), 34–49. https://doi.org/10.1080/10926771.2016.1228020

Pommier, E., Neff, K. D., & Tóth-Király, I. (2020). The development and validation of the compassion scale. *Assessment, 27*(1), 21–39. https://doi.org/10.1177/1073191119874108

Porges, S. W. (2007). The polyvagal perspective. *Biological Psychology, 74*(2), 116–143. https://doi.org/10.1016/j.biopsycho.2006.06.009

Raab, K. (2014). Mindfulness, self-compassion, and empathy among health care professionals: A review of the literature. *Journal of Health Care Chaplaincy, 20*(3), 95–108. https://doi.org/10.1080/08854726.2014.913876

Raab, K., Sogge, K., Parker, N., & Flament, M. F. (2015). Mindfulness-based stress reduction and self-compassion among mental healthcare professionals: A pilot study. *Mental Health, Religion & Culture, 18*(6), 503–512. https://doi.org/10.1080/13674676.2015.1081588

Ramseyer, F., & Tschacher, W. (2011). Nonverbal synchrony in psychotherapy: Coordinated body movement reflects relationship quality and outcome. *Journal of Consulting and Clinical Psychology, 79*(3), 284–295. https://doi.org/10.1037/a0023419

Ramseyer, F., & Tschacher, W. (2014). Nonverbal synchrony of head-and body-movement in psychotherapy: Different signals have different associations with outcome. *Frontiers in Psychology, 5*, 979. https://doi.org/10.3389/fpsyg.2014.00979

Raque-Bogdan, T. L., Ericson, S. K., Jackson, J., Martin, H. M., & Bryan, N. A. (2011). Attachment and mental and physical health: Self-compassion and mattering as mediators. *Journal of Counseling Psychology, 58*(2), 272–278. https://doi.org/10.1037/a0023041

Razzaque, R., Okoro, E., & Wood, L. (2015). Mindfulness in clinician therapeutic relationships. *Mindfulness, 6*(2), 170–174. https://doi.org/10.1007/s12671-013-0241-7

Reich, C. M., Berman, J. S., Dale, R., & Levitt, H. M. (2014). Vocal synchrony in psychotherapy. *Journal of Social and Clinical Psychology, 33*(5), 481–494. https://doi.org/10.1521/jscp.2014.33.5.481

Reiner, I., Bakermans-Kranenburg, M., Van IJzendoorn, M., Fremmer-Bombik, E., & Beutel, M. (2016). Adult attachment representation moderates psychotherapy treatment efficacy in clinically depressed inpatients. *Journal of Affective Disorders, 195*, 163–171. https://doi.org/10.1016/j.jad.2016.02.024

Rizzolatti, G., Fadiga, L., Gallese, V., & Fogassi, L. (1996). Premotor cortex and the recognition of motor actions. *Cognitive Brain Research, 3*(2), 131–141. https://doi.org/10.1016/0926-6410(95)00038-0

Robins, C. J., Schmidt, H., III, & Linehan, M. M. (2004). Dialectical behavior therapy: Synthesizing radical acceptance with skillful means. In S. Hayes, V. Follette, & M. Linehan (Eds.), *Mindfulness and acceptance*. Guilford Press.

Rockliff, H., Gilbert, P., McEwan, K., Lightman, S., & Glover, D. (2008). A pilot exploration of heart rate variability and salivary cortisol responses to compassion-focused imagery. *Clinical Neuropsychiatry: Journal of Treatment Evaluation, 5*(3), 132–139.

Roemer, L., Orsillo, S. M., & Salters-Pedneault, K. (2008). Efficacy of an acceptance-based behavior therapy for generalized anxiety disorder: Evaluation in a randomized controlled trial. *Journal of Consulting and Clinical Psychology, 76*(6), 1083–1089. https://doi.org/10.1037/a0012720

Rogers, C. R. (1951). *Client-centered therapy: Its current practice, implications, and theory*. Houghton Mifflin.

Rogers, C. (1980). *A way of being*. Houghton Mifflin Company.

Ross, N. D., Kaminski, P. L., & Herrington, R. (2019). From childhood emotional maltreatment to depressive symptoms in adulthood: The roles of self-compassion and shame. *Child Abuse & Neglect, 92*, 32–42. https://doi.org/10.1016/j.chiabu.2019.03.016

Rowe, A. C., Shepstone, L., Carnelley, K. B., Cavanagh, K., & Millings, A. (2016). Attachment security and self-compassion priming increase the likelihood that first-time engagers in mindfulness meditation will continue with mindfulness training. *Mindfulness, 7*(3), 642–650. https://doi.org/10.1007/s12671-016-0499-7

Ryan, A., Safran, J. D., Doran, J. M., & Muran, J. C. (2012). Therapist mindfulness, alliance and treatment outcome. *Psychotherapy Research, 22*(3), 289–297. https://doi.org/10.1080/10503307.2011.650653

Saarela, M. V., Hlushchuk, Y., Williams, A. C. D. C., Schürmann, M., Kalso, E., & Hari, R. (2007). The compassionate brain: Humans detect intensity of pain from another's face. *Cerebral Cortex, 17*(1), 230–237. https://doi.org/10.1093/cercor/bhj141

Safran, J. D., & Muran, J. C. (2000). Resolving therapeutic alliance ruptures: Diversity and integration. *Journal of Clinical Psychology, 56*(2), 233–243. https://doi.org/10.1002/(sici)1097-4679(200002)56:2<233::aid--jclp9>3.0.co;2-3

Samstag, L. W., & Norlander, K. (2019). Characteristics of trainees' early sessions: A naturalistic process-outcome study tribute to Jeremy D. Safran. *Psychoanalytic Psychology, 36*(2), 148–158. https://doi.org/10.1037/pap0000239

Schanche, E., Stiles, T. C., McCullough, L., Svartberg, M., & Nielsen, G. H. (2011). The relationship between activating affects, inhibitory affects, and self-compassion in patients with Cluster C personality disorders. *Psychotherapy, 48*(3), 293–303. https://doi.org/10.1037/a0022012

Schoenherr, D., Paulick, J., Strauss, B. M., Deisenhofer, A. K., Schwartz, B., Rubel, J. A., Lutz, W., Stangier, U., & Altmann, U. (2019). Nonverbal synchrony predicts premature termination of psychotherapy for social anxiety disorder. *Psychotherapy, 56*(4), 503–513. https://doi.org/10.1037/pst0000216

Schore, A. (1994). *Affect regulation and the origin of the self: The neurobiology of emotional development*. Lawrence Erlbaum Associates.

Schore, A. N. (1998). Early shame experiences and infant brain development. In P. Gilbert & B. Andrews (Eds.), *Shame: Interpersonal behavior, psychopathology, and culture* (pp. 57–77). Oxford University Press.

Schwartz, R. C. (1995). *Internal family systems*. Guilford.

Schwartz, R. C., & Sweezy, M. (2019). *Internal family systems therapy*. Guilford.

Scoglio, A. A., Rudat, D. A., Garvert, D., Jarmolowski, M., Jackson, C., & Herman, J. L. (2018). Self-compassion and responses to trauma: The role of emotion regulation. *Journal of Interpersonal Violence, 33*(13), 2016–2036. https://doi.org/10.1177/0886260515622296

Segal, Z. V., & Teasdale, J. (2018). *Mindfulness-based cognitive therapy for depression*. Guilford Publications.

Segal, Z., Williams, J., & Teasdale, J. (2012). *Mindfulness-based cognitive therapy for depression*. The Guilford Press.

Serpell, L., Amey, R., & Kamboj, S. K. (2020). The role of self-compassion and self-criticism in binge eating behaviour. *Appetite, 144*, 104470. https://doi.org/10.1016/j.appet.2019.104470

Shahar, B. (2020). New developments in emotion-focused therapy for social anxiety disorder. *Journal of Clinical Medicine, 9*(9), 2918. https://doi.org/10.3390/jcm9092918

Shahar, B., Carlin, E. R., Engle, D. E., Hegde, J., Szepsenwol, O., & Arkowitz, H. (2012). A pilot investigation of emotion-focused two-chair dialogue intervention for self-criticism. *Clinical Psychology & Psychotherapy, 19*(6), 496–507. https://doi.org/10.1002/cpp.762

Shapiro, S. L., & Carlson, L. E. (2009). *The art and science of mindfulness: Integrating mindfulness into psychology and the helping professions*. American Psychological Association.

Shapiro, S. L., Schwartz, G. E., & Bonner, G. (1998). Effects of mindfulness-based stress reduction on medical and premedical students. *Journal of Behavioral Medicine, 21*(6), 581–599. https://doi.org/10.1023/a:1018700829825

Shapiro, S. L., Astin, J. A., Bishop, S. R., & Cordova, M. (2005). Mindfulness-based stress reduction for health care professionals: Results from a randomized trial. *International Journal of Stress Management, 12*(2), 164. https://doi.org/10.1037/1072-5245.12.2.164

Shaver, P. R., & Mikulincer, M. (2007). Adult attachment strategies and the regulation of emotion. *Handbook of Emotion Regulation, 446*, 465.

Shaver, P. R., Mikulincer, M., Sahdra, B. K., & Gross, J. T. (2017). Attachment security as a foundation for kindness toward self and others. In K. W. Brown & M. R. Leary (Eds.), *The Oxford handbook of hypo-egoic phenomena* (pp. 223–242). Oxford University Press.

Shonin, E., & Van Gordon, W. (2015). Practical recommendations for teaching mindfulness effectively. *Mindfulness, 6*(4), 952–955. https://doi.org/10.1007/s12671-014-0342-y

Shorey, H. S., & Snyder, C. (2006). The role of adult attachment styles in psychopathology and psychotherapy outcomes. *Review of General Psychology, 10*(1), 1–20. https://doi.org/10.1037/1089-2680.10.1.1

Sick, K., Pila, E., Nesbitt, A., & Sabiston, C. M. (2020). Does self-compassion buffer the detrimental effect of body shame on depressive symptoms? *Body Image, 34*, 175–183. https://doi.org/10.1016/j.bodyim.2020.05.012

Siegel, D. J. (1999). *The developing mind: Toward a neurobiology of interpersonal experience*. Guilford Press.

Siegel, R. D. (2009). *The mindfulness solution: Everyday practices for everyday problems*. Guilford Press.

Siegel, D. J. (2010). *Mindsight: The new science of personal transformation*. Bantam.

Singer, T., & Klimecki, O. M. (2014). Empathy and compassion. *Current Biology, 24*(18), R875–R878. https://doi.org/10.1016/j.cub.2014.06.054

Singer, T., & Lamm, C. (2009). The social neuroscience of empathy. *Annals of the New York Academy of Sciences, 1156*(1), 81–96. https://doi.org/10.1111/j.1749-6632.2009.04418.x

Sirois, F. M., Bögels, S., & Emerson, L.-M. (2019). Self-compassion improves parental Well-being in response to challenging parenting events. *The Journal of Psychology, 153*(3), 327–341. https://doi.org/10.1080/00223980.2018.1523123

Skelton, W. M., Cardaciotto, L., O'Hayer, C. V., & Goldbacher, E. (2020). The role of self-compassion and shame in persons living with HIV/AIDS. *AIDS Care, 33*(6), 818–826. https://doi.org/10.1080/09540121.2020.1769836

Stanley, S., Reitzel, L. R., Wingate, L. R., Cukrowicz, K. C., Lima, E. N., & Joiner, T. E., Jr. (2006). Mindfulness: A primrose path for therapists using manualized treatments? *Journal of Cognitive Psychotherapy, 20*(3), 327–335. https://doi.org/10.1891/jcop.20.3.327

Steffen, P. R., Foxx, J., Cattani, K., Alldredge, C., Austin, T., & Burlingame, G. M. (2020). Impact of a 12-week group-based compassion focused therapy intervention on heart rate variability. *Applied Psychophysiology and Biofeedback, 46*(1), 61–68. https://doi.org/10.1007/s10484-020-09487-8

Steindl, S. R., Matos, M., & Creed, A. K. (2018). Early shame and safeness memories, and later depressive symptoms and safe affect: The mediating role of self-compassion. *Current Psychology, 40*, 761–771. https://doi.org/10.1007/s12144-018-9990-8

Stern, D. N. (2018). *The interpersonal world of the infant: A view from psychoanalysis and developmental psychology*. Routledge.

Stevens, L., Gauthier-Braham, M., & Bush, B. (2018). The brain that longs to care for itself: The current neuroscience of self-compassion. In L. Stevens & C. C. Woodruff (Eds.), *The neuroscience of empathy, compassion, and self-compassion* (pp. 91–120). Elsevier.

Stipek, D. J. (1983). A developmental analysis of pride and shame. *Human Development, 26*(1), 42–54. https://doi.org/10.1159/000272869

Svendsen, J. L., Osnes, B., Binder, P. E., Dundas, I., Visted, E., Nordby, H., Schanche, E., & Sørensen, L. (2016). Trait self-compassion reflects emotional flexibility through an association with high vagally mediated heart rate variability. *Mindfulness, 7*(5), 1103–1113.

Svendsen, J. L., Schanche, E., Osnes, B., Vøllestad, J., Visted, E., Dundas, I., Nordby, H., Binder, P. E., & Sørensen, L. (2020). Is dispositional self-compassion associated with psychophysiological flexibility beyond mindfulness? An exploratory pilot study. *Frontiers in Psychology, 11*, 614. https://doi.org/10.3389/fpsyg.2020.00614

Sweezy, M. (2013). Emotional cannibalism: Shame in action. In M. Sweezy & E. Ziskind (Eds.), *Internal family systems therapy: New dimensions* (pp. 24–34). Routledge.

Symons, C. S., & Johnson, B. T. (1997). The self-reference effect in memory: A meta-analysis. *Psychological Bulletin, 121*(3), 371. https://doi.org/10.1037/0033-2909.121.3.371

Taylor, B. L., Strauss, C., Cavanagh, K., & Jones, F. (2014). The effectiveness of self-help mindfulness-based cognitive therapy in a student sample: A randomised controlled trial. *Behaviour Research and Therapy, 63*, 63–69. https://doi.org/10.1016/j.brat.2014.09.007

Thompson, B. L., & Waltz, J. (2008). Self-compassion and PTSD symptom severity. *Journal of Traumatic Stress, 21*(6), 556–558. https://doi.org/10.1002/jts.20374

Thoresen, S., Aakvaag, H. F., Strøm, I. F., Wentzel-Larsen, T., & Birkeland, M. S. (2018). Loneliness as a mediator of the relationship between shame and health problems in young people exposed to childhood violence. *Social Science & Medicine, 211*, 183–189. https://doi.org/10.1016/j.socscimed.2018.06.002

Toole, A. M., & Craighead, L. W. (2016). Brief self-compassion meditation training for body image distress in young adult women. *Body Image, 19*, 104–112. https://doi.org/10.1016/j.bodyim.2016.09.001

Trompetter, H. R., de Kleine, E., & Bohlmeijer, E. T. (2017). Why does positive mental health buffer against psychopathology? An exploratory study on self-compassion as a resilience mechanism and adaptive emotion regulation strategy. *Cognitive Therapy and Research, 41*(3), 459–468. https://doi.org/10.1007/s10608-016-9774-0

Uliaszek, A. A., Al-Dajani, N., Ferguson, A., & Segal, Z. V. (2020). Third-wave psychotherapies. In *Landmark papers in psychiatry* (pp. 215–227). Oxford University Press. https://doi.org/10.1093/med/9780198836506.003.0013

Valdesolo, P., & DeSteno, D. (2011). Synchrony and the social tuning of compassion. *Emotion, 11*(2), 262. https://doi.org/10.1037/a0021302

Valk, S. L., Bernhardt, B. C., Trautwein, F. M., Böckler, A., Kanske, P., Guizard, N., Collins, D. L., & Singer, T. (2017). Structural plasticity of the social brain: Differential change after socio-affective and cognitive mental training. *Science Advances, 3*(10), e1700489. https://doi.org/10.1126/sciadv.1700489

Van Dam, N. T., Sheppard, S. C., Forsyth, J. P., & Earleywine, M. (2011). Self-compassion is a better predictor than mindfulness of symptom severity and quality of life in mixed anxiety and depression. *Journal of Anxiety Disorders, 25*(1), 123–130. https://doi.org/10.1016/j.janxdis.2010.08.011

Van den Brink, E., & Koster, F. (2015). *Mindfulness-based compassionate living: A new training programme to deepen mindfulness with heartfulness*. Routledge.

Vasconcelos, P., Oliveira, C., & Nobre, P. (2020). Self-compassion, emotion regulation, and female sexual pain: A comparative exploratory analysis. *The Journal of Sexual Medicine, 17*(2), 289–299. https://doi.org/10.1016/j.jsxm.2019.11.266

Vettese, L. C., Dyer, C. E., Li, W. L., & Wekerle, C. (2011). Does self-compassion mitigate the association between childhood maltreatment and later emotion regulation difficulties? A preliminary investigation. *International Journal of Mental Health and Addiction, 9*(5), 480. https://doi.org/10.1007/s11469-011-9340-7

Wadsworth, L. P., Forgeard, M., Hsu, K. J., Kertz, S., Treadway, M., & Björgvinsson, T. (2018). Examining the role of repetitive negative thinking in relations between positive and negative aspects of self-compassion and symptom improvement during intensive treatment. *Cognitive Therapy and Research, 42*(3), 236–249. https://doi.org/10.1007/s10608-017-9887-0

Wang, J., Lloyd-Evans, B., Giacco, D., Forsyth, R., Nebo, C., Mann, F., & Johnson, S. (2017). Social isolation in mental health: A conceptual and methodological review. *Social Psychiatry and Psychiatric Epidemiology, 52*(12), 1451–1461. https://doi.org/10.1007/s00127-017-1446-1

Wang, Y., Fan, L., Zhu, Y., Yang, J., Wang, C., Gu, L., Zhong, S., Huang, Y., Xie, X., Hui, Z., Luo, S., & Zhou, H. (2019). Neurogenetic mechanisms of self-compassionate mindfulness: The role of oxytocin-receptor genes. *Mindfulness, 10*(9), 1792–1802. https://doi.org/10.1007/s12671-019-01141-7

Webb, J. B., Fiery, M. F., & Jafari, N. (2016). "You better not leave me shaming!": Conditional indirect effect analyses of anti-fat attitudes, body shame, and fat talk as a function of self-compassion in college women. *Body Image, 18*, 5–13. https://doi.org/10.1016/j.bodyim.2016.04.009

Wei, M., Liao, K. Y. H., Ku, T. Y., & Shaffer, P. A. (2011). Attachment, self-compassion, empathy, and subjective well-being among college students and community adults. *Journal of Personality, 79*(1), 191–221. https://doi.org/10.1111/j.1467-6494.2010.00677.x

Wei, M., Liu, S., Ko, S. Y., Wang, C., & Du, Y. (2020). Impostor feelings and psychological distress among Asian Americans: Interpersonal shame and self-compassion. *The Counseling Psychologist, 48*(3), 432–458. https://doi.org/10.1177/0011000019891992

Weibel, D. T., McClintock, A. S., & Anderson, T. (2017). Does loving-kindness meditation reduce anxiety? Results from a randomized controlled trial. *Mindfulness, 8*(3), 565–571. https://doi.org/10.1007/s12671-016-0630-9

Werner, K. H., Jazaieri, H., Goldin, P. R., Ziv, M., Heimberg, R. G., & Gross, J. J. (2012). Self-compassion and social anxiety disorder. *Anxiety, Stress & Coping, 25*(5), 543–558. https://doi.org/10.1080/10615806.2011.608842

Werner, A. M., Tibubos, A. N., Rohrmann, S., & Reiss, N. (2019). The clinical trait self-criticism and its relation to psychopathology: A systematic review–update. *Journal of Affective Disorders, 246*, 530–547. https://doi.org/10.1016/j.jad.2018.12.069

Westphal, M., Leahy, R. L., Pala, A. N., & Wupperman, P. (2016). Self-compassion and emotional invalidation mediate the effects of parental indifference on psychopathology. *Psychiatry Research, 242*, 186–191. https://doi.org/10.1016/j.psychres.2016.05.040

Wetterneck, C. T., Lee, E. B., Smith, A. H., & Hart, J. M. (2013). Courage, self-compassion, and values in obsessive-compulsive disorder. *Journal of Contextual Behavioral Science, 2*(3–4), 68–73. https://doi.org/10.1016/j.jcbs.2013.09.002

Wheatley, T., Kang, O., Parkinson, C., & Looser, C. E. (2012). From mind perception to mental connection: Synchrony as a mechanism for social understanding. *Social and Personality Psychology Compass, 6*(8), 589–606. https://doi.org/10.1111/j.1751-9004.2012.00450.x

Williams, J. C., & Lynn, S. J. (2010). Acceptance: An historical and conceptual review. *Imagination, Cognition and Personality, 30*(1), 5–56. https://doi.org/10.2190/IC.30.1.c

Wilson, A. C., Mackintosh, K., Power, K., & Chan, S. W. (2019). Effectiveness of self-compassion related therapies: A systematic review and meta-analysis. *Mindfulness, 10*(6), 979–995. https://doi.org/10.1007/s12671-018-1037-6

Winders, S. J., Murphy, O., Looney, K., & O'Reilly, G. (2020). Self-compassion, trauma, and posttraumatic stress disorder: A systematic review. *Clinical Psychology & Psychotherapy, 27*(3), 300–329. https://doi.org/10.1002/cpp.2429

Wisener, M., & Khoury, B. (2020). Specific emotion-regulation processes explain the relationship between mindfulness and self-compassion with coping-motivated alcohol and marijuana use. *Addictive Behaviors, 112*, 106590. https://doi.org/10.1016/j.addbeh.2020.106590

Wolf, C., & Serpa, J. G. (2015). *A clinician's guide to teaching mindfulness: The comprehensive session-by-session program for mental health professionals and health care providers*. New Harbinger Publications.

Wong, C. C. Y., & Yeung, N. C. (2017). Self-compassion and posttraumatic growth: Cognitive processes as mediators. *Mindfulness, 8*(4), 1078–1087. https://doi.org/10.1007/s12671-017-0683-4

Woods, H., & Proeve, M. (2014). Relationships of mindfulness, self-compassion, and meditation experience with shame-proneness. *Journal of Cognitive Psychotherapy, 28*(1), 20–33. https://doi.org/10.1891/0889-8391.28.1.20

Xavier, A., Pinto-Gouveia, J., & Cunha, M. (2016). The protective role of self-compassion on risk factors for non-suicidal self-injury in adolescence. *School Mental Health, 8*(4), 476–485. https://doi.org/10.1007/s12310-016-9197-9

Yadavaia, J. E., Hayes, S. C., & Vilardaga, R. (2014). Using acceptance and commitment therapy to increase self-compassion: A randomized controlled trial. *Journal of Contextual Behavioral Science, 3*(4), 248–257. https://doi.org/10.1016/j.jcbs.2014.09.002

Yela, J. R., Gómez-Martínez, M. Á., Crego, A., & Jiménez, L. (2020). Effects of the Mindful Self-Compassion programme on clinical and health psychology trainees' well-being: A pilot study. *Clinical Psychologist, 24*(1), 41–54. https://doi.org/10.1111/cp.12204

Young, K. S., van der Velden, A. M., Craske, M. G., Pallesen, K. J., Fjorback, L., Roepstorff, A., & Parsons, C. E. (2018). The impact of mindfulness-based interventions on brain activity: A systematic review of functional magnetic resonance imaging studies. *Neuroscience & Biobehavioral Reviews, 84*, 424–433. https://doi.org/10.1007/s10880-018-9548-9

Zadra, J. R., & Clore, G. L. (2011). Emotion and perception: The role of affective information. *Wiley Interdisciplinary Reviews: Cognitive Science, 2*(6), 676–685.

Zhang, H., Carr, E. R., Garcia-Williams, A. G., Siegelman, A. E., Berke, D., Niles-Carnes, L. V., Patterson, B., Watson-Singleton, N. N., & Kaslow, N. J. (2018). Shame and depressive symptoms: Self-compassion and contingent self-worth as mediators? *Journal of Clinical Psychology in Medical Settings, 25*(4), 408–419.

Zhang, X., Noah, J. A., Dravida, S., & Hirsch, J. (2020). Optimization of wavelet coherence analysis as a measure of neural synchrony during hyperscanning using functional near-infrared spectroscopy. *Neurophotonics, 7*(1), 015010. https://doi.org/10.1117/1.NPh.7.1.015010

Compassion Focused Therapy – What It Is, What It Targets, and the Evidence

23

James N. Kirby and Nicola Petrocchi

Introduction

There is an increasing recognition within mental health and psychotherapy of the crucial role that compassion has in helping alleviate suffering and in promoting well-being (Forster & Kanske, 2021). There are many different approaches that can help cultivate a compassionate mindset, with some of the most evaluated being mindful self-compassion (Neff & Germer, 2013), cognitively based compassion training (Pace et al., 2009), compassion cultivation training (Jazaieri et al., 2013), and mindfulness-based compassionate living (Schuling et al., 2021). These programmes all aim to explicitly cultivate compassion, are time-limited, usually spanning between 8 and 12 sessions, and include various experiential exercises, imagery exercises, and meditations to help foster a compassionate mindset. These programmes were initially developed to help those in the community struggling with stress and self-criticism; however, these programmes are now being applied to clinical settings as well, such as those experiencing depression (Schuling et al., 2021). In contrast, one approach to cultivating compassion that began as a therapy and is beginning to be applied more commonly to community settings is compassion-focused therapy (CFT). This chapter will (1) discuss the definition of compassion from a CFT perspective, (2) review the origins of CFT and compassionate mind training (CMT), (3) discuss social mentality theory which underpins both CFT and CMT, (4) describe a case formulation using CFT, and (5) describe a new 12 module CFT intervention.

Compassion Definition

A defining feature of CFT is the definition of compassion, which is *"the sensitivity to suffering in self and others, with a commitment to alleviate and prevent it"* (Gilbert, 2014, p. 19). As the definition suggests, compassion can be directed to others, it can be directed towards the self (self-compassion), and we can also receive compassion from others. Interestingly, many compassionate approaches do not emphasise the importance of receiving compassion from others to the same degree, despite research suggesting that receiving compassion from others is possibly more important than self-compassion in buffering against depression (Hermanto et al., 2016). Compassion is contextual and interactional, and the emotions, reasoning, and behaviours recruited will be dependent on the nature of the suffering

J. N. Kirby (✉)
Compassionate Mind Research Group, School of Psychology, The University of Queensland, St Lucia, QLD, Australia
e-mail: j.kirby@psy.uq.edu.au

N. Petrocchi
Department of Economics and Social Sciences, John Cabot University, Rome, Italy

A. Finlay-Jones et al. (eds.), *Handbook of Self-Compassion*, Mindfulness in Behavioral Health,
https://doi.org/10.1007/978-3-031-22348-8_23

encountered. This definition conceptualises compassion as a motivation, which is rooted in mammalian caregiving. This evolutionary approach argues that compassion evolved from caring behaviour, as a result of mammalian *K*-selected reproductive parental investment strategies (Kirby et al., 2017a, b). *K*-selected reproductive strategies are those where there is live birth, small numbers of young, and heavy parental investment post-birth (e.g. elephants, chimpanzees, humans). Thus, parents are highly attuned to signals of distress from their young, with attachment bonds forming, with parents providing important sources of growth and physiological regulation of the infant (Mayseless, 2016).

According to this definition, compassion can be conceptualised as an evolved algorithm that operates on a stimulus detection/stimulus response system (Gilbert, 2020, and see Chap. 4). Specifically for the compassion algorithm, the stimulus detection involves being sensitive to signals of suffering and distress (e.g. threats, danger), and the stimulus response system enables behaviours which are congruent to the context of suffering that help to alleviate it. Therefore, the actions enabled in the stimulus response will vary – there is no one compassionate behaviour. For example, in contexts of sadness, the compassionate response could be comfort, whereas in the context of anxiety the compassionate response could be validation, encouragement, and gradual exposure. In CFT, there are specific skills taught to help those engage with distress that is often avoided (e.g. anxiety). In addition, skills are also taught to help with stimulus response, which will enable actions that help alleviate the distress, for example, soothing rhythm breathing or compassionate imagery.

In CFT, a core emphasis is on the qualities of *courage* and *wisdom*. That is, the therapist emphasises to the client that it takes courage to engage with life difficulties and suffering which cause pain. Equally, it takes wisdom to engage in actions to help effectively alleviate the suffering in the long term. This definition of compassion is shared with the client, and it is discussed collaboratively, specifically how the aims of the CFT approach will be to help the client to engage with their own suffering and learn wise and effective ways to help reduce their suffering.

The Origins of CFT

CFT is an integrated and evolutionary-informed biopsychosocial and contextual model of therapy developed by Paul Gilbert (2020). The origins of CFT date back to the late 1980s when Gilbert was studying and using evolutionary models of psychology with traditional cognitive behaviour therapy (CBT) to help those with clinical depression. A core insight during that period, which led to his book, Depression: From Psychology to Gilbert, P. (1984), was that the emotional tone (i.e. hostile) and not just the content of the inner thoughts mattered, and this would relate to internal physiology. Specifically, Gilbert theorised that when challenging your dysfunctional automatic thoughts and examining alternative ones, the tone of voice and the feelings associated with that experience mattered. The content could be identical, but how that content was experienced in terms of emotion and feeling was critical. If the approach to challenging thoughts was critical, attacking, or hostile, this would result in a dominate–subordinate internal relating style, which would maintain the depressive brain state. In contrast, Gilbert hypothesised that if one was to use a more compassionate, friendly, reassuring tone when challenging inner thoughts, this could lead to a different experience, which would have a different physiological pattern in the body, thus began the compassion-focused approach to therapy, specifically to help with depression. Importantly, there is now research supporting the need for such an approach. For example, Whelton and Greenberg (2005) examined the emotional tone when engaging in self-criticism. The researchers found that those who were high self-critics adopted tones with high levels of contempt and disgust compared to controls. Moreover, those with high levels of self-criticism were less resilient, less assertive, and more submissive than controls. They concluded that when it comes to self-criticism, the emotional tones of contempt and disgust are critical in the genesis of depressive mind states, which is what Gilbert was attempting to shift in the 1980s.

Compassion-Focused Therapy and Compassionate Mind Training

One of the key aspects of the origin of CFT is the emphasis on "motivational *focus*"; that is, the approach to CFT is to orient the therapeutic approach in compassion. Using this compassionate mindset, one can then engage in the therapeutic tasks necessary to help the specific difficulty (e.g. anxiety, depression). As a result, CFT is an integrated model, which includes a range of different techniques and processes common among almost all therapies, for example, Socratic dialogues, guided discovery, inference chaining, psychoeducation, mindfulness, exposure, behavioural practice, which are common among many of the therapy schools (e.g. cognitive-behavioural therapy, acceptance and commitment therapy, dialectical behaviour therapy; Gilbert, 2020). There is a growing recognition by scholars in mental health and psychotherapies for the need to move to a more process-based approach to therapy, focusing on identifying the processes that give rise to better mental health, rather than staying fixed in separate schools of therapy (Gilbert & Kirby, 2019; Hoffman & Hayes, 2020). As such, there should be increased convergence among the many schools of therapy, as the science indicates which factors are more and less supportive of good mental health. Indeed, in many ways this is already happening, with almost all therapies including identified therapeutic active ingredients such as exposure, mindfulness, breathing, imagery, behaviour activation, and the inclusion of homework (Gilbert & Kirby, 2019). CFT includes these active ingredients as part of its approach, as the science indicates that they are effective in helping with life challenges (e.g. emotional difficulties, tragedies). However, where CFT might be unique relative to other therapeutic models is its definition of compassion as a motivation, and its emphasis on an evolutionary functional approach to compassion and emotions, and the development of the "compassionate-self," which is used as a mindset to help soothe and encourage the individual when working with life difficulties.

CFT is an open-ended therapeutic approach, based on the evolutionary model of social mentality theory, which Gilbert describes in Chap. 4 of this book (see also Gilbert, 2020). The aims of CFT are to address the key targets of self-criticism, shame, threat, fear, and trauma that underpin so many mental health disorders by using a compassion-focused approach characterised by affiliative relating (both to ourselves and others) and behaviours to help alleviate suffering and improve quality of life and well-being. In CFT, core therapeutic tasks such as assessments and case formulations, along with the therapeutic bond, guide the therapy process. CFT has been used for a range of different clinical presentations, including, but not limited to, depression (Falconer et al., 2016; Noorbala et al., 2013; Savari et al., 2021), obsessive compulsive disorder (Petrocchi et al., 2021), psychosis (Braehler et al., 2013), personality disorder (Lucre & Corten, 2013), anxiety (Cuppage et al., 2018; Gharraee et al., 2018), posttraumatic stress disorder (Daneshvar et al., 2020; Lawrence & Lee, 2014), eating disorder (Kelly & Carter, 2015; Steindl et al., 2017), substance use disorder (Carlyle et al., 2019), chronic pain (Dhokia et al., 2020), problematic sleep (Eslamian et al., 2019), and intellectual disability (Clapton et al., 2018). The evidence in support of this approach is constantly growing, with a recent systematic review indicating most support for its approach being in a group format delivery with at least 12 sessions (Craig et al., 2020). The systematic review also identified that CFT led to reductions in symptoms among difficult-to-treat clinical populations, such as forensic populations, eating disorders, and personality disorders.

CFT also has a specific skill-based programme called compassionate mind training (CMT). The aim of CMT is the same as CFT to help address core issues of self-criticism and shame by using a set series of compassion-focused strategies and exercises. CMT is time-limited and can be delivered as briefly as using a 15-minute audio-guided exercise (Kim et al., 2020), to a 2-hour brief seminar (Matos et al., 2017a, b), to four module sessions (Maratos et al., 2019), and to a longer eight-session programme (Irons & Heriot-Maitland, 2021). CMT is aimed for those in the community dealing with stress and self-criticism, but not necessarily meeting criteria for a clinical disorder. Unlike CFT, CMT does not include individual assessments and case formulations to guide the therapeutic process. Rather, CMT is a manualised programme, which delivers set skills to the target group that aim to cultivate a compassionate mindset. In the delivery of CFT, these specific CMT exercises are used when needed, according to the CFT formulation. CMT/CFT are highly related, with the key point of difference being that the former is a manualised skill-based programme aimed specifically at developing the compassionate-self to support well-being, whereas the latter is an open-ended therapy guided by assessment and formulation, using an arrangement of active ingredients to help alleviate suffering for the specific life challenge presented in therapy.

A Brief Overview of Social Mentality Theory

Social mentality theory has been outlined elsewhere (Gilbert, 2020) and in another chapter in this volume (Gilbert, Chap. 4). Essentially, it is an evolutionary model that suggests that individual patterns of cognition, affect, physiology, and behaviours are determined by the contextual social roles in which the individual is engaging (Gilbert, 2020). CFT is based on social mentality theory. We have many social motives (e.g. competitive, cooperative, sexual, and compassionate) that can lead to different role relationships with others (e.g. parent–child, therapist–client, employer–employee) and ourselves. Gilbert (2020) suggests that each social mentality has its own feature detector and that environments can shape how these feature detectors are tuned. Feature detectors enable animals to identify and pay attention to (take an interest in) different kinds of stimuli in the environment and then respond to those stimuli in appropriate ways (Gilbert, 2020). For example, in hostile environments "*threat detectors*" tend to be sensitised, whereas in safe, caring environments, detectors of opportunities for affiliative (play, curiosity, trust) are sensitised.

In CFT, there is a central focus placed on two of these motives: competitive and compassionate motives. The competitive motive posits that individuals monitor social relationships and engage in social comparisons to determine their relative rank or status (e.g. *"am I superior or inferior to others"*), which then can lead to feelings of external shame (e.g. *"others see me as less than or inferior"*), and this results in submissive or aggressive behaviours (e.g. hiding, avoidance, or perfectionistic over-compensation). The key vulnerability factor in a competitive motive is when individuals are competing to avoid inferiority, and thus, rejection or exclusion by others becomes central. A recently published meta-analysis on competitive motives and social rank has supported this model in understanding depression, anxiety, and other clinical disorders (Wetherall et al., 2019).

The other motive is compassion, which is focused on alleviating distress, and can facilitate self-reassurance and encouragement. Thus, rather than activating a rank social mentality (both with ourselves and others), and subsequently being self-critical when triggered by threats, one can consciously activate a compassionate mentality and be self-reassuring, thus buffering themselves from the deleterious effects of depression and anxiety (Petrocchi et al., 2021). This theoretical model was recently tested with 1695 participants when examining body weight shame, with results finding support for how social mentality theory predicts the associations between social rank, self-relating, and mental health (Carter et al., 2021b). The aim of CFT is to shift people from relying on competitive social mentalities that feature self-criticism and shame, to that of a compassionate mind.

Three Affect Regulation Systems

Motives have evolved to perform important life tasks, such as avoiding harm, cooperating with others, and finding reproductive partners, all of which aid in survival and reproduction (Gilbert, 2020). Importantly, emotions help guide the enactment of the motive, providing the energy or information signals as to whether one is being successful or if the motive is being impeded. A defining feature in CFT is its model of affect regulation. The model is referred to as the three affect regulation systems of emotion. It is depicted in Fig. 23.1 and is informed by affective neuroscience research into the evolutionary functions of emotion (Depue & Morrone-Strupinsky, 2005; LeDoux, 1998; Panksepp, 2010).

These three emotion regulation systems interact and include the following: (a) the threat/self-protect system, (b) the drive/reward system, and (c) the affiliative/soothing system. The function of the threat system is to detect and respond to threats (physical and social) in order to provide self-protection and to avoid harm and danger. This system is responsible for the physiological fight/flight response (sympathetic system) when faced with threat (Gilbert, 2014). As such, the threat system is the most dominant emotion regulation system and is prioritised over the other systems, given it is focused on harm avoidance (Gilbert, 2014). The drive system refers to seeking out, obtaining resources (e.g. food, status, sexual opportunities, and friendships), and achieving desired goals (Gilbert, 2009). The emotions elicited in the drive system include positive emotions of excitement, pleasure, and happiness. Finally, the function of the soothing/affiliative system is to slow the organism to allow for important physiological actions such as recovery and rest and digest, and emotions connected to this system include feeling calm, content, and peaceful. According to Gilbert's model, the soothing system is linked to the parasympathetic nervous system which activates a slowing down of the autonomic nervous system. It is also linked to the experience of intra- and interpersonal safeness, which, for humans, is not necessarily derived from the absence of threats, but more specifically from the presence of "affiliative and helping others," and our own ability to compassionately self-soothe when stressed. Indeed, from an evolutionary point of view, it is crucial to consider the sense of safeness in the context of interpersonal motivations, that is, what makes the internal and external world safe for humans (Petrocchi & Cheli, 2019). Gilbert (2014) proposes that these three emotion regulation systems are influenced by motivation. How these systems

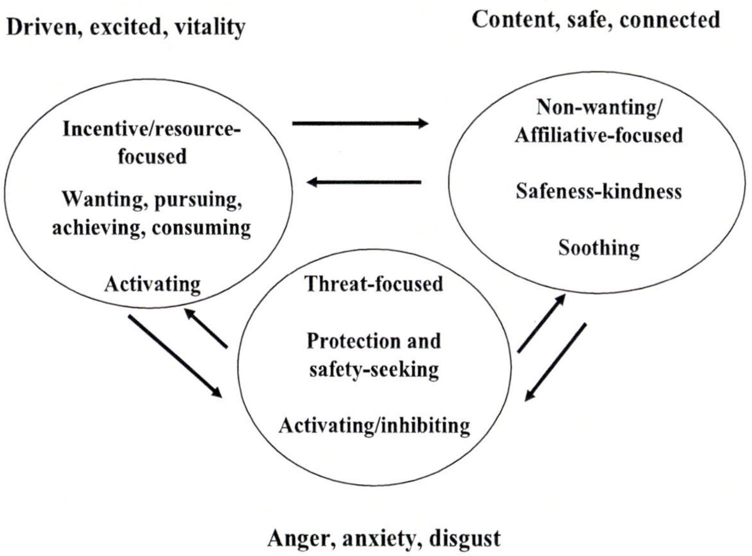

Fig. 23.1 The interaction between the three major emotion regulation systems. From Gilbert, The Compassionate Mind (2009), reprinted with permission from Constable & Robinson Ltd.

operate is contextual, and in some contexts, it is important to have more dominant threat-drive system (e.g. responding to a crisis). The key, however, is to examine how these systems are operating across all contexts, and the theory postulates that if one is compassionately motivated there is greater balance between these systems. Conversely, if one is competitively motivated, there is an over-reliance on the threat-drive systems, which is theorised to be a vulnerability factor for distress and mental health difficulties (Gilbert, 2009). How these systems are activated can be triggered by external (immediate threat in front of you) and internal factors (judgements we make). Therefore, a balance of these three affect regulation systems is required to obtain successful emotion regulation.

The aim of CMT/CFT therefore is to cultivate the compassionate motive to bring about a greater balance of these three systems, with emerging research supporting this approach. For example, Matos et al. (2017a, b) examined a brief 2-hour CMT seminar using a randomised controlled trial design with 97 participants and found it led to improvements in positive affect, soothing affect, and reductions in negative affect. In addition, at post-intervention, heart rate variability (HRV), a physiological index of parasympathetic activation, increased compared to a control condition, indicating CMT leads to improvements in the soothing system, theorised to be connected to the parasympathetic system, as proposed by Gilbert's three emotion regulation system. This was replicated in a recent study using a brief 15-minute CMT exercise, the compassionate-self, where Kim et al. (2020) found that this practice improved the parasympathetic system, as measured by HRV.

Fears of Compassion

Not all individuals respond favourably to compassion-focused approaches; in the first ever evaluated group delivery of CMT, Gilbert and Procter (2006) found that people with high self-criticism and shame found self-compassion and receiving compassion extremely difficult, and many reported fearing compassion. Just like any motivation, fears associated with the expression of compassion may lead to its downregulation (Gilbert et al., 2011; Gilbert & Mascaro, 2017). Specifically, fears of compassion relate to the avoidance or fear response that individuals can have to all three flows of compassion (for the self, from others, and to others). These might include beliefs that compassion is a weakness or self-indulgent or that compassionate efforts will be seen as incompetent, unhelpful, rejected, or in some way aversive (Gilbert & Mascaro, 2017). A reluctance to engage in compassion can also stem from the fear that one might suffer too much personal distress (Vitaliano et al., 2003), or the fear that compassion will be perceived by others as a form of manipulative self-interest (Gilbert & Mascaro, 2017). In the first group evaluation of CMT, Gilbert and Procter found those with high self-criticism and shame were fearful of compassion, believing it would lead them to become overwhelmed or too deeply saddened, or even that they were undeserving of compassion. Moreover, the participants indicated that they did not like to receive compassion from others, fearing that if they let others in, they would see the "bad in them"; thus, it was safer to keep people at an "arm's distance," so that they could not be hurt (Gilbert & Procter, 2006).

To understand how fears of compassion can manifest, Gilbert (2020) draws upon evolutionary models, attachment theory, and classical conditioning. Specifically, Gilbert postulated that a fear of compassion can develop in those who have learned associations between prosocial actions and aversive outcomes. For example, children can be punished by parents or other "authoritative" figures (e.g. teachers) for being too generous, kind, and compassionate, as it might be seen as naïve or inappropriate (Matos et al., 2017a, b). As a result, the punishment (e.g. being yelled at, anger, criticism), is paired with compassion, and thus, individuals can become fearful or avoidant of compassion in the future. In addition, if one is raised in an environment where there is little modelling of compassionate behaviour and high levels of shame, this too can lead to fears of expressing compassion (Matos et al.,

2017a, b). Meta-analytic research has since found the importance that fears of compassion have as a unique contributor to general mental health and well-being (Kirby et al., 2019). Specifically, Kirby et al. (2019) meta-analysed data from over 5000 participants and found medium to large effect sizes between both fears of self-compassion and fears of receiving compassion, and adverse outcomes such as shame, self-criticism, depression, and anxiety. Moreover, these effects were significantly greater for those with clinical disorders compared to those without. As a result, one of the core tasks in CFT is to explore the client's fears of compassion as part of the therapeutic journey. Indeed, CFT interventions have been found to be effective in reducing fears of compassion (e.g. Kelly et al., 2017; Savari et al., 2021).

Compassion and Physiology

The evolution of caring and the "hyper-affiliative" motivations of human beings were paralleled by the evolution of a range of physiological processes, including the interplay of hormones oxytocin and vasopressin, and the development of a branch of the parasympathetic system, the myelinated vagus nerve (Porges, 2007). The result of the regulatory activity exerted by the vagus nerve is the dynamic balancing of the sympathetic and parasympathetic nervous systems that gives rise to the variability in heart rate (Porges, 2007). This in turn enables a calm physiological state that enables social affiliations, caring, and sharing and inhibits sympathetically driven threat-defensive behaviours (e.g. fight/flight). In support of this, research has shown that vagal tone, as measured by HRV, is critically important for compassion (Stellar et al., 2015). Higher HRV (i.e. more variability of the inter-beat intervals) is related with greater ability to regulate stress and arousal (Thayer & Lane, 2009), which, for humans, mostly stem from perceiving social safeness both in their "external world" (e.g. in social environments dominated by compassionate dynamics) and "internal world" (when we create internal relationships with ourselves based on the intention to be supportive, helpful, and reassuring towards our inner struggles). The significance of this is that compassion has a vagal component that regulates autonomic physiology, which is postulated to be the key to the successful self-regulation of daily setbacks in everyday life (Di Bello et al., 2020). The role of the vagus nerve is to help downregulate our physiological responses in times of threat or suffering (acting as a "vagal brake"), hence the clear links to compassion (Kirby et al., 2017a, b). This is why scholars have suggested that HRV can be considered a key physiological measurement of compassion (Di Bello et al., 2020).

HRV is a robust predictor of improved mental and physical health (Di Bello et al., 2020; Kirby et al. 2017a, b). A recent meta-analysis found significant moderate effect sizes ($g = 0.54$) between higher baseline HRV and compassion (Di Bello et al., 2020). It is typical within CFT/CMT research to examine baseline HRV, with evidence supporting increases to baseline HRV at post-intervention (Matos et al., 2017a, b; Kim et al., 2020). HRV, however, is dynamic, and if a person is under threat, it will drop temporarily. A better measure of HRV is "*vagal flexibility*" also known as HRV reactivity (Balzarotti et al., 2017). Vagal flexibility refers to how the HRV signal responds after a threat disappears. A recent evaluation of a 12-session protocol of CFT found that those with secure attachment styles had greater HRV reactivity or vagal flexibility than those with insecure attachment styles (Steffen et al., 2021). This was assessed by measuring the HRV signal during a threat exercise, through self-critical writing, then examining the HRV signal during a recovery period. The researchers found that participants with secure attachment styles were better able to engage with the threat task, as evidenced by a significant decrease in HRV. Moreover, during the recovery period, the HRV signal recovered to baseline more quickly for those with secure attachment than for those with insecure attachment styles.

Compassion should not be confused with chronic inflexible state of soothing positive emotions and the "removal or denial" of the suffering, which acts by simply reducing the threat or anxiety experienced. Indeed, in order to fully engage in actions aimed to alleviate (self or others') distress, the pain should resonate with, and empathic sensitivity should be experienced first. This observation has led Di Bello, Ottaviani and Petrocchi (2021) to conduct a study with the intention to shed light on the complex nature of the relationship between compassion and HRV. They measured HRV in students exposed first to a video inducing empathic sensitivity (the first component of compassion) and then to another video eliciting compassionate actions (the second component of compassion). They found that HRV decreased after the first video but significantly increased after the second video, suggesting that it is simplistic to link compassion with higher vmHRV. Indeed, compassion-focused interventions might often produce increased sensitivity to emotional pain, which is naturally associated with lower HRV. However, such interventions are also associated with a concomitant increase in willingness to take actions to alleviate one's suffering, which is ultimately associated with increased vmHRV. Thus, CFT adopts a nuanced perspective on the complex physiological regulation that underlies compassionate responding to suffering.

Applying CFT to a Case: The Basics

When applying CFT to a case, the target population will inform how the intervention itself is delivered. For example, what is offered to parents of children with chronic illness might be different to that of teen parents, where there is often a lot of social shame, which will be different again to an adult client with depression. Thus, it is important to be guided by the science, that is, what the research suggests to be most helpful when tailoring CFT to that specific population. Given CFT attempts to engage in motivational shifting from competitive to compassionate motives, a core target addressed in the model is self-criticism and shame, as these are central in competitive motives. Therefore, when using a CFT approach, a shame-based formulation may prove to be helpful. The formulation may help the client make sense of current difficulties and to also provide options in terms of intervention target points. Fig. 23.2 is an example of a shame-based formulation for an example client, in this instance a parent struggling in their role who is highly self-critical of their parenting, and fears being judged as a bad parent.

The shame-based formulation links parenting to past experiences, to key fears, safety behaviours, unintended consequences, and self-to-self-relating. A core aspect of this shame-based case formulation model is differentiating the source of fear (or threat) or safety as being from an external or internal source. External sources are the minds of others, that is, *"how do others see me or judge me?"*. Internal sources are one's own views and judgements. External sources of shame are more powerful in impacting depression and anxiety, compared to internal sources of shame (Kim et al., 2011). External sources of shame can be internalised, such that our parents (external source) can judge us as annoying, which we can then adopt and believe that we are annoying (internal source). It is important to differentiate between the sources of shame, threat, and safety as being either external or internal, as the interventions to address them will differ. In the case of external threats and shame, assertiveness training may become important, whereas in the case of internal shame, developing self-reassuring self-relating styles may be more critical.

To aid in the development of this formulation, the client can be asked to complete measures such as the Early Life Events Scale, which assesses for the emotional memories of childhood, including whether you recalled being devalued or frightened (Gilbert et al., 2003). In addition, the Others as Shamer Scale can be used to assess for external shame (Goss et al., 1994), with the External and Internal Shame Scale for internal shame (Ferreira et al., 2020). The Forms of Self-Criticism and Self-Reassurance Scale (Gilbert et al., 2004) can be used to assess different forms of self-criticism, with two key forms of inadequacy and self-hatred being linked to mental health difficulties. Finally, the Fears of Compassion Scale (Gilbert et al., 2014) can also be used to determine how the client views different experiences of compassion. Identifying fears of compassion have been shown to be important for treatment, with recent research reporting that fears of compassion can interfere with treatment progress for social anxiety disorder, obsessive compulsive disorder, generalised anxiety disorder, and depression (Merrit & Purdon, 2020). Like other therapy models, there are many ways to develop a case formulation, with the shame-based formulation being one such example in CFT. An alternative CFT case formulation could be to use the three affect regulation systems model with a client, with interventions aimed to develop the soothing/affiliative system (e.g. soothing rhythm breathing), or alternatively develop the drive system (e.g. behaviour activation). After assessment and case formulation, interventions can be introduced, and in partnership with the client, specific exercises and strategies can be implemented to help shift from competitive to compassionate motives.

CFT Intervention Components

There can be many different intervention components within a CFT approach. Often specific CMT exercises are used and applied to the case formulation developed with the client. For example, in relation to the client case formulation depicted in Fig. 23.2, the self-critical parent may be guided in how to ground in the body and to slow down when feeling overwhelmed, and then how to switch to a compassionate motivation by using imagery, such as the compassionate-self or the ideal compassionate other exercises. This can help shift the self-relating style from being one characterised by attacking self-criticism, to a relating style that is validating, encouraging, and self-reassuring. To help researchers in their continued evaluation efforts of CFT, Kirby, Gilbert and Petrocchi have recently attempted to manualise the CFT process in a 12-module intervention. The manualised protocol can be delivered over any required number of sessions. For example,

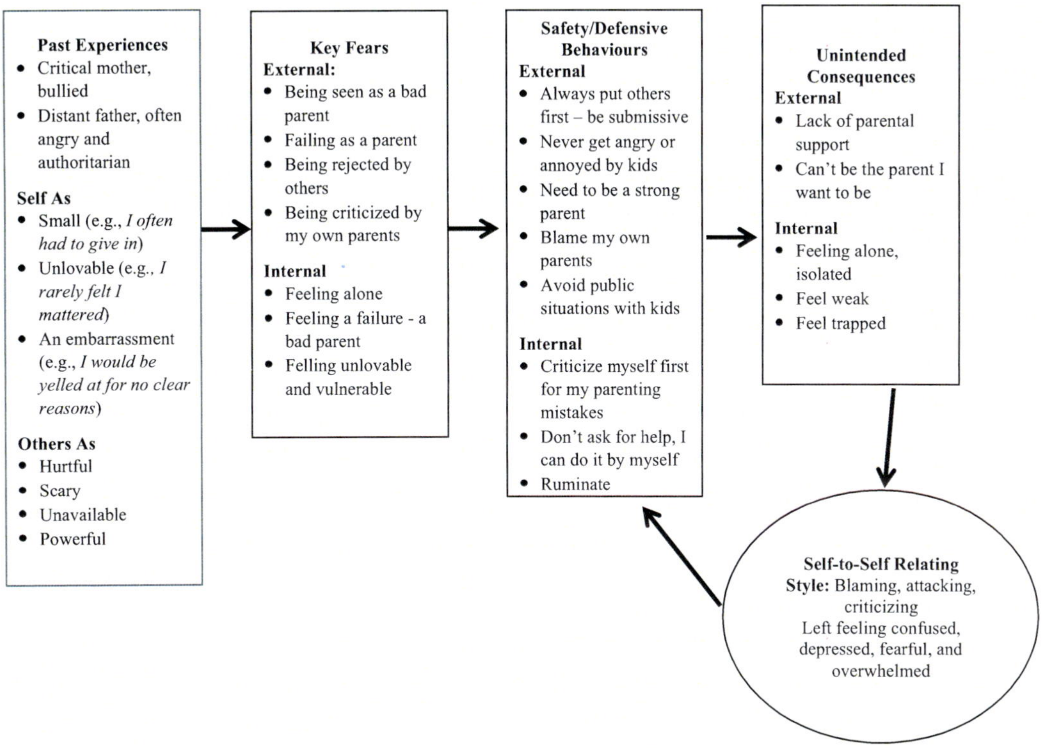

Fig. 23.2 An example of a shame-based CFT formulation for a self-critical parent

you may deliver one module for one two-hour session; conversely, you may deliver a module across two or three one-hour sessions. The number of sessions will be influenced by several factors, including the target population, severity of symptoms, as well as external pressures such as number of sessions available for access under health system options. The structure of the 12 modules is described in Table 23.1, which outlines the aim of each module and a brief overview of some of the module exercises.

This 12-module protocol was evaluated in a recent randomised controlled trial (RCT) examining its effectiveness at helping with body weight shame (Carter et al., 2021a). The CFT intervention was delivered in weekly 2-hour modules. Each module included (a) didactic presentation of material by facilitator; (b) guided meditation or imagery exercises; (c) pair exercises; and (d) large group discussions. As outlined in Table 23.1, the first five sessions of the programme aimed at developing the participants "compassionate-self." Thus, the programme started with psychoeducation on defining compassion, how the mind works, and the three affect regulation systems that underpin CFT. The programme sessions then transitioned to active skills training by focusing on body posture, soothing rhythm breathing, attention and mindfulness, and the development of the compassionate-self. The remaining seven sessions used the participant's compassionate-self to work on difficult emotions, specifically self-criticism, shame, self-compassion, assertiveness, and forgiveness. In each session, there was a guided meditation or visual imagery exercise which was recorded and sent to the participants as part of personal practice between sessions. Each participant also received a workbook with the key content covered across the 12 sessions. After each session, the participants received a standardised email summarising the core components of the session, along with the live audio recording of the in-session meditation or imagery exercise. Results from the RCT found that the CFT intervention compared to a waitlist control group significantly reduced shame, self-criticism, as well as signifi-

Table 23.1 A 12 module compassion-focused therapy intervention

Module	Content
1: Introduction to Compassion and the tricky brain	**Aim:** To understand how individuals experience compassion, fears they have towards compassion and beginning psychoeducation of the evolved mind, and how it functions. **Module exercises:** Large group discussions, pair exercises, compassionate imagery, and the realities of life meditation.
2: Three types of emotion	**Aim:** To introduce evolutionary functional analysis of emotion, assist individuals to understand the nature and function of threat-based emotions (anger, anxiety, and disgust), drive-based emotions happiness and excitement), and soothing-based emotions (safeness and contentment) and help individuals clarify compassion as a motive. **Module exercises:** Large group discussions, pair exercises, and soothing rhythm breathing exercise.
3: Attention training and mindfulness	**Aim:** To introduce individuals to the nature and function of attention (how to pay attention to attention), with introductions to mindfulness-based practices. **Module exercises:** Large group discussions, pair exercises, connecting mindfulness skills with breathing, grounding, and body awareness skills such as use of body posture, facial expressions, and voice tones
4: Safety/ safeness and Compassion from others	**Aim:** To introduce individuals to the concept of safeness (affiliative and exploratory focus) and how that differs to safety (threat focused). To explore how it feels to experience compassion from others. Discussions on how and why our relationships are important to us and support a range of physiological processes within us. **Module exercises:** Large group discussions, pair exercises, breathing exercises, compassionate imagery, and safe place imagery.

(continued)

Table 23.1 (continued)

Module	Content
5: The compassionate-self	**Aim:** To introduce individuals to the nature and concept of, the compassionate-self. The compassionate-self includes three key qualities: Wisdom, strength, and commitment. These qualities are described and explored. **Module exercises:** Large group discussions, pair exercises, breathing exercises, and cultivating the compassionate-self
6: Multiple selves	**Aim:** To introduce group members to the concept of multiple selves, with a particular focus on threat-based emotions, examining specifically angry-self, anxious-self, and sad-self. **Session exercises:** Large group discussion, pair exercises, exploration and experiential practice of multiple selves and responses, compassionate imagery, and breathing exercises.
7: Self-criticism	**Aim:** To help individuals understand the forms and functions of self-criticism and how to use the compassionate-self to work with disappointments, setbacks, and rejections. **Module exercises:** Large group discussion, pair exercises, experiential exercises (self-monitoring), breathing exercises, compassionate imagery, and breathing exercises.
8: Shame and guilt	**Aim:** To help individuals understand the evolution of the threat, drive, and soothing systems in social relationships. Exploration of social rank systems and emotions: Shame (external, internal), humiliation, and guilt. **Module exercises:** Large group discussion, pair exercises, experiential exercises (experiencing shame and how the compassionate-self works with shame), and breathing exercises

(continued)

Table 23.1 (continued)

Module	Content
9: Deepening Compassion for self	**Aim:** To help individuals deepen compassion for the self by facilitating broader abilities to opening and tolerating emotional and motivational experiences **Module exercises:** Large group discussion, pair exercises, experiential exercise (directing compassion towards others and self), compassionate letter writing, making compassion-focused flashcards, and breathing exercises.
10: Compassionate assertiveness	**Aim:** To help individuals understand assertiveness and how assertiveness is linked to strength and authority of compassion. Thus, allowing for individuals to express themselves confidently not aggressively or passively. **Module exercises:** Large group discussion, pair exercises (unhelpfulness of aggressive, passiveness, and passive–aggressive responses), refection of compassionate-self-identity, and breathing exercises.
11: Forgiveness	**Aim:** To introduce individuals to exploring how to engage in the flow of compassion for others as well as forgiveness and how the suffering of others can influence us and what we feel in our body and what we do. **Module exercises:** Large group discussion, pair exercises, practising of perspective taking and empathy, and breathing exercises
12: Envisioning a compassionate future	**Aim:** To revisit the journey the group has been on and invite individuals to consider prevention and emergency strategies for future difficulties and envision what a compassionate future would involve. **Module exercises:** Large group discussion, pair exercises (how to cultivate and strengthen compassion), acknowledgement of challenges, self-gratitude letter, compassionate imagery (compassionate wishes), and breathing exercises

cantly improved levels of self-compassion and self-reassurance.

This 12-session protocol was also used in a recent pre-post evaluation of 31 participants who were attending university counselling centre (Steffen et al., 2021). Steffen et al. (2021) found the intervention improved levels of compassion and reduced self-criticism and shame. They also examined HRV, and while they found no significant pre-post-changes in baseline HRV at the group level, they did find that those who reliably improved in self-compassion (as measured by the Compassion Engagement and Action Scale) also had a significant increase in HRV from baseline to post-intervention. Furthermore, they found those with secure attachment styles had greater HRV reactivity compared to those with insecure attachment styles. Steffen et al. (2021) suggested one method to help support continued practice of the exercises designed to increase HRV is to include HRV biofeedback within the sessions. As the study did not have a control group, it is recommended that future research integrating HRV biofeedback include an active comparator condition so that the effects can be attributed to the CFT protocol.

Future Work for Compassion Focused Therapy

Over the past years, many studies have begun to explore the effectiveness of CFT and CFT-based approaches on several mental health outcomes. In his review of the literature, Kirby (2016) identified at least five randomised controlled trials, yielding promising preliminary evidence for the effectiveness of CFT, both as group-based intervention or as unguided self-help, in clinical and nonclinical samples. However, since that time there have been additional RCTs examining CFT with populations which have included youth forensic populations (Ribeiro da Silva et al., 2020), those who experience body weight shame (Carter et al., 2021a), those with chronic pain (Dhokia et al., 2020), those transitioning to motherhood (Kelman et al., 2018), those with paranoid ideation (Ascone et al., 2017), and those with PTSD (Daneshvar et al., 2020). In a recent systematic review of effectiveness and acceptability of CFT in clinical populations, Craig et al. (2020) found that CFT shows promise for a range of mental health problems, especially when delivered in a group format over at least 12 h. The authors also found that CFT was well accepted by clients and clinicians and that a standard manual and protocol might facilitate future research, promoting the widespread implementation into mainstream clinical practice. Indeed, the 12-module intervention by Kirby et al. discussed previously will allow researchers to compare a standardised CFT protocol across samples with appropriate treatment fidelity checks. Moreover, it will allow for assessment of the specific psychophysiological impact of each component of the intervention. For example, researchers will be more able to conduct dismantling trials to monitor changes produced by each module of the protocol, and potentially isolate the "active ingredients" of CFT, with the intention of delivering an intervention that is powerful and parsimonious at the same time.

There are two major future research endeavours for CFT to target over the next 10 years. First, researchers should continue to examine the efficacy of CFT using RCTs with clinical populations and comparing its effectiveness to other active therapies. Second, attention should be given to examining the active processes within CFT. There are many components within CFT that are used as part of the therapeutic change process. For example, the psychoeducation process of the evolved "*Tricky Brain*" is a core component that aims to de-shame the difficulties the client is experiencing. However, whether this component acts as an active ingredient within the therapy process is unknown. When we refer to active ingredients within a therapy, we are referring to those aspects of the intervention that help drive a positive impact, are conceptually well defined, and related to a proposed mechanism of action (Sebastian et al., 2021). The question of which components are effective for which populations remain an ongoing scientific discovery. There are many studies that have evaluated isolated CFT components such as the compassionate-self (Kim et al., 2020) or the ideal compassionate-other (Rockliff et al., 2013) or a form of compassionate imagery (Carlyle et al., 2019). However, these interventions are examined by themselves, not as a component within a larger therapy intervention. We hope that the manualised CFT protocol developed by Kirby et al. (2021) will usher in a new wave of evaluations, allowing for specific examination of components and active ingredients within the therapy process.

Conclusion

This chapter has examined the origins of CFT, how it aims to target the processes of self-criticism and shame that can arise due to having a dominant competitive social mentality, which can lead to many different mental health difficulties. The aim of CFT is to help cultivate the compassionate social mentality, which enables the individual to become an internal source of self-reassurance, validation, and encouragement. There have been several CFT evaluations using RCTs with many different clinical populations. Future work should continue to examine the efficacy of CFT for different clinical populations, along with examining the specific active ingredients within the approach that help bring about positive change.

References

Ascone, L., Sundag, J., Schlier, B., & Lincoln, T. M. (2017). Feasibility and effects of a brief compassion-focused imagery intervention in psychotic patients with paranoid ideation: A randomized experimental pilot study. *Clinical Psychology and Psychotherapy, 24*(2), 348–358. https://doi.org/10.1002/cpp.2003

Balzarotti, S., Biassoni, F., Colombo, B., & Ciceri, M. R. (2017). Cardiac vagal control as a marker of emotion in healthy adults: A review. Biological Psychology, 130, 54–66. https://doi.org/10.1016/j.biopsycho.2017.10.008

Braehler, C., Gumley, A., Harper, J., Wallace, S., Norrie, J., & Gilbert, P. (2013). Exploring change processes in compassion focused therapy in psychosis: Results of a feasibility randomized controlled trial. *The British Journal of Clinical Psychology, 52*(2), 199–214. https://doi.org/10.1111/bjc.12009

Carlyle, M., Rockliff, H., Edwards, R., Ene, C., Karl, A., Marsh, B., Hartley, L., & Morgan, C. J. (2019). Investigating the feasibility of brief compassion focused therapy in individuals in treatment for opioid use disorder. *Substance Abuse: Research and Treatment, 13*, 1178221819836726. https://doi.org/10.1177/1178221819836726

Carter, A., Gilbert, P., & Kirby, J. N. (2021a). Compassion-focused therapy for body weight shame: A mixed methods pilot trial. *Clinical Psychology and Psychotherapy, 28*, 93–108. https://doi.org/10.1002/cpp.2488

Carter, A., Gilbert, P., & Kirby, J. N. (2021b). A systematic review of compassion-based interventions for individuals struggling with body weight shame. *Psychology & Health*, 1–31. https://doi.org/10.1080/08870446.2021.1955118

Clapton, N. E., Williams, J., Griffith, G. M., & Jones, R. S. (2018). 'Finding the person you really are ... on the inside': Compassion focused therapy for adults with intellectual disabilities. *Journal of Intellectual Disabilities: JOID, 22*(2), 135–153. https://doi.org/10.1177/1744629516688581

Craig, C., Hiskey, S., & Spector, A. (2020). Compassion focused therapy: A systematic review of its effectiveness and acceptability in clinical populations. *Expert Review of Neurotherapeutics, 20*(4), 385–400. https://doi.org/10.1080/14737175.2020.1746184

Cuppage, J., Baird, K., Gibson, J., Booth, R., & Hevey, D. (2018). Compassion focused therapy: Exploring the effectiveness with a transdiagnostic group and potential processes of change. *British Journal of Clinical Psychology, 57*(2), 240–254. https://doi.org/10.1111/bjc.12162

Daneshvar, S., Shafiei, M., & Basharpoor, S. (2020). Group-based compassion-focused therapy on experiential avoidance, meaning-in-life, and sense of coherence in female survivors of intimate partner violence with PTSD: A randomized controlled trial. *Journal of Interpersonal Violence, 37*(7–8), NP4187–NP4211. https://doi.org/10.1177/0886260520958660

Depue, R. A., & Morrone-Strupinsky, J. V. (2005). A neurobehavioral model of affiliative bonding. *Behavioral and Brain Sciences, 28*(3), 313–395. https://doi.org/10.1017/S0140525X05000063

Di Bello, M., Ottaviani, C., & Petrocchi, N. (2020). Compassion is not a benzo: Distinctive associations of heart rate variability with its empathic and action components. *Frontiers in Neuroscience, 15*, 617443. https://doi.org/10.3389/fnins.2021.617443

Dhokia, M., Elander, J., Clements, K., & Gilbert, P. (2020). A randomized-controlled pilot trial of an online compassionate mind training intervention to help people with chronic pain avoid analgesic misuse. *Psychology of Addictive Behaviors, 34*(7), 726–733. https://doi.org/10.1037/adb0000579

Eslamian, A., Moradi, A., & Salehi, A. (2019). Effectiveness of compassion focused group therapy on sleep quality, rumination and resilience of women in Isfahan city suffering from depression in summer 2018. *International Journal of Medical Investigation, 8*(4), 41–50.

Falconer, C. J., Rovira, A., King, J. A., Gilbert, P., Antley, A., Fearon, P., Ralph, N., Slater, M., & Brewin, C. R. (2016). Embodying self-compassion within virtual reality and its effects on patients with depression. *BJPsych open, 2*(1), 74–80. https://doi.org/10.1192/bjpo.bp.115.002147

Forster, K., & Kanske, P. (2021). Exploiting the plasticity of compassion to improve psychotherapy. Current Opinion in Behavioral Sciences, 39, 64–71. https://doi.org/10.1016/j.cobeha.2021.01.010

Gharraee, B., Tajrishi, K. Z., Farani, A. R., Bolhari, J., & Farahani, H. (2018). A randomized controlled trial of compassion focused therapy for social anxiety disorder. *Iranian Journal of Psychiatry and Behavioral Sciences, 12*(4), e80945.

Gilbert, P. (1984). *Depression: From psychology to brain state*. Routledge.

Gilbert, P., Cheung, M.S.P., Grandfield, T., Campey, F. & Irons, C. (2003). Recall of threat and submissiveness in childhood: Development of a new scale and its relationship with depression, social comparison and shame. Clinical Psychology and Psychotherapy, 10, 108–115.

Gilbert, P., Clark, M., Hempel, S., Miles, J. N. V., & Irons, C. (2004). Criticising and reassuring oneself: An exploration of forms, styles and reasons in female students. British Journal of Clinical Psychology, 43(1), 31–50. https://doi.org/10.1348/014466504772812959

Gilbert, P. (2009). Evolved minds and compassion-focused imager in depression. In L. Stopa (Ed.). Imagery and the threatened self in cognitive therapy (pp. 206–231). London: Routledge.

Gilbert, P., McEwan, K., Matos, M., & Rivis, A. (2011). Fears of compassion: Development of three self-report measures. Psychology and Psychotherapy: Theory, Research and Practice, 84, 239–255. https://doi.org/10.1348/147608310X526511.

Gilbert, P. (2014). The origins and nature of compassion focused therapy. *British Journal of Clinical Psychology, 53*, 6–41. https://doi.org/10.1111/bjc.12043

Gilbert, P., & Mascaro, J. (2017). Compassion fears, blocks, and resistances: An evolutionary investigation. In E. Sappla, & J. Doty (Eds.). Handbook of compassion (pp. 399–416). New York, NY: Oxford University Press.

Gilbert, P. (2020). Compassion: From its evolution to a psychotherapy. *Frontiers in Psychology, 11*, 586161. https://doi.org/10.3389/fpsyg.2020.586161

Gilbert, P., & Kirby, J. N. (2019). Building an integrative science for psychotherapy for the 21st century: Preface and introduction. *Psychology and Psychotherapy: Theory, Research and Practice, 92*, 151–163. https://doi.org/10.1111/papt.12225

Gilbert, P., McEwan, K., Matos, M., & Rivis, A. (2010). Fears of compassion: Development of three self-report measures. *Psychology and Psychotherapy: Theory, Research and Practice, 84*, 239–255. https://doi.org/10.1348/147608310X526511

Gilbert, P., & Procter, S. (2006). Compassionate mind training for people with high shame and self-criticism: Overview and pilot study of a group therapy approach. *Clinical Psychology & Psychotherapy, 13*(6), 353–379. https://doi.org/10.1002/cpp.507

Goss K, Gilbert P, Allan S (1994) An exploration of shame measures - I: the other as Shamer scale. Personality & Individual Differences, 17, 713–717

Hayes, S.C., Hoffman, S.G., & Ciarrochi, J. (2020). A process-based approach to psychological diagnosis and treatment: The conceptual and treatment utility of an extended evolutionary meta model. *Clinical Psychology Review, 82*, https://doi.org/10.1016/j.cpr.2020.101908

Hermanto, N., Zuroff, D. C., Kopala-Sibley, D. C., Kelly, A. C., Matos, M., Gilbert, P., & Koestner, R. (2016). Ability to receive compassion from others buffers the depressogenic effect of self-criticism: A cross-cultural multi-study analysis. *Personality and Individual Differences, 98*, 324–332. https://doi.org/10.1016/j.paid.2016.04.055

Irons, C., & Heriot-Maitland, C. (2021). Compassionate mind training: An 8-week group for the general public. Psychology and Psychotherapy: Theory, Research and Practice, 94(3), 443–463. https://doi.org/10.1111/papt.12320

Jazaieri, H., McGonigal, K., Jinpa, T., Doty, J. R., Gross, J. J., & Goldin, P. R. (2013). A randomized controlled trial of compassion cultivation training: Effects on mindfulness, affect, and emotion regulation. *Motivation and Emotion, 38*, 23–35. https://doi.org/10.1007/s11031-013-9368-z

Kelly, A. C., & Carter, J. C. (2015). Self-compassion training for binge eating disorder: A pilot randomized controlled trial. *Psychology and Psychotherapy: Theory, Research and Practice, 88*(3), 285–303. https://doi.org/10.1111/papt.12044

Kelly, A. C., Wisniewski, L., Martin-Wagar, C., & Hoffman, E. (2017). Group-based compassion-focused therapy as an adjunct to outpatient treatment for eating disorders: A pilot randomized controlled trial. *Clinical Psychology & Psychotherapy, 24*(2), 475–487. https://doi.org/10.1002/cpp.2018

Kelman, A. R., Evare, B. S., Barrera, A. Z., Munoz, R. F., & Gilbert, P. (2018). A proof-of-concept pilot randomized comparative trial of brief internet-based compassionate mind training and cognitive- behavioral therapy for perinatal and intending to become pregnant women. *Clinical Psychology & Psychotherapy, 25*, 608–619. https://doi.org/10.1002/cpp.2185

Kim, S., Thibodeau, R., & Jorgensen, R. S. (2011). Shame, guilt, and depressive symptoms: A meta-analytic review. *Psychological Bulletin, 137*, 68–96. https://doi.org/10.1037/a0021466

Kim, J. J., Parker, S., Henderson, T., & Kirby, J. N. (2020). Physiological fractals: Visual and statistical evidence across timescales and experimental states. Journal of The Royal Society Interface, 17(167), Article 20200334. https://doi.org/10.1098/rsif.2020.0334

Kirby, J.N. (2017). Compassion interventions: The programmes, the evidence, and implicaitons for research and practice. Psychology and Psychotherapy: Theory, Research and Practice, 90 (3), 432-455. https://doi.org/10.1111/papt.12104

Kirby, J. N., Day, J., & Sagar, V. (2019). The 'flow' of compassion: A meta-analysis of the fears of compassion scales and psychological functioning. *Clinical Psychology Review, 70*, 26–39. https://doi.org/10.1016/j.cpr.2019.03.001

Kirby, J. N., Doty, J. R., Petrocchi, N., & Gilbert, P. (2017a). The current and future role of heart rate variability for assessing and training compassion. *Frontiers in Public Health, 5*(SUPPL.3), 393–395. https://doi.org/10.3389/fpubh.2017.00040

Kirby, J. N., Petrocchi, N., & Gilbert, P. (2021). *Compassion focused therapy: Manual*. Unpublished Therapy Manual.

Kirby, J. N., Tellegen, C. L., & Steindl, S. R. (2017b). A meta-analysis of compassion-based interventions: Current state of knowledge and future directions. *Behavior Therapy, 48*(6), 778–792. https://doi.org/10.1016/j.beth.2017.06.003

LeDoux, J. (1998). The Emotional Brain. London: Weidenfeld and Nicolson.

Lucre, K., & Corten, N. (2013). An exploration of group compassion focused therapy for personality disorder. *Psychology and Psychotherapy, 86*(4), 387–400. https://doi.org/10.1111/j.2044-8341.2012.02068.x

Lawrence, V. A., & Lee, D. (2014). An exploration of people's experiences of compassion-focused therapy for trauma, using interpretative phenomenological analysis. *Clinical Psychology & Psychotherapy, 21*(6), 495–507. https://doi.org/10.1002/cpp.1854

Maratos, F. A., Montague, J., Ashra, H., Welford, M., Wood, W., Barnes, C., Sheffield, D., & Gilbert, P. (2019). Evaluation of a compassionate mind training intervention with school teachers and support staff. *Mindfulness, 10*(11), 2245–2258. https://doi.org/10.1007/s12671-019-01185-9

Matos, M., Duarte, J., & Pinto-Gouveia, J. (2017a). The origins of fears of compassion: Shame and lack of safeness memories, fears of compassion and psychopathology. *The Journal of Psychology, 151*, 804–819. https://doi.org/10.1080/00223980.2017.1393380

Matos, M., Duarte, C., Duarte, J., Pinto-Gouveia, J., Petrocchi, N., Basran, J., & Gilbert, P. (2017b). Psychological and physiological effects of compassionate mind training: A pilot randomised controlled study. *Mindfulness, 8*(6), 1699–1712. https://doi.org/10.1007/s12671-017-0745-7

Mayseless, O. (2016). *The caring motivation: An integrated theory*. Oxford University Press.

Merritt, O. A., & Purdon, C. L. (2020). Scared of compassion: Fear of compassion in anxiety, mood, and non-clinical. *British Journal of Clinical Psychology, 59*, 354–368. https://doi.org/10.1111/bjc.12250

Neff, K. D., & Germer, C. K. (2013). A pilot study and randomized controlled trial of the mindful self-compassion program. *Journal of Clinical Psychology, 69*, 28–44. https://doi.org/10.1002/jclp.21923

Noorbala, F., Borjali, A., Ahmadian-Attari, M. M., & Noorbala, A. A. (2013). Effectiveness of compassionate mind training on depression, anxiety, and self-criticism in a group of Iranian depressed patients. *Iranian Journal of Psychiatry, 8*(3), 113–117.

Pace, T. W., Negi, L. T., Adame, D. D., Cole, S. P., Sivilli, T. I., Brown, T. D., Issa, M. J., & Raison, C. L. (2009). Effect of compassion meditation on neuroendocrine, innate immune and behavioral responses to psychosocial stress. *Psychoneuroendocrinology, 34*(1), 87–98. https://doi.org/10.1016/j.psyneuen.2008.08.011

Panksepp, J. (2010). Affective neuroscience of the emotional brainmind: Evolutionary perspectives and implications for understanding depression. Dialogues in Clinical Neuroscience, 12, 383–399.

Petrocchi, N., & Cheli, S. (2019). The social brain and heart rate variability: Implications for psychotherapy. *Psychology & Psychotherapy: Theory, Research & Practice, 92*(2), 208–223. https://doi.org/10.1111/papt.12224

Petrocchi, N., Cosentino, T., Pellegrini, V., Femia, G., D'innocenzo, A., & Mancini, F. (2021). Compassion focused group therapy for treatment-resistant OCD: Initial evaluation using a multiple baseline design. *Frontiers in Psychology, 11*, 594277. https://doi.org/10.3389/fpsyg.2020.594277

Porges, S. W. (2007). The polyvagal perspective. *Biological Psychology, 74*, 116–143. https://doi.org/10.1016/j.biopsycho.2006.06.009

Ribeiro da Silva, D., Rijo, D., Salekin, R. T., Paulo, M., Miguel, R., & Gilbert, P. (2020). Clinical change in psychopathic traits after the PSYCHOPATHY.COMP program: Preliminary findings of a controlled trial with male detained youth. *Journal of Experimental Criminology, 17*, 397–421. https://doi.org/10.1007/s11292-020-09418-x

Rockliff, H., Gilbert, P., McEwan, K., Lightman, S., & Glover, D. (2008). A pilot ex- ploration of heart rate variability and salivary cortisol responses to compassion-fo- cused imagery. Journal of Clinical Neuropsychiatry, 5(3), 132–139.

Savari, Y., Mohagheghi, H., & Petrocchi, N. (2021). A preliminary investigation on the effectiveness of compassionate mind training for students with major depressive disorder: A randomized controlled trial. *Mindfulness, 12*, 1159–1172. https://doi.org/10.1007/s12671-020-01584-3

Schuling, R., Huijbers, M. J., van Ravesteijn, H., Kuyken, W., & Speckens, A. E. M. (2021). Mindfulness-based compassionate living (MBCL): A qualitative study into the added value of compassion in recurrent depression. *Mindfulness, 12*(9), 2196–2206. https://doi.org/10.1007/s12671-021-01675-9

Sebastian, C. L., Pote, I., & Wolpert, M. (2021). Searching for active ingredients to combat youth anxiety and depression. *Nature Human Behavior, 5*, 1266–1268. https://doi.org/10.1038/s41562-021-01195-5

Steindl, S. R., Buchanan, K., Goss, K., & Allan, S. (2017). Compassion focused therapy for eating disorders: A qualitative review and recommendations for further applications. *Clinical Psychologist, 21*(2), 62–73. https://doi.org/10.1111/cp.12126

Stellar, J. E., Cohen, A., Oveis, C., & Keltner, D. (2015). Affective and physiological responses to the suffering of others: Compassion and vagal activity. Journal of Personality and Social Psychology, 108(4), 572–585. https://doi.org/10.1037/pspi0000010

Steffen, P. R., Foxx, J., Cattani, K., Alldredge, C., Austin, T., & Burlingame, G. M. (2021). Impact of a 12-week group-based compassion focused therapy intervention on heart rate variability. *Applied Psychophysiology and Biofeedback, 46*(1), 61–68. https://doi.org/10.1007/s10484-020-09487-8

Thayer, J. F., & Lane, R. D. (2009). Claude Bernard and the heart- brain connection: Further elaboration of a model of neurovisceral integration. *Neuroscience and Biobehavioral Review, 33*, 81–88. https://doi.org/10.1016/j.neubiorev.2008.08.004

Vitaliano, P. P., Zhang, J., & Scanlan, J. M. (2003). Is caregiving hazardous to one's health: A meta-analysis. Psychological Bulletin, 129, 946–972. https://doi.org/10.1037/0033-2909.129.6.946.

Wetherall, K., Robb, K. A., & O'Connor, R. C. (2019). Social rank theory of depression: A systematic review of self-perceptions of social rank and their relationship with depressive symptoms and suicide risk. *Journal of Affective Disorders, 246*, 300–319. https://doi.org/10.1016/j.jad.2018.12.045

Whelton, W. J., & Greenberg, L. S. (2005). Emotion in self-criticism. *Personality and Individual Differences, 38*(7), 1583–1595. https://doi.org/10.1016/j.paid.2004.09.024

24

A House with Many Doors – Toward a More Nuanced Self-Compassion Intervention Science

Amy Finlay-Jones

Introduction

As demonstrated throughout the chapters in this Handbook, there is now a robust body of evidence demonstrating the associations between self-compassion and adaptive outcomes across a range of populations and contexts. From a translational perspective, the true value of this research is what it means for prevention and intervention science; if self-compassion can be effectively cultivated through intervention, it has the potential to deliver a broad range of benefits across the spectrum of health promotion through to clinical intervention. In support of this proposition, the rapidly burgeoning compassion intervention literature has demonstrated that self-compassion can be effectively cultivated, using a variety of different techniques. As this literature advances, we are able to start asking more nuanced questions: Which self-compassion interventions are most effective for which outcomes? What are the active ingredients of self-compassion interventions that drive outcome change? Who do such interventions work (and not work) for, and under which circumstances? And finally, what are the mechanisms underpinning treatment effects that can explain why these treatment works?

A. Finlay-Jones (✉)
Telethon Kids Institute, Nedlands, WA, Australia
e-mail: amy.finlay-jones@telethonkids.org.au

This chapter aims to provide some insight into the state of the science in the context of these questions, with a focus on the application of self-compassion interventions within transdiagnostic approaches to mental health problems. Transdiagnostic frameworks of psychopathology attempt to highlight common factors implicated in different psychological syndromes that may serve as efficient intervention targets across a range of presenting issues and their comorbidities (Sauer-Zavala et al., 2016; Fusar-Poli et al., 2019). This chapter first summarizes the observational and experimental literature on self-compassion and mental health outcomes, including depression, anxiety, eating disorders, and posttraumatic stress disorder. The range of methods that are available to cultivate self-compassion are then discussed alongside evidence of their effectiveness in clinical populations and in the context of head-to-head trials that have sought to determine relative effectiveness of different self-compassion interventions. Furthermore, evidence regarding predictors, moderators, and mediators of treatment outcome is presented with a view to elucidating what is known about whom self-compassion interventions work for, under which circumstances, and how. Putative mechanisms underpinning these interventions, including self-compassion, emotion regulation, and negative self-relational processes, are discussed.

A. Finlay-Jones et al. (eds.), *Handbook of Self-Compassion*, Mindfulness in Behavioral Health,
https://doi.org/10.1007/978-3-031-22348-8_24

Self-Compassion and Risk and Resilience to Psychopathology

Research examining the associations between self-compassion and both adaptive and maladaptive mental health outcomes indicates that in general, self-compassion is associated with greater positive mental health and resilience and reduced risk of psychopathology and associated outcomes. A substantial proportion of this work has documented cross-sectional relationships between self-compassion and internalizing pathology, including depression and anxiety symptoms. However, self-compassion has also been considered longitudinally and in the context of early risk factors to explore its protective role in the development of symptoms of psychopathology over time. Such studies have found that self-compassion both mediates and moderates the relationship between risk factors for adverse mental health outcomes and their symptoms. For example, the relationship between child maltreatment and adult depression is mediated by self-compassion (Tao et al., 2021), while self-compassion appears to weaken the link between stress and depression and anxiety over time (Stutts et al., 2018). Such associations suggest that reductions in self-compassion are one reason that early risk factors might lead to adverse outcomes, but also that the practice of self-compassion can buffer against the deleterious effects of such risk factors.

Self-Compassion and Mood and Anxiety Disorders

An early meta-analysis of 20 studies found that self-compassion was inversely related to symptoms of depression, anxiety, and stress in adults, with a large effect size (MacBeth & Gumley, 2012). Similarly, a recent meta-analysis of 19 studies with adolescents (10–19 years; $N = 7049$) found a large effect size for the negative relationships between self-compassion and anxiety, depression, and stress (Marsh et al., 2018). Relationships between self-compassion and depressive symptoms have been observed in both community (Körner et al., 2015) and clinical populations, as well as in a range of specific populations, including adolescents (Pullmer et al., 2019), individuals with chronic illness (Hughes et al., 2021), and women in the perinatal period (Carona et al., 2022). While largely correlational, these findings are supported by longitudinal research, which has demonstrated that self-compassion represents a protective factor in the development of depression over time (Krieger et al., 2016; Raes, 2011; Tobin & Dunkley, 2021). Additionally, depressed patients report less self-compassion than those who have never been depressed (Krieger et al., 2013), while those living with depression report that their condition undermines their ability to be self-compassionate (Pauley & McPherson, 2010). Together, these findings indicate that a lack of self-compassion may act as both a precipitating and maintaining factor in depressive episodes. Moreover, pooled effect sizes from meta-analyses (d = 0.40–0.66) indicate that the effect of self-compassion interventions on depression symptoms is moderate (Kirby et al., 2017; Ferrari et al., 2019; Wilson et al., 2018).

Self-compassion also shares inverse associations with various facets of anxiety disorder symptomology, including social anxiety (Gill et al., 2018), obsessive-compulsive disorder (Leeuwerik et al., 2019), and generalized anxiety (Hoge et al., 2013). Self-compassion also negatively correlates with specific mechanisms implicated in the onset and perpetuation of these disorders, such as anxiety sensitivity and worry (Hoge et al., 2013; Raes, 2010). These correlational studies are supported by longitudinal findings demonstrating that self-compassion predicts anxiety over time (Ştefan, 2019; Zhu et al., 2019). Further, self-compassion is associated with important aspects of functioning among people living with mood and anxiety disorders. For example, in a help-seeking community sample with mixed anxiety and depression, Van Dam et al. (2011) found that self-compassion was associated with unique variance in symptom severity and quality of life, over and above that accounted for mindfulness. Pooled effect sizes from the intervention literature suggest that the impact of self-compassion interventions on anxiety (d = 0.46–0.57) is moderate (Kirby et al., 2017; Ferrari et al., 2019; Wilson et al., 2018)

Self-Compassion and Eating Disorders

Self-compassion is inversely associated with several dimensions of eating disorder pathology, in both clinical and non-clinical samples. This includes disordered eating (Ferreira et al., 2013; Webb & Forman, 2013; Adams & Leary, 2007), eating-related guilt (Adams & Leary, 2007), drive-for-thinness (Ferreira et al., 2013), and body image concerns (for a review, see Chap. 11). Self-compassion is also positively associated with factors that appear protective for eating disorders, such as intuitive eating (Schoenefeld & Webb, 2013) and positive body image (Siegel et al., 2020). A recent meta-analysis of 59 studies found medium-large pooled effects for the relationship between self-compassion and eating pathology, body image concerns, and positive body image (r = −0.34, r = −0.45, r = 0.52, respectively), while the pooled effects of self-compassion interventions on eating pathology and body image were superior to controls, g = 0.58, g = 0.39 (Turk & Waller, 2020). Longitudinal studies demonstrating the relationship between self-compassion and onset of eating disorder pathology (e.g., Linardon, 2021; Fresnics et al., 2019) further underscore the potential of self-compassion as a target for eating disorders prevention. Importantly, self-compassion explains more variance in eating disorder pathology than related constructs, such as mindfulness (Fresnics et al., 2019; Messer et al., 2021).

Self-Compassion and Posttraumatic Stress

Posttraumatic stress disorder (PTSD) is the term given to a cluster of persistent symptoms that emerge following trauma exposure – including hypervigilance, avoidance of trauma-related stimuli, reexperiencing the traumatic event, and angry outbursts. A systematic review including 35 studies reporting associations between self-compassion and posttraumatic stress found largely consistent evidence of an inverse association between these outcomes in both univariate and multivariate models, although the strength of these associations varied across studies (Winders et al., 2020). Self-compassion is also positively associated with posttraumatic growth, a term used to refer to the phenomenon of experiencing personal development following trauma, to the point that one achieves a higher level of functioning than before the trauma occurred (Wong & Yeung, 2017). Furthermore, a recent meta-analysis of self-compassion interventions for posttraumatic stress included 12 studies, eight of which were conducted with clinical samples (Luo et al., 2021). Several of these studies were uncontrolled, and in the overall sample, a medium pooled effect of self-compassion interventions on PTSD symptoms was reported. In a subgroup analysis of three controlled interventions delivered by a therapist, Luo et al. (2021) reported a large pooled effect on PTSD symptoms.

Cultivating Self-Compassion: A House with Many Doors

In addition to the pooled effects on depression and anxiety reported above, prior meta-analyses of self-compassion interventions have reported moderate effect size improvements in self-compassion (d = 0.52–.0.75), mindfulness (d = 0.54–0.62), and well-being (d = 0.51) (Kirby et al., 2017; Ferrari et al., 2019; Wilson et al., 2018). These findings demonstrate the benefit of such interventions for promoting resilience as well as for reducing adverse mental health outcomes. In many ways, self-compassion cultivation can be considered a house with many doors. There are multiple ways that the principles, skills, and benefits of self-compassion can be accessed and cultivated, including through modeling (for example, through the therapeutic relationship; see Chaps. 20 and 22, or via parent–child relationships, see Chap. 16), scalable public health efforts (Mak et al., 2019; Sommers-Spijkerman et al., 2018), informal self-compassion practices, and meditation training on its own (Galante et al., 2014), or as part of a multi-component program (e.g., Neff & Germer, 2013). These latter approaches are described in further detail in this section.

Mindful Self-Compassion Training Mindful Self-Compassion (MSC) is an 8-week group intervention that integrates meditation training with interpersonal exercises and informal practices to support the cultivation and application of self-compassion. MSC was developed for general populations, and there have been formal adaptations of the program for adolescents (Bluth et al., 2015), as well as for healthcare professionals (Neff et al., 2020). Informal adaptations of the program have also been trialed, including brief formats, and online and video-conferencing delivery (Eriksson et al., 2018). Together, findings from experimental and quasi-experimental studies have demonstrated sustained intervention effects of MSC and its derivatives on self-compassion, depression, anxiety, and stress, in general adult (Neff & Germer, 2013), adolescent (Bluth et al., 2015), clinical (Torrijos-Zarcero et al., 2021), and at-risk populations (Knier et al., 2020; Neff et al., 2020), as well as across cultures (Finlay-Jones et al., 2017).

Compassion Focused Therapy Compassion Focused therapy (CFT; Gilbert, 2010) was the first explicitly compassion-focused program designed for clinical populations. As documented in Chap. 23, CFT is an individual or group-based approach that involves imagery-based and experiential exercises aimed at increasing capacity for self-compassion, safeness, and soothing, and reducing fears of receiving compassion from self or others. A recent review of 29 studies concluded that CFT is acceptable and effective for a range of clinical populations (Craig et al., 2020), including mixed psychiatric samples (Stroud & Griffiths, 2021), patients with psychosis (Braehler et al., 2013), and as an adjunctive (Gale et al., 2014; Kelly et al., 2017) and standalone (Kelly & Carter, 2015) therapy for people with eating disorders.

Compassionate Mind Training Compassionate mind training (CMT) includes psychoeducation on the nature of compassion, fears of compassion, Gilbert's social mentality theory (see Chap. 4), and the different affect regulation systems that drive various mood states. Similar to CFT, CMT includes imagery-based and experiential exercises designed to training attention, reduce physiological arousal, and cultivate the "compassionate self" (Gilbert, 2009). Quasi-experimental studies of CMT have reported pre- and post-improvements in self-compassion, well-being, and distress in community samples (Irons & Heriot-Maitland, 2021), while a recent RCT of CMT among students with major depressive disorder reported significant intervention effects for self-compassion, depression, and some aspects of rumination relative to no-treatment controls (Savari et al., 2021).

Compassion Cultivation Training Compassion Cultivation Training (CCT) is an 8- to 9-week structured protocol that comprises weekly group classes and daily compassion-focused meditation practice. The protocol integrates psychoeducation on mindfulness, compassion, and loving-kindness, and related socio-cognitive processes, but is distinguished from MSC, CFT, and CMT in the frequency and duration of the mediation practices included. Further, unlikely other self-compassion-based programs, most of the time spent in CCT focuses on other-focused compassion, although this is grounded in an understanding of the interconnectedness between self and other. Accordingly, cultivating compassion for others is seen to benefit the cultivation of self-compassion, and vice versa. RCTs of CCT have been conducted with general community samples (Jazaieri et al., 2012, 2013; Brito-Pons et al., 2018) and caregivers of people with mental illness (Hansen et al., 2021) and have consistently reported treatment effects for self-compassion alongside several facets of other-focused compassion, well-being, and distress.

Cognitively Based Compassion Training Cognitively Based Compassion Training (CBCT) is framed as a secular protocol that follows an 8-week format, involving 2-h group sessions once a week. Sessions involve psychoeducation, group discussion, and guided meditation practice derived from Tibetan Buddhist mind training (lojong) practices. These practices involve both attention and awareness (mindfulness) practices, as well as "analytical" practices focused on compassion for self and others. Quasi-experimental studies have documented pre- and post-improvements in biomarkers of stress responses among adolescents in foster care (Pace et al., 2012). Experimental and quasi-experimental studies have shown that participation in CBCT improves veteran's PTSD symptoms in both within-groups and between-groups comparisons relative to active control (Lang et al., 2019, 2020). Relative to waitlist or treatment-as-usual, clinical trials have demonstrated the efficacy of CBCT for improving self-compassion, depression, stress, and avoidance in cancer survivors (Gonzalez-Hernandez et al., 2018; Dodds et al., 2015), although no group differences were observed for quality of life (Gonzalez-Hernandez et al., 2018). Interestingly, a recent trial of CBCT has also explored program effects on children when their parents undertake training. In small, waitlist-controlled RCT, Poehlmann-Tynan et al. (2020) found significant decreases in cortisol were observed for children of parents in the CBCT group, despite there being no group differences in parent cortisol or parenting interactions (Engbretson et al., 2020).

Mindfulness-Based Compassionate Living Mindfulness-Based Compassionate Living (MBCL) was developed as a follow-on intervention for people who have completed standardized mindfulness-based interventions such as Mindfulness-Based Cognitive Therapy (MBCT) or Mindfulness-Based Stress Reduction (MBSR) but who are still experiencing residual symptoms of distress (Van den Brink & Koster, 2015). Like other self-compassion-based interventions, MBCL involves psychoeducation, meditation and experiential practice, and dyadic and group exercises, and it follows a similar format to MBCT or MBSR. Bi-weekly sessions of 2.5 h are delivered for 8 weeks, with an additional half-day silent session, and daily meditation practices of between 45–60 min. Unlike MBCT or MBSR, however, these practices focus explicitly on the cultivation of self-compassion and compassion toward others. MBCL has an emerging evidence base supporting its acceptability, feasibility (Bartels-Velthuis et al., 2016; Schuling et al., 2018), and efficacy (Schuling et al., 2020; ter Avest et al., 2021) with clinical populations. For example, among 104 patients with recurrent depression, an RCT of MBCL with 100 found significant treatment effects on depressive symptoms, self-compassion, mindfulness, and quality of life that were maintained over six months (Schuling et al., 2020).

Mindfulness-Based Interventions There are several mindfulness-based interventions (MBIs), including Mindfulness-Based Cognitive Therapy (MBCT) and Mindfulness-Based Stress Reduction (MBSR) in which compassion is implicitly taught as an attitudinal foundation (see Chap. 2), even if not taught explicitly. The impact of such programs on self-compassion is supported by a recent meta-analysis ($k = 26$), which found a medium pooled effect for pre- and post-changes in self-compassion in MBIs compared to control conditions ($g = 0.60$, 95% CI = 0.41 to 0.80, $p < 0.001$) (Golden et al., 2021). Demonstrating the overlap between self-compassion and MBIs, another recent review included MBIs as part of the group of self-compassion interventions for people with serious mental illness and found that MBCT was the most frequently researched intervention for people with depressive disorders (Kurebayashi & Sugimoto, 2022). In four non-randomized controlled trials, significant pre- and post-improvements in self-compassion were observed in patients with major depressive disorder or bipolar disorder undertaking MBCT (Geurts et al., 2020; Schoenberg & Speckens, 2015; Williams et al., 2020; Hanssen et al., 2019).

Toward a More Nuanced Self-Compassion Intervention Science: What, for Whom, Under Which Circumstances, and Why?

Despite overlaps in goals, principles, and practices, self-compassion interventions differ from each other, and from mindfulness-based interventions in several important ways. These include differences in where attention is focused (present-moment versus past/future; intrapersonal versus interpersonal), how experiences and affective states are related to (open monitoring and acceptance of "what is" versus active cultivation or savoring of specific experiences or affective states), and the extent to which meditation practice is considered a necessary element of the program. The focus of psychoeducational content can also vary between programs and may be tailored to the specific needs of the population receiving the program. For example, Finlay-Jones et al. (2020) worked with a group of young people with chronic conditions to adapt and tailor self-compassion exercises specifically for that population.

The What: Differential Outcomes Across Treatment Conditions

While the experimental literature emphasizes the promise of self-compassion interventions, key questions that need to be addressed are (a) self-compassion interventions superior to other types of interventions for improving a specific outcome? (b) of the available self-compassion interventions, are some more effective for improving specific outcomes than others? and (c) what are the "core components" of self-compassion interventions that are most closely linked with intervention effects? Exploring the first question, one meta-analysis found no difference in outcomes between treatment conditions when self-compassion interventions are compared to active controls (Wilson et al., 2018), suggesting that self-compassion interventions may not confer specific benefits relative to other forms of intervention. Despite some noted limitations of this review such as the heterogeneity in the interventions categorized as "self-compassion interventions" (Kirby & Gilbert, 2019), some studies published since have aligned with this conclusion. For example, no differences were found between CFT and meta-cognitive therapy for reducing anxiety in mothers of children with cerebral palsy (Negin et al., 2021), and Haukaas et al. (2018) found that a brief (3-week) mindful self-compassion intervention was comparable to attention training in reducing symptoms of anxiety and depression among a non-clinical student sample.

Other head-to-head studies have demonstrated mixed results, with some showing that self-compassion-based interventions are superior to active controls for some outcomes. For example, in a clinical sample of patients with chronic pain, Torrijos-Zarcero et al. (2021) found that MSC treatment effects were superior to cognitive behavioral therapy for self-compassion and that outcomes for pain acceptance, pain interference, catastrophizing, and anxiety also favored the MSC group. Studies have also reported head-to-head trials of self-compassion and mindfulness-based programs, providing some insight into specific effects of self-compassion practice. For example, one study found that MBSR and CCT had comparable effects for improving psychological well-being and mindfulness; however, effects on self-compassion, empathic concern, and common humanity were greater in the CCT group (Brito-Pons et al., 2018). Similarly, in a study comparing CBCT and a support group for people who had attempted suicide, effects on depression, suicidal ideation, and mindfulness were comparable; however, only the CBCT group report improvements in self-compassion (LoParo et al., 2018). Moreover, differential changes in proposed treatment mechanisms have been demonstrated in head-to-head studies (Roca et al., 2021; LoParo et al., 2018; Hildebrandt et al.,

2017); together, these findings highlight the likelihood of both common and practice-specific treatment effects. Other studies have underscored the importance of examining engagement metrics in addition to effectiveness outcomes when trying to determine which interventions are likely to be most efficient under real-world conditions. For example, one study comparing MSC, MBSR, and a waitlist control among psychology trainees reported no differences between the active conditions in terms of effects on mindfulness, self-compassion, and anxiety and found that while MBSR was superior for reducing depression, adherence rates were higher in MSC (Jiménez-Gómez et al., 2022).

While the question of "which intervention, for which outcomes?" remains pertinent, there is also a shift away from distinct therapy "packages" and toward elucidating core components or active ingredients germane to effective approaches. Insights into core components of effective self-compassion interventions can be provided by dismantling trials that compare, for example, self-compassion interventions that are purely psychoeducational versus those that also involve a meditation training component. Within the mindfulness literature, such studies have successfully identified differential mechanisms and outcomes specific to particular meditation practices (Britton et al., 2018), as well as common factors such as participant ratings of instructor and group (Canby et al., 2021). While dismantling studies are yet to be conducted in the self-compassion intervention literature, emerging research has considered whether component interventions focused on individual dimensions of Neff's (2003) self-compassion construct (i.e., mindfulness, common humanity, self-kindness) have differential effects. In a four-group randomized experimental design with adults with Major Depression Disorder, Ceclan and Nechita (2021) found no differences between groups in pre- and post-changes in depression and shame-proneness and found that outcome improvements were also observed in the no-treatment control group. A limitation of this study was that the target constructs (mindfulness, common humanity, self-kindness) were not measured, making it difficult to conclude whether the conditions effectively targeted these outcomes. Further, as the components of self-compassion are thought to operate synergistically, there may be risks associated with focusing on one component at the expense of others. A more pertinent question remains: What are the active ingredients needed to optimize self-compassion interventions?

For Whom and Under Which Circumstances? Predictors and Moderators of Treatment Effects

Insights from the mindfulness literature have highlighted the importance of considering not just "what" is taught, but also "how" and "in what context" teaching occurs. For example, in a mixed-methods study, Canby et al. (2021) found that participant's ratings of instructor characteristics (such as empathy) predicted changes in depression and stress, while their ratings of group processes (including hope, safety, social learning, and interpersonal dynamics) predicted changes in stress and mindfulness. Furthermore, they found that while degree of formal meditation predicted changes in anxiety and stress, informal mindfulness practice did not predict outcomes. Qualitative data further supported the finding that bonding, feelings of hope, and safe emotional expression were integral to participants' change processes. These findings emphasize that our understanding of active ingredients cannot be reduced to a set of content or practice-driven mechanisms but must also include relational processes. This, in turn, has important implications for facilitator training, as well as for the development of interventions where relational processes must be explicitly designed and integrated ahead of time (e.g., self-guided online interventions) rather than emerge as part of an interpersonal dynamic.

At present, the impact of therapist and group variables on outcome in self-compassion training is unknown. However, there is some evidence regarding the contextual circumstances that influence self-compassion intervention outcome, including intervention dose, setting, and delivery.

In a moderator analysis of the treatment effects reported in their meta-analysis of self-compassion interventions for reducing self-criticism, Wakelin et al. (2021) found that treatment effects were greater when self-compassion interventions were longer. Similarly, in Luo et al.'s (2021) meta-analysis, longer self-compassion interventions were associated with better treatment effects for posttraumatic stress. Intervention delivery and setting were not significant moderators (Wakelin et al., 2021).

Studies have also examined participant characteristics as predictors and moderators of treatment outcome, providing some insight into the question of "for whom" these interventions are most effective. For example, prior work has found no moderating effect of sociodemographic characteristics (Sommers-Spijkerman et al., 2018; Goldin & Jazaieri, 2017), or exposure to positive or negative life events (Sommers-Spijkerman et al., 2018) on treatment outcome. Findings concerning participants' psychological profiles are mixed. For example, in a meta-analysis of self-compassion training for posttraumatic stress, baseline self-compassion was not associated with treatment outcomes for posttraumatic stress (Luo et al., 2021). Similarly, Sommers-Spijkerman et al. (2018) found that baseline or post-treatment psychological well-being did not impact treatment outcomes of CFT, while Finlay-Jones et al. (2017) found that perfectionism did not moderate MSC outcomes. However, in a trial of CCT, Goldin and Jazaieri (2017) found that people with more emotion regulation difficulties, lower mindfulness, and greater perceived stress at baseline reported greater improvements in fears of compassion for self, although these variables did not moderate treatment effects on other outcomes, including self-compassion and other fears of compassion. Interestingly, one study found that while MBCT only increased mindfulness for people with high levels of baseline rumination, CFT enhanced this outcome regardless of baseline scores (Frostadottir & Dorjee, 2019). Furthermore, among African American people who had attempted suicide, Sun et al. (2019) found that CBCT was more effective that a support group for participants with higher levels of reactivity at baseline.

The How: Mechanisms of Change

Understanding the theoretical mechanisms that putatively underlie treatment effects supports the capacity to refine interventions and optimize their impact (Kazdin, 2007). Furthermore, recent qualitative work has demonstrated that understanding "how" self-compassion interventions work can influence participation engagement. For example, in a qualitative study with youth, Egan et al. (2022) found that youth were more willing to engage in self-compassion interventions because they reduce self-criticism than because they increase self-kindness. In this section, the theoretical basis and empirical evidence for three "clusters" of potential mediators that have been proposed to underpin the relationship between self-compassion interventions and improved mental health outcomes: namely, increases in self-compassion, improvements in emotion regulation, and reductions in negative self-relational processes such as self-criticism, perfectionism, and shame-proneness. As the focus of this chapter is weighted more toward the relevance of self-compassion for clinical populations, additional mechanisms (such as increased resilience, positive affect, and self-efficacy) that are more relevant to well-being outcomes are not discussed. It should be noted, however, that the literature surrounding such mechanisms is relatively underdeveloped. A key goal for future research is thus to unpack the different pathways by which self-compassion might lead to positive well-being outcomes in both clinical and non-clinical populations.

Self-Compassion Tests of mediation effects in self-compassion intervention studies generally support the proposition that these programs improve psychological outcomes by increasing self-compassion, reducing fears of self-compassion, and making people more open to receiving compassion from others (Matos et al., 2022). One exception is a recent meta-analysis of self-compassion interventions for posttraumatic stress, which found that while self-compassion interventions had a small effect on self-compassion in clinical populations, there was no effect in non-clinical samples (Luo et al., 2021). Nevertheless, changes in self-compassion appear to have meaningful implications for clinical outcomes. For example, Geurts et al. (2020) reported that greater pre- and post-changes in self-compassion during MBCT were associated with greater improvements in depression, while in Schuling et al.'s (2020) RCT, changes in self-compassion preceded improvements in negative affect, indicating that self-compassion likely plays a mechanistic role in alleviating depressive symptoms (ter Avest et al., 2021).

Emotion Regulation Emotion regulation refers to the processes that influence the nature, intensity, and duration of an individual's emotional experiences (Gross, 2015a, b). In turn, these processes can influence the physiological and behavioral consequences of emotional experiences. From a clinical perspective, the emotion *dys*regulation model of mood and anxiety disorders suggests that the ways in which individuals experience and respond to different emotional states play a key role in precipitating and maintaining psychopathology (Hofmann et al., 2012). Similarly, according to the affect regulation model of eating disorders, binge eating and purging can develop as a form of maladaptive emotion regulation designed to manage negative cognitive-affective experiences (Stice et al., 1996, 1998; Cardi et al., 2015), while non-suicidal self-injury is also a maladaptive means of emotion regulation (see Chaps. 6 and 21). Given that emotion dysregulation is a transdiagnostic predictor of psychopathology, understanding the association between self-compassion and emotion regulation provides insight into how self-compassion might act as a transdiagnostic *resilience* mechanism, as well as generate understanding regarding the utility of self-compassion in clinical populations.

According to Gross' (1998, 2015b) process model, there are five interconnected stages that influence emotion *generation* and *regulation*: (1) selecting which situations to engage in, and which to avoid; (2) attempting to modify a situation once engaged; (3) deciding where to deploy, or focus, one's attention; (4) changing the way one interprets or appraises the situation; and (5) modulating one's responses to the situation. In addition to influencing normative emotional experiences and responses, each of these stages has implications for emotion dysregulation and, in turn, for psychopathology. For example, visual biases toward threat (a process related to the third stage that involves increased automatic attention to potentially threatening stimuli, such as angry faces) are thought to play a causal role in anxiety disorders (Van Bockstaele et al., 2014), while maladaptive cognitive appraisals (stage four) and deployment of unhelpful emotion regulation strategies (stage five) are associated with a range of psychopathological outcomes (Mennin et al., 2007; Mehu & Scherer, 2015).

Self-compassion appears to influence affective styles, both in terms of how individuals experience, process, and respond to their emotions on both the cognitive-affective (Finlay-Jones, 2017) and physiological levels (Guan et al., 2021). Prior work has supported an emotion regulation model of self-compassion (Finlay-Jones et al., 2015), in which the relationship between self-compassion and depression or distress is mediated by a lower propensity toward self-reported emotion regulation difficulties (Carona et al., 2022; Finlay-Jones, 2017; Bakker et al., 2019; Inwood & Ferrari, 2018; Eichholz et al., 2020). These include difficulties with emotional awareness, clarity, and acceptance, as well as problems deploying adaptive emotion regulation strategies and challenges maintaining impulse control and goal regulation in the face of difficult emotions (Finlay-Jones, 2017). An important finding from studies with clinical samples is that emotion regulation difficulties mediate the relationship between self-compassion and symptoms of anxiety (Eichholz et al., 2020) and depression (Bakker et al., 2019). These findings provide useful guidance for mechanistic testing in intervention studies, particularly those with a clinical application.

As with the broader emotion regulation literature (Sheppes et al., 2015), a limitation of existing studies is that most have focused on the fifth stage, that is, the implementation of specific regulatory strategies that are associated with adverse outcomes. These have either examined emotion regulation difficulties using composite measures, or the use of specific emotion regulation *strategies*, such as rumination or avoidance, as mediators between self-compassion and a range of different mental health outcomes. Additionally, because these studies tend to use self-report trait-based measures at a single time point, there is less clarity around the specific ways in which self-compassion might influence the experience of emotions at different aspects of the emotion generation and regulation process. Accordingly, there is a need for further exploration of how self-compassion supports the upregulation of positive affect and downregulation of negative affect at other stages of the process model. Figure 24.1 provides some examples of potential ways in which self-compassion might contribute to more adaptive emotion regulation at each of these stages.

Furthermore, in an extension to the process model, Sheppes et al. (2015) highlighted four specific points at which difficulties in emotion regulation can be linked to psychopathology: (i) identifying the need to regulate emotions, (ii) selecting from available regulatory strategies, (iii) implementation of regulatory strategies, and (iv) monitoring implemented strategies over time. Understanding how self-compassion influences behavior at each of these points, as well as the original stages highlighted in the process model, can provide insights into some of the mechanisms potentially underlying the relationship between self-compassion and psychopathology. Testing these proposed pathways using experience-sampling methods with clinical and non-clinical populations would help to provide a more complete understanding of how self-compassion contributes to emotion regulation, and whether these pathways vary according to type or severity of clinical presentation. Potential ways in which self-compassion may support adaptive emotion regulation at each of the stages outlined in the extended process model are outlined in Fig. 24.2 and discussed further below.

Identifying the Need to Regulate Emotions As outlined by Sheppes et al. (2015), difficulties at the identification stage – including the overrepresentation and underrepresentation of emotional states - are implicated in a range of mental health problems. By definition, self-compassion involves mindful awareness of emotions, and evidence indicates that individuals with higher self-compassion report greater emotional awareness and clarity regarding their emotional states. This is reflected in studies documenting associations between self-compassion and cognitive awareness of affective states, as well as studies that have shown that individuals with higher self-compassion have greater interoceptive awareness (Barker, 2019).

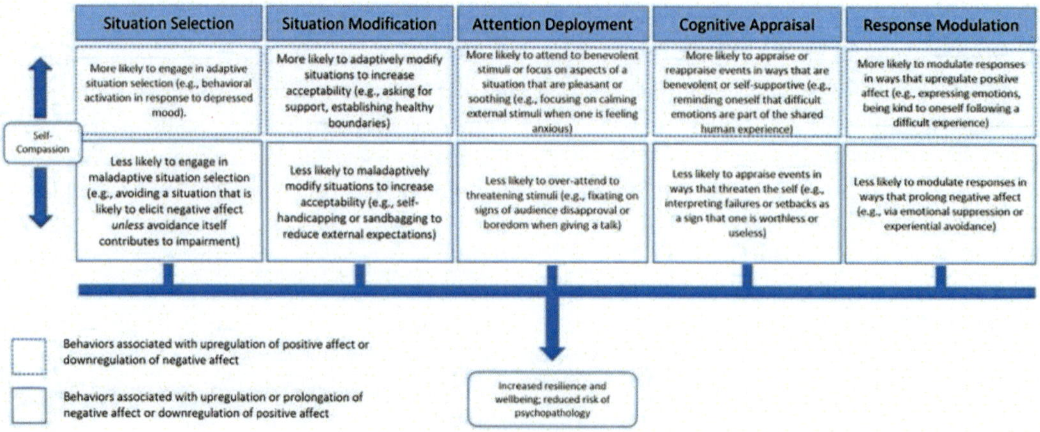

Fig. 24.1 Examples of ways in which self-compassion might contribute to upregulation of positive affect and downregulation of negative affect at each of the stages articulated in Gross' process model

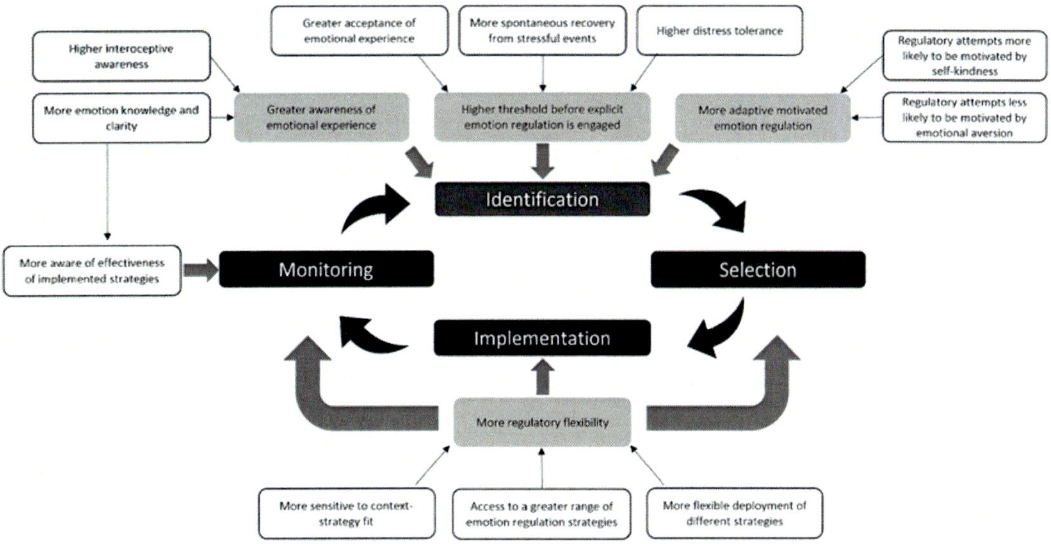

Fig. 24.2 Examples of ways in which self-compassion might contribute to more adaptive emotion regulation at each stage of Gross' extended process model

Interestingly, the implication of this appears to be that individuals with higher self-compassion are more sensitive to shifts in mood and therefore may be more likely to report drops in mood following a negative affective stimulus (Beshai et al., 2018). However, the relationship between self-compassion and the *identified need* to regulate emotions is likely more complex.

First, not all emotions need to be regulated, and attempts to regulate emotions are generally enacted when one's emotional state conflicts with one's desired state or goals (see, e.g., Tamir et al., 2020). As self-compassion is thought to facilitate a more equanimous response to difficult emotions, including increased distress tolerance, it may therefore allow for a much wider range of emotional states to be accepted without explicit regulatory attempts. Prior work examining the relationship between self-compassion and depressive symptoms in people with unipolar depression found that those with higher self-compassion had greater capacity to tolerate negative emotions and, in turn, reported lower depressive symptoms (Diedrich et al., 2017). Other research has indicated that individuals with higher self-compassion are less likely to appraise stressful events as threatening (Chishima et al., 2018), suggesting that those with higher self-compassion might have a greater window of tolerance for stressful events before explicit regulatory attempts are activated. Further, it appears that self-compassion is associated with spontaneous recovery from negative moods (Beshai et al., 2018), indicating that some of the benefits of self-compassion for emotion regulation (and by extension, mental health) may not occur as the result of motivated behavior.

Such findings are supported by functional magnetic resonance imaging work indicating that individuals with higher self-compassion require less cognitive effort to regulate their emotions in response to negative affective stimuli (Liu et al., 2022). Additionally, several studies have found that self-compassion is associated with reduced dorso-lateral prefrontal cortex (DLPFC) activity (Berry et al., 2020; Liu et al., 2022; Williams et al., 2020) and volume (Guan et al., 2021), under both experimental and non-experimental conditions. The DLPFC is an area of the brain associated with executive function and emotion regulation, and while greater volume in this area has traditionally been associated with better functioning, Guan et al.'s (2021) findings challenge this assumption. In addition to finding that self-compassion was associated with lower DPFC volume, Guan et al. (2021) also found that self-criticism was associated with increased volume in this area. They thus concluded that higher self-criticism and lower self-compassion predispose people to more frequent activation of cognitive emotion regulation. This may reflect (a) attempts to regulate negative affect elicited by self-criticism itself; (b) attempts to regulate negative affect elicited by external events that are more likely to be appraised as threatening; or (c) a propensity toward more frequent cognitive emotion regulation that co-occurs with self-criticism but is not driven by it.

On the other hand, it is plausible that, given that self-compassion reflects a kind and nurturing form of self-relation, individuals with higher self-compassion may be more motivated to regulate difficult emotions where they are appraised as overwhelming. Additionally, it is possible that the motivations underpinning emotion regulation attempts are qualitatively different for those with higher versus lower self-compassion. For example, people with high self-compassion may be more likely to regulate their emotions out of a desire for self-nurturance and goal regulation, rather than being driven by aversion or avoidance of difficult emotions. Insights into how self-compassion relates to this facet of emotion regulation are limited; however, one study found that in the context of behavioral activation (an emotion regulation strategy that is effective for reducing depressive symptoms), individuals with higher self-compassion were more likely to engage in values-based behavior (Takagaki et al., 2021). Additionally, there is some suggestion that individuals with greater self-compassion are more likely to have adaptive beliefs about emotion (Sydenham et al., 2017) although more exploration of how such beliefs influence emotion regulation motivation is required. Further research, using experience-sampling methods (English & Eldesouky, 2020), may provide useful insights into the relationship between self-compassion and the motivation to regulate emotions once specific affective thresholds have been reached. Another opportunity for future research is to determine whether self-compassion intervention influences the thresholds at which individuals identify the need to regulate their emotions.

Selecting from Available Regulatory Strategies Early taxonomies of emotion regulation tending to focus on distinguishing broadly "adaptive" (i.e., associated with more helpful outcomes) from "maladaptive" (i.e., associated with unhelpful outcomes) strategies. Aligned with this perspective, work on self-compassion and specific emotion regulation strategies has highlighted that in general, people with higher levels of self-compassion are less likely to report to use of maladaptive strategies such as rumination and worry (Raes, 2010), and more likely to engage adaptive strategies such as acceptance (Bakker et al., 2019). However, more recent advances in the emotion regulation literature have emphasized that the function of different emotion regulation strategies varies according to context and that individual strategies are not necessarily adaptive or maladaptive (Troy et al., 2013). Accordingly, what is most adaptive may be access to a range of potential emotion regulation strategies, regulatory flexibility, and the capacity to fit the appropriate strategy to the situation and context (Haines et al., 2016; Bonanno & Burton, 2013).

Studies are yet to directly test whether self-compassion supports the capacity to flexibly select and match emotion regulation strategies to one's context; however, there are several findings that suggest this might be the case. For example, positive associations between trait self-compassion and vagally mediated heart rate variability suggest that individuals with higher self-compassion also have greater emotional flexibility (Svendsen et al., 2016, 2020). This is reinforced by findings that self-compassion is associated with more self-reported psychological flexibility in community (Marshall & Brockman, 2016) and clinical (Davey et al., 2020) populations. To advance our understanding of the self-compassion-emotion regulation nexus, future research should identify whether self-compassion is also associated with access to a greater range of emotion regulation strategies, particularly among individuals with mental health problems, who typically report restricted emotion regulation repertoires. Additionally, there is a need to examine whether cultivating self-compassion supports the capacity to flexibly select appropriate emotion regulation strategies across different contexts.

Implementation of Regulatory Strategies Selection and implementation of emotion regulation strategies can play a role in psychopathology where maladaptive strategies are positively valued, such as the role of worry in anxiety disorders, or where people struggle to activate adaptive strategies (such as behavioral activation in depression). As noted above, several authors have suggested that increased implementation of adaptive emotion regulation strategies and decreased use of maladaptive emotion regulation strategies is likely a prominent mechanism driving treatment effects for self-compassion interventions (Diedrich et al., 2017; Finlay-Jones, 2017). Additionally, findings from one study of individuals with major depressive disorder indicate that priming participants with self-compassion *prior* to engaging cognitive reappraisal enhanced the efficacy of emotion regulation following a sad mood induction (Diedrich et al., 2016). To date, intervention studies testing implementation of emotion regulation strategies as a mediator of intervention effects are relatively few. One study found that reductions in experiential avoidance, alongside increases in self-compassion, mediated the link between meditation and mental health (Yela et al., 2020). In contrast, however, one study found that changes in emotion regulation and emotional suppression did not mediate treatment effects of CCT on depression and stress in caregivers of people with mental illness (Hansen et al., 2021).

Monitoring Implemented Strategies Over Time Understanding the effect of implemented strategies and knowing when to alter or discontinue them is a crucial aspect of risk and resilience to psychopathology. The ability to notice when a deployed emotion regulation strategy is ineffective and try an alternative strategy is crucial for more efficient emotion regulation. It is also closely linked to the other stages of the extended process model (i.e., being aware of the need to regulate emotions, having a range of strategies to choose from, making wise choices about those strategies, and deploying them effectively). Research has found that individual's monitoring decisions tend to align with their established regulatory preferences (Ilan et al., 2019), which has important implications for psychopathology risk. Specifically, it has been proposed that limited regulatory flexibility – including insensitivity to context, limited availability of different regulatory strategies, and low responsiveness to feedback, plays a key role in precipitating and maintaining psychopathology (Bonanno & Burton, 2013; Sheppes et al., 2015). It may be expected that people with greater self-compassion are both more aware of their context and able to use this information to inform regulatory decisions, have a greater range of emotion regulation strategies at their disposal, and are more able to switch regulatory tactics following feedback. Some support for this proposition is provided by studies linking self-compassion to psychological flexibility and adaptive goal disengagement and reengagement processes (Beshai et al., 2018; Davey et al., 2020; Miyagawa et al., 2021; Svendsen et al., 2016). However, there is a need for further study using experience-sampling methods to assess variety and change in strategy use across contexts.

Negative Self-Relational Processes: Self-Criticism, Perfectionism, and Shame-Proneness Another proposed reason for the transdiagnostic utility of self-compassion in clinical contexts is its relationship with negative self-directed cognitive and affective processes such as self-criticism, perfectionism, and shame-proneness (Gilbert & Procter, 2006). These constructs are unique but highly interconnected transdiagnostic processes in psychopathology because of their associations with the onset and maintenance of a range of different mental health problems (Schanche, 2013; Egan et al., 2011). For example, shame-proneness is associated with a range of psychopathological outcomes, including depression (Kim et al., 2011), anxiety (Swee et al., 2021; Szentágotai-Tătar et al., 2020), eating disorders (Nechita et al., 2021), and personality disorders (Rüsch et al., 2007). Self-compassion is often framed as an "antidote" to self-criticism, perfectionism, and shame, and available research has tested models in which self-compassion mediates the relationship between these processes and adverse mental health outcomes (Fletcher et al., 2019), as well as models in which the relationship between self-compassion and mental health is mediated by these processes. Given the likely bidirectionality of the relationship between self-compassion and self-criticism, perfectionism, and shame, longitudinal studies are needed to parse these relationships further.

Other research has suggested that self-compassion can attenuate the adverse impacts of negative self-relational processes on mental health outcomes. Self-compassion has been found to attenuate the links between self-criticism (Kaurin et al., 2018), perfectionism (Ferrari et al., 2018), and early shame experiences (Farr et al., 2021), such that individuals with these risk factors are less likely to experience distress and depression the more self-compassionate they are. These processes also overlap with emotion regulation mechanisms described above. For example, Farr et al. (2021) found that the relationship between early shame and distress was mediated by experiential avoidance, such that those with early shame experiences were more likely to avoid potentially difficult experiences, and experience greater distress as a result. However, self-compassion moderated this relationship such that those with higher levels of self-compassion were less likely to experience distress, regardless of their level of experiential avoidance.

Findings from the intervention literature also support the role of negative self-relational processes in mediating the treatment effects of self-compassion training. For example, a meta-analysis of 19 RCTs in both clinical and non-clinical populations found that compared to control participants, individuals in self-compassion-based interventions report significant improvements in self-criticism, with a medium effect size (Wakelin et al., 2021). One study found that a brief self-compassion training was associated with both pre- and post-reductions in shame-proneness and social anxiety in a primarily female undergraduate sample with high levels of social anxiety (Cândea & Szentágotai-Tătar, 2018), while another found that a brief self-compassion writing intervention led to reductions in state shame, shame-proneness, and depressive symptoms in a student sample, when compared to an expressive writing control condition (Johnson & O'Brien, 2013). Further, among patients with eating disorders, Kelly et al. (2014) found that increases in self-compassion early in treatment predicted decreases in shame, which in turn predicted faster increases in eating disorder symptoms. Similarly, Kelly and Tasca (2016) reported cyclical relationships between shame and eating disorder symptoms such that eating pathology increased following periods of elevated shame, while shame decreased following a period of fewer eating disorder symptoms. Moreover, they found that self-compassion appeared to interrupt this cycle, such that shame was lower after a period of increased self-compassion (Kelly & Tasca, 2016).

Future Directions

In addition to the opportunities for extending the self-compassion intervention science described above, there are several general recommendations that can be made to improve the state of the science and optimize its impact. These include increasing methodological rigor and generalizability, considering scalability through economics and implementation science, testing theoretical models and mechanisms, and exploring predictors and moderators of treatment effects.

Increasing Methodological Rigor and Generalizability

Across prior reviews and meta-analyses, authors have noted that the capacity to reliably estimate the impact of self-compassion studies is limited by the heterogeneity across studies and the number of studies that are only moderate quality (e.g., Wakelin et al., 2021). With several sites around the world now actively researching intervention science, it is now time for more collaborative efforts to generate multi-site trials, align outcome measures, standardize trial reporting (including measurement and reporting of variables believed to influence treatment outcome, such as therapeutic alliance and group process variables), and ensure that compassion interventions are tested against active conditions with sufficient sample sizes. Additionally, a consistent limitation of the self-compassion intervention literature is its overrepresentation of Western, educated, industrialized, rich, and democratic (WEIRD) populations. While this is not unique to the self-compassion literature, it immensely limits the generalizability, accessibility, and equity of the available evidence base.

Considering Scalability Through Economics and Implementation Science

One of the undersold benefits of self-compassion interventions is that they generally do not require qualified mental health practitioners to deliver them and they are relatively inexpensive to implement., Further, many of the established manualized interventions are supported by robust, train-the-trainer teaching models, and thriving communities of practice. This has important implications for scalability and fidelity of implementation. Future work should therefore focus on measuring and testing important facets of program implementation (such as differences in training models and facilitator experience) as well as establishing evidence for the cost-effectiveness of self-compassion interventions for improving health outcomes.

Testing Theoretical Models and Mechanisms

While recent work by Ash et al. (2019) has provided a useful, testable model mapping CBCT components to proposed mechanisms and outcomes, in general, self-compassion intervention science is absent a coherent theoretical framework. Accordingly, little is known about how self-compassion interventions work or how they should be refined to optimize outcomes. It is plausible that several theoretical frameworks are necessary to explain the diverse impacts of self-compassion interventions; however, the development of testable frameworks that include some of the mechanisms reviewed in this chapter (including self-compassion, emotion regulation, and negative self-relational processes) is a warranted direction for future research.

Understanding Predictors and Moderators of Treatment Effects

Given that fear of self-compassion is particularly elevated in clinical populations, including those with anxiety disorders (Merritt & Purdon, 2020) and personality disorders (Ebert et al., 2018), determining the effect of baseline fear of compassion on treatment outcome is an important extension of work to date. Understanding which variables influence treatment outcome and why they influence treatment can help to guide decisions about sequential staging of interventions (for example, offering mindfulness-based intervention prior to self-compassion training), as well as help guide the therapeutic process (for example, gradually "titrating" self-compassion into therapy).

Conclusion

Self-compassion is a malleable, transdiagnostic risk, and resilience variable that has the potential to promote more adaptive functioning across clinical and non-clinical groups. As the literature reviewed in this chapter demonstrates, self-compassion intervention science is generating compelling evidence documenting the benefits of self-compassion-based interventions in both clinical and non-clinical populations. Synthesis of this literature highlights promising directions for expanding the science and translating this evidence into more effective, targeted, and scalable intervention approaches. Specifically, this encompasses a shift away from the simple question of "do self-compassion interventions work?" and toward a more nuanced enquiry regarding whom they work for, under which circumstances, and how.

References

Adams, C. E., & Leary, M. R. (2007). Promoting elf-compassionate attitudes toward eating among restrictive and guilty eaters. *Journal of Social and Clinical Psychology, 26*(10), 1120–1144. https://doi.org/10.1521/jscp.2007.26.10.1120

Ash, M., Harrison, T., Pinto, M., DiClemente, R., & Negi, L. T. (2019). A model for Cognitively-Based Compassion Training: Theoretical underpinnings and proposed mechanisms. *Social Theory & Health, 19*(1), 43–67. https://doi.org/10.1057/s41285-019-00124-x

Bakker, A. M., Cox, D. W., Hubley, A. M., & Owens, R. L. (2019). Emotion regulation as a mediator of self-compassion and depressive symptoms in recurrent depression. *Mindfulness, 10*(6), 1169–1180. https://doi.org/10.1007/s12671-018-1072-3

Barker, E. (2019). *Exploring the association between interoceptive awareness, self-compassion and emotional regulation*. Honours dissertation, University of Adelaide, School of Psychology.

Bartels-Velthuis, A. A., Schroevers, M. J., van der Ploeg, K., Koster, F., Fleer, J., & van den Brink, E. (2016). A Mindfulness-Based Compassionate Living Training in a heterogeneous sample of psychiatric outpatients: A feasibility study. *Mindfulness, 7*(4), 809–818. https://doi.org/10.1007/s12671-016-0518-8

Berry, M. P., Lutz, J., Schuman-Olivier, Z., Germer, C., Pollak, S., Edwards, R. R., Gardiner, P., Desbordes, G., & Napadow, V. (2020). Brief self-compassion training alters neural responses to evoked pain for chronic low back pain: A pilot study. *Pain Medicine (Malden, Mass.), 21*(10), 2172–2185. https://doi.org/10.1093/pm/pnaa178

Beshai, S., Prentice, J. L., & Huang, V. (2018). Building blocks of emotional flexibility: Trait mindfulness and self-compassion are associated with positive and negative mood shifts. *Mindfulness, 9*(3), 939–948. https://doi.org/10.1007/s12671-017-0833-8

Bluth, K., Gaylord, S. A., Campo, R. A., Mullarkey, M. C., & Hobbs, L. (2015). Making Friends with Yourself: A mixed methods pilot study of a Mindful Self-Compassion program for adolescents. *Mindfulness, 7*(2), 479–492. https://doi.org/10.1007/s12671-015-0476-6

Bonanno, G. A., & Burton, C. L. (2013). Regulatory flexibility: An individual differences perspective on coping and emotion regulation. *Perspectives on Psychological Science, 8*(6), 591–612. https://doi.org/10.1177/1745691613504116

Braehler, C., Gumley, A., Harper, J., Wallace, S., Norrie, J., & Gilbert, P. (2013). Exploring change processes in compassion focused therapy in psychosis: Results of a feasibility randomized controlled trial. *British Journal of Clinical Psychology, 52*(2), 199–214. https://doi.org/10.1111/bjc.12009

Brito-Pons, G., Campos, D., & Cebolla, A. (2018). Implicit or explicit compassion? Effects of compassion cultivation training and comparison with mindfulness-based stress reduction. *Mindfulness, 9*(5), 1494–1508. https://doi.org/10.1007/s12671-018-0898-z

Britton, W. B., Davis, J. H., Loucks, E. B., Peterson, B., Cullen, B. H., Reuter, L., Rando, A., Rahrig, H., Lipsky, J., & Lindahl, J. R. (2018). Dismantling Mindfulness-Based Cognitive Therapy: Creation and validation of 8-week focused attention and open monitoring interventions within a 3-armed randomized controlled trial. *Behaviour Research and Therapy, 101*, 92–107. https://doi.org/10.1016/j.brat.2017.09.010

Canby, N. K., Eichel, K., Lindahl, J., Chau, S., Cordova, J., & Britton, W. B. (2021). The contribution of common and specific therapeutic factors to mindfulness-based intervention outcomes. *Frontiers in Psychology, 11*, 603394. https://doi.org/10.3389/fpsyg.2020.603394

Cândea, D.-M., & Szentágotai-Tătar, A. (2018). The impact of self-compassion on shame-proneness in social anxiety. *Mindfulness, 9*(6), 1816–1824. https://doi.org/10.1007/s12671-018-0924-1

Cardi, V., Leppanen, J., & Treasure, J. (2015). The effects of negative and positive mood induction on eating behaviour: A meta-analysis of laboratory studies in the healthy population and eating and weight disorders. *Neuroscience and Biobehavioral Reviews, 57*, 299–309. https://doi.org/10.1016/j.neubiorev.2015.08.011

Carona, C., Xavier, S., Canavarro, M. C., & Fonseca, A. (2022). Self-compassion and complete perinatal mental health in women at high risk for postpartum depression: The mediating role of emotion regulation difficulties. *Psychology and Psychotherapy, 95*, 561–574. https://doi.org/10.1111/papt.12388

Ceclan, A. A., & Nechita, D. M. (2021). The effects of self-compassion components on shame-proneness in individuals with depression: An exploratory study. *Clinical Psychology and Psychotherapy, 28*(5), 1103–1110. https://doi.org/10.1002/cpp.2560

Chishima, Y., Mizuno, M., Sugawara, D., & Miyagawa, Y. (2018). The influence of self-compassion on cognitive appraisals and coping with stressful events. *Mindfulness, 9*(6), 1907–1915. https://doi.org/10.1007/s12671-018-0933-0

Craig, C., Hiskey, S., & Spector, A. (2020). Compassion focused therapy: A systematic review of its effectiveness and acceptability in clinical populations. *Expert Review of Neurotherapeutics, 20*(4), 385–400. https://doi.org/10.1080/14737175.2020.1746184

Davey, A., Chilcot, J., Driscoll, E., & McCracken, L. M. (2020). Psychological flexibility, self-compassion and daily functioning in chronic pain. *Journal of Contextual Behavioral Science, 17*, 79–85. https://doi.org/10.1016/j.jcbs.2020.06.005

Diedrich, A., Hofmann, S. G., Cuijpers, P., & Berking, M. (2016). Self-compassion enhances the efficacy of explicit cognitive reappraisal as an emotion regulation strategy in individuals with major depressive disorder. *Behaviour Research and Therapy, 82*, 1–10. https://doi.org/10.1016/j.brat.2016.04.003

Diedrich, A., Burger, J., Kirchner, M., & Berking, M. (2017). Adaptive emotion regulation mediates the relationship between self-compassion and depression in individuals with unipolar depression. *Psychology and Psychotherapy, 90*(3), 247–263. https://doi.org/10.1111/papt.12107

Dodds, S. E., Pace, T. W., Bell, M. L., Fiero, M., Negi, L. T., Raison, C. L., & Weihs, K. L. (2015). Feasibility of Cognitively-Based Compassion Training (CBCT) for breast cancer survivors: A randomized, wait list controlled pilot study. *Supportive Care in Cancer, 23*(12), 3599–3608. https://doi.org/10.1007/s00520-015-2888-1

Ebert, A., Edel, M. A., Gilbert, P., & Brüne, M. (2018). Endogenous oxytocin is associated with the experience of compassion and recalled upbringing in Borderline Personality Disorder. *Depression & Anxiety, 35*(1), 50–57. https://doi.org/10.1002/da.22683

Egan, S. J., Wade, T. D., & Shafran, R. (2011). Perfectionism as a transdiagnostic process: A clinical review. *Clinical Psychology Review, 31*(2), 203–212. https://doi.org/10.1016/j.cpr.2010.04.009

Egan, S. J., Rees, C. S., Delalande, J., Greene, D., Fitzallen, G., Brown, S., Webb, M., & Finlay-Jones, A. (2022). A review of self-compassion as an active ingredient in the prevention and treatment of anxiety and depression in young people. *Administration and Policy in Mental Health, 49*(3), 385–403. https://doi.org/10.1007/s10488-021-01170-2

Eichholz, A., Schwartz, C., Meule, A., Heese, J., Neumüller, J., & Voderholzer, U. (2020). Self-compassion and emotion regulation difficulties in obsessive–compulsive disorder. *Clinical Psychology and Psychotherapy, 27*(5), 630–639. https://doi.org/10.1002/cpp.2451

Engbretson, A. M., Poehlmann-Tynan, J. A., Zahn-Waxler, C. J., Vigna, A. J., Gerstein, E. D., & Raison, C. L. (2020). Effects of Cognitively-Based Compassion Training on parenting interactions and children's empathy. *Mindfulness, 11*(12), 2841–2852. https://doi.org/10.1007/s12671-020-01495-3

English, T., & Eldesouky, L. (2020). Emotion regulation flexibility: Challenges and promise of using ecological momentary assessment. *European Journal of Psychological Assessment, 36*(3), 456–459. https://doi.org/10.1027/1015-5759/a000581

Eriksson, T., Germundsjö, L., Åström, E., & Rönnlund, M. (2018). Mindful Self-Compassion training reduces stress and burnout symptoms among practicing psychologists: A randomized controlled trial of a brief web-based intervention. *Frontiers in Psychology, 9*. https://doi.org/10.3389/fpsyg.2018.02340

Farr, J., Ononaiye, M., & Irons, C. (2021). Early shaming experiences and psychological distress: The role of experiential avoidance and self-compassion. *Psychology and Psychotherapy, 94*(4), 952–972. https://doi.org/10.1111/papt.12353

Ferrari, M., Yap, K., Scott, N., Einstein, D. A., & Ciarrochi, J. (2018). Self-compassion moderates the perfectionism and depression link in both adolescence and adulthood. *PLoS One, 13*(2), e0192022. https://doi.org/10.1371/journal.pone.0192022

Ferrari, M., Hunt, C., Harrysunker, A., Abbott, M. J., Beath, A. P., & Einstein, D. A. (2019). Self-compassion interventions and psychosocial outcomes: A meta-analysis of RCTs. *Mindfulness, 10*(8), 1455–1473. https://doi.org/10.1007/s12671-019-01134-6

Ferreira, C., Pinto-Gouveia, J., & Duarte, C. (2013). Self-compassion in the face of shame and body image dissatisfaction: Implications for eating disorders. *Eating Behaviors: An International Journal, 14*(2), 207–210. https://doi.org/10.1016/j.eatbeh.2013.01.005

Finlay-Jones, A. (2017). The relevance of self-compassion as an intervention target in mood and anxiety disorders: A narrative review based on an emotion regulation framework. *Clinical Psychologist, 21*(2), 90–103. https://doi.org/10.1111/cp.12131

Finlay-Jones, A., Rees, C. S., & Kane, R. T. (2015). Self-compassion, emotion regulation and stress among Australian psychologists: Testing an emotion regulation model of self-compassion using structural equation modeling. *PloS One, 10*(7), e0133481–e0133481. https://doi.org/10.1371/journal.pone.0133481

Finlay-Jones, A., Xie, Q., Huang, X., Ma, X., & Guo, X. (2017). A pilot study of the 8-Week Mindful Self-Compassion training program in a Chinese community sample. *Mindfulness, 9*(3), 993–1002. https://doi.org/10.1007/s12671-017-0838-3

Finlay-Jones, A., Boyes, M., Perry, Y., Sirois, F., Lee, R., & Rees, C. (2020). Online self-compassion training to improve the wellbeing of youth with chronic medical conditions: Protocol for a randomised control trial. *BMC Public Health, 20*(1), 106. https://doi.org/10.1186/s12889-020-8226-7

Fletcher, K., Yang, Y., Johnson, S. L., Berk, M., Perich, T., Cotton, S., Jones, S., Lapsley, S., Michalak, E., & Murray, G. (2019). Buffering against maladaptive perfectionism in bipolar disorder: The role of self-compassion. *Journal of Affective Disorders, 250*, 132–139. https://doi.org/10.1016/j.jad.2019.03.003

Fresnics, A. A., Wang, S. B., & Borders, A. (2019). The unique associations between self-compassion and eating disorder psychopathology and the mediating role of rumination. *Psychiatry Research, 274*, 91–97. https://doi.org/10.1016/j.psychres.2019.02.019

Frostadottir, A. D., & Dorjee, D. (2019). Effects of Mindfulness Based Cognitive Therapy (MBCT) and Compassion Focused Therapy (CFT) on symptom change, mindfulness, self-compassion, and rumination in clients with depression, anxiety, and stress. *Frontiers in Psychology, 10*, 1099–1099. https://doi.org/10.3389/fpsyg.2019.01099

Fusar-Poli, P., Solmi, M., Brondino, N., Davies, C., Chae, C., Politi, P., Borgwardt, S., Lawrie, S. M., Parnas, J.,

& McGuire, P. (2019). Transdiagnostic psychiatry: A systematic review. *World psychiatry: Official Journal of the World Psychiatric Association (WPA), 18*(2), 192–207. https://doi.org/10.1002/wps.20631

Galante, J., Galante, I., Bekkers, M. J., & Gallacher, J. (2014). Effect of kindness-based meditation on health and well-being: A systematic review and meta-analysis. *Journal of Consulting and Clinical Psychology, 82*(6), 1101–1114. https://doi.org/10.1037/a0037249

Gale, C., Gilbert, P., Read, N., & Goss, K. (2014). An evaluation of the impact of introducing Compassion Focused Therapy to a standard treatment programme for people with eating disorders. *Clinical Psychology & Psychotherapy, 21*(1), 1–12. https://doi.org/10.1002/cpp.1806

Geurts, D. E. M., Compen, F. R., Beek, M. H. C. T., & Speckens, A. E. M. (2020). The effectiveness of Mindfulness-Based Cognitive Therapy for major depressive disorder: Evidence from routine outcome monitoring data. *BJPsych Open, 6*(6), e144–e144. https://doi.org/10.1192/bjo.2020.118

Gilbert, P. (2009). *The compassionate mind: A new approach to facing the challenges of life*. Constable Robinson.

Gilbert, P. (2010). *Compassion-focused therapy: Distinctive features*. Routledge.

Gilbert, P., & Procter, S. (2006). Compassionate mind training for people with high shame and self-criticism: Overview and pilot study of a group therapy approach. *Clinical Psychology & Psychotherapy, 13*(6), 353–379. https://doi.org/10.1002/cpp.507

Gill, C., Watson, L., Williams, C., & Chan, S. W. Y. (2018). Social anxiety and self-compassion in adolescents. *Journal of Adolescence, 69*, 163–174. https://doi.org/10.1016/j.adolescence.2018.10.004

Golden, H. L., Vosper, J., Kingston, J., & Ellett, L. (2021). The impact of mindfulness-based programmes on self-compassion in nonclinical populations: A systematic review and meta-analysis. *Mindfulness, 12*(1), 29–52. https://doi.org/10.1007/s12671-020-01501-8

Goldin, P. R., & Jazaieri, H. (2017). Investigating moderators of compassion meditation training in a community sample. *Mindfulness, 11*(1), 75–85. https://doi.org/10.1007/s12671-017-0857-0

Gonzalez-Hernandez, E., Romero, R., Campos, D., Burychka, D., Diego-Pedro, R., Baños, R., Negi, L. T., & Cebolla, A. (2018). Cognitively-Based Compassion Training (CBCT®) in breast cancer survivors: A randomized clinical trial study. *Integrative Cancer Therapies, 17*(3), 684–696. https://doi.org/10.1177/1534735418772095

Gross, J. J. (1998). The emerging field of emotion regulation: An integrative review. *Review of General Psychology, 2*(3), 271–299. https://doi.org/10.1037/1089-2680.2.3.271

Gross, J. J. (2015a). Emotion regulation: Current status and future prospects. *Psychological Inquiry, 26*(1), 1–26. https://doi.org/10.1080/1047840X.2014.940781

Gross, J. J. (2015b). The extended process model of emotion regulation: Elaborations, applications, and future directions. *Psychological Inquiry, 26*(1), 130–137. https://doi.org/10.1080/1047840X.2015.989751

Guan, F., Liu, G., Pedersen, W. S., Chen, O., Zhao, S., Sui, J., & Peng, K. (2021). Neurostructural correlates of dispositional self-compassion. *Neuropsychologia, 160*, 107978. https://doi.org/10.1016/j.neuropsychologia.2021.107978

Haines, S. J., Gleeson, J., Kuppens, P., Hollenstein, T., Ciarrochi, J., Labuschagne, I., Grace, C., & Koval, P. (2016). The wisdom to know the difference: Strategy-situation fit in emotion regulation in daily life is associated with well-being. *Psychological Science, 27*(12), 1651–1659. https://doi.org/10.1177/0956797616669086

Hansen, N. H., Juul, L., Pallesen, K.-J., & Fjorback, L. O. (2021). Effect of a Compassion Cultivation Training program for caregivers of people with mental illness in Denmark: A randomized clinical trial. *JAMA Network Open, 4*(3), e211020–e211020. https://doi.org/10.1001/jamanetworkopen.2021.1020

Hanssen, I., Dord, M. I., Compen, F. R., Geurts, D. E. M., Schellekens, M. P. J., & Speckens, A. E. M. (2019). Exploring the clinical outcome of Mindfulness-Based Cognitive Therapy for bipolar and unipolar depressive patients in routine clinical practice: A pilot study. *International Journal of Bipolar Disorders, 7*(1), 1–10. https://doi.org/10.1186/s40345-019-0153-0

Haukaas, R. B., Gjerde, I. B., Varting, G., Hallan, H. E., & Solem, S. (2018). A randomized controlled trial comparing the attention training technique and Mindful Self-Compassion for students with symptoms of depression and anxiety. *Frontiers in Psychology, 9*(827). https://doi.org/10.3389/fpsyg.2018.00827

Hildebrandt, L. K., McCall, C., & Singer, T. (2017). Differential effects of attention-, compassion-, and socio-cognitively based mental practices on self-reports of mindfulness and compassion. *Mindfulness, 8*(6), 1488–1512. https://doi.org/10.1007/s12671-017-0716-z

Hofmann, S. G., Sawyer, A. T., Fang, A., & Asnaani, A. (2012). Emotion dysregulation model of mood and anxiety disorders. *Depression and Anxiety, 29*(5), 409–416. https://doi.org/10.1002/da.21888

Hoge, E. A., Hölzel, B. K., Marques, L., Metcalf, C. A., Brach, N., Lazar, S. W., & Simon, N. M. (2013). Mindfulness and self-compassion in generalized anxiety disorder: Examining predictors of disability. *Evidence-Based Complementary and Alternative Medicine: eCAM, 2013*, 576258. https://doi.org/10.1155/2013/576258

Hughes, M., Brown, S. L., Campbell, S., Dandy, S., & Cherry, M. G. (2021). Self-compassion and anxiety and depression in chronic physical illness populations: A systematic review. *Mindfulness, 12*(7), 1597–1610. https://doi.org/10.1007/s12671-021-01602-y

Ilan, S. D., Shafir, R., Birk, J. L., Bonanno, G. A., & Sheppes, G. (2019). Monitoring in emotion regulation: behavioral decisions and neural consequences. Social cognitive and affective neuroscience, 14(12), 1273–1283. https://doi.org/10.1093/scan/nsaa001

Inwood, E., & Ferrari, M. (2018). Mechanisms of change in the relationship between self-compassion, emotion regulation, and mental health: A systematic review. *Applied Psychology: Health and Well-Being, 10*(2), 215–235. https://doi.org/10.1111/aphw.12127

Irons, C., & Heriot-Maitland, C. (2021). Compassionate Mind Training: An 8-week group for the general public. *Psychology and Psychotherapy, 94*(3), 443–463. https://doi.org/10.1111/papt.12320

Jazaieri, H., Jinpa, G. T., McGonigal, K., Rosenberg, E. L., Finkelstein, J., Simon-Thomas, E., Cullen, M., Doty, J. R., Gross, J. J., & Goldin, P. R. (2012). Enhancing compassion: A randomized controlled trial of a compassion cultivation training program. *Journal of Happiness Studies, 14*(4), 1113–1126. https://doi.org/10.1007/s10902-012-9373-z

Jazaieri, H., McGonigal, K., Jinpa, T., Doty, J. R., Gross, J. J., & Goldin, P. R. (2013). A randomized controlled trial of compassion cultivation training: Effects on mindfulness, affect, and emotion regulation. *Motivation and Emotion, 38*(1), 23–35. https://doi.org/10.1007/s11031-013-9368-z

Jiménez-Gómez, L., Yela, J. R., Crego, A., Melero-Ventola, A. R., & Gómez-Martínez, M. Á. (2022). Effectiveness of the Mindfulness-Based Stress Reduction (MBSR) vs. the Mindful Self-Compassion (MSC) programs in clinical and health psychologist trainees. *Mindfulness, 13, 584–599*. https://doi.org/10.1007/s12671-021-01814-2

Johnson, E. A., & O'Brien, K. A. (2013). Self-compassion soothes the savage ego-threat system: Effects on negative affect, shame, rumination, and depressive symptoms. *Journal of Social and Clinical Psychology, 32*(9), 939–963. https://doi.org/10.1521/jscp.2013.32.9.939

Kaurin, A., Schönfelder, S., & Wessa, M. (2018). Self-compassion buffers the link between self-criticism and depression in trauma-exposed firefighters. *Journal of Counseling Psychology, 65*(4), 453–462. https://doi.org/10.1037/cou0000275

Kazdin, A. E. (2007). Mediators and mechanisms of change in psychotherapy research. *Annual Review of Clinical Psychology, 3*, 1–27. https://doi.org/10.1146/annurev.clinpsy.3.022806.091432

Kelly, A. C., & Carter, J. C. (2015). Self-compassion training for binge eating disorder: A pilot randomized controlled trial. *Psychology and Psychotherapy, 88*(3), 285–303. https://doi.org/10.1111/papt.12044

Kelly, A. C., & Tasca, G. A. (2016). Within-persons predictors of change during eating disorders treatment: An examination of self-compassion, self-criticism, shame, and eating disorder symptoms. *The International Journal of Eating Disorders, 49*(7), 716–722. https://doi.org/10.1002/eat.22527

Kelly, A. C., Carter, J. C., & Borairi, S. (2014). Are improvements in shame and self-compassion early in eating disorders treatment associated with better patient outcomes? *The International Journal of Eating Disorders, 47*(1), 54–64. https://doi.org/10.1002/eat.22196

Kelly, A. C., Wisniewski, L., Martin-Wagar, C., & Hoffman, E. (2017). Group-based Compassion-Focused Therapy as an adjunct to outpatient treatment for eating disorders: A pilot randomized controlled trial. *Clinical Psychology and Psychotherapy, 24*(2), 475–487. https://doi.org/10.1002/cpp.2018

Kim, S., Thibodeau, R., & Jorgensen, R. S. (2011). Shame, guilt, and depressive symptoms: A meta-analytic review. *Psychological Bulletin, 137*(1), 68–96. https://doi.org/10.1037/a0021466

Kirby, J. N., & Gilbert, P. (2019). Commentary regarding Wilson et al. (2018) "Effectiveness of 'self-compassion' related therapies: A systematic review and meta-analysis." All is not as it seems. *Mindfulness, 10*, 1006–1016. https://doi.org/10.1007/s12671-018-1088-8

Kirby, J. N., Tellegen, C. L., & Steindl, S. R. (2017). A meta-analysis of compassion-based interventions: Current state of knowledge and future directions. *Behavior Therapy, 48*(6), 778–792. https://doi.org/10.1016/j.beth.2017.06.003

Knier, S., Watson, J., & Duffy, J. (2020). The effects of Mindful Self-Compassion (MSC) training on increasing self-compassion in healthcare professionals. *The American Journal of Occupational Therapy, 74*(S1), 7411515371–7411515371p1. https://doi.org/10.5014/ajot.2020.74S1-PO2900

Körner, A., Coroiu, A., Copeland, L., Gomez-Garibello, C., Albani, C., Zenger, M., & Brähler, E. (2015). The role of self-compassion in buffering symptoms of depression in the general population. *PLoS One, 10*(10), e0136598. https://doi.org/10.1371/journal.pone.0136598

Krieger, T., Altenstein, D., Baettig, I., Doerig, N., & Holtforth, M. G. (2013). Self-compassion in depression: Associations with depressive symptoms, rumination, and avoidance in depressed outpatients. *Behavior Therapy, 44*(3), 501–513. https://doi.org/10.1016/j.beth.2013.04.004

Krieger, T., Berger, T., & Holtforth, M. G. (2016). The relationship of self-compassion and depression: Cross-lagged panel analyses in depressed patients after outpatient therapy. *Journal of Affective Disorders, 202*, 39–45. https://doi.org/10.1016/j.jad.2016.05.032

Kurebayashi, Y., & Sugimoto, H. (2022). Self-compassion and related factors in severe mental illness: A scoping review. *Perspectives in Psychiatric Care*. https://doi.org/10.1111/ppc.13017

Lang, A. J., Malaktaris, A. L., Casmar, P., Baca, S. A., Golshan, S., Harrison, T., & Negi, L. (2019). Compassion meditation for posttraumatic stress disorder in veterans: A randomized proof of concept Study. *Journal of Traumatic Stress, 32*, 299–309. https://doi.org/10.1002/jts.22397

Lang, A. J., Casmar, P., Hurst, S., Harrison, T., Golshan, S., Good, R., Essex, M., & Negi, L. (2020). Compassion meditation for veterans with posttraumatic stress disorder (PTSD): A nonrandomized study. *Mindfulness, 11*(1), 63–74. https://doi.org/10.1007/s12671-017-0866-z

Leeuwerik, T., Cavanagh, K., & Strauss, C. (2019). The association of trait mindfulness and self-compassion with obsessive-compulsive disorder symptoms: Results from a large survey with treatment-seeking adults. *Cognitive Therapy and Research, 44*(1), 120–135. https://doi.org/10.1007/s10608-019-10049-4

Linardon, J. (2021). Positive body image, intuitive eating, and self-compassion protect against the onset of the core symptoms of eating disorders: A prospective study. *The International Journal of Eating Disorders, 54*(11), 1967–1977. https://doi.org/10.1002/eat.23623

Liu, G., Zhang, N., Teoh, J. Y., Egan, C., Zeffiro, T. A., Davidson, R. J., & Quevedo, K. (2022). Self-compassion and dorsolateral prefrontal cortex activity during sad self-face recognition in depressed adolescents. *Psychological Medicine, 52*(5), 864–873. https://doi.org/10.1017/S0033291720002482

LoParo, D., Mack, S. A., Patterson, B., Negi, L. T., & Kaslow, N. J. (2018). The efficacy of Cognitively-Based Compassion Training for African American suicide attempters. *Mindfulness, 9*(6), 1941–1954. https://doi.org/10.1007/s12671-018-0940-1

Luo, X., Che, X., Lei, Y., & Li, H. (2021). Investigating the influence of self-compassion-focused interventions on posttraumatic stress: A systematic review and meta-analysis. *Mindfulness, 12*(12), 2865–2876. https://doi.org/10.1007/s12671-021-01732-3

MacBeth, A., & Gumley, A. (2012). Exploring compassion: A meta-analysis of the association between self-compassion and psychopathology. *Clinical Psychology Review, 32*(6), 545–552. https://doi.org/10.1016/j.cpr.2012.06.003

Mak, W. W. S., Wong, C. C. Y., Chan, A. T. Y., & Lau, J. T. F. (2019). Mobile self-compassion programme for promotion of public mental health: A randomised controlled trial. *Hong Kong Medical Journal, 25*, 15–17.

Marsh, I. C., Chan, S. W. Y., & MacBeth, A. (2018). Self-compassion and Psychological Distress in Adolescents-a Meta-analysis. Mindfulness, 9(4), 1011–1027. https://doi.org/10.1007/s12671-017-0850-7

Marshall, E.-J., & Brockman, R. N. (2016). The relationships between psychological flexibility, self-compassion, and emotional well-being. *Journal of Cognitive Psychotherapy, 30*(1), 60–72. https://doi.org/10.1891/0889-8391.30.1.60

Matos, M., Duarte, C., Duarte, J., Pinto-Gouveia, J., Petrocchi, N., & Gilbert, P. (2022). Cultivating the compassionate self: An exploration of the mechanisms of change in Compassionate Mind Training. *Mindfulness, 13*(1), 66–79. https://doi.org/10.1007/s12671-021-01717-2

Mehu, M., & Scherer, K. R. (2015). The appraisal bias model of cognitive vulnerability to depression. *Emotion Review, 7*(3), 272–279. https://doi.org/10.1177/1754073915575406

Mennin, D. S., Holaway, R. M., Fresco, D. M., Moore, M. T., & Heimberg, R. G. (2007). Delineating components of emotion and its dysregulation in anxiety and mood psychopathology. *Behaviour Therapy, 38*(3), 284–302. https://doi.org/10.1016/j.beth.2006.09.001

Merritt, O. A., & Purdon, C. L. (2020). Scared of compassion: Fear of compassion in anxiety, mood, and non-clinical groups. *British Journal of Clinical Psychology, 59*(3), 354–368. https://doi.org/10.1111/bjc.12250

Messer, M., Anderson, C., & Linardon, J. (2021). Self-compassion explains substantially more variance in eating disorder psychopathology and associated impairment than mindfulness. *Body Image, 36*, 27–33. https://doi.org/10.1016/j.bodyim.2020.10.002

Miyagawa, Y., Tóth-Király, I., Knox, M. C., Taniguchi, J., & Niiya, Y. (2021). Development of the Japanese Version of the State Self-Compassion Scale (SSCS-J). Frontiers in Psychology, 12, 779318–779318. https://doi.org/10.3389/fpsyg.2021.779318

Nechita, D. M., Bud, S., & David, D. (2021). Shame and eating disorders symptoms: A meta-analysis. *The International Journal of Eating Disorders, 54*(11), 1899–1945. https://doi.org/10.1002/eat.23583

Neff, K. D. (2003). Self-Compassion: An Alternative Conceptualization of a Healthy Attitude Toward Oneself. Self and identity, 2(2), 85-101. https://doi.org/10.1080/15298860309032

Neff, K. D., & Germer, C. K. (2013). A pilot study and randomized controlled trial of the Mindful Self-Compassion program. *Journal of Clinical Psychology, 69*(1), 28–44. https://doi.org/10.1002/jclp.21923

Neff, K. D., Knox, M. C., Long, P., & Gregory, K. (2020). Caring for others without losing yourself: An adaptation of the Mindful Self-Compassion program for healthcare communities. *Journal of Clinical Psychology, 76*(9), 1543–1562. https://doi.org/10.1002/jclp.23007

Negin, K., Farah, N., Sahar, S., & Marjan, A. (2021). Comparison of the effects of Metacognitive Therapy and Compassion- Focused Therapy on anxiety in the mothers of children with cerebral palsy. *Women's Health Bulletin, 8*(1), 1–9. https://doi.org/10.30476/whb.2020.88585.1087

Pace, T. W., Negi, L. T., Dodson-Lavelle, B., Ozawa-de Silva, B., Reddy, S. D., Cole, S. P., Danese, A., Craighead, L. W., & Raison, C. L. (2012). Engagement with Cognitively-Based Compassion Training is associated with reduced salivary C-reactive protein from before to after training in foster care program adolescents. *Psychoneuroendocrinology, 38*(2), 294–299. https://doi.org/10.1016/j.psyneuen.2012.05.019

Pauley, G., & McPherson, S. (2010). The experience and meaning of compassion and self-compassion for individuals with depression or anxiety. *Psychology and Psychotherapy: Theory, Research and Practice, 83*(Pt 2), 129–143. https://doi.org/10.1348/147608309X471000

Poehlmann-Tynan, J., Engbretson, A., Vigna, A. B., Weymouth, L. A., Burnson, C., Zahn-Waxler, C., Kapoor, A., Gerstein, E. D., Fanning, K. A., & Raison, C. L. (2020). Cognitively-Based Compassion Training for parents reduces cortisol in infants and young children. *Infant Mental Health Journal, 41*(1), 126–144. https://doi.org/10.1002/imhj.21831

Pullmer, R., Chung, J., Samson, L., Balanji, S., & Zaitsoff, S. (2019). A systematic review of the relation between self-compassion and depressive symptoms in adolescents. *Journal of Adolescence, 74*(C), 210–220. https://doi.org/10.1016/j.adolescence.2019.06.006

Raes, F. (2010). Rumination and worry as mediators of the relationship between self-compassion and depression and anxiety. *Personality and Individual Differences, 48*(6), 757–761. https://doi.org/10.1016/j.paid.2010.01.023

Raes, F. (2011). The effect of self-compassion on the development of depression symptoms in a non-clinical sample. *Mindfulness, 2*(1), 33–36. https://doi.org/10.1007/s12671-011-0040-y

Roca, P., Vazquez, C., Diez, G., Brito-Pons, G., & McNally, R. J. (2021). Not all types of meditation are the same: Mediators of change in mindfulness and compassion meditation interventions. *Journal of Affective Disorders, 283*, 354–362. https://doi.org/10.1016/j.jad.2021.01.070

Rüsch, N., Lieb, K., Göttler, I., Hermann, C., Schramm, E., Richter, H., Jacob, G. A., Corrigan, P. W., & Bohus, M. (2007). Shame and implicit self-concept in women with borderline personality disorder. The American journal of psychiatry, 164(3), 500–508. https://doi.org/10.1176/ajp.2007.164.3.500

Sauer-Zavala, S., Gutner, C. A., Farchione, T. J., Boettcher, H. T., Bullis, J. R., & Barlow, D. H. (2016). Current definitions of "transdiagnostic" in treatment development: A search for consensus. *Behavior Therapy, 48*(1), 128–138. https://doi.org/10.1016/j.beth.2016.09.004

Savari, Y., Mohagheghi, H., & Petrocchi, N. (2021). A preliminary investigation on the effectiveness of Compassionate Mind Training for students with major depressive disorder: A randomized controlled trial. *Mindfulness, 12*(5), 1159–1172. https://doi.org/10.1007/s12671-020-01584-3

Schanche, E. (2013). The transdiagnostic phenomenon of self-criticism. *Psychotherapy, 50*(3), 316–321. https://doi.org/10.1037/a0032163

Schoenberg, P. L., & Speckens, A. E. M. (2015). Multi-dimensional modulations of alpha and gamma cortical dynamics following mindfulness-based cognitive therapy in major depressive disorder. *Cognitive Neurodynamics, 9*(1), 13–29. https://doi.org/10.1007/s11571-014-9308-y

Schoenefeld, S. J., & Webb, J. B. (2013). Self-compassion and intuitive eating in college women: Examining the contributions of distress tolerance and body image acceptance and action. *Eating Behaviors, 14*(4), 493–496. https://doi.org/10.1016/j.eatbeh.2013.09.001

Schuling, R., Huijbers, M., Jansen, H., Metzemaekers, R., Den Brink, E. V., Koster, F., Van Ravesteijn, H., & Speckens, A. (2018). The co-creation and feasibility of a compassion training as a follow-up to Mindfulness-Based Cognitive Therapy in patients with recurrent depression. *Mindfulness, 9*(2), 412–422. https://doi.org/10.1007/s12671-017-0783-1

Schuling, R., Huijbers, M. J., van Ravesteijn, H., Donders, R., Cillessen, L., Kuyken, W., & Speckens, A. (2020). Recovery from recurrent depression: Randomized controlled trial of the efficacy of mindfulness-based compassionate living compared with treatment-as-usual on depressive symptoms and its consolidation at longer term follow-up. *Journal of Affective Disorders, 273*, 265–273. https://doi.org/10.1016/j.jad.2020.03.182

Sheppes, G., Suri, G., & Gross, J. J. (2015). Emotion regulation and psychopathology. *Annual Review of Clinical Psychology, 11*, 379–405. https://doi.org/10.1146/annurev-clinpsy-032814-112739

Sommers-Spijkerman, M. J., Trompetter, H. R., Schreurs, K. G., & Bohlmeijer, E. T. (2018). Compassion-focused therapy as guided self-help for enhancing public mental health. *Journal of Consulting and Clinical Psychology, 86*(2), 101–115. https://doi.org/10.1037/ccp0000268

Ştefan, C. A. (2019). Self-compassion as mediator between coping and social anxiety in late adolescence: A longitudinal analysis. *Journal of Adolescence, 76*, 120–128. https://doi.org/10.1016/j.adolescence.2019.08.013

Stice, E., Nemeroff, C., & Shaw, H. E. (1996). Test of the dual pathway model of bulimia nervosa: Evidence for dietary restraint and affect regulation mechanisms. *Journal of Social and Clinical Psychology, 15*(3), 340–363. https://doi.org/10.1521/jscp.1996.15.3.340

Stice, E., Shaw, H., & Nemeroff, C. (1998). Dual pathway model of bulimia nervosa: Longitudinal support for dietary restraint and affect-regulation mechanisms. *Journal of Social and Clinical Psychology, 17*(2), 129–149. https://doi.org/10.1521/jscp.1998.17.2.129

Stroud, J., & Griffiths, C. (2021). An evaluation of compassion-focused therapy within adult mental health inpatient settings. *Psychology and Psychotherapy, 94*(3), 587–602. https://doi.org/10.1111/papt.12334

Stutts, L. A., Leary, M. R., Zeveney, A. S., & Hufnagle, A. S. (2018). A longitudinal analysis of the relationship between self-compassion and the psychological effects of perceived stress. *Self and Identity, 17*(6), 609–626. https://doi.org/10.1080/15298868.2017.1422537

Sun, S., Pickover, A. M., Goldberg, S. B., Bhimji, J., Nguyen, J. K., Evans, A. E., Patterson, B., & Kaslow, N. J. (2019). For whom does Cognitively Based Compassion Training (CBCT) work? An analysis of predictors and moderators among African American suicide attempters. *Mindfulness, 10*(11), 2327–2340. https://doi.org/10.1007/s12671-019-01207-6

Svendsen, J. L., Osnes, B., Binder, P. E., Dundas, I., Visted, E., Nordby, H., Schanche, E., & Sørensen, L. (2016). Trait self-compassion reflects emotional flexibility through an association with high vagally mediated heart rate variability. *Mindfulness, 7*(5), 1103–1113. https://doi.org/10.1007/s12671-016-0549-1

Svendsen, J. L., Schanche, E., Osnes, B., Vøllestad, J., Visted, E., Dundas, I., Nordby, H., Binder, P. E., & Sørensen, L. (2020). Is dispositional self-compassion associated with psychophysiological flexibility beyond mindfulness? An exploratory pilot study. *Frontiers in Psychology, 11*, 614. https://doi.org/10.3389/fpsyg.2020.00614

Swee, M. B., Hudson, C. C., & Heimberg, R. G. (2021). Examining the relationship between shame and social anxiety disorder: A systematic review. *Clinical*

Psychology Review, 90, 102088–102088. https://doi.org/10.1016/j.cpr.2021.102088

Sydenham, M., Beardwood, J., & Rimes, K. A. (2017). Beliefs about emotions, depression, anxiety and fatigue: A mediational analysis. *Behavioural and Cognitive Psychotherapy, 45*(1), 73–78. https://doi.org/10.1017/S1352465816000199

Szentágotai-Tătar, A., Nechita, D.-M., & Miu, A. C. (2020). Shame in anxiety and obsessive-compulsive disorders. *Current Psychiatry Reports, 22*(4), 16–16. https://doi.org/10.1007/s11920-020-1142-9

Takagaki, K., Yokoyama, S., & Kambara, K. (2021). Do self-compassion factors affect value-based behavior promotion for adolescents? The context of behavioral activation. *Frontiers in Psychology, 12*, 566181–566181. https://doi.org/10.3389/fpsyg.2021.566181

Tamir, M., Vishkin, A., & Gutentage, T. (2020). Emotion regulation is motivated. *Emotion, 20*(1), 115–119. https://doi.org/10.1037/EMO0000635

Tao, J., He, K., & Xu, J. (2021). The mediating effect of self-compassion on the relationship between childhood maltreatment and depression. *Journal of Affective Disorders, 291*, 288–293. https://doi.org/10.1016/j.jad.2021.05.019

ter Avest, M. J., Schuling, R., Greven, C. U., Huijbers, M. J., Wilderjans, T. F., Spinhoven, P., & Speckens, A. E. M. (2021). Interplay between self-compassion and affect during Mindfulness-Based Compassionate Living for recurrent depression: An Autoregressive latent trajectory analysis. *Behaviour Research and Therapy, 146*, 103946. https://doi.org/10.1016/j.brat.2021.103946

Tobin, R., & Dunkley, D. M. (2021). Self-critical perfectionism and lower mindfulness and self-compassion predict anxious and depressive symptoms over two years. *Behaviour Research and Therapy, 136*, 103780. https://doi.org/10.1016/j.brat.2020.103780

Torrijos-Zarcero, M., Mediavilla, R., Rodríguez-Vega, B., Del Río-Diéguez, M., López-Álvarez, I., Rocamora-González, C., & Palao-Tarrero, Á. (2021). Mindful Self-Compassion program for chronic pain patients: A randomized controlled trial. *European Journal of Pain, 25*(4), 930–944. https://doi.org/10.1002/ejp.1734

Troy, A. S., Shallcross, A. J., & Mauss, I. B. (2013). A person-by-situation approach to emotion regulation: Cognitive reappraisal can either help or hurt, depending on the context. *Psychological Science, 24*(12), 2505–2514. https://doi.org/10.1177/0956797613496434

Turk, F., & Waller, G. (2020). Is self-compassion relevant to the pathology and treatment of eating and body image concerns? A systematic review and meta-analysis. *Clinical Psychology Review, 79*, 101856. https://doi.org/10.1016/j.cpr.2020.101856

Van Bockstaele, B., Verschuere, B., Tibboel, H., De Houwer, J., Crombez, G., & Koster, E. H. W. (2014). A review of current evidence for the causal impact of attentional Bias on fear and anxiety. *Psychological Bulletin, 140*(3), 682–721. https://doi.org/10.1037/a0034834

Van Dam, N. T., Sheppard, S. C., Forsyth, J. P., & Earleywine, M. (2011). Self-compassion is a better predictor than mindfulness of symptom severity and quality of life in mixed anxiety and depression. *Journal of Anxiety Disorders, 25*(1), 123–130. https://doi.org/10.1016/j.janxdis.2010.08.011

Van den Brink, E., & Koster, F. (2015). *Mindfulness-based compassionate living: Living with heart*. Routledge.

Wakelin, K. E., Perman, G., & Simonds, L. M. (2021). Effectiveness of self-compassion-related interventions for reducing self-criticism: A systematic review and meta-analysis. *Clinical Psychology and Psychotherapy, 29*, 1–25. https://doi.org/10.1002/cpp.2586

Webb, J. B., & Forman, M. J. (2013). Evaluating the indirect effect of self-compassion on binge eating severity through cognitive–affective self-regulatory pathways. *Eating Behaviors: An International Journal, 14*(2), 224–228. https://doi.org/10.1016/j.eatbeh.2012.12.005

Williams, K., Elliott, R., McKie, S., Zahn, R., Barnhofer, T., & Anderson, I. M. (2020). Changes in the neural correlates of self-blame following mindfulness-based cognitive therapy in remitted depressed participants. *Psychiatry Research. Neuroimaging, 304*, 111152–111152. https://doi.org/10.1016/j.pscychresns.2020.111152

Wilson, A. C., Mackintosh, K., Power, K., & Chan, S. W. Y. (2018). Effectiveness of self-compassion related therapies: A systematic review and meta-analysis. *Mindfulness, 10*(6), 979–995. https://doi.org/10.1007/s12671-018-1037-6

Winders, S.-J., Murphy, O., Looney, K., & O'Reilly, G. (2020). Self-compassion, trauma, and posttraumatic stress disorder: A systematic review. *Clinical Psychology & Psychotherapy, 27*(3), 300–329. https://doi.org/10.1002/cpp.2429

Wong, C. C. Y., & Yeung, N. C. Y. (2017). Self-compassion and posttraumatic growth: Cognitive processes as mediators. *Mindfulness, 8*(4), 1078–1087. https://doi.org/10.1007/s12671-017-0683-4

Yela, J. R., Crego, A., Gómez-Martínez, M. Á., & Jiménez, L. (2020). Self-compassion, meaning in life, and experiential avoidance explain the relationship between meditation and positive mental health outcomes. *Journal of Clinical Psychology, 76*(9), 1631–1652. https://doi.org/10.1002/jclp.22932

Zhu, L., Yao, J., Wang, J., Wu, L., Gao, Y., Xie, J., et al. (2019). The predictive role of self-compassion in cancer patients' symptoms of depression, anxiety, and fatigue: A longitudinal study. *Psycho-Oncology, 28*(9), 1918–1925. https://doi.org/10.1002/pon.5174

Index

A. Finlay-Jones et al. (eds.), *Handbook of Self-Compassion*, Mindfulness in Behavioral Health,
https://doi.org/10.1007/978-3-031-22348-8

T

V

W